Comparative Inorganic Chemistry

Comparative Inorganic Chemistry

Third Edition

Bernard Moody, MA, CChem, FRSC

Edward Arnold
A division of Hodder & Stoughton
LONDON NEW YORK MELBOURNE AUCKLAND

The preparation of this 3rd edition led me to recall my many former students, however distinguished or otherwise they have become as chemists or biochemists (or other forms of life), in candid and grateful recognition of what they did for me by their efforts, albeit sometimes reluctantly, and their enterprising laboratory work which has made the study of chemistry such a stimulating discipline (yet an art in some of its preparatory aspects) and let me say, great fun.

© 1991 Bernard Moody

First published in Great Britain 1965
First edition 1965
Second edition 1969
Third edition 1991

Distributed in the USA by Routledge, Chapman and Hall, Inc.
29 West 35th Street, New York, NY 10001

British Library Cataloguing in Publication Data
Moody, Bernard
    Comparative inorganic chemistry. – 3rd. ed.
    1. Inorganic high pressure chemistry
    I. Title
    546

    ISBN 0–7131–3679–0

Typeset in 10/11pt Times by Blackpool Typesetting Services Ltd. Printed and bound in Great Britain by Thomson Litho Ltd, East Kilbride for Edward Arnold, a division of Hodder and Stoughton Limited, Mill Road, Dunton Green, Sevenoaks, Kent TN13 2YA.

Contents

Preface to first edition

This book is an attempt to present in a systematic way the principles and facts needed to give a sound, coherent foundation to the study of Inorganic Chemistry. It has been written with the needs of students preparing for the Advanced Level Examinations of the General Certificate of Education in mind, and also for candidates attempting examinations for university and college Open Awards, and examinations for University Entrance. The treatment is comparative and reflects the shift in emphasis which has appeared in Inorganic Chemistry in recent years. The principal aim is to teach the generalizations, many of them subtle, which lie at the heart of Inorganic Chemistry. To do this properly and with intellectual honesty, factual detail must also be included. The amount of Physical Chemistry included has been cut to a minimum. It is hoped that the book will complement an experimental approach to the subject, both in the laboratory and lecture room.

Answers to the fundamental questions of classical chemistry—why we accept the existence of atoms and molecules—appear in the introductory chapter. Classification of the elements has been related to atomic structure. Evidence for the concept of energy levels has been included, a fundamental notion taken on trust without thought by so many of our pupils. Questions raised by the synthesis of the transuranic elements have been included to illustrate the way in which Seaborg and his co-workers have predicted the properties of elements in the best traditions of Mendeléeff. Valency has been developed in detail to show the evolution from its conception in the last century to the electronic theory, with a mention of the spin theory. Many bond diagrams are given and the transition from ionic to covalent bond type is emphasized and related both to physical properties and to the position of elements in the Period Table, the long form of which is preferred. This is followed by an electronic treatment of certain aspects of oxidation and reduction to provide another unifying thread to the book, making the deduction of many equations possible.

Principles governing the extraction of elements and some details of the processes, to show the essential chemistry and give some idea of the magnitude and type of operation used, are grouped together. Metals of growing importance, such as beryllium, titanium and zirconium, are included. The general treatment followed removes a formidable quantity of detail out of chapters on families of elements, a barrier which reduces the stamina of so many readers of conventional books. Likewise, the usual odd collections of data about alloys have been grouped in one chapter and treated generally. Industrial processes linked with air, coal, water, salt and sulphuric acid have also been grouped together.

The study of each periodic group commences with an overall survey of the gradation of properties and of the bonding encountered. While the emphasis lies on the more important elements, other elements of the family and their compounds are introduced where a generalization is well illustrated. Compounds are given in the periodic order of the other principal element combining, and a cross-reference system avoids repetition. Each chapter includes qualitative tests, including those for elements showing anomalous behaviour. Enough detail is included to support generalizations made and to link the book with laboratory work, essential to the stimulating teaching of Inorganic Chemistry.

The treatment of transition elements is designed to bring out their similar, yet graded, properties. Short accounts of the chemistry of titanium, vanadium and cobalt and certain principal valency states are included not only for the benefit of the scholarship candidate but also to justify the general trends mentioned.

Many industrial organizations were most helpful. Their generous help is most gratefully acknowledged and if any are omitted from the following list, the author offers his sincere apologies:

Albright and Wilson Ltd
The Aluminium Development Association
The Associated Ethyl Company Ltd
The Association of British Chemical
   Manufacturers
Borax Consolidated Ltd
BDH Chemicals Ltd
The British Iron and Steel Federation
The British Metal Corporation Ltd
The British Oxygen Company Ltd
The Chilean Iodine Educational Bureau
The Copper Development Association
The Dead Sea Works Ltd
The Dow Chemical Company (UK) Ltd
Hoechst Chemicals Ltd
The Gas Council
Henry Wiggin and Company Ltd
Imperial Chemical Industries Ltd
Imperial Smelting Corporation Ltd
Johnson, Matthey and Company Ltd
The Lead Development Association

Laporte Industries Ltd
Magnesium Elektron Ltd
Marchon Products Ltd
The Mond Nickel Company Ltd
Monsanto Chemicals Ltd
N. V. Norit-Vereeniging Verkoop Centrale
Northern Aluminium Company Ltd
North Thames Gas Board
Streetley Magnesite Company Ltd
The Tin Research Institute
Union Oxide and Chemical Company Ltd
The Zinc Development Association

Care has been taken to include a wide range of questions from the various examination boards and scholarship groups. A full acknowledgement is given with the questions.

I offer my sincere thanks to Mr. F. P. Dunn who read the entire book and suggested numerous improvements in content and style and to Mrs. Pauline Woodhouse who most competently deciphered and typed a difficult manuscript. Finally, I wish to record my appreciation of the generous and patient help given to me by my Publishers.

1964   B.J.M.

Preface to second edition

This book was written to project a personal view of what Inorganic Chemistry should be like at about VIth form level. The motivation came, I like to think, from a genuine dissatisfaction with the books then available and not from a desire to burden the market. The contents are based on the chemistry which had been taught under the direction of the author at Latymer Upper School and later, at Bristol Grammar School. That the first printing sold out within 3–4 months was all the more gratifying because a conscious attempt had been made to present an individual approach, in the early infancy of the Nuffield developments. I would like to express my warmest thanks to the many chemists who have either written to me or sought me out to discuss what merit the book has, to offer criticisms and constructive advice, and if I have fallen short of their standards in this edition, the fault is entirely mine.

The 1960s have seen a rapid re-organization of some parts of the chemical and metallurgical industries, which in turn has required some reshaping of certain chapters in this book. Although the production of town gas from coal has almost completely disappeared in this country, the outline of the coal gas industry has been retained in this edition. The production of ammonia is now based on the steam re-forming of naphtha and will in the not too distant future, if the price is right, be based on North Sea gas. It is a pleasure to record the help given by the Agricultural Division of ICI Ltd at Billingham in supplying information and in answering my questions prior to the rewriting of this section. To Dr. M. J. Rhydderch for giving me permission to draw on his account of Spray Steelmaking in the July 1968 edition of *Science Journal*, and to the Editor of that journal, I extend my warmest thanks. I am also pleased to acknowledge the help given to me by my youngest brother, Stuart Moody, BSc, PhD, of the University of Edinburgh, in keeping me abreast of certain recent developments.

The Stock Notation for nomenclature was a feature of the first edition and the underlying plan of the book was organized on it. In this edition the nomenclature has been completely revised to comply with the *IUPAC* recommendations (*The Nomenclature of Inorganic Chemistry* (1957), *IUPAC, London: Butterworth Scientific Publications*, 1959) and with the suggestions put forward by the Nuffield Foundation. A rationalized system of metric units, the *Système International d'Unités* (*SI*) is coming into international use and the units in this book are based on that system. But for both nomenclature and units the established, or *trivial*, names and units are also given where it seemed realistic and helpful to do so. Attention is drawn to the Royal Institute of Chemistry Monograph no. 15: *M. L. McGlashan, Physico-chemical Quantities and Units* (1968). In the first edition, values for ionic radii were based on those of V. M. Goldschmidt (1926). In the current edition the values adopted are those in the Nuffield Book of Data, and recommended by Ahrens and Yatsimirskii, to avoid confusion to those pursuing Nuffield courses, but it must be emphasized that while internuclear distances are often known precisely, how far each ion extends is arguable.

In addition to minor corrections and improvements, sections have been added on the shape of covalent molecules, based on simple theory of electrical repulsion, on the concept of lattice energy and its relation to the transition from ionic to covalent bonding and to the solubility of ionic compounds, and on the importance of coordination number. The terms ionic and co-ionic are still preferred to their alternatives (although these are given) largely for the reasons put forward by W. G. Palmer, my former Director of

Studies at Cambridge, in his book on *Valency* (Cambridge University Press) and used also in his university text, *Experimental Inorganic Chemistry* (Cambridge University Press).

Finally, I record my continued appreciation of the friendly and intelligent help given to me by my Publishers.

1969   B.J.M.

Preface to third edition

The experienced reader will see numerous altera-tions in this edition including the rearrangement of some of the early chapters. Units have been brought in line with current internationally accepted recommendations, some features of the text have been expanded and others modified. The alterations reflect changes in emphasis of the author's thinking and teaching, based on many years experience as a Chief Examiner at Advanced level in chemistry to two Examination Boards and are in turn a reflection of evolutionary change, or perhaps tidal change would be a more apposite metaphor, in the subject. In particular it is hoped that while the prime purpose of the book will be enhanced by the changes, the book will also fulfil another function: it will continue to act as a foun-dation for early college and university courses, being a reliable infill for those aspects of chemistry which have been glossed over fleetingly in some modern courses as the treasure-house of knowledge and achievement is plundered.

The opening chapters are historical in a double sense: the route passes some of the important land-marks of classical chemistry but more importantly, the chapters trace the development of thinking in the subject for they were written with many of the source papers to hand. Currently, and in the same vein, such books as Francis Crick's *What Mad Pursuit* (Basic Books, 1988), a personal view of scientific discovery by the Nobel Prize winner (with James Watson and Maurice Wilkins) who dis-covered the double helix structure of DNA, should be prescribed reading for all students. Therein lies the cultural value of science.

Two early chapters have been expanded and reorganized into four chapters. The properties of elements in relation to atomic structure has been separated from the matter of electronic configura-tion. Mass spectrometry has been promoted in this edition from being a side issue. While simple octet theory is the springboard to what follows, the spin theory and the idea of electron-pairs is developed, and where it is helpful, molecular orbitals are not overlooked in the main text. In this edition there is greater emphasis on the quantitative aspects of ionization energies, electron affinity and exchange energy. As before, however, the criterion for inclu-sion of theory is strictly its relevance to the com-parative theme of the book. Bonding and structure merits a separate chapter after the electronic theory of valency. Bonding found in complex ions involves hybridization, inner and outer orbital complexes, and a later discussion of the associated characteristic colours. Atomic, ionic, metallic and van der Waals' radii are dealt with in some detail and the significance of lattice energies, the theory, the relationship to oxidation states, solubility, interionic distance and physical properties, extended. Much of the foregoing is developed further in the introductory remarks at the beginning of each chapter on the respective Periodic families of elements. The treatment is always comparative, concerned with finding rela-tionships for use as a teaching aid as much as anything else. Here, comparisons of ionization energies, various radii, electronegativities and standard electrode potentials are illustrated in simple graphical form. Crystallization and the crystalline state, formerly a middle order topic, has been promoted to support the opening chapters and now, in an expanded form to show the theoretical building up of close-packed (and not so close-packed) metal structures, it follows the critical examination of bonding, at the simple level. It remains the contention of the author that appar-ently quite simple concepts can be used to acquire a basic understanding and intuitive grasp of what happens during chemical reactions, and an insight into structural chemistry. The problem is age-old: to prepare the way for work at a higher level in such a way that the subject unfolds like a flower, without implanting in the student the misunderstandings and old-fashioned prejudices of the teacher. The problem is well known and has to be accepted since it is inescapable.

While expanding the section on nuclear energy, the opportunity has been taken to discuss the hazards associated with the generation of nuclear power and the chemical processing of the products from nuclear reactors (and other sources). Current medical disquiet about the effects of radiation is discussed along with other related issues including politics and economics.

Redox processes are again considered from first principles but the use of equivalents has been omitted, with calculations in titrimetry (volumetric analysis) indicated along molarity lines. The *Stock Notation*, always a feature, has been extended as an alternative nomenclature for the oxoacids of the halogens, to avoid stress for students following the Nuffield courses. Here, and later, the perceptive reader will appreciate the experimental nature of the underlying scheme upon which this chapter, and later chapters is based.

The discussion of Gibbs' Free Energy has been taken further in looking at the principles governing the extraction of metals. In the associated chapters there are minor changes in emphasis and the Midrex process for iron is outlined. The retention of certain processes largely displaced in the United Kingdom is explained by the wider readership of the book.

The section on acids and bases, which already ranged from the initial concept right through to non-protonic systems has been enhanced by inclusion of the Lewis theory, with its link to electron-pairs and bonding. Perhaps it is appropriate here to mention that while extra headings have been introduced to facilitate reference, the temptation to subdivide various discourses has been resisted on the assumption that the student is quite capable of absorbing the study of (say) carbon dioxide, carbonic acid and its salts in one digestible whole.

There are a series of minor alterations to the main text and for up-to-date interest, short topics touching on such matters as metal clusters (for organometallic catalysis) and so-called one dimensional metals (replacing the silicon chip) are added, to give, it must be admitted, a whiff of current research. The text has been considerably enhanced by many illustrations of molecular and ionic structures, with a continuing emphasis on simple concepts like the electrical repulsion theory which accounts for (as an example) the octahedral structure from which the square planar shape of the nickel cyanide complex, $Ni(CN)_4^{2-}$ may be derived. Certain aspects of pollution are also discussed in this edition including the contamination of water supplies from the excessive use of chemical fertilizers, the European concern over acid rain and the world problem of ozone depletion and the greenhouse effect.

Examination questions are included again but not for cramming! They have been vetted, pruned and extended, not primarily for examination preparation but as a useful and surprisingly interesting way of assessing how much of what has been read, has penetrated and been retained.

I would like to acknowledge David Kitchen (ICI, Billingham) and Geoff Daniels (ICI, Runcorn) for their help with Chapters 12 and 14, respectively. In addition to the organizations listed in the prefaces to the first and second editions, I would like to thank the following:

The Fertiliser Manufacturers Association
The National Sulphuric Acid Association
The Royal Mint
The Royal Society of Chemistry
The Science and Engineering Research Council

Finally I would like to acknowledge the help of my Publishers.

1991   B.J.M.

A Note of Prefixes for SI Units

The following are the approved decimal fractions or multiples of the basic SI Units or of the SI specially named derived units.

Fraction	Prefix	Symbol	Multiple	Prefix	Symbol
10^{-1}	deci	d	10	deka	da
10^{-2}	centi	c	10^2	hecto	h
10^{-3}	milli	m	10^3	kilo	k
10^{-6}	micro	μ	10^6	mega	M
10^{-9}	nano	n	10^9	giga	G
10^{-12}	pico	p	10^{12}	tera	T
10^{-15}	femto	f			
10^{-18}	atto	a			

Example $\text{Mg km}^2\,\mu\text{s}^{-2}$

$$= (10^3\,\text{kg})(10^3\,\text{m})^2(10^{-6}\,\text{s})^{-2}$$
$$= 10^3\,\text{kg} \times 10^6\,\text{m}^2 \times 10^{12}\,\text{s}^{-2}$$
$$= 10^{21}\,\text{kg m}^2\text{s}^{-2}$$

Note that the prefix must be written immediately adjacent to the unit, i.e. touching it in normal writing and separated from the next unit by a small space.

e.g. kg mm, Mg, nm but $\text{kg m}^2\text{s}^{-2}\text{K}^{-1}\text{mol}^{-1}$ ($= \text{J K}^{-1}\text{mol}^{-1}$)

Further cm^2 always means $(0.01\,\text{m})^2$ and never $0.01\,\text{m}^2$ which may trap the unwary when faced with, say, km^2 or μs^{-2} as above.

The development of fundamental ideas in 19th century chemistry

Science and chemistry

The Latin word *scientia* was Anglicized to 'science' and applied to any formal branch of knowledge until the beginning of the 17th century when its use became gradually restricted to the natural sciences. At the beginning of the last century the term philosophy was usually applied to all branches of knowledge, science having the same meaning. With the rapid advances in science as we know it, came a separation of the two terms. Philosophy was divided at first into moral philosophy and natural philosophy and science into moral science and natural science. Moral refers to man and his ways: thus, moral law is concerned with the requirements to which right action must conform, and moral sense, the power of distinguishing between right and wrong. Natural in this context has a diffuse meaning: it concerns everything external to man and his conduct. Thus, natural philosophy and natural science were equivalent in meaning but the former has declined in use and only survives in the 'Chairs of Natural Philosophy' at the older universities, philosophy being retained nowadays for the older branches of formal knowledge. The adjective, natural, is usually dropped from Natural Science although its use is retained in the 'Schools of Natural Science' serving to distinguish them from the 'Schools of Moral Science' of the older universities.

The divisions of Natural Science, or the Natural Sciences, are well known: fundamentally, Physics is the general study of matter, heat, light, electricity, etc., Chemistry involves the separation and detailed examination of selected kinds of matter, while Biology is the study of living organisms, Geology, the study of the Earth, and Astronomy, the Universe. However, it is extremely difficult to define science itself and the term is still sometimes used in its earlier sense, as in 'he is a scientific bowler', i.e. skilful. Scientists follow a highly

individual profession but it is generally agreed that any definition of science must indicate the methods by which scientific knowledge is obtained as well as the knowledge itself. Sir Richard Gregory defined science as organized and formulated knowledge of natural objects and phenomena derived from verifiable observations and experiments.

The concept of hypothesis and the crucial experiment lie at the heart of any understanding of science. Starting from the collection of verifiable observations, classification may lead to general statements called laws for which explanations are sought. To this end, hypotheses are formed. A hypothesis is the principal intellectual tool of the scientist. Hypotheses are intelligent guesses and require the possession of imaginative and intuitive powers. They enable the planning of observations and experiments so that predictions based on one or more rival hypotheses may be tested. Experiments are designed to ask specific questions of nature. A crucial experiment enables a choice to be made between rival hypotheses. Science advances by the formulation of hypotheses, with the prediction from them of results which may be tested by experiment. Theories to account for observed phenomena are then propounded. Hypotheses which seem well established become theories. But these must be modified or discarded as fresh evidence is uncovered which proves at variance with them. Of course, chance plays a role in scientific discovery and a clear sequence of events is not always apparent. Most hypotheses by their nature will prove wrong, but by the critical examination of such concepts, science progresses. With hypotheses established, an organized structure based on reason is erected.

In the study of Chemistry, the naturally occurring mixtures of the Earth's crust must be separated and purified so that an examination of the characteristic properties of the components of the mixtures and their relationships with each other may be undertaken. During the interaction of different kinds of matter, the relative quantities of matter and associated energy are measured.

Chemistry may be roughly divided into the organic and inorganic branches. The former describes the chemistry of the countless compounds of carbon, although the behaviour of the element and a few simple compounds is left to Inorganic Chemistry, into which the study of all other elements and compounds falls. Inorganic Chemistry as a science may easily become choked by facts. After critical selection and examination of observations concerning the characteristics of elements and their compounds, a periodic pattern emerges. Order appears in the classification of the elements and the study of quantitative and qualitative trends. Physical Chemistry involves the study of chemical phenomena in general terms of Physics and of Chemistry. In what follows it will be assumed that much of the Physical Chemistry required to understand Inorganic Chemistry is known, or is readily available elsewhere.

The purification and identification of substances in the laboratory

Substance is a very vague term. The simplest forms of matter, which cannot be broken down any further, are termed chemical elements. By chemical action, involving the release or absorption of energy by the system, and described as exothermic or endothermic reactions accordingly, compounds are formed. A substance may be an element, a compound or any kind of physical mixture. The detection and identification of the element present in a compound is called analysis and the building up of the compound from simpler units, synthesis.

The details for purification techniques belong properly to laboratory manuals. Gases must be chemically separated into the constituents, dried and collected. This involves the use of wash bottles, absorption bulbs and U-tubes, with collection in air, over water or mercury, or into an evacuated vessel. Mixtures of gases are homogeneous; they consist of a single phase of uniform molecular distribution. Systems involving liquids or solids, and their vapour, consist of a number of phases, and are described as heterogeneous, each phase being of uniform composition and defined by boundary surfaces. Liquids may be separated by means of a separating funnel if immiscible, but their purification usually requires distillation. This may be accomplished generally under atmospheric pressure, but where thermal decomposition would occur, distillation under reduced pressure may be used. The apparatus may consist of a simple retort or involve a condenser, which may be water- or air-cooled. Mixtures of liquids may require fractional distillation if they are miscible and their boiling-points are very similar. Liquids immiscible with water and otherwise difficult to isolate may be distilled in steam. Bromine, nitric acid, phosphorus trichloride and sulphuryl chloride require distillation, preferably in a glass-jointed apparatus. This is cleaner, easier to arrange and avoids the corrosion problems encountered with the use of corks. Hypochlorous acid and hydrogen peroxide are concentrated under reduced pressure. Hydrochloric acid requires fractionation. The separation of solids may require extraction with solvents. This will be followed by filtration and evaporation, possibly under reduced pressure. Filtration may be replaced by the use of a centrifuge, or in simple cases by decantation. Purification may depend on crystallization or fractional crystallization. Sometimes, solids may be sublimed. Drying may be accomplished in a desiccator, either at ordinary pressure or under reduced pressure in the presence of a desiccating agent, such as concentrated sulphuric acid.

For inorganic solids, crystallization is the most important process of purification and the formation of well-formed crystals from a given solvent may be sufficient to render identification possible. With the covalent solids of organic chemistry, their comparatively low melting-points make the practice of melting-point and mixed melting-point determination a major factor in their identification. Usually inorganic solids have comparatively high melting-points. But there are no homologous series to complicate matters and qualitative analysis may be used. The use of the flame test and spectroscope gives a ready means of identifying metals by their characteristic atomic spectra. The inert gases were identified by the spectra obtained by electrical discharge in them under reduced pressure.

Separation of mixtures may be accomplished by chromatographic techniques. They are of great importance in research and in the routine analytical work on complex mixtures in industry.

Atoms: classical arguments for their existence

Dalton's Atomic Theory was first described in Thomas Thomson's *System of Chemistry*, Volume

III (1807) and by John Dalton himself in *A New System of Chemical Philosophy* (1808). The concept that matter could not be divided indefinitely but would be left as ultimately indivisible particles was not new. It was apparent in Greek thought, as shown in the writings of Leukippos (480 B.C.?) and Demokritos of Abdera (468–370 B.C.). The word 'atom' comes from the Greek, *atomos*, meaning indivisible. The Roman, Lucretius, wrote about an atomic concept in his poem concerning *The Nature of the Universe*. Boyle, Newton and others revived the idea of ultimate particles. Dalton's contribution was to create a working hypothesis out of speculation. He developed the idea that atoms of the same kind of matter would be alike, especially in weight and shape, and would differ from the atoms of other kinds of matter. However, the main step forward came with his explanation of chemical combination in terms of the rearrangement of atoms: 'Chemical analysis and synthesis go no further than to the separation of particles from one another and to their union. No new creation or destruction of matter is within the reach of chemical agency.'

In the absence of evidence to the contrary, Dalton assumed that atoms united in the simplest ratios possible; for one compound, 1:1, and for more than one compound being formed between two elements, 1:1, 1:2, 2:1, etc. While the arbitrary assignment of combining ratios proved incorrect the hypothesis that chemical combination involved the regrouping of atoms into the same pattern for a given compound gave a simple explanation to the previously observed relationships between the (equivalent) combining weights of elements.

Dalton stressed the importance of the weight of atoms. Nowadays we distinguish between the mass of a body (the quantity of matter present) and its weight (the force of attraction exerted by gravity towards the earth). The quantity of matter in a body is the same wherever the body is while the weight may change. A popular example will be the weightless astronaut in space whose body (and therefore, his mass) is very much in evidence. Within the confines of a chemical balance, the directional pull of gravity on the unknown mass is compared, or balanced, against that of a known mass, the magnitude of the gravitational force cancelling so that masses are compared.

We now know that atoms of a given element may differ slightly in mass, the different kinds being called isotopes. It has been known since 1895 that

atoms are not indivisible and various sub-atomic particles have been described. Further, matter is not indestructible but its loss is associated with an enormous release of energy. But in the context of 'chemical agency', Dalton's Atomic Theory is still substantially correct. As a hypothesis it was based on quantitative laws and tested by predicting another law, that of multiple proportions. **The Atomic Theory** may be summarized:

'All matter is composed of very small discrete particles.

For elements the particles are called atoms and are indivisible and indestructible in chemical reactions.

The atoms of a given element are all identical and differ from the atoms of other elements. Dalton stressed weight as a property distinguishing atoms of different elements.

A chemical compound is formed from its elements by the combination of the different atoms in a fixed ratio for that compound.'

During the course of his researches into the role of oxygen in combustion, Lavoisier formulated the **Law of Conservation of Mass** (1774).

In a chemical reaction, the mass of the products is equal to the mass of the reactants.

He weighed a sealed retort containing tin and air, before and after heating. The experimental statement of the law is sometimes discarded for the statement that 'matter is neither created nor destroyed in the course of a chemical reaction'. An accuracy of 1 in 10^7 was achieved by Landolt in further experiments during the period 1891–1908 and Morley in 1912 recorded an accuracy of 1 in 10^8. Until 1801, the constancy of chemical composition of pure compounds had been assumed. Indeed, without it, chemistry would be very different. In that year the law, formally stated two years before by Proust, was challenged by Berthollet who drew the wrong conclusions from some experimental work on oxides of lead. However, Proust did further work and using compounds of copper he established the **Law of Constant Composition (or Definite Proportions)**.

All pure samples of a particular chemical compound contain the same elements combined in the same proportions by mass.

By 1808, after further work, the law was accepted. It will be seen later that elements usually contain a fixed proportion of isotopes or atoms of slightly different mass but where the formation of some specimens by radioactive decay has occurred a variation in composition by mass of certain compounds will be found. Interstitial compounds formed between the transition metals and the lighter elements (hydrogen, boron, carbon, nitrogen) and one or two sulphides and oxides show slight variations in composition. These are known as Berthollides as distinct from the much more usual Daltonide compounds. On the basis of the atomic hypothesis and these quantitative laws, Dalton predicted a relationship which would be found to hold where more than one compound was formed between two elements. This was then established experimentally and known as the **Law of Multiple Proportions**.

When two or more elements combine together to form more than one compound the several masses of one element combining with a fixed mass of the other element are in a simple ratio.

The concept that chemical interaction results in the regrouping of atoms in patterns peculiar to the compounds concerned, together with the notion that the mass of an atom is a characteristic property of the element of which it is the smallest particle explains the fourth quantitative law, the **Law of Equivalent (or Reciprocal) Proportions** established by Richter between 1791 and 1802.

The ratio of the mass of two elements which separately combine with a given mass of a third element is the ratio in which the two elements react with each other, or a simple multiple of this ratio.

A direct consequence of the laws of chemical combination and not depending on theory in any way is the importance attached to the idea of equivalent mass. This is the mass of the element which will combine with a standard mass of a standard element. Where direct combination does not occur, or difficulties arise, an indirect determination may be made by combination with an element of previously determined equivalent mass. The standard chosen will be that which proves most convenient, readily forming a wide range of compounds. The standard value should not be too cumbersome nor should values based on it be less than unity. Dalton chose hydrogen, which is the element with the lowest equivalent mass, as the standard (equivalent = 1) but relatively few elements combine with it. Berzelius favoured oxygen and this is usually used today in the school laboratory. The value selected (equivalent = 8 exactly) makes the lowest equivalent mass, that of hydrogen, just greater than unity. Therefore, **the equivalent mass of an element is**

the number of parts by mass of that element which will combine with or displace, directly or indirectly, 8 parts by mass of oxygen exactly.

Dalton emphasized the importance of determining the relative mass of atoms and adopted arbitrary combination ratios for atoms uniting. The mass of individual atoms is very small indeed and it is much more convenient to define a scale in terms of the mass of one of the elements. Clearly, the scale should be based chemically on oxygen. The table of relative atomic masses so devised gives values near to whole numbers for most elements. When hydrogen and oxygen combine to form steam, the volumes uniting are in the ratio of $2:1$. That the atoms of hydrogen and oxygen unite in groups of three, two of hydrogen to one of oxygen, will be deduced later. In round numbers, the equivalent mass of hydrogen = 1 when that of oxygen is defined as 8. Briefly, for steam the combining ratios of hydrogen and oxygen are: by atoms, $\frac{2}{1}$; by mass, $\frac{1}{8}$. By adopting the atomic mass standard based on 16 exactly as the value for oxygen, the atomic mass of hydrogen is equal to the equivalent mass and just greater than unity (1.0080): it has the smallest atomic mass. The atomic mass standard, and therefore the atomic mass scale for other elements, is arbitrary. **Until 1960, atomic masses were based on the scale showing the mass of an atom of oxygen as 16 exactly**. The masses of individual atoms may be calculated using Avogadro's constant (formerly Avogadro's Number), which determined experimentally, shows that there are 6.023×10^{23} atoms in one gram atom (i.e. the mole, p. 8).

The alchemists used a sort of shorthand to represent substances. Thus, silver was shown by a crescent moon, and silver nitrate is still sometimes called lunar caustic. Dalton first used symbols in a quantitative, as well as a qualitative way, in the development of the Atomic Theory. To him, a symbol represented a definite quantity of an element, namely, one atom. He pictured atoms as circles with shading and signs to distinguish different kinds of atoms so that a compound could be shown as a cluster of touching atoms, showing a definite number of atoms of each of the combining elements. Berzelius invented the present system in 1811. He suggested that it was easier to write and print an abbreviated word than a diagram. He selected the initial letter of the Latin name of each element and where several had the same initial letter he added the next letter or if this failed, he used the first consonant not in common. He also stressed the need for a symbol to represent an atom of the element. It now represents that quantity of the element equal to the relative atomic mass measured in grams, to use the obsolescent definition.

The combining capacity of an atom of an element is called the valency, from the Latin, *valentia*, vigour or capacity. Because two elements may unite to form more than one compound, the valency may assume more than one value. Since equivalents are mass-combining ratios experimentally determined with atoms combining in small whole number ratios and atomic masses are ratios, both being based usually for our purposes on O = 16 exactly, there must be a simple relationship between the atomic mass and equivalent of an element referred to oxygen:

$$\frac{\text{Atomic mass}}{\text{Equivalent mass}} = \text{a small integer, which is the}$$

valency of the element in the compound in which it exhibits the equivalent mass shown.

The valency of an element may be deduced in another way. This requires the use of test elements of indivisible valency, or univalency. The valency of an element equals the number of atoms of the test element with which one atom of the element unites. Hydrogen is accepted as the universal test element. The evidence in support of this will be given later. But this method is not so different from that used before. In counting the number of atoms of hydrogen uniting with the atom of the element, the number of gram atoms of hydrogen, or gram equivalents, combining with the gram atom of the element is being determined. The number of hydrogen atoms counted represents the number of gram equivalents in the gram atom of the element.

The **valency** of an element is equal to the number of hydrogen atoms with which one atom of the element will unite.

This classical view will be examined in detail again.

Until 1960, the standard for atomic mass was O = 16 exactly. Formerly, the term relative atomic mass was used, but nowadays the molar mass is preferred, which has the unit g mol^{-1} of the species quoted. The occurrence of isotopes, atoms of an element with different atomic masses, has been mentioned. The chemical scale of atomic masses was based on oxygen being, in this context, the ordinary mixture of isotopes. The nature of isotopes will be described in detail later. Atmospheric oxygen contains 99.758% of the isotope† ^{16}O, 0.0373% of ^{17}O, and 0.2039% of ^{18}O. The weighted mean is 16.004462. The physical scale of atomic masses was based on the most abundant isotope ^{16}O taken as 16. The prefixes refer to the mass number of each isotope, the integer nearest to the atomic mass. Atomic masses on the physical scale were larger by a factor of 1.000272 than those on the chemical scale. **A scale based on $^{12}C = 12.0000$ was proposed in 1960**; it differs from the chemical scale (O = 16.0000) by 42 p.p.m. and the physical scale ($^{16}O = 16.0000$) by 317 p.p.m.

The Table of relative atomic mass is given as Table 1.1, p. 6.

Molecules: classical arguments for their existence

In his conception of the Atomic Theory, Dalton postulated that atoms united in small whole numbers. A few months later in 1808, Gay–Lussac announced his **Law of Combining Volumes**:

Gases combine together in volumes which bear a simple ratio to each other and to the volume of the product if it be gaseous and all measurements are made under fixed conditions of temperature and pressure.

Three years before Gay–Lussac and von Humboldt had published *Experiments on the Ratio of the constituents of the Atmosphere*. One method of analysis for oxygen involved exploding measured volumes of air with hydrogen and measuring the volume of the product. Clearly, data for the combination of oxygen and hydrogen was required but there proved to be nothing reliable at hand. As a preliminary, they measured the combining ratio of hydrogen and oxygen by volume. As a mean of 24 experiments, they calculated that 100 parts of oxygen by volume reacted with 199.89 parts of

† *spoken:* oxygen-16

Table 1.1 Relative atomic masses of the elements
Based on the relative atomic mass of $^{12}C = 12$

The values for relative atomic mass given in the Table apply to elements as they exist in nature, without artificial alteration of their isotopic composition, and further to natural mixtures that do not include isotopes of radiogenic origin. If the term **molar mass** is preferred, the unit is $g\,mol^{-1}$.
Alphabetical Order in English

Name	Symbol	Atomic number	Atomic mass	Name	Symbol	Atomic number	Atomic mass
Actinium	Ac	89	227.028	Mercury	Hg	80	200.59
Aluminium	Al	13	26.9815	Molybdenum	Mo	42	95.94
Americium*	Am	95	243.061	Neodymium	Nd	60	144.24
Antimony	Sb	51	121.75	Neon	Ne	10	20.179^b
Argon	Ar	18	39.948	Neptunium*	Np	93	239.053
Arsenic	As	33	74.9216	Nickel	Ni	28	58.71
Astatine*	At	85	209.987	Niobium	Nb	41	92.906
Barium	Ba	56	137.34	Nitrogen	N	7	14.0067
Berkelium*	Bk	97	247.070	Nobelium*	No	102	255.093
Beryllium	Be	4	9.0122	Osmium	Os	76	190.2
Bismuth	Bi	83	208.980	Oxygen	O	8	15.9994^a
Boron	B	5	10.811^a	Palladium	Pd	46	106.4
Bromine	Br	35	79.904^b	Phosphorus	P	15	30.9738
Cadmium	Cd	48	112.40	Platinum	Pt	78	195.09
Caesium	Cs	55	132.905	Plutonium*	Pu	94	239.052
Calcium	Ca	20	40.08	Polonium	Po	84	210.000
Californium*	Cf	98	252.082	Potassium	K	19	39.102
Carbon	C	6	12.01115^a	Praseodymium	Pr	59	140.907
Cerium	Ce	58	140.12	Promethium*	Pm	61	144.9126
Chlorine	Cl	17	35.453^b	Protactinium	Pa	91	231.036
Chromium	Cr	24	51.996	Radium	Ra	88	226.025
Cobalt	Co	27	58.9332	Radon	Rn	86	222.018
Copper	Cu	29	63.546^a	Rhenium	Re	75	186.2
Curium*	Cm	96	247.070	Rhodium	Rh	45	102.905
Dysprosium	Dy	66	162.50	Rubidium	Rb	37	85.47
Einsteinium*	Es	99	254.088	Ruthenium	Ru	44	101.07
Erbium	Er	68	167.26	Rutherfordium*	Rf	104	260
Europium	Eu	63	151.96	Samarium	Sm	62	150.35
Fermium*	Fm	100	253.086	Scandium	Sc	21	44.956
Fluorine	F	9	18.9984	Selenium	Se	34	78.96
Francium*	Fr	87	223.020	Silicon	Si	14	28.086^a
Gadolinium	Gd	64	157.25	Silver	Ag	47	107.868^b
Gallium	Ga	31	69.72	Sodium	Na	11	22.9898
Germanium	Ge	32	72.59	Strontium	Sr	38	87.62
Gold	Au	79	196.967	Sulphur	S	16	32.064^a
Hafnium	Hf	72	178.49	Tantalum	Ta	73	180.948
Hahnium*	Ha	105	—	Technetium*	Tc	43	98.906
Helium	He	2	4.0026	Tellurium	Te	52	127.60
Holmium	Ho	67	164.930	Terbium	Tb	65	158.924
Hydrogen	H	1	1.00797^a	Thallium	Tl	81	204.37
Indium	In	49	114.82	Thorium	Th	90	232.038
Iodine	I	53	126.9044	Thulium	Tm	69	168.934
Iridium	Ir	77	192.2	Tin	Sn	50	118.69
Iron	Fe	26	55.847^b	Titanium	Ti	22	47.90
Krypton	Kr	36	83.80	Tungsten	W	74	183.85
Lanthanum	La	57	138.91	Uranium	U	92	238.03
Lawrencium*	Lr	103	257	Vanadium	V	23	50.942
Lead	Pb	82	207.19	Xenon	Xe	54	131.30
Lithium	Li	3	6.939	Ytterbium	Yb	70	173.04
Lutetium	Lu	71	174.97	Yttrium	Y	39	88.905
Magnesium	Mg	12	24.305	Zinc	Zn	30	65.37
Manganese	Mn	25	54.9380	Zirconium	Zr	40	91.22
Mendelevium*	Md	101	257.096				

Rutherfordium, Rf, 104 is also called kurchatovium, Ku

^{a} Elements with a relative atomic mass so designated are known to be variable because of natural variations in isotopic composition. The observed ranges are:

Boron	±0.003	Hydrogen	±0.00001	Silicon	±0.001
Carbon	±0.00005	Oxygen	±0.0001	Sulphur	±0.003

^{b} Elements with a relative atomic mass so designated are believed to have the following experimental uncertainties:

Bromine	±0.001	Copper	±0.001	Neon	±0.003
Chlorine	±0.001	Iron	±0.003	Silver	±0.001

[Table adapted from Comptes Rendues of the XXIV Conference of IUPAC, Prague, September 1967, with additions]
*Not found naturally on Earth

hydrogen by volume, concluding that with an appropriate allowance for experimental error, the ratio would become 100 : 200. This surprisingly simple result stimulated the intellectual curiosity of Gay–Lussac who investigated other gaseous combinations and derived his law. The same surprising simplicity is apparent in other gas laws: Boyle's Law, Charles' Law and Dalton's Law of Partial Pressures are obeyed by gases of widely diverse chemical nature.

In seeking the link between Gay–Lussac's Law of Combining Volumes, and the Atomic Theory of Dalton, the proposition that equal volumes of all gases under the same physical conditions contain equal numbers of atoms must be examined. Dalton conjectured on this idea and rejected it but his hostility to Gay–Lussac's work must also have been due to the conflicting results obtained by other experimenters with combining gases. Consider hydrogen and oxygen uniting above 100°C and at constant pressure to form steam:

2 volumes hydrogen + 1 volume oxygen →
2 volumes steam

which will reduce to

2 atoms hydrogen + 1 atom oxygen →
2 'compound atoms' of steam

The 'compound atoms' of steam would be identical and each would consist of 1 atom of hydrogen and $\frac{1}{2}$ atom of oxygen. But the atom is indivisible by chemical agency and the Atomic Theory was strongly supported. The hypothesis must be discarded. This difficulty always arises where the volume of the product exceeds that of either of the reactants.

In 1811, Avogadro supplied the necessary modification to the idea of ultimate particles of matter. In addition to the smallest chemical entity he postulated the molecule as the smallest particle of a gas with independent existence. Molecules of gaseous elements consist of one or more identical atoms while molecules of compounds consist of clusters of different atoms united. **Avogadro's Hypothesis** may be stated:

Under the same conditions of temperature and pressure, equal volumes of all gases contain the same number of molecules.

A **molecule** is the smallest particle of a gaseous element or compound which can exist by itself in the free state.

An **atom** is the smallest indivisible particle of an element which can take part in a chemical change.

Nearly 50 years elapsed before Avogadro's Hypothesis was accepted finally and its value understood. This is why the term *hypothesis* is retained here, being preferred to *theory*, which is becoming fashionable. Relatively few gaseous compounds were known until the rapid development of Organic Chemistry supplied many more for testing the hypothesis and attention was directed towards the consolidation of the Atomic Theory. The molecule seemed even more abstract than the imaginary atom.

Avogadro died in 1856. In that year Cannizzaro wrote: *A sketch of a Course of Theoretical Chemistry held at the Royal University of Genoa*, in which he showed that the atomic and molecular theories were not incompatible. In 1860, at Karlsruhe, an International Congress met to attempt the preparation of agreed tables of formulae and atomic masses for universal use. Nothing was achieved apparently but Cannizzaro's paper was circulated privately. Its value was recognized immediately and provided methods for the determination of relative molecular masses and relative atomic masses using densities of volatile elements and compounds. The molecular mass of an element or compound is naturally based on the atomic mass scale, ^{12}C† = 12.000 and refers to the molecular state of the element or compound. The term molar mass is also used, with the unit $g\,mol^{-1}$ of the species quoted.

Avogadro's Hypothesis enables a coherent body of chemical theory to be constructed. Two aspects of this will be briefly examined.

The atomicity of hydrogen may be deduced. Suppose a large number of gaseous compounds of hydrogen are prepared and analysed. Their densities ($g\,dm^{-3} = g\,l^{-1}$, here) under fixed conditions of temperature and pressure are measured and the percentage by mass of hydrogen in each determined. The mass of combined hydrogen in one dm^3 of each of the gases is now calculated. From a series of gaseous compounds, the mass of hydrogen is found to be related simply to the mass of one dm^3 of pure hydrogen under these conditions: each is either one half of the density of hydrogen or a simple multiple of one half of the density. Assuming Avogadro's Hypothesis, equal numbers of

† *spoken:* carbon-12

molecules under the given conditions are being considered.

It is reasonable to suppose, though not at all certain, that the minimum mass of hydrogen recorded corresponds to one hydrogen atom per molecule. Thus, three times the minimum denotes three atoms of hydrogen. The more compounds that are examined, the more probable it will be that the minimum value mentioned corresponds to one hydrogen atom. But the minimum is one half the mass of the volume of pure hydrogen equal to the volume of the gaseous compound under the given conditions. Therefore, if Avogadro's Hypothesis is true, the hydrogen molecule is diatomic, H_2. This method may be applied to determine the atomicity of any gaseous element forming a range of volatile compounds.

Returning to the original problem raised by Gay–Lussac's Law of Combining Volumes, Avogadro's Hypothesis may be applied to the volumes recorded when gases react to form gaseous products. Hydrogen and chlorine may be sparked to form hydrogen chloride:

1 volume hydrogen + 1 volume chlorine →
2 volumes hydrogen chloride

Assuming Avogadro's Hypothesis, 2 volumes of hydrogen chloride contain twice as many molecules as 1 volume of hydrogen and twice as many as 1 volume of chlorine. It must be remembered that the volumes here do not represent the actual volumes of the molecules but the space in which they move: the ease of compression of a gas makes this point clear. Hydrogen and chlorine may be shown to be diatomic, and by application of Avogadro's Hypothesis to the experimental measurements:

1 molecule hydrogen + 1 molecule chlorine →
2 molecules hydrogen chloride

∴ $H_2 + Cl_2 →$ 2 (hydrogen chloride), i.e. 2HCl

Therefore, hydrogen chloride is HCl. Because no volume change occurs, one difficulty is not made clear by this example. Consider the synthesis of steam where a reduction in volume occurs:

2 volumes hydrogen + 1 volume oxygen →
2 volumes steam

As before,

$2H_2 + O_2 →$ 2 molecules steam, i.e. $2H_2O$

Therefore, steam (water) has the formula H_2O. The loss in volume is due to the regrouping of atoms bringing increasing complexity to the molecule.

Now, the molecular mass of an element or compound is the mass of one molecule of that element or compound relative to the atomic mass standard, $^{12}C = 12.0000$, but from our viewpoint not significantly different from the value based on O = 16.0000.

By experiment, there are 6.023×10^{23} molecules of an element or compound in the gram molecule (mole) or the quantity in grams represented by the relative molecular mass, the molar mass. This is a fundamental physical quantity called the **Avogadro constant** (formerly Avogadro's Number).

Avogadro constant
$$= (6.022\ 52 \pm 0.000\ 28) \times 10^{23}\ mol^{-1}$$

The term gram molecule has now been superseded by the *IUPAC* term, mole.

The mole is the amount of substance which contains as many elementary units as there are atoms in 0.012 kilogramme of carbon-12. The elementary unit must be specified and may be an atom, a molecule, an ion, a radical, an electron, a photon, etc., or a specified group of such entities.

One mole of oxygen has a mass equal to 32.00 g and the density of oxygen at STP is found by experiment to be $1.429\ g\ dm^{-3}$. The volume of one mole of oxygen is therefore $32.00/1.429\ dm^{-3}$, i.e. $22.4\ dm^3$ at STP. This volume of all gases at STP will contain the same number of molecules, which equals the number in one mole. It is called the **Molar Volume**†.

For all gases at 0°C (273 K) and 760 mmHg pressure ($101\ kNm^{-2}$, 101 kPa), i.e. at STP the molar volume is $22.4\ dm^3$.

Further evidence for the existence of molecules will not be given here. Enough has been outlined to establish the classical reasoning leading to the acceptance of atoms as chemical units, and molecules as the physical units of gases. The number of atoms in a molecule is termed the atomicity. Although liquids and solids may also consist of molecules of the type envisaged in Avogadro's Hypothesis the

† Formerly known as the gram molecular volume

uncritical transfer of what has been well-established for gases to the other states of matter, must be avoided.

Following the dissemination of Cannizzaro's ideas on atomic and molecular mass, two men returned home to write books during the following decade. Both came to far-reaching conclusions regarding the classification of elements based on relative atomic mass. They were Mendeléeff, going home to Russia, and Lothar Meyer, to Germany.

To bring the developments up to date, attention must be drawn to the term **amount of substance**; given the symbol, n. It is one of seven independent basic physical quantities chosen by international agreement.† The amount of substance is proportional to the number of specified particles of that substance, the proportionality constant universally being the Avogadro constant, L. Because it is basic, its dimension is simply the amount of substance. The definition has nothing to do with any choice of unit; it is wrong to call n the number of moles just as it is wrong to call m the number of kilogrammes. The amount of substance is not the same as its mass although for any single substance it will be proportional to it. Thus, for chemists equal amounts of various substances are those quantities which contain the same number of specified particles (e.g. atoms, molecules, ions, radicals) and to compare equal amounts of different substances it is necessary to weigh out, not equal masses, but masses of the substances in the ratio of the masses of the specified elementary particles present. Much of this will have been realised already in the discussion of relative atomic mass.

Amount of substance × Avogadro constant
= number of specified particles

Amount of substance
$$= \frac{\text{number of specified particles}}{\text{Avogadro constant}}$$

Since the unit for the Avogadro constant is mol^{-1} (the mole has been arbitrarily related to a certain mass of carbon-12), and numbers are dimensionless, the amount of substance is therefore expressed here in the unit, mole.

The Kinetic Theory

The discrete particles of which matter consists are in a perpetual state of motion. The notion that above absolute zero, the molecules of a gas display incessant random motion is the basis of the Kinetic Theory. By adopting certain assumptions it is possible to devise a simple model from which very good predictions of the behaviour of (ideal) gases may be made. A simple mental picture of the structure of matter emerges which may be used to explain the mechanism of chemical reactions. A chemical reaction is essentially the rearrangement of atoms and the accompanying energy change.

The most obvious differences between the three states of matter concern their shape and the influence of pressure and temperature changes upon it. A gas has no definite shape, filling the container, being readily compressed by increase of pressure and expanding rapidly on heating. Liquids merely fill the container into which they are poured but there is now a surface between liquid and atmosphere. Solids have a definite shape. Liquids and solids are affected only slightly by altering conditions of temperature and pressure. There are no simple universal laws similar to those for gases.

Because a gas is highly compressible it is assumed to consist of a very large number of extremely small particles (molecules), relatively far apart. Further, it is assumed that the molecules are spherical in shape and move in straight lines between collisions with each other or the walls of the container. The molecules are without influence on each other except during impact. The collisions undergone by molecules account for the pressure exerted by the gas. Motion is random and therefore molecular velocities are spread according to the normal distribution curve. A rise in temperature is associated with increased mean kinetic energy and mean molecular velocity. Collisions are assumed to be perfectly elastic so that no energy is lost; otherwise, both pressure and temperature would fall. The diameter of a molecule is approximately 2×10^{-10} m and calculation shows that the average distance travelled between collisions may be 2×10^{-7} m. If the number of molecules contained in a given volume of gas is increased, the rate of collision is increased, bringing an increased rate of change of momentum and a rise in pressure. Reducing the volume of a given mass of gas at constant temperature has the same effect. Heating the gas increases the kinetic energy of the molecules which move faster and have a higher collision rate with a consequent rise in pressure. From the simple kinetic picture, **Boyle's Law**, discovered experimentally, may be derived:

† The others are length, mass, time, electric current, thermodynamic temperature and luminous intensity

The volume of a given mass of gas at constant temperature is inversely proportional to the pressure.

Using Charles' Law, it may be shown that the absolute temperature is proportional to the mean kinetic energy per molecule of the gas.

However, real gases do not obey the gas laws exactly. For gases which are not easily liquefied, the discrepancies are 5% or less below 10 atmospheres pressure. Contrary to our assumption, the volume of molecules may not be neglected in comparison with the volume occupied by the whole gas when pressure becomes great. Further, sufficient lowering of the temperature brings molecules closer together with a resulting increase in their mutual attraction. Liquefaction may occur as the result of one or both of these factors under the proper conditions. For complex molecules, which in this context means all but monatomic molecules, collisions are not perfectly elastic: upon collision, rotational and vibrational energy changes may occur. The principle of the conservation of energy will apply to the sum total of energies of translation, rotation and vibration. In the liquid state, there is translational motion but to a much lesser extent than in the gaseous state. For solids, the particles are in a condition of vibration, each about a mean position. Increase of temperature leads to an increase in the frequency of vibration and eventually to a breakdown of the structure: fusion has taken place. Evaporation occurs from the surface of both liquids and solids to an extent depending on their natures. Again, the loss of particles with higher kinetic energies and the consequent cooling (latent heat) effect is easily pictured in terms of the Kinetic Theory.

A kinetic view of chemical reactions must take into account energy changes. Collision between molecules must occur for a reaction between gases to take place. It has been shown that for collisions to be effective, molecules must have the necessary Activation Energy for the rearrangement of atoms to occur. The proportion of activated molecules increases rapidly with rise in temperature although the rate of collision rises only comparatively slowly. For simple reactions the predicted increase in the reaction rate of gases has agreed with experiment. Generally for the same initial concentrations of reactants, the reaction rate is approximately doubled for a rise in temperature of 10 K. At a given temperature, the proportion of activated molecules will be constant and the reaction rate will be proportional to the molecular concentration of the reactants for ideal gases. This is the **Law of Mass Action**.

At a constant temperature the rate of a homogeneous reaction at a given instant is proportional to the active masses of the reactants at that instant.

This law was stated by Guldberg and Waage in 1864–7. For homogeneous processes, active mass may be equated to molecular concentration or to partial pressure.

Electrolytes, ions and electrons

In 1800, Volta generated electricity by means of a pile of zinc and copper plates, arranged alternately and each separated from the next by a felt pad moistened with brine. This arrangement, known as a voltaic pile, produced relatively small electric currents but considerable progess resulted from the discovery. Nicholson and Carlisle decomposed water into its elements in 1800 and Davy isolated various metals from 1807 onwards by electrical decomposition of fused hydroxides and salts.

When a difference of electrical potential is applied to metals, alloys or graphite, an electric current passes but no chemical effect is seen. However, certain compounds when fused or dissolved in a suitable solvent, usually water, will conduct electricity and chemical action is observed, localized at the poles, or electrodes of metal or carbon, by which the electric current enters and leaves the liquid. Such compounds are alkalis, salts and in water, acids; they are called electrolytes and the process of conduction and chemical action is called electrolysis. Electrolytes may be classed as strong or weak according to their conducting characteristics. Pure water is a very poor electrical conductor and is therefore a very weak electrolyte. Mineral acids, alkalis and most salts are strong electrolytes. Among inorganic compounds, ammonia, carbonic acid, hydrogen sulphide, lead(II) acetate and mercury(II) chloride are weak electrolytes. Compounds which do not conduct electricity are called non-electrolytes; such is carbon tetrachloride. Sometimes, conduction occurs because of interaction with the solvent: phosphorus trichloride is hydrolysed, i.e. reacts with water, and electrolytes are formed:

$$PCl_3 + 3H_2O \rightarrow H_3PO_3 + 3HCl$$

A detailed study of electrolysis and theories of electrical conduction will not be attempted although certain features are relevant here.

In 1803, Berzelius and Hisinger demonstrated that non-metals, except hydrogen, were liberated at the positive electrode during the electrolysis of solutions while hydrogen and metals appeared at the negative electrode. During electrolysis the products may originate from electrolyte, solvent or electrode. The terminology was devised by Faraday. The positive pole is called the anode and the negative, the cathode. He called the particles of electrolyte, ions; those moving to the cathode are cations and those going to the anode, anions. Faraday investigated the chemical effects of an electric current and stated two laws, the first in 1832 and the second a year later. **Faraday's Laws** are:

The mass of any product liberated in electrolysis is directly proportional to the quantity of electricity which has passed.

When the same quantity of electricity is passed through a number of electrolytes in series, the masses of products liberated are in the ratio of their equivalent masses.

The gram equivalent mass of an element is liberated by $(9.64870 \pm 0.00016) \times 10^4$ C mol^{-1}, usually taken to 3 significant figures as 96 500 coulombs per mole and called 1 Faraday. In 1834, Faraday showed that his laws applied to fused electrolytes. Fundamental studies of the electrical conductance of electrolytes were accomplished by Hittorf between 1853 and 1859, and by Kohlrausch, who had established by 1876 that each ion contributes a definite amount to the total conductance in dilute solution (infinite dilution) irrespective of the nature of the other ion. Clausius (1857) rejected the earlier theory of Grotthus that 'molecules' of electrolytes were orientated in an electric field and split by the electrical forces of attraction when adjacent to electrodes because this mechanism would require large electrical forces which in fact were not present. Because electrolytes obey Ohm's Law, Clausius suggested that the 'molecules' undergo slight dissociation into charged atoms, or ions:

$$NaCl \rightleftharpoons Na^+ + Cl^-$$

Molecular collisions caused dissociation and recombinations. Ions discharged in electrolysis would be replaced by further dissociation. Van't Hoff showed that electrolytes deviated widely from his osmotic pressure equation. This was explained by the Ionic Theory of Arrhenius (1883–7) in which he proposed that strong electrolytes are largely dissociated and weak electrolytes, feebly dissociated into ions. We now believe that strong electrolytes, which will conduct electricity in the fused state, are completely ionized.

From Faraday's Laws, the quantity of electricity associated with one gram equivalent is a fixed quantity, conveniently called the Faraday (F). If ions are electrically charged atoms, the charge associated with a gram ion will be nF where n is the valency exerted by the element (or group). Dividing by the Avogadro constant, L, the number of atoms (etc.) in the gram atom (etc.), the charge associated with one atom is $n(F/L)$ coulomb. A univalent ion would carry (F/L) coulomb, a bivalent ion, $2(L/n)$ coulomb and a tervalent ion, $3(F/L)$ coulomb. (F/L) was seen by Johnstone Stoney (1874) and Helmholtz (1881) as a fundamental 'atom' of electricity. For this, Stoney proposed the term 'electron' in 1891. In 1897, J. J. Thomson performed quantitative experiments on the 'cathode rays' emanating from the cathode during the passage of an electrical discharge through gases at reduced pressure and calculated the ratio of charge : mass. The constant value under differing conditions, and the results of other experiments, lead to the conclusion that the electron is a fundamental unit of electricity. The electronic charge = 1.60×10^{-19} coulomb.

2

The properties of elements and the structure of their atoms

The method of study

In any branch of the natural sciences, classification of facts collected by accurate observations is very important because regularities and trends in properties may be seen. Then, by induction, general statements or laws summarizing this information may be propounded. Classification is an essential stage in the development of theories. The simple (or apparently simple) division of elements into metals and non-metals is useful but not very productive: the boundary between the two groups is diffuse and some elements are neither markedly metals nor non-metals and are called metalloids. The system which is used to classify the elements must group together elements which are similar in as many physical and chemical properties as possible.

Classification of the elements in the 19th century

Dalton, Berzelius and Döbereiner

In the first decade of the 19th century, Dalton published his Atomic Theory with its concept of the regrouping of small whole numbers of atoms of elements reacting. The atoms, as their name suggests, were indivisible and of masses characteristic of the elements involved.

The most valuable experimental work done in establishing the Atomic Theory was by Berzelius. He determined equivalent masses and, having deduced the composition by atoms of many common compounds, he was able to publish the first table of accurate relative atomic masses in 1826; these were based on oxygen as the standard with $O = 100$. The values were often remarkably accurate. A knowledge of the chemical properties of many substances was also being gathered.

In 1817 and in a detailed paper published in 1829, Döbereiner was the first to group similar elements together and link this with their atomic masses. Here are some of the analogies which he noted: calcium, strontium and barium form a sequence of elements with similar properties and strontium ($= 88$) has an atomic mass which is the arithmetical mean of those of calcium (40) and barium (137). Chlorine (35), bromine (80), iodine (127); and sulphur (32), selenium (79), tellurium (128) are two more sequences. He stressed that similar arithmetical relationships, such as between the atomic masses of carbon (12), nitrogen (14) and oxygen (16), were meaningless without chemical similarities. However, some elements which are similar do not form these sequences, or 'triads', and Döbereiner suspended judgment here until further data were available. Such a group is manganese (55), iron (56), cobalt (59), nickel (59), copper (64) and zinc (65), which are elements with nearly the same atomic masses.

Cannizzaro, de Chancourtois and Newlands

At the Congress of Karlsruhe in 1860, the paper written by Cannizzaro linking Avogadro's Hypothesis and the Atomic Theory of Dalton was circulated, with the result that values of relative atomic masses could be calculated with certainty when the corresponding equivalents had been determined accurately. Much of this was done by Stas. The way was now open for the complete classification of the elements in terms of relative atomic masses.

The first real attempt to correlate chemical properties and atomic mass was made by de Chancourtois, who plotted atomic masses in a 45° spiral or helix on the surface of an upright cylinder. One turn round the cylinder was 16 atomic mass units, the value of the atomic mass of oxygen, and each value was located by the vertical distance from the

Table 2.1 Classification of the elements: Newlands (1863–6)

Order of increasing atomic mass	H Li Be B C N O	F Na Mg Al Si P S	Cl K Ca Cr Ti Mn Fe	Co, Ni Cu Zn Y In As Sc	Br Rb Sr etc.	Horizontal similarities

Table 2.2 Mendeléeff (1869): Periodic Table

				Ti = 50	Zr = 90	? = 180
				V = 51	Nb = 94	Ta = 182
				Cr = 52	Mo = 96	W = 186
				Mn = 55	Rh = 104.4	Pt = 197.4
				Fe = 56	Ru = 104.4	Ir = 198
			Ni = Co = 59	Pd = 106.6	Os = 199
H = 1				Cu = 63.4	Ag = 108	Hg = 200
	Be = 9.4	Mg = 24	Zn = 65.2	Cd = 112	
	B = 11	Al = 27.4	? = 68	Ur = 116	Au = 197?
	C = 12	Si = 28	? = 70	Sn = 118	
	N = 14	P = 31	As = 75	Sb = 122	Bi = 210?
	O = 16	S = 32	Se = 79.4	Te = 128?	
	F = 19	Cl = 35.5	Br = 80	I = 127	
Li = 7	Na = 23	K = 39	Rb = 85.4	Cs = 133	Tl = 204
		Ca = 40	Sr = 87.6	Ba = 137	Pb = 207
		? = 45	Ce = 92		
		?Er = 56	La = 94		
		?Yt = 60	Di = 95		
		In = 75.6	Th = 118?		

base and rotation of the cylinder through the same distance. Similar elements were in the same vertical straight lines and were separated by 16 units of atomic mass.

Another system was developed between about 1863 and 1866 by Newlands. Listing elements in increasing order of the (at that time) new relative atomic masses, he noted that similar elements were separated by intervals of eight elements. He called this observation the 'Law of Octaves': it was received with ridicule. Comparison, remembering that the inert gases had not been discovered at the time, shows that his table bears a striking similarity to the Periodic Classification although the order of the elements was disorganized by some incorrect values of atomic masses and by undiscovered elements. Part of the scheme advanced by Newlands is shown in Table 2.1.

Mendeléeff and Lothar Meyer

Attending the Congress of Karlsruhe in 1860 and influenced by the ideas of Cannizzaro were two men who would shortly write textbooks requiring the collation of the physical and chemical properties of elements and a detailed examination of atomic mass values: both were struck by the periodicity of properties with atomic mass. Neither seemed aware of the work of others in this field.

Lothar Meyer was primarily interested in the relationships obtained by plotting the magnitude of physical properties against atomic masses. We are chiefly indebted to him for the atomic volume curve of which a modern form, substantially the same as the original, is shown in Fig. 3.1 (p. 31). The periodic value of the property is clearly displayed.

The **atomic volume** of a solid element is the volume of the molar mass.

$$\text{Atomic volume} = \frac{\text{molar mass}}{\text{density}}$$

The unit for atomic volume is $cm^3\,mol^{-1}$. Many other physical properties such as melting-point, boiling-point, coefficient of thermal expansion, and malleability, to name but a few, show similar types of curve. From these studies Meyer drew up a periodic table.

In 1869, just prior to the publication of Meyer's conclusions, Mendeléeff announced his system of

Table 2.3   The modern/short (Mendeléeff) Periodic Table

Period	Series	I a	I b	II a	II b	III a	III b	IV a	IV b	V a	V b	VI a	VI b	VII a	VII b	VIII	VIII	VIII	0
1	(1)	– H 1.00797																	1 He 4.0026
2	(2)	3 Li 6.939		4 Be 9.0122			5 B 10.811		6 C 12.0112		7 N 14.0067		8 O 15.9994		9 F 18.9984				10 Ne 20.179
3	(3)	11 Na 22.9898		12 Mg 24.305			13 Al 26.9815		14 Si 28.086		15 P 30.9738		16 S 32.064		17 Cl 35.453				18 Ar 39.948
4	(4)	19 K 39.102		20 Ca 40.08		21 Sc 44.956		22 Ti 47.90		23 V 50.942		24 Cr 51.996		25 Mn 54.9380			26 Fe 55.847	27 Co 58.9332	28 Ni 58.71	
	(5)		29 Cu 63.546		30 Zn 65.37		31 Ga 69.72		32 Ge 72.59		33 As 74.9216		34 Se 78.96		35 Br 79.904				36 Kr 83.80
5	(6)	37 Rb 85.47		38 Sr 87.62		39 Y 88.905		40 Zr 91.22		41 Nb 92.906		42 Mo 95.94		43 Tc 98.906			44 Ru 101.07	45 Rh 102.905	46 Pd 106.4	
	(7)		47 Ag 107.868		48 Cd 112.40		49 In 114.82		50 Sn 118.69		51 Sb 121.75		52 Te 127.60		53 I 126.904				54 Xe 131.30
6	(8)	55 Cs 132.905		56 Ba 137.34		57 La 138.91 LANTHANIDES		72 Hf 178.49		73 Ta 180.948		74 W 183.85		75 Re 186.2		76 Os 190.2	77 Ir 192.2	78 Pt 195.09	
	(9)		79 Au 196.967		80 Hg 200.59		81 Tl 204.37		82 Pb 207.19		83 Bi 208.98		84 Po 210.000		85 At 209.987				86 Rn 222.018
7	(10)	87 Fr 223.020		88 Ra 226.025		89 Ac 227.028 ACTINIDES		104 Rf* (260)		105 Ha –		(106)		(107)		(108)	(109)	(110)	

Group

Note: The Lanthanides and Actinides are not shown. The long form of the complete Periodic Table is shown as Table 3.5, p. 35.

*Rutherfordium, Rf is also known as kurchatovium, Ku.

classification. Basing his reasoning on chemical characteristics he listed elements in order of atomic masses and showed the periodicity of properties. By making predictions about the existence of hitherto undiscovered elements, which were confirmed within a few years, he quickly got his ideas accepted. His observations are summarized in the next section.

Mendeléeff's Periodic Law

The properties of the elements are a periodic function of their atomic masses.

Mendeléeff published his first Periodic Table of 63 elements in 1869. This is shown in Table 2.2. An improved version was published in 1872. The periodic system of 1872 brought up to date by the inclusion of elements undiscovered at that time is shown in Table 2.3. It is called the *short form* of the Periodic Table. In the table illustrated, each element has been assigned a number called the *atomic number* which describes the position of the element in the series of increasing atomic masses. The horizontal rows are called *periods* and vary in the number of elements included. In the Mendeléeff Table, one row of elements was called a *series* so that the long periods contained two *series*. These have been retained in the table shown. The first three periods are termed *short periods* and the others, *long periods*. The vertical columns are called *groups* and are divided into *subgroups*. Elements appearing directly underneath each other are closely similar, except for hydrogen which is unique. Complications which arise in the use of this table may be resolved by the use of the *long form* described later.

In reaching his general conclusion which we have called the **Periodic Law**, Mendeléeff noted that similar elements had either almost the same atomic masses, e.g. platinum (195), iridium (192), osmium (190), or values which showed a uniform increase, e.g. potassium (39), rubidium (85), caesium (133), the modern values having been given here, to the nearest integer. Mendeléeff stressed that atomic mass determines the character of an element and the arrangement of groups brings out the differences in chemical nature and valency along each series. The elements with light atoms were found to be generally widely distributed and to have clearly defined properties. He called these *the typical elements*. They are located in the three short periods. Hydrogen was recognized as unique, or 'typical of itself'.

The group similarities are so striking that it tends to be forgotten that Mendeléeff also stressed the similarity of some adjacent elements. Group VIII contains three sequences of triads to which Mendeléeff gave the name transitional or *transition elements*. This term has been extended to cover the

Table 2.4 Germanium: a comparison of properties predicted by Mendeléeff with those discovered by Winkler

	Eka-silicon (Es) Prediction: Mendeléeff 1871	Germanium (Ge) Discovery: Winkler 1886
1	Relative atomic mass 72	Relative atomic mass 72.6
2	Density 5.5 g cm^{-3}	Density 5.47 g cm^{-3}
3	Atomic volume 13 cm^3 mol^{-1}	Atomic volume 13.2 cm^3 mol^{-1}
4	Dark grey element which will give a white powder EsO$_2$ on calcination	Greyish-white metal which on calcination gives a white powder, GeO$_2$
5	Metal will decompose steam with difficulty	Metal does not react with water
6	Resistant to acids, less so to alkalis	Metal not attacked by HCl but dissolves in aqua regia. Molten, but not aqueous, KOH reacts
7	Element will be isolated by the action of sodium on EsO$_2$ or K$_2$EsF$_6$	Isolated by reduction of GeO$_2$ with C or of K$_2$GeF$_6$ with Na
8	Oxide EsO$_2$ will be refractory, specific gravity 4.7, with properties less basic than those of TiO$_2$ or SnO$_2$, but more marked than those of SiO$_2$	GeO$_2$ refractory oxide, specific gravity 4.703 and very feebly basic although indications of oxosalts found
9	Hydroxide will react with acids but solutions will be hydrolysed readily to a precipitate of meta-hydroxide	Acids do not precipitate hydroxide from dilute alkaline solutions but from concentrated solutions, acids precipitate GeO$_2$ or meta-hydroxide
10	Chloride EsCl$_4$ will be a liquid, b.p. less than 100°C, specific gravity 1.9 at 0°C	GeCl$_4$ is a liquid, b.p. 86.5°C, specific gravity 1.887 at 18°C
11	Fluoride EsF$_4$ will not be gaseous	GeF$_4$3H$_2$O is a white crystalline solid
12	Organo-metallic compounds will be formed e.g. Es(C$_2$H$_5$)$_4$, b.p. 160°C, specific gravity 0.96	Ge(C$_2$H$_5$)$_4$, b.p. 160°C, specific gravity just less than that of water

sequence scandium (21) to nickel (28) in period 4 (and sometimes copper and zinc) and the corresponding elements of the following periods. Lanthanides and actinides are discussed later.

Mendeléeff left gaps in his table for elements which were at that time undiscovered where otherwise his group similarities would be spoilt. From a study of adjacent elements and their compounds he was able to make predictions which proved amazingly accurate. Eka-aluminium (Sanskrit: *eka* = one) was discovered in 1874 (gallium), eka-boron in 1879 (scandium) and eka-silicon in 1885 (germanium). Table 2.4 shows a comparison of the properties of germanium as predicted by Mendeléeff and determined by Winkler, who isolated the element. Of recent years, Seaborg and his co-workers have synthesized transuranic elements by nuclear reactions and they claim that the prediction of the properties of these elements is very straightforward.

In the original Mendeléeff Periodic Table, Group O was missing because none of these elements was known. When the work of Rayleigh and Ramsay resulted in the discovery of a gas, which was chemically inert and quite unlike any known element, and which was therefore called argon (Greek: *argos* = inert or idle), other members of the family were immediately sought and found.

For some elements the positions allotted in the table and those apparently indicated by the study of chemical properties were not in agreement using the values of the atomic masses assigned to the elements at that time. Mendeléeff considered that the wrong multiple of the equivalent had been used in fixing the atomic masses of beryllium, indium and uranium. He adjusted the values so that the elements fitted into his scheme and subsequent experiments duly confirmed his predictions.

Minor adjustments of atomic masses have also been made with success but a major weakness of the Mendeléeff classification is made manifest in this connection. In the modern form of the Mendeléeff Table it will be seen that four pairs of elements, positioned by their chemical characteristics, are in the reverse order of atomic mass. These elements with atomic numbers and atomic masses are listed in Table 2.5. The root of the problem lies in the fact that the atomic mass of an element is not a fundamental quantity by which to classify the element. The difficulty may only be resolved by the study of atomic structure when a table of rather different shape will be preferred. In addition to the lack of emphasis on the unique nature of hydrogen,

some elements in the short form of the Periodic Table appear in positions which are at variance with their chemical properties. Because of the double series in the long period, transition metals in the wider sense, such as copper, silver and gold (in this book called 'associated with transition elements'), are placed alongside the metals of Group I, potassium, rubidium and caesium, to which they bear no resemblance. The original transitional elements of Mendeléeff were placed in triads in Group VIII. The term 'transition element' will be defined in terms of atomic structure later. The rare earth elements, all very similar chemically, and now usually called the lanthanides, proved even more difficult to place on the basis of atomic masses and are seen to be confined to one place in the Mendeléeff Short Table.

Table 2.5   Atomic mass anomalies in Mendeléeff's classification

Element	Symbol	Atomic number	Atomic mass
Argon	Ar	18	39.948
Potassium	K	19	39.102
Cobalt	Co	27	58.9332
Nickel	Ni	28	58.71
Tellurium	Te	52	127.60
Iodine	I	53	126.9044
Thorium	Th	90	232.038
Protactinium	Pa	91	231.036

We will now review the main discoveries and theories which have resulted in a knowledge of atomic structure and finally the classification of elements by the electronic configuration of their atoms.

The divisible atom

Natural radioactivity

According to Dalton's Atomic Theory, the ultimate particle of matter was the atom, indivisible and characterized by its mass, all atoms of a given element being identical and different from atoms of other elements. It had been suggested by Prout in 1815–16 that the hydrogen atom was the fundamental unit from which all other atoms were made. While current values of atomic masses seemed to support this hypothesis, the more accurate determinations, made later in the 19th century, notably by Stas, led to its rejection. Experiments now show that the concept of indivisibility and uniqueness of mass of atoms must be modified. Yet the effectiveness of the Classical Atomic Theory in stressing the fundamental nature of the atom in chemical reactions has not been diminished with the greater understanding of atomic phenomena.

Atoms of certain elements undergo spontaneous disintegration with the emission of particles and rays. This is known as radioactivity and was discovered by Becquerel in 1896.

When cathode rays, which are high-speed electrons, collide with a solid object, X-rays are produced. These have characteristic properties—notably, they produce fogging of photographic plates wrapped in black paper to exclude ordinary light, excite fluorescence in certain compounds, render gases electrically conducting by causing ionization of their molecules, and pass through thin sheets of substances opaque to ordinary light. Investigating the connection between X-rays and fluorescence, Becquerel discovered that uranium and its compounds, without previous treatment of any kind, produce the same effects as X-rays. He concluded that the element uranium spontaneously emits radiation. The intensity of the radioactivity, which could be measured by the rate of discharge of a charged electroscope due to ionization of the contained air, depended only on the quantity of element, uranium, present. The radioactive decay of an element proceeds spontaneously and is independent of changes in the physical conditions. Temperature change has no effect on radioactive decay which is in marked contrast to the influence seen on the rates of ordinary chemical reactions.

Thorium was shown to be radioactive by Madame Curie and Schmidt working independently in 1898. In the same year, because the observed radioactivity of certain uranium minerals, notably pitchblende (U_3O_8), exceeded that calculated for its uranium content, Marie Curie with her husband, P. Curie, looked for other radioactive elements in these minerals. This led to the discovery of polonium and radium. During the next year (1899), Debierne, another colleague of Marie Curie, discovered actinium also in uranium minerals. In addition, many other elements have been discovered which exhibit radioactivity to some degree besides those cases of artificial radioactivity produced in laboratories and atomic piles.

The Becquerel rays are of three different kinds, two named by Rutherford in 1899, α (alpha) rays and β (beta) rays, and the third γ (gamma) rays, following the work of P. Curie (1900) and Villard (1900). The three types of ray may be separated by using a powerful magnetic field, when β-rays are deflected in accordance with their negative charge, α-rays in the opposite direction and to a much less extent, and γ-rays pass undeflected. Ramsay and Soddy in 1903 showed that helium is continuously produced during the decay of radium. During the period 1903–9, Rutherford and his co-workers proved that α-rays are positively charged helium atoms, He^{2+}, of mass 4 (on the relative atomic mass scale) and each bears a double positive charge, i.e. twice the magnitude of the charge of the electron but positive. β-rays have been shown to be fast-moving electrons and are also called β^--particles. γ-rays are similar to X-rays, but with wave-lengths in the range approximately 10^{-12} m to 10^{-10} m, just shorter than those for X-rays, endowing them with even greater penetrating power.

The Theory of Radioactive Disintegration was advanced by Rutherford and Soddy in 1902–3. The spontaneous disintegration of atoms with the emission of radiation results in the formation of atoms of new elements, which may also be radioactive and break down further. The rate of decay of radioactivity of each element will be characteristic of that element.

The rate of decay follows an exponential law which means that at any instant the rate of decay is proportional to the intensity of the activity at that instant. The intensity of the radioactivity will be reduced with time but will not disappear. The rate of decay, being independent of physical and chemical conditions, and being a characteristic property of the element undergoing radioactive changes, is usually expressed in terms of the 'half-life period' which is the time that elapses for the radioactivity to decay to one half of its value from a given instant. A high decay rate is associated with a short half-life and a slow rate with a long half-life. This is illustrated in Fig. 2.1.

Mathematically, such a (first-order) process may be expressed:

$$\text{Radioactivity, } A = \frac{-\,dN}{dt} = \lambda N$$

(where N = number, t = time, λ = decay constant)

$$\text{Re-arranging, } \frac{+\,dN}{N} = -\lambda\,dt$$

and integrating, $\ln(N_t/N_0) = -\lambda t$

(where N_0 = number at $t = 0$ and N_t at time, t)

$$\text{or } N_t = N_0 \, e^{-\lambda t}$$

and if $t_{1/2}$ represents the half-life period of the process†

$$\ln 1/2 = -\lambda t_{1/2}$$

$$-2 = -\lambda t_{1/2}$$

$$t_{1/2} = \frac{\ln 2}{\lambda} = \frac{0.693}{\lambda}$$

In calculations, N_t represents the number of atoms which have not disintegrated at time, t. The unit for measuring

$$A\left(= \frac{dN}{dt} \right)$$

is $s^{-1}(= \text{Hz})$.

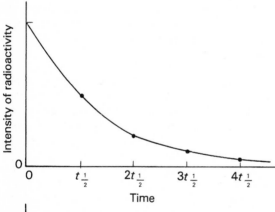

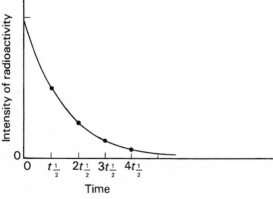

Fig. 2.1   Radioactive decay illustrated by examples in which one half-life is twice the other. For clarity, the initial radioactivity has been the same in each case

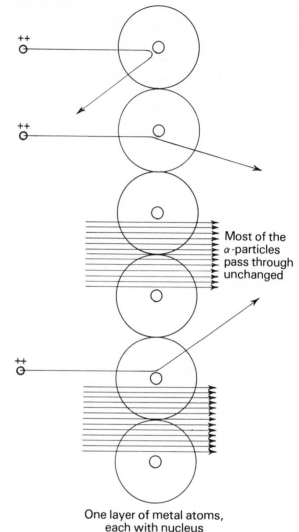

One layer of metal atoms, each with nucleus

Fig. 2.2   Rutherford's experiment on scattering of α-particles by atomic nuclei (schematic)

The atomic nucleus

When a stream or pencil of α-particles, travelling at high speed from a suitable source, such as radium, is directed towards a thin piece of metal foil, a limited number of particles are deflected through wide angles while the majority suffer no appreciable deflection at all. In the first experiment, copper, silver, gold and platinum of thickness about 0.0005 mm (5×10^{-7} m) were used and

† Remember that the basis for this is statistical, some atoms disintegrating immediately, others taking time, up to infinite time

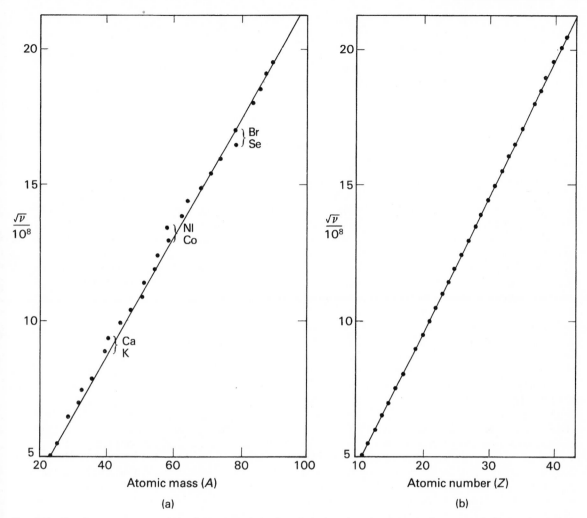

Fig. 2.3 Moseley: square root of the frequency of the $K\alpha_1$, line plotted against (a) atomic mass and (b) atomic number, showing the precise relationship with the latter quantity (units ignored)

the α-particles were detected by the scintillations of light which each produced on collision with a zinc sulphide screen. As an illustration, using gold foil 0.0004 mm (4×10^{-7} m) thick, the average deflection of the α-particles was less than 7°, but one α-particle in 20 000 was deflected through 90° or more.

The experiments on the scattering of α-particles were directed by Rutherford (1911). He concluded that the reflection of each α-particle was due to a single collision with an atom of metal. Because atoms contain (negative) electrons there must also be a positive charge to give electrical neutrality. That relatively few α-particles are deflected widely indicates that this charge and the greater part of the mass is concentrated in a small nucleus. Electrons

are much too light to cause deflections and most of the α-particles travel straight on, so it is assumed that the greater part of an atom consists of empty space. A picture emerges of a central, dense, small, positively charged nucleus surrounded at a relatively great distance by a sort of planetary system of electrons. Rutherford supposed that their high speed prevented electrons from spiralling into the nucleus but this is not so. A more accurate treatment requires the use of wave, or quantum, mechanics. The charge on the nucleus is about one half of the atomic mass. The scattering of α-particles is shown schematically in Fig. 2.2.

In 1913, Moseley bombarded various elements, or compounds where it proved difficult to use the

Table 2.6   A comparison of the three main sub-atomic particles

	Relative charge	Mass relative to that of the electron	Mass on the relative atomic mass scale	Mass as convenient approximation
proton	+1	1836	1.008	1
neutron	0	1839	1.009	1
electron	−1	1	0.00055	0

charge of electron = $(1.602\,10 \pm 0.000\,07) \times 10^{-19}$ C
mass of electron   = $(9.109\,1 \pm 0.000\,4) \times 10^{-31}$ kg

elements, with cathode rays in a discharge tube and showed that the resulting X-rays had characteristic frequencies. The square root of the frequency of a characteristic line in each spectrum was directly proportional to the atomic number of the element. Results are plotted in Fig. 2.3. The atomic number, with few exceptions, describes the position of an element when listed with the other elements in order of increasing atomic mass as in the original periodic classification. In 1913 van de Broek suggested the hypothesis that the atomic number equalled the nuclear charge as determined by scattering experiments. From his experiments, Moseley concluded that the atomic number was a fundamental quantity for an element and probably equalled the positive charge on the nucleus, the unit being the charge of an electron. There was no such precise relationship as indicated above when atomic mass was substituted for atomic number. This shows the decline in importance of atomic mass in element classification and the exceptions in the order based on atomic mass are shown in a new light. In 1920, Chadwick determined the nuclear charge of copper, silver and platinum directly using a refinement of the scattering technique and the results were in excellent agreement with the atomic numbers.

Now, the α-particle is a positively charged helium atom and helium has a higher atomic mass than hydrogen. Therefore, a positively charged particle which is smaller than the α-particle was expected. The hydrogen atom was thought to consist of a nucleus of unit positive charge and one electron. The nucleus of the hydrogen atom was called the **proton** (Greek: *protos* = first) by Rutherford in 1920 following his transmutation experiments of the previous year. He had produced these particles during the bombardment of nitrogen by α-particles. Protons could also be produced by the similar bombardment of other atoms and were assumed to be a constituent of all atoms. Protons were also detected in the positive rays of

the discharge tube. Hence, the atomic number of an element is equal to its proton number, the terms being interchangeable. Also in 1920, Rutherford postulated the existence of an uncharged particle, which he called the **neutron**, of the same mass as the proton and forming part of the nucleus. This particle was identified by Chadwick in 1932. A comparison of the electron, proton and neutron is shown in Table 2.6.

The structure of the atom

An atom consists of a nucleus, which is relatively small and dense, surrounded at relatively great distances by electrons, which are of negligible mass. The simplest atom, that of hydrogen, consists of one proton, which has a diameter of approximately 10^{-15} m and one electron moving at a distance around it. The diameter of the atom is approximately 10^{-10} m. More complex atoms have diameters of the same order with the diameter of the nuclei being between 10^{-15} m and 10^{-14} m. As the diameter of an atom is approximately 10^5 times the diameter of the nucleus, the atom is seen to contain mostly empty space. If the nucleus were magnified to 20 mm diameter, the electron would be at a distance of about one kilometre from it! The detailed pattern taken up by the electrons, upon which the properties of an element depend, will be described shortly. A simple picture rather like a miniature solar system in which the sun represents the nucleus and the planets are the electrons moving in orbits, is sometimes very helpful but not accurate. The atom should be visualized as a small, dense, positively charged core surrounded by a diffuse cloud of moving electrons, the particle nature of which cannot be distinguished in the picture while the density of negative charge represented by this cloud falls off with increasing distance from the nucleus. It is difficult to define an exact boundary for the atom.

The nucleus consists of protons and neutrons sometimes collectively known as *nucleons*, which account for nearly the whole of the mass of the atom. The nucleus has a positive charge due to the protons and in the neutral atom the number of external electrons is equal to the number of protons, i.e. the proton number or atomic number. Different elements may be successively constructed from the simplest atom, hydrogen, which consists of one proton and one electron, by adding one proton to the nucleus and one external electron at each step. Neutrons provide the ballast and make up the mass of the atom. For convenience, the atomic mass is sometimes rounded off to the nearest integer and called the Mass Number or Nucleon Number.

The Atomic Number (Z) of an element is equal to the following:

1   The value of the positive charge on the nucleus of an atom of the element, the unit of charge being the charge of one electron.
2   The order of the element in the Periodic Classification.
3   The number of electrons in a neutral atom of the element.

Z is also called the Proton Number.

The Mass Number (A) of an element is the relative atomic mass rounded off to the nearest whole number. A also represents the Nucleon Number.

An Element is a substance having the same number of protons in the nucleus of each atom of the element, the number being characteristic of the element.

By subtracting the proton number (Z) from the nucleon number (A), the number of neutrons (N) in the respective nucleus is obtained,

$$N = A - Z$$

We have seen that the atomic mass of an element is much less important than was formerly believed, but why are the values sometimes so far from being approximately integral as would be expected from structures composed of neutrons and protons, each of mass approximately unity? Why is the atomic mass of element No. 17 (chlorine) 35.453?

As early as 1910, Soddy suggested that the atoms of a given element could differ in mass. This came from the study of radioactive decay where successive elements are formed from uranium and thorium. The non-radioactive end-products are chemically identical with lead but of atomic masses 206 and 208 respectively while the atomic mass of lead is 207.1. The different forms of an element were called *isotopes* (Greek: *isos* and *topos* = equal place, i.e. in the Periodic Table). As a result of the preliminary experiments by J. J. Thomson and the later work of great precision by Aston, who had developed the mass spectrograph by 1919, it was clear that elements not associated with radioactive decay could also consist of atoms of different masses. Neon was shown to consist mainly of atoms of mass number 20 with some of mass number 22, and chlorine of mass numbers 35 and 37. The chemical atomic mass is seen to be an average for the various atoms making up the sample used. The number of naturally occurring isotopes of successive elements fluctuates from as many as ten for tin down to just one for iodine, sodium, arsenic and others.

Isotopes are forms of the same element which differ in relative atomic mass.

At one time it was thought that except for elements resulting from radioactive decay, the isotopic ratio for an element would be constant because of the thorough mixing of samples during the cosmic creation of this planet. Precise measurements show that while this is correct for some elements, there is a smaller isotopic variation for others. Chlorine and silicon show no variations but this is not the case for sulphur, boron, carbon, silicon, hydrogen, krypton, xenon or nitrogen, and a slight variation in the isotopic composition of atmospheric oxygen and oxygen combined in water has been demonstrated.

In 1918, Stewart pointed out that different elements may have isotopes of the same mass and he called these *isobares* or *isobars* (Greek: same weight). Thus, there are naturally occurring isotopes of argon, potassium and calcium all with mass number 40.

We seem to be very near the early hypothesis advanced by Prout because now that we believe the different isotopes of an element are composed of neutrons and protons, the mass of an atom should equal the total mass of the constituent particles. However, for a given isotope there is a small but significant difference in mass. This is ascribed to the *packing effect*. When isolated nucleons are assembled to form the atomic nucleus of an atom, the energy of formation comes from the mass change recorded. Einstein (1905) explained this equivalence of mass (m) and energy (E) in terms of the equation, $E = mc^2$, where c = velocity of light (3×10^8 m s^{-1}) so that a minute loss of mass results in an enormous release of energy because the c^2 term introduces a factor of 9×10^{16}. The effect is small and ($A - Z$) still gives the number of neutrons. This may be illustrated. Helium is virtually 100% $^4_2$He and may be built up with two hydrogen atoms, each containing 1 proton and 1 electron, and 2 neutrons:

Relative atomic mass calculated
$$= (2 \times 1.008) + (2 \times 1.009) = 4.034$$

Relative atomic mass by experiment = 4.003

Sodium is 100% $^{23}_{11}$Na and may be built up from 11 H atoms, each 1 proton and 1 electron, and 12 neutrons.

Calculated relative atomic mass
$$(11 \times 1.008) + (12 \times 1.009) = 23.196$$
Relative atomic mass by experiment = 22.990

In the symbols $^4_2$He and $^{23}_{11}$Na, the upper figure refers to the nucleon (mass) number (A) and the lower figure to the proton (atomic) number (Z) defined earlier. In Chapter 1, ^{16}O, ^{17}O, ^{18}O and ^{12}C were mentioned: the upper figure is sometimes placed after the symbol.

The determination of atomic masses by the mass spectrometer

Isotopic masses and the abundance ratio of the isotopes of elements may be determined, and hence the *weighted mean* unitless relative atomic mass calculated (or molar mass/g mol^{-1}), taking into account the proportions of the atoms of different mass present. A mass *spectograph* generally used for individual mass determinations, delivers a photographic record while a mass *spectrometer* incorporates a meter to measure ion current and is used to determine isotopic abundance ratios. Today mass spectrometry is concerned with the determination of relative molecular mass (molar mass/ g mol^{-1}) and the deduction of molecular structures, particularly of organic compounds. The principle of the mass spectrometer is described here.

Positive ions are commonly produced by electron bombardment,

$$M + {}^0_{-1}e^- \rightarrow M^+ + 2{}^0_{-1}e^-$$

but ultraviolet irradiation, sparking or field emission from a surface upon which the sample has been absorbed may be used. The element, or a suitable compound, is introduced as a gas under very low pressure (1×10^{-4} Pa (N m^{-2})) to avoid interference from other entities as far as possible. Molecules may ionize to ions with one or more charges, or simply disintegrate. The positive ions are accelerated by a powerful electric field, a narrow beam is selected and this enters the field of an electromagnetic analyser. The ions sweep round a semi-circular curved path, the radius of curvature of which is related to the magnitude of the accelerating voltage ($\propto \sqrt{E}$) and to the strength of the magnetic field ($\propto (1/B)$). The lighter ions are deflected most, and the heaviest, least. Paths are altered so as to bring successive beams through the slit of the collector. On neutralization of the positive ions, suitable electrical amplification produces a signal which can be recorded.

Suppose the accelerating field, E, acts on positive ions, mass m, to produce a velocity v and hence a kinetic energy of $\frac{1}{2}mv^2$ for an ion assumed to be of single charge, e.

$$Ee = \tfrac{1}{2}mv^2$$

Ions, equivalent to an electric current of ev arriving at the collector have experienced a magnetic field B which caused them to move in a curve of radius r.

$$Bev = \frac{mv^2}{r} \quad \text{i.e. } Be = \frac{mv}{r}$$

The radius is fixed during manufacture of the instrument, i.e. by elimination of v

$$\frac{m}{e} = \frac{B^2r^2}{2E}\dagger$$

Thus ions may be selectively collected according to their m/e ratio by altering E or B or both E and B. The isotopic composition of the gas neon could be investigated using the simple mass spectrometer shown in Fig. 2.4.

Neon has three stable isotopes of masses, 20, 21 and 22, of relative abundance 90.92, 0.257 and 8.82 respectively. Ignoring the least abundant isotope, pathway Y in Fig. 2.4 will be pursued by the heavier ^{22}Ne$^+$ ions of higher momentum, mv, and pathway X by the lighter ^{20}Ne$^+$ ions ($r \propto \sqrt{m/e}$ for given values of B and E and $e = 1$). Ions with a double charge would be deflected into a curve of smaller radius, since the charge is doubled in the previous expression. Decreasing the voltage applied in the accelerating field (between P and Q in Fig. 2.4)

$\dagger$ m/kg, e/C, r/m, E/Vm^{-1}, B/Vsm^{-2} or Wbm^{-2}

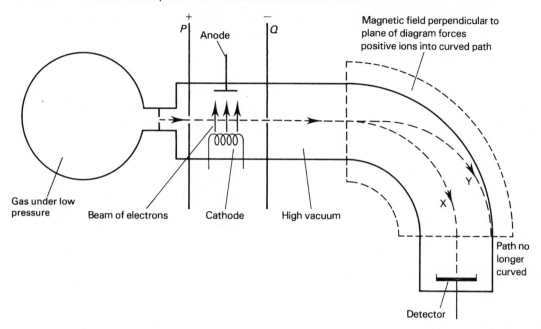

Acceleration of positive ions in electric field between P and Q

Fig. 2.4   Mass spectrometry

brings the heavier isotope into the collector ($r \propto \sqrt{E}$). On the chart record (Fig. 2.5), the height of each peak is proportional to the rate of collection of ions. If a photographic plate is used a series

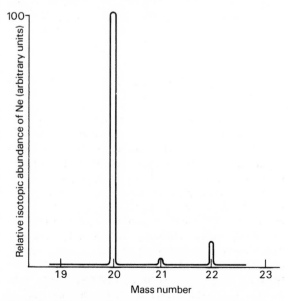

Fig. 2.5   Diagrammatic representation by mass spectrum of neon recorded by pen on chart

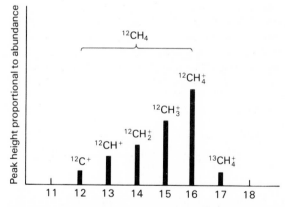

Fig. 2.6   Stick diagram for CH_4 showing ^{12}C species and a trace of $^{13}CH_4$ (as ion): not to scale

of lines is recorded, their positions depending on the radius of curvature of the pathway. By use of a known standard substance in the spectrometer, the recorded mass spectrum may be calibrated and ions identified.

Usually the spectrum is drawn as a *stick diagram* with the most abundant species given a height of 100 units (the *base peak*) and each successive peak height proportional to species abundance. Stick diagrams are used during the identification of ions formed during the breakdown of molecules of an organic compound whose structure is under investigation by mass spectrometry. A simple illustration shows the fragmentation pattern of methane, the mass spectrum in Fig. 2.6 not being drawn to scale. The stick diagram for tetrachloromethane (carbon tetrachloride, CCl_4) is much more complicated. Various peaks for $^{12}CCl_4^+$, $^{12}CCl_3^+$, $^{12}CCl_2^+$, $^{12}CCl^+$, Cl^+, $^{12}C^+$ might be expected but each of the species containing chlorine shows multiple peaks due to the isotopes ^{35}Cl and ^{37}Cl. Thus, there are four peaks corresponding to CCl_3^+

117	$^{35}Cl\ ^{35}Cl\ ^{35}Cl\ ^{12}C^+$
119	$^{35}Cl\ ^{35}Cl\ ^{37}Cl\ ^{12}C^+$
121	$^{35}Cl\ ^{37}Cl\ ^{37}Cl\ ^{12}C^+$
123	$^{37}Cl\ ^{37}Cl\ ^{37}Cl\ ^{12}C^+$

The relative abundance of the isotopes of chlorine may be calculated from the height of the peaks.

Nuclear reactions

We are now in a position to examine the phenomena of radioactivity further. A naturally occurring radioactive isotope may be placed in one of three series of radioactive elements. In the thorium series, in which all mass numbers of elements are multiples of 4 (i.e. mass numbers = $4x$), after ten changes thorium, $^{232}_{90}Th$, forms an isotope of lead, $^{208}_{82}Pb$. The changes involve the formation and disintegration of other elements, including another isotope of thorium:

$$^{228}_{88}Ra \rightarrow\ ^{228}_{89}Ac \rightarrow\ ^{228}_{90}Th \rightarrow\ ^{224}_{88}Ra \rightarrow\ ^{220}_{86}Rn$$

$$\rightarrow\ ^{216}_{84}Po \rightarrow\ ^{212}_{82}Pb\ (or\ ^{216}_{85}At) \rightarrow\ ^{212}_{83}Bi$$

$$\rightarrow\ ^{212}_{84}Po\ (or\ ^{208}_{81}Tl) \rightarrow\ ^{208}_{82}Pb$$

In the uranium series (mass numbers = $4x + 2$), uranium, $^{238}_{92}U$, forms a different isotope of lead, $^{206}_{82}Pb$, after fourteen reactions. The first member of the actinium series (mass numbers = $4x + 3$) is now known to be an isotope of uranium, $^{235}_{92}U$, which forms yet another isotope of lead, $^{207}_{82}Pb$, after eleven changes. Another series, the neptunium series (mass numbers = $4x + 1$), has been discovered for certain artificially prepared radioactive elements; it ends with an isotope of bismuth, $^{209}_{83}Bi$. At each stage in a radioactive series, an unstable atomic nucleus loses energy by the emission of an α-particle (helium nucleus, ^4_2He) and/or β^--particle (electron, $^{\ 0}_{-1}e$) and probably γ-radiation.

Thus, the emission of a (negative) β^--particle (electron) requires the conversion of one neutron into one proton in the nucleus, the atomic number rising by one unit.

$$^1_0n \rightarrow\ ^1_1p +\ ^{\ 0}_{-1}e$$

Emission of energy as γ-radiation leaves the nucleus in a more stable state, with less energy but with the same nuclear charge. Loss of an α-particle (helium nucleus) means that the nucleus loses two protons and two neutrons.

In 1919, Rutherford bombarded nitrogen with α-particles, hoping to cause transmutation of nitrogen. He was able to demonstrate that protons were formed. In 1925, Blackett showed that the collision of a fast-moving helium nucleus with a nitrogen atom yields a proton and an atom of oxygen of mass number 17 as distinct from the more usual isotope of mass number 16:

$$^{14}_7N +\ ^4_2He \rightarrow\ ^1_1H +\ ^{17}_8O$$

A more compact notation, in which α represents ^4_2He and p, ^1_1H expresses this change as

$$^{14}_7N\ (\alpha, p)\ ^{17}_8O$$

or generally

$$\text{initial nuclide} \left(\begin{array}{ccc} & & \text{outgoing} \\ \text{incoming} & & \text{particle(s)} \\ \text{particle(s)} & , & \text{or} \\ & & \text{quanta} \end{array} \right) \text{final nuclide}$$

If n represents the neutron and γ the photon (of radiation), write full equations for the following changes and then read on

$$^{59}_{27}Co\ (n, \gamma)\ ^{60}_{27}Co$$
$$^{23}_{11}Na\ (\gamma, 3n)\ ^{20}_{11}Na$$
$$^{27}_{13}Al\ (\alpha, p)\ ^{30}_{14}Si$$

In nuclear reactions the total (mass + energy) of the nuclei is conserved. In addition the total kinetic energy of the moving particles is unaltered. The upper figures in the equations show the mass numbers, atomic masses to the nearest integer, and the lower figures, the atomic numbers or charges. The sum of the mass numbers on one side of the equation must equal the sum of mass numbers on the other side. Similarly, the atomic (proton) numbers or charges must balance. Energy–mass changes ($E = mc^2$) have too small an influence on the mass to affect the equation although they may be calculated. Returning to natural radioactivity, a series of radioactive changes may be summarized in an equation. The decay of uranium-238, $^{238}_{92}U$, involving fourteen changes, has been mentioned:

$$^{238}_{92}U \rightarrow\ ^{206}_{82}Pb + 8\,^{4}_{2}He + 6\,^{0}_{-1}e$$

The decay of radium, found in the actinium series, yields a different isotope of lead after six changes:

$$^{223}_{88}Ra \rightarrow\ ^{207}_{82}Pb + 4\,^{4}_{2}He + 2\,^{0}_{-1}e$$

In both cases helium nuclei (α-particles) and electrons (β^--particles) are emitted. Most changes are accompanied by γ-radiation.

Artificial radioactivity and nuclear transmutations

Artificial radioactivity was discovered in 1932 by Irene Curie and Joliot (the Joliot-Curies). The bombardment of magnesium nuclei with helium nuclei (α-particles) produces an isotope of silicon and a neutron which may be expressed

$$^{24}_{12}Mg(\alpha, n)\,^{27}_{14}Si$$

The silicon isotope has a half-life of 4 seconds and decays, emitting a positive β-particle (a positron) and leaving aluminium:

$$^{24}_{12}Mg + \,^{4}_{2}He \rightarrow\ ^{1}_{0}n + \,^{27}_{14}Si$$

$$^{27}_{14}Si \rightarrow\ ^{0}_{+1}e(\beta^+) + \,^{27}_{13}Al$$

The existence of the positive electron (positron) had been predicted by Dirac in 1930 and discovered in 1932 by Anderson. Many artificially radioactive isotopes may be produced by disintegration of atomic nuclei under the bombardment of high-energy protons, neutrons and deuterons ($^{2}_{1}H$), etc., using various accelerating machines. Protons cause the transmutation of lithium to beryllium with the loss of a neutron,

$$^{7}_{3}Li + \,^{1}_{1}H \rightarrow\ ^{1}_{0}n + \,^{7}_{4}Be$$

and neutrons convert cobalt into manganese,

$$^{59}_{27}Co + \,^{1}_{0}n \rightarrow\ ^{4}_{2}He + \,^{56}_{25}Mn$$

and sulphur to phosphorus,

$$^{32}_{16}S + \,^{1}_{0}n \rightarrow\ ^{1}_{1}H + \,^{32}_{15}P$$

while deuterons ($^{2}_{1}H$, d) convert chromium to manganese

$$^{53}_{24}Cr + \,^{2}_{1}H \rightarrow\ ^{1}_{0}n + \,^{54}_{25}Mn$$

These changes may be briefly set out:

$$^{7}_{3}Li\ (p, n)\ ^{7}_{4}Be \quad ^{59}_{27}Co\ (n, \alpha)\ ^{56}_{25}Mn \quad ^{32}_{16}S\ (n, p)\ ^{32}_{15}P$$

$$^{53}_{24}Cr\ (d, n)\ ^{54}_{25}Mn$$

The quantities involved are relatively very small.

Nuclear fission and nuclear energy

Nuclear fission was discovered in 1939. The uranium-235 nucleus was shown to break down upon capturing a slow (i.e. a low-energy) neutron with the emission of a large quantity of energy, produced by the annihilation of mass ($E = mc^2$). From 1934, Fermi in Rome had been obtaining confusing results from the bombardment of heavy nuclei (including uranium) with neutrons. In 1939, Hahn and Strassmann in Germany identified an isotope of barium in the products from uranium bombardment. Frisch and Meitner, working in the Copenhagen laboratory of Bohr, interpreted the results of these experiments as showing the splitting or fission of the uranium nucleus. 'Atomic Energy', nuclear energy, has developed from this discovery. A number of processes occur during fission of uranium, one of which may be:

$$^{235}_{92}U + \,^{1}_{0}n \rightarrow\ ^{147}_{60}Nd + \,^{87}_{32}Ge + 2\,^{1}_{0}n + \gamma$$

Notice that two neutrons are formed. They may be lost or absorbed by other materials, but if at least one is used in another fission process, a chain reaction results with the continuous production of energy. If the escape of neutrons produced by fission of uranium-235, or their absorption by impurities, is reduced to a minimum the chain reaction may get out of control with a consequent enormous release of energy, and a nuclear explosion: this can happen when the sample reaches a certain *critical size*. The atomic, or nuclear, bomb requires two pieces of purified uranium-235 (or plutonium-239), each of sub-critical size, to be brought together suitably to form a mass of larger than critical size. The chain reaction then escalates with horrific, and unforgettable results, as seen at Hiroshima and Nagasaki. The net number of neutrons formed is rarely a whole number, as seen in the following fission processes of uranium and plutonium.

$$^{235}_{92}U + \,^{1}_{0}n\ (slow)\dagger \rightarrow\ ^{236}_{92}U + \gamma$$
$$\rightarrow \text{fission fragments}\dagger\dagger + 2.42\,^{1}_{0}n$$

$$^{235}_{92}U + \,^{1}_{0}n\ (fast) \rightarrow\ ^{236}_{92}U + \gamma$$
$$\rightarrow \text{fission fragments} + 2.58\,^{1}_{0}n$$

$$^{233}_{92}U + \,^{1}_{0}n\ (slow) \rightarrow\ ^{234}_{92}U + \gamma$$
$$\rightarrow \text{fission fragments} + 2.49\,^{1}_{0}n$$

$$^{233}_{92}U + \,^{1}_{0}n\ (fast) \rightarrow\ ^{234}_{92}U + \gamma$$
$$\rightarrow \text{fission fragments} + 2.59\,^{1}_{0}n$$

$\dagger$ Usually with energy less than 1 eV (1.6×10^{-19} J)
$\dagger\dagger$ e.g. $^{95}_{42}Mo$, $^{139}_{57}La$ and 32 other elements

$$^{239}_{94}\text{Pu} + ^1_0\text{n (slow)} \rightarrow ^{240}_{92}\text{U} + \gamma$$

$$\rightarrow \text{fission fragments} + 2.88 \, ^1_0\text{n}$$

$$^{239}_{94}\text{Pu} + ^1_0\text{n (fast)} \rightarrow ^{240}_{92}\text{U} + \gamma$$

$$\rightarrow \text{fission fragments} + 3.0 \, ^1_0\text{n}$$

A nuclear reactor (*atomic pile*) has at its core a matrix of fissionable material (uranium) permeated by a medium (e.g. heavy water, deuterium oxide) which is chosen to slow down the neutrons and with an arrangement by which the chain reaction can be controlled, or moderated, by absorbing surplus neutrons in rods of a suitable material (e.g. cadmium, or boron). The heat generated may be used to generate electricity in the normal way by using steam turbines. The intense neutron field may be used to prepare artificial radioactive isotopes as above or as one way to build up nuclei in the search for new elements. Thus, an isotope of plutonium may be converted into americium

$$^{242}_{94}\text{Pu} + ^1_0\text{n} \rightarrow ^0_{-1}\text{e} + ^{243}_{95}\text{Am}$$

or

$$^{242}_{94}\text{Pu (n, e)} \, ^{243}_{95}\text{Am}$$

the initial stage involving neutron capture and formation of $^{243}_{94}\text{Pu}$.

Nuclear fusion

Nuclear reactions involving the aggregation of light nuclei to form heavier nuclei are called fusion reactions. Energy is evolved in these processes and it is believed that the energy of the sun is provided by fusion of hydrogen nuclei:

$$^1_1\text{H} + ^1_1\text{H} \rightarrow ^2_1\text{H} + ^0_{+1}\text{e} + \text{light}$$

$$^2_1\text{H} + ^1_1\text{H} \rightarrow ^3_2\text{He} + \gamma$$

$$^3_2\text{He} + ^3_2\text{He} \rightarrow ^4_2\text{He} + 2 \, ^1_1\text{H}$$

Helium is the product.

Nuclear power

Calder Hall (and the nuclear/radioactive processing plant at Windscale) near Sellafield in Cumbria opened in 1956 as the first industrial scale nuclear electricity power station in the world. There was early optimism over prospects for generating abundant, comparatively cheap electrical power as a peaceful application of the nuclear fission process, and as it turned out, wild optimism over

the illusive fusion phenomenon. It would be idle to pretend that military applications were not associated with the research and development work on the chemical treatment of nuclear materials.

The engineering problems associated with the construction of nuclear reactors were formidable and pioneering considering that malfunctions needed to be anticipated and health hazards avoided when nobody could forsee entirely what would happen as reactor materials aged under continuous running conditions. To the original construction costs must be added those of decommissioning such stations after their 20–30 year economic lifespan and management of the highly radioactive materials remaining. The Magnox stations come to mind in this context.

The United Kingdom has encountered commercial disappointments over the running of the more advanced gas cooled reactors since their performance has rarely reached expectations. In addition, recent calculations of the 'time cost' of generating electricity in this way has made such stations seem uneconomic in comparative terms. Much depends, however, on what is meant by 'time cost'. The time cost of electricity generated by combustion of fossil fuels (coal, oil) in an ideal world would include a factor to cover environmental protection from the sulphurous products causing 'acid rain' and respiratory problems in those people prone to such weakness. As an alternative to the (wasteful) use of fossil fuels, nuclear power has many attractions especially where supplies of such fuels are not abundant. Most of the electricity generated in France comes from nuclear sources and the commercial price, when supplies are exported via cables to adjacent countries (including England at peak times), is very competitive. Medium sized nuclear power stations with their characteristic hemispherical reactor shape can be seen in West Germany, and there is much to be said for such smaller stations.

There are alternative ways of generating electrical power which are becoming popular. Hydro-electric schemes have been developed for many years and at last the possibility of harnessing the energy of wave motion and tides is being explored. Along the North Sea coast, in Holland, and in Schleswig-Holstein and Denmark especially, tall, elegant two or three bladed 'windmills' (2-4 times the height of traditional windmills) dot the landscape, their propellor-like blades twirling in the breeze.

Hazards of nuclear power and processing

Design faults, inadequate engineering techniques and the so-called human factor have resulted in various nuclear power station incidents in the USA and in Europe, causing public alarm. When coupled with a growing cynicism about how forthcoming governments are over such incidents (the Macmillan government was less than candid over an emission of radioactivity from Windscale) it is hardly surprising that people can develop an antipathy towards the nuclear power industry.

An international catastrophe such as the explosion at the Russian nuclear power station at Chernobyl caused serious loss of life and lowered life-expectancy in the area. The effects were also experienced in other countries, as seen in the serious blow to the livelihoods of sheep farmers as far away as Cumbria when their meat was barred from the market.

The discovery that the Greifswald nuclear power station, near to the military airfield at Peenemünde (with an inherent risk of aircraft crashes) in East Germany, was without a concrete shield over the reactor led to its closure. The closure followed inspections by West German experts during preliminary unification meetings and makes one wonder about the quality of design and construction there and elsewhere. Numerous incidents at reactors (many of little significance) and agitation over the location of radioactive processing plants have cast a political shadow which could blight an industry which should have enjoyed an assured future. An example is in West Germany where the construction of a plant at Wackendorf was abandoned.

The biological damage caused by nuclear irradiation is insidious and the long term effects of exposure, even to seemingly safe radiation levels, tragic. There is, of course, the other side of the coin—the beneficial effects of radiation treatment for cancer. However, it remains that the cost of Chernobyl still has to be paid for by some with their lives. In the United Kingdom, a television programme in 1983, *Windscale: The Nuclear Laundry* first suggested that there were unusually high incidents of blood cancer leukaemia around Sellafield, the former Windscale plant. The *Black Report* (1984) concluded that the increased incidence of leukaemia was real but that there was nothing in the environment (including radioactive discharges) to account for them.

The Committee on Medical Aspects of Radiation was set up and in 1988 found six cases of leukaemia in children where only one would have been expected. The normal chance of contracting the disease is 1 in 2000. In the previous year *The British Medical Journal* had highlighted other cases. At nuclear plants at Aldermaston and Burghfield there were 41 cases of leukaemia among children under 14 years of age where only 28 would have been expected. Other data were collected but no investigation could explain the cause, even taking the local carcinogenic bracken into account. The *Garden Report* (1990) suggested a genetic link between parental occupational exposure and the appearance of leukaemia, suggesting that a father experiencing higher than normal radiation levels may carry genetically damaged sperm cells. The report stated, 'It is something we suggest is biologically possible, rather than proven'. Sellafield is situated in a bleak, beautiful region that has limited economic prospects. British Nuclear Fuels Ltd (BNFL), a major employer, processes some of the UK nuclear power waste there. Together with its workers it has to face health hazard problems which exceed the range and depth of those faced by other industrial companies by far.

Finally, it must be remembered that we all experience background radiation, the intensity of which depends on where we live and our occupation. We are all at risk to various degrees, however slight this may be.

The detection of radioactivity

The image produced on a photographic plate by a crystal of a uranium salt led Becquerel in 1896 to the discovery of radioactivity. The plate was protected from the action of visible light. The action of radiation on a photographic emulsion remains one of the simplest tests for the detection of radiation from radioactive sources and X-ray equipment. In radiation laboratories and industrial plants where there is the risk of exposure to harmful radiation, workers carry suitably prepared pieces of film which are developed to check that exposure has not been excessive. There is a necessary time lag for development and another principle is used for the immediate detection of radiation.

Ionization of a gas is caused by charged particles (α, β^-, β^+), and by particles (β^-) resulting from the interaction of X- and γ-radiation with the gas. Collision of a charged particle with a gas molecule may result in the ejection of an electron, leaving behind a positively charged ion. The Wilson cloud chamber is used in laboratory experiments, and the

Geiger–Müller counter ('Geiger counter') may be used in the laboratory and as a portable detector.

In the Wilson cloud chamber, dust-free air is saturated with water vapour and cooled suddenly by adiabatic expansion, becoming supersaturated. Condensation occurs on the ions produced and the track of each sub-atomic particle is revealed as a trail of fine droplets. With suitable illumination, the track may be photographed.

In the Geiger–Müller counter, a small cylindrical ionization chamber contains an ionizable gas (air or argon) at low pressure, the wall acting as the negative electrode. A central axial wire acts as the positive electrode and is insulated from the wall of the chamber. A potential difference in the range 1000–2000 V is maintained between the electrodes. This must be insufficient to cause electrical discharge but enough to drive all ions produced to the electrodes. When a charged particle enters the gas, ionization occurs. The ions produced are accelerated sufficiently to cause further ionization. Each particle causes a burst of ionization. By suitable amplification an electric current may be measured, the bursts of ionization counted in a counting device or made audible through a loudspeaker. The circuit is arranged so that the discharge lowers the potential between the electrodes and ionization ceases until the next particle enters the chamber.

Radioactive isotopes

The course of a chemical process may be followed by substituting a radioactive isotope of an element for the normal non-radioactive form. Chemically they are identical and their isotopic ratio remains essentially unchanged. By determining the intensity of the radioactivity due to the isotope, the concentration of the element may be calculated, even in very small concentrations. This is seen at once to be especially useful in biological systems. Photosynthesis has been studied using radioactive carbon, $^{14}_{6}C$, by Calvin. The process of (say) phosphate absorption by plants may be followed by using radioactive phosphorus in the phosphate supplied.

Of special interest is the estimation of the date at which animals or vegetation died by measuring the intensity of radioactivity due to the carbon-14 isotope. During their lifetime plants and trees take up small quantities of $^{14}_{6}C$ as carbon dioxide, the quantity reaching an equilibrium value. Plants are eaten by animals which also acquire carbon-14. At death, the intake is stopped and radioactive decay continues. The half-life period is 5720 years. By determining their radioactivity the age of various carbon products has been measured, covering a period of 1000–30 000 years, by Libby. $^{14}_{6}C$ is produced in the upper atmosphere by the action of neutrons (produced by cosmic rays) on nitrogen:

$$^{14}_{7}N + {}^{1}_{0}n \rightarrow {}^{14}_{6}C + {}^{1}_{1}H$$

Radioactive isotopes used in 'labelling' ions, atoms and molecules during the study of chemical, industrial, biochemical and biological processes are called isotopic tracers, or radiotrace elements.

3

The properties of elements and the electronic configuration of their atoms

The discovery of Argon and its significance

Towards the end of the last century, just at the time when ideas concerning atomic structure were being developed, following the discovery of radioactivity and the electron, elements belonging to a new group of the Periodic Table were isolated. These elements, all gases, which appeared to be devoid of chemical reactivity were called the inert gases, rare gases or, noble gases (see p. 424). They provide the key to an understanding of chemical reactions and the periodic classification in terms of atomic structure.

In 1784, Cavendish repeatedly passed electric sparks through a measured volume of air and absorbed the brown fumes (nitrogen dioxide) produced in alkali until the volume of air was reduced no further. Small quantities of oxygen were introduced and sparking continued until the volume reached a minimum when excess oxygen was absorbed by a solution of potassium polysulphides (liver of sulphur). A small bubble of gas remained which was neither nitrogen nor oxygen and to quote Cavendish, 'we may safely conclude that it is not more than $\frac{1}{120}$ part of the whole.' Cavendish had isolated a mixture of inert gases. Here, the matter rested for over 100 years.

The names and some physical characteristics of the inert gases are listed in Table 3.1, and the composition of the Earth's atmosphere, which contains all of the inert gases except radon, is given in Table 3.2. Radon is an isotopic mixture of radon, thoron and actinon and was once called *emanation* (1900) because these gases are respectively emanations of

Table 3.2 Composition of the Earth's atmosphere/parts in 10^6 of dry air

By volume	Gas	By mass
780 800	Nitrogen	755 200
209 500	Oxygen	231 500
9 320	Argon	12 800
300	Carbon dioxide	460
18	Neon	12
5.2	Helium	0.72
1.5	Methane	0.80
1.1	Krypton	3.0
0.5	Dinitrogen oxide	0.8
0.5	Hydrogen	0.04
0.4	Ozone	0.7
0.09	Xenon	0.4

Table 3.1 Some physical constants for the noble (or inert) gases

	Helium	Neon	Argon	Krypton	Xenon	Radon
Symbol	**He**	**Ne**	**Ar**	**Kr**	**Xe**	**Rn**
Atomic number	2	10	18	36	54	86
Relative atomic mass†	4.002 6	20.179	39.948	83.80	131.30	222.018
Melting-point/°C (101 kPa)	− 271.4 (3 MPa)	− 248.7	− 189.2	− 157	− 112	− 71
Boiling-point/°C (101 kPa)	− 268.9	− 246.1	− 185.9	− 153.2	− 108.1	− 62
Density/g dm^{-3} (standard conditions*)	0.1785	0.8999	1.784	3.743	5,896	9.96
Ratio specific heats c_p/c_v (= 1.67 for monatomic gases)	1.65	1.64	1.65	1.69	1.67	—

exactly: 273.15 K, 101 325 Pa (Nm^{-2})
†Molar mass/g mol^{-1} of the species quoted

radium, thorium, and actinium, i.e. produced during the radioactive decay of these elements. All are radioactive. The half-life of thoron is 54.5 s, of actinon, 3.9 s, and of radon, 3.8 days.

It has been estimated that c. 6% of lung cancer deaths (41 000 every year in the UK) are caused by radon, the single most important source of radiation exposure, produced by radioactive decay in the granite of SW England and NE Scotland.

The National Radiological Protection Board recommend that the annual dose of radiation received at work should not exceed 15 mSv (note: mSv = millisievert). It is estimated that ten times as many people experience more radiation at home every year than do people working in industry.

In 1892, Lord Rayleigh, the physicist, determined the density of nitrogen prepared from air by removal of all other known constituents, and of nitrogen prepared from chemical sources: the respective values were $1.2572 \, g \, dm^{-3}$ and 1.2505 $g \, dm^{-3}$, under standard conditions, 273 K and 101 kPa. The density of atmospheric 'nitrogen' was the higher by about 1 part in 200 which was more than the degree of experimental error.

In 1894, Sir William Ramsay, the chemist, believing that the discrepancy was due to an unknown impurity in atmospheric nitrogen, also started investigating the problem. He started with dry air, from which carbon dioxide had been removed, and removed oxygen by repeated passage over heated copper, which was converted to the oxide. He passed the remaining gas over heated magnesium, which combined with the nitrogen to form magnesium nitride. This was continued until the residual gas had a constant relative density (19.41, H = 1). The gas had a spectrum which was new and quite unlike that of nitrogen. It was chemically inert. Lord Rayleigh had continued his experiments of sparking air with oxygen and had obtained small quantities of the gas. They named the gas argon (Greek: *argos* = inert or idle) and published their results in a joint paper.

Ramsay now looked for other sources of argon and his attention was drawn to a mineral clèveite from which Hildebrand had released what seemed to be nitrogen in 1889. Having removed the nitrogen, Ramsay found that samples of the gas were composed mostly of a gas with a spectrum different from that of argon or nitrogen. This was identified with spectrum lines detected by Janssen in the solar spectrum during an eclipse in 1868 and interpreted by Lockyer as being due to an element unknown on the Earth, which he had named helium (Greek:

helios = sun), the suffix -*ium* indicating that a metallic nature was assumed. The identification was made in 1895. Helium is associated with many radioactive minerals and may be released by heating them. It is formed by radioactive decay and a helium nucleus is identical with an α-particle.

With the appearance of a group 0 (zero valency) in the Periodic Table to accommodate new elements with unusual characteristics, Ramsay realized there was probably an undiscovered element of similar properties with an intermediate atomic mass and probably others which had larger atomic masses. At about this time, liquid air had been made in bulk by Hampson in England and Linde in Germany. By controlled evaporation of liquid air, various components could be obtained.

With M. W. Travers in 1898, Ramsay continued the investigation of liquid air. From the residual few millilitres of liquid left when nearly all of about one and a half litres of liquid air had been allowed to evaporate, they isolated krypton (Greek: *kryptos* = hidden), identifying it by its distinct spectrum. Afterwards, neon (Greek: *neos* = new) and xenon (Greek: *xenos* = stranger) were also isolated and identified spectroscopically.

In 1907, up to 1.84% by volume of helium was discovered in natural gas by Cady and McFarland in Kansas, USA. Although up to 7–8% by volume has been found, quantities nearer 1% are more usual. The natural gas is associated with the petroleum regions of the USA and consists mostly of methane, the helium having (presumably) collected as a result of radioactivity. Except for radon, the other inert gases are produced from the atmosphere.

Because of their inert nature, the gases of this family have interesting and specialized uses. Deep-sea divers require the supply of oxygen under considerable pressure. If air is supplied at high pressure, large quantities of nitrogen are dissolved in the blood stream and the sudden release of this by bringing the diver up rapidly would result in blockage of the circulation by bubbles of nitrogen. A mixture of helium, which is much less soluble, and oxygen is used instead of air and the diver may be raised quickly. Helium (b.p. $-268.9\,°C$) is used in low-temperature studies and in providing an inert atmosphere for certain welding processes. Discharge tubes filled with neon are used for illumination and for advertisement signs, the red-orange colour being modified by additions of other rare gases, mercury, or the use of coloured glass as desired. With 15% nitrogen, argon is used in gas-filled electric lamps and in fluorescent lighting tubes. It is also used as an inert atmosphere for welding. Radon has been used in the treatment of cancer because of its radioactivity.

The classification of the elements in the 20th century

The periodic pattern of electron groups

We have seen that an atom is believed to be composed of a relatively small, dense, positively charged nucleus, which is formed from protons and neutrons, surrounded at a distance by sufficient negatively charged electrons to leave the whole atom electrically neutral. In theory, elements may be formed by constructing atoms of the appropriate atomic number by using this number of protons in the nucleus with the required number of neutrons as a sort of nuclear ballast and surrounding the nucleus at a distance with enough electrons to preserve electrical neutrality, the number of electrons being equal to the atomic number. Nothing has been said about the configuration adopted by these electrons.

Many physical properties depend to some degree on the size of the atoms of the element under review. The size or volume of an atom will be determined by the distance at which the outermost electrons lie from the nucleus. The atomic volume is a

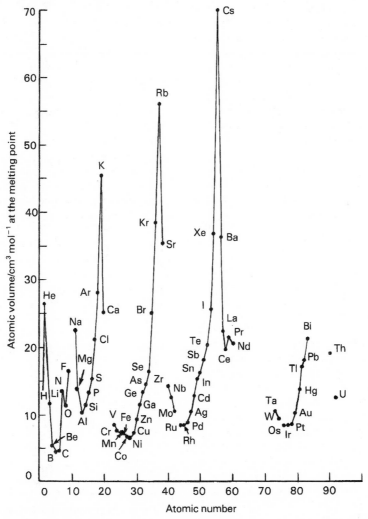

Fig. 3.1 Atomic volume curve: atomic volume is plotted against atomic number. In the original curve, Lothar Meyer plotted atomic volume against atomic weight

convenient but rough way of measuring an atom's size. It equals the volume occupied by the molar mass of the element and depends on the geometrical packing of atoms in the solid, especially where polyatomic entities occur, and on the conditions under which the density of the solid is determined.

Lothar Meyer showed that when plotted against atomic mass, the atomic volumes of successive elements form a periodic curve. The fundamental quantity for an atom is now known to be the nuclear charge, proton number or atomic number, not the atomic mass. The graph showing atomic volume plotted against atomic number for the elements is shown in Fig. 3.1. It is essentially the same as that produced by Lothar Meyer. It is observed that corresponding positions on successive peaks belong to the successive elements of each periodic family. Although atomic volume, as defined and measured here, depends on so many factors, it seems reasonable to suppose that this periodicity is related to the electron distribution in the atoms selected.

When the light from a sodium vapour lamp is examined in a spectroscope a bright yellow line is observed. On narrowing the slit by which light enters the spectroscope, two lines close together may be observed. This spectrum, produced by excitation of sodium atoms, is called a line spectrum or an atomic spectrum. Some are very much more complex. Atomic spectra may be produced by electrical discharge, arcing or by heating and are emitted by atoms in the vapour state. On the other hand, light from an incandescent solid, on spectroscopic analysis, forms a continuous spectrum. Some metals may be identified by the characteristic colours, or, if necessary, their spectral lines, given under suitable conditions in qualitative analysis. Chlorides of the metals are used. They are usually relatively volatile and easily obtained by mixing a powdered solid compound with concentrated hydrochloric acid. A platinum wire, moistened with the mixture and held in a (non-luminous) Bunsen flame may impart a characteristic coloration to the flame. Molecules give a different type of spectrum, in which successive bands appear, called a band spectrum. On resolution, each band appears as a number of close-packed lines. Spectra may appear in the infrared, ultra-violet and visible parts of the spectrum. At present, we confine ourselves to the knowledge that suitably excited atoms give line spectra. Figure 3.2 shows line spectra for lithium, sodium and potassium. Wave-lengths are expressed in nanometres ($nm = 10^{-9}$ m) and picometres

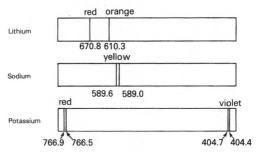

Fig. 3.2 Line spectra for lithium, sodium and potassium showing colours and wave-length/nm

($pm = 10^{-12}$ m)[†]. The emission of lines of characteristic wave-lengths must be associated with definite energy changes. Further, it is found that elements from the same family often have similar atomic spectra. As would be expected, the atomic spectrum for hydrogen has proved to be relatively the most simple and the wave-lengths (λ) of the multiplicity of lines in ultra-violet, visible and infrared regions are found to obey the simple formula:

$$\frac{1}{\lambda} = R_H \left(\frac{1}{n_1^2} - \frac{1}{n_2^2} \right)$$

where n_1 and n_2 are integral and n_1 assumes a fixed value for all lines of a given series. Table 3.3. gives the values of n_1 and n_2 for the different series, each of which is named after the discoverer, together with the region to which the series belongs. R_H is the Rydberg constant for hydrogen—observed value = $10\,967\,818$ m^{-1}.

The continuous spectrum of an incandescent solid is in no way dependent on the nature of the solid. The distribution of energy or intensity of radiation is characteristic of the temperature. For a perfect radiator, called a black body, the total energy emitted per second is proportional to the fourth power of the absolute temperature (Stefan's Law) and the maximum value of the radiation for

Table 3.3   The atomic spectrum of hydrogen

Series	Year of discovery	Region	n_1	n_2
Lyman	1906	Ultra-violet	1	2, 3, 4, …
Balmer	1885	Visible	2	3, 4, 5, …
Paschen	1908	Infra-red	3	4, 5, 6, …
Brackett	1922	Infra-red	4	5, 6, 7, …
Pfund	1924	Infra-red	5	6

[†] The unit of measurement formerly used was the Ångström (10^{-10} m), but the use of this unit of measurement is discouraged in the SI system

unit wave-length interval is at a wave-length which is inversely proportional to the absolute temperature (Wien's Law). In order to explain these observations the **Quantum Theory** was advanced by Planck in 1900. Planck suggested that energy is emitted or absorbed in discrete units or quanta: each quantum is a quantity of energy (E) which is proportional to the frequency v (greek letter, nu) of the radiation and is given by the formula:

$$E = hv$$

where $h = 6.6256 \times 10^{-34}$ J s (Planck's constant). This hypothesis has developed into a theory which is one of the most important in physics and chemistry.

Thus, quanta (hv) emitted as infra-red radiation from a hot body will be smaller than those emitted as the body reaches incandescence. In 1905, Einstein extended these ideas to the transmission of energy by atoms and suggested that radiation travelled in quanta. Each quantum is called a photon and represents a small increment of energy ($E = hv$) travelling with the velocity of light ($c = 3 \times 10^8$ m s^{-1}).

The Rutherford atom had been pictured as a miniature solar system upon which was imposed a distribution of electricity resulting in a positively charged nucleus exerting a force of attraction on the negatively charged electrons revolving in their orbits. Rapid motion was the best explanation afforded for the apparent stability of electrons remaining in orbit against the pull of attraction from the heavier nucleus. However, this brought in another difficulty since such an arrangement would require, from a classical theory, *continuous* radiation with consequent spiralling of the electrons into the nucleus as energy was lost.

However, novel suggestions, which have resulted in the Quantum Theory being extended to atomic structure, were put forward in 1913 by Bohr. It was postulated that electrons may occupy orbits (at a distance from the nucleus) from which no energy is absorbed or emitted. These were called stationary states and are now referred to as energy levels. A further postulate required that absorption of a suitable quantum of energy results in the virtually *instantaneous* promotion of an electron from an inner to an outer orbit, from the normal or ground state of least energy to a higher level. The energy absorbed is related to the frequency of radiation absorbed by $\Delta E = hv$.

When an electron returns from this energy level to the original level, the same quantity of energy is emitted and appears as radiation of the same frequency as before ($\Delta E = hv$). By having a number of energy levels into which an electron can be promoted or fall without necessarily returning directly to its original level, Bohr developed a theory of the hydrogen atom which explains the series of lines observed in the atomic spectrum of the gas. He deduced the formula:

$$\frac{1}{\lambda} = \frac{2\pi^2 e^4 \mu_H}{h^3 c} \left(\frac{1}{n_1^2} - \frac{1}{n_2^2} \right)$$

where

$$\frac{1}{\mu_H} = \frac{1}{m_e} + \frac{1}{m_p}$$

in which μ_H is called the reduced mass, m_e being the mass of an electron and m_p that of a proton, e the numerical value of the electrical charge on an electron and on a proton, c the velocity of light and λ the wave-length of the spectral line given by integral values of n_1 and n_2. Comparison shows that this equation is very similar to the previous equation and substitution of the values of the symbols, in appropriate units gives $R_H = 10\,967\,758$ m^{-1}, which is very close indeed to the observed value of $10\,967\,818$ m^{-1}. The concept of energy levels is thus based on atomic spectra.

Figure 3.3 shows a pictorial representation of the formation of the various series making up the full atomic spectrum of hydrogen. Electrons are found in quantum shells or electron shells, associated with energy levels. Each energy level is numbered with

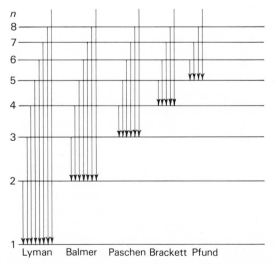

Fig. 3.3 The formation of the various series of lines belonging to the atomic spectrum of hydrogen

a principal *quantum number*. Alternatively, the letters, used originally to describe X-ray spectra, may be used instead: K, L, M, ...

Other experiments have confirmed the concept of energy levels. The position with other atoms is much more complex and is treated in the subject of wave mechanics or quantum mechanics. The treatment regards the atom as a mathematical concept without regard for a simple pictorial representation because it is now believed that the electron cannot be regarded as a conventional material particle. The measurement of ionization energies, providing firm evidence for the orderly expansion of the electronic configuration of atoms and necessary associated theory is described later in this chapter after the Periodic Table has been established according to the principles set out so far. Events in the gas discharge tube were interpreted in terms of a particle concept (= the electron) while later developments in theory and experiment have resulted in the need to explain some further phenomena in terms of a wave character. Just as with light, the electron is the name given to a group of linked phenomena which are best explained sometimes in particle terms and in other instances in wave terms. Now, it is quite meaningless to ascribe physical properties to things unless measurements can be made. The electron is so small that any attempt to measure its speed, or the related quantities, momentum or energy (by using suitable radiation perhaps) would result in the disturbance of the electron (because of collision with at least one photon of radiation, photon being the name given to one quantum of light) and uncertainty about its position. This is an aspect of the Heisenberg *Uncertainty Principle*; it is unconnected with ordinary experimental errors in any way, and is implicit in the study of sub-atomic phenomena. However, in chemistry, we may safely visualize the electron as a particle if we do not insist on defining exactly its speed and position in orbit. The mathematical treatment involves wave equations and the wave function may best be interpreted as giving the probability of finding an electron at a particular position. While we find it useful to regard electrons as tiny particles, the overall picture of an atom, when built from the main sub-atomic units, should be of a positively charged nucleus surrounded by a diffuse cloud of negative electric charge ('electron-cloud'), gradually thinning out to the periphery of the atom. The probability of finding an electron is greatest where the cloud is densest. Having dwelt on the inadequacy of physical models, we now turn

Table 3.4 Electronic configurations of the noble (or inert) gases

Name	Symbol	Atomic number	Designation of electron shell					
			1	2	3	4	5	6
			K	**L**	**M**	**N**	**O**	**P**
Helium	**He**	2	2					
Neon	**Ne**	10	2	8				
Argon	**Ar**	18	2	8	8			
Krypton	**Kr**	36	2	8	18	8		
Xenon	**Xe**	54	2	8	18	18	8	
Radon	**Rn**	86	2	8	18	32	18	8

to the question of configurations adopted by the electrons in atoms.

Soon after the discovery of the electron the link between chemical properties and electronic configuration was apparent and the idea grew that the lack of chemical reactivity shown by inert gases was due to completed shells of electrons. Starting here, a scheme was developed to explain the position of elements in the Periodic Table in terms of electronic configurations. Associated with this theory are Langmuir (1919), Bury (1921), Bohr (1921) and Pauli (1925). The electronic configurations of the inert gases are listed in Table 3.4. Helium contains two electrons, when in its ground state, occupying the first or K electron shell which cannot expand further. Neon has eight electrons in the second or L electron shell besides the two in the K shell. Argon has a further eight electrons in the third or M shell. This straightforward expansion is now modified. In passing from argon to krypton, the M shell is expanded to contain eighteen electrons while the fourth or N shell is built up to eight electrons. For xenon, the outer shell, the fifth or O shell, contains eight electrons but the penultimate N shell has now eighteen electrons. Radon is formed by changes in three shells: the N shell takes in fourteen more electrons and the O shell, ten electrons, while the sixth or P shell contains eight electrons. Except for helium, each inert gas has an outer shell of eight electrons. The family is known as the noble, or inert, gases. A group of eight electrons forming a closed shell is called an *octet* and the electronic configurations shown in Table 3.4 are considered to be endowed with a special stability. All but the K and L shells may be later expanded to hold 18 (M), 18 or 32 (N) and 18 (O) electrons. The maximum number of electrons in a shell of quantum number n is seen to be $2n^2$ up to and including the fourth shell.

Table 3.5 The long form of the Periodic Table

Period	Ia	IIa	IIIa	IVa	Va	VIa	VIIa	VIII			Ib	IIb	IIIb	IVb	Vb	VIb	VIIb	O
1	1 H 1.00797																	2 He 4.0026
2	3 Li 6.939	4 Be 9.0122											5 B 10.811	6 C 12.0112	7 N 14.0067	8 O 15.9994	9 F 18.9984	10 Ne 20.179
3	11 Na 22.898	12 Mg 24.305											13 Al 26.9815	14 Si 28.086	15 P 30.9738	16 S 32.064	17 Cl 35.453	18 Ar 39.948
4	19 K 39.102	20 Ca 40.08	21 Sc 44.956	22 Ti 47.90	23 V 50.942	24 Cr 51.996	25 Mn 54.9380	26 Fe 55.847	27 Co 58.9332	28 Ni 58.71	29 Cu 63.546	30 Zn 65.37	31 Ga 69.72	32 Ge 72.59	33 As 74.9216	34 Se 78.96	35 Br 79.904	36 Kr 83.80
5	37 Rb 85.47	38 Sr 87.62	39 Y 88.905	40 Zr 91.22	41 Nb 92.906	42 Mo 95.94	43* Tc 98.906	44 Ru 101.07	45 Rh 102.905	46 Pd 106.4	47 Ag 107.868	48 Cd 112.40	49 In 114.82	50 Sn 118.69	51 Sb 121.75	52 Te 127.60	53 I 126.904	54 Xe 131.30
6	55 Cs 132.905	56 Ba 137.34	57 La 138.91	72 Hf 178.49	73 Ta 180.948	74 W 183.85	75 Re 186.2	76 Os 190.2	77 Ir 192.2	78 Pt 195.09	79 Au 196.967	80 Hg 200.59	81 Tl 204.37	82 Pb 207.19	83 Bi 208.980	84 Po 210.000	85* At 209.987	86 Rn 222.018
7	87 Fr 223.020	88 Ra 226.025	89 Ac 227.028	104* Rf (260)	105* Ha	(106)	(107)	(108)	(109)	(110)	(111)	(112)						

LANTHANIDES

ACTINIDES

Lanthanide Series	57 La 138.91	58 Ce 140.12	59 Pr 140.907	60 Nd 144.24	61* Pm 144.913	62 Sm 150.35	63 Eu 151.96	64 Gd 157.25	65 Tb 158.924	66 Dy 162.50	67 Ho 164.93	68 Er 167.26	69 Tm 168.934	70 Yb 173.04	71 Lu 174.97
Actinide Series	89 Ac 227.028	90 Th 232.038	91 Pa 231.036	92 U 238.03	93* Np 239.053	94* Pu 239.052	95* Am 243.061	96* Cm 247.070	97* Bk 247.070	98* Cf 252.082	99* Es 254.088	100* Fm 253.086	101* Md 257.096	102* No 255.093	103* Lr (257)

*Not found naturally on Earth
Rutherfordium, Rf, 104 is also called kurchatovium, Ku

Except for the K shell, each shell is made up of two or more *orbits* or *sub-shells*, each energy level being subdivided into a number of closely grouped different levels.

A new basis for classification

As we have seen, atomic masses are no longer regarded as fundamental qualities for elements and have been superseded by atomic numbers, which represent not only the order of the element in the periodic table but also the nuclear charge and the number of electrons in the electron shells. Shells are also called energy levels and of course have no material existence themselves.

The Periodic Law

The properties of elements are a periodic function of their atomic numbers.

The Periodic Table used today is a long form produced by extending the long periods of the Mendeléeff Table: it is very similar to the Bohr-Thomsen version, which is shown in Table 3.5 and inside the back cover. Each horizontal period corresponds to the quantum number assigned to the outer shell which is completed with an octet of electrons (two for the first period) for the inert gas seen at the end of the period. The term, *series*, as used by Mendeléeff, may now be discarded.

Notations for subgroups   In the long form of the Periodic Table, the original designation of A and B subgroups shown in the Mendeléeff Table has been retained. A minority of authors prefer to list transitional elements exclusively as B subgroups and therefore the normal elements as A subgroups. Thus IIIa to VIIa inclusive become IIIb to VIIb respectively and vice versa. However, this is not very important because it is customary to indicate the composition of a family of elements by giving at least the first-named element. Finally, it may be mentioned that some have argued for 18 groups, there being eighteen places (excluding lanthanides and actinides) in the longest periods of the Periodic Table.

The construction of the Periodic Table from electronic configurations

Hydrogen has the simplest atom, the most abundant isotope having a nucleus of one proton with one electron outside it. From this simplest of structures, by the addition at each step of one proton to the nucleus and one external electron, with the appropriate number of neutrons to provide its isotopic mass, the atom of any element may be formed in theory. Chemical character depends on electronic configuration.

Periods 1, 2 and 3

The K shell may have one or two electrons only. It is expanded during the formation of atoms of the elements hydrogen and helium, the only members of period 1. Hydrogen, which is quite unlike any other element, has sometimes been placed by itself above the table, or even included above fluorine, F, as well as above lithium, Li.

The lithium atom, which has three electrons, has a completed K shell of two and a new shell, L, starting with one electron. The elements beryllium to fluorine, of atomic numbers 4–9 inclusive, have progressively extra electrons in the L shell until neon is reached. At this element, both K and L shells are full.

The M shell commences with sodium, which has eleven electrons, one more than can be accommodated in the K and L shells of its atoms. This shell is built up steadily until argon is reached, all elements being placed in the third period. The development of the first three periods is shown in Table 3.6.

Table 3.6   Periods 1, 2 and 3

1	1 **H** 1								2 **He** 2
2	3 **Li** 2 1	4 **Be** 2 2		5 **B** 2 3	6 **C** 2 4	7 **N** 2 5	8 **O** 2 6	9 **F** 2 7	10 **Ne** 2 8
3	11 **Na** 2 8 1	12 **Mg** 2 8 2		13 **Al** 2 8 3	14 **Si** 2 8 4	15 **P** 2 8 5	16 **S** 2 8 6	17 **Cl** 2 8 7	18 **Ar** 2 8 8

Period 4

The electronic configuration of the last member of this period, which will be the inert gas krypton, is 2, 8, 18, 8 and is seen to involve the expansion of the M and N shells. The energy levels filled in the start of shell N and the further filling of shell M are about the same. The potassium atom contains nineteen electrons, one starting the N shell, the remainder having the argon configuraton 2, 8, 8. Two electrons are in the N shell for the calcium atom, otherwise the pattern is that of potassium. Starting with the next element, No. 21, scandium, the penultimate M shell is expanded to contain a further ten electrons while, except for chromium and copper, two electrons remain in the outermost shell. At this stage, it may be assumed that the transference of an electron from the outermost N shell to the penultimate M shell gives a symmetrical distribution, with lower energy, for the five electrons of chromium and ten electrons of copper, which have been added to the penultimate shell. With zinc comes the completion of the M shell and the restoration of two electrons in the N shell. The outermost shell then takes in extra electrons to form the electronic configurations of gallium, germanium, arsenic, selenium, bromine and krypton in succession. Certain metals in Period 4 are called transition metals or transition elements.

Transition elements are characterized by atoms in which an inner electron shell (or level) is in process of expansion but is not yet complete.

In this period, by definition, the transition series comprises scandium to nickel inclusive. However, because of some similarities, copper and zinc are sometimes classed with the transition elements. In this book they will be referred to as 'associated with transition elements'. Chapter 24 is devoted to the transition and associated elements.

Elements other than transition elements and inert gases are called **normal elements** or **representative elements**.

Period 4 is shown in Table 3.7.

Period 5

Broadly, this follows closely the sequence in Period 4, except that in the transition series the absorption of an outermost electron by the penultimate shell

Table 3.7 Period 4

	19 K 2 8 8 1	20 Ca 2 8 8 2					31 Ga 2 8 18 3	32 Ge 2 8 18 4	33 As 2 8 18 5	34 Se 2 8 18 6	35 Br 2 8 18 7	36 Kr 2 8 18 8
4												

	21 Sc 2 8 9 2	22 Ti 2 8 10 2	23 V 2 8 11 2	24 Cr 2 8 13 1	25 Mn 2 8 13 2	26 Fe 2 8 14 2	27 Co 2 8 15 2	28 Ni 2 8 16 2	29 Cu 2 8 18 1	30 Zn 2 8 18 2

Table 3.8 Period 5

	37 Rb 2 8 18 8 1	38 Sr 2 8 18 8 2					49 In 2 8 18 18 3	50 Sn 2 8 18 18 4	51 Sb 2 8 18 18 5	52 Te 2 8 18 18 6	53 I 2 8 18 18 7	54 Xe 2 8 18 18 8
5												

	39 Y 2 8 18 9 2	40 Zr 2 8 18 10 2	41 Nb 2 8 18 12 1	42 Mo 2 8 18 13 1	43 Tc 2 8 18 14 1	44 Ru 2 8 18 15 1	45 Rh 2 8 18 16 1	46 Pd 2 8 18 18 0	47 Ag 2 8 18 18 1	48 Cd 2 8 18 18 2

occurs more often. By strict definition, the transition series starts with yttrium ($Z = 39$) and ends with rhodium ($Z = 45$), although palladium, silver and cadmium may be included. Period 5 is shown in Table 3.8.

Period 6

The first element, caesium ($Z = 55$) has the electronic configuration of xenon for fifty-four of its electrons and the last electron is placed in the P

Table 3.9   The electronic configurations of isolated atoms of the elements listed in order of increasing atomic number

Z	El					
1	H	1				
2	He	2				
3	Li	2	1			
4	Be		2			
5	B		3			
6	C	He	4			
7	N	core	5			
8	O		6			
9	F		7			
10	Ne		8			
11	Na	2	8	1		
12	Mg			2		
13	Al			3		
14	Si	Ne		4		
15	P	core		5		
16	S			6		
17	Cl			7		
18	Ar			8		
19	K	2	8	8	1	
20	Ca	2	8	8	2	
21	Sc			9	2	
22	Ti			10	2	
23	V	Ar		11	2	
24	Cr	con-		13	1	
25	Mn	figuration		13	2	
26	Fe	expanding		14	2	
27	Co			15	2	
28	Ni			16	2	
29	Cu			18	1	
30	Zn			18	2	
31	Ga	2	8	18	3	
32	Ge	Zn			4	
33	As	configuration			5	
34	Se	expanding			6	
35	Br				7	
36	Kr				8	
37	Rb	2	8	18	8	1
38	Sr	2	8	18	8	2
39	Y				9	2
40	Zr	Kr			10	2
41	Nb	configuration			12	1
42	Mo	expanding			13	1
43	Tc				14	1
44	Ru				15	1
45	Rh				16	1
46	Pd				18	0
47	Ag				18	1
48	Cd				18	2

Z	El						
49	In	2	8	18	18	3	
50	Sn					4	
51	Sb	Cd				5	
52	Te	configuration				6	
53	I	expanding				7	
54	Xe					8	
55	Cs	2	8	18	18	8	1
56	Ba	2	8	18	18	8	2
57	La	2	8	18	18	9	2
58	Ce	2	8	18	20	8	2
59	Pr				21	8	2
60	Nd	Formation			22	8	2
61	Pm	of			23	8	2
62	Sm	Lanthanides			24	8	2
63	Eu				25	8	2
64	Gd				25	9	2
65	Tb				27	8	2
66	Dy				28	8	2
67	Ho				29	8	2
68	Er				30	8	2
69	Tm				31	8	2
70	Yb				32	8	2
71	Lu				32	9	2
72	Hf	2	8	18	32	10	2
73	Ta					11	2
74	W	Outer				12	2
75	Re	Transition				13	2
76	Os	Series				14	2
77	Ir	now being				15	2
78	Pt	completed				17	1
79	Au					18	1
80	Hg					18	2
81	Tl	2	8	18	32	18	3
82	Pb						4
83	Bi	Hg					5
84	Po	configuration					6
85	At	expanding					7
86	Rn						8

Z	El							
87	Fr	2	8	18	32	18	8	1
88	Ra	2	8	18	32	18	8	2
89	Ac	2	8	18	32	18	9	2
90	Th	2	8	18	32	18	10	2
91	Pa					20	9	2
92	U	Formation				21	9	2
93	Np	of				22	9	2
94	Pu	Actinides				24	8	2
etc.								

shell. Similarly, the next element, barium ($Z = 56$) has two electrons in the P shell. The first member of the transition series, lanthanum, now appears with the expansion of the penultimate shell. However, a new phenomenon appears with the expansion of the O shell from eighteen electrons in successive stages to thirty-two electrons. The fourteen new elements are called *lanthanons, lanthanides* or *rare earths*. They are very similar to one another. This is a *transition series within another transition series*, or an *inner transition series*. Hafnium ($Z = 72$) is the next element of the transition series which by rigid definition finishes with platinum ($Z = 78$) followed by gold ($Z = 79$) and mercury ($Z = 80$). The outermost shell is then expanded with the appearance of the sequence, thallium ($Z = 81$) to radon ($Z = 86$). The lanthanides are shown separately below the Periodic Table.

Period 7

Before the 1939 war, no transuranic elements were known. As the result of the work of Seaborg and his associates at the University of California, a whole series from $Z = 93$ onwards has been produced by transmutation starting from uranium. In the best Mendeléeff tradition of prediction, it has been established by these workers that an inner transition series, the actinide series, is formed. This is expected to contain fourteen elements like the lanthanides and so the second member of the outer transition series should be element $Z = 104$.

Table 3.9 lists the elements in order of atomic number together with the electronic configurations assigned to them and of which their physical and chemical properties are a function. It refers to free atoms in their normal states, not to atoms in a solid or liquid where electronic interaction occurs.

Atomic orbitals and quantum numbers

The concept of quantum shells (electron shells) associated with energy levels is based on the interpretation of atomic spectra. However, single shells present too simple a picture to account for the complexities seen. It has already been noted that the simple mechanical model of the atom must be abandoned, the result of *Wave (Quantum) Mechanical* treatment and application of the Heisenberg *Uncertainty (Indeterminacy) Principle*. Each quantum shell is replaced by a group of atomic orbitals, except for the first shell which is replaced by one type of orbital. The atomic orbital represents the boundaries of the space outside the nucleus in which there is the greater probability of finding a particular electron. In the mathematical terms of wave mechanics, the wave functions describing single electrons are called orbitals. Each orbital may hold only two electrons, each spinning in the opposite sense to the other. Electrons are assigned to orbitals so that those associated with the lowest energy levels are filled first. For orbitals of a given type as many single electrons as possible are accommodated before pairing starts. The number of unpaired electrons determines the valency.

Orbitals of a given type occur in sets in successive energy levels:

$$1s \quad 2s \quad 3s \quad 4s \quad \cdots$$
$$2p \quad 3p \quad 4p \quad \cdots$$
$$3d \quad 4d \quad \cdots$$
$$4f \quad \cdots$$

The number refers to the *Principal Quantum Number* (n) already mentioned, the energy level nearest to the nucleus (K shell) having $n = 1$. The letters refer to the *Subsidiary Quantum Number* (l) which serves to distinguish the types of orbital in each energy level. The letters originate in the spectroscopic terms for the series: sharp, principal, diffuse, fundamental. For a given value of n, the subsidiary quantum number may assume values up to $(n - 1)$:

$$l \quad 0 \quad 1 \quad 2 \quad 3 \quad \cdots \quad (n - 1)$$
$$\text{or} \quad \text{or} \quad \text{or} \quad \text{or}$$
$$s \quad p \quad d \quad f$$

The subsidiary quantum number is also called the *Angular Momentum Quantum Number*. The first energy level ($n = 1$) has only one type of orbital (s), the second, two types (s, p), the third, three (s, p, d), and so on. The p orbital differs from the s orbital in the same energy level by having an energy distribution which depends on the angle and radial distance from the nucleus: it has a directional character. By symmetry, the three p orbitals in a given level have the same energy, unless an electric or magnetic field is applied to the atom.

In a powerful magnetic field the lines of an atomic spectrum are split up, the new lines being associated with only slightly different energies from those seen in the absence of the field. This is

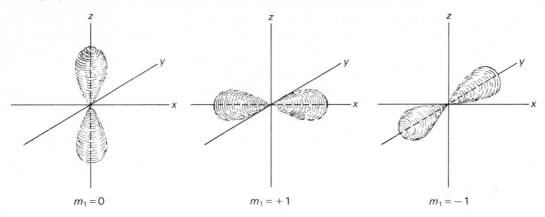

Fig. 3.4 Pictorial representation of 2p orbitals

the *Zeeman Effect*. A similar happening in an electric field is called the *Stark Effect*. The application of a strong magnetic field leads to different energies being associated with orbitals which were previously identical. The magnetic quantum number (m_1) is assigned to distinguish between orbitals of the same type in a given level. It may assume any integral value between $+l$ and $-l$ including zero:

l	m
$0 (\equiv s)$	0
$1 (\equiv p)$	$+1, 0, -1$
$2 (\equiv d)$	$+2, +1, 0, -1, -2$
$3 (\equiv f)$	$+3, +2, +1, 0, -1, -2, -3$

The fourth energy level ($n = 4$) comprises one s orbital, three p orbitals, five d orbitals and seven f orbitals. The three quantum numbers given are found in the *Schrödinger wave equation*, which has only been precisely solved for hydrogen.

A pictorial representation of orbitals, or rather, the electron-cloud probability pattern they represent, may be attempted. The 1s orbital is spherically symmetrical about the nucleus, rather like a fluffy ball of wool, though with no material existence, thinning out at the periphery as the probability of finding the electron rapidly diminishes. The 2s orbital is similarly spherically symmetrical but less dense and at a greater distance and separated from the 1s orbital by a region (spherical nodal surface) where the probability of finding the electron is low. In the higher quantum states, the electron-cloud spreads over a much larger region of space, and since the probability refers to finding one electron only, the density of the cloud is much reduced. There are three 2p orbitals, 'dumb bell' in shape placed along axes at right angles to each

other. See Fig. 3.4. 3p, 4p, ... orbitals are similar to the 2p orbitals. The z axis is defined by reference to an external magnetic field and the x and y axes by reference to additional external fields, as created (for example) by adjacent atoms in a crystal. Notice the nodal plane (virtual zero probability of locating an electron) at the origin, passing through the atomic nucleus.

Each orbital may contain two electrons. A fourth quantum number, the *Spin Quantum Number* (m_s) is assigned to distinguish between these electrons. From studies of the fine structure of atomic structure, it is considered that electrons behave as if spinning about their centres. An electron is given a value of m_s equal to $+\frac{1}{2}$ or $-\frac{1}{2}$. The difference between the energies of the electrons which differ only in m_s is very small, but increases with atomic number.

The state of an electron is defined by its four quantum numbers, no two electrons in the same atom having exactly the same four quantum numbers. This is the *Pauli Exclusion Principle* and was put forward empirically. When the electrons in an atom are in the lowest possible energy levels consistent with this principle, the atom is said to be in its *ground state*. The *Rule of Maximum Multiplicity (the Hund Rule)* states that in filling a p, d or f orbital, as many unpaired electrons as possible are placed before coupling of electrons with opposed spins is allowed.

Electrons are placed in orbitals of the lowest energy level (Table 3.10) and orbitals filled in order of increasing energy, the inert gas at the end of the period resulting from completion of the orbitals being shown in parentheses. However, energy levels change relative to each other with change in

Table 3.10   The general order of energy levels (see text)

1s (He) lowest energy	2s, 2p (Ne)	3s, 3p (Ar)	4s, 3d, 4p (Kr)	5s, 4d, 5p (Xe)	6s, 5d, 4f, 6p (Rn) highest energy
		electrons placed in orbitals of lowest energy first, e.g. 4s before 3d and 3d before 4p →			

Table 3.11   Completion of successive energy levels

	+3	+2	+1	0	−1	−2	−3	Number of electrons	
1s				↑↓				2	
2s				↑↓				2	
2p			↑↓	↑↓	↑↓			6 } 8	
3s				↑↓				2	Arrows indicate the sense (clockwise or anti-clockwise) in which the electrons are spinning.
3p			↑↓	↑↓	↑↓			6	
3d		↑↓	↑↓	↑↓	↑↓	↑↓		10 } 18	
4s				↑↓				2	
4p			↑↓	↑↓	↑↓			6	
4d		↑↓	↑↓	↑↓	↑↓	↑↓		10	
4f	↑↓	↑↓	↑↓	↑↓	↑↓	↑↓	↑↓	14 } 32	

atomic number, while above atomic number 60, overlap of principal shells no longer occurs.

The completion of successive energy levels may be considered further (Table 3.11). Numbers below the coupled electrons refer to the magnetic quantum numbers. The other numbers are principal quantum numbers and the letters, subsidiary numbers, ($s \equiv 0$, $p \equiv 1$, $d \equiv 2$, $f \equiv 3$). Spin quantum numbers ($\pm \frac{1}{2}$) have not been shown. The electrons in the $3p$ orbitals have (for example) the following quantum numbers:

$3, 1, +1, +\frac{1}{2}$ $3, 1, 0, +\frac{1}{2}$ $3, 1, -1, +\frac{1}{2}$
$3, 1, +1, -\frac{1}{2}$ $3, 1, 0, -\frac{1}{2}$ $3, 1, -1, -\frac{1}{2}$

The numbers 2, 8, 18, 32 are the numbers of elements in successive periods of the periodic table.

By use of a small superscript number to represent the number of electrons in the orbital, or group of similar orbitals, indicated, electronic configurations may be listed:

H	$(1s)^1$
He	$(1s)^2$
Li	$(1s)^2(2s)^1$
Be	$(1s)^2(2s)^2$

B	$(1s)^2(2s)^2(2p)^1$
C	$(1s)^2(2s)^2(2p)^2$
N	$(1s)^2(2s)^2(2p)^3$
O	$(1s)^2(2s)^2(2p)^4$
F	$(1s)^2(2s)^2(2p)^5$
Ne	$(1s)^2(2s)^2(2p)^6$

This building-up is sometimes called the *aufbau process*.

Argon, the next inert gas, has the electronic configuration:

$$\text{Ar}\quad (1s)^2(2s)^2(2p)^6(3s)^2(3p)^6$$

In the next period a transition series starts with scandium and ends with nickel. This corresponds to the appearance of $3d$ orbitals:

K	$(1s)^2(2s)^2(2p)^6(3s)^2(3p)^6$	$(4s)^1$
Ca		$(4s)^2$

Tran-	Sc		$(4s)^2(3d)^1$
sition	Ti		$(4s)^2(3d)^2$
series	V		$(4s)^2(3d)^3$
	Cr	Argon	$(4s)^1(3d)^5$
	Mn	core	$(4s)^2(3d)^5$
	Fe		$(4s)^2(3d)^6$

Co	$(4s)^2(3d)^7$
Ni	$(4s)^2(3d)^8$
Cu	$(4s)^1(3d)^{10}$
Zn	$(4s)^2(3d)^{10}$

Krypton, the next inert gas, has the electronic configuration:

Kr $(1s)^2(2s)^2(2p)^6(3s)^2(3p)^6(4s)^2(3d)^{10}(4p)^6$

with 36 electrons in all.

Returning to the second period, the $2p$ electrons are shared between $2p_x$, $2p_y$, $2p_z$ orbitals, singly at first, and then in pairs:

C	$(1s)^2(2s)^2(2p_x)^1(2p_y)^1$
N	$(1s)^2(2s)^2(2p_x)^1(2p_y)^1(2p_z)^1$
O	$(1s)^2(2s)^2(2p_x)^2(2p_y)^1(2p_z)^1$
F	$(1s)^2(2s)^2(2p_x)^2(2p_y)^2(2p_z)^1$
Ne	$(1s)^2(2s)^2(2p_x)^2(2p_y)^2(2p_z)^2$

A mutally paired set of electrons in an outer shell is called a *lone pair*, e.g. $(2s)^2$ and $(2p_x)^2$ for oxygen, $(2s)^2$ for nitrogen, etc. The number of singly occupied orbitals determines the valency of an element. Nitrogen is tervalent:

$$
\begin{array}{cccccc}
 & 1s & 2s & 2p_x & 2p_y & 2p_z \\
\text{N} & \uparrow\downarrow & \uparrow\downarrow & \uparrow & \uparrow & \uparrow
\end{array}
$$

When electrons are transferred to form ions or shared to form covalent bonds, electrons are coupled in pairs with opposing spin.

The 's, p, d, f blocks' of elements

The modern Periodic Table reflects the order of energy levels in the atoms of elements. Sometimes elements are described as belonging to a particular 'block' according to whether s, p, d, or f levels are being filled, and the number of elements in a period will depend on the number of 'blocks' traversed, each holding sets of 2, 6, 10 and 14 corresponding to s, p, d and f orbitals. A block diagram of the Periodic Table in Fig. 3.5 summarises the position.

Note that the d block is split by the f block, and helium, $(1s)^2$, is isolated on the right.

Quantitative evidence for assigning electronic configurations to atoms

Ionization energy, ionization potential

The amount of energy required to eject one electron from an isolated atom of an element, leaving both the resulting ion and electron without surplus kinetic energy, is called the first ionization energy (or ionization potential) of the element. Successive removal of electrons requires input of the 1st, 2nd, 3rd, ... ionization energies: see Table 3.12.

Table 3.12 Ionization energy

Ionization	Process	Ionization energy/ kJ mol^{-1} (273 K)
1st	Ca $\rightarrow$ Ca$^+$ + e$^-$	590
2nd	Ca$^+$ $\rightarrow$ Ca^{2+} + e$^-$	1148
3rd	Ca^{2+} $\rightarrow$ Ca^{3+} + e$^-$	4940

Energy is always required to draw the electron away from the nuclear charge of the atom or ion and is a measure of the bonding energy of the electron removed. The value is always positive. There is always a surge in value after removal of the valency electrons, as shown by the 3rd ionization energy in the table.

The first ionization energy may be measured by the potential difference required to give electrons, which are being used to bombard gaseous atoms, enough energy to eject the first electron from these atoms. Prior to the ejection of the electrons, new spectral lines appear as electrons return to lower energy levels (shells) within the atoms after promotion to higher levels (outer shells). As the potential is raised, ionization is reached. The first ionization energy is therefore a measure of the energy required

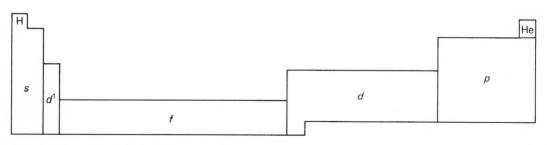

Fig. 3.5 Orbitals and the Periodic Table: s, p, d, f blocks

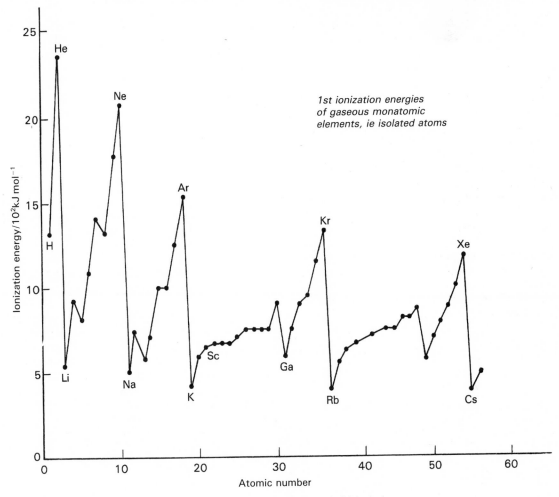

Fig. 3.6 Ionization potential plotted against atomic number for elements 1–56 inclusive

to remove the first electron from an isolated atom (in the gaseous state). It is a periodic property of atoms and is illustrated in Fig. 3.6 for elements 1–56 (hydrogen to barium) inclusive. Ionization energy increases along a period with increasing proton number because the effective nuclear charge increases while electrons are being accommodated in the same quantum shell i.e. at the same energy level; it decreases down the groups because electrons in outer shells tend to be shielded more from the nucleus. Mutual repulsion between electrons of opposite spin in the same quantum level accounts for minor dips in the curve. Values for ionization energies are given with individual group studies later.

Close examination of the curve provides evidence for the arrangements of electrons known as sub-shells or orbitals. In building up atoms from lithium to neon, the first peak corresponds to completion of the $2s$ orbital, the next illustrates that the incoming $2p$ electrons remain unpaired until four electrons have to be accommodated. Finally, the stability of the noble gas electronic configuration should be noted. The ns^2np^6 ($1s^2$ for He) configurations are the most stable in the Periodic Classification, all requiring a high input of energy for removal of an electron.

	$1s$	$2s$	$2p$		Configuration
Li	↑↓	↑			$(1s)^2(2s)^1$
Be	↑↓	↑↓			$(1s)^2(2s)^2$
B	↑↓	↑↓	↑		$(1s)^2(2s)^2(2p_x)^1$
C	↑↓	↑↓	↑ ↑		$(1s)^2(2s)^2(2p_x)^1(2p_y)^1$
N	↑↓	↑↓	↑ ↑ ↑		$(1s)^2(2s)^2(2p_x)^1(2p_y)^1(2p_z)^1$
O	↑↓	↑↓	↑↓ ↑ ↑		$(1s)^2(2s)^2(2p_x)^2(2p_y)^1(2p_z)^1$
F	↑↓	↑↓	↑↓ ↑↓ ↑		$(1s)^2(2s)^2(2p_x)^2(2p_y)^2(2p_z)^1$
Ne	↑↓	↑↓	↑↓ ↑↓ ↑↓		$(1s)^2(2s)^2(2p_x)^2(2p_y)^2(2p_z)^2$

In the later periods it will be seen that the d^5 and f^7 half-filled shell configurations have enhanced stabilities similar to that for p^3 above, an illustration of *Hund's Rule of maximum multiplicity* and the idea of *exchange energy*.

Successive ionization energies and electronic configuration

Further evidence for the orderly build up of the electronic configuration of atoms comes from a study of the energy required to remove electrons one by one. When the measured values are plotted against the *order of ionization*, or *ionization number* which equals the number of electrons removed, this is apparent. However, because of the wide range of values of ionization energies it is usual to plot the logarithm of successive ionization energies against ionization numbers, i.e. expressing ionization energies as powers of 10 and plotting

the index e.g. instead of (say) plotting 1000, this is expressed as 10^3 and plotted as 3; 100 000 would be plotted as 5, rather like pH plots except that here minus the logarithm is used to convert negative indices into positive numbers. Tables 3.13 and 3.14 show the successive ionization energies of four elements (of different groups) from the early periods of the Periodic Table, and their logarithms, and Fig. 3.7 the respective plots for the latter.

Each sequence of points lies on a rising curve showing that successively larger ionization energies are required to eject an electron from the atom. As each electron goes, so some screening is lost and the positive nuclear charge exerts a more powerful attraction for the electrons left. Steep rises between certain successive (logarithms of) ionization energies are seen, showing that proportionately more energy is suddenly required to shift an electron. We may deduce that this electron was located nearer to the nucleus than the previous one because the effect is more pronounced than what might have been expected from a progressive rise in effective proton number. The broken vertical lines therefore denote a change in quantum shell. In the light of what has been discussed before, for potassium the first electron ejected has come from the $4s$ orbital and not from the $3d$ orbital. Note the disposition of the curves: because of its higher effective nuclear charge for a given number of residual electrons, the curve for potassium lies above that for sulphur, and so on.

Table 3.13   Successive ionization energies/kJ mol^{-1} (273 K)

Number of electrons removed	Atomic number and symbol of element			
	9 **F**	12 **Mg**	16 **S**	19 **K**
1	1679	737	1003	419
2	3377	1447	2258	3068
3	6040	7728	3377	4435
4	8413	10544	4564	5858
5	10999	13640	6995	(7741)
6	15147	18033	8490	9623
7	17849	21757	27111	11380
8	92042	25690	30890	14937
9	107200	28200	36564	
10		28500		
11		331000		

Table 3.14   Log$_{10}$ (ionization energy/kJ mol^{-1})

Ionization number	9 **F**	12 **Mg**	16 **S**	19 **K**
1	3.25	2.87	3.00	2.62
2	3.53	3.16	3.35	3.49
3	3.78	3.89	3.53	3.65
4	3.92	4.02	3.66	3.75
5	4.04	4.14	3.84	(3.89)
6	4.18	4.26	3.93	3.98
7	4.25	4.34	4.43	4.06
8	4.96	4.41	4.49	4.17
9	5.03	4.50	4.56	
10		4.55		
11		5.20		

Electron affinity

Traditionally this is the energy released when an electron unites with an isolated atom, ion or molecule: it is the energy associated with the reverse of ionization. However, assigning a positive sign to the release of heat reverses normal practice. The resulting confusion is resolved by giving the equation for the process:

$$X + e^- \rightarrow X^-; \Delta H = \text{electron affinity}$$

Theoretically at least the electron-cloud around the nucleus of an atom is not confined to the immediate proximity of that nucleus so the nuclear charge is never fully balanced by the electrons. This leaves some residual attraction by atoms for electrons, with a consequent release of energy. On the other hand, second (and higher) electron affinities are necessarily large and endothermic, energy being

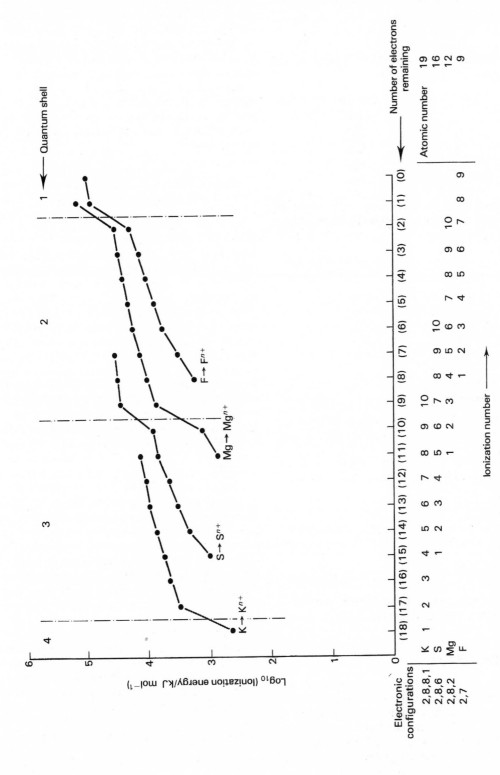

Fig. 3.7 Electron configurations and successive ionization energies

required to overcome electrical repulsion from the charged ion:

$$e.g. \; O^- + e^- \rightarrow O^{2-}$$

Electron affinities for Group VII (halogens) are given below.

Process	Electron affinity/ $kJ\,mol^{-1}$ (273 K)
$F\,(g) + e^- \rightarrow F^-(g)$	-349
$Cl\,(g) + e^- \rightarrow Cl^-(g)$	-364
$Br\,(g) + e^- \rightarrow Br^-(g)$	-343
$I\,(g) + e^- \rightarrow I^-(g)$	-314

These values may be compared with those of the ionization energies of Group 1 (alkali metals) elements:

Process	Ionization energy/ $kJ\,mol^{-1}$ (273 K)
$Li\,(g) \rightarrow Li^+(g) + e^-$	$+520$
$Na\,(g) \rightarrow Na^+(g) + e^-$	$+496$
$K\,(g) \rightarrow K^+(g) + e^-$	$+419$
$Rb\,(g) \rightarrow Rb^+(g) + e^-$	$+403$
$Cs\,(g) \rightarrow Cs^+(g) + e^-$	$+375$

Thus the reaction

$$M\,(g) \rightarrow M^+(g) + e^- \quad \Delta H_1 \text{ positive}$$
$$\underline{X\,(g) + e^- \rightarrow X^-(g) \quad \Delta H_2 \text{ negative}}$$
$$X\,(g) + M\,(g) \rightarrow M^+X^-(g); \quad \Delta H = \Delta H_1 + \Delta H_2$$

(in which M and X belong to Groups I and VII respectively) is endothermic and requires an input of energy

e.g.
$$Cl\,(g) + e^- \rightarrow Cl^-(g); \qquad \Delta H = -364\,kJ\,mol^{-1}$$
$$\underline{Na\,(g) \rightarrow Na^+(g) + e^-; \quad \Delta H = +496\,kJ\,mol^{-1}}$$
$$\therefore Na\,(g) + Cl\,(g)$$
$$\rightarrow Na^+Cl^-(g); \qquad \Delta H = +132\,kJ\,mol^{-1}$$

The energy required is provided by the lattice energy. Refer to the Born-Haber cycle on p. 85.

Exchange energy

This concept comes from quantum mechanics and arises from the impossibility of distinguishing between electrons during the interaction of atoms. It is a function of the number of pairs of electrons with parallel spins (when clockwise and anticlockwise spins of electron pairs, ↑↓, are treated

independently) and different from the energy associated with the electrostatic interaction between nucleus and electrons, and between electrons. In the mathematical study of the hydrogen molecule, H_2, the total energy for the molecule appears to be lowered by allowing either electron to be near to either nucleus, the electrons being identical and subject to simultaneous attraction to the two positive nuclei (instead of one, as in the atom). Each electron is assumed to have a separate 'identity' so that

↑↑ = 2 electrons with parallel spins: 1 pair
↑↑↑ = 3 electrons with parallel spins: 3 pairs†
↑↑↑↑ = 4 electrons with parallel spins: 6 pairs
↑↑↑↑↑ = 5 electrons with parallel spins: 10 pairs

Thus ↑↑↑↓↓ depicts 5 electrons with $3 + 1 = 4$ pairs with parallel spins. This is the basis of *Hund's Rule* that in orbitals of equal energy electrons are assigned with parallel spins until coupling is unavoidable.

As an illustration the number of pairs of electrons with parallel spins may be compared for maximum pairing and minimum pairing (Hund's Rule obeyed). Table 3.15 shows that the exchange energy (directly dependent on the number of parallel spins) is greatest for p^2, p^3, p^4 configurations when Hund's Rule is obeyed. Without attempting to explore the mathematical treatment, we may accept that an extra stability is conferred on an atom where the exchange energy factor appears.

Similarly in Table 3.16 a comparison can be made for half-filled shells, the configurations p^3, d^5, and f^7, and in Table 3.17 for where shells are completed apparently prematurely, as for copper. Here the gain in exchange energy more than compensates for the loss due to transferring an electron between energy levels, from the $4s$ orbital to the $3d$ orbital, i.e. $(4s)^1(3d)^{10}$ is preferred to $(4s)^2(3d)^9$. Refer to Table 3.10. This is distinct from the configuration adopted by nickel in its ground state where the reverse is true, i.e. $(4s)^2(3d)^8$ is preferred to $(4s)^0(3d)^{10}$.

Finally, the term exchange energy must *not* be associated with an oscillation of electrons or any other picture. It arises from the mathematical equations of quantum mechanics and represents an extra stabilization factor in studying the energy of atoms, with no classical analogy to be drawn upon. Indeed, if we do not know which electron is which, to say that they change places is meaningless.

† Combinations may be *ab, ac, bc for abc*

Table 3.15 Indication of the difference in exchange energies for two possible configurations adopted by p electrons (see Difference column) assuming that exchange energy is proportional to the number of pairs of electrons with parallel spins

Number of p electrons	Maximum pairing	Number of pairs of parallel spins	Minimum pairing	Number of pairs of parallel spins	Difference
1	↑	0	↑	0	0
2	↑↓	0	↑ ↑	↑↑ = 1	1
3	↑↓ ↑	↑↑ = 1	↑ ↑ ↑	↑↑↑ = 3	2
4	↑↓ ↑↓	↑↑↓↓ = 2	↑↓ ↑ ↑	↑↑↑↓ = 3	1
5	↑↓ ↑↓ ↑	↑↑↑↓↓ = 4	↑↓ ↑↓ ↑	↑↑↑↓↓ = 4	0
6	↑↓ ↑↓ ↑↓	↑↑↑↓↓↓ = 6	↑↓ ↑↓ ↑↓	↑↑↑↓↓↓ = 6	0

Table 3.16 Exchange energy ($\propto$ number of pairs of parallel spins) compared for p^3, d^5, f^7 electron configurations

Number of electrons	Maximum pairing	Number of pairs of parallel spins	Minimum pairing	Number of pairs of parallel spins	Difference
p^3 : 3	↑↓ ↑	↑↑ = 1	↑↑↑	3	2
d^5 : 5	↑↓ ↑↓ ↑	↑↑↑↓↓ = 4	↑↑↑↑↑	10	6
f^7 : 7	↑↓ ↑↓ ↑↓↑	↑↑↑↑↓↓↓ = 9	↑↑↑↑↑↑↑	21	12

Using the long form of the Periodic Table

Clearly, by suitable compression this form of the Periodic Classification may be converted into the Mendeléeff Table. However, the creation of the subgroups leads to paradoxes such as the inclusion of Cu (copper), Ag (silver) and Au (gold) in juxtaposition to K (potassium), Rb (rubidium) and Cs (caesium), although these elements are quite dissimilar: further, Mn (manganese) and Br (bromine) provide another example of unlike elements placed together. The very real success of the Mendeléeff Table in other ways leads the unwary to seek resemblances where none exist. The long form avoids this. It shows the grouping of similar elements more clearly, both in the same vertical families and in the long period transition series. The gradation in properties and trends across the table is also more obvious.

Groups Ia and Ib are widely separated in the table which is now in current use. The elements of Group Ia are a well-defined family named the *alkali metals*: lithium, sodium, potassium, rubidium and caesium. Melting-points range from 186°C for lithium down to 30°C for caesium which has the highest atomic number. No data are available for francium but there is no doubt about its belonging to the alkali metals family. Of low density and soft enough to be easily cut with a knife, these metals are good conductors of electricity and fairly good conductors of heat, are soon corroded in air and react vigorously with water forming hydrogen and the metal hydroxide. With a dilute acid, the liberation of hydrogen proceeds with violence and the resulting salts are very similar to each other. With no physical or chemical resemblance to Group Ia are the so-called coinage metals of Group Ib: copper, silver and gold. Excellent conductors of electricity and heat, they are hard and show a high lustre with surface tarnishing for copper and silver. They have high melting-points: copper (1083°C), silver (961°C) and gold (1063°C), and have high densities. These metals do not react with water or non-oxidizing acids. The valency shown by Group Ia is 1 while one valency state of Group Ib is also 1. This is the only similarity.

Although by strict definition the elements of Group Ib are not transition elements it is often instructive to compare their characteristics with those of the true transition series. This is in accordance

Table 3.17 Exchange energy ($\propto$ number of pairs of parallel spins) compared for two possible electronic configurations for copper, ignoring the argon core of electrons

Configuration	Electron pairs ($3d$)	Number of pairs of parallel spins
$(4s)^1(3d)^{10}$	↑↓ ↑↓ ↑↓ ↑↓ ↑↓	↑↑↑↑↑↓↓↓↓↓ = 10 + 10 = 20
$(4s)^2(3d)^9$	↑↓ ↑↓ ↑↓ ↑↓ ↑	↑↑↑↑↑↓↓↓↓ = 10 + 6 = 16

with Mendeléeff's observations. Elements of the transition series are all metals. In varying degrees they show the ability to combine in various valency (and oxidation) states, form coloured ions and complex compounds readily and frequently, as elements or compounds and behave as catalysts. Corresponding salts sometimes form isomorphous crystals.

Quite different from the alkali metals except in the degree to which they hold their properties are the elements of Group VIIb: fluorine, chlorine, bromine, iodine and astatine (the *halogens*). These elements are all strongly reactive non-metals although there are little data available about the short-lived radioactive astatine. Non-conductors of electricity, they have melting-points ranging from $-223°C$ for fluorine to $+114°C$ for iodine, and boiling-points from $-118°C$ for fluorine to $+184°C$ for iodine. They form very similar compounds, with a well-marked gradation in reactivity, fluorine being the most reactive non-metal known. Each member of the halogens is separated from the nearest alkali metal by an inert gas.

The great value of the Periodic Table results from the way in which the detailed memorization of facts may be reduced. By the grouping of elements to show similarities, contrasts and trends in characteristics, a wider grasp and deep understanding of the subject may be attained. Although the properties of an element and its compounds may be generally predicted from a knowledge of the behaviour of nearby elements in the table there are pitfalls: some elements do show anomalous properties, chemical and physical, and predictions should be checked by use of suitable reference books or, if feasible, by direct experiment. The most useful form of the Periodic Table, designed for ease of reference, will be two-dimensional and the long form is the simplest: electrons are systematically added in successive steps along any period working from left to right and elements fall into vertical family groups. Admittedly, hydrogen is by itself and for this reason is sometimes included twice, above the alkali metals (as here) and above the halogens, but no geometrical solution will show the unique nature of hydrogen—unless it is set aside by itself! There seems to be no simple way in which the lanthanides and actinides can be directly included in the table. Splitting the (outer) transition series so as to include these elements would reduce the value of the rest of the table because it would become too wide and cumbersome.

The gradation of properties shown in using the Periodic Classification will be considered further in the next chapter, which deals with the subject of valency, with which chemical nature is firmly linked.

Further predictions using the Periodic Table—the transuranium elements

Before the Second World War, the elements thorium, protactinium and uranium were usually positioned below hafnium, tantalum and tungsten respectively in the Periodic Table. The synthesis of neptunium, a new element of atomic number 93, by McMillan and Abelson in 1940 altered the position radically. Neptunium was discovered during the investigation of the fission products (i.e. breakdown products) of uranium induced by neutrons. $^{239}_{93}Np$ results from the β decay of $^{239}_{92}U$ formed by neutron capture in $^{238}_{92}U$. Now the position immediately to the right of tungsten is occupied by rhenium. McMillan and Abelson showed that the properties of neptunium resembled those of uranium rather than rhenium, the implication being that a new transition series was being built up. This is called the *actinide series* and is shown at the bottom of the Periodic Table. The preliminary studies of the new element were done by radiotrace techniques on quantities of the order of 10^{-10} g, the radioactive character of the atoms being studied. Thorium, protactinium and uranium are no longer placed below hafnium, tantalum and tungsten.

Later in 1940, McMillan, Kennedy, Wahl and Seaborg bombarded uranium with deuterons and produced $^{238}_{93}Np$, another isotope of neptunium, which decayed to a new element, plutonium, $^{238}_{94}Pu$. Its military potential led to its production in 'visible quantity'. Cunningham and Werner prepared about a microgram of $^{239}_{94}Pu$ in 1943. Ultra-microchemical techniques in which volumes of liquids in the range of 10^{-1} to 10^{-5} cm^3 were measured to within 1% were developed and chemical work was performed under the microscope. Using atomic (or nuclear) reactors, plutonium has been produced in kilogram quantities. The synthesis of other elements followed rapidly. Americium ($^{241}_{95}Am$) and curium ($^{242}_{96}Cm$) were identified largely on the basis of prediction of properties from scrutiny of the Periodic Table in 1944–5. Berkelium ($^{243}_{97}Bk$) and californium ($^{245}_{98}Cf$) were synthesized in 1949–50. It must be realized that the quantities formed were very small (e.g. about 5000 atoms of californium) and the isotopes are generally very short-lived. The tervalent state (oxidation number $+3$) is generally the principal state from americium onwards, although available information is sparse. Solid CmF_4, curium (IV) fluoride has been prepared but the oxidation state does not survive aqueous solution. Berkelium (III) can be oxidized in solution to oxidation state $+4$. Einsteinium ($^{254}_{99}Es$) and fermium ($^{253}_{100}Fm$) were discovered in 1952 in the debris of the first test thermonuclear explosion. Mendelevium ($^{256}_{101}Md$), nobelium ($^{255}_{102}No$) and lawrencium ($^{257}_{103}Lr$) were then prepared by suitable particle bombardment methods and established, while more recently rutherfordium ($^{260}_{104}Rf$), which has also been named kurchatovium (Ku), and hahnium (105) have been claimed. In these experiments nuclei of light elements have been used, e.g. $^{11}_{5}B$, $^{12}_{6}C$, while the use of elements of even higher proton number, e.g. Se, has been proposed. By using the techniques of nuclear physics new elements have been synthesized by transmutation.

Ultra-microchemical work has shown that the actinide series resembles the lanthanides. The series is therefore expected to end with lawrencium, element No. 103. This is evident from scrutiny of the long form of the Periodic Table, Table 3.5, p. 35. The general family similarity with the actinides means that the investigation of one element gives information about the remainder.

The synthesis of elements beyond lawrencium should see the start of an outer transition series: rutherfordium (104) (under hafnium), hahnium (105) (under tantalum), 106 (under tungsten) and the end of the period with an element of inert gas configuration (118). In writing about his work in this field Seaborg predicted that element 102 would have a stable tervalent state and somewhat unstable bivalent state, with an unusually low density and relatively high volatility. Element 103 was expected to be only tervalent. Element 104 would probably be quadrivalent in solution only and resemble hafnium of the previous period. Element 105 should resemble niobium and tantalum. Element 106 should fit into the Cr, Mo and W sequence with valencies 3, 4, 5 and 6.

The synthesis of further elements is expected to become increasingly difficult because the predicted half-lives of the new isotopes are expected to be very short, of the order of seconds or minutes although there could be a sequence around elements 114 and 126 with somewhat higher stability. This means the development of fresh techniques for the identification of the new elements, and these will, no doubt, be evolved. To reflect on the choice of names of scientists and places used in naming these (and other) elements, and of certain SI units, is to embark on a fascinating study of the history of science.

4

The electronic theory of valency and the periodic classification

Simple definitions of valency

The idea of describing the valency, or combining power, of an element in terms of an element of indivisible combining power was introduced in Chapter 1. Hydrogen is believed to be such an element of indivisible combining power.

Two related definitions were stated. The valency of an element is equal to the number of hydrogen atoms with which one atom of the element will unite.

$$\frac{\text{Atomic mass}}{\text{Equivalent mass}} = \text{a small integer}$$

This small integer is usually the valency of the element in the compound in which it has the equivalent used. Some elements have more than one equivalent and, hence, more than one valency.

The valencies may be described by the adjectives univalent, bivalent, tervalent, quadrivalent, quinquevalent, sexvalent, septivalent and octavalent, in which Latin prefixes (1–8) are used with the Latin word (*valentia* = strength, capacity). Sometimes, less desirably, the Greek prefixes mono-, di-, tri-, tetra-, penta-, hexa-, etc. are used. Chlorine in HCl is univalent, oxygen in H_2O is bivalent and nitrogen in NH_3, tervalent. Valencies may often be deduced by considering other compounds. Sodium is univalent in NaCl and Na_2O as would follow from the hydride, NaH. Aluminium forms $AlCl_3$ and Al_2O_3, being tervalent in these compounds, while tin is bivalent in SnO and $SnCl_2$ but quadrivalent in SnO_2 and $SnCl_4$. With a knowledge of valency states and sometimes of chemical nature, the formulae of common compounds may be written.

While this short-cut to formulae is very useful, it in no way helps to the understanding of the forces resulting in combination nor does it explain the differences in reactivity of compounds of similar formulae.

The development of the concept of valency

Dalton's Atomic Theory (1807) gives a picture of a chemical reaction as a process involving a regrouping of atoms. When more than one compound is formed between the atoms of two elements, the Law of Multiple Proportions results as a direct consequence of assigning characteristic masses to the atoms involved. Because of the vast number of atoms reacting, the existence of isotopes does not make this law invalid as long as the isotopic ratio in the reactants is always the same. Dalton listed rules governing multiple combinations between pairs of elements and any chance that characteristic combining powers would be assigned to atoms at that time disappeared with his rejection of Avogadro's Hypothesis.

In 1800, an elementary form of electric cell was devised by Volta and is now known as the Voltaic Pile. With this, Nicholson and Carlisle electrolysed water in 1800 and Davy isolated sodium and potassium by electrolysis of their fused hydroxides in 1807. Chemical combinations were seen to be reversed by electrolysis and Berzelius published his *Electrochemical* or *Dualistic Theory* to account for this. Following the observation that during electrolysis oxygen and the acids are formed at the positive pole, he supposed that particles of compound were formed by the electrostatic attraction of the reactants: a salt particle, on this theory, would consist of a negatively charged acidic portion associated with a positively charged basic portion. In turn, these portions could be further subdivided into electropositive and electronegative elements. That chemical affinity was not linked to the magnitude of the electrical charges was made clear by Faraday's Laws of Electrolysis (1832–3) which showed the link between equivalent masses and electrical charge. Other difficulties included the overall neutrality of matter and multiple combination. The Electrochemical Theory survived until 1838 when Dumas discovered that chlorine (electronegative) could replace hydrogen (electropositive) in ethanoic acid (acetic acid) leaving a product, chloroethanoic acid (chloroacetic acid), which was fundamentally similar to acetic acid in chemical nature. (The compounds belonged to the same organic type: *The Theory of Types*.) Although Berzelius' theory had failed the quantitative test of Faraday's Laws and Dumas' discoveries, the idea of electrostatic attraction within molecules, with the classification of elements into electropositive and electronegative, has proved useful.

Edward Frankland in 1852 noted that there was a general symmetry about many formulae of inorganic compounds. He was impressed especially by compounds of nitrogen, phosphorus, arsenic and antimony; using modern notation, these were:

$$NH_3 \quad PH_3 \quad AsH_3 \quad SbH_3$$
$$NI_3$$
$$PCl_3 \quad AsCl_3 \quad SbCl_3$$
$$N_2O_3 \quad P_2O_3 \quad As_2O_3 \quad Sb_2O_3$$
$$N_2O_5 \quad P_2O_5$$
$$NH_4I \quad PH_4I$$

Frankland continued, 'without offering any hypothesis regarding the cause of this symmetrical grouping of atoms, it is sufficiently evident, from the examples just given, that such a tendency or law prevails, and that, no matter what the character of the uniting atoms may be, the combining-power of the attracting element, if I may be allowed the term, is always satisfied by the same number of these atoms.' By assuming a fixed valency for carbon

$(= 4)$ and certain other elements, hydrogen $(= 1)$, oxygen $(= 2)$, and nitrogen $(= 3)$, Kekulé developed the idea of combining power and introduced order thereby in organic chemistry. This led to rapid progress. The idea of fixed valencies was not so fruitful with other elements and Mendeléeff considered that some elements could assume more than one valency, a view held ever since. He recognized valency as a periodic property.

The periodicity of valency

The number describing a periodic group is also the value of a valency shown by the elements in that group. When there is more than one valency, the group number does not necessarily correspond to the most important value. The group number is equal to the minimum valency for Group Ib (Cu, Ag, Au) and the maximum for some others, for example, Group Vb (N, P, As, Sb, Bi). In the Mendeléeff Table, the only resemblance between the subgroups of a particular group lies in a common valency, the differences being most marked at each side of the table and least marked in the middle, for Group IV.

As an illustration of the periodicity of valency, fluorides, hydrides and oxides are listed for Periods 2, 3 and 4 (excluding transition and similar metals) in Table 4.1. When more than one compound is formed, the maximum valency has been selected because both fluorine and oxygen have the power to induce an element to exhibit its maximum combining capacity. Formulae in square brackets indicate compounds not fitting into the general periodic pattern while oxides of bromine are omitted because of their extreme instability. Notice

Table 4.1 The periodicity of valency illustrated by the hydrides, oxides and fluorides of the typical elements of Periods 2, 3 and 4

LiH	BeH_2		B_2H_6	CH_4	NH_3	OH_2	FH	Ne
Li_2O	BeO		B_2O_3	CO_2	N_2O_5	—	$[F_2O]$	
LiF	BeF_2		BF_3	CF_4	$[NF_3]$	$[OF_2]$	—	
NaH	$(MgH_2)_x$		$(AlH_3)_x$	SiH_4	PH_3	SH_2	ClH	Ar
Na_2O	MgO		Al_2O_3	SiO_2	P_4O_{10}	SO_3	Cl_2O_7	
NaF	MgF_2		AlF_3	SiF_4	PF_5	SF_6	$[ClF_3]$	
KH	CaH_2		Ga_2H_6	GeH_4	AsH_3	SeH_2	BrH	Kr
K_2O	CaO		Ga_2O_3	GeO_2	As_4O_{10}	$[SeO_2]$	†	
KF	CaF_2		GaF_3	GeF_4	AsF_5	SeF_6	$[BrF_5]$	

The valencies adopt the following general pattern:

Hydrides	1	2	3	4	3	2	1	0
Oxides	1	2	3	4	5	6	7	0
Fluorides	1	2	3	4	5	6	—	0

† Br_2O, BrO_2, Br_3O_5 are all very unstable.

that in certain cases, addition of valency towards oxygen and hydrogen for an element gives eight. This holds for Group IV onward and is an aspect of the 'rule-of-eight' proposed by Abegg in 1904.

The basic assumptions of the electronic theory

The magnitude of the energy change involved in a chemical reaction is only sufficient to disturb the outer electrons of the atoms interacting. For this reason, electrons are usually divided into *core electrons*, composing the inner electrons which are not affected, and *valence or valency electrons*, which are those regrouped.

The chemical nature of the inert or noble gases was attributed, in the previous chapter, to possession of completed shells of electrons although some of these could be expanded later to explain the existence of the various transition series.

The basis of present theories about chemical combination may be traced back to propositions advanced independently in 1916 by W. Kossel and G. N. Lewis. The former was struck by the fact that in the Periodic Table strongly electronegative elements are separated from strongly electropositive elements by the inert gases. It was suggested that:

during a chemical reaction, the atoms of an element adjust their electronic configuration to that of the nearest inert gas.

While a few elements close to helium have two electrons (the helium configuration) after the reaction, many others finally have eight electrons in the valency shell. They are said to obey 'the octet rule'. However, *there are certain other configurations which result from chemical combination* and these will be described in their proper context.

The ionic bond and the formation of ions

In 1916, Kossel suggested that

an atom acquired the electronic configuration of an inert gas by the loss or gain of electrons.

The result is the formation of ions. This valency bond, which is an electrostatic attraction between ions, is called the ionic bond or electrovalent bond. There is no molecule in the kinetic theory sense and during electrolysis of either the fused compound or solution, ions will migrate towards oppositely charged poles.

Consider a hypothetical reaction between atoms of lithium and fluorine to form lithium fluoride. Inserting the outer shell of valency electrons, we may describe the formation of lithium fluoride as follows:

$$\text{Li}^{\circ} \;+\; {}_{\circ}^{\circ\circ}\text{F}_{\circ}^{\circ} \longrightarrow \big[\text{Li}\big]^{+} \; \Big[{}_{\circ}^{\circ\circ}\text{F}_{\circ}^{\circ}\Big]^{-}$$

Protons (Z)	3	9	3	9
Electrons	2, 1	2, 7	2 (He)	2, 8 (Ne)
	lithium atom	fluorine atom	lithium ion	fluoride ion

lithium fluoride

The transference of one electron from the atom of lithium to that of fluorine, to form respectively lithium and fluoride ions, leaves the former ion with a net positive charge of one unit and the latter with an equal net negative charge. While the charge of each nucleus remains the same, the electronic configurations are those of inert gases. Lithium assumes the electronic configuration of helium and fluorine assumes that of neon. The ions assume positions of equilibrium, about which they vibrate, in the crystal lattice. The bond diagram for lithium fluoride may be written:

$$\big[\text{Li}\big]^{+} \Big[{}_{\circ}^{\circ\circ}\text{F}_{\circ}^{\circ}\Big]^{-} \quad \text{or simply, Li}^{+}\text{F}^{-}$$

since F^- has a completed shell of electrons, or LiF.

Consider a similar reaction between atoms of elements two places from inert gases. For the synthesis of calcium sulphide:

$$\text{Ca}_{\circ}^{\circ} \;+\; {}^{\circ\circ}_{\circ\circ}\text{S}_{\circ}^{\circ} \longrightarrow \big[\text{Ca}\big]^{2+} \; \Big[{}^{\circ\circ}_{\circ\circ}\text{S}_{\circ}^{\circ}\Big]^{2-}$$

Protons (Z)	20	16	20	16
Electons	2, 8, 8, 2	2, 8, 6	2, 8, 8 (Ar)	2, 8, 8 (Ar)
	calcium atom	sulphur atom	calcium ion	sulphide ion

calcium sulphide

Each atom assumes the electronic configuration of argon, the charge on the nucleus remaining unchanged. Omitting the octet of electrons in the outer shell of the sulphide ion, calcium sulphide may be written $Ca^{2+}S^{2-}$ or CaS.

Aluminium oxide is probably an ionic compound. A scheme in which the atoms assume the electronic configuration of neon may be devised:

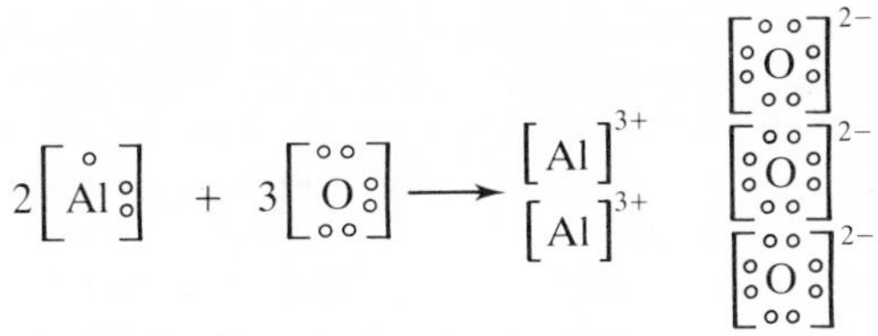

$$2\left[\text{Al}\right] + 3\left[\text{O}\right] \longrightarrow \begin{matrix}\left[\text{Al}\right]^{3+} \\ \left[\text{Al}\right]^{3+}\end{matrix} \quad \begin{matrix}\left[\text{O}\right]^{2-} \\ \left[\text{O}\right]^{2-} \\ \left[\text{O}\right]^{2-}\end{matrix}$$

This may be written

$$(\text{Al}^{3+})_2(\text{O}^{2-})_3 \quad \text{or} \quad \text{Al}_2\text{O}_3$$

	13	8	13	8
Protons (Z)	2, 8, 3	2, 6	2, 8 (Ne)	2, 8 (Ne)
Electrons	aluminium atom	oxygen atom	aluminium ion	oxygen (oxide) ion

aluminium oxide

By arranging for the electrons shed by the metal in forming the positive ion, or cation, to be taken by the non-metal in forming negative ions or anions, the *bond diagrams*, as these structures are called, may be written for any compound formed by this method. The ionic charge equals the valency.

The chlorides and bromides of sodium and strontium may be written:

$$\text{Na}^+\text{Cl}^- \qquad \text{Sr}^{2+}(\text{Cl}^-)_2$$
$$\text{Na}^+\text{Br}^- \qquad \text{Sr}^{2+}(\text{Br}^-)_2$$

or simply

$$\text{NaCl} \qquad \text{SrCl}_2$$
$$\text{NaBr} \qquad \text{SrBr}_2$$

$$\text{Na}^\circ \longrightarrow \text{Na}^+ + e^-$$

	11	11	—	
Protons (Z)				
Electrons	2, 8, 1	2, 8 (Ne)	1	neon electronic configuration

* sodium atom $\longrightarrow$ sodium ion $+$ one electron

$$\text{Sr} \longrightarrow \text{Sr}^{2+} + 2e^-$$

	38	38	—	
Protons (Z)				
Electrons	2, 8, 18, 8, 2	2, 8, 18, 8 (Kr)	2	krypton electronic configuration

strontium atom $\longrightarrow$ strontium ion $+$ two electrons

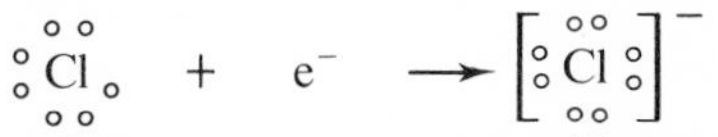

$$\text{Cl} + e^- \longrightarrow \left[\text{Cl}\right]^- \quad \text{or simply, Cl}^-$$

	17	—	17	
Protons (Z)				
Electrons	2, 8, 7	1	2, 8, 8 (Ar)	argon electronic configuration

chlorine atom $+$ one electron $\longrightarrow$ chloride ion

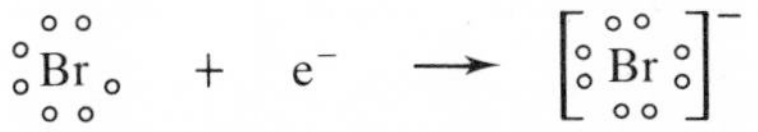

$$\text{Br} + e^- \longrightarrow \left[\text{Br}\right]^- \quad \text{or simply, Br}^-$$

	35	—	35	
Protons (Z)				
Electrons	2, 8, 18, 7	1	2, 8, 18, 8 (Kr)	krypton electronic configuration

bromine atom $+$ one electron $\longrightarrow$ bromide ion

* Normally the symbol of an element etc. represents that quality (in grams) equal to one mole, i.e. the relative atomic mass in this case. The electron would need to be replaced by one mole of electrons in the sodium equation, i.e. one Faraday $(F = Le)$ of negative electrical charge, and two moles $(2F)$ of negative electrical charge in the second.

Hydrogen forms two ions:

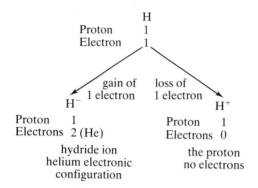

	H
Proton	1
Electron	1

gain of 1 electron   loss of 1 electron

H⁻

Proton	1
Electrons	2 (He)

hydride ion
helium electronic
configuration

H⁺

Proton	1
Electrons	0

the proton
no electrons

The proton, H^+, is found in the discharge tube while the negative ion, H^-, occurs in the hydrides of most metals of Groups Ia and IIa:

e.g. Li^+H^- $Ca^{2+}(H^-)_2$

In solution, the hydrogen ion (H^+) is attached to a solvent molecule. In aqueous solution the ion is H_3O^+ and will be described later. In liquid ammonia, the hydrogen ion becomes NH_4^+ (ammonium ion).

Extending the octet rule

Transition elements and those elements just after a transition series in the Periodic Table clearly can-

not form ions with inert gas electronic configurations. If they could, there would be a Co^{9+} ion for cobalt to give but one example to show the high charges involved, from which electrons would not be able to escape. Examples of the formation of ions without the electronic configuration of an inert gas will be chosen from Period 4.

The metals vanadium to nickel inclusive form bivalent ions of greatly differing stability, described in detail in Chapter 24. In Table 4.2, the ions are shown under atoms with the same number of electrons.

Of special interest is the formation of *the 2, 8, 18 configuration* or *the 18 electron group configuration*. This term also describes the corresponding configurations of the two following periods, 2, 8, 18, 18 and 2, 8, 18, 32, 18. The numbers of electrons held but not their configurations, correspond to the numbers held by nickel, palladium and platinum respectively. The grouping is called a *nickel group configuration* and sometimes, the *pseudo inert gas electronic configuration*. In examples, the 18 electron group may conveniently be abbreviated to *(core), 18* which will apply to all three periods. The 18 electron group pattern is created in the formation of copper(I) ions, Cu^+, of zinc ions, Zn^{2+}, and gallium(III) ions, Ga^{3+} from the metals. The ion Ge^{4+} occurs in a natural form of germanium dioxide, GeO_2. Arsenic forms no

Table 4.2 Bivalent ions of some transition metals

Protons	21 Sc	22 Ti	23 V	24 Cr	25 Mn	26 Fe	27 Co	28 Ni
Electrons (isolated atoms)	2 8 9 2	2 8 10 2	2 8 11 2	2 8 13 1	2 8 13 2	2 8 14 2	2 8 15 2	2 8 16 2

Protons		23 V^{2+}	24 Cr^{2+}	25 Mn^{2+}	26 Fe^{2+}	27 Co^{2+}	28 Ni^{2+}
Electrons		2 8 11	2 8 12	2 8 13	2 8 14	2 8 15	2 8 16

vanadium (II)   manganese (II)   cobalt (II)
chromium (II)   iron (II)   nickel (II)

Table 4.3   The formation of 'pseudo inert gas electronic configuration' in Cu^+, Zn^{2+} and Ga^{3+}

Protons	**28** **Ni**	**29** **Cu**	**30** **Zn**	**31** **Ga**	**32** **Ge**	**33** **As**
Electrons (isolated atoms)	2 8 16 2	2 8 18 1	2 8 18 2	2 8 18 3	2 8 18 4	2 8 18 5

Protons	29 **Cu**$^+$
Protons	30 **Zn**$^{2+}$
Protons	31 **Ga**$^{3+}$
Electrons	2, 8, 18

All have the same number of electrons as Ni, but a different configuration.

Table 4.4   Ions with the 'zinc group electronic configuration': tin(II) and lead(II)

					32 **Ge**
Protons					
Electrons					2 8 18 4

Protons	50 **Sn**$^{4+}$		50 **Sn**$^{2+}$		50 **Sn**
Electrons	(core), 18		(core), 18, 2		2 8 18 18 4
	$\left(\begin{array}{c}46\\ \textbf{Pd}\end{array}\right)$	$\left(\begin{array}{c}47\\ \textbf{Ag}\end{array}\right)$	$\left(\begin{array}{c}48\\ \textbf{Cd}\end{array}\right)$	$\left(\begin{array}{c}49\\ \textbf{In}\end{array}\right)$	

Protons	82 **Pb**$^{4+}$		82 **Pb**$^{2+}$		82 **Pb**
Electrons	(core), 18		(core), 18, 2		2 8 18 32 18 4
	$\left(\begin{array}{c}78\\ \textbf{Pt}\end{array}\right)$	$\left(\begin{array}{c}79\\ \textbf{Au}\end{array}\right)$	$\left(\begin{array}{c}80\\ \textbf{Hg}\end{array}\right)$	$\left(\begin{array}{c}81\\ \textbf{Tl}\end{array}\right)$	

cations. This pattern of electrons results in a greater stability than that conferred by the irregular configuration of the transition elements, which are characterized by their multiplicity of valency states. On the other hand, it is less stable, i.e. less readily formed, than the simple ions of true inert gas electronic configuration. Table 4.3 shows the formation of the ions, Cu^+, Zn^{2+} and Ga^{3+}.

An interesting phenomenon appears when the ions formed within a vertical group of elements in this part of the table are compared. As the atomic number increases, the electronic configuration *(core), 18* gives way to *(core), 18, 2* as the most stable configuration adopted by atoms of these elements. The electrons which are retained are called an *inert pair*. The configuration *(core), 18, 2* which occurs in the three Periods 4, 5 and 6, is sometimes called a *zinc group configuration*. This is illustrated in Tables 4.4 and 4.5. In Group IVb simple ions are generally only found in compounds formed by tin and lead, although germanium may form them in some compounds in addition to Ge^{4+} in GeO_2 mentioned above. While tin(II) ions, Sn^{2+}, readily form the more stable tin(IV) ions, Sn^{4+}, lead(II) ions, Pb^{2+}, are more stable than lead(IV) Pb^{4+}, ions. Lead compounds containing Pb^{4+} may readily be converted into those of the bivalent metal. In Group Vb, antimony and bismuth but not arsenic form tervalent ions. The bismuth ion is more stable than that formed by antimony.

Anions of oxoacids

Ionic compounds are formed between the anions of oxoacids and the metals. The structure of the anion itself involves another type of bonding. Examples are:

$$(K^+)_2SO_4^{2-} \qquad Ca^{2+}CO_3^{2-} \qquad (Na^+)_3PO_4^{3-}$$

potassium sulphate   calcium carbonate   sodium phosphate

The hydration of ions

In aqueous solution the ionic charges attract water molecules. Cations draw to themselves the oxygen atoms of water molecules and form hydrated ions. This has a bearing on the stability of the ion in the presence of water. Examples are $Al(H_2O)_6^{3+}$, (the hexaquo-aluminium ion), $Fe(H_2O)_6^{2+}$, (the hexaquo-iron(II) ion), and the irregular structure

Table 4.5   Ions with the 'zinc group electronic configuration': antimony(III) and bismuth(III)

Protons					33 As
Electrons					2 8 18 5
Protons	51 Sb³⁺				51 Sb
Electrons	(core), 18, 2 ←				2 8 18 18 5
	(48 Cd)	(49 In)	(50 Sn)		
Protons	83 Bi³⁺				83 Bi
Electrons	(core), 18, 2 ←				2 8 18 32 18 5
	(80 Hg)	(81 Tl)	(82 Pb)		

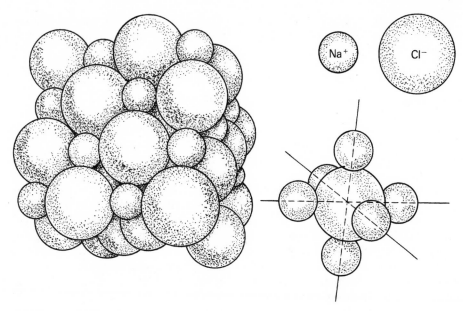

Fig. 4.1   The sodium chloride structure

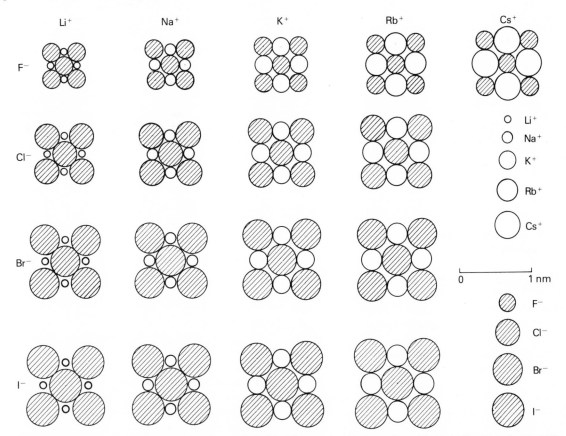

Fig. 4.2   The arrangement of ions in cube-face layers of alkali halide crystals with the sodium chloride lattice. Note that for LiCl, LiBr and LiI the anions are in mutual contact (from Pauling, 1945, *Nature of the Chemical Bond*, Corneil)

$Cu(H_2O)_6^{2+}$, (the hexaquo-copper(II) ion). Alternatively, the formulae may be given as $Al(OH_2)_6^{3+}$, etc, which suggests the bonding.

The solid structure of ionic compounds

With the condition that electrical neutrality must be preserved, the structures of ionic compounds are largely determined by the geometrical packing together of the ions.

X-ray analysis has shown that the halides of the alkali metals, except CsCl, CsBr and CsI, have the same *lattice*. This is called the sodium chloride structure (Figs 4.1 and 4.2). Ions of sodium and chlorine are packed together alternately so that each sodium ion is surrounded by six chloride ions and each chloride ion by six sodium ions. No conventional molecule exists. Table 4.6 lists ionic radii of alkali metals and halide ions.

Table 4.6   Ionic radii of some alkali metal and halide ions/nm

Li^+ 0.060	F^- 0.136
Na^+ 0.095	Cl^- 0.181
K^+ 0.133	Br^- 0.195
Rb^+ 0.148	I^- 0.216

The covalent bond and the formation of molecules

The idea of complete transference of electrons between atoms which are uniting cannot be used to explain the formation of molecules between atoms of the same element, such as nitrogen, N_2, oxygen, O_2, and hydrogen, H_2, and of non-electrolytes, such as phosphorus trichloride, PCl_3.

In 1916, Lewis suggested that

'the inert gas electronic configuration could be attained by the sharing of electrons between atoms'.

A pair of electrons held in a common region between two atoms, and shared by them, constitutes one covalent bond and can be conveniently represented by a single line drawn between the two atoms. According to modern valency theory, electrons may be visualized as coupled in pairs, the members of each pair spinning in opposite senses, clockwise and anticlockwise, and localized between the atoms concerned.

The shared electrons will be held between the linked atoms at some position which will depend on the electrical nature of the atoms: the pair being attracted nearer to the atom with the higher affinity for electrons, i.e. the more electronegative element. This results in a small electrical dipole. A dipole is a system of two equal and opposite electrical charges separated by a short distance. The product of the charge and the distance between the charges is called the *dipole moment*.

Inserting the outer shell of valency electrons, formation of covalent molecules may be shown schematically, the total number of electrons remaining unaltered. Each shared pair of electrons represents one covalent bond. A pair of electrons not used for bonding is called a *lone pair*.

Simple examples of covalent molecules are depicted opposite (p. 59). In these structures the signs ○, • and × have been used to show the origin of the bonding electrons. This is for convenience. The electrons cannot be distinguished from each other in the final structure.

Hydrides of some Period 2 elements are shown in Table 4.7 and the trichlorides of the elements of Group Vb in Table 4.8.

Extending the octet rule

It is generally accepted that the elements in Period 2, lithium to fluorine, obey the octet rule rigorously.

However, there are other structures which have more electrons than can be accommodated in the valency octets. Ingenious use of the Theory of Resonance (see below) may enable some compounds to obey the rule but the balance of opinion favours the view that the valency octet may sometimes be expanded.

Hydrogen, H₂

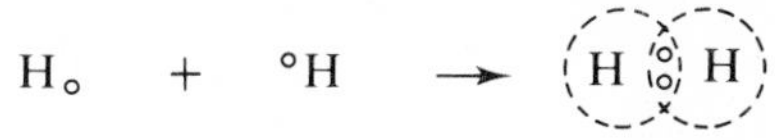

H₀ + °H ⟶ (H ⦂ H)   or simply, H—H

Protons (Z)	1	1	1 1	Each atom assumes
Electrons	1	1	(2)	the helium electronic
			one shared pair	configuration: 2

hydrogen   hydrogen   hydrogen
atom   atom   molecule

Hydrogen is univalent

Chlorine, Cl₂

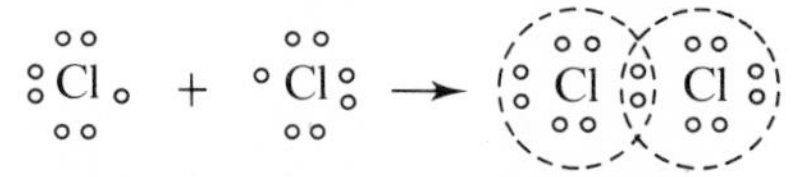

⦂Cl₀ + °Cl⦂ ⟶ (⦂Cl ⦂ Cl⦂)   or simply, Cl—Cl

Protons (Z)	17	17	17 17	Each atom assumes
Electrons	2, 8, 7	2, 8, 7	2, 8, 6 (2) 6, 8, 2	the argon electronic
			one shared pair	configuration: 2, 8, 8

chlorine   chlorine   chlorine
atom   atom   molecule

Chlorine is univalent

Nitrogen, N₂

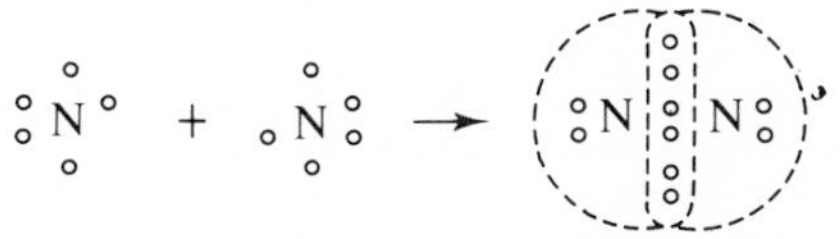

⦂N° + ₀N⦂ ⟶ (⦂N ⦂⦂⦂ N⦂)   or simply, N≡N

Protons (Z)	7	7	7 7	Each atom assumes
Electrons	2, 5	2, 5	2, 2 (6) 2, 2	the neon electronic
			three shared pairs	configuration: 2, 8

nitrogen   nitrogen   nitrogen
atom   atom   molecule

Nitrogen is tervalent

Carbon dioxide, CO₂

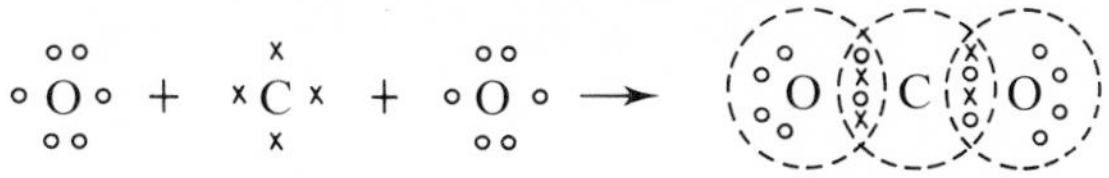

°O° + ×C× + °O° ⟶ (°O× C ×O°)   or simply, O=C=O

Protons (Z)	8	6	8	8 6 8	Each atom assumes
Electrons	2, 6	2, 4	2, 6	2, 4 (4) 2 (4) 4, 2	the neon electronic
				two shared two shared	configuration: 2, 8
				pairs pairs	

oxygen   carbon   oxygen   carbon dioxide
atom   atom   atom   molecule

Carbon is quadrivalent
Oxygen is bivalent

Hydrogen cyanide, HCN

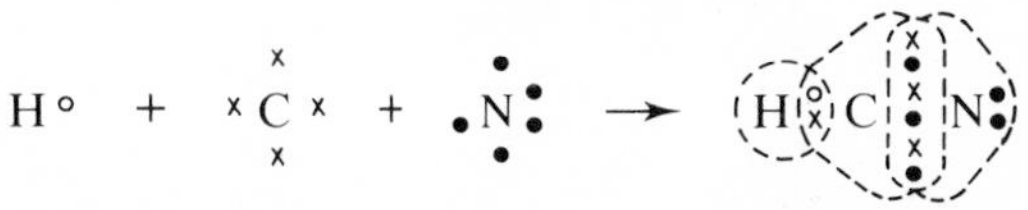

H° + ×C× + •N⦂ ⟶ (H×C×N⦂)   or simply, H—C≡N

Protons (Z)	1	6	7	1 6 7	Hydrogen assumes the
Electrons	1	2, 4	2, 5	(2) 2(6) 2,2	helium electronic
					configuration: 2
				one three	Carbon and nitrogen both
				shared shared	assume the neon electronic
				pair pairs	configuration: 2, 8

hydrogen   carbon   nitrogen   hydrogen cyanide
atom   atom   atom   molecule

Hydrogen is univalent
Carbon is quadrivalent
Nitrogen is tervalent

Table 4.7   The hydrides of carbon, nitrogen, oxygen and fluorine

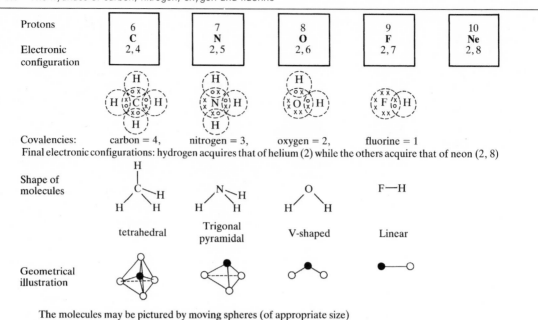

Protons	6 **C** 2, 4	7 **N** 2, 5	8 **O** 2, 6	9 **F** 2, 7	10 **Ne** 2, 8
Electronic configuration					

Covalencies: carbon = 4,   nitrogen = 3,   oxygen = 2,   fluorine = 1

Final electronic configurations: hydrogen acquires that of helium (2) while the others acquire that of neon (2, 8)

Shape of molecules: tetrahedral   Trigonal pyramidal   V-shaped   Linear

Geometrical illustration

The molecules may be pictured by moving spheres (of appropriate size) towards each other along lines (which then disappear) until they partly coalesce to give a bulbous mass with a diffuse outside

Table 4.8   The trichlorides of the Group Vb elements

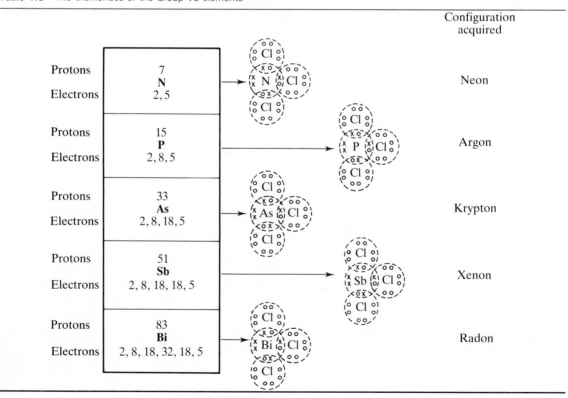

			Configuration acquired
Protons Electrons	7 **N** 2, 5		Neon
Protons Electrons	15 **P** 2, 8, 5		Argon
Protons Electrons	33 **As** 2, 8, 18, 5		Krypton
Protons Electrons	51 **Sb** 2, 8, 18, 18, 5		Xenon
Protons Electrons	83 **Bi** 2, 8, 18, 32, 18, 5		Radon

In Group IVb there is a series of stable ions of formulae SiF_6^{2-}, GeF_6^{2-}, SnF_6^{2-}; in Group Vb, PF_6^-, AsF_6^-, SbF_6^-; and in Group VI, stable compounds, SF_6, SeF_6, TeF_6 although the last-named is more reactive than the others. These ions and molecules have similar bond diagrams. Using silicon, phosphorus and sulphur as examples:

14	15	16
Si	**P**	**S**
2, 8, 4	2, 8, 5	2, 8, 6

Sulphur hexafluoride will have the structure:

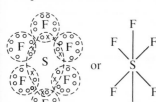

Fluorine assumes the neon electronic configuration: 2, 8

Sulphur assumes the electronic configuration: 2, 8, 12

Sulphur ends up with twelve electrons in its outer shell. The ions SiF_6^{2-} and PF_6^- may be formed by adding (in theory) six F atoms to Si^{2-} and P^- respectively. The electronic configurations of these last ions are the same as that of sulphur.

$$Si^{2-} \qquad P^-$$
$$2, 8, 6 \qquad 2, 8, 6$$

The bond diagrams may be constructed similarly:

For all of these structures, the disposition of fluorine atoms about the central atom or ion may be depicted as

which, though not drawn to scale, reveals an octahedral shape. Of course, the ions are met as salts, such as $K_2(SiF_6)$ and $K(PF_6)$, and are quite stable. Presumably the extra electrons occupy energy levels normally reached with the transition metals. Silicon and phosphorus assume the electronic configuration: 2, 8, 12. The derivation of the last two structures by substituting Si^{2-} and P^- in turn for S illustrates the **Isoelectronic Principle**:

'Simple structures containing the same number of valency electrons are represented by the same bond diagram.' (Penney and Sutherland, 1936.)

Covalency achieved by donation: the co-ionic bond

A pair of electrons shared between two atoms constitutes a single covalent bond. So far, each atom has contributed equally to each shared pair, but both electrons may come from the same atom. The covalent bond formed in this way is given a special name to indicate its origin. It may be called a dative bond or a co-ionic bond. The term *dative bond*, with its arrow sign, emphasizes the method of formation but disguises the essential covalent character of the link formed. For this reason, the term *co-ionic* will be preferred here and the link shown as a covalent bond between ions.

The term co-ordinate bond has also been used because this type of linkage explains the formation of certain complex compounds or *co-ordination compounds*, some of which were investigated and their structure elucidated by Werner from 1893 onwards. A valency problem arises because stable compounds are formed by the combination of molecules and ions which themselves have a separate, stable existence. A similar difficulty appears with some elements which form more than a single oxide. The structure of carbon dioxide may apparently be described quite simply in terms of the octet theory. But what about the structure of carbon monoxide?

Bond diagrams for various complex compounds are listed below. Note that the electrons used in creating a co-ionic bond originate from a lone pair. A molecule or ion which attaches itself to an atom in this way is known as a *ligand*. An ion formed in this way is called a *complex ion*.

Compounds formed from atoms and/or molecules

Boron trichloride forms very stable ammines with ammonia and its organic derivatives, amines. The ammonia complex is $BCl_3 . NH_3$.

The simplest bond diagram for BCl_3 leaves the boron atom with only a sextet of electrons, while in ammonia, the nitrogen atom has a lone pair of electrons:

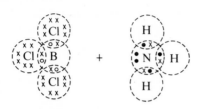

+

The electronic configuration of boron in combination is 2, 6, i.e., two short of the electronic configuration of neon

nitrogen has a lone pair of electrons

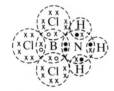

Both boron and nitrogen have the electronic configuration of neon: 2, 8

The electron notation shown is used to facilitate reference.

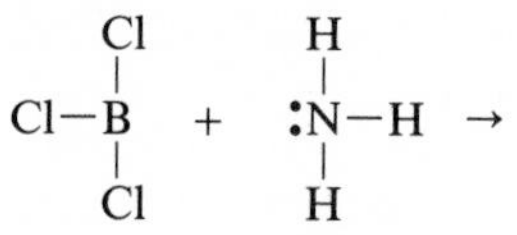

The boron–nitrogen linkage is by two electrons which belonged solely to nitrogen before the reaction. In a concise way, it may be shown either as a dative bond or a co-ionic bond:

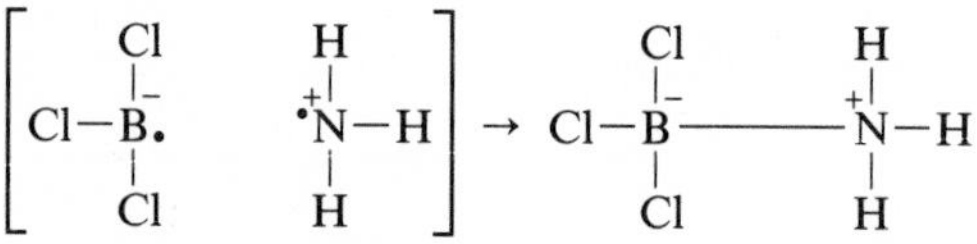

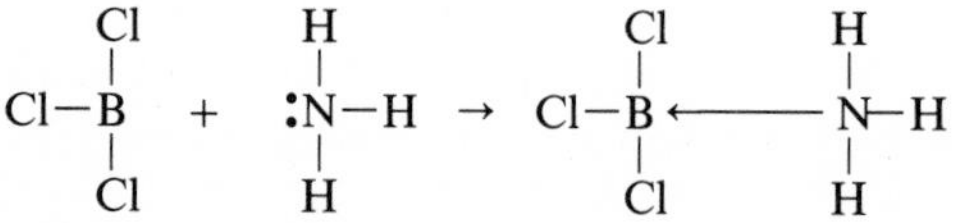

$$
\begin{array}{ccc}
\overset{\displaystyle Cl}{\underset{\displaystyle Cl}{|}}\!\!-\!B & + & \overset{\displaystyle H}{\underset{\displaystyle H}{|}}\!\!N\!-\!H \rightarrow \\
\end{array}
$$

Cl—B + :N—H → Cl—B ←——— N—H

B (2, 6)   N (2, 8)   B (2, 8)   N (2, 8)

The arrow represents a dative bond and points to the **acceptor** of electrons from the **donor** of the electrons. The original lone pair is now associated with both atoms.

Cl—B + :N—H →

B (2, 6)   N (2, 8)

$$
\left[Cl-\overset{Cl}{\underset{Cl}{|}}B\bullet \qquad \bullet\overset{H}{\underset{H}{|}}\overset{+}{N}-H \right] \rightarrow Cl-\overset{Cl}{\underset{Cl}{|}}\overset{-}{B}\text{——}\overset{H}{\underset{H}{|}}\overset{+}{N}-H
$$

B (2, 7)   N (2, 7)   B (2, 8)   N (2, 8)

transference of one electron from nitrogen to boron

The simplest bond diagram for boron trichloride is more correctly written with *back co-ordination*

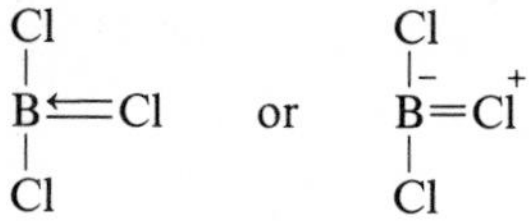

$$
\overset{Cl}{\underset{Cl}{|}}B\text{⇐}Cl \quad \text{or} \quad \overset{Cl}{\underset{Cl}{|}}\overset{-}{B}\text{=}\overset{+}{Cl}
$$

This leaves boron with an octet and is indicated by physical evidence (interatomic distances). The distribution of electric charge would be spread to leave all B–Cl distances equal.

Metallic nickel will react with carbon monoxide when warmed to form nickel carbonyl, a volatile, unstable, lethal compound. Carbon monoxide contains a co-ionic bond:

C   O
2, 4   2, 6

transfer one electron and compare with nitrogen

$\overset{-}{C}$ $\overset{+}{O}$ N   N
2, 5   2, 5   2, 5   2, 5

whereupon

$\overset{-}{C}$≡≡$\overset{+}{O}$ N≡≡≡N
2, 8   2, 8   2, 8   2, 8

Alternatively,

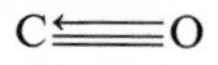

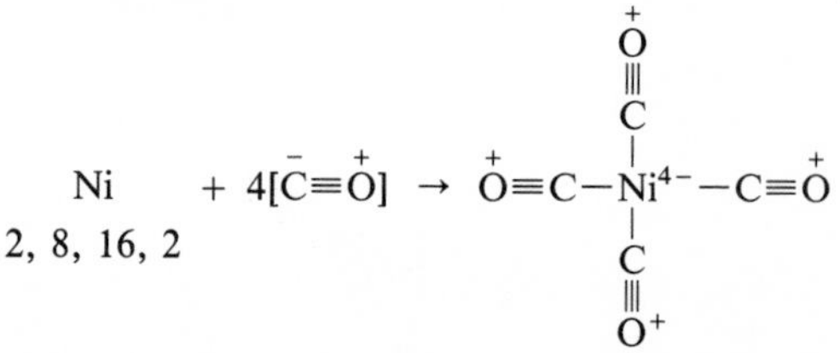

C≡≡O

$$
Ni + 4[\overset{-}{C}\text{≡}\overset{+}{O}] \rightarrow \overset{+}{O}\text{≡}C-\overset{Ni^{4-}}{}-C\text{≡}\overset{+}{O}
$$

Ni
2, 8, 16, 2

Nickel acquires the electronic configuration of krypton (2, 8, 18, 8) while carbon and oxygen both attain the electronic configuration of neon (2, 8)

Ammonia and hydrogen chloride gases interact with the formation of solid ammonium chloride:

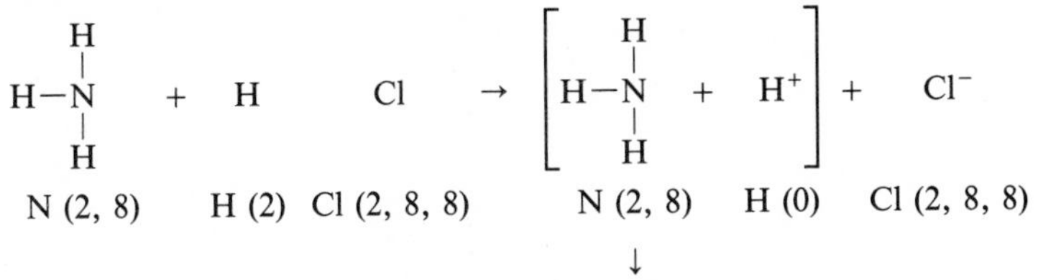

$$
\begin{array}{c}
\text{H} \\
| \\
\text{H}-\text{N} \\
| \\
\text{H}
\end{array}
\; + \; \text{H} \qquad \text{Cl} \;\rightarrow\;
\left[
\begin{array}{c}
\text{H} \\
| \\
\text{H}-\text{N} \\
| \\
\text{H}
\end{array}
\; + \; \text{H}^+
\right]
+ \quad \text{Cl}^-
$$

Note: The shape of the molecule is tetrahedral, like methane, CH_4.

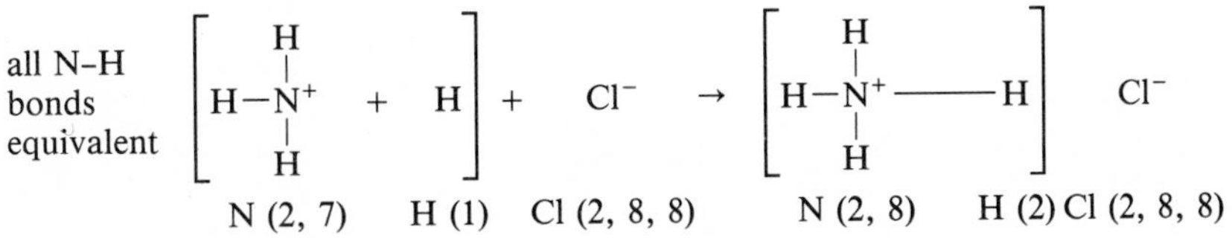

N (2, 8)   H (2) Cl (2, 8, 8)   N (2, 8)   H (0)   Cl (2, 8, 8)

↓

all N–H bonds equivalent
$$
\left[
\begin{array}{c}
\text{H} \\
| \\
\text{H}-\text{N}^+ \\
| \\
\text{H}
\end{array}
\; + \; \text{H}
\right]
+ \quad \text{Cl}^- \;\rightarrow\;
\left[
\begin{array}{c}
\text{H} \\
| \\
\text{H}-\text{N}^+\!\!-\!\!-\!\!-\text{H} \\
| \\
\text{H}
\end{array}
\right]
\quad \text{Cl}^-
$$

N (2, 7)   H (1)   Cl (2, 8, 8)   N (2, 8)   H (2) Cl (2, 8, 8)

The transference of the proton to ammonia confers a net positive charge on the ammonium ion. Each bond linking hydrogen atoms with the central nitrogen atom must be regarded as equivalent. The formation of ammonium chloride may be written concisely:

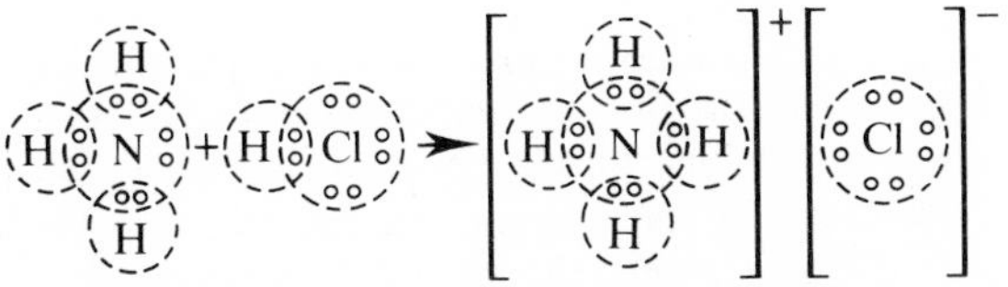

Hydrogen assumes the electronic configuration of helium (2), nitrogen assumes that of neon (2, 8) and chlorine, that of argon (2, 8, 8)

Hydrogen chloride gas reacts with water similarly:

$$HCl + H_2O \rightarrow H_3O^+ + Cl^-$$

The ion, H_3O^+, is a hydrated proton (H^+) and is called the hydroxonium ion, oxonium ion or hydronium ion. The ionization of hydrochloric acid, previously written in the manner of Arrhenius as

$$HCl \rightleftharpoons H^+ + Cl^- \quad \text{or} \quad HCl \rightarrow H^+ + Cl^-$$

when highly dissociated, is more correctly written as above. The very small proton is drawn close to the oxygen atom of a water molecule and away

from its chlorine atom:

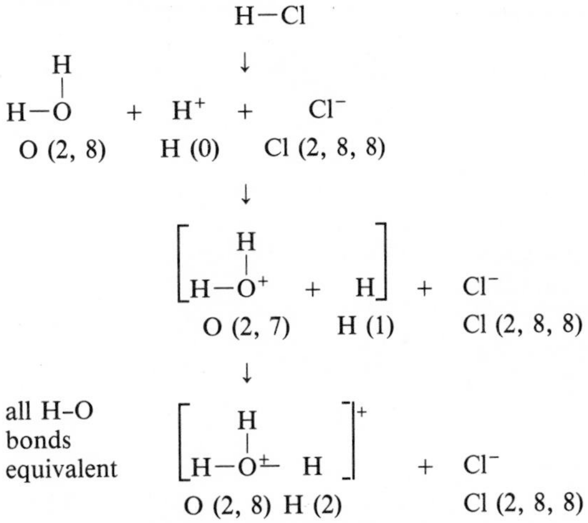

$$\text{H}-\text{Cl}$$
↓

$$
\begin{array}{c}
\text{H} \\
| \\
\text{H}-\text{O}
\end{array}
\; + \; \text{H}^+ \; + \quad \text{Cl}^-
$$

O (2, 8)   H (0)   Cl (2, 8, 8)

↓

$$
\left[
\begin{array}{c}
\text{H} \\
| \\
\text{H}-\text{O}^+ \; + \; \text{H}
\end{array}
\right]
+ \quad \text{Cl}^-
$$

O (2, 7)   H (1)   Cl (2, 8, 8)

↓

all H–O bonds equivalent
$$
\left[
\begin{array}{c}
\text{H} \\
| \\
\text{H}-\text{O}^{\pm} \;\; \text{H}
\end{array}
\right]^{+}
+ \quad \text{Cl}^-
$$

O (2, 8) H (2)   Cl (2, 8, 8)

Hydrogen assumes the electronic configuration of helium, oxygen assumes that of neon (2, 8) and chlorine, that of argon (2, 8, 8)

Compounds formed from ions and molecules

Copper(II) sulphate reacts with an excess of ammonia in aqueous solution with the formation of the deep blue tetramminecopper(II) ion, $Cu(NH_3)_4^{2+}$. This is one example of a large group of compounds, called 'ammines', formed between metal ions and nitrogen compounds, such as ammonia and amines, in which the lone pair of electrons previously held by the nitrogen is used as the linkage. They are examples of complex cations. $Cu(NH_3)_4^{2+}$ is loosely associated with two molecules of water as $Cu(NH_3)_4(H_2O)_2^{2+}$. The hexa-ammine, $Cu(NH_3)_6^{2+}$, sometimes quoted exists only in liquid ammonia. In both cases the extra two ligands create an irregular structure.

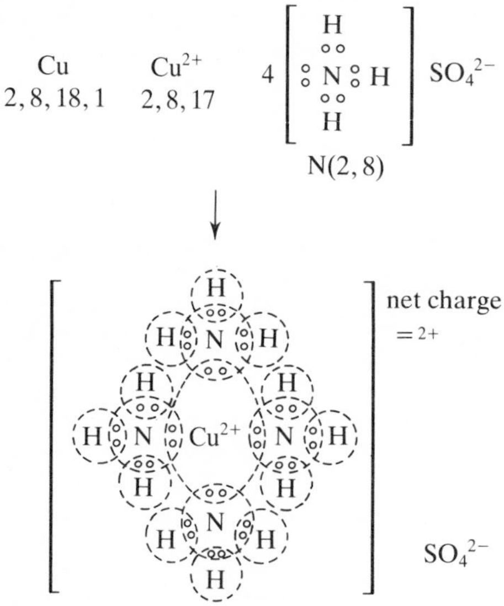

$$\begin{array}{cc} Cu & Cu^{2+} \\ 2,8,18,1 & 2,8,17 \end{array} \quad 4 \left[\begin{array}{c} H \\ {}^{\circ\circ} \\ {}^{\circ}_{\circ} N {}^{\circ}_{\circ} H \\ {}^{\circ\circ} \\ H \end{array} \right] SO_4^{2-}$$

N(2, 8)

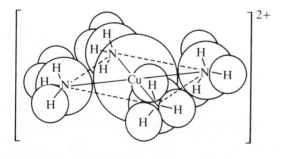

$$\begin{array}{c} \text{net charge} \\ = 2+ \end{array}$$

$$SO_4^{2-}$$

The copper atom now has the electronic configuration 2, 8, 18, 7. Hydrogen assumes the electronic configuration of helium (2) and nitrogen that of neon (2, 8).

The complex cation adopts a square planar structure:

The central copper atom has not achieved the electronic configuration of an inert gas. This is not unusual. The stability of these ions is associated with either an inert gas electronic configuration or a configuration very near to it. Tervalent cobalt forms a wide range of ammines of which one of the simplest is $[Co(NH_3)_6]Cl_3$. The bond diagram may be represented:

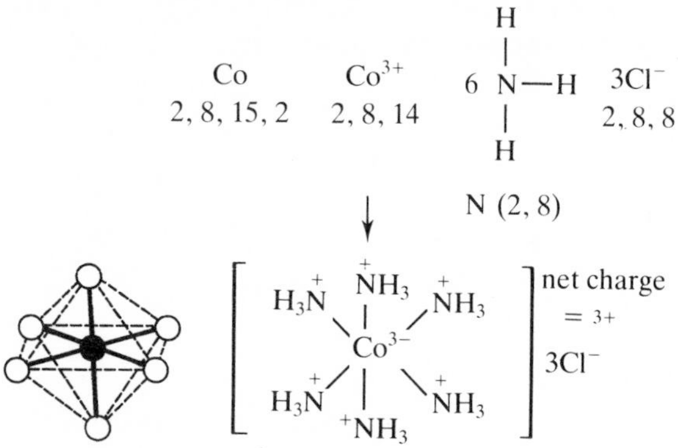

$$\begin{array}{ccc} Co & Co^{3+} & \overset{H}{\underset{H}{\overset{|}{N}}-H} \quad 3Cl^- \\ 2,8,15,2 & 2,8,14 & 6 \quad \begin{array}{c} | \\ H \end{array} \quad 2,8,8 \end{array}$$

N (2, 8)

octahedral shape

Cobalt requires the electronic configuration of krypton (2, 8, 18, 8), nitrogen assumes that of neon (2, 8) and hydrogen, that of helium (2). Chlorine has the argon electronic configuration (2, 8, 8) in the chloride ion.

Compounds formed from ions

Although potassium hexacyanoferrate(II) contains iron in the bivalent condition it does not give a positive test for iron(II). The iron may be shown to be present in the anion. The formula is written $K_4[Fe(CN)_6]$ where the square brackets enclose a complex anion which may be constructed from iron(II) and cyanide ions by the following scheme:

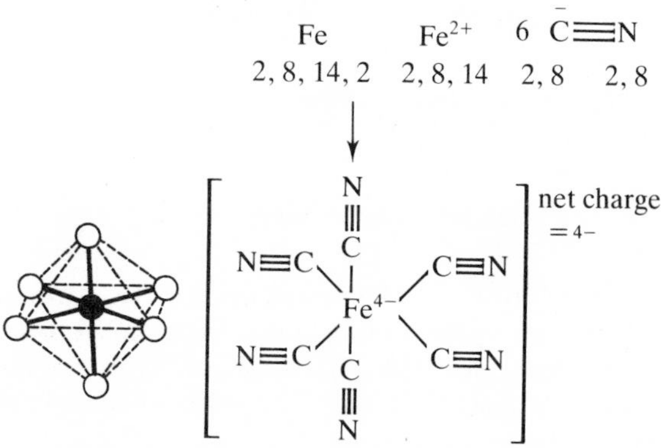

$$\begin{array}{cccc} Fe & Fe^{2+} & 6 & \bar{C}\equiv N \\ 2,8,14,2 & 2,8,14 & & 2,8 \quad 2,8 \end{array}$$

net charge = 4−

Iron acquires the electronic configuration of krypton (2, 8, 18, 8). Carbon and nitrogen both assume the electronic configuration of neon (2, 8).

Potassium hexacyanoferrate(III), $K_3[Fe(CN)_6]$, contains the complex hexacyanoferrate(III) ion and gives no test for simple iron(III) ions. Building the ion from iron(III) ions and cyanide ions, the structure becomes:

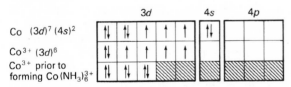

$$
\left[
\begin{array}{c}
N \\
\parallel\parallel \\
C \\
N\equiv C \diagdown \ \ \mid \diagup C\equiv N \\
Fe^{3-} \\
N\equiv C \diagup \ \ \mid \diagdown C\equiv N \\
C \\
\parallel\parallel \\
N
\end{array}
\right]
\begin{array}{l}
\text{net charge} \\
= 3-
\end{array}
$$

octahedral
shape

Iron may be given the electronic configuration 2, 8, 17, 8, being one electron short of the krypton configuration

The final electronic configuration of the central atom of a complex ion, where it is not that of an inert gas, is decided by a more detailed study of electronic configuration and is intimately connected with the shape of the complex ion.

The bonding in complex ions

There are several ways of approaching this highly complicated subject.

Pauling developed the valence bond theory route but to some degree this has been superseded. Consider the $Co(NH_3)_6^{3+}$ ion shown earlier. Hybridization (see p. 81) of the immediately available orbitals of the central atom gives octahedrally disposed orbitals available for bonding, using the lone pair electron-clouds of the *ligand*, in this case, ammonia. The outer electronic configuration of the cobalt atom in terms of electron spin may be represented:

The next *six* orbitals are 3d, 3d, 4s, 4p, 4p, 4p and when these are occupied by the electron pairs of incoming molecules of ammonia they combine to give an octahedral disposition of electron density lobes, the levels being termed *hybridized* [d^2sp^3] *orbitals*.

The spin and orbital motion of unpaired electrons gives rise to magnetic properties of the compound which contains them. The number of unpaired electrons can often be determined experimentally. Information about the number of unpaired *d* electrons leads to a further development of the Pauling approach. Consider complexes of iron(III) and their unpaired *d* electrons

species	$Fe(H_2O)_6^{3+}$	$Fe(CN)_6^{3-}$	FeF_6^{3-}
unpaired electrons	5	1	5

outer electronic configurations are shown at the bottom of the page.

Orbitals required to link the ligand ions or molecules are:

$Fe^{3+} \rightarrow Fe(CN)_6^{3-}$ 3d, 3d, 4s, 4p, 4p, 4p i.e. d^2sp^3

$Fe^{3+} \rightarrow FeF_6^{3-},$
$Fe(H_2O)_6^{3+}$ 4s, 4p, 4p, 4p, 4d, 4d i.e. sp^3d^2

giving alternative modes of hybridization and accounting for the number of unpaired electrons discovered.

The complexes were classified according to the style of hybridization:

inner orbital complex   e.g. $Fe(CN)_6^{3-}$

outer orbital complex   e.g. FeF_6^{3-}, $Fe(H_2O)_6^{3+}$

	3d	4s	4p
Co $(3d)^7 (4s)^2$	↑↓ ↑↓ ↑ ↑ ↑	↑↓	
Co^{3+} $(3d)^6$	↑↓ ↑ ↑ ↑ ↑		
Co^{3+} prior to forming $Co(NH_3)_6^{3+}$	↑↓ ↑↓ ↑↓		

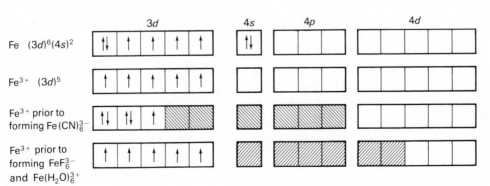

	3d	4s	4p	4d
Fe $(3d)^6(4s)^2$	↑↓ ↑ ↑ ↑ ↑	↑↓		
Fe^{3+} $(3d)^5$	↑ ↑ ↑ ↑ ↑			
Fe^{3+} prior to forming $Fe(CN)_6^{3-}$	↑↓ ↑↓ ↑			
Fe^{3+} prior to forming FeF_6^{3-} and $Fe(H_2O)_6^{3+}$	↑ ↑ ↑ ↑ ↑			

This theory has given way to the *ligand field theory*, itself a development of crystal field theory and the theory of molecular orbitals.

Coordination results from the electrostatic effect of the lone pair electrons of the ligands on the *d* orbitals of the central atom. An isolated iron atom which has lost 3 electrons will have five electrons in the five *d* orbitals, all equivalent in energy and an over-all positive charge ($+3$). When fluoride ions approach through electrostatic attraction these electrons increase in potential energy. Because of their electrical repulsion of each other, the incoming fluoride ions move towards the iron atom along axes at right angles to each other and an octahedral positioning occurs. But the increase in energy is greater for two of the orbitals than the other three because of the spatial distribution of $3d$ orbitals. The extent of this splitting of energy levels can be measured spectroscopically. The ligands which cause the widest splitting of levels are those which have the capacity to re-accept some control over electrons by back coordination (see p. 62). Thus the cyanide ion but not the fluoride ion (because of available electron orbitals) causes wider splitting and electrons occupy the 3 lower modified $3d$ orbitals ($\uparrow\downarrow\ \uparrow\downarrow\ \uparrow$) of a comparatively much lower energy level (in forming $Fe(CN)_6^{3-}$) than when all five modified $3d$ orbitals are used for iron(III) in forming FeF_6^{3-}. The former is termed a *low spin complex* (corresponding to an inner orbital complex) and the latter a *high spin complex* (corresponding to an outer orbital complex).

For most purposes a simple electrostatic picture of the formation of a complex ion will explain the shape adopted.

The splitting of energy levels in transition metal atoms is discussed further in relation to the colours of their simple and complex ions (p. 452).

A note on formulae

Square brackets [] are used to enclose complex structures and bond diagrams. In one or two cases they are used to make some particular point clear, such as analogous structures. The net charge is shown outside the brackets. Square brackets are also used in formulae where parentheses ( ) are also required, e.g.

$[Cr(H_2O)_6]^{3+}(Cl^-)_3$ and $[Cu(NH_3)_4]SO_4 . H_2O$

the charge not necessarily being shown. The charge shown in formulae may be placed upon an atom inside the brackets if this makes the bonding clearer.

In equations for reaction kinetics and equilibria, square brackets refer to molar concentrations.

The positions occupied by metals and non-metals in the Periodic Table

The classification of elements into metals and non-metals according to their physical and chemical properties is very convenient, especially when a correlation with the type of chemical bond favoured is sought, but the distinction proves to be not as clear cut as might be expected. The vast majority of elements are metallic.

At room temperature, metals are usually solids while non-metals may be solids, liquids or gases. In fact, only mercury (m.p. $-38.9°C$) is a liquid among the metals although caesium (m.p. $29°C$) and gallium (m.p. $30°C$) only just qualify as solids. Non-metals are usually easily volatilized, exceptions including boron, carbon and silicon, while the majority of metals have high melting-points and boiling-points, the highest being those of the transition elements.

Metals are characterized by a distinctive lustre and as a class are denser than non-metals. For a reactive metal, such as potassium, quickly corroded by exposure to the atmosphere, the lustre may only appear when it is sliced with a knife. On the other hand, the lustre shown by copper, silver and gold may be preserved and enhanced by polishing.

Non-metals, except graphite, are poor conductors of electricity, but metals are well known for high electrical conductivity, which decreases in magnitude with increase in temperature. Metals, as distinct from non-metals, are good heat conductors.

In mechanical properties, metals are noted for high tensile strength, ductility (may be drawn into wire) and malleability (may be hammered into thin sheets).

The chemical properties of metals are also distinctive: they are usually soluble in mineral acids and form salts in combination with oxoacids, while non-metals do neither. Metals form basic oxides and non-metals form acidic oxides. The hydrides of metals are either non-volatile or very unstable volatile substances as are most, but not all, of the halides. Hydrides and most of the halides of non-metals are readily volatile.

It will be realized that the characteristics of metals and non-metals are not mutually exclusive. Elements which quite definitely belong to one class may have properties suggestive of the other—for example, the low melting-points of the alkali metals (lithium to caesium) and the electrical conductance of graphite which is due to its internal bonding. However, it proves impossible to classify some elements as metals or non-metals and these are separately classified as *metalloids* (e.g. germanium and arsenic).

Linked with the metallic properties listed here is the electropositive nature of metals, their ability to throw off electrons leaving positively charged ions. Non-metals show an affinity for electrons, forming negative ions and are electronegative. Non-metals also form many covalent compounds and indeed some never form ions. Metals form few covalencies.

A **metal** is an element which forms simple positive ions (cations) by the loss of electrons.

In the metallic crystal the valency electrons are shared between the atoms so that the structure is essentially an array of positive ions immersed in a common pool of delocalized electrons.

Before locating metals, metalloids and non-metals in the Periodic Table, it will greatly simplify matters if the transitional elements are removed. All are typical metals in the general sense and together with elements of the copper and zinc groups have certain characteristics more or less in common. These properties are linked with the incomplete nature of their electron shells. The transition metals form coloured ions, have a variable valency and several oxidation states, form complex compounds and act, either as elements or combined, as catalysts. Oxides in the higher valency states are usually covalent and acidic while in the lower valency states they are basic and simple cations are formed, although the oxide itself may still be covalent, and possibly macromolecular in structure. Intermediate valency states may have amphoteric oxides. The intricacies of the relationship of $3d$ electron levels and physical and chemical properties have been traced elsewhere (p. 446).

The periodic relationships displayed by such properties as electronegativity, ionization energy, ionic and atomic radii etc. when plotted against atomic number have been noted.

Going down a family of the Periodic Table, the tendency to form positive ions increases. There is an increase in metallic nature or a decrease in non-metallic character. With increase in atomic number in a family, the atomic radius increases with the increasing number of electron shells. Therefore, peripheral electrons are less firmly held by the nucleus and are screened from it increasingly by intermediate electron shells. The tendency towards cation formation increases. For non-metals, easier polarization of large anions to form covalent molecules is seen. In the middle groups of the Periodic Table these effects result in the transition from non-metal to metal. A valency drop occurs with the appearance of positive ions. A pair of electrons becomes inert, with the result that high charges are avoided. The valency (oxidation state) is reduced by two units. Group I shows the progressive increase in metallic nature from lithium to caesium, all being metallic elements. In Group VII, there is a reduction in non-metallic character, from fluorine to iodine, all being non-metallic elements. In Group IV the gradual change from the non-metallic element carbon to the metal, lead, is seen.

In working along Period 2, it will be observed that the elements become progressively more electronegative. Lithium is a metal and fluorine the most electronegative non-metallic element. There are corresponding changes in the other periods. A general decrease in the volume of the atom occurs along a period due to increasing the nuclear charge without the addition of further electron shells. The increased forces of attraction result in a more compact atom. Periods containing transition metals show little change in the size of atoms while inner electron shells are being expanded. Decrease in the size of atoms with increase in nuclear charge brings a stronger control of the valency electrons by the nucleus and the power to attract and hold further electrons to form negative ions. This results in the observed change from metallic to non-metallic nature along the period.

Acquisition of the electronic configuration of a nearby inert gas by atoms during ionic combination does not mean that the ions formed have the same stability as the atoms of inert gas. The difference in nuclear charge ensures a difference in size and in the capacity to attract or repel electrons.

Families in which there is a progressive change from non-metal to metal, with increase in atomic number, contain elements with very weak metallic characteristics. These are called *metalloids*. They lie in the general direction of a diagonal drawn from lithium (top left) to iodine (bottom right) across the Periodic Table. This is shown in Table 4.9. Metalloids behave as weak metals. They will be discussed in the appropriate chapters. Chemists do not always agree on what should be a metalloid and

Table 4.9   Metals, metalloids and non-metals in the Periodic Table

I **H**								**2** **He**
3 **Li**	**4** **Be**	**5** **B**	**6** **C**	**7** **N**	**8** **O**	**9** **F**		**10** **Ne**
11 **Na**	**12** **Mg**	**13** **Al**	**14** **Si**	**15** **P**	**16** **S**	**17** **Cl**		**18** **Ar**
19 **K**	**20** **Ca**	**31** **Ga**	**32** **Ge**	**33** **As**	**34** **Se**	**35** **Br**		**36** **Kr**
37 **Rb**	**38** **Sr**	**49** **In**	**50** **Sn**	**51** **Sb**	**52** **Te**	**53** **I**		**54** **Xe**
55 **Cs**	**56** **Ba**	**81** **Tl**	**82** **Pb**	**83** **Bi**	**84** **Po**			**86** **Rn**
87 **Fr**	**88** **Ra**							

metalloids   non-metals   metals

transition elements

what should not. The line dividing metals from non-metals is drawn from between beryllium and boron to between polonium and astatine, although it is probably better to give the successive pairs of elements of each period lying on the diagonal as being on the borderline between the metallic and non-metallic elemental state. Taken vertically each pair lies in the region where the drop in electro-negativity brings the change from non-metal to weak metal.

The first member of each family shows slight variations in properties when compared with the remainder of the family. These differences are sometimes described as anomalous. The effect is probably caused by there being only two electrons between the nucleus and the valency electrons for elements of the second period. Non-metallic elements exert a stronger attraction for electrons and metals show an unexpected reluctance to form cations. This is made clear by a reference to the electronegativity (p. 79) and ionization energy (p. 43) curves above where it will be seen that the effect is more pronounced as the atomic number increases along the period. One aspect of this is revealed as the *diagonal relationship*, which refers to the slight resemblance between the general properties of compounds of the pairs: Li and Mg, Be and Al, B and Si.

5

Bonding and the structures displayed by elements and their compounds

The general physical properties of compounds related to bond type

The nature of the bonding in a compound and of the geometrical pattern adopted by the ions or molecules in a solid, will largely determine the physical properties of that substance. While distinctive properties associated with ionic and covalent bonding may be discerned, there is a gradual merging of characteristics when the compounds of a large number of elements are compared. This gradual transition is not altogether unexpected.

When fused or dissolved in water, an ionic compound will conduct electricity. The current is carried through the liquid by the ions which gain their mobility when the compound is melted or dispersed in a solvent. Unless a reaction occurs with the solvent, covalent substances yield non-conducting liquids.

In the crystal lattice of an ionic compound, each ion is surrounded by oppositely charged ions, the number depending on the particular pattern adopted in the crystal. Strong electrical forces hold the ions in position although each atom oscillates by virtue of its thermal energy. Considerable energy is required to overcome the forces of attraction and ionic compounds usually melt at high temperatures and are non-volatile. On the other hand, covalent molecules, each electrically neutral, are held by much weaker intermolecular forces. Therefore, fusion, boiling and sublimation are relatively easy to accomplish. To illustrate this point, the melting-points of the fluorides formed by the elements of Period 3, sodium–sulphur, are shown in Table 5.1.

Ionic bonding is usually assigned to the first three fluorides on account of their high melting-points although Pauling considers that AlF_3 is only slightly different from SiF_4 and the high melting-point is due to the crystal structure.

The force between two electrical charges Q_1 and Q_2, separated by a distance l, in a medium of dielectric constant D, is given by $F = Q_1Q_2/Dl^2$. In a vacuum, $D = 1$. High values of the dielectric constant will reduce the magnitude of the electrical attraction. Solvents which dissolve ionic compounds usually, but not necessarily, have high dielectric constants. Aggregates of ions dissociate in these solvents. While ionic compounds are soluble in relatively few solvents, covalent compounds are soluble in many. Liquefied ammonia, sulphur dioxide, hydrogen fluoride, hydrogen peroxide and water are ionizing solvents. Covalent compounds dissolve in a wide range of organic solvents, including benzene, tetrachloromethane (carbon tetrachloride), turpentine, paraffin, petroleum spirit, etc. Dielectric constants (20°C) are listed in Table 5.2 for some solvents.

The molecules of a covalent compound have a definite shape because the constituent atoms,

Table 5.1 Melting-points of the fluorides of the Period 3 elements/°C

NaF	MgF$_2$	AlF$_3$	SiF$_4$	PF$_5$	SF$_6$
992	1260	1291 (sublimes)	− 90	− 83	− 51

Table 5.2 Dielectric constants of some common solvents at 20°C

Paraffin	2.1	Acetone[4]	21.3
Turpentine	2.6	Carbon tetrachloride[5]	2.3
Benzene	2.3	Chloroform[6]	5.0
Amyl acetate[1]	5.1	Ethanol	25.7
Diethyl ether[2]	4.3	Water	78
Hexane	1.9	Ammonia	15.5
Toluene[3]	2.4	Methanol	33.1

Systematic names are:
1. amyl ethanoate
2. ethoxyethane
3. methylbenzene
4. propanone
5. tetrachloromethane
6. trichloromethane

although they are in a state of continuous vibration, adopt mean positions which are fixed in distance and direction from each other. This leads to the possibility of *structural isomerism*, being the occurrence of compounds with the same molecular formula but different structures, and *stereoisomerism* which means that the structures are the same, but more than one configuration of atoms is possible. While structural isomerism belongs chiefly to organic chemistry and is due to the covalent chain-forming properties of the carbon atom, examples of stereoisomerism also occur in inorganic chemistry.

Structural isomerism

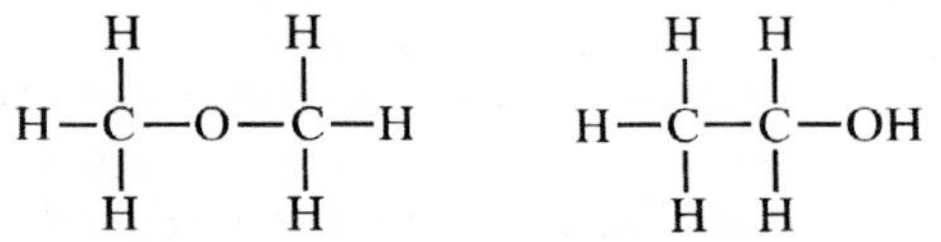

H−C−O−C−H   H−C−C−OH

methoxymethane   hydroxyethane
(dimethyl ether)   (ethanol)

Stereoisomerism: geometrical type

cis-   trans-

diammineplatinum(II) chloride

Both molecules are planar

Stereoismerism: geometrical type

N=N C₆H₅   C₆H₅ N=N C₆H₅

C₆H₅

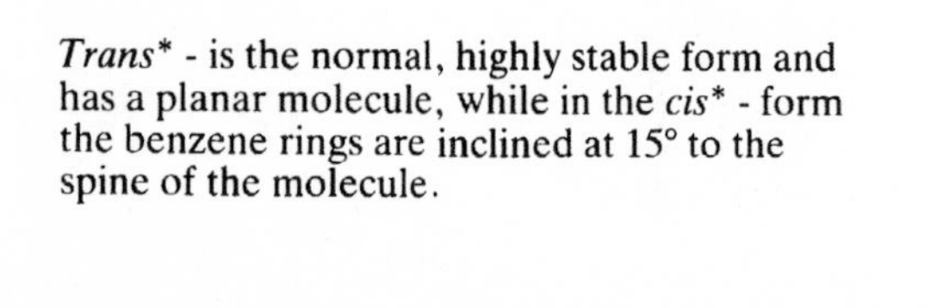

trans -azobenzene   *cis*-azobenzene

*Trans** - is the normal, highly stable form and has a planar molecule, while in the *cis** - form the benzene rings are inclined at 15° to the spine of the molecule.

Stereoisomerism: geometrical type

$trans$-dinitrogen difluoride cis -dinitrogen difluoride

*Trans** – is the more stable form.

Stereoisomerism: geometrical type

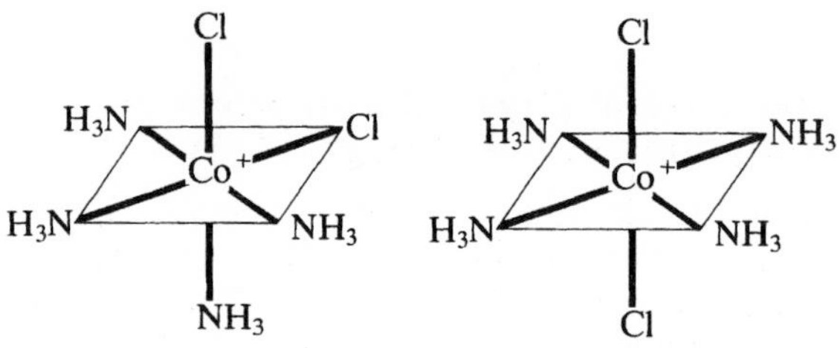

cis-   trans-

$[Co(NH_3)_4Cl_2]^+$

A three dimensional Co(III) complex anion with Cl and NH_3 as ligands.

Note that the sides of the squares (viewed obliquely) have no significance chemically: they are added to enhance the three dimensional effect of what might be called the equitorial plane of the octagonal structure.

Optical isomerism is not confined to organic chemistry, in which the early emphasis is placed on the *asymmetric carbon grouping* (not the asymmetric carbon atom)

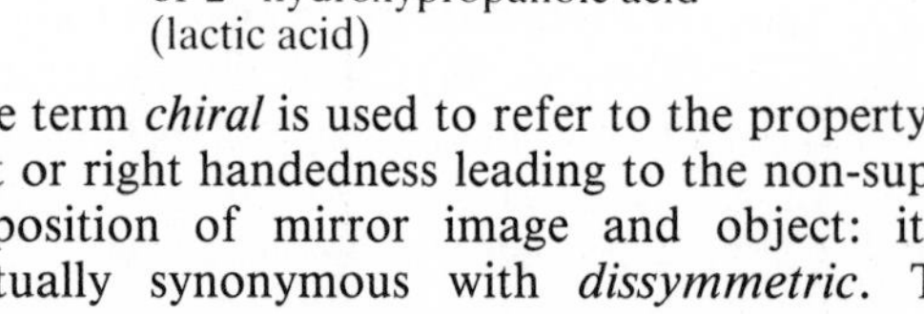

d-(dextro) and l-(laevo) forms of 2– hydroxypropanoic acid (lactic acid)

The term *chiral* is used to refer to the property of left or right handedness leading to the non-superimposition of mirror image and object: it is virtually synonymous with *dissymmetric*. The

* *cis*-(Latin) means 'on this side of', *trans*-(Latin) means 'on the other side of', the corresponding Greek prefixes, used in the elegant investigations into the stereochemistry of oximes are respectively *syn*- and *anti*-.

above molecules are not superimposable and in solution rotate the plane of polarization of plane–polarized light in opposite senses. (The actual configuration of each molecule, as distinct from its effect on polarized light and certain physical properties, can be determined by reference to an accepted standard molecule.) *Chirality* is not confined to the stereoisomerism of organic compounds but relates to three-dimensional structures in general. The optical isomers of $[Co(en)_2Cl_2]^+$ in which pairs of ammonia molecules linked to the cobalt atom in $[Co(NH_3)_4Cl_2]^+$ above are replaced by molecules of ethane-1, 2-diamine (ethylenediamine), $NH_2CH_2CH_2NH_2$ ($=$ en), may be represented:

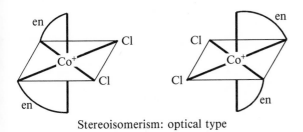

Stereoisomerism: optical type

$NH_2CH_2CH_2NH_2$ which 'bites' the cation is an example of a *chelate* ligand, and since two co-ordination positions are used, a *bidentate ligand*. The tetranegatively charged anion of ethylenediaminetetra-acetic acid,† abbreviated *EDTA*, a useful complexing agent and an example of a polydentate ligand, links with Co^{3+} in all six positions.

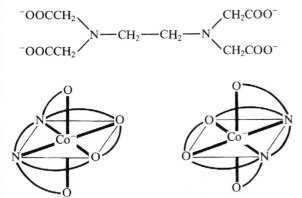

Stereoisomerism: *d* and *l* optical enantiomorphs
optical type   of Co(EDTA)⁻ complex

The nomenclature adopted for the naming of aqua (aquo), chloro and ammino complexes of Cobalt(III) is stated on pp. 473–4. An example of a *bidentate ligand* producing both *cis-trans* and *optical* stereoisomerism in chromium is shown on p. 460.

The shapes of simple inorganic molecules

A polyatomic molecule, like any other object, has three moments of inertia about three axes, at right angles to each other and intersecting at the centre of gravity of the molecule. It is customary to choose the axes of symmetry of the molecule as axes of rotation and according to the degree of symmetry, there may be one, two or three numerical values for the moments of inertia. For a simple covalent molecule, the number of moments of inertia, and their value, may be determined by investigation of the infra-red spectrum of the vapour of the compound. Alternatively, electron diffraction studies of the vapour may yield information about the structure of the molecule of the compound.

The shapes of isolated simple covalent molecules, AX_n, where n assumes values of 1–6 are shown in Fig. 5.1. The shapes may be explained by assuming that electrical repulsion occurs between the various $A–X$ bonds, that is, between the electron pairs forming these several bonds, so that each $A–X$ takes up a position in a direction as far away from the other $A–X$ bonds in the same molecule, AX_n, as is possible. Examples of binary compounds are listed in Table 5.3.

The same basic shapes are found for molecules of the type AX_nY_m, but the configuration of the molecule is less symmetrical. Maximum repulsion occurs between the shorter, stronger bonds (of higher electron density) at the expense of that between the weaker bonds in the same molecule. An extreme example of this type of molecule is where the central atom has a lone pair, or more than one lone pair, of valency electrons available. A molecule which is apparently AX_2 (for example) but which has one lone pair of electrons, may assume a shape which is based on the shape of AX_2Y, or with two lone pairs of electrons the shape of AX_2Y_2, and so on, where each Y atom has replaced one lone pair of electrons. In predicting the probable basic shape of a molecule, replace each lone pair of electrons on the central atom, A, by an atom, Y, and choose the shape associated

†Bis [bis(carboxymethyl)amino] ethane.

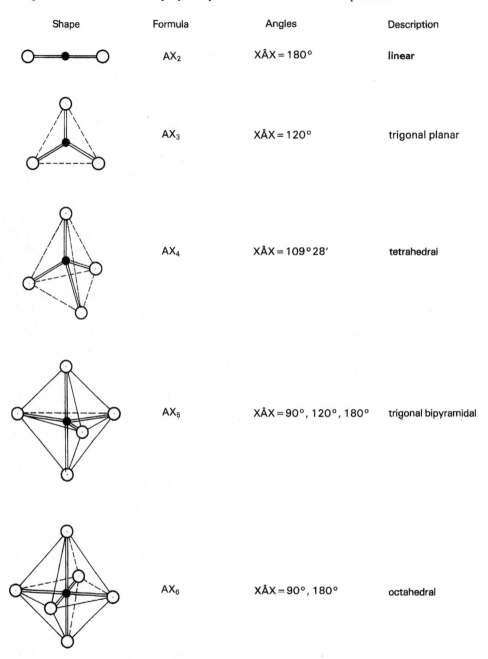

Shape	Formula	Angles	Description
	AX_2	$X\hat{A}X = 180°$	linear
	AX_3	$X\hat{A}X = 120°$	trigonal planar
	AX_4	$X\hat{A}X = 109°28'$	tetrahedral
	AX_5	$X\hat{A}X = 90°, 120°, 180°$	trigonal bipyramidal
	AX_6	$X\hat{A}X = 90°, 180°$	octahedral

Fig. 5.1 The shape of isolated simple covalent molecules, AX_n

with the molecule so derived; then, eliminate Y and remember that a lone pair of electrons exerts a maximum of electrical repulsion on other electron pairs already used in bonding in the same molecule, thereby causing a closing-up of the angles between these bonds. This idea is developed in Table 5.4 and in Fig. 5.2. Notice particularly how a square planar structure falls out of an octahedron when two of the atoms (top and bottom) are omitted and replaced by two electron-pairs, exerting maximum repulsion.

The effect of lone pair repulsion may be illustrated by water and ammonia, where the

Table 5.3 The shape of simple covalent molecules, AX_n

General formula type	Shape associated with type	Examples
AX_2	linear	$BeCl_2$ beryllium chloride
AX_3	trigonal planar	BCl_3 boron trichloride
AX_4	tetrahedral	$SiCl_4$ silicon tetrachloride
AX_5	trigonal bipyramidal	PCl_5 phosphorus pentachloride
AX_6	octahedral	SF_6 sulphur hexafluoride

Table 5.4 The shape of covalent molecules: lone pair effect

Apparent type AX_n	Number of lone pairs on A m	Actual type AX_nY_m	Shape associated with AX_nY_m	Example: formula AX_n		Periodic Group of A in example	Actual shape (by experiment)
AX_2	1	AX_2Y	trigonal planar	$PbCl_2$ (g)	Lead(II) chloride	IV	V-shaped
	2	AX_2Y_2	tetrahedral	H_2O	water	VI	V-shaped
	3	AX_2Y_3	trigonal bipyramidal	XeF_2	xenon difluoride	0	linear
AX_3	1	AX_3Y	tetrahedral	NH_3	ammonia	V	trigonal pyramid
	2	AX_3Y_2	trigonal bipyramidal	ClF_3	chlorine trifluoride	VII	T-shaped
AX_4	2	AX_4Y_2	octahedral	XeF_4	xenon tetrafluoride	0	square planar

(geometrically predicted) tetrahedral angle 109° 28' (109.5°) is reduced to $\widehat{HOH} = 104.5°$ and $\widehat{HNH} = 107.3°$ respectively. (Bond angles measured for common molecules are usually accurate to within ±0.5°.) Of the type AX_3Y_2, there are three theoretically possible configurations for the molecule AX_3: only ClF_3 has been investigated and the molecule is found to be T-shaped, with lone pair repulsion reducing $\widehat{FClF}$ to 87.5°. In comparing NF_3 and OF_2 with NH_3 and OH_2 respectively, the angles between bonds in the fluorides are closed up more than the corresponding angles in the hydrides. The bonding electrons lie

closer to the fluorine atoms, away from the central atom, and are less resistant to the repulsion of the lone pair(s).

The configuration adopted for atoms such as boron and carbon may be reconciled with the spin theory of valency by adopting the notion that valency orbitals are *hybridized*. For boron s^2p becomes $[sp^2]$ and for carbon, s^2p^2 becomes $[s^1p^3]$: where the [] sign means hybridization. Similarly for beryllium, s^2 hybridizes to $[sp]$. This leads to the respective shapes trigonal planar (BCl_3), tetrahedral (CCl_4) and linear ($BeCl_3$). For the remaining elements of the period (N, O, F) there is at least one

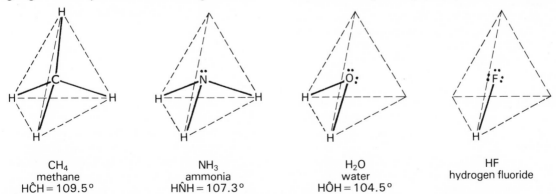

CH_4	NH_3	H_2O	HF
methane	ammonia	water	hydrogen fluoride
$\widehat{HCH} = 109.5°$	$\widehat{HNH} = 107.3°$	$\widehat{HOH} = 104.5°$	

Fig. 5.2 The shapes of covalent molecules of common hydrides of Period 2

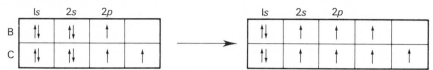

occupant of all possible orbitals in the ground state, so the ground state and the state of maximum valency are identical.

An input of energy is required for the uncoupling of electrons but this is recovered as part of the enthalpy of formation of the compound. The $[s^1p^2]$ and $[s^1p^3]$ orbitals are now 'welded' in a mathematical operation which shows that the equivalent orbits are stable. Only in the discharge tube can enough energy be supplied to uncouple the electron pair, $1s^2$, of helium. Interestingly an alternative hybrid for carbon has three equivalent bonding orbitals directed to the corners of an equilateral triangle with another weaker bond at right angles to this plane, corresponding to multibond formation. The subtle nature of the chemical bond is already emerging.

Atomic and ionic (including metallic) radii

At first sight it would seem that the determination of atomic and ionic radii involves only the adoption of a method of refined measurement for use with a well-defined, discrete entity. Yet the electron-cloud, or probability electron density pattern, in an atom does not end abruptly at a sharply defined distance from the nucleus. Further, the radius measured for an isolated atom (in the gaseous state) will differ from that found in the solid state where there is interaction with electron-clouds of other atoms, and in combination where it may differ, according to the nature of the element(s) with which it is combined, from one compound to the next. Finally, the measurements made can only be those of the distance *between* the centres of atoms and from such measurements (of bond lengths) the radii of the atoms (or ions) have to be deduced. The radii remain to a large extent empirical and are adjusted to reproduce measured bond lengths: predictions are usually within 10% of the observed value. The values depend on the nature of the bonding.

The easiest investigations are those where a univalent element forms simple molecules, e.g. Cl_2, when the interatomic distance is conveniently halved to give a radius for the atom, as illustrated

in Fig. 5.3 and Table 5.5. Such a radius is sometimes called the *covalent atomic radius* to distinguish it from the *metallic radius* and the *ionic radius*, and to indicate that the value relates to a *covalent compound*.

Table 5.5 Atomic radii of diatomic covalent elements (Group VII)

Element	Molecule	Bond length/nm	Atomic radius/nm
Fluorine	F	0.146	0.073*
Chlorine	Cl	0.198	0.099
Bromine	Br	0.228	0.114
Iodine	I	0.267	0.134

* 0.064 nm (from a large number of F compounds is preferred)

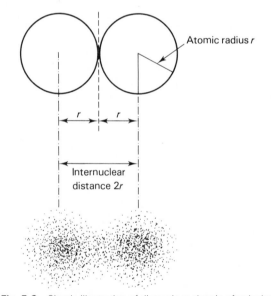

Fig. 5.3 Simple illustration of diatomic molecule of univalent element e.g. Cl_2

Where multiple bonds are formed between atoms, the internuclear distance (and hence the apparent atomic radius) decreases as the number of bonds increases. Carbon compounds (Table 5.6) provide a good example. A corollary is that halving the internuclear distance for oxygen in O_2 and nitrogen in N_2 will not give the respective atomic radii because of the multiple bonding in the

Table 5.6 Multiple bonds and internuclear distance: carbon compounds

Bond	Internuclear distance/nm	Apparent atomic radius/nm
C—C	0.154	0.077
C=C	0.134	0.067
C≡C	0.120	0.060

molecules of these elements. Similarly the atomic radius of a metallic element in the vapour state (e.g. Li_2 molecules) will differ from that in the solid metallic state (Li atoms/ions with some pooled electrons). The *metallic radius* is defined as half the internuclear distance between adjacent atoms in the metal structure. However, the atomic radii of such elements as oxygen and nitrogen may be deduced after measurement of bond lengths in molecules of compounds formed with elements of known atomic radii. In Table 5.7 values for the atomic radius of carbon are deduced and may be

Table 5.7 Atomic radius of carbon (from tetrahalides)

Compound	Bond	Experimental value/nm	Atomic radius of halogen/nm	Subtraction gives atomic radius of carbon/nm
CCl_4	C—Cl	0.177	0.099	0.078
CBr_4	C—Br	0.191	0.114	0.077
CI_4	C—I	0.211	0.134	0.077

compared with the value obtained by halving the interatomic distance (0.154 nm) in diamond, an allotrope of carbon of macromolecular structure. In this way a coherent self-consistent system of atomic radii may be tabulated. Pronounced deviations from expected values point to unexpected bonding and related phenomena. Fluorine and hydrogen are among those elements which deviate.

As electrolytes, ionic compounds necessarily contain at least two elements, e.g. Na^+Cl^-. The interatomic, or interionic distance in the solid state at its simplest is the sum of the radii of adjacent, touching ions. The lattice arrangement depends on ionic sizes and the stoichiometric requirements of the compound. Credence to the touching sphere notion of ionic species is lent by such observations as those shown in Table 5.8 in which the radius of the common ion is eliminated by subtraction to give the difference in radius of the other ions, often but by no means always the same or nearly the same.

Table 5.8 Ions as spheres of constant radius

	Interionic distance of univalent ions/nm				
Halide	Sodium salt		Potassium salt		Rubidium salt
M^+Cl^-	0.281	(0.033)*	0.314	(0.015)	0.329
	(0.017)		(0.015)		(0.014)
M^+Br^-	0.298	(0.031)	0.329	(0.014)	0.343
	(0.025)		(0.024)		(0.023)
M^+I^-	0.323	(0.030)	0.353	(0.013)	0.366

* radius K^+ − radius Na^+ = 0.314−0.281 = 0.033 nm
Similarly, radius Br^- − radius Cl^- = 0.298 − 0.281 = 0.017 nm. Compare with values in Table 5.10

Following the investigations of Landé (1920) and Wasastjerna (1923), Goldschmidt (1926) compiled a table of *ionic radii* by starting with the assumption that for electrolytes with large anions and comparatively small cations, the anions would be in contact so that their ionic radius could be deduced from the simple geometry of the structure (Fig. 5.4, 4.2 and 6.4). The radii of cations of comparable size can be deduced from measured interionic distances and in turn, the radii of other anions.

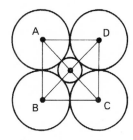

AC = 2× interionic distance (measured)
AB = 2× radius of anion = BC
and $AC^2 = AB^2 + BC^2$ (Pythagorus)
Hence, radius of I^- known

Fig. 5.4 Deduction of radius of large anion in M^+X^-, e.g. Li^+I^-

Pauling (1927) assumed that for salts with isoelectronic ions e.g. K^+Cl^- the interionic distance could be apportioned to the ions in inverse ratio of their effective nuclear charges and with this start compiled another self-consistent sequence of radii, but not compatible with the Goldschmidt values. Pauling adopted the term *crystal radius* when dealing with multivalent ions. Detailed X-ray analysis of a few salts has given electron density maps of ions which support neither approach to the problem entirely, although it must be remembered

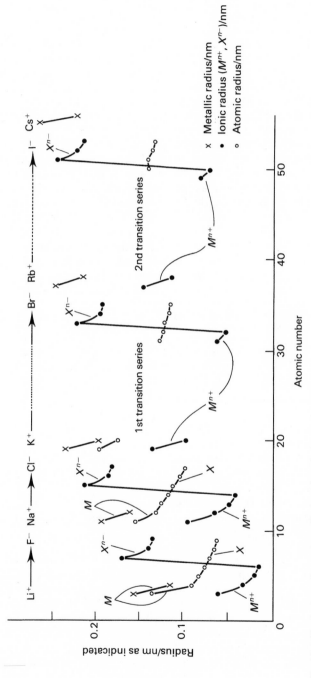

Fig. 5.5　Metallic, ionic and atomic radii plotted against atomic number

Table 5.9   Hypothetical ionic radii in crystals (Pauling)/nm (Note size variation across period and down group)

Li^+	0.060	Be^{2+}	0.031	B^{3+}	0.020	C^{4+}	0.015	N^{3-}	0.071	O^{2-}	0.140	F^-	0.136
Na^+	0.095	Mg^{2+}	0.065	Al^{3+}	0.050	Si^{4+}	0.041	P^{3-}	0.212	S^{2-}	0.184	Cl^-	0.181
K^+	0.133	Ca^{2+}	0.099	Ga^{3+}	0.062	Ge^{4+}	0.053	As^{3-}	0.222	Se^{2-}	0.198	Br^-	0.195
Rb^+	0.148	Sr^{2+}	0.113	In^{3+}	0.081	Sn^{4+}	0.071	Sb^{3-}	0.245	Te^{2-}	0.221	I^-	0.216
Cs^+	0.169	Ba^{2+}	0.135	Tl^{3+}	0.095	Pb^{4+}	0.084						

Small, highly charged cations M^{n+} and large, highly charged anions, X^{n-} are hypothetical (see Fajans' Rules, pp. 82–3)

that the ionic size may vary according to the chemical environment.

Ionic metallic, and atomic radii are listed in Tables 5.9, 5.10 and 5.11. When plotted against atomic number (Fig. 5.5), these values indicate a periodic variataion in the radii of atoms and ions, another illustration of the Periodic Law and clearly linked to the electronic structure. Comparison with

Table 5.10   Group I and Group II elements: metallic radius/nm

Li	0.155	Be	0.112
Na	0.191	Mg	0.160
K	0.235	Ca	0.197
Rb	0.248	Sr	0.215
Cs	0.267	Ba	0.222

the plot of ionization energy against atomic number (Fig. 3.6) illustrates what might be expected, that elements with low ionization energies have large internuclear distances and hence atomic radii. The plot of ionic and atomic radii against atomic number confirms the shrinkage effect of increasing the nuclear charge on the surrounding electron shells of the elements in each period. From the above tables and the graph of radii *versus* atomic number, the following observations may be made.

1   A cation e.g. Na^+ has a smaller radius than its atom, the loss of its electron(s) rendering the nuclear charge more effective and thereby causing shrinkage.
2   An anion e.g. Cl^- is larger than its atom, the extra electron(s) reducing the effective nuclear charge and thereby allowing expansion.
3   Increase of atomic number, i.e. nuclear charge, across a period brings a shrinkage of both

atoms and ions, and especially cations, e.g. O^{2-}, F^-, Na^+, Mg^{2+} are isoelectronic and the interionic distances of NaF and MgO may be measured and compared:

$$Na^+F^-\ 0.227\ nm \qquad Mg^{2+}O^{2-}\ 0.211\ nm$$

4   Although nuclear charge increases down a group of the Periodic Table, extra electron shells bring about expansion of both atoms and ions. When using tables of ionic radii, it must be remembered that the distance between the centres of ions is measured experimentally, and then an assumption is made about the nature of and charge on the ions. Reference to Fajans' Rules (pp. 82–3) makes the issue clear.
5   Finally, remember that values quoted by different researchers can vary according to the assumptions made in executing the calculations. Differences of the order of 0.005 nm may be ignored for our purposes.

van der Waals' radii

Finally, another set of radii may be distinguished in which each value relates to the external influence an atom has on another with which it is *not* combined. In a crystal of chlorine, $Cl_2(s)$, the molecules adopt equilibrium positions (about which they vibrate) in relation to one another. This distance halved, would give the *van der Waals' radius* of chlorine. Since there are two electron pairs rather than one between the neighbouring atom of adjacent molecules, the van der Waals' radius of chlorine is expected to be more than the atomic (covalent) radius (0.099 nm) and much nearer the

Table 5.11   Atomic radii in covalent compounds/nm (Note size variations in period and group)

Li	0.134	Be	0.090	B	0.082	C	0.077	N	0.070	O	0.066	F	0.064
Na	0.154	Mg	0.130	Al	0.125	Si	0.117	P	0.110	S	0.104	Cl	0.099
K	0.196	Ca	0.174	Ga	0.126	Ge	0.122	As	0.121	Se	0.117	Br	0.114
Rb	0.216	Sr	0.191			Sn	0.141	Sb	0.141	Te	0.137	I	0.133
Cs	0.235	Ba	0.198			Pb	0.154	Bi	0.152				

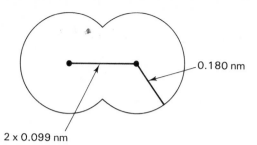

0.180 nm

2 x 0.099 nm

Fig. 5.6   Representation of Cl_2 molecule to show its effective volume

atomic radius	0.099 nm
van der Waals' radius	0.180 nm
ionic radius	0.181 nm

ionic radius (0.181 nm) and this is found. This is illustrated in Figure 5.6.

The values shown in Table 5.12 were determined by investigation of internuclear distances in suitable compounds of the elements listed. Molecules may be sketched, using van der Waals' radii rather than atomic radii, to show the effective volume in which their influence would be experienced. For more complicated structures, one part of a molecule may, by the same mode, exert a steric influence on another part of the same molecule.

Table 5.12   van der Waals' radii/nm

N	0.15	O	0.140	F	0.135
P	0.19	S	0.185	Cl	0.180

The force of attraction between the adjacent chlorine molecules is termed the *van der Waals' force* and arises from fluctuation of charge density and it increases in magnitude, as would be expected, with increase in the size of the molecule. Increase of melting-point and boiling-point with atomic number increase is accounted for in this way, except that elements in the middle groups of the Periodic Table have enough valency electrons to permit the formation of giant structures (or macromolecules) while the metals also differ in forming lattices with shared, mobile (delocalized) electrons which conduct electricity.

The absence of powerful bonding forces between atoms of the noble (or inert) gases allows a comparatively large size for the atoms in the liquid and solid states.

Except in very rare cases, these elements form no compounds (see XeF_4 etc.) and so each cannot fulfil the prerequisites for defining and determining an atomic radius as normally understood. When atomic radii of the Group O (noble gas) elements are required, the van der Waals' radii are used, thereby defining the sphere of influence of the atoms.

	van der Waals' radius/nm
Neon	0.160
Argon	0.192
Krypton	0.197
Xenon	0.217

The transition from ionic to covalent bond type

That the electron defies physical description has already been noted. In drawing bond diagrams we have found it convenient to show electrons as particles although it was realized that a sub-atomic particle cannot be described in the material terms which are familiar to our senses. The qualitative experiments of J. J. Thomson gave results best explained in corpuscular terms, while the experiments of his son, G. P. Thomson, with diffraction patterns produced by a beam of electrons impinging on thin metal foil, required a wave theory explanation. These are different aspects of what we call electrons and no one picture can represent them. Further, a simple convention involving + and − signs and lines has been used—and will continue to be used—to cover the complexities of chemical bonding in order to show simple models of united atoms. The ionic bond required the complete transference of electrons and the covalent bond, their sharing. Regarding electrons as diffuse electrical clouds, we may visualize degrees of bonding, from the complete separation of electron shells for ions to their overlapping in covalency. For a particular compound, the actual structure may be intermediate in nature and only described in simple, conventional terms by fusing together the structures represented by alternative ionic and covalent bond diagrams in the required proportion. If the final result requires more of the ionic form than the covalent, then, for general purposes, the ionic structure is used. When a structure is more than 50% covalent, a covalent bond diagram is drawn. At the same time the intermediate character of the bonding must be noted. We must recognize that simple diagrams are not always adequate.

The strong electrical forces between ions have been noted, but those between covalent molecules, while smaller, are not always negligible. Even the atoms of inert gases must exert some forces of attraction for each other because they may be liquefied. If r is the distance between the centres of the nuclei of two atoms, the potential energy produced is proportional to $-1/r^6$, being zero when the atoms are at an infinite distance apart.

The term van der Waals' forces is used to describe these attractions which are believed due to a fluctuation in electrical moment, or dipole moment, brought about by the motion of electrons about the atomic nuclei although the whole molecule will have no net charge. In addition to these weak forces, covalent molecules are often polarized because of the unequal sharing of electron pairs between atoms with differing electron affinity. The shared pair of electrons in hydrogen chloride is located between the two nuclei but nearer to that of chlorine. This is shown in the bond diagram. The resultant structure has an electrical dipole and this may be depicted by indicating partial changes, $+\delta\varepsilon$ and $-\delta\varepsilon$, above the atoms:

The hydrogen chloride molecule is a hybrid structure between those represented by:

$$H-Cl \quad \text{and} \quad H^+Cl^-$$

But covalency predominates and the former bond diagram is used. Chlorine has a greater affinity for electrons than hydrogen. It is said to be more electronegative.

Electronegativity

Pauling has calculated values of electronegativity from a study of bond energies. Values for typical elements are given in Table 5.13. The most highly electronegative elements, with their avid affinity for electrons, are comparatively small and have highly effective nuclear charges due to inadequate screening by electron shells. They are found in the upper right of the Periodic Table. On the other hand, strong metals in the opposite corner tend to eject electrons, not attract them. In Fig. 5.7, the value of electronegativity is plotted against atomic number for elements 1–56 (barium) inclusive. A periodic curve results.

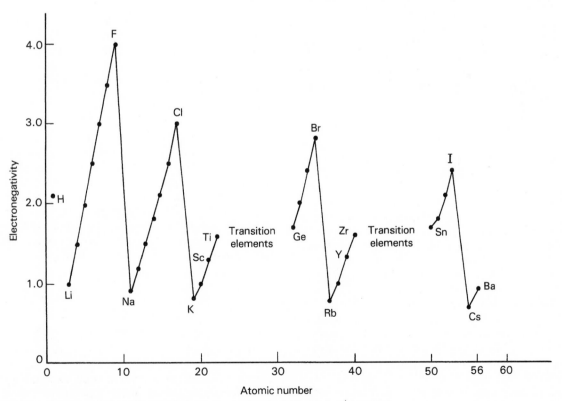

Fig. 5.7 Electronegativity plotted against atomic number for elements 1–56 inclusive

Table 5.13 Electronegativities (calculated by Pauling) for typical elements

H 2.1						
Li 1.0	Be 1.5	B 2.0	C 2.5	N 3.0	O 3.5	F 4.0
Na 0.9	Mg 1.2	Al 1.5	Si 1.8	P 2.1	S 2.5	Cl 3.0
K 0.8	Ca 1.0		Ge 1.7	As 2.0	Se 2.4	Br 2.8
Rb 0.8	Sr 1.0		Sn 1.7	Sb 1.8	Te 2.1	I 2.4
Cs 0.7	Ba 0.9					

Table 5.14 Electronegativities (calculated by Allred and Rochow) for typical elements

H 2.1						
Li 0.95	Be 1.5	B 2.0	C 2.5	N 3.05	O 3.5	F 4.1
Na 1.0	Mg 1.25	Al 1.45	Si 1.75	P 2.05	S 2.45	Cl 2.85
K 0.9	Ca 1.05	Ga 1.8	Ge 2.0	As 2.2	Se 2.5	Br 2.75
Rb 0.9	Sr 1.0	In 1.5	Sn 1.7	Sb 1.8	Te 2.0	I 2.2
Cs 0.85	Ba 0.95	Tl 1.45	Pb 1.55	Bi 1.65		

Fluorine is the most electronegative element and caesium the least. Electronegativity decreases down any family because the valency electrons are situated further from the positive nucleus and there is an increasing amount of screening by inner shells of electrons. This means that the attraction for these outer valency electrons decreases. The bond type may be predicted from electronegativities. Elements of widely different electronegativity (i.e. affinity for electrons) form ionic bonds, while those of nearly the same electronegativity form covalent bonds. A difference of 1.7 units represents roughly equal contributions of ionic and covalent bonds.

The original scale was defined in 1932. An alternative approach involving calculations based on atomic radii and effective nuclear charge has been made by Allred and Rochow. The values are given in Table 5.14. They are based on an electronegativity of 2.1 for H to correspond to Pauling's values. Comparison shows that the values are not very different, especially when it is remembered that *differences* in electronegativities are used in estimating the degree of polarization in a chemical bond. This polarization does not necessarily correspond to the dipole moments of a molecule because of the overriding influence of lone pairs of electrons on the latter.

Polar covalent molecules, such as hydrogen chloride described above, will tend to associate because of the attraction of electrical charges of opposite signs. Thermal energy will be required to break down aggregates of molecules so that an abnormally high boiling-point and latent heat of evaporation are encountered. Ammonia, water and hydrogen fluoride have very much higher boiling-points than might be expected from a study of the boiling-point data of the hydrides formed by other members of their respective families (see Fig. 15.1).

Dipole moments

When a molecule has an electrical distribution which may be described in terms of partial charges, $+\delta Q$ and $-\delta Q$, and their distance apart, l, the electric dipole moment, p, is given by $p = l \cdot \delta Q$. Dipole moments were measured in units called Debyes (related to the obsolete electrostatic unit: $1D = 1 \times 10^{-18}$ cm e.s.u.) but under the S.I. system revert to Cm ($1D = \frac{1}{3} \times 10^{-29}$ Cm). A dipole moment is a vector quantity. Some molecules may have zero dipole moments yet have bonds with polar characteristics. Thus, tetrachloromethane (carbon tetrachloride), CCl_4 has a resultant dipole of zero because of the tetrahedral arrangement of C—Cl bonds. The zero dipole moment here indicates symmetry. Similarly, carbon dioxide has a linear molecule with a zero dipole moment but water ($p = 6.13 \times 10^{-30}$ Cm) is not linear:

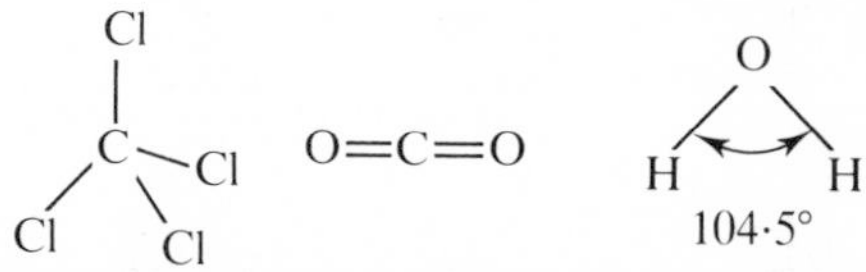

Dipole moments of a few common elements and compounds, measured in the gaseous or vapour phase, are given in Table 5.15.

Table 5.15 Dipole moments/10^{-30} C m

H_2O	6.13	H_2	0.0	$SnCl_4$	0.0
CO	0.40	Cl_2	0.0	CCl_4	0.0
CO_2	0.0	N_2	0.0	BCl_3	0.0
N_2O	0.47	HCl	3.43	CH_3Cl	6.17
SO_2	5.37	NH_3	4.87	$CHCl_3$	3.83

The Theory of Resonance

Strictly, the structure of hydrogen chloride is not

$$H-Cl$$

i.e. 100% covalent, although it is usually written so. We have seen that it is a hybrid structure, corresponding neither to $H-Cl$ nor $H^+ Cl^-$ but being intermediate. It may be calculated from experimental data that hydrogen chloride owes about 80% to the covalent structure and 20% to the ionic. The separate formulae represent *resonance structures*.

$$\{H-Cl \leftrightarrow H^+Cl^-\}$$

Hydrogen chloride has a fixed structure. It does not alternate between the structures shown. Hydrogen chloride cannot be accurately represented by simple bond diagram conventions (e.g.—represents a shared pair of electrons) but the correct picture is obtained by fusing the separate structures given above in the requisite proportions. The concept of *resonance* or *mesomerism*† comes from the mathematical treatment of bonding by electrons in wave mechanics. Hybrids are well known in everyday life—a mule is a hybrid between a horse and an ass. Notice that the structures, in terms of which the real structure is described, have no separate existence. Thus, on seeing a rhinoceros for the first time, a mediaeval traveller would perhaps relate its appearance to that of the unicorn and the dragon, neither of which leads an independent existence outside mythology—like bond diagrams!

Resonance may occur between covalent structures: thus the bond lengths (interatomic distances) in ozone are both equal, which, since bond lengths are usually characteristic of bond type, is not apparent from a single bond diagram. The ozone molecule is a hybrid of the structures represented by the resonance forms:

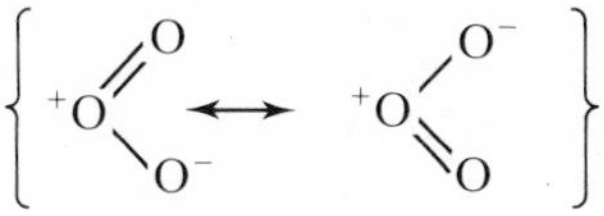

The bonding electrons are equally distributed round the molecule—as might be anticipated by symmetry.

Probably the best known example of mesomerism is shown by benzene, C_6H_6, which is a flat, regular hexagonal molecule.

Kekulé (1865) postulated the six carbon-ring, thereby laying the foundations of aromatic organic chemistry. Two structures apparently co-existed:

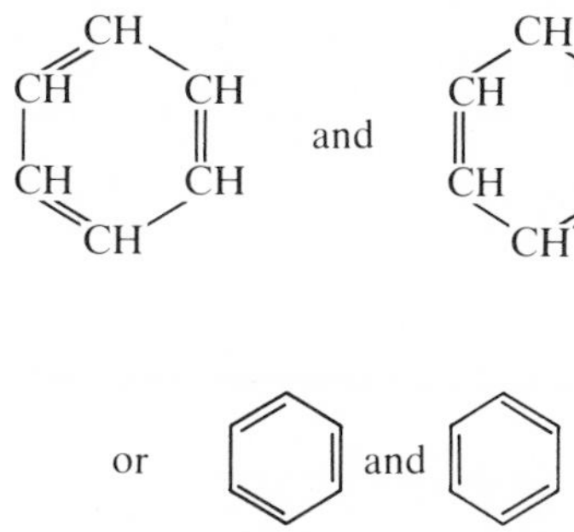

Classical and later arguments about the structure need not concern us here and the tide of opinion has ebbed and flowed. It is sufficient to say that recent studies reveal that the structures of benzene and other aromatic hydrocarbons investigated, are best explained by a wave mechanical treatment of structures resembling those of Kekulé with localized electron double bonds, rather than those of the molecular orbital theory of Hückel, which is the basis of the widespread, popular delocalized electron model. That structures akin to Kekulé's postulate of 1865 are lower in energy than the conventionally used model serves as a reminder that all scientific theories must be watched critically.

While a detailed study lies outside the scope of this book, various physical methods (spectroscopy, X-rays and electron diffraction) indicate that the interatomic distance for a given bond between a given pair of atoms in various compounds is very nearly constant. It follows that if an experimental value for the bond in a compound under examination is intermediate between the distance expected for single and double covalent bonds, resonance very probably occurs. Thermochemical data may also be indicative because resonance is associated with extra stability.

† Ingold, C. K., J. Chem. Soc., 1120 (1933).

Here are a few examples of resonance, the shapes of the entities as determined by experiment and the structures which are ordinarily used:

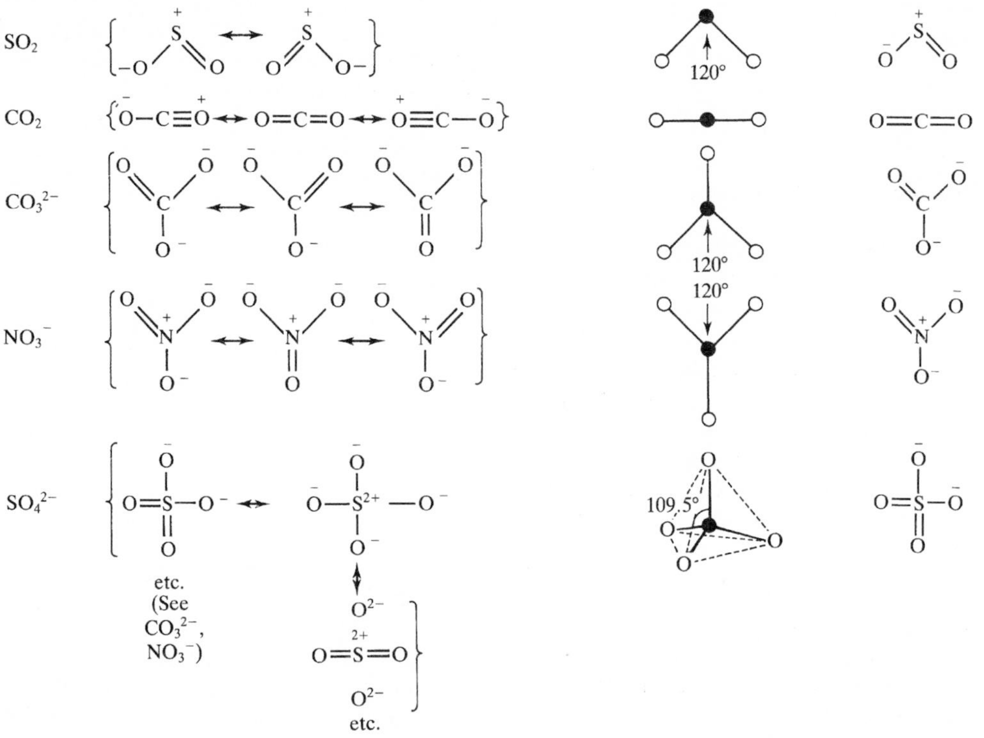

SO_2

CO_2 $\{\bar{O}-C\equiv \overset{+}{O} \leftrightarrow O=C=O \leftrightarrow \overset{+}{O}\equiv C-\bar{O}\}$

CO_3^{2-}

NO_3^-

SO_4^{2-}

etc.
(See CO_3^{2-}, NO_3^-)

etc.

The existence of certain stable molecules which contain an odd number of electrons, so that they are unable to form octets of electrons, is explained by resonance. Nitrogen oxide (formerly nitric oxide) and nitrogen dioxide are such *odd electron molecules*.

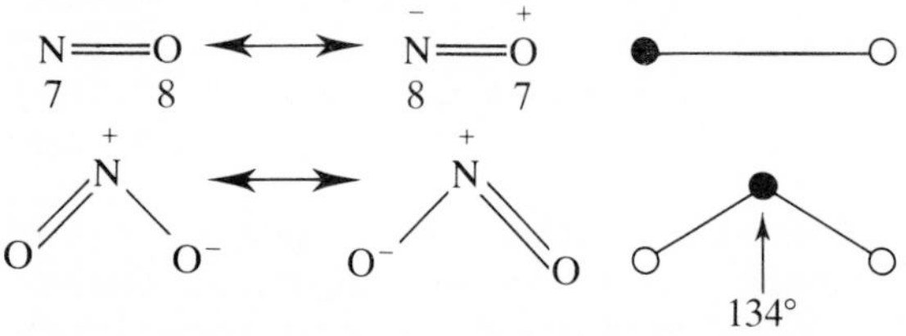

$N=O \leftrightarrow N=O$
7   8   8   7

All resonance structures are assigned with a full consideration of necessary experimental data. *The Theory of Resonance* requires the separate resonance structures to have the same, or very nearly the same nuclear framework, the same number of unpaired electrons and the same, or nearly the same stability. The heat of formation (enthalpy change of formation) of the actual mole formed is more (negative) than that calculated for the resonance hybrid obtained by fusing the various structures

together. i.e. more heat is expelled from the system than would be expected. Where the theory of resonance is required to explain a structure, it is always found that the structure has a greater stability than expected. This resonance stability is very important.

Fajans' Rules

The characteristics of a compound which enable a classification into ionic or covalent to be made are electrical conductance, melting-point, boiling-point, volatility and solubility. Other physical measurements (bond lengths, etc.) are useful. The findings usually agree with predictions that can be made on the basis of rules formulated by Fajans in 1923–4. *Fajans' Rules* enable the clearest visual picture of the transition in bond type to be made. Starting with the assumption that ions are initially present, certain conditions suggest that polarization may occur with sufficient distortion of electronic configuration to bring about the formation of covalency. The ions may be purely hypothetical.

High ionic charges are unlikely; for cations the successive loss of electrons becomes increasingly difficult as the retarding attraction of the oppositely charged ion being left behind increases. For anions, the successive acquisition of electrons results eventually in repulsion. Na^+, Ca^{2+}, Cl^- and O^{2-} are representative of most ions but higher charges are not so usual; there are some like Al^{3+} but very few like Pb^{4+}, which probably only occurs in the oxides PbO_2 and Pb_3O_4.

The size of the atom or ion is very important. In large ions, valency electrons are comparatively far from the positively charged nucleus and consequently the force holding the electron ($F = Q_1Q_2/Dl^2$, p. 69) is weak. This means that the electron clouds of an anion of large volume will be readily distorted to produce a covalent molecule. On the other hand, an atom of large volume will, if other conditions are favourable, readily form a cation by ejecting the necessary electrons. X-ray analysis of the crystal structure of the silver halides (see Table 5.16) shows that the iodide exists as a macromolecular covalent structure (wurtzite structure) while the other halides are ionic.

Table 5.16   Radii of ions forming silver halides (AgX)/nm

Ag^+	1.26	F^-	0.136
		Cl^-	0.181
		Br^-	0.195
		I^-	0.216

While the fluoride is soluble in water, the other salts are insoluble but there is a gradation in their solubilities in aqueous ammonia to form the diammine, $Ag(NH_3)_2^+$. The chloride dissolves readily, the bromide sparingly, while the iodide is insoluble. Presumably, solubility is connected with the presence of silver ions and supports the concept of a change in bond type. The precipitation of covalent silver iodide from the interaction of ions may be imagined as shown in Fig. 5.8 to illustrate the idea of distortion of the electron clouds of ions to covalent bonding.

The colours exhibited by the silver halides have interest. The anhydrous fluoride is yellow, but when hydrated is colourless. Silver chloride is white, the bromide cream and the iodide yellow. The colour change is not necessarily linked with change in bond type. A small cation will be more effective in polarizing an anion because of the concentration of charge attracting the electrons, so

that an excellent set of conditions for covalent bond formation involves the meeting of a large anion and small cation.

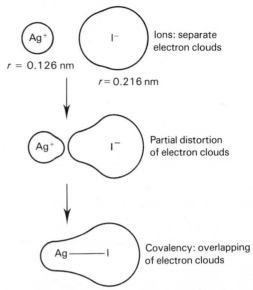

Fig. 5.8   The formation of covalent molecules from ions $Ag^+ + I^- \rightarrow AgI$ (covalent)

In the vapour phase, the tendency to covalency increases. Whereas in the solid state the effect of the cations surrounding an anion will to some extent be evened out (see Fig. 4.1: the sodium chloride structure), the association of an ion-pair in the vapour leads to a more pronounced unilateral distortion. Silver chloride is covalent in the vapour phase.

The electronic configuration in the ions is also important. It appears that cations with the electronic configurations of inert gases are less effective in their polarizing effect than those without. This is not unexpected and may be illustrated by comparison of sodium chloride (m.p. 801°C) and copper(I) chloride (m.p. 422°C). The cations have roughly the same radius: Na^+ (0.095 nm), Cu^+ (0.096 nm). The electronic configurations are respectively, 2, 8, 8 and 2, 8, 18. Sodium chloride is ionic while copper(I) chloride has a macromolecular covalent structure.

To summarize, covalency is favoured when there are

1 high ionic charges,
2 a small cation and large anion,
3 cations with a non-inert gas electronic configuration.

These are **Fajans' Rules**.

Table 5.17   Selected values for lattice energies

| Compound | Structure assumed | Lattice energy $kJ\,mol^{-1}$ | | Percentage difference (approx.) (sign ignored) |
		Experimental	Theoretical	
LiF	NaCl	−1007	−1021	1
NaF	NaCl	−903	−899	<1
KF	NaCl	−803	−807	<1
CaF_2	CaF_2	−2579	−2583	<1
SrF_2	CaF_2	−2458	−2458	<1
BaF_2	CaF_2	−2321	−2324	<1
AgF	NaCl	−953	−920	4
AgCl	NaCl	−903	−833	8
AgBr	NaCl	−895	−815	9

Lattice energy and crystal energy

The discussion of the transition from ionic to covalent bonding may be taken further by the study of the energy changes that occur on forming a crystal lattice.

The energy associated with a regular assembly of anions and cations in a lattice will depend on the charge on the ions, the distance between adjacent ions and on their distribution pattern, i.e. the geometry of the lattice. The electrostatic energy of a pair of point charges Q_1, Q_2, separated by a distance l, is Q_1Q_2/l.

For $A^{2+}B^{2-}$, the electrostatic energy will be numerically four times that of A^+B^-, if other variables remain constant. The numerical value of the energy will fall as the point charges are moved away from each other, so that when comparing the chlorides of the alkali metals of similar crystal structure, where ionic charges and the geometry of the lattice are the same, the numerical value of the energy of the lattice per mole is lowered with increase in cationic size. Both remarks are illustrated in Table 5.17. Knowledge of the geometrical characteristics of the lattice enable an adjustment to be made to this simple approach to allow for the complexities of electrical interaction between the various ions in the neighbourhood.

The **Lattice Energy** of a crystal is the energy evolved when the isolated ions, separated from each other by an infinite distance, are brought together to their equilibrium positions in the stable, solid lattice

It may be regarded as the heat of formation of one mole of crystal from its gaseous ions: for an ionic crystal, the lattice energy equals the standard enthalpy change of formation of the crystal lattice from its constituent ions in the gaseous phase $(\Delta H^{\ominus}_{298})$†. The process is exothermic and the value has a negative sign. (The term is nowadays preferred to that of *crystal energy* which has a positive sign attached to the same numerical value, being effectively the energy (under standard conditions) required to disperse the lattice.)

$$Na^+(g) + Cl^-(g) = NaCl(s); \quad \Delta H^{\ominus}_{298} = -765\,kJmol^{-1}$$

∴   Lattice energy, $U_L = -765\,kJmol^{-1}$

while, crystal energy, $U_C = +765\,kJmol^{-1}$

Now the heat of formation of a crystalline salt from its elements in their normal (standard) states, the standard enthalpy change of formation, is usually known accurately:

$$Na(s) + \tfrac{1}{2}Cl_2(g) = NaCl(s); \quad \Delta H^{\ominus}_{298} = -410\,kJmol^{-1}$$

This may be regarded as a multistage process, summarized

$$Na(s) + \tfrac{1}{2}Cl_2(g) = Na^+(g) + Cl^-(g)$$

$$Na^+(g) + Cl^-(g) = NaCl(s)$$

Computing the energy change accompanying the first stage involves knowledge of the enthalpy changes associated with the sublimation of sodium and its ionization, and the dissociation into atoms of chlorine and its electron affinity. Then, using the known standard enthalpy change of formation of crystalline salt from gaseous molecular chlorine

† A convenient standard state adopted is that in which the substance is stable at 298 K (strictly, 298.15 K) and 101 kPa (1 atmosphere), temperature and pressure respectively. $\Delta H^{\ominus}_{298}$ is read $\Delta H_{standard}$.

and solid sodium, the lattice energy, accompanying the second stage, may be derived. This is the 'experimental' value. If the lattice is not too complex, the lattice energy may be predicted for a given geometrical model of solid, undistorted spheres rigidly held in position, using the introductory electrostatic argument. Substantial agreement between the two values supports the assumption of ionic bonding in the structure: otherwise, polarization towards the covalent condition is probably indicated. Table 5.17 also shows data illustrating this approach to the bonding problem. Small differences may be assumed to lie within the bounds of experimental error, although there is a wide range of values for each compound in the literature, emphasizing the difficulties inherent in the determination of the various quantities used. Compare the perfectly acceptable value for NaF here with that quoted in Table 5.18.

The Born–Haber thermochemical cycle

The foregoing argument, devised by Born and Haber (1919), may be summarized diagrammatically in the following cycle† (based on Hess's Law of Constant Heat Summation) for potassium chloride at 298 K and one atmosphere pressure (1.01 kPa).

would be distinguished from true lattice energy, although the relationship is very close.) Ionization energies are obtained from atomic spectra, dissociation energies from molecular spectra. The other quantities, except electron affinity (p. 44), may be determined by measurement of change of heat content, ΔH, for the process. Until recently, electron affinities could not be determined experimentally with confidence so the Born–Haber cycle was used with the calculated value of the lattice energy to give the electron affinity of the non-metallic element involved. Sherman obtained consistent results for the halogens using the alkali metal halides.†

The basis of modern lattice theory

The advent of X-ray analysis (1912) made the development of lattice theory possible. The pioneers were Born and Lande (1918), who considered the potential energy of ions in a lattice, the attraction of oppositely charged ions for each other and the balancing forces of repulsion which must operate at close approach to confer on ions their characteristic radii. The missing geometrical dimension came from X-ray analysis, the sodium

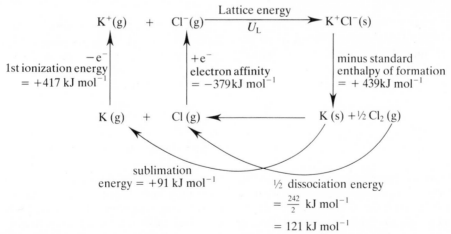

∴ energy required to disperse lattice
  = crystal energy
  = $+439 + (\frac{1}{2} \times 242) + 91 + 417 - 379$
  = $+689$ kJ mol^{-1}
  i.e. Lattice energy = -689 kJ mol^{-1}

(Note that this is, strictly, the lattice enthalpy and that in a rigorous thermodynamical treatment it

chloride lattice providing a comparatively simple structure for mathematical treatment.

Arising from this exploratory work, the lattice energy, U_L i.e. the enthalpy change for a process producing a sodium chloride type of lattice

$$A^{z+} + B^{z-} = A^{z+}B^{z-}$$

† Data from Sherman, *Chem. Rev.* **11**, 93 (1982).

Table 5.18   Lattice energies (calculated values) and ionic charge

	M^+X^- (Mayer and Helmholtz, 1932)				$M^{2+}X^{2-}$ (Mayer and Maltbie, 1934)			
M^+F^-	Lattice energy/kJ mol^{-1}	$M^{2+}O^{2-}$	Lattice energy/kJ mol^{-1}	M^+Cl^-	Lattice energy/kJ mol^{-1}	$M^{2+}S^{2-}$	Lattice energy/kJ mol^{-1}
NaF	-893	MgO	-3929	NaCl	-766	MgS	-3347
KF	-794	CaO	-3477	KCl	-692	CaS	-3084
RbF	-760	SrO	-3205	RbCl	-672	SrS	-2870
CsF	-727	BaO	-3042	CsCl	-637	BaS	-2707

may be computed as

$$U_L = -\frac{LAz^2e^2}{r_0}\left(1 - \frac{1}{n}\right)$$

in which

L = the Avogadro constant (= the number of ion pairs per mole)

r_0 = the equilibrium distance between adjacent ions (i.e. for minimum potential energy)

ze = ionic charge (see equation)

n = term evaluated from experimentally measured compressibility ($n \sim 10\%$) for the structure

A = the *Madelung* constant, obtained by complex calculations for a particular structure ($A \sim 1$)

It is seen that the lattice energy depends on the magnitude of the ionic charges, the interionic distance adopted and through the Madelung constant to the geometrical configuration of the lattice.

$$U_L = -Q\frac{z^2e^2}{r_0} \quad \text{where } Q = NA\left(1 - \frac{1}{n}\right)$$

By far the most important term is that involving ionic charge: the lattice energy of $A^{2+}B^{2-}$ (s) is four times that of A^+B^- (s) if the lattice geometry of the solids is the same. On the other hand the Madelung constant has only a minor effect, changing only very slightly for different structures of the same stoichiometry, e.g. NaCl (s), $A = 1.784$; CsCl (s), $A = 1.763$. The effect of ionic charge is illustrated in Table 5.18 in which Group I metal fluorides are compared with Group II metal oxides, and the corresponding chlorides with sulphides, to show the approximately fourfold ratio. Notice that adjacent pairs of isoelectric structures are compared.

The decline in magnitude of lattice energy with increase of interionic distance is also apparent. The product of interionic distance and lattice energy for these compounds is shown in Table 5.19, which again illustrates two of the factors under discussion, the effect of charge and interionic distance.

Table 5.19   Product of lattice energy and interionic distance/ kJ mol^{-1} nm

NaF	-206	MgO	-805	NaCl	-211	MgS	-833
KF	-214	CaO	-831	KCl	-217	CaS	-873
RbF	-216	SrO	-811	RbCl	-221	SrS	-852
CsF	-222	BrO	-837	CsCl	-223	BaS	-864

The likelihood that an ionic structure will be formed when two elements react is high when the calculated lattice energy is large. As we have seen this is favoured by high ionic charges and short interionic distances.

Lattice energy and oxidation states

In combination the elements of Groups I, II and III have only one oxidation state, positive and equal to the group number. The Born–Haber thermochemical cycle may be used with the calculated value of lattice energies, where compounds are hypothetical, to explore this matter further. If calculations based on the cycle are performed for a range of chlorides of calcium CaCl, $CaCl_2$, $CaCl_3$, ..., $CaCl_n$, it is found that the dominant energy factors are the appropriate ionization energy and the lattice energy, and only $CaCl_2$ turns out to have a negative standard enthalpy change of formation.

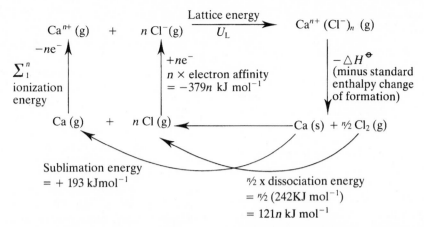

By Hess's Law, the energy required to disperse the lattice may calculated directly and indirectly:

$$-U_L = -\Delta H^\ominus + (n/2 \times 242) + 193 - 379n$$
$$+ \text{ ionization energy required}$$

Re-arranging and simplifying,

$$\Delta H^\ominus = 193 - 258n + (\text{ionization energy}$$
$$+ \text{ lattice energy})$$

By substitution, for $n = 1, 2, 3$:

n	1	2	3
$\Delta H^\ominus/\text{kJ mol}^{-1}$	$+148$	-832	$+1913$

Table 5.20 Ionization and lattice energies for CaCl, CaCl$_2$ and CaCl$_3$

n	Total ionization energy/kJ mol^{-1} $Ca \rightarrow Ca^{n+} + ne$	Corresponding lattice energy/ kJ mol^{-1}	Sum of ionization and lattice energies/ kJ mol^{-1}
1	590	$-$ 377 (calc)	$+$ 213
2	590 + 1148 = 1738	-2247 (exptl)	$-$ 509
3	1738 + 4940 = 6678	-4184 (calc)	$+2494$

Lattice energy and the solubility of ionic compounds

Although water is a poor solvent for covalent compounds, even if their molecules have electrical dipoles, many ionic compounds dissolve and their ions are hydrated in solution, the cation attracting the oxygen atom, with its two lone pairs of electrons, and the anion attracting one of the hydrogen atoms of the water molecule, itself already polarized. While a comprehensive theory explaining the range and extent of solubility has not emerged, an examination of the energy changes that occur on dissolving an ionic compound in water (or any solvent) is profitable.

Usually, but not always, heat is evolved when a crystal lattice disperses in water, the energy of vibration of the ions initiating the process. From the energy standpoint in deciding whether a process is potentially possible, the breakdown and dispersal of the lattice structure may be regarded as a two-stage process. The first is endothermic, requiring an input of energy equal in magnitude to the lattice energy, but with a positive sign to indicate absorption of energy (see definition), and resulting in isolated, gaseous anions and cations. Secondly, two exothermic processes involve the hydration of these gaseous ions, the effect being greatest for small ions of high electrical charge. The heat of solution is therefore the balance between the energies of dispersal and hydration, both relatively large and comparable in magnitude but opposite in sign.

The lattice energy has been seen to depend on the ionic charge, the size of the ions and the geometry of the lattice. Increase of charge, the other factors remaining constant, raises the lattice energy. The stability of the lattice is increased and its dispersal is rendered more difficult. Increase of charge also increases the heat of hydration of the ion but the effect is less marked. Compounds of high ionic charge (e.g. $A^{2+}B^{2-}$) tend to be less soluble than of lower charge (e.g. A^+B^-) with the same structure.

The sum of the heats of hydration of the isolated gaseous ions may be calculated from the values of the lattice energy and heat of solution for a given compound. For a pair of salts with a common cation, the difference between these sums should be independent of the nature of the cation. This has been verified, and supports the underlying principles of this treatment.

Lattice energy, interionic distance and general physical properties

Many physical properties depend essentially on lattice energy. They include heat of fusion, heat of sublimation, boiling-point, melting-point, solubility etc. In turn, the value of the lattice energy is inversely related to the interionic distance. Therefore regularities are expected in the measured values of the physical properties mentioned. A mathematical study has been made for the alkali metal halides which predicts accurately distances between the ions. The picture is complicated where anions are in contact and allowance must be made for double repulsion (Fig. 4.2, page 57). After correcting for these effects the deviations shown by real alkali metal halides when compared with their hypothetical forms can be explained.

Thus, of the postulated versions of calcium chloride, only that for $n = 2$ has a negative enthalpy change of formation; calcium chloride is $CaCl_2$. Similar arguments may be mounted for covalent compounds for which lattice energy is replaced by the heat of formation of the covalent bond from gaseous ions.

Table 5.20 illustrates another general point. As the magnitude of the charge on the cation increases, so do both the aggregated ionization energies and the corresponding lattice energy (obtained by calculation, making reasonable assumptions), with opposite signs, the dominant factors in the calculation. Their different rates of increase determine the value of the enthalpy change of formation. Since the entropy change of formation of the compound is invariably small, $\Delta H^{\ominus}$ effectively determines whether or not a solid compound is stable ($\Delta G = \Delta H - T\Delta S$, p. 135). The maximum possible value of the oxidation state occurs first immediately prior to an escalation in the totalled ionization energies.

Physical properties of the elements in relation to their structures and to their positions in the Periodic Table

Reference has been made (pp. 13 and 31) to the pioneering work done by Lothar Meyer (1869) on establishing the Periodic Law, using physical properties, notably atomic volume, the molar volume/cm^3 mol^{-1} of the element. A periodic curve (Fig. 3.1) results when atomic volume is plotted against atomic number. When density is plotted against atomic number for elements in the solid or liquid state a reflection of this curve is obtained in the atomic number axis, with troughs where there were peaks. The density of an element (mass/volume) will depend on the atomic mass and on factors affecting the volume, namely the volume of the atoms, their geometrical arrangement in the crystal and the bonding. Atomic and ionic radii have been plotted against atomic number elsewhere (p. 76) to show a periodic relationship.

In moving across a period of typical elements the number of electrons involved in bonding increases. The metals (e.g. Na, Mg, Al) have structures with closed-packed atoms (or ions) and delocalized electrons. The early Group IV elements (e.g. C, Si) have giant, macromolecular structures in which all four valency electrons are used in bonding adjacent atoms into a massive three dimensional structure. The first members of the next groups form simple molecules, X_2 but other member of Groups V and VI form polyatomic covalent molecules, e.g. P_4, S_8. There are comparatively weak forces between the molecules which means that the crystals are less rigid than in Group IV and of lower density.

The position is clearly illustrated by plotting[†] melting-points of the elements against atomic number, as shown in Fig. 5.9. Melting-points and boiling-points are listed in Table 5.21. The highest melting-points are shown by Group IV elements. The noble gases occupy the troughs with the halogens and the alkali metals on either side. The units of which a crystal is made, vibrate with increasing frequency as the crystal is heated. Energy is absorbed in quanta. The stronger the bonding, the higher the frequency of vibration and the higher the energy input required to achieve the collapse of the crystal lattice, i.e. melting. Bonding in metals and macromolecules is much stronger than that between the discrete molecules of molecular crystals (e.g. I_2, P_4, S_8), held together by the weak forces usually known as van der Waals' forces. The van der Waals' forces are responsible for holding together the close-packed structures of the inert gases when solidified. This means that the melting-point against atomic number curve is similar to that for density. Across a period of typical elements the melting-point rises with increase in the number of electrons available for bonding and reaches a sharp peak for the macromolecular structures, falling abruptly for

† There is no substitute for actually plotting these graphs, and those for other qualities of the elements to appreciate the astonishing rhythm of the Periodic Law for so many apparently diverse elements.

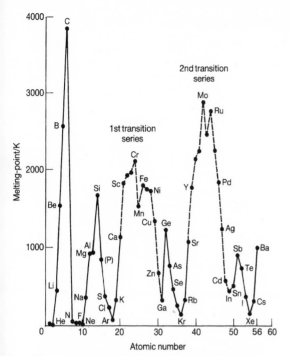

Fig. 5.9 Melting-point plotted against atomic number

Table 5.21 Melting-points and boiling-points of elements 1–56 inclusive*

Atomic number	Symbol	Melting-point/K	Boiling-point/K
1	H	14	20
2	He	(1)	4
3	Li	452	1609
4	Be	1556	3243
5	B	2573	2823
6	C	3843	5100
7	N	63	77
8	O	54	90
9	F	53	85
10	Ne	24	27
11	Na	371	1156
12	Mg	924	1380
13	Al	933	2543
14	Si	1687	2628
15	P	(862)	—
16	S	392	718
17	Cl	172	239
18	Ar	84	87
19	K	337	1031
20	Ca	1124	1460
21	Sc	1814	3104
22	Ti	1933	3560
23	V	1993	3653
24	Cr	2103	2533
25	Mn	1515	2173
26	Fe	1800	3023
27	Co	1763	3143
28	Ni	1728	3003
29	Cu	1356	2840
30	Zn	692	1180
31	Ga	303	2676
32	Ge	1232	3103
33	As	886	(906)
34	Se	490	958
35	Br	266	332
36	Kr	116	120
37	Rb	312	973
38	Sr	1073	1639
39	Y	1773	3611
40	Zr	2130	4650
41	Nb	2223	5015
42	Mo	2895	5833
43	Tc	2445	5150
44	Ru	2583	4173
45	Rh	2239	—
46	Pd	1830	3243
47	Ag	1234	2425
48	Cd	594	1037
49	In	429	2373
50	Sn	505	2635
51	Sb	904	1598
52	Te	723	1663
53	I	387	460
54	Xe	161	165
55	Cs	302	943
56	Ba	998	1810

* Some variation in reference books

the molecular crystals of Group V and onwards. The periodic variation will be shown by the molar heat of fusion of the elements, this being a measure of the energy required to cause disintegration of the crystal lattice at its melting-point, although for metals the liquid phase still has atoms, or ions, in a mutually held pool of delocalized electrons and thus essentially retains its bonding.

When the liquid element boils, the atoms or molecules must detach themselves from the main bulk of the liquid. Molar heats of vaporization are larger than the corresponding heats of fusion because of this. For molecular crystals the difference between melting-point and boiling-point will be small, reflecting the very weak forces between the molecules at boiling. The graph of boiling-point against atomic number appears in Fig. 5.10. It will be seen at once that the alkali metals have much higher boiling-points than might have been expected, hence the idea of mobile atoms, or ions, in a pool of delocalized bonding electrons being retained in the liquid state. The macromolecular structures in the centre of the Periodic Table do not entirely disintegrate into atoms on melting and the dissolved liquid structure still needs a considerable input of energy for boiling to occur, thus accounting for the extra high peaks for elements such as silicon ($Z = 14$).

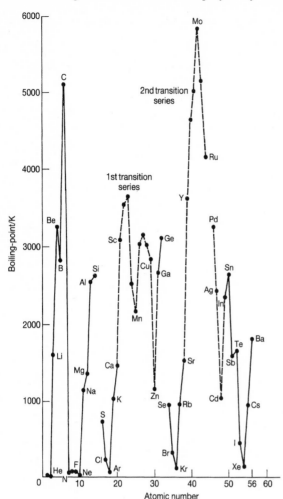

Fig. 5.10   Boiling-point plotted against atomic number

Semiconductors

One way of classifying elements (and other materials) is in terms of their electrical conductance, or alternatively their resistance to the passage of an electric current. Metals are electrical conductors with low resistance to the flow of electrons because of their lattice arrangement, but this resistance increases with rise in temperature. At the other extreme are the solid non-metals with covalent bonding which are classed as insulators. Between these groups the metalloids provide elements of great importance in electronics, the semiconductors. The electrical resistance of these decrease with rising temperature as electrons are sufficiently excited to move from energy levels associated with bonding to those beyond which merge, so providing electrical conduction. The electrical characteristics of a crystal depend on the size of the energy gap between the filled valence band, or quantum level, and the higher, hitherto empty conduction bonds. To illustrate this consider some elements with the silicon structure for which the band energy gap/$kJ\ mol^{-1}$ has the following values

C (diamond)	Si	Ge	Sn (grey)
502	105	58	8

Diamond is a splendid insulator. Tin will form the metallic allotrope on heating. For silicon and germanium, conduction electrons will be released by normal thermal excitation. These elements are called *intrinsic semiconductors*. Promotion of an electron to the conductance band leaves behind a hole in the valence band, or level. This means that there is room for another electron to be promoted to the empty state created. This hole is effectively a positive charge. Herein lies the essentials of the *band theory* of semiconductance.

The conducting property of a semiconductor is greatly enhanced by the controlled addition of certain impurities, when the semiconductor is said to be doped. Elements from adjacent groups of the Periodic Table are used for this purpose thereby introducing either a surplus or a shortage of electrons. If phosphorus atoms are introduced into a silicon lattice, four of the valence electrons of the phosphorus are needed in the lattice but the fifth is expelled to a higher level. The energy gap is thereby reduced to $1.2\ kJ\ mol^{-1}$. The semiconductor silicon doped with phosphorus (as donor) is of the *negative type* or *n*-type, conduction being stimulated by negative (electron) charges, introduced by the Group V element.

Very many physical properties reflect the periodicity of the properties of the elements. These can be traced back to the external features of the atom, the size of the nucleus and its charge, and the number and configuration of the attendant electrons, or 'electron-clouds'. In the foregoing the salient feature has been the way in which the orders of magnitude of melting-points (and to a large extent boiling-points) reflect the type of crystal structure adopted by the element, from the metallic macromolecular structures with delocalized electrons in the early elements of a period, the distinctive rigid macromolecular crystals of the middle groups, the molecular crystals of later groups, and the molecules of the noble gases which are single atom entities not to mention the various transition series.

On the other hand if an element of Group III is introduced into the silicon lattice, an electron is missing. If boron were to be used, there would be a hole (effectively a positive charge) in the tetrahedral bonding superimposed on the boron atoms. Thus, a new energy level appears for the migration of electrons. The energy gap is $0.96 \, kJ \, mol^{-1}$. Conductance is considerably enhanced and silicon doped with boron is now a semiconductor of the p-type, or *positive type*. Boron is an acceptor impurity.

From this outline it will be seen that whether extra electrons (n-type) or extra 'holes' (p-type) appear in the metalloid crystal, the electrical properties of the semiconductor material are markedly altered. In visualizing applications of this in the use of transistors, one may think of electrons moving one way or of positively charged holes moving the other. The diode provides a simple example.

A diode is a device which permits the conduction of electrical current (or charge) in one direction. It is manufactured by forming regions of n-type and p-type semiconductors *within the same crystal*. At the junction electrons diffuse from the n-type and holes diffuse from the p-type, which absorb the electrons so creating a depletion (or barrier) layer of high resistance because the conduction units have disappeared. The n-type material develops a positive potential with respect to the p-type material, eventually causing migration to cease. If an applied potential difference reduces this, a current will flow; if it reinforces it, there will be no conduction.

Silicon and germanium are used as semiconductors, doped with phosphorus, arsenic and antimony of Group V, or boron, iridium and gallium of Group III.

6

Crystallization and the crystalline state

The characteristics of crystals, their structure and bonding

Crystallization from aqueous solution

On being shaken with water, a solid may dissolve to form a solution. The solid is usually called the solute and the liquid, not necessarily water, the solvent. Continued addition of solid will lead to *saturation*, when no more solid will dissolve in the presence of excess solid. However, on warming, the undissolved solid will probably dissolve and more solid will have to be added to achieve saturation at the higher temperature. The mass of solute required to saturate a specified convenient mass or volume of a solvent at a given temperature is called the *solubility* of the solute: the mass of solvent chosen is usually 100 g although g cm^{-3}, g dm^{-3} or percentage by weight of solute in the solution are also used. Where a compound contains water of crystallization, the solubility usually quoted is that of the anhydrous compound. It is most important that units, temperature and anhydrous condition (or otherwise) be stated with numerical data. The solubility of a solid in water—and we are concerned chiefly with this solvent here—varies with temperature: in most cases, a rise in temperature results in a rise in solubility although some compounds show a decrease in solubility with increased temperature. The relationship between solubility and temperature is best expressed in the form of *solubility curves*, some examples of which are shown in Fig. 6.1. Some compounds have very small solubilities in water. Calcium hydroxide has a retrograde solubility curve, the solubility decreasing with rise in temperature: 0.19 (0°C), 0.17 (15°C), 0.077 (100°C) g 100 g^{-1} water. The solubility of calcium sulphate dihydrate† (CaSO$_4$.2H$_2$O) rises and then falls: 0.18 (0°C), 0.28 (40°C), 0.26 (60°C) g 100 g^{-1}water. Both are said to be sparingly

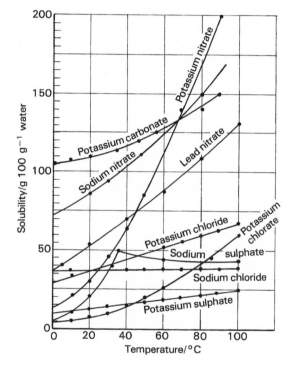

Fig. 6.1 Solubility curves

soluble. There are many compounds which have such low solubilities that for many purposes they are classified as insoluble. The solubility curve for sodium sulphate shows a sharp discontinuity at 32.38°C, indicating a change in composition of the solid phase in equilibrium with the saturated solution: below the transition point, it is Na$_2$SO$_4$.10H$_2$O and above, Na$_2$SO$_4$.

At a given temperature, a saturated solution is considered to be in equilibrium with any undissolved solid present. The equilibrium is dynamic, the rate at which solid passes into solution

† Greek prefix, di = 2; Greek root, *hūdor*, water.

equalling the rate of deposition. Until saturation is reached, the solution is said to be unsaturated. Under certain circumstances a solution may contain more dissolved solid than the value of its solubility at that temperature would suggest as possible, but the addition of a little more solid results in precipitation until saturation holds. Such a solution is *supersaturated*. A good example, probably encountered already, is shown by the ease with which sodium thiosulphate pentahydrate ('hypo') $Na_2S_2O_3.5H_2O$, forms supersaturated solutions. The hydrated crystals melt on gentle warming to form a solution which usually does not yield crystals immediately on cooling, especially if dust is excluded. Addition of a tiny crystal of sodium thiosulphate pentahydrate, called 'seeding' or inoculating, results in crystallization of the whole solution. Heat is evolved. Addition of a little solid enables a quick test to be made for the degree of saturation in a given solution: disappearance indicates unsaturation, while further deposition indicates supersaturation. Some degree of supersaturation occurs widely and is not confined to spectacular cases like that of sodium thiosulphate.

The solubility of a solid in a liquid will depend on the chemical nature of the solute and solvent, on the temperature and perhaps on the physical state of the solid. Water is the common solvent in inorganic work but there are simple examples of other solvents being used. Rhombic sulphur is usually crystallized from solution in carbon disulphide, while monoclinic sulphur may be crystallized from boiling toluene. White phosphorus is soluble in carbon disulphide. The red allotrope is insoluble. Both in industry and the laboratory, the final stage in the preparation of inorganic salts, usually ionic compounds, frequently involves re-crystallization, the practice of which depends on the variation of solubility with temperature.

Crystallization may be effected by cooling a saturated solution, if the solubility decreases with fall in temperature, or in a few cases where the solubility curves are retrograde, by raising the temperature. Removal of solvent may be preferable or complementary to the above, by careful evaporation on a water bath or sometimes at room temperature by use of a vacuum desiccator or by leaving a very concentrated aqueous solution in a desiccator over concentrated sulphuric acid which will absorb water vapour. Tin(II) chloride is crystallized over concentrated sulphuric acid. Occasionally the addition of a miscible organic liquid to the aqueous solution results in precipitation of the solute. Organic compounds used for this purpose include ethanoic acid (acetic acid), methanol, ethanol and propanone (acetone). The purpose of crystallization may be either purification or the preparation of well-formed specimen crystals.

For crystallization to occur, it seems certain that the unit of pattern for the crystal must be present in the solution. Where solutions prove difficult to crystallize, the unit, or nucleus, is provided by deliberate inoculation ('seeding') with the addition of a tiny crystal to the supersaturated solution. The same intention is behind the practice of scratching the inside glass walls of the vessel containing the solution with a glass rod. Crystallization often appears to be spontaneous and this is attributed to 'invisible atmospheric dust'. Rapid crystallization results in many small crystals being formed and these are usually purer than larger crystals because less mother liquor, the liquid remaining after crystallization, is trapped inside. This cannot be removed by washing. Titrimetric (volumetric) analysis of suitably grown hydrated crystals will illustrate this point (ammonium iron(II) sulphate by potassium permanganate, copper(II) sulphate by iodine-thiosulphate). Rapid crystallization results from the rapid cooling of hot (super)saturated solutions or, sometimes, by suitable double decomposition reactions where supersaturation occurs momentarily. Large crystals are developed under quite different conditions. Very slow crystallization of saturated solutions at an even temperature, conveniently provided by the warmth from a closely adjacent electric lamp, yields large crystals. Large single specimen crystals may be grown by suspending a suitable crystal by a thread in a saturated solution.

Crystal habit may be altered by the presence of dissolved impurities in the solution. Large crystals which are well formed may usually be obtained by crystallizing salts from solutions which also contain the corresponding acid in moderate concentration. Thus, the addition of nitric acid produces better-formed crystals of lead(II) nitrate, and sulphuric acid improves the shape of copper(II) sulphate crystals.

During the preparation of an inorganic salt, the initial product will probably contain impurities due to the reactants and other products formed. Purification will be by re-crystallization, if the solubilities of product and impurities are sufficiently different or show a sufficient variation of solubility with temperature change. Fortunately, many compounds show a pronounced variation of solubility with change in temperature.

With the minimum quantity of boiling distilled water, or other suitable solvent, such as ethanoic acid (acetic acid), methanol, ethanol, isopropanol or propanone (acetone), a saturated solution is made in a conical flask. Care is taken to use the minimum permissible quantity of solvent to avoid diminishing the final yield by leaving behind more saturated solution than is necessary. The solution is filtered rapidly through a fluted filter in a short-stemmed filter funnel to remove insoluble impurities and then cooled. This is done conveniently

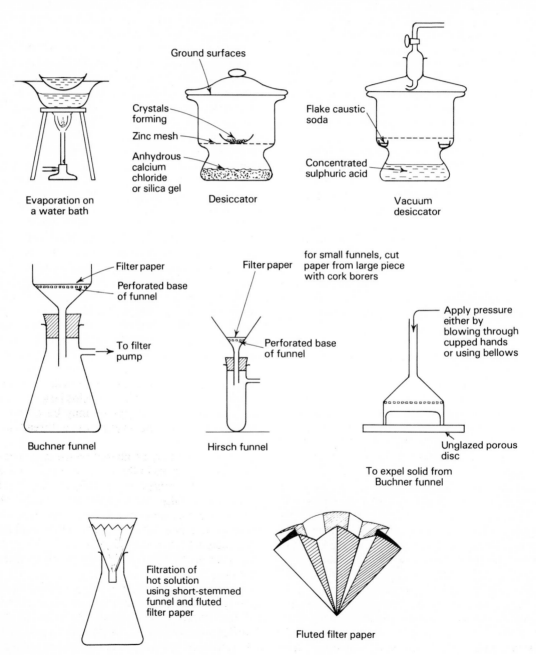

Fig. 6.2   Apparatus used in purification by crystallization

by holding the flask with its side in a stream of cold water from the tap and swirling the contents. The use of an ice-water mixture may be desirable for a maximum yield, reducing the quantity remaining inevitably in the saturated solution.

The crystals are filtered on a Buchner or Hirsch filter by suction and rammed down on the filter paper, conveniently by using an inverted reagent bottle stopper, until as much adhering solution as possible has been drained away. Crystals left in the conical flask must be washed into the filtration funnel with filtrate solution. A metal filter pump with a non-return valve is recommended. In place of the traditional funnels of porcelain or polythene, which require filter papers, sintered glass funnels are sometimes used.

After turning off the pump, or disconnecting the pressure tubing if a non-return valve is not fitted, the solid is moistened with a little cold solvent. This is sucked through and allowed to drain away. The procedure is repeated. Washing in a suitable volatile organic liquid miscible with the main solvent greatly facilitates drying, which is accomplished by placing the filtration funnel and its contents in an oven or in a warm place in the open laboratory. The pure, dry solid with the dry filter paper may be removed from the inverted funnel by blowing down the stem through a cupped hand. The use of bellows may be preferred. Sometimes, gentle tapping is sufficient to remove the contents of the funnel.

In Fig. 6.2 various pieces of apparatus used for crystallization are shown.

The shape and internal structure of crystals

A crystal is a piece of solid substance which is homogeneous and bounded by plane faces, meeting with sharp edges at fixed angles characteristic of that substance. Crystals may occur naturally or be produced artificially. Examination of a batch of crystals of a particular substance will reveal that they are certainly not all the same size or shape. Faces which meet in some crystals may be cut across by new faces or facets in others, conditions during crystallization leading to unequal rates of growth. However, there is one characteristic common to all of these crystals.

The angles between corresponding faces of the crystals of the same substance have exactly the same value, which is characteristic of that substance.

This law was established by Steno (1669), Guglielmini (1688–1705) and de l'Isle (1772–83) and is called the **Law of Constancy of Angle**.

Choice of solvent, impurities, temperature conditions and the rate of crystallization all influence the final crystal *habit* or characteristic shape. Thus, sodium chloride normally crystallizes as simple cubes but forms cubic crystals with octagonal faces in the presence of urea. All six faces of the cubic crystals are of the same *form* but the new crystals have faces of two different forms, octagonal and equilateral triangular faces. Each crystal is said to have a *combination form*. Crystallographically speaking, these crystals are identical. Shapes produced by successively removing portions from the eight corners or *coigns* of a cube are shown in Fig. 6.3.

That the regular external shape of a crystal is an outward manifestation of orderliness within remained an intelligent conjecture until confirmed by X-ray analysis from 1912 onwards. Ions, atoms or molecules take up positions in a regularly repeating pattern. Crystallography has become the scientific study of the internal crystalline state through the development of the techniques of X-ray analysis. The term *solid* is restricted nowadays to crystalline substances: apparent solids like pitch and glass are classified as supercooled liquids, being *amorphous, i.e. without a shape of their own*. Over the years, pitch left in a filter funnel will be observed to flow away and old soda glass will sometimes crystallize or undergo devitrification when heated.

There are certain properties which are characteristic of the crystalline state and possession of these is an indication of inner symmetry whatever the outward appearance. Many crystalline substances may be split, fractured or *cleaved* easily in directions usually parallel to external faces of the original crystal, but not always. Internal forces must, therefore, be weakest along these cleavage planes. Knowledge of the four cleavage planes of diamond is applied by jewellers to 'cut' these extremely hard crystals. For calcite, there are three planes, mica has one while quartz has none. In addition, many crystals have physical properties which vary with the direction of measurement. For example, thermal conductance may depend on the direction in which the heat flow is measured. The thermal decomposition of lead(II) nitrate is accompanied in the early stages by decrepitation, which is the audible shattering of crystals produced by unequal expansion in different directions. Some crystals are doubly refracting and are capable of rotating the plane of polarization of plane polarized light. Where properties vary with the direction of measurement, the substance is said to be *anisotropic* (Greek, *anisos* = unequal, *tropos* = manner) while uniformity of behaviour classes the substance as *isotropic* (Greek, *isos* = equal). Elasticity, hardness and magnetic properties have all been found to vary with the direction of measurement in a crystal.

Crystals may be deposited by cooling a liquid or vapour, or from a saturated solution. Crystals of metals may be grown during electrolytic deposition or electrochemical displacement by other metals. Some salts form hydrated crystals which are quite dry in appearance but contain an integral number of molecules of *water of crystallization* combined with each unit of the crystal but readily expelled on heating. Sometimes, water is lost to the atmosphere, because the vapour pressure over the crystal is more than that of the water vapour in the atmosphere at that temperature. Such crystals are said to *effloresce*. Washing soda crystals (sodium carbonate decahydrate, $Na_2CO_3.10H_2O$) soon become coated with a white film of monohydrate, rendering the translucent crystals opaque. Some crystalline compounds take up water vapour from the atmosphere, forming a solution. They are said to *deliquesce*. The process starts because the water vapour pressure over the crystal is less than that in the atmosphere at that temperature and continues because the saturated vapour pressure of water over the resulting solution is less than that in the atmosphere. The process continues until the saturated water vapour pressure over the dilute solution equals that of the atmosphere.

The geometry of crystals

The variation of crystal habit when crystallization occurs under different conditions for a given pure element or compound means that any classification of crystals cannot be made by external shape. The simplest classification comes from the study of their degree of symmetry. *Seven major crystal systems* may be defined in terms of *axes of symmetry*. For a description of crystal systems a textbook of crystallography should be consulted.

From observations of the cleavage of crystals, which started with the accidental breakage of a calcite crystal, the Abbé Haüy suggested in 1784 that a crystal was composed of very many sub-microscopic units, solid and of a definite shape not necessarily that of the final crystal, packed together in a regular manner. Clearly, cleavage planes exist between successive layers of 'bricks'. It is easy to produce a solid crystal with faces parallel to the faces of the individual bricks but Haüy went on to show that other final shapes could arise from the omission of successive layers where necessary. Taking a small cube as the unit, the various shapes seen in Fig. 6.3 can be derived and since the units are sub-microscopic, the final crystals will appear to have smooth edges and faces.

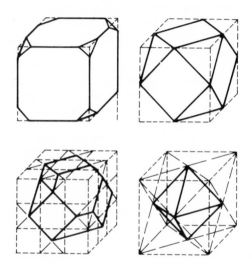

Fig. 6.3 Shapes obtained from a cube by removing portions from the corners

However, some crystals cannot be broken by cleavage or show cleavage in one direction only or do not yield an obvious simple structural unit. The concept of solid structural units has evolved into the idea of a repeating geometrical pattern, without thinking in terms of material units. Imagine hanging a wallpaper with a pronounced pattern. The pattern on adjacent pieces must be matched. Imaginary straight lines criss-crossing the paper, connecting corresponding points all over the pattern, would produce a network in each 'cell' of which the repeating pattern is contained. The actual point chosen does not matter; the whole is still made up of repeating units. A crystal is three-dimensional and a series of intersecting planes, separated by a distance depending on the appropriate axis length and parallel to each face, will produce a system of points in space. The result is called a *space lattice* and the smallest unit enclosed, the *unit cell*. In 1848, Bravais showed that there are fourteen different types of arrangement for lattices, which may be further divided into 230 general types.

The concept of a space lattice is geometrical and while these considerations have been confirmed by experiment, the points of a lattice are not necessarily to be thought of as representing atoms. The whole purpose of a lattice is to explain crystal shapes in terms of a repeating geometrical unit.

Direct evidence for the orderly arrangement within the solid state has come from the development of X-ray techniques for the examination of crystals. By analogy with the diffraction of visible light, it was suggested that X-rays would produce diffraction patterns with crystals because wavelengths and interatomic distance were considered comparable. In 1912, von Laue, Friedrich and Knipping demonstrated this: a beam of X-rays with a range of wave-lengths was passed through a crystal and received on to a photographic plate upon which was recorded the position of the main beam and surrounding it, a symmetrical arrangement of spots. In 1913, Sir William Bragg and his son, (Sir) Lawrence Bragg, invented the X-ray spectrometer in which a beam of X-rays of (nearly) a single known wave-length was 'reflected' from a crystal and the angles at which strong beams of X-rays were diffracted were measured. Nowadays, the rotating crystal method is used where crystals are available but otherwise, a fine powder will suffice. Interpretation of the photographs is involved and becomes increasingly difficult with the less symmetrical crystal systems. However, modern physical methods have brought a clear understanding of the structure of inorganic compounds. Under the direction of Sir Lawrence Bragg at Cambridge this powerful experimental technique has been used to unravel the complexities of certain proteins (haemoglobin by Perutz and myoglobin by Kendrew) and of the genetic material, the celebrated double helix of deoxyribonucleic acid, DNA (by Crick and Watson, and in parallel investigations at King's College by Wilkins and the young Rosalind Franklin, who was sadly dead by 1962, when the other five received their Nobel Prizes).

Simple ionic crystals: the sodium chloride, caesium chloride, fluorite and rutile structures

The first crystal analysis was that of rock salt by Sir William Bragg in 1913. The solid is an assembly of ions: each sodium ion is surrounded by six chloride ions and each chloride ion by six sodium ions. The whole is held together by electrostatic forces of attraction. No simple NaCl molecule exists, the formula merely showing the ratio of the number of different ions forming the structure. The whole may be regarded as one molecule. All the halides of the alkali metals crystallize in this way, except caesium chloride, bromide and iodide. Structures, not to scale, are shown in Figs 4.1 and 4.2 (p. 57). Dissimilar ions are 'in contact' except for the chloride, bromide and iodide of lithium where the larger anions touch. The form of the crystal lattice depends on the ratio of the radii of the ions in the structure. The electrostatic forces of attraction and repulsion between ions are non-directional and do not determine the form of the lattice. On fusing or dissolving the crystal in water, the ions dissociate from each other.

The sodium chloride lattice is one of two commonly found structures for ionic compounds of the general formula AB, being formed by the oxide and sulphides of the Group II elements, magnesium, calcium, strontium and barium, in addition to the alkali metal halides listed above. The other structure frequently adopted by ionic compounds of the type AB is that of caesium chloride, where the caesium cation, larger than the other cations of Groups I and II but smaller than the chloride anion, is surrounded by eight chloride ions, and to maintain the stoichiometric requirements of the formula CsCl (there being no space limitation), each chloride anion is surrounded by eight caesium ions. The cubical unit cells of NaCl and CsCl are shown in Fig. 6.4, which should be studied in conjunction with Fig. 4.1 and 4.2 mentioned earlier. The body-centred cubical cell of caesium chloride has effectively one ion-pair while the less simple sodium chloride structure has four ion-pairs in each unit cell.

Structures commonly adopted by ionic compounds of general formula AB_2 are those of fluorite (calcium fluoride) and rutile (titanium(IV) oxide). These are also shown in Fig. 6.4 where the 'anti-fluorite' structure of Li_2O (i.e. A_2B) is also indicated. The fluorides of strontium and barium have the fluorite structure, while magnesium fluoride and tin(IV) oxide adopt that of rutile. In fluorite, the calcium ions are arranged in a cubic face-centred lattice and the associated fluoride ions are at the centres of the eight cubelets shown as making up the unit cell. In rutile, six oxygen ions, placed at six corners of a slightly distorted regular octahedron surround each titanium ion, while each oxygen lies in the plane of a nearly equilateral triangle formed by three adjacent titanium ions.

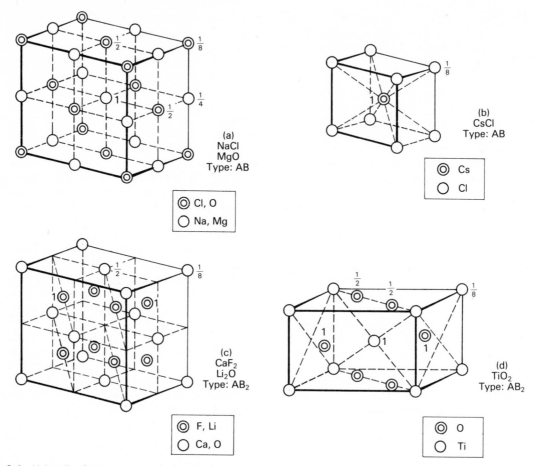

Fig. 6.4 Unit cells of some common ionic crystals
(a)   Cubical unit cell of NaCl or MgO (sodium chloride structure)
(b)   Cubical unit cell of CsCl (caesium chloride structure)
(c)   Cubical unit cell of CaF$_2$ (fluorite structure) and of Li$_2$O ('anti-fluorite' structure)
(d)   Tetragonal unit cell of TiO$_2$ (rutile structure)
(Adapted from Palmer (1945), *Valency: Classical and Modern*, Cambridge University Press)

In assessing the net number of ions in the unit cell, remember that ions at the corner, edge or in the face of the cell are shared with the adjacent cells. Reference to Table 6.1 and Fig. 6.4 will make this clear.

Table 6.1 Interpretation of unit cell diagrams

Position of atom or ion	Number of cells sharing atom or ion	Net contribution of atom or ion to unit cell
Inside	1	$\therefore$ 1
Face	2	$\therefore \frac{1}{2}$
Edge	4	$\therefore \frac{1}{4}$
Corner	8	$\therefore \frac{1}{8}$

For sodium chloride:

$$Na^+ \quad (12 \times \tfrac{1}{4}) + 1 \qquad = 4$$

$$Cl^- \quad (6 \times \tfrac{1}{2}) + (8 \times \tfrac{1}{8}) = 4$$

$\therefore$ sodium chloride has 4 Na$^+$Cl$^-$ in each cell.
For calcium fluoride (fluorite):

$$Ca^{2+} \quad (8 \times \tfrac{1}{8}) + (6 \times \tfrac{1}{2}) = 4$$

$$F^- \qquad\qquad\qquad = 8$$

$\therefore$ calcium fluoride has 4 Ca^{2+} (F$^-$)$_2$ in each unit cell.

Co-ordination number

The simplest way of indicating the environment of an atom of an element in a crystal structure lies in the concept of co-ordination number. This is determined from the structure alone and does not depend on whether the bonding is covalent or ionic, and while it applies to both elements (e.g. metal structures) and compounds, it will be illustrated by reference in this section to ionic compounds only.

A cation, being formed by loss of electrons from an atom is smaller than that atom, while an anion, formed by gain of electrons is larger than the corresponding atom. Further, some common anions are charged groups of atoms united together. Therefore, cations are generally smaller than anions and it is usual to speak of the limiting number of anions which can be packed round a cation in a structure, although the terms can be interchanged where necessary.

> The **Co-ordination Number** of an ion in the solid structure of a specified compound equals the number of nearest neighbouring ions which surround it; for a covalently bound atom, the co-ordination number of the atom equals the number of neighbouring atoms to which it is directly linked.

The co-ordination number of the smaller ion having been determined, that of the larger ion, for which, of course, there is no overcrowding problem, is obtained from the stoichiometric requirements of the formula. For compound A^+B^-, the co-ordination number of B^- equals that of A^+; for $A^{2+}(B^-)_2$, the co-ordination number of B^- is one half of that of A^{2+}.

The ions in a structure are subjected to forces of electrical attraction by oppositely charged ions and electrical repulsion by similar ions; these are a minimum when the ions are arranged as regularly as possible. A structure may be visualized as made up of regular polyhedra. The co-ordinating anions are situated at the corners of a polyhedron, at the centre of which lies the cation. This is illustrated in Fig. 6.5.

Simple molecular crystals: iodine, sulphur, ice

Crystals formed by covalent elements and compounds either show the individual molecules or appear to be one large molecule in which all atoms are covalently bound. Since intermolecular forces are relatively small, crystals containing individual

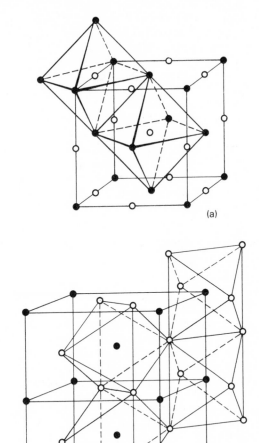

Fig. 6.5 The co-ordinating octahedra of anions around the cations in (a) sodium chloride, NaCl, (b) rutile, TiO_2. (Adapted from Evans (1946), *An Introduction to Crystal Chemistry*, Cambridge University Press)

molecules will be soft and easily fused: iodine, sulphur and sulphur dioxide provide examples, in which the molecular units have the formulation I_2, S_8, SO_2 respectively. However, hydrogen bonding plays an important role in the structure of ice. Water molecules have electron-pairs arranged tetrahedrally around oxygen with two protons attached, giving an angled molecule $H\widehat{O}H$ (Fig. 5.2). Each water molecule may be linked to four other molecules by hydrogen bonding, the hydrogen atoms being attracted to the lone pairs of oxygen atoms of other molecules while the lone pairs of the oxygen atom attract hydrogen atoms of a pair of adjacent water molecules. X-ray investigations reveal that ice has a structure similar to that of wurtzite (see below) in which each oxygen atom is surrounded tetrahedrally by four other oxygen atoms at a distance of 0.276 nm. The result is a very open structure and accounts for the lower density of ice, which floats on water, and the rise in density of water to 4°C is due to a change in equilibrium between hydrogen-bonded structures of different configuration. Also of great interest, however, are the giant molecules or macro-molecules. In the face of evidence acquired by the use of modern physical methods many preconceived notions about molecular individuality have had to be abandoned.

The covalent macromolecular crystals of carbon and zinc sulphide

Zinc sulphide forms two crystal structures: it is dimorphous. In each, a zinc atom is surrounded by four sulphur atoms in a tetrahedral configuration and each sulphur atom similarly by four atoms of zinc. Each atom is bound covalently. No single ZnS molecule is distinguishable and so the whole must be regarded as a giant molecule or macro-molecule. The two types of zinc sulphide are zinc blende and wurtzite, names which are now given to the type of lattice formed. Alternatively, the crystal system may be quoted after the name or formula. Zinc blende belongs to the cubic system and wurtzite to the hexagonal crystal system. The lattices are shown in Fig. 6.6. The diagram shows the cubical

cell unit of zinc blende, consisting of eight cubelets. Zinc atoms occur at the corners of the cube and face centres, being shared with adjacent cells, in what is called a face-centred cubic lattice. Sulphur atoms are at the centres of alternate cubelets, each being linked to the four zinc atoms at the corners of its cubelet. Sulphur atoms also lie in a face-centred cubic lattice but this is not apparent in the diagram, only a small portion being shown for simplicity. Zinc and sulphur are identically situated.

It will have been noted that β-cristobalite (SiO_2) and diamond (C) form the zinc blende structure. Other elements of Group IV: silicon, germanium and tin, and silicon carbide, SiC, also form this structure as do the copper(I) halides: chloride, bromide and iodide.

The geometrical relationship between zinc blende and wurtzite is well brought out by the diagrams. Beryllium oxide and aluminium nitride form the wurtzite structure. It is interesting to note that silver iodide exists in both zinc blende and wurtzite structures.

Diamond has the zinc blende structure. Replacing both zinc and sulphur atoms by carbon and allowing the lattice framework to 'fade' into the background, emphasizes the tetrahedral configuration in the diamond structure. Each carbon atom lies at the centre of a tetrahedron, at the corners of which lie four other carbon atoms. The internuclear distance is 0.154 nm, which is nearly equal to the distance between the centres of adjacent

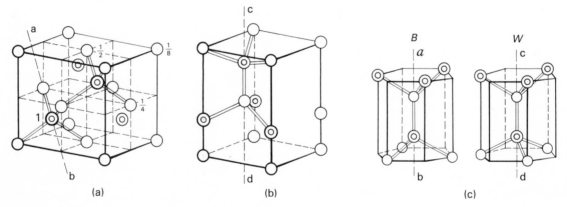

Fig. 6.6

(a)   The cubical unit of zinc blende (ZnS) and diamond for which S = Zn = C
      (For SiO_2 β-cristobalite put C = Si, and inter-pose O at midpoint between each pair of Si atoms)
(b)   The hexagonal unit cell of wurtzite (ZnS)
(c)   The closely-related structures of zinc blende (*B*), and wurtzite (*W*): axes *ab* and *cd* are shown in (a) and (b)
(From Palmer (1945), *Valency: Classical and Modern*, Cambridge University Press)

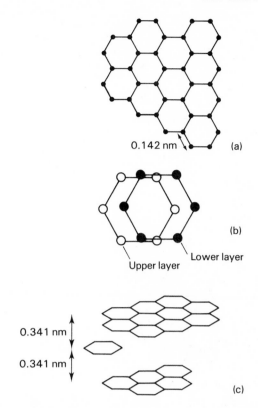

0.142 nm   (a)

Upper layer   Lower layer   (b)

0.341 nm

0.341 nm   (c)

Fig. 6.7   The structure of graphite
(a)   Individual layer of carbon atoms showing honey-comb structure
(b)   Showing relative position of atoms in adjacent layers of atoms in natural graphite
(c)   General picture showing three layers

carbon atoms in saturated aliphatic compounds. The special hardness of diamond comes from this rigid giant structure.

Graphite, the other form of carbon, is quite different. The atoms lie in flat honeycomb layers which are relatively far apart. The atom is linked covalently to three other atoms in the same layer, the distance between the centres of adjacent atoms being 0.142 nm. The layers are 0.341 nm apart, being loosely held together by van der Waals' forces. The use of graphite as a lubricant depends on the sliding of layer upon layer and on the hardness of the element at right angles to the layers, the sheets resisting compression. Each layer may be regarded as one giant molecule. The structure is shown in Fig. 6.7. X-ray analysis has shown that in natural graphite alternate planes of atoms are usually similarly disposed to each other. There are other forms of graphite, including synthetic graphite, in which every third plane of atoms is similarly disposed. In some industrially produced graphite, and in coal, the graphite structure has jagged holes in it where large numbers of atoms have been omitted.

The structure of metal crystals: face-centred cubic and hexagonal close-packing, body-centred cubic

The elements on the left of the Periodic Table, unlike those on the right which tend to form covalent bonds by electron pair sharing, have been seen to favour ionic bonding in combining with

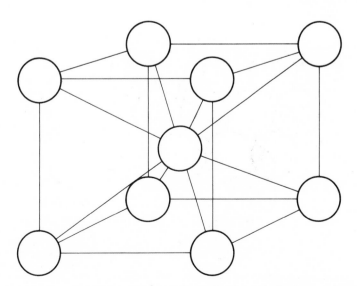

Fig. 6.8   Metal structure: body-centred cubic arrangement unit cell showing co-ordination number of 8

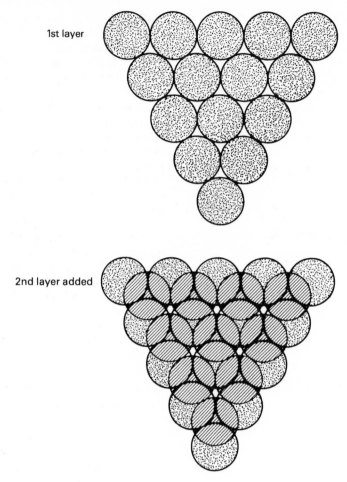

1st layer

2nd layer added

Fig. 6.9   Building up close-packed structures in layers

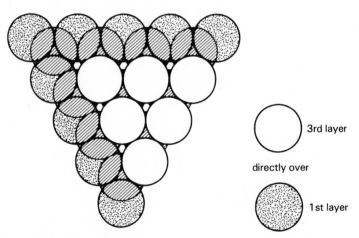

3rd layer

directly over

1st layer

Fig. 6.10   Building up close-packed structures in layers:
3rd layer added to form hexagonal close-packing (*hcp*): $A, B, A, B, A, B \ldots$ with co-ordination number of 12

other elements, involving the transfer of electrons. In the solid state the atoms of these elements relinquish control of their outer electrons which become delocalized. The picture emerges of a geometrical array of cations through which percolates a 'sea of electrons', accounting for the high electrical and thermal conductance of these elements. The bonding, metallic bonding, is electrostatic and non-directional. Nearly all metals adopt one of three common structures, also adopted by metal alloys and by the so-called interstitial hydrides, borides, carbides and nitrides of transition elements, in which the smaller early-period atoms fit into a metal structure different from that of the original metal and some valency electrons remain freely available for electrical conduction.

No close packing

The *body-centred cubic structure*, in which the co-ordination number is eight, which is not close-packed, is shown in Fig. 6.8. Examples of metals forming this atomic arrangement are the alkali metals (Group I). The descriptive term is conveniently shortened to *bcc*.

Alternative ways of close packing

There are alternative ways of packing identical spheres as tightly as possible into a space. The resulting close-packed structures are described as *hexagonal close-packing (hcp)* and *face-centred cubic close-packing (fcc)*.

The first two layers of spheres can fit together in only one way (Fig. 6.9). However, there are two possible ways in which the third layer may be added. In the first, the new layer is directly over the base layer, so that the successive layers alternate *A, B, A, B, A, B*... This creates a hexagonal close-packed structure, with the co-ordination number of 12 (Fig. 6.10). Magnesium and zinc display such a *hcp* structure, the hexagonal basis for which is readily seen (Fig. 6.11). Here, the central sphere is in contact with six spheres in its own layer, three from above and three from below, making twelve in all. On exploding the structure to show the unit

Fig. 6.11   The hexagonal unit in *hcp*

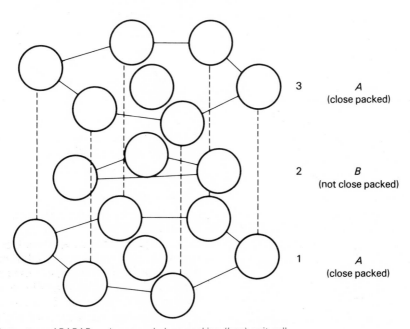

3   *A*
(close packed)

2   *B*
(not close packed)

1   *A*
(close packed)

Fig. 6.12   Metal structure: *ABABAB* ... hexagonal close-packing (*hcp*) unit cell

cell (Fig. 6.12), the three layers may be seen. In the second arrangement, the third layer fits over the holes which remain in the two layer structure. The layers are successively *A, B, C, A, B, C, A, B, C*... Inspection shows this to be a face-centred cubic close-packed structure, also with a co-ordination number of 12 (Fig. 6.13). Calcium, copper, silver and gold exhibit this *fcc* structure, in which the six spheres shown in the top layer form the diagonal plane of a cube (Fig. 6.14). The structure is completed by adding two spheres to the upper empty corners of the cube shown, three more to the base corners and three to the unoccupied faces. This is

shown in the unit cell of the exploded structure (Fig. 6.15), and in the *fcc* structure itself (Fig. 6.16).

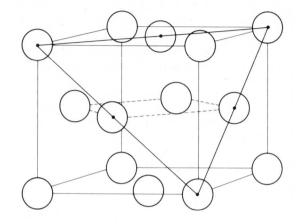

Fig. 6.15 Metal structure: face-centred cubic close-packing (*fcc*) unit cell

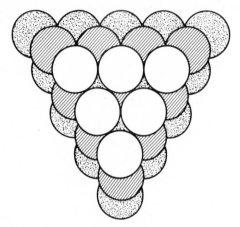

Fig. 6.13 Building up close-packed structures in layers: 3rd layer added to form face-centred cubic close-packing (*fcc*): *A, B, C, A, B, C,* ... also with co-ordination number of 12

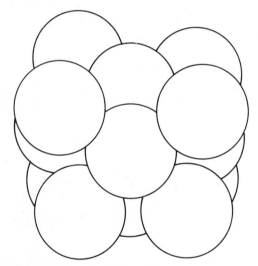

Fig. 6.16 Metal structure: face-centred cubic close-packing (*fcc*) unit cell

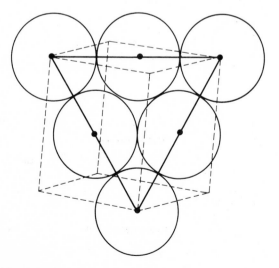

Fig. 6.14 Face-centred cubic close-packing: the diagonal plane

Just as with the other example of close-packing each sphere is in contact with twelve other spheres, six arranged hexagonally in its own plane, three from the plane above and three from below. This is made clear by a sketch of adjacent unit cells in which only the spheres actually used in the arrangement described are shown (Fig. 6.17). The face-centred cubic structure is close-packed across diagonal planes but not close-packed across its faces.

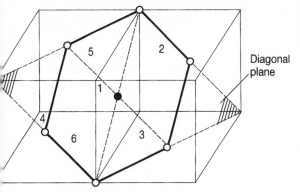

Fig. 6.17 Face-centred cubic close-packed structure: three atoms located above (1, 2, 3) and three below (4, 5, 6) the hexagon to give the central atom co-ordination number = 12

Allotropy, polymorphism and isomorphism

Different substances of very similar crystalline form are described as **isomorphous** (Greek, *isos morphē* = equal form).

Such are the alums. The phenomenon of isomorphism (Mitscherlich, 1819) was originally applied to compounds of similar chemical constitution, but this qualification is usually waived and the term is used in the general sense indicated above. It has already been seen that some compounds and elements may exist in more than one crystalline form (zinc sulphide, silver iodide, carbon): they exhibit *polymorphism* (Greek, *poly morphē* = many shapes; Mitscherlich, 1821). The substance is said to be polymorphic; dimorphic where two forms exist, trimorphic for three, etc. Polymorphism is not confined to Inorganic Chemistry. For elements,

the term *allotropy* (Greek, *allos tropos*, another shape; Berzelius, 1841), is used. The different forms are called allotropes or allotropic modifications. However, allotropy extends beyond polymorphism and crystal shape, covering the appearance of different forms of an element in the liquid state, as with the so-called different allotropic modifications of molten sulphur, and in the gaseous, or vapour state, illustrated by oxygen and ozone.

An element which exists in more than one form in the same state exhibits **allotropy**.

Two types of polymorphism, and allotropy of solid elements, may be distinguished by the relative stability of the different forms to each other. They are enantiotropic and monotropic polymorphism or allotropy. In addition, a third type of allotropy, dynamic allotropy, will be described.

A comparison of enantiotropy and monotropy using vapour pressure curves

Shown by both elements and compounds, these types of allotropy and polymorphism may be defined:

Enantiotropy (or Enantiotropic Allotropy, etc.) occurs when one solid form of the substance on heating changes into another solid form at a definite transition temperature, the reverse happening on cooling.

Monotropy (or Monotropic Allotropy, etc.) occurs when one form of the substance is stable over the whole temperature range for which the allotropes exist, all other forms being metastable and passing, sometimes very slowly indeed, into the other form.

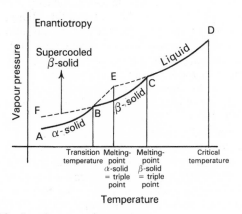

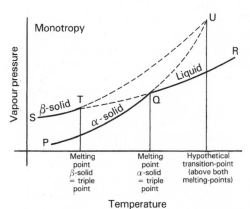

Fig. 6.18 Enantiotropy and monotropy

Table 6.2   Examples of enantiotropy and monotropy

	Enantiotropy		
Element or compound	Form below	Transition temperature/°C	Form above
Sulphur	Rhombic	95.5	Monoclinic
Tin	Grey	13.2	White
Mercury(II) iodide	Tetragonal red	126	Orthorhombic yellow
Ammonium chloride	'CsCl' structure	184	'NaCl' structure
Copper(I) mercury(II) iodide	Scarlet (β)	70	Chocolate-purple (α)

	Monotropy	
Element or compound	Metastable	Stable
Phosphorus	White (yellow)	Red
Carbon	Diamond	Graphite
Sulphur	All other forms	Rhombic or monoclinic
Iodine monochloride	Brown (β)	Red (α)

A **Transition Temperature** is the temperature at which two crystalline forms of a substance co-exist in equilibrium, whereas below this temperature, one form, and above it, the other form, is stable.

Vapour pressure curves are drawn in Fig. 6.18, while Table 6.2 shows some examples of enantiotropy and monotropy. Remember that transition temperatures refer to closed systems where solid and vapour are in equilibrium, and the atmosphere is excluded.

Consider enantiotropy first. On warming α solid, the vapour pressure follows curve AB until the transition-point, B, is reached. If equilibrium is maintained, the temperature rise is arrested and β solid is formed. Interconversion of solids is slow. The vapour pressure of solid β, on warming, follows the curve BC, ending in the melting-point, C. The molten (liquid) form has the vapour pressure curve CD, ending in the critical-point, D. Slow cooling, so that equilibrium is maintained, reverses the process. However, rapid heating of the α form or rapid cooling of the liquid results in the vapour pressure curves AE and DE respectively being recorded. E is the triple-point of the α form, which is metastable above the transition temperature. The β form does not appear. These triple-points are very close indeed to the melting-points under normal atmospheric pressure, because changes in pressure have relatively little effect. It will be appreciated that a substance exhibiting enantiotropy will melt at the temperature corresponding to the solid form actually melting. Thus sulphur melts at 114°C (rhombic) or 119°C (monoclinic).

Transition-points are definite and may be used, where appropriate, for thermometric fixed points. There are some important differences between transition-points and melting-points. Both are affected by changes in pressure, the effect depending on the density change. Expansion on melting or on raising the temperature through the transition-point, leads to a raising of the melting- or transition-point when the pressure is increased. This follows from **Le Châtelier's Principle**:

Whenever a system in equilibrium is subjected to a constraint, that change takes place which tends to annul the effects of the constraint.

There is also an evolution of heat when one form is cooled and changes into the other at the transition-point. For the solidification of a molten element or compound, the latent heat of fusion is evolved. On heating the solids, melting is easier to accomplish than the solid–solid change. The former requires the breakdown of the solid lattice from the vibrations of increasing amplitude of the lattice units while a transition involves a rearrangement in the solid state. Transition-points may be missed by superheating; melting-points, never.

In the illustration of monotropy, the stable form will have a lower vapour pressure than the metastable forms at any chosen temperature at which they exist. The α form, which is the stable form, has the vapour pressure curve PQ and it melts under its own vapour pressure at Q and then the molten element or compound has the vapour pressure curve QR, ending in the critical-point, R. Slow cooling might be expected to reverse this but the usual rapid cooling results in the appearance of the β solid at the triple-point, T. This is metastable and may change spontaneously to the α form although this change, from solid to solid, will be very slow and may not even be apparent at all. Under no circumstances can the stable form pass directly to the metastable form, the only ways of interconversion being by rapid cooling of the liquid or vapour forms of the substance. The vapour pressure curves of the solid α and β forms do intersect, but only when extrapolated beyond their triple-points, so that the transition-point is not realized in practice. The melting-points under normal atmospheric pressure are very close to the triple-points.

Generally, it is not the stable, but a metastable form, which appears when a substance is prepared chemically, or formed from the liquid or vapour. For enantiotropic systems, it has been seen that supercooling beyond the transition temperature leaves a form which is metastable with respect to the normal stable form. Monoclinic sulphur persists although it is only stable above the transition temperature 95.5°C. White (yellow) phosphorus is always the form produced on cooling phosphorus vapour, and changes only very slowly to the red, stable form. Precipitation of mercury(II) iodide by mixing aqueous solutions of mercury(II) chloride and potassium iodide yields the crimson tetragonal form but the metastable yellow rhombic form is seen momentarily. Using dilute solutions containing 1% gelatin, mercury(II) iodide is formed in colloidal suspension and in very dilute solutions, in the colloidal state. In a series of experiments with increasing dilution, the yellow form persists for increasingly long periods with greater dilution, and may persist for several weeks, gradually changing in colour through various shades from yellow to crimson, as the metastable form at room temperature changes to the stable form. These are illustrations of a more general law, called **Ostwald's Law of Successive Reactions:**

A substance passes from a less stable condition (of higher energy content) to the most stable condition (of lowest energy content) at a given temperature through intermediate conditions of progressively greater stability where these exist.

Each form of an element or compound at a given temperature is associated with a definite quantity of energy for a given mass. Where a metastable form exists, this will have more energy than the stable form, resulting in a higher vapour pressure

at this temperature, and greater solubility. When the same quantity of each form separately undergoes the same reaction with another element or compound, the heats of reaction (enthalpy changes of reaction) will differ by that quantity of heat associated with the interconversion of the given quantity of the substance. This illustrates Hess' Law of Constant Heat Summation (1840).

Dynamic allotropy

This type of allotropy is found in liquid systems and gaseous or vapour systems, both of which are homogeneous.

Dynamic allotropy occurs when different forms of the element in a given physical state co-exist in equilibrium, the relative concentrations of each depending on the physical conditions.

Oxygen and ozone (p. 254), formerly quoted as examples in this context, are better described as showing monotropy. The colour and viscosity changes of molten sulphur on heating may be explained in terms of liquid allotropes, λ-S and μ-S.

An examination of some allotropic systems

While the phenomenon of allotropy is wide-spread, a detailed study of sulphur, tin, phosphorus and carbon only will be attempted. Other allotropes will be mentioned in the text.

Orthorhombic and monoclinic sulphur

The element sulphur is a yellow non-metal, usually found as orthorhombic crystals, and commonly called rhombic sulphur. The molecules, S_8, consist of puckered rings of atoms. In each ring the atoms lie in two parallel planes, with the angles between the bonds all equalling 105° and the bond lengths, 0.212 nm. The molecule, S_8, is shown in Fig. 6.19.

Bond angles 105°
Bond lengths 0.212 nm

Fig. 6.19   The S_8 molecule

When rhombic sulphur (α-S) is heated slowly, it changes into monoclinic sulphur (β-S) at the transition temperature, 95.5°C. Monoclinic sulphur is the stable form up to the melting-point, 119°C. Sulphur is exhibiting enantiotropy. Rapid heating results in a melting-point of 114°C being recorded, that of the rhombic form metastable above 95.5°C. Crystallization from suitable solvents below the transition temperature yields the rhombic form, and above, the monoclinic form. The crystals are shown in Fig. 6.20. Rhombic sulphur is usually prepared from a solution of sulphur in carbon disulphide, the solvent readily evaporating on standing. Transparent amber crystals, density 2.07 g cm^{-3}, are formed. Monoclinic sulphur may be obtained by very slowly cooling a saturated solution of sulphur in boiling toluene. Alternatively, molten sulphur is carefully melted in an evaporating dish and allowed to cool but as soon as a crust forms, any fluid sulphur is poured away through holes pierced in the crust. Long prismatic crystals are seen, density 1.96 g cm^{-3}. After a few days the monoclinic sulphur crystals become powdery due to reversion to the stable, rhombic form.

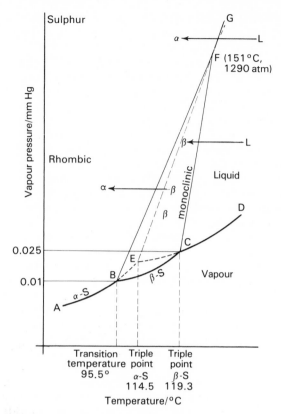

Fig. 6.21   The phase diagram for sulphur*
α-S = rhombic sulphur β-S = monoclinic sulphur

*mmHg = 13.5951 × 9.80665 N m^{-2}
atm = 101325 N m^{-2}

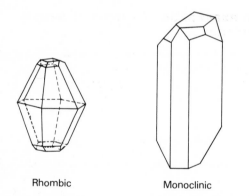

Rhombic   Monoclinic

Fig. 6.20   Crystals of rhombic and monoclinic sulphur

There is a decrease in density during the change from rhombic sulphur to the monoclinic form, and on melting both rhombic and monoclinic allotropes. The phase diagram, not to scale, is shown in Fig. 6.21 where slopes of BF, EF, CF have been exaggerated to show the effect of increased pressure, raising the transition-point and melting-points. B is the transition-point, C the melting-point of monoclinic sulphur and E the melting-point of rhombic sulphur, under the vapour pressure of the system. These temperatures are,

therefore, triple-points. As the pressure is increased, points B and C assume positions on BF and CF respectively. The areas are labelled with the form of sulphur stable under the conditions appertaining within these areas. Area FBC may be divided to show the metastable forms: metastable rhombic sulphur may exist in FBE, metastable liquid in FEC and metastable vapour in EBC. Monoclinic sulphur is stable within the closed area FBC. FG is the continuation of EF. At a pressure above F, cooling molten sulphur produces the stable rhombic allotrope, which probably accounts for the appearance of large crystals of rhombic sulphur in nature.

There are four triple-points, B, C, E and F. The co-ordinates of F, for rhombic solid, liquid and vapour are (151°C, 1290 atmospheres).

Other solid forms of sulphur are relatively unimportant. Several have been prepared but are metastable with respect to rhombic sulphur below the transition-point, and to monoclinic sulphur, above. Sulphur exhibits monotropy in these cases.

Sulphur melts at either 114°C or 119°C to a mobile amber liquid, which deepens in colour as the temperature rises to give a deep red at about 180°C when the viscosity reaches a maximum and the liquid cannot be poured. The dark colour persists but the viscosity decreases on further heating so that mobility is regained by the boiling-point, 444.6°C. The almost black liquid gives a reddish-orange vapour. The increase in viscosity on heating is contrary to normal experience: usually, the extra energy absorbed results in greater mobility of the molecules. The abnormal behaviour of sulphur is due to the formation of very long chains of atoms. The light yellow mobile liquid has been called λ-sulphur, and the dark red liquid, μ-sulphur. Calculations based on viscosity measurements give an estimate of the lengths of the chains of sulphur atoms. Just above the melting-point, the chains formed from broken rings will be S_8. At 180°C, they are about S_{20000} and at 300°C, about S_{1000}. S_8 molecules will re-form on cooling slowly, but rapid cooling by pouring molten sulphur into water produces plastic sulphur, a supercooled liquid. Its fibrous nature is confirmed by X-ray studies. After a few days, the rubbery, elastic nature disappears and the S_8 rings of the rhombic form of sulphur are re-formed.

Molten sulphur may be regarded as exhibiting dynamic allotropy, although the two forms, λ-S and μ-S do not correspond to definite structures but in general to short and long chains of sulphur atoms.

$$\alpha\text{-S} \underset{\substack{95.5°C \\ \text{transition}}}{\overset{95.5°C}{\rightleftarrows}} \beta\text{-S} \underset{\text{fusion}}{\rightleftarrows} \lambda\text{-S} \underset{\substack{\text{dynamic} \\ \text{equilibrium}}}{\rightleftarrows} \mu\text{-S}$$

White (or yellow) and red (or violet) phosphorus

The allotropy of phosphorus is complex and not fully elucidated. At least four forms have been reported: α-white, β-white (transition-point −77°C), violet and black. In the laboratory, two forms, white and red, are used. The discussion is limited to these.

In the final stage of phosphorus manufacture the vaporized element is condensed under water. It is strikingly white and translucent but on exposure to light becomes tinged with yellow within a few hours, and more like the allotrope usually found in the laboratory. This form may be called white or yellow phosphorus. It is metastable and forms traces of the stable modification, commonly called red phosphorus, although again the colour of this modification varies. The pure stable allotrope is said to be violet in colour. Red is thought to be either very finely divided violet or a solid solution of white in violet. However, for practical purposes, we are interested in the metastable white and stable red laboratory forms. They constitute a monotropic system. Some physical data are given in Table 6.3. The formula of phosphorus is P_4.

On cooling molten phosphorus, the freezing-point at which the red form might be expected is passed and the metastable white solid allotrope appears on solidification. The same colourless liquid is obtained on fusing both red and white allotropes in the absence of air and under pressure to avoid sublimation in the first case. The formation of the white modification illustrates Ostwald's Law of Successive Reactions. The metastable white form is so reactive that special precautions must be taken. White phosphorus inflames at about 35°C in air and must not be handled because the warmth of the hand is enough to cause it to catch fire. It is stored under water. Sticks of white phosphorus are cut under water, each stick being held in a pair of tongs. Red phosphorus is supplied as what is apparently a reddish-violet amorphous powder, but actually microcrystalline. White phosphorus is also crystalline. Red phosphorus has neither taste nor odour, is not poisonous and becomes slightly moist in air. White phosphorus is very poisonous (lethal dose = 0.05 g), giving off a poisonous vapour associated usually with garlic. It readily inflames in

Table 6.3 Some physical properties of the allotropes of phosphorus

	White phosphorus	Red phosphorus
Density at room temperature	1.82 g cm^{-3}	$2.15\text{–}2.35 \text{ g cm}^{-3}$
Melting-point	44.25°C	589.5°C (under pressure)
Boiling-point (101 kPa)	280°C	Sublimes at 400°C
Ignition temperature in air	35°C (approx.)	260°C (approx.)

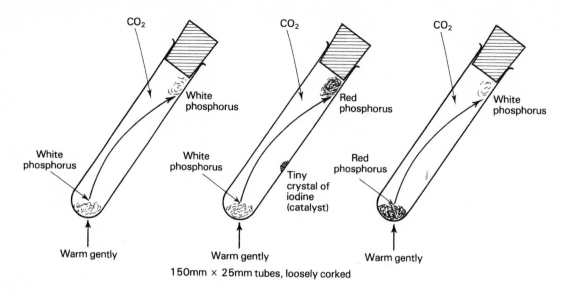

Fig. 6.22   Interconversion of red and white phosphorus

air. Combusion of both allotropes yields white fumes of phosphorus pentoxide in air or oxygen, equal masses of the allotropes yielding the same mass of the oxide.

The metastable white allotrope, unlike the red, is easily soluble in carbon disulphide, benzene and other organic solvents. Evaporation yields the finely divided white allotrope, which is spontaneously flammable. In a comparison of their chemical reactions, white phosphorus is easily the more reactive. White phosphorus catches fire spontaneously in chlorine, while the red allotrope usually requires heating, although it does inflame if sprinkled into chlorine gas. White phosphorus is more vigorous in reacting with bromine. Unlike red phosphorus, the white modification reacts with hot aqueous alkali to form phosphine.

When exposed to air, light or heat in the region 240–260°C, in an inert atmosphere, the red allotrope is formed from white phosphorus. The higher temperature accelerates the interconversion. A series of simple experiments is shown in Fig. 6.22. These are nearly self-explanatory. Air is displaced from each tube by carbon dioxide and replenished as necessary during the manipulations. Each tube (150 mm × 25 mm) is loosely corked and clamped during heating. A small pea-size piece of white phosphorus, on warming, melts and volatilizes to deposit drops which solidify to the white allotrope in the cool upper regions of the tube. Repetition but with a tiny crystal of iodine

placed just above the piece of phosphorus results in the stable red allotrope being formed. Red phosphorus, in the last experiment, sublimes on heating to give a yellowish vapour and the white allotrope. Where phosphorus condenses on to hot regions of the tube, the red allotrope is seen because the increased temperature accelerates the change from the metastable to the stable condition. Use of a catalyst means that a lower temperature is effective in promoting the interconversion.

Grey and white tin

Because of the importance of tin through the ages, its allotropic modifications have been known since classical times. In intense cold, statues made of tin collapsed because of a profound change in the nature of the metal, noted by Aristotle (384–322 B.C.) and Plutarch (c. A.D. 46–c. 120). In the severe winter of 1850, tin pipes of the church organ at Zeit in Germany were ruined and in St Petersburg (Leningrad) during 1867–8, blocks of tin decayed in the Customs House. The crumbling of tin buttons used for military uniforms during severe cold was also known. This so-called tin plague may sometimes be seen on medallions and coins stored in museums, appearing as a sort of fungoid growth.

Tin exhibits enantiotropic allotropy with a transition-point of 13.2°C. Ordinary metallic tin is the white allotrope, passing into the less dense grey form:

$$\underset{\substack{\text{diamond}\\\text{structure}\\\text{density}\\5.80\,\text{g cm}^{-3}}}{Grey\ tin} \overset{13.2\,°C}{\rightleftharpoons} \underset{\substack{\text{tetragonal}\\\text{structure}\\\text{density}\\7.29\,\text{g cm}^{-3}}}{White\ tin} \overset{232\,°C}{\rightleftharpoons} Molten\ tin$$

The change is extremely sluggish when white tin is cooled just below 13.2°C and is fastest at about −50°C. This is a good illustration of the slowness of solid→solid change. The grey allotrope has the lower density and is bulkier than the original white allotrope from which it was formed, giving the powdery, wart-like growths associated with 'tin plague'. Because the prevailing temperature is often below 13.2°C in many countries, metallic white tin is frequently in the metastable condition. It will be appreciated that utensils made of tin and tin solders for petrol and oil drums may not be used in polar regions. The conversion of grey tin into the white form may be quickly accomplished by pouring hot water over the grey allotrope. It appears that a third allotrope, γ-tin, previously reported with transition-point 161°C, does not exist.

Graphite and diamond

There are just two allotropes of carbon. X-ray analysis has shown that apart from diamond, all varieties of carbon are fundamentally graphite, probably in a fine microcrystalline form. The crystal structures of diamond and graphite are shown in Figs 6.6 and 6.7 (pp. 100, 101).

The hardness of diamond is attributed to the rigid tetrahedral configuration of the covalent bonding while the hexagonal honeycomb layers of graphite slide upon each other, bringing its important lubricating properties. Diamond has a structural configuration rather like the aliphatic hydrocarbons and is colourless, while graphite, except in very thin layers which are transparent, is black.

In forming diamond and graphite, carbon exhibits monotropy. Diamond is the metastable form but has never been observed to change into graphite. This is not altogether surprising because of the strength of the bonding in diamond and the close values of the heats of combustion (enthalpy changes of combustion) of each allotrope, pointing to a similar energy content. The enthalpy changes of combustion for diamond and graphite are respectively 395 kJ mol^{-1} and 393 kJ mol^{-1}.

It has been suggested that diamond was formed during the crystallization of carbon from molten rock inside the Earth at temperatures exceeding 4000°C, under enormous pressures, probably in excess of $23 \times 10^9\,\text{N m}^{-2}$. At Kimberley, South Africa, diamonds are mined in a region of extinct volcanic activity, where they have been brought near to the surface by volcanic eruption. Synthetic diamonds were made in 1936 by cooling graphite in molten iron from 4000–5000°C to ice-salt temperatures. Moissan claimed to have made diamonds in 1893, but it is considered that his temperatures and pressures were inadequate. Late in 1959, the General Electric Company of the USA reported that very small diamonds had been made by dissolving carbon in excess molten iron containing iron sulphide with an added catalyst. Catalysts incorporating chromium, nickel, manganese, cobalt, platinum and tantalum have been used. With a nickel catalyst, diamonds were formed at pressures exceeding 5.5 kN mm^{-2} and temperatures in excess of 1500°C. The reaction vessel was approximately 2.5 mm diameter and 12 mm long. The diamonds were small (a fraction of a mm across) and coloured, being suitable for industrial cutting and abrading machinery.

Isomorphism

The Law of Isomorphism was discovered by Mitscherlich when working as a student in the laboratory of Berzelius, who did so much to establish the Atomic Theory of Dalton. It was published in 1819. The principle was subsequently used by Berzelius in fixing certain chemical formulae and obtaining values of 'atomic weights' for the tables published in 1826. Following up the notion that matter was composed of atoms which united in definite small whole numbers, Mitscherlich examined the hypothesis that the same crystalline form could be shown by compounds of different elements if the compounds had similar chemical constitutions. The crystal form would be determined solely by the combination of the same numbers of atoms of the corresponding elements and not by their chemical natures. Similar formulae would reflect similar crystal form. This seemed to be true at first, but later, as more examples were investigated, only partially true. When grouped on account of their similar chemical constitution, some compounds appeared in the same crystalline form and Mitscherlich referred to the elements belonging to a group of similar salts, such as the sulphates, as *isomorphs*. Nowadays the term isomorphism is applied to the crystalline form. In his preliminary work, Mitscherlich discovered that disodium hydrogenorthophosphate and the corresponding arsenate, $Na_2HPO_4.12H_2O$ and $Na_2HAsO_4.12H_2O$, were isomorphous. Further, not only were monosodium dihydrogenorthophosphate and the corresponding arsenate, $NaH_2PO_4.H_2O$ and $NaH_2AsO_4.H_2O$, isomorphous but each existed in two forms. He had discovered polymorphism.

In the original form, for isomorphous compounds the law of isomorphism stated: 'An equal number of atoms, if they are bound in the same way, produce similar crystal forms and the crystal form depends not on the nature of the atoms, but on the number and method of combination.'

Isomorphous substances will have identical space lattices.

According to the Abbé Haüy, a fundamental postulate of crystallography advanced in 1784 was that every definite chemical element or compound had its own crystal form. Very precise measurement on the isomorphous crystals of different substances reveals that minute differences exist. This was first shown by Wollaston in 1812: corresponding angles in calcite, $CaCO_3$, dolomite $(Ca,Mg)CO_3$, spathic iron ore, $FeCO_3$, are respectively 74° 55′, 73° 45′ and 73° 0′, all three carbonates being isomorphous. The work of Tutton and others, during the period 1893–1925, has shown that this state of affairs is general except for the most symmetrical system, the cubic. In some cases, the difference for corresponding angles in other systems may amount to several degrees.

The formation of crystals which are isomorphous is not confined to compounds with similar chemical formulae, those with the same number of atoms in combination. **The Law of Isomorphism** is now written:

Substances of similar chemical constitution frequently crystallize in very nearly the same form.

In the strictest sense of the word, isomorphism is restricted to substances with analogous chemical formulae. As such it has been used very successfully to fix atomic masses, although care must be exercised because very similar crystal form does not always mean similar chemical constitution. Isomorphous series include the sulphate, selenate and chromate of potassium, K_2SO_4, K_2SeO_4 and K_2CrO_4 respectively, in which the first may be regarded as a derivative of sulphur trioxide SO_3, so the corresponding oxides of selenium and chromium become SeO_3 and CrO_3, and the atomic

mass of each element unites with 'three atomic masses' of oxygen. The alums form another series,

$$(M^+)_2SO_4.(M^{3+})_2(SO_4)_3.24H_2O$$

or $$M^+M^{3+}(SO_4)_2.12H_2O$$

where M^+ is a univalent metal (Na, K, Rb, Cs, etc.) and M^{3+}, tervalent (Al, V, Cr, Mn, Fe, Co, etc.). The general term alum originates with potash alum, $K_2SO_4.Al_2(SO_4)_3.24H_2O$, the whole structure consisting of ions and water molecules in the proportions given. Iron(III) ammonium alum contains the same number of ions but more atoms, due to the ammonium ion, NH_4^+:

$$(NH_4)_2SO_4.Fe_2(SO_4)_3.24H_2O$$

The original hypothesis of Mitscherlich could therefore be amended to refer to ions and not atoms. The alums are double salts. Another well-known series of double sulphates contains metals in the uni- and bivalent states:

$$(M^+)_2SO_4.M^{2+}SO_4.6H_2O$$

Mohr's salt, ammonium iron(II) sulphate, provides an example, $(NH_4)_2SO_4.FeSO_4.6H_2O$, but iron(II) may be replaced by V, Cr, Mn, Co, Ni, Cu and Zn in the bivalent state without change of crystal form. Continuing with sulphates, there is an isomorphous series of general formula $MSO_4.7H_2O$ where M may be V, Cr, Mn, Fe, Co, Ni, Zn, etc., in the bivalent state, although chromium(II) sulphate is more usually isomorphous with copper(II) sulphate: $CrSO_4,5H_2O$ and $CuSO_4.5H_2O$. In all of these examples, elements, or the ammonium group, of similar valency occupy corresponding positions and the valency of any one can be deduced. However, there are substances which form isomorphous crystals although not of similar chemical constitution and there are some with similar chemical constitutions which are not isomorphous. Argentite, Ag_2S, and galena, PbS, mineral sulphides of silver and lead respectively, are isomorphous. So are calcite, $CaCO_3$, and sodium nitrate, $NaNO_3$, although from the standpoint of the electronic threory, they may be regarded as analogous: $Ca^{2+}CO_3^{2-}$ and $Na^+NO_3^-$. The isomorphism of potassium perchlorate, $KClO_4$, and barium sulphate, $BaSO_4$, may be similarly interpreted. Yet such similar compounds as potassium and sodium nitrates are not isomorphous and neither are magnesium and strontium carbonates. Although in its original conception, isomorphism required not only closely similar crystal form but also analogous chemical composition, crystallographers have extended the use of the term so that it applies to analogous formulae in the wide sense where atoms or ions are of comparable size, as for example with lead sulphide, PbS, and sodium chloride, NaCl.

Isomorphous compounds of closely similar chemical constitutions usually have interesting characteristics connected with crystal growth in common, notably the ability to form *solid solutions* and *overgrowths*. The formation of solid solutions or overgrowths is indicative of isomorphism although not all isomorphous compounds necessarily show these characteristics. Not only must the crystal lattices be the same but the repeating crystal unit of structure must be of the same, or very nearly the same, volume for the successful formation of overgrowths and solid solutions.

Potash alum, $K_2SO_4.Al_2(SO_4)_3.24H_2O$, which is colourless, and the purple chrome alum, $K_2SO_4.Cr_2(SO_4)_3.24H_2O$, both crystallize in octahedral form. Mixed solutions of various relative concentrations of these alums will crystallize, each solution yielding crystals of similar form but with a colour intermediate between deep purple and colourless depending on the proportion of chrome alum. These crystals are homogeneous and are examples of *solid solutions*, although the term *mixed crystals* has been applied. The latter is not favoured because the final crystal is not composed of a mixture of the crystals of the separate alums. Isomorphous compounds of similar chemical constitution cannot usually be separated by crystallization. There are many interesting examples of solid solution formation. Mixed solutions of the colourless potassium perchlorate, $KClO_4$, and dark purple potassium permanganate, $KMnO_4$, will yield crystals ranging from colourless through pink to purple, depending on the proportion of permanganate. The complementary colours of ammonium nickel(II) sulphate (green) and ammonium cobalt(II) sulphate (pink), of respective formulae,

$$(NH_4)_2SO_4.NiSO_4.6H_2O$$
and $$(NH_4)_2SO_4.CoSO_4.6H_2O$$

enable the proportions of the components to be adjusted to give crystals of a neutral tint. Where solid solutions are not formed, the compounds crystallize separately from a mixed solution.

A crystal of chrome alum suspended in a saturated solution of potash alum by a thread will continue to grow, the latter forming a colourless overgrowth over the violet nucleus of chrome alum. Green nickel(II) sulphate, $NiSO_4.7H_2O$, may similarly be covered by an overgrowth of colourless zinc sulphate, $ZnSO_4.7H_2O$. The formation of overgrowths is not entirely confined to isomorphous compounds in the Mitscherlich sense: sodium nitrate, $NaNO_3$, will form growths parallel to the faces of crystals of calcite, $CaCO_3$, but they are, in fact, of analogous ionic constitution. The formation of crystals from a solution may sometimes be induced by using crystals of an isomorphous compound as 'seeds'.

Crystals of similar form result from a similarity in size and proportions of the units out of which the identical crystal lattices are built. Although crystals may be closely similar, the variation in angle noted previously will be due to slight differences in the size of the unit. Incompatibility of the unit cells, due to a difference in volume, will prevent the formation of solid solutions and overgrowths. In the crystallographic interpretation of isomorphism, the size of the ion or atom is of paramount importance. Isomorphous compounds need not be related chemically.

Finally, both polymorphism and isomorphism may appear together. Arsenic trioxide and antimony trioxide both exist in octahedral and rhombic forms. They are *isodimorphous*. A few cases of isotrimorphism are known. The general phenomenon is called *isopolymorphism*.

7

Oxidation, reduction and electrochemical processes

Oxidation and reduction

When sulphur is heated in air, it melts, catches fire and burns with a blue flame. The product is a choking gas, sulphur dioxide, and the combination of sulphur with oxygen is an example of *oxidation*,

$$S + O_2 \rightarrow SO_2$$

Sulphur has been oxidized. The converse of oxidation is *reduction*.

Many metals unite with oxygen and they do so with varying degrees of vigour. Magnesium burns brilliantly in oxygen and a white ash, magnesium oxide, falls,

$$2Mg + O_2 \rightarrow 2MgO$$

On gentle heating, fused lead forms a yellow scum of oxide:

$$2Pb + O_2 \rightarrow 2PbO$$

When a very small quantity of a mixture of lead(II) oxide (lead monoxide) and magnesium filings or powder is cautiously heated in a crucible, a sudden, violent reaction occurs. Oxygen is transferred from lead to magnesium:

$$Mg + PbO \rightarrow MgO + Pb$$

Lead monoxide has been reduced to lead and magnesium oxidized to magnesium oxide. Reduction is the loss of oxygen. In this example lead monoxide supplies oxygen to magnesium and is called the oxidizing agent or oxidant while magnesium is the reducing agent or reductant. Under the conditions of this experiment, magnesium has a greater affinity for oxygen than lead.

Consider another example. Hydrogen burns in oxygen, or air, with a pale blue flame to form water,

$$2H_2 + O_2 \rightarrow 2H_2O$$

When finely powdered copper is heated in a rapid stream of oxygen, black copper(II) oxide is produced:

$$2Cu + O_2 \rightarrow 2CuO$$

On passing dry hydrogen over the heated copper(II) oxide, the black powder is reduced to pink copper and steam passes out of the end of the tube:

$$CuO + H_2 \rightarrow Cu + H_2O$$

Under these conditions hydrogen is a stronger reducing agent than copper and has the higher affinity for oxygen.

Experiments like these are purely qualitative in nature and the conditions in them vary so much that valid comparisons cannot be made. When comparisons are to be made, the conditions must be rigorously defined.

When iron powder is heated in a combustion tube through which steam is passed, tri-iron tetroxide remains and hydrogen issues from the end of the tube. However, reversing the flow of hydrogen and stopping the flow of steam reverses the reaction, iron and steam being formed:

$$3Fe + 4H_2O \rightleftharpoons Fe_3O_4 + 4H_2$$

Sweeping away the gaseous product in each case results in more of that product being formed, otherwise an equilibrium mixture would result. The conditions applied to this balanced system dictate the products.

The concept of oxidation and reduction may be extended to include compounds. Carbon monoxide burns in air with a blue flame to form carbon dioxide,

$$2CO + O_2 \rightarrow 2CO_2$$

and iron(II) oxide on exposure to air is rapidly oxidized, sometimes with incandescence,

$$4FeO + O_2 \rightarrow 2Fe_2O_3$$

Hydrogen can so often be used to remove oxygen from compounds that the terms oxidation and reduction were extended to include this reducing agent specifically. Reduction now includes the addition of hydrogen to elements or compounds and oxidation the removal of hydrogen from a compound. Hydrogen bromide is oxidized by chlorine to bromine and by the same reaction it may be said that chlorine is reduced to hydrogen chloride:

$$2HBr + Cl_2 \rightarrow 2HCl + Br_2$$

Oxidation and reduction occur at the same time and are complementary aspects of the same overall process, sometimes called a redox process.

At this stage we have defined oxidation as the addition of oxygen to an element or compound or the removal of hydrogen from a compound and reduction as the removal of oxygen from a compound or the addition of hydrogen to an element or compound. The oxidizing agent brought about oxidation and the reducing agent, reduction.

An extension of the original ideas

When dry chlorine is passed over heated anhydrous iron(II) chloride, iron(III) chloride is formed and sublimes away. In aqueous solution also, iron(II) chloride is converted into iron(III) chloride by chlorine. This is analogous to the oxidation of iron(II) oxide to iron(III) oxide by oxygen in the air already mentioned. The iron(II) oxide could have been prepared from iron(II) chloride, which would only involve converting a salt into the corresponding base, and the base, iron(III) oxide, produced by oxidation, could readily be converted into the corresponding salt, iron(III) chloride. The net change is oxidation and this must describe also the direct formation of iron(III) chloride from the iron(II) salt. This is shown schematically in Fig. 7.1. In this way, oxidation may be further extended to include the addition of chlorine, which has a high affinity for electrons and is therefore an electronegative element. Further studies along similar lines lead to defining oxidation as a process involving the addition of any electronegative element or group (of elements). A group of elements (e.g. SO$_4$) is sometimes conveniently called a radical. But oxidation involves relative affinities, to oxygen in the simplest examples, and must therefore describe the addition of a more electronegative element. Some examples will illustrate these points. Carbon

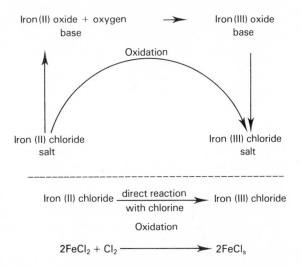

Fig 7.1 Extending the definition of oxidation and reduction

burns vigorously in fluorine and is oxidized to carbon tetrafluoride,

$$C + 2F_2 \rightarrow CF_4$$

An intimate mixture of iron filings and powdered sulphur reacts with incandescence when heated, iron being oxidized to the iron(II) sulphide,

$$Fe + S \rightarrow FeS$$

Lead(II) sulphide, which may be synthesized from the elements, is oxidized to the sulphate by hydrogen peroxide,

$$PbS + 4H_2O_2 \rightarrow PbSO_4 + 4H_2O$$

Iron(II) sulphate solution, acidified with sulphuric acid, is oxidized to iron(III) sulphate by hydrogen peroxide, with an increase of the electronegative proportion,

$$2FeSO_4 + H_2SO_4 + H_2O_2 \rightarrow Fe_2(SO_4)_3 + 2H_2O$$

Consider the oxidation of aqueous iron(II) chloride by chlorine in more detail. Iron(II) ions are converted into the iron(III) state and chlorine molecules become ions, electrons (e$^-$) being transferred. Hydration of ions may be ignored:

$$Fe^{2+} - e^- \rightarrow Fe^{3+}$$

$$Cl_2 + 2e^- \rightarrow 2Cl^-$$

Fe^{2+} represents one gram ion (i.e. one mole) and e$^-$, not one electron in this context but one mole of electrons which equals one Faraday (96 500 coulombs) of negative charge, which is the quantity of electricity associated with one gram equivalent (Faraday's Second *Law*, Chapter 1, page 11). The use of F would be ambiguous. In the oxidation of acidified iron(II) sulphate by hydrogen peroxide, the essential oxidation process is again the conversion of iron(II) ions into iron(III). This change may be reversed by use of a suitable reducing agent, e.g. zinc amalgam.

Oxidation may be regarded as the loss of electrons and reduction as the gain of electrons. The electrons are transferred from the substance oxidized to that reduced, which again emphasizes that oxidation and reduction occur simultaneously. During electrolysis, the discharge of a cation at the cathode may be classed as reduction while the anode process is oxidation. In the electrolysis of fairly concentrated copper(II) chloride solution, the electrode processes are

$$\text{reduction at cathode: } Cu^{2+} + 2e^- \rightarrow Cu$$

$$\text{oxidation at anode: } 2Cl^- - 2e^- \rightarrow Cl_2$$

Electrons flow along the circuit. However, many reactions involving oxidation occur with covalent (or co-ionic) compounds and the ionic extension will not apply. Sometimes both the formation and/or rupture of covalent bonds and the transference of electrons occur. It is convenient to distinguish between two kinds of oxidation and reduction, namely, covalent and ionic. In these definitions and subsequent examples look for the valency changes that occur.

By definition (p. 67), metals are elements which form simple positive ions (cations) by loss of electrons: these are electropositive elements.

Final general definitions

Covalent oxidation is the addition of a more electronegative element or radical or the removal of hydrogen and involves covalent (or co-ionic) bonds.

Covalent reduction is the removal of the more electronegative element or radical from a compound or the addition of hydrogen and involves covalent (or co-ionic) bonds.

Ionic oxidation is the loss of electrons by atoms, radicals, ions or molecules.

Ionic reduction is the gain of electrons by atoms, radicals, ions or molecules.

An **oxidizing agent** (oxidant) supplies the electronegative element or removes hydrogen in covalent oxidation or takes up electrons in ionic oxidation.

A **reducing agent** (reductant) contains the element which is oxidized in covalent oxidation or surrenders electrons in ionic reduction.

The oxidation–reduction reaction is conveniently called a **redox** process.

For covalent oxidation and reduction, reacting (equivalent) masses can be related to the standard element by proportion. For ionic systems the link is indirect through the quantity of electricity representing one mole of electrons, the Faraday.

For ionic oxidation and reduction:

The gram equivalent (mass) of an oxidizing agent (oxidant) is the mass in grams of the oxidizing agent which takes up one Faraday of (negative) electrical charge.

The gram equivalent (mass) of a reducing agent (reductant) is the mass in grams of the reducing agent which gives up one Faraday of (negative) electrical charge.

Fashions change even in chemistry! For many the term equivalent will seem almost archaic yet the concept supplies a simple method of calculating quantities resulting from redox reactions, as an alternative to using the equation for the reaction. This is especially true for redox reactions involving ions in solution.

In outline only, we will now consider how different redox systems are compared.

Electrode potentials and redox potentials

When a metal is immersed in a solution containing its ions, a potential difference is set up between the metal and solution. The magnitude of this potential depends on the nature of the metal selected, the temperature and the molar concentration (strictly, activity) of the metal ions in solution. A metal placed in a solution of one of its salts for electrochemical purposes is called an electrode. Strictly, the metal rod is part of the electrode.

The metal consists of a lattice of ions which share a common pool of valency electrons, or 'electron gas' as it is sometimes called. The positive ions are therefore already present as the structural units of metal. When the metal acquires a net negative charge with respect to the solution, metal ions have escaped into solution leaving surplus electrons behind. Such an element, having a tendency to furnish positive ions in solution is said to be electropositive and the higher the tendency, the more electropositive it is. A positive charge, on the other hand, results from the deposition of metal ions on to the electrode surface. The transfer of matter is too small to be detected by weighing and soon halts through the potential difference created.

When metallic zinc is placed in copper(II) sulphate solution, a displacement reaction occurs: copper is deposited, zinc passes into solution and heat energy is lost to the surroundings:

$$Zn + Cu^{2+} \rightarrow Zn^{2+} + Cu$$

Electrons pass from zinc to copper; the former is oxidized and the latter reduced. The Daniell cell is an arrangement by which these electrons pass along a connecting wire as an electric current. The cell consists essentially of a zinc rod dipping into zinc sulphate solution, or dilute sulphuric acid, in a porous pot, which is placed in saturated copper(II) sulphate solution contained in a copper can, acting as the other electrode.

Zinc dissolves,

$$Zn \rightarrow Zn^{2+} + 2e^-$$

and electrons flow along the external circuit to the copper. Copper(II) ions from solution are now discharged,

$$Cu^{2+} + 2e^- \rightarrow Cu$$

The electromotive force of this galvanic cell is approximately 1.1 V. When zinc reacts directly with copper(II) sulphate, the electrical energy is dissipated as heat.

At this stage the electrode terms, anode and cathode, may be defined for both galvanic cells and cells in which an externally applied potential difference causes electrolysis.

The **anode** is the electrode at which oxidation occurs; electrons flow from the anode along the external circuit to the cathode, and negative ions (anions) in solution migrate towards it.

The **cathode** is the electrode at which reduction occurs; electrons flow to the cathode along the external circuit from the anode, and positive ions (cations) in solution migrate towards it.

When generating electricity by means of a Daniell cell, the zinc rod is the anode and copper the cathode. Electrons flow along the external circuit from zinc to copper. **The electric potential difference ΔV is equal in sign and magnitude to the electric potential of a metallic conducting lead on the right minus that of a similar lead on the left.** The cell is represented by the diagram

$$Zn|Zn^{2+}|Cu^{2+}|Cu$$

and the reaction of the cell taken overall is

$$\tfrac{1}{2}Zn + \tfrac{1}{2}Cu^{2+} = \tfrac{1}{2}Zn^{2+} + \tfrac{1}{2}Cu$$

so that while electrons flow from left to right (zinc to copper) along the external circuit, and from copper to zinc effectively through the solution, the 'positive current' flows from left to right through the solution. Under normal conditions this is what happens when the cell is short-circuited, and the electrical potential difference is positive. The electromotive force of the cell is the limiting value of the electric potential difference for zero current through the cell. (By convention in Physics, the 'positive current', which is fictitious, flows in the opposite direction to the flow of electrons.) If a current is driven through the Daniell cell by an externally applied potential difference, the chemical reaction may be reversed. Zinc is then the cathode and copper the anode.

Values of individual electrode potentials cannot be determined because the electrical circuit must be connected to the solution and another potential is thereby introduced, although ingenious ways of circumventing this problem appear from time to time. A standard reference electrode is therefore used, called the *standard hydrogen electrode*, consisting of a piece of platinum partially immersed in a molar solution of hydrogen ions under an atmosphere of hydrogen at standard pressure (101 kPa) and at a standard quoted temperature. The platinum is previously platinized by electrolysis of a solution of platinum(IV) chloride (1%) containing a trace of lead(II) acetate with frequent reversal of polarity to give a coating of finely divided platinum. Hydrogen is adsorbed by the platinum black surface which catalyses the change:

$$H_2 \rightleftharpoons 2H^+ + 2e^-$$

The half-cell reaction is written

$$\tfrac{1}{2}H_2 \rightleftharpoons H^+ + e^-$$

(of course, H^+ ions exist in aqueous solutions as H_3O^+ but this need not detain us).

Table 7.1   Standard electrode potentials at 298K (25°C)

Implied electrode	Electrode	Electrode process	$E^{\ominus}_{298}/V$
	Li$^+$; Li	Li$^+$ + e$^-$ $\rightleftharpoons$ Li	−3.05
	K$^+$; K	K$^+$ + e$^-$ $\rightleftharpoons$ K	−2.93
	Na$^+$; Na	Na$^+$ + e$^-$ $\rightleftharpoons$ Na	−2.71
	Mg^{2+}; Mg	$\frac{1}{2}$Mg^{2+} + e$^-$ $\rightleftharpoons$ $\frac{1}{2}$Mg	−2.37
	Al^{3+}; Al	$\frac{1}{3}$Al^{3+} + e$^-$ $\rightleftharpoons$ $\frac{1}{3}$Al	−1.66
	Zn^{2+}; Zn	$\frac{1}{2}$Zn^{2+} + e$^-$ $\rightleftharpoons$ $\frac{1}{2}$Zn	−0.763
	Fe^{2+}; Fe	$\frac{1}{2}$Fe^{2+} + e$^-$ $\rightleftharpoons$ $\frac{1}{2}$Fe	−0.440
	Co^{2+}; Co	$\frac{1}{2}$Co^{2+} + e$^-$ $\rightleftharpoons$ $\frac{1}{2}$Co	−0.277
	Ni^{2+}; Ni	$\frac{1}{2}$Ni^{2+} + e$^-$ $\rightleftharpoons$ $\frac{1}{2}$Ni	−0.250
	Sn^{2+}; Sn	$\frac{1}{2}$Sn^{2+} + e$^-$ $\rightleftharpoons$ $\frac{1}{2}$Sn	−0.140
The standard	Pb^{2+}; Pb	$\frac{1}{2}$Pb^{2+} + e$^-$ $\rightleftharpoons$ $\frac{1}{2}$Pb	−0.126
hydrogen	H$^+$; H$_2$, (Pt)	H$^+$ + e$^-$ $\rightleftharpoons$ H	0.000
electrode	Cu^{2+}; Cu	$\frac{1}{2}$Cu^{2+} + e$^-$ $\rightleftharpoons$ $\frac{1}{2}$Cu	+0.337
Pt, H$_2$|H$^+$	I$^-$; I$_2$, (Pt)	$\frac{1}{2}$I$_2$ + e$^-$ $\rightleftharpoons$ I$^-$	+0.536
	Ag$^+$; Ag	Ag$^+$ + e$^-$ $\rightleftharpoons$ Ag	+0.799
	Br$^-$; Br$_2$, (Pt)	$\frac{1}{2}$Br$_2$ + e$^-$ $\rightleftharpoons$ Br$^-$	+1.06
	Cl$^-$; Cl$_2$, (Pt)	$\frac{1}{2}$Cl$_2$ + e$^-$ $\rightleftharpoons$ Cl$^-$	+1.36
	HF aq; F$_2$, (Pt)	H$^+$ + $\frac{1}{2}$F$_2$ + e$^-$ $\rightleftharpoons$ HF	+3.06

The physical states are implied by the standard conditions

At 25°C, the standard hydrogen electrode is assigned an arbitrary electrode potential of zero volt. It may be coupled to another electrode and the electromotive force of the resulting cell may be determined by means of a potentiometer as required by the definition. Usually a more robust standard electrode (of known standard electrode potential on the hydrogen scale) is used. Details are in the practical Physical Chemistry books.

Using a molar (1.0 M) concentration of metal ions at 25°C, the electrode potential is called the standard electrode potential. For gases the pressure is 101 kPa (one atmosphere). Standard electrode potentials are listed in Table 7.1. Each equilibrium process may be used as half of a cell, two forming a whole cell when suitably connected. The sign convention adopted shows metals with an overall tendency to throw off ions and thus acquire a negative potential with respect to the solution with a negative standard electrode potential. Thus the standard hydrogen electrode ($E^{\ominus}_{298}$ = 0 V by convention) is placed on the left (also by convention) and this is indicated in the table. The reversible arrows indicate that equilibrium between element and solution is necessary.

Here are two examples of cell reactions in which the short vertical lines indicate interphase boundaries within the cell:

Pt, H$_2$|H$^+$|Zn^{2+}|Zn implies
$$\tfrac{1}{2}H_2 + \tfrac{1}{2}Zn^{2+} \rightarrow H^+ + \tfrac{1}{2}Zn$$

Pt, H$_2$|H$^+$|Cl$^-$|Cl$_2$, Pt implies
$$\tfrac{1}{2}H_2 + \tfrac{1}{2}Cl_2 \rightarrow H^+ + Cl^-$$

In both, the electrode on the left is the standard hydrogen electrode, so that the measured electric potential of each cell may be called a relative electrode potential, or simply, electrode potential. This is what is meant when chemists speak of the potentials of half-cells such as

$$Zn^{2+}|Zn \text{ and } Cl^-|Cl_2, Pt$$

which are abbreviated forms of the cell reactions already given. If on the other hand, the half cells are depicted as Zn|Zn^{2+} and Pt, Cl$_2$|Cl$^-$ the reactions within the cells are the reverse of those shown above and the potential differences with the standard hydrogen electrode on the right differ in sign and should not be called electrode potentials, standard or otherwise. For metals reacting with water, indirect methods of determination are used.

The value of the standard electrode potential is a measure of the tendency of the metal to form (hydrated) cations, the half-cell equilibrium moving towards reduction under the stated conditions or, in other words, with the standard hydrogen electrode on the left. At the upper end of the table the tendency towards the donation of electrons is greatest and these elements are readily oxidized. For other concentrations, the value of the electrode potential may be calculated from the equation:

$$E = E^{\ominus}_{298} + \frac{RT}{zF}\ln m$$

where $E^{\ominus}_{298}$ = the standard electrode potential at 298 K, z = valency of (charge on) the ion, T =

absolute temperature, $F = 96\,500\,\mathrm{C\,mol^{-1}}$ (the Faraday), and $\ln m = \log_e$ (molar concentration of ions). Strictly, m refers to the activity of the ions, taken as molar concentration here. For zinc, Zn^{2+}, this expression at 298 K becomes

$$E/V = -0.763 + \frac{0.0591}{2} \log_{10} m_{Zn^{2+}}\,\mathrm{mol\,dm^{-3}}$$

The order of elements in the table of standard electrode potentials is called the *electrochemical series*. A metal high in the series is a stronger reducing agent than a metal lower down, being readier to shed electrons.

To consolidate the foregoing, the following convention has been adopted in calculating the electromotive force of a cell and its sign. The Daniell cell mentioned above should be represented by a diagram

$$Zn \mid Zn^{2+} \mid Cu^{2+} \mid Cu$$

whereupon the electromotive force is equal in sign and magnitude to the electrical potential of the metallic conducting lead on the right when that of the similar lead on the left is taken as zero, the circuit being open. The so-called positive current flows through the cell from left to right, the reaction of the cell being

$$\tfrac{1}{2}Zn + \tfrac{1}{2}Cu^{2+} \rightarrow \tfrac{1}{2}Zn^{2+} + \tfrac{1}{2}Cu$$

because the ions are doubly charged and the equation must relate to the passage of one Faraday of electric charge.

If the equation had been written the other way round

$$\tfrac{1}{2}Cu + \tfrac{1}{2}Zn^{2+} \rightarrow \tfrac{1}{2}Cu^{2+} + \tfrac{1}{2}Zn$$

this would imply the diagram

$$Cu \mid Cu^{2+} \mid Zn^{2+} \mid Zn$$

and the electromotive force of the cell would be negative (this relates to driving a current through the cell to reverse the reaction and to 're-charge' it). When a half cell is written as $Zn^{2+} \mid Zn$, the electromotive force of this half cell means the electromotive force of the cell

$$Pt, H_2 \mid H^+ \mid Zn^{2+} \mid Zn$$

for which the reaction is

$$\tfrac{1}{2}H_2 + \tfrac{1}{2}Zn^{2+} \rightarrow H^+ + \tfrac{1}{2}Zn$$

the standard hydrogen electron being on the left. By the convention above, the electromotive force is thus the relative electrode potential, the standard electrode potential.

On the other hand a half cell written as $Zn \mid Zn^{2+}$ means the electromotive force of the cell

$$Zn \mid Zn^{2+} \mid H^+ \mid H_2, Pt$$

implying the reaction

$$\tfrac{1}{2}Zn + H^+ \rightarrow \tfrac{1}{2}Zn^{2+} + \tfrac{1}{2}H_2$$

with the standard hydrogen electrode on the right. The sign is the opposite to that above. Thus the electromotive force (positive, the reaction goes left to right) is *not* the electrode potential.

When an inert electrode, such as a platinum wire, is immersed in a solution containing both iron(II) and iron(III) ions, a difference of potential develops between the metal and solution. The magnitude of this potential depends on the net tendency of the system to absorb electrons from the platinum electrode or to donate them to it,

$$Fe^{3+} + e^- \rightleftharpoons Fe^{2+}$$

The potential is called a redox potential. At 25°C, using concentrations of iron(II) and iron(III) salts which are both 1.0 M, the standard redox potential

Table 7.2 Standard electrode potentials (ctd.): standard redox potentials at 298K (25°C)

Pt electrode in	Electrode process	$E^{\ominus}_{298}/V$
H^+; H_2	$H^+ + e^- \rightleftharpoons \tfrac{1}{2}H_2$	0.000
Sn^{4+}; Sn^{2+}	$\tfrac{1}{2}Sn^{4+} + e^- \rightleftharpoons \tfrac{1}{2}Sn^{2+}$	+0.15
Cu^{2+}; Cu^+	$Cu^{2+} + e^- \rightleftharpoons Cu^+$	+0.153
$Fe(CN)_6^{3-}$; $Fe(CN)_6^{4-}$	$Fe(CN)_6^{3-} + e^- \rightleftharpoons Fe(CN)_6^{4-}$	+0.360
Fe^{3+}; Fe^{2+}	$Fe^{3+} + e^- \rightleftharpoons Fe^{2+}$	+0.771
Hg^{2+}; Hg_2^{2+}	$Hg^{2+} + e^- \rightleftharpoons \tfrac{1}{2}Hg_2^{2+}$	+0.920
$Cr_2O_7^{2-}$; Cr^{3+}	$\tfrac{1}{6}Cr_2O_7^{2-} + \tfrac{7}{3}H^+ + e^- \rightleftharpoons \tfrac{1}{3}Cr^{3+} + \tfrac{7}{6}H_2O$	+1.33
Ce^{4+}; Ce^{3+}	$Ce^{4+} + e^- \rightleftharpoons Ce^{3+}$	+1.45
MnO_4^-; Mn^{2+}	$\tfrac{1}{5}MnO_4^- + \tfrac{8}{5}H^+ + e^- \rightleftharpoons \tfrac{1}{5}Mn^{2+} + \tfrac{4}{5}H_2O$	+1.52

The physical states are implied by the standard conditions

may be determined on the standard hydrogen scale. Standard redox potentials are listed in Table 7.2. They are separated from those given in Table 7.1 purely for convenience: both tables show redox potentials. The relative positions of two systems indicate the possibility of a reaction between them but give no information about the rate of reaction, the influence of catalysts, or the other variables such as complex ion or insoluble product formation. Note that the expression, electrode potential or redox potential of the half cell written

$$Fe^{2+}, Fe^{3+} \mid Pt$$

means the electric potential difference of the whole cell

$$Pt, H_2 \mid H^+ \mid Fe^{2+}, Fe^{3+} \mid Pt$$

implying the reaction

$$\tfrac{1}{2}H_2 + Fe^{3+} \rightarrow H^+ + Fe^{2+}$$

For conditions other than standard, the following expression may be deduced for the redox potential, in its general form

$$\underset{\text{form}}{\text{Oxidized}} + ne^- \rightleftharpoons \underset{\text{form}}{\text{Reduced}}$$

$$E = E^{\ominus}_{298} + \frac{RT}{nF} \ln \frac{[\text{Oxidized form}]}{[\text{Reduced form}]}$$

In which the square brackets indicate activities for which we may substitute molar concentrations, so that for the Fe^{3+}, Fe^{2+} equilibrium ($E^{\ominus} = +0.771$ V)

$$E/V = +0.771 + 0.0591 \log_{10} \left(\frac{m_{Fe^{3+}}/\text{mol dm}^{-3}}{m_{Fe^{2+}}/\text{mol dm}^{-3}} \right)$$

where $m_{Fe^{3+}}/m_{Fe^{2+}}$ is the ratio of the respective molar concentrations and $n = 1$.

The sign of the potential is that of the electrode. A positive electrode potential is the result of the removal of electrons from the electrode by the system and since the gain of electrons results in reduction, the system must have an oxidizing tendency. Therefore, high positive potentials represent systems with a high oxidizing potential (see Table 7.1 also) while a negative potential indicates reducing potential. The relative position of any two systems shows the system with the higher (positive) potential as the oxidant and the other as the reductant in a possible reaction between them. Reference to Tables 7.1 and 7.2 reveals that the standard electrode potential of the Cl^-, Cl_2(Pt) system is

between those of $Cr_2O_7^{2-}$, Cr^{3+} (Pt) and MnO_4^-, Mn^{2+} (Pt), with the last named the more powerful oxidizing agent under the standard conditions. This has application in titrimetric (volumetric) analysis in the determination of iron in the presence of chloride ions where dichromate is preferred to permanganate which would oxidize chloride ions to chlorine.

Some examples of oxidation and reduction

Oxidizing or reducing power is relative. However, a general qualitative approach may be made to the diagnosis of the oxidizing or reducing nature of an element or compound.

Tests for reducing agents

1. The magenta colour of potassium permanganate solution acidified with dilute sulphuric acid is discharged. Warming may be necessary. For example, hot oxalic acid (ethanedioic acid) is oxidized:

$$2KMnO_4 + 5 \begin{array}{c} COOH \\ | \\ COOH \end{array} + 3H_2SO_4 \rightarrow$$
$$K_2SO_4 + 2MnSO_4 + 10CO_2 + 8H_2O$$

Decolorization is a sufficient indication that the reaction has occurred. In this example, the reaction is autocatalysed by manganese(II) ions and the removal of colour is not immediate.

2. The orange colour of potassium dichromate solution acidified with dilute sulphuric acid or hydrochloric acid, and warmed, if necessary, turns to the green of chromium(III) ions. Using iron(II) chloride:

$$K_2Cr_2O_7 + 6FeCl_2 + 14HCl \rightarrow$$
$$2KCl + 2CrCl_3 + 6FeCl_3 + 7H_2O$$

Iron(III) chloride is formed. The colour change is enough to demonstrate that the reaction has taken place.

Test for oxidizing agents

1. Iodine is liberated from potassium iodide solution acidified with dilute hydrochloric acid or sulphuric acid, with warming if necessary. It appears as a yellow or reddish-brown colour which forms a deep blue coloration with starch solution. Using potassium bromate,

$$KBrO_3 + 6KI + 6HCl \rightarrow$$
$$KBr + 6KCl + 3I_2 + 3H_2O$$

In an alternative procedure, carbon tetrachloride is added to form an immiscible lower layer in which iodine dissolves to give a violet solution. This is enough to show that the redox reaction has occurred.

2. A freshly prepared solution of an iron(II) salt, usually the sulphate, but the chloride may be used, is oxidized to the iron(III) salt, on warming if necessary. The colour change from pale water-green to yellow-brown may be observed but a chemical test is desirable to show that a reaction has occurred. An immediate deep blue precipitate with potassium hexacyanoferrate(II) solution indicates an iron(III) salt. Because iron(II) salts are readily oxidized in air, a check on the original reagent should be performed.

Using potassium peroxodisulphate, on warming:

$$K_2S_2O_8 + 2FeSO_4 \rightarrow K_2SO_4 + Fe_2(SO_4)_3$$

Potassium permanganate in titrimetric analysis†

When potassium permanganate* solution is run into acidified iron(II) sulphate solution, the intense magenta colour of the former is discharged until all of the iron(II) has been oxidized. Qualitative tests show that manganese(II) and iron(III) ions are produced. Separate ion-electron equations, where e^- represents one Faraday,†† may be written:

$$MnO_4^- + 8H^+ + 5e^- \rightleftharpoons Mn^{2+} + 4H_2O$$
oxidant   (K^+ and SO_4^{2-} ions unchanged)

$$Fe^{2+} - e^- \rightleftharpoons Fe^{3+}$$
reductant   (SO_4^{2-} ions unchanged)

Electrons are transferred during the reaction of the two systems:

$$MnO_4^- + 8H^+ + 5Fe^{2+} \rightarrow$$
$$5Fe^{3+} + Mn^{2+} + 4H_2O$$

The transfer of one Faraday of electric charge requires the interaction of $\frac{1}{5}$ mole (0.2 mole) of $KMnO_4$ as oxidant††† and 1 mole of Fe^{2+} which could have been weighed out as 1 mole of $FeSO_4$, or 1 mole of $FeSO_4 7H_2O$, or 1 mole of $FeSO_4$ $(NH_4)_2SO_4 6H_2O$ or simply as 1 mole of Fe (to be dissolved in dilute sulphuric acid) as reductant. This equivalence is the key to any calculation which has to be made.

The iron(II) sulphate solution is acidified with sulphuric acid. Hydrochloric acid reacts with permanganate, unless used under very carefully controlled conditions, while nitric acid is an oxidant itself.

The ion-electron equations allow much simpler equations to be written than the cumbersome ones used before and for these equations to be derived from first principles.

Potassium permanganate is used widely as an oxidizing agent in the presence of acid. The slightest excess imparts the characteristic magenta colour to the solution titrated. However, it is not a primary standard and solutions must therefore be standardized. For this purpose, oxalates** (or oxalic acid***) or iron(II) salts, especially Mohr's salt, ammonium iron(II) sulphate, may be used. Sodium oxalate is ideal: it is anhydrous and may be prepared in a high degree of purity with no problem concerning the amount of hydration. Oxalates must be heated to about 60°C for titration against permanganate:

$$\begin{array}{l} COO^- \\ | \quad\quad\quad - 2e^- \rightleftharpoons 2CO_2 \\ COO^- \end{array}$$

$$MnO_4^- + 8H^+ + 5e^- \rightleftharpoons Mn^{2+} + 4H_2O$$

$$\therefore 5\begin{array}{l} COO^- \\ | \\ COO^- \end{array} + 2MnO_4^- + 16H^+ \rightarrow$$
$$10CO_2 + 2Mn^{2+} + 8H_2O$$

The transfer of one Faraday of electric charge requires the interaction of $\frac{1}{5}$ mole of $KMnO_4$ as oxidant and $\frac{1}{2}$ mole of $C_2O_4^{2-}$.

Common sources of the latter are:

$\frac{1}{2}$ mole of oxalic acid $(COOH)_2 2H_2O$
$\frac{1}{2}$ mole of sodium oxalate $(COONa)_2$
$\frac{1}{2}$ mole of potassium oxalate $(COOK)_2 H_2O$
$\frac{1}{2}$ mole of ammonium oxalate $(COONH_4)_2 H_2O$

† The traditional term is volumetric analysis, used in the restricted sense of applying to the quantitative estimation of substances in solution, and not to gas analysis. Titrimetry seems a better description of titration techniques

†† The use of $F(=Le)$ might be confused with F. Traditionally we speak of ions, atoms and electrons in these equations

††† From the equations it is seen that both MnO_4^- and H^+ are necessary for the oxidation to occur as described

*Alternative UPAC name: Potassium tetraoxomanganate(VII)

** ethanedioates

*** ethanedioic acid

but

$\frac{1}{4}$ mole of potassium tetroxalate (or quadroxalate)

$$\begin{bmatrix} \text{COOK} & \text{COOH} \\ | & \cdot & | & \cdot & 2H_2O \\ \text{COOH} & \text{COOH} \end{bmatrix}$$

Again, this is the basis of any calculation demanded.

Potassium permanganate may be used in solutions which are not acidic but the reactions are relatively unimportant:

in neutral or alkaline solution (using sodium carbonate):

$$MnO_4^- + 2H_2O + 3e^- \rightleftharpoons MnO_2(s) + 4OH^-$$

and in strongly alkaline solution:

$$MnO_4^- + e^- \rightleftharpoons MnO_4^{2-}$$

In the first reaction gelatinous manganese(IV) oxide (manganese dioxide) is precipitated and in the second, a green solution of potassium manganate(VI) is formed. The transfer of one Faraday of electron charge is associated with the reaction of $\frac{1}{3}$ mole of KMnO$_4$ and 1 mole of KMnO$_4$ respectively.

Potassium dichromate in titrimetric analysis

When potassium dichromate† solution is added to acidified iron(II) sulphate solution, the orange colour of the former changes to green with the production of chromium(III) ions; iron(III) ions are also formed. The end-point is determined by use of an oxidation–reduction indicator because the development of the green makes judging the colour change impossible.

The ion-electron equations may be written:

$$Cr_2O_7^{2-} + 14H^+ + 6e^- \rightleftharpoons 2Cr^{3+} + 7H_2O$$
$$\text{oxidant}$$

$$Fe^{2+} - e^- \rightleftharpoons Fe^{3+}$$
$$\text{reductant}$$

The transfer of one Faraday of electric charge is associated with the interaction of $\frac{1}{6}$ mole K$_2$Cr$_2$O$_7$ and one mole of Fe^{2+} (whatever the source). By eliminating the charge transferred in these equations, we obtain the net reaction:

$$Cr_2O_7^{2-} + 14H^+ + 6Fe^{2+} \rightarrow$$
$$6Fe^{3+} + 2Cr^{3+} + 7H_2O$$

The fuller equation may be derived by including potassium and sulphate ions, but the result is much less clear than the ionic equation which does emphasize the mechanism of the change,

$$6FeSO_4 + K_2Cr_2O_7 + 7H_2SO_4 \rightarrow 3Fe_2(SO_4)_3$$
$$+ K_2SO_4 + Cr_2(SO_4)_3 + 7H_2O$$

Another example: tin may be estimated as tin(II) chloride in the presence of hydrochloric acid by dichromate:

$$Cr_2O_7^{2-} + 14H^+ + 3Sn^{2+} \rightarrow$$
$$2Cr^{3+} + 3Sn^{4+} + 7H_2O$$

Once again, the calculations based upon the equivalence relationships implied by the equations are straightforward.

A detailed discussion of redox indicators is beyond the scope of the present book, but an idea of their mode of action may be given. They are usually organic compounds which have reduced and oxidized states of different colour, and are selected because of these colours, and the redox potential of the system. For example, a solution of diphenylamine in phosphoric acid is used as an indicator in iron(II) ion versus dichromate titrations. The iron(II) ion is oxidized preferentially, but when it is all oxidized to iron(III), further addition of dichromate oxidizes the diphenylamine from colourless to deep blue.

Why should potassium dichromate be used when permanganate is available, and acts as its own indicator giving such an easily seen end-point? Permanganate solutions tend to give off oxygen because of their high redox potential and the manganese(IV) oxide produced by this and the oxidation of any organic matter present accelerates the change. Such solutions require frequent standardization and are not made by direct weighing. Furthermore, in the acidic medium, chloride ions will be oxidized to chlorine with the result that titration readings will be too high. Potassium dichromate being a weaker oxidizing agent does not do this and is perfectly stable. In addition, it is a salt which can be prepared in a very high state of purity and is not decomposed by the moderate heat used in drying ovens. Potassium dichromate is a most valuable redox standard and is used especially for the estimation of iron in the presence of chloride ion. Other uses are in the estimation of tin, and manganese in steel or manganese ore.

† *Alternative IUPAC name:* potassium μ-oxobistrioxochromate(VI)

Iodine in titrimetric (or volumetric) analysis

The oxidizing power, as measured by the tendency of the system,

$$I_2 + 2e^- \rightleftharpoons 2I^-$$

to take up electrons from a platinum wire under standard conditions, is less than that of most other oxidizing agents. Its standard redox potential is $+0.54\,V$ while that of dichromate in acid is $+1.33\,V$ and of permanganate in acid $+1.52\,V$. Stronger oxidizing agents ought to liberate iodine from a solution of iodide ions. This happens, and in an excess of iodide ions (usually a potassium iodide solution) the iodine, which has a very low solubility in water ($0.28\,g\,dm^{-3}$ at $18°C$), remains in solution as a complex tri-iodide ion,

$$I_2 + I^- \rightleftharpoons I_3^-$$

When potassium permanganate solution is run into acidified potassium iodide solution, iodine is liberated. Potassium dichromate will also liberate iodine. The ion-electron equation for iodine is:

$$2I^- - 2e^- \rightleftharpoons I_2$$

One Faraday of electric charge is associated with $\frac{1}{2}(2I^-)$. i.e. 1 mole of I^- as KI.

The final equations are:

$$2MnO_4^- + 16H^+ + 10I^- \rightarrow 2Mn^{2+} + 5I_2 + 8H_2O$$

$$Cr_2O_7^{2-} + 14H^+ + 6I^- \rightarrow 2Cr^{3+} + 3I_2 + 7H_2O$$

The liberated iodine may be estimated by titration against standard sodium thiosulphate† solution:

$$I_2 + 2S_2O_3^{2-} \rightarrow 2I^- + S_4O_6^{2-}$$
$$\text{thiosulphate} \qquad\qquad \text{tetrathionate}$$
$$\text{ion} \qquad\qquad\qquad \text{ion}$$

By using the principle of liberation of iodine which is then determined in this way, many oxidizing agents may be estimated volumetrically. Standard solutions of sodium thiosulphate are not usually made by weighing alone because there is some doubt about the degree of hydration of the (usually) hydrated salt and the solutions do not keep indefinitely. Dichromate could be used as a primary standard for thiosulphate solutions. For the titrimetric iodine–thiosulphate reaction, one Faraday of electric charge is associated with the reaction of $\frac{1}{2}$ mole of I_2 and 1 mole of $Na_2S_2O_35H_2O$, deduced from the respective ion-electron equations:

$$I_2 + 2e^- \rightleftharpoons 2I^-$$

$$2S_2O_3^{2-} - 2e^- \rightleftharpoons S_4O_6^{2-}$$

The products of the reaction are iodide and tetrathionate ions.

On running sodium thiosulphate solution into an iodine solution (in potassium iodide solution), the reddish-brown colour is gradually discharged through yellow to colourless. Just before this stage a few drops of a freshly prepared starch solution impart an intense blue colour to the iodine solution. At the end-point, this is discharged. Alternatively, carbon tetrachloride is added to the solution in a stoppered bottle. Iodine is about 85 times more soluble in the organic solvent than in water at room temperature. A violet coloured solution is formed which persists down to very low concentrations of iodine. Thiosulphate may be added with shaking until the colour of iodine is no longer visible.

A standard solution of iodine may be used to estimate various reducing agents. Weaker reducing agents, such as arsenites, sulphites, antimonates(III) and hexacyanoferrates(II) tend to react reversibly unless the acid is neutralized:

$$AsO_2^- + I_2 + 2H_2O \rightarrow AsO_4^{3-} + 2I^- + 4H^+$$
$$\text{meta-arsenite} \qquad\qquad\qquad \text{arsenate}$$
$$\text{ion} \qquad\qquad\qquad\qquad \text{ion}$$

$$SO_3^{2-} + I_2 + H_2O \rightarrow SO_4^{2-} + 2H^+ + 2I^-$$

Addition of oxygen as reduction

Oxygen is not the most electronegative element. It is second to fluorine. With the extension of the definition of oxidation, the formation of fluorine monoxide (oxygen difluoride) presents a problem. This compound is formed as a colourless gas by passing fluorine very slowly through a 2% solution of sodium hydroxide,

$$2F_2 + 2OH^- \rightarrow 2F^- + F_2O + H_2O$$

Although oxygen has been added to fluorine, for a consistent general view we must conclude that fluorine has been reduced. Fluorine always undergoes reduction on combination.

Addition of hydrogen as oxidation

The alkali metals react with hydrogen on heating: sodium forms colourless crystals of sodium hydride,

$$2Na + H_2 \rightarrow 2NaH$$

† Alternative IUPAC name: sodium trioxothiosulphate(VI)

Sodium hydride is an ionic compound and two separate ion-electron equations may be distinguished:

$$2Na - 2e^- \rightarrow 2Na^+$$

$$H_2 + 2e^- \rightarrow 2H^-$$

Sodium has undergone ionic oxidation, and hydrogen, ionic reduction. In the formation of ionic hydrides, hydrogen is reduced to the H^- ion and the metal oxidized to the cation.

Oxidation number and valency

The concept of oxidation number (O.N.) was developed by Johnson in 1880 to calculate the molar combining ratios of oxidizing agents. It follows from the recognition that oxidation and reduction are accompanied by changes in valency.

Oxidation number is a number assigned to an element in a particular compound as a measure of the amount of oxidation or reduction required to convert one atom of the element from the free state into the state of the atom in the compound.

It is the electric charge which is carried by the atom of the element when the electrons are assigned according to an agreed convention. The number is positive if oxidation is necessary and negative if reduction is necessary. The rules used are as follows.

1. The oxidation number (i.e. electric charge) of a compound is zero. The oxidation number of each element is multiplied by the number of atoms of that element in the formula of the compound and the resulting numbers, when added algebraically, total zero.
2. The oxidation number of a free or uncombined element is zero, e.g. Ba, O_2, O_3, S_8, P_4. Note that the oxidation number does not equal the valency here although it usually does so in the following.
3. The oxidation number of an ion is the ionic charge, e.g. $Ca^{2+}(+2)$, $S^{2-}(-2)$.
4. The oxidation number of hydrogen is $+1$ except where it is present as the hydride ion (H^-) when it is -1 (the ionic charge).
5. The oxidation number of oxygen is -2 except in peroxides (O_2^{2-}) where it is -1, the ionic charge being shared between two atoms.
6. In a compound, the more electronegative element (See Fig. 5.7 and Table 5.13) is assigned a negative oxidation number (electric charge) and the

relatively electropositive element, a positive charge. Examples follow.

BrF_3:
   If $F_3 \rightarrow 3F^-$, then $Br \rightarrow Br^{3+}$
   O.N. $F(-1)$, $Br(+3)$

H_2S:
   If $H_2 \rightarrow 2H^+$, then $S \rightarrow S^{2-}$
   O.N. $H(+1)$ (convention), $S(-2)$

SO_2:
   If $O_2 \rightarrow 2O^{2-}$, then $S \rightarrow S^{4+}$
   O.N. $O(-2)$ (convention), $S(+4)$

SO_3:
   If $O_3 \rightarrow 3O^{2-}$, then $S \rightarrow S^{6+}$
   O.N. $O(-2)$, $S(+6)$

H_2SO_4:
   If $H_2SO_4 \rightarrow 2H^+ + SO_4^{2-}$
   and $4O$ in $SO_4^{2-} \rightarrow 4O^{2-}$, then $S \rightarrow S^{6+}$ to leave a charge of minus two units on SO_4^{2-}
   O.N. $H(+1)$, $O(-2)$, $S(+6)$

MnO_4^-:
   If $O_4 \rightarrow 4O^{2-}$, then $Mn \rightarrow Mn^{7+}$
   O.N. $O(-2)$, $Mn(+7)$

$FeCl_2$:
   If $Cl_2 \rightarrow 2Cl^-$, then $Fe \rightarrow Fe^{2+}$
   (ionic charge)
   O.N. $Cl(-1)$, $Fe(+2)$.

For various compounds of nitrogen the oxidation numbers of nitrogen may be listed:

+5	N_2O_5	+1	N_2O
+4	N_2O_4	0	N_2
+3	N_2O_3	−1	NH_2OH
+2	NO	−2	N_2H_4
		−3	NH_3

and similarly for sulphur:

+7	$H_2S_2O_8$	+2.5	$H_2S_4O_6$
+6	H_2SO_4	+2	$H_2S_2O_3$
+5	$H_2S_2O_6$	+1	—
+4	H_2SO_3	0	S_8
+3	$H_2S_2O_4$	−1	—
		−2	H_2S

Oxidation is a process which results in an increase in the (positive) oxidation number of an element.

When an element has a particular oxidation number in a series of compounds, it is said to be in that particular **oxidation state** in those compounds.

Using three compounds of sulphur listed above, the following reaction sequence may be effected.

$$H_2S \rightarrow SO_2 \rightarrow SO_3$$

The oxidation numbers of sulphur are respectively -2, $+4$ and $+6$. Sulphur has been oxidized and passes from oxidation state -2, to $+4$, and on to $+6$. Note that the maximum oxidation state of a typical element corresponds to the number of its group in the Periodic Table, which is also the number of electrons available for transfer in chemical reactions. The minimum oxidation state corresponds to the group number minus eight, or zero (as in the pure element).

A knowledge of change in oxidation number enables the calculation of equivalent masses in which oxidizing and reducing agents interact. For instance in the reduction of potassium permanganate in dilute sulphuric acid by iron(II) sulphate:

$$K^+MnO_4^- \rightarrow Mn^{2+}SO_4^{2-}$$
$$Fe^{2+}SO_4^{2-} \rightarrow (Fe^{3+})_2(SO_4^{2-})_3$$

For manganese the change in oxidation number is from $+7$ to $+2$, by 5 units of reduction. For iron the change in oxidation number is from $+2$ to $+3$, by 1 unit of oxidation.

Thus $\frac{1}{5}$ mole of $KMnO_4$ reacts with 1 mole of $FeSO_4$ and the equation may be derived:

$$KMnO_4 + 5FeSO_4 + acid \rightarrow$$
$$MnSO_4 + K_2SO_4 + \tfrac{5}{2}Fe_2(SO_4)_3 + water$$

whereupon by inspection,

$$2KMnO_4 + 10FeSO_4 + 8H_2SO_4 \rightarrow$$
$$2MnSO_4 + K_2SO_4 + 5Fe_2(SO_4)_3 + 8H_2O$$

This equation is cumbersome and the ionic equation given earlier is to be preferred:

$$MnO_4^- + 5Fe^{2+} + 8H^+ \rightarrow Mn^{2+} + 5Fe^{3+} + 4H_2O$$

When iron(III) chloride is reduced by hydrogen sulphide, sulphur is precipitated and iron(II) chloride remains:

$$FeCl_3 \rightarrow FeCl_2$$
$$H_2S \rightarrow 2H^+ + S$$

For iron the change in oxidation number is from $+3$ to $+2$, by 1 unit of reduction.

H_2S is oxidized to free S (oxidation number zero), so that the change in oxidation number for sulphur is from -2 to 0, or by 2 units of oxidation.

Thus $2FeCl_3$ reacts with $1H_2S$, and the equation becomes

$$2FeCl_3 + H_2S \rightarrow 2FeCl_2 + 2HCl + S$$

although the ionic equation is to be preferred:

$$2Fe^{3+} + H_2S \rightarrow 2Fe^{2+} + 2H^+ + S$$

Chemical nomenclature

The concept of oxidation number is useful in chemical nomenclature.† Fe in $FeCl_3$ is tervalent but this is not apparent from Fe_2Cl_6 in the vapour state. Fe in $K_3Fe(CN)_6$ is tervalent and in $K_4Fe(CN)_6$ is bivalent, but this is not readily seen. Ni in $Ni(CO)_4$ is assigned zero valency yet the bonds are not weak. In the traditional nomenclature ferrous and ferric have symbols Fe^{2+} and Fe^{3+} respectively but the unwary might be caught out by stannous and stannic where the symbols are respectively Sn^{2+} and Sn^{4+}. Further, N in NH_3 and in NO_2^- is properly described as tervalent yet the former has to be oxidized to form the latter.

Now, valency, as has been seen earlier, describes the bonding in a compound and requires rather more to be understood than straightforward chemical analysis and application of an agreed set of rules: it sets out to express more than oxidation number, or oxidation state. Thus the central metal atom in a complex ammine, e.g. $[Co(NH_3)_6]^{3+}$, has a covalency of 6 (from its co-ionic bonds) and the entity exhibits an ionic valency of 3, while the state of affairs within the complex cation may be open to argument. These problems of fine detail are avoided by using the concept of oxidation state or number, which is a cruder, though exact measure.

In the *Stock Notation* the oxidation number is inserted immediately after the name of the principal atom. Roman numerals are most frequently used after the name or symbol of the element. Consider examples:

Ferrous chloride	$FeCl_2$	Iron(II) chloride
Ferric chloride	$FeCl_3$	Iron(III) chloride
Stannous chloride	$SnCl_2$	Tin(II) chloride
Stannic chloride	$SnCl_4$	Tin(IV) chloride

Consider anions containing iron, using the term ferrate as the basis for nomenclature:

† *International Union of Pure and Applied Chemistry: Nomenclature of Inorganic Chemistry* (Butterworth Scientific Publications)

FeO_3^{3-} ferrate: Fe^{3+} united with $3O^{2-}$ to give FeO_3^{3-} so the anion becomes ferrate(III),
FeF_6^{3-} hexafluoroferrate(III),
$Fe(CN)_6^{3-}$ hexacyanoferrate(III) instead of ferricyanide,
$Fe(CN)_6^{4-}$ hexacyanoferrate(II) instead of ferrocyanide.

The latter seem cumbersome but have these advantages:

(a) the name and number of ligands is stated,
(b) the central atom is that of ferrate, i.e. iron,
(c) the oxidation number of iron is stated,
(d) the conversion of hexacyanoferrate(II) to hexacyanoferrate(III) is readily seen to be oxidation.

With certain exceptions, a polyatomic electronegative group is designated by modifying the name of the characteristic or central atom to end in *-ate*. In fully systematic nomenclature, sulphate, for example, becomes the general name of an electronegative group:

sodium sulphate $Na_2[SO_4]$
                     sodium *tetraoxosulphate*
sodium sulphite $Na_2[SO_3]$
                     sodium *trioxosulphate*
sodium thiosulphate $Na_2[S_2O_3]$
                     sodium *trioxothiosulphate*

Applicable to all types of compound but especially suitable for binary compounds of non-metals, is the system whereby the relative proportion of constituents is given:

nitrous oxide N_2O *dinitrogen oxide*
'sulphur
   monochloride' S_2Cl_2 *disulphur dichloride*
ferroso-ferric oxide Fe_3O_4 *tri-iron tetroxide*

In the Stock Notation, the relative proportion of constituents is indicated indirectly by showing the oxidation number. The idiosyncrasies of the older system of nomenclature will be explained later as examples arise.

Thus the Stock Notation can be applied to both cations and anions, but should preferably *not* be applied to compounds between non-metals. In this book, in accordance with international convention the right superscript position is used to indicate states of valency, *viz* $KMn^{VII}O_4$ while oxidation numbers are included in the names as explained earlier.

The action at electrodes during electrolysis

We have already seen that certain compounds, called electrolytes, conduct electricity when fused or dissolved in water and that chemical action occurs at the electrode surfaces only, where the quantities of substances liberated are in accordance with Faraday's Laws of Electrolysis. The electric charge passes through the electrolyte by virtue of the migration of ions. Strong electrolytes are completely ionic even in the solid state (e.g. sodium chloride) or largely ionized in aqueous solution (e.g. sulphuric acid), while weak electrolytes are feebly ionized. The action at the electrodes involves the transfer of electrons and is therefore either oxidation or reduction. Oxidation occurs at the anode: negative ions (anions) migrate towards it and on being discharged, surrender their electrons which flow along the external circuit. Reduction occurs at the cathode: the positive ions (cations) migrate towards it and on receiving electrons, are discharged. In the electrolysis of fused sodium chloride with carbon rod electrodes, sodium appears at the cathode and chlorine at the anode:

$$Na^+ + e^- \rightarrow Na$$

$$Cl^- - e^- \rightarrow Cl \qquad 2Cl \rightarrow Cl_2$$

For each Faraday of electric charge passing:

$$Na^+Cl^- \rightarrow Na + \tfrac{1}{2}Cl_2$$

When dilute aqueous solutions of sulphuric acid, sodium hydroxide or sodium sulphate are electrolysed with platinum (inert) electrodes, the gaseous products are identical. At the cathode, hydrogen is liberated and at the anode, oxygen, in the volume ratio 2:1 respectively. In the electrolysis of a solution of hydrogen chloride (hydrochloric acid), the cathode product is hydrogen yet the product at the anode depends on the concentration: very dilute solutions yield oxygen and more concentrated solutions yield chlorine while at intermediate concentrations, a mixture of the two gases results. Electrolysis of sodium chloride solutions with various inert electrodes in a range of concentrations gives the same products unless a mercury cathode is used, when sodium is discharged into the mercury. In the production of aluminium, sodium and aluminium cations are both present but only aluminium is produced.

A detailed treatment will be found in text-books on Physical Chemistry. The following topics are relevant:

(a) discharge potentials,
(b) decomposition voltage,
(c) decomposition of water, oxoacids, alkalis and oxosalts by electrolysis,
(d) overvoltage (overpotential), especially at a mercury cathode.

The importance of hydrogen overvoltage (overpotential)

For the deposition of a metal during electrolysis the cathode must be at a sufficient negative potential with respect to the solution for the discharge potential of the metal to be reached. Discharge of hydrogen may interfere with this process. However, the fact that a hydrogen overvoltage is required at certain surfaces enables a high yield of some metals to be obtained at cathodes during electrolysis if the conditions are carefully controlled. Zinc, iron, cadmium, nickel, tin and lead, which lie between the very reactive highly electro-positive metals and the noble metals in the table of standard electrode potentials, are those to be considered in this context. Taking zinc as an illustration, in an aqueous solution which contains zinc sulphate and sulphuric acid, each cation being in molar (1.0M) concentration, the standard electrode potentials of zinc and hydrogen at 25°C (298K) are respectively -0.76 V and 0.00 V. The overpotential of hydrogen at a zinc cathode, or a cathode coated with zinc, is -0.70 V. For discharge of hydrogen the negative potential difference between cathode and solution must just exceed -0.70 V and for zinc, -0.76 V. The values are so close that from molar solutions both zinc and hydrogen are liberated.

When a solution of a salt of an alkali metal is electrolysed using a mercury cathode, no hydrogen is evolved from the surface of the mercury. In the industrial electrolysis of brine, a solution of sodium in mercury, sodium amalgam, is formed. As the concentration of sodium in the mercury cathode increases, hydrogen may eventually be evolved. Sodium is discharged from brine at a mercury cathode in preference to hydrogen because of the high hydrogen overpotential (-0.78 V) at a mercury surface.

Overpotential at anodes

When brine is electrolysed between a mercury cathode and a graphite anode, the anode product is chlorine, although under certain circumstances oxygen may also be evolved. However, oxygen has a high overpotential at a carbon anode which leads to the discharge potential of chlorine being the lowest in saturated brine (6M). If the chloride ion concentration drops to below molar (1M) concentration, the required discharge potential for chlorine rises and the gaseous anode product is mainly oxygen. At intermediate concentrations, a mixture of chlorine and oxygen is formed. Oxygen may also be formed on increasing the hydroxide ion concentration by addition of alkali or by very rapid electrolysis with a high current density when a limit is reached for the diffusion and discharge of chloride ions. The proportion of chlorine obtained from a solution of sodium chloride of intermediate concentration may be increased by reducing the concentration of hydroxide ions—by the addition of acid. Chlorine is the sole anode gas from the electrolysis of fairly concentrated hydrochloric acid.

Electroplating

The deposition of metals by electrolysis is of great technical importance. Both metallic and, in certain cases, non-metallic articles, the latter previously rendered conducting by a surface coating of graphite, are used as cathodes in suitable cells. The deposit might serve as decoration, as in the silver-plating of tableware, or to provide resistance to corrosion, as does rhodium plating on silver-plated propelling pencils. For engineering purposes, hard chromium surfaces may be deposited on to the surface of special tools, or worn components may be built up again. Printing surfaces for books and textiles are formed by copper-plating.

In electroplating, the new surface must adhere to the article, giving a fine-grained coherent deposit, possibly with a lustrous finish. Complex ions are frequently used, reducing the concentration of metal ions to avoid coarse deposition and altering the discharge potential. Various chemical agents are added. By suitably controlling the composition of the electrolyte and conditions, mixtures of metals may be deposited. Thus, brass may be deposited from a solution containing the complex cyanides of zinc and copper. The metal for deposition may come from the electrolyte, as in chromium-plating, or from the anode, as in copper-plating. In this connection, it may be pointed out that articles to be chromium-plated are usually previously plated with copper—an extremely important application of both of these metals.

Electrolytic corrosion

When a piece of pure zinc is placed in dilute sulphuric acid, it dissolves gradually and hydrogen is evolved slowly. Less pure zinc dissolves faster, with a brisker evolution of hydrogen. The metal throws off zinc ions into the solution and is left with surplus electrons and is therefore at a negative potential with respect to the solution. Zinc ions remain near to the piece of metal unless the residual charge is dissipated. Discharge of hydrogen ions and evolution of gas is slow because of the relatively large hydrogen overpotential at a zinc cathode.

However, touching the zinc with a piece of platinum below the surface of the acid has a marked effect: hydrogen is briskly evolved from the platinum and zinc dissolves, forming zinc sulphate solution. The platinum is unchanged. This arrangement is really a voltaic cell in which the current passes across the short-circuit of the two metal electrodes held in contact with each other. After ejection of zinc ions, electrons pass from zinc to platinum, which has a lower hydrogen overpotential, and hydrogen ions are discharged at its surface:

$$Zn \rightarrow Zn^{2+} + 2e^-$$

$$2e^-(Zn) \rightarrow 2e^-(Pt)$$

$$2e^-(Pt) + 2H^+ \rightarrow (Pt) + 2H$$

$$2H \rightarrow H_2$$

Other metals, less electropositive than zinc i.e. with less of a tendency to furnish positive ions in solution and with relatively low hydrogen overpotentials, may be used similarly. Drops of copper(II) sulphate solution added to the sulphuric acid and zinc mixture result in some electrochemical displacement of copper by zinc. The copper adheres to the zinc and short-circuited cells are created. Hydrogen is evolved briskly from the copper. Similarly, impure zinc reacts faster with acid than pure specimens because the impurities are less electropositive than zinc.

The reaction between zinc and sulphuric acid, accelerated by the presence of metals less electropositive than zinc, illustrates the principle of electrolytic corrosion. The pairs of metals involved are called *galvanic couples*. In the zinc/platinum galvanic couple, electrons pass from zinc to platinum so that zinc is the anode and platinum is the cathode. The adjectives anodic and cathodic are frequently used and in the case of impure zinc

reacting with acid, zinc dissolves from anodic areas and hydrogen is evolved from those which are cathodic. The driving force for this type of reaction comes from the tendency of the baser metal in the couple to form ions. Corrosion will only occur where the base metal is at a higher negative potential than that at which hydrogen is discharged from the other metal, so that electrons flow in the required direction.

Electrolytic corrosion is slow with pure zinc because of its relatively high hydrogen overvoltage. Hydrogen ions will migrate towards the more noble, i.e. less base, metal of the couple. Addition of alkali reduces the hydrogen ion concentration and impedes corrosion. On the other hand, removal of hydrogen gas as soon as it is formed with a chemical depolarizer, such as dissolved oxygen, increases the rate of discharge of hydrogen ions and the rate of corrosion. The presence of depolarizers will reduce the protection afforded against corrosion by a high hydrogen overpotential.

Differential oxygenation, occurring where the concentration of oxygen depolarizer is not uniform, is very important in corrosion and especially in the rusting of iron and steel. The heterogeneous nature of iron and steel, due to alloying constituents, leads to internal potential differences. Abnormally high corrosion rates have been found in the steel structures erected in the North Sea by those involved in the oil and natural gas industry, due no doubt to the enormous stresses and strains such structures experience: differences of potential will occur where the metal undergoes strain. Differential oxygenation results in easier hydrogen discharge because the gas is oxidized by the depolarizer. The areas from which hydrogen ions accept electrons are cathodic. They receive electrons from the anodic areas cut off from the air by rust or paintwork. Electrons remain in the anodic areas when iron(II) ions pass into solution and then pass to the cathodic areas. Deep pitting in ironwork may be ascribed to differential oxygenation, because oxygen concentration is lowest at the bottom of each pit which constitutes the anodic region. Pitting therefore goes deeper. It may already have been inferred that rust is not a primary product in the electrolytic corrosion of iron and steel.

Iron, in our climate, is usually moist. The layer of liquid water on the surface of the metal dissolves oxygen and carbon dioxide. To these must be added traces of sulphur dioxide downwind from the polluted atmospheres of industrial areas, and on ships, or near the sea, salt-spray. Iron is soon in contact with a solution of electrolyte and chemical depolarizer. In the initial stages of rusting, iron(II) ions pass into solution at anodic areas and hydrogen ions are discharged at the cathodic areas. The hydrogen ions originate from the water or suitable electrolytes, such as acids, present. Other ions migrate to the electrode areas and promote electrochemical action by producing a potential difference between solution and metal. Dissolved oxygen oxidizes hydrogen to water, acting as a depolarizer, and iron(II) ions to the iron(III) state. Hydroxide ions are present and iron(III) hydroxide or hydrated iron(III) oxide is formed as a powdery deposit of rust. Rust is a secondary product and affords no protection for the metal. It does not adhere to the metal surface and thereby impede corrosion but, by excluding oxygen, promotes

further rusting. Many other metals, such as aluminium and zinc, are protected by a surface coating of oxide. Iron and steel are so widely used that the fight against rusting is of paramount importance. Rusting is depicted schematically in Fig. 7.2.

Iron and steel are protected in many ways. Rusting may be inhibited, as in the motor-car industry, by the application of solutions containing phosphoric acid and orthophosphate which form an insoluble, tenacious film of iron phosphate. Films of oil and grease may be used on tools and machinery. Silicone grease, highly water repellant, is used in reels fitted to fishing rods. Ironware is sometimes protected by vitreous enamel which may also be decorative. Iron and steel structures are painted: red lead or zinc chromate paints being used as priming coats. But the application of metal coatings is particularly instructive in the study of

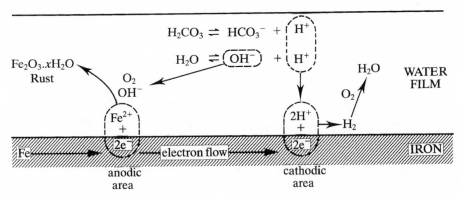

Fig. 7.2   Rusting

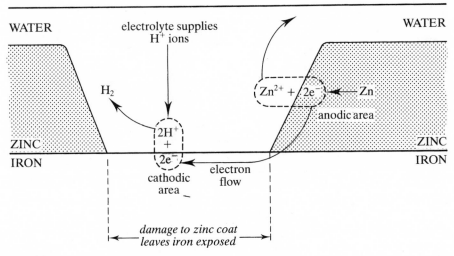

Fig. 7.3   Galvanized iron

electrolytic corrosion. A protective coat of zinc may be applied by galvanizing or sherardizing. Galvanized iron is obtained by immersing pickled (i.e. acid-cleaned) metal in molten zinc at 450°C. Sherardizing is accomplished by heating articles in zinc dust to about 350°C. Tin-plate is formed by immersion of the sheet steel in molten tin. What happens when the protective film of zinc or tin breaks down?

Consider galvanized iron which has been imperceptibly damaged but with the underlying iron exposed to the action of a solution of electrolyte as shown in Fig. 7.3. Zinc passes into solution and hydrogen is discharged from the surface of the iron. This happens because zinc is the more electropositive metal and the hydrogen overvoltage at an iron surface is relatively low. Dissolved gases provide the electrolyte. Zinc is thus sacrificed in the protection of iron. Where the surface coating is undamaged, zinc is protected from corrosion by a firmly adhering layer of zinc oxide. When tin-plate is damaged, rusting occurs. Iron, being the more electropositive metal, passes into solution and forms rust. Hydrogen ions, originating from the solution of electrolyte, are discharged at the adjacent tin coating, where dissolved oxygen acts as a depolarizer. Electrons pass from iron to tin which are therefore anodic and cathodic areas respectively. Tin-plate is used in canning food. When the coating is undamaged, the tin surface, having a high hydrogen overpotential (0.53 V), withstands the attack of acids present in fruit juice (etc.).

When damage to the surface has occurred, as shown in Fig. 7.4, iron is sacrificed and poisonous tin compounds still do not pass into solution. Zinc cannot be used for the food industry because of the poisonous nature of its salts although traces of zinc (in common with other metals) are required in the diet for healthy living.

In the examples considered, the anodic parts of the galvanic couples corrode away while the cathodic areas undergo protection. Such cathodic protection may be used to protect underground pipelines made of steel. Pieces of magnesium alloy are buried at intervals along a pipeline and connected to it by insulated conducting wires. The soil contains water and electrolyte. Magnesium dissolves and hydrogen is discharged at the iron. In oil refineries, magnesium alloys are used to protect the inside of distillation and cracking plant. The pieces of alloy, bolted in position, dissolve preferentially in sulphuric acid produced from the breakdown and oxidation of sulphur compounds. Such a refining plant is in continuous use and even a very low sulphur content could build up into an awkward corrosion problem between the annual maintenance inspections.

The electrochemical series and the properties of metals

The order in which elements are listed in the table of standard electrode potentials (Table 7.1) indicates the relative tendency of the elements to

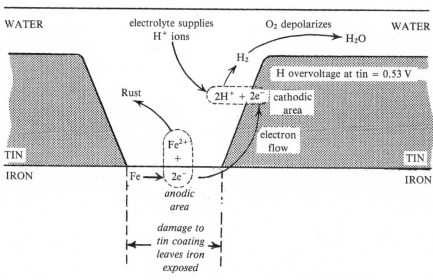

Fig. 7.4   Tin-plate

form ions in aqueous solution of molar (1.0 M) concentration of the ions at 298K (25°C). These ions are often hydrated. The list of elements in order of standard electrode potentials is called the electrochemical series, electromotive series or activity series.

In a convenient qualitative manner, the electrochemical series helps to summarize certain very important properties of elements involving the formation of ions. Metals form ionic compounds chiefly and the discussion will be limited to reactions involving ions. The electrochemical series shows only the tendency for ions to form. Many factors influence the rate of a reaction and the nature of the final product, including the temperature, the relative energies (strictly, free energies) of formation of reactants and products, lattice energies and the size of ions. The relative tendency to lose electrons (reducing power) in reactions involving solids or gases is roughly the same as in aqueous solution in spite of the wide differences in conditions.

Displacement of metal by metal

The displacement of copper from copper(II) sulphate solution by zinc has been interpreted already in terms of electrode potentials and the Daniell cell:

$$Zn + Cu^{2+} \rightarrow Zn^{2+} + Cu$$

The value of each electrode potential differs from the standard electrode potential by a term involving the absolute temperature and logarithm of the ionic concentration; at 298 K (25°C),

$$E = E_{298}^{\ominus} + \frac{0.0591}{2} \log_{10} m_{M^{2+}} \, mol \, dm^{-3}$$

As displacement proceeds the value of the electrode potential changes with change in concentration. Although copper(II) ions are never completely displaced from solution, displacement in general will be effectively complete from the analytical standpoint when the difference between standard electrode potentials is not less than 0.4 V. Selective electrodeposition of copper from a dissolved alloy (using concentrated nitric acid, etc.) for which a rotating platinum cathode may be conveniently used exploits this difference. Simple displacement experiments may be devised to establish the order in which metals displace each other from solutions of salts. Metals very high in the series displace

hydrogen from water anyway and cannot be arranged in this simple way. The activity of aluminium is suppressed by its thin surface layer of extremely impervious oxide, even after cleaning with emery paper. A metal high in the series displaces one lower down:

$$Cu(s) + 2Ag^+ \rightarrow Cu^{2+} + 2Ag(s)$$

$$Fe(s) + Cu^{2+} \rightarrow Fe^{2+} + Cu(s)$$

$$Zn(s) + Fe^{2+} \rightarrow Zn^{2+} + Fe(s)$$

Displacement reactions are actually reversible but the equilibrium position lies well to the right. However, under exceptional circumstances the reactions may be reversed. In the presence of excess aqueous potassium cyanide, which forms complex cyanides with copper and zinc, $Cu(CN)_3^{2-}$ and $Zn(CN)_4^{2-}$ respectively, the former being a copper(I) complex, copper will displace zinc. The ionic concentration of copper is reduced to an extremely small value by the formation of the cyanocuprate(I) ion.

In the solid state, sodium will displace aluminium from the chloride on heating,

$$3Na + AlCl_3 \rightarrow Al + 3NaCl$$

and aluminium will displace iron from iron(III) oxide in the thermite process,

$$2Al + Fe_2O_3 \rightarrow Al_2O_3 + 2Fe$$

Displacement of hydrogen from dilute acids and water

Metals above hydrogen in the electrochemical series displace it from solutions of dilute acids. The difference of electrical potential between metal and solution is an indication of the vigour of the reaction: the higher the metal and the larger the concentration of hydrogen ions, the more vigorous the reaction. Metals high in the series (e.g. magnesium) are very active. Intermediate metals (e.g. lead) just above hydrogen are very slow in action. (Magnesium is highly electropositive while lead is much less so.) The reaction rate will also depend on the solubility of the salt formed. An insoluble coating of lead(II) sulphate soon stops the reaction between lead and dilute sulphuric acid. Metals below hydrogen do not displace it from acids except in one or two exceptional circumstances, which emphasize the equilibrium nature of the displacement reaction. Thus, hydrocyanic acid, which is very weakly dissociated (ionized) in solution, reacts with copper to form the cyanocuprate(I) complex, reducing the copper(I) ion concentration to a very low value indeed:

$$2Cu + 2H^+ \rightleftharpoons 2Cu^+ + H_2$$

$$Cu^+ + 3CN^- \rightleftharpoons Cu(CN_3)^{2-}$$

removal of Cu^+ ions leads
to copper reacting and
hydrogen being liberated

When a metal is immersed in water, cations are flung off from the metal lattice until the potential difference established halts further net transfer. Without overpotential, the potential difference between a cathode and a chemically neutral solution (i.e. in which $[H^+] = [OH^-] = 1 \times 10^{-7} \, mol \, dm^{-3}$) must be at least -0.41 V, the calculated value of the electrode potential of hydrogen in neutral solution, for discharge of hydrogen. Therefore, only metals in the upper part of the table will displace hydrogen. Potassium and sodium react vigorously, the former catching fire, while calcium reacts quietly. The products are soluble hydroxides, alkalis:

$$2K + 2H_2O \rightarrow 2KOH + H_2$$

$$2Na + 2H_2O \rightarrow 2NaOH + H_2$$

$$Ca + 2H_2O \rightarrow Ca(OH)_2 + H_2$$

Magnesium reacts only very slowly with cold water, because of the film of insoluble magnesium hydroxide which is formed. However, the reaction is faster in hot water, especially with the metal in a powdered form. Aluminium is unreactive until the surface layer of oxide is penetrated. This can be done by amalgamation: immersion of the metal in hot aqueous mercury(II) chloride† leads to displacement of mercury which is removed from the surface of the aluminium by rinsing, wiping and rubbing with emery paper. Aluminium, cleaned in this way, reacts with water to give filaments of gelatinous aluminium hydroxide. Iron reacts with steam reversibly on heating to form tri-iron tetroxide and hydrogen,

$$3Fe + 4H_2O \rightleftharpoons Fe_3O_4 + 4H_2$$

Oxide and hydroxide formation

Metals high in the series have a high affinity for oxygen. Cold potassium and sodium are oxidized by dry air in a matter of minutes. Aluminium, cleaned as above, becomes very warm within minutes and loose white flakes of oxide cover the surface: a thin piece, e.g. a milk bottle top, soon becomes too hot to hold. Zinc and lead are only superficially oxidized. The values of heat of formation (enthalpy change of formation) of oxides roughly follows the order in the electrochemical series. Heating increases the rate of oxidation of the metal: the higher metals burn: the lower ones form a layer of oxide.

The highly electropositive metals are difficult to recover from their oxides. Extraction is usually by electrolysis of fused salts. Intermediate metals may be recovered by reduction with hydrogen or other reducing agents:

$$Fe_3O_4 + 4H_2 \rightleftharpoons 3Fe + 4H_2O$$

$$ZnO + C \rightarrow Zn + CO$$

$$PbO + H_2 \rightarrow Pb + H_2O$$

The noble metals form oxides which are readily decomposed by heat:

$$2Ag_2O \rightarrow 4Ag + O_2$$

The hydroxides of the alkali metals and alkaline earths are very stable and difficult to decompose thermally. Those of the intermediate metals are fairly readily decomposed, some being hydrated oxides and only nominally hydroxides, e.g. iron(III) hydroxide, hydrated iron(III) oxide. Hydroxides are not formed by metals very low in the electrochemical series, the oxide and water resulting from their attempted formation. Silver is such a metal:

$$2Ag^+ + 2OH^- \rightarrow Ag_2O(s) + H_2O$$

† Care—very poisonous

8

The principles governing the extraction of elements

The nature of the problem

Very few elements occur uncombined in the Earth's crust. The estimated relative abundance of the more frequently occurring elements is listed in Table 8.1, but the economic desirability of extraction of a particular element will depend on its localized abundance. The method employed will be determined by the nature of the compound recovered, the element extracted and other minerals associated with it.

An **ore** is a naturally occurring substance from which an element may be extracted.

The process in which the ore is used must be economically worth while, although with strategic materials simple business profit may not be the motive for the extraction. Ores are usually mined.

A **mineral** is a substance obtained by mining but the term is often extended to inorganic substances generally. It may be a single element or compound or a complex mixture.

Oxygen (50%) and silicon (26%), both non-metals, head Table 8.1. Hydrogen, unique among the elements, occurs abundantly as the oxide,

water. Other non-metals, even if they occur in quantities large enough to be included in the table, are very much less abundant. Broadly, metals are found united as oxides, sulphides, chlorides and to widely different extents, as oxosalts, including silicates, phosphates, carbonates, sulphates and nitrates. The less reactive metals, to some extent called noble metals to distinguish them from the more reactive base metals, are sometimes found natively. They include copper, gold, mercury, and also tin. Because oxygen occurs in such abundance and in electronegative character is second only to fluorine, oxides may be readily formed from other compounds. Metals are often extracted by reduction of their oxides.

The extraction of a metal from its ore by a process involving melting is called **smelting**.

Because there are far more metals than non-metals, smelting assumes a paramount position in the extraction of elements. In a general way, the extraction of a metal M from a compound MX may be written:

$$MX + Y \rightleftharpoons M + XY$$

where the left to right process is smelting.

The art of working metals and of extracting metals from ores forms part of the science of **Metallurgy**.

A comparison of chemical reducing agents: chemical affinities

$$MX + Y \rightleftharpoons M + XY$$

The possibility of this reaction proceeding to the right and to completion, with recovery of the metal M, depends on the relative avidities of M and Y for X. The discussion need not be confined to the extraction of metals.

Table 8.1 The relative abundance of elements in the Earth's crust (estimated weight per cent)

Oxygen	50	Potassium	2.4
Silicon	26	Magnesium	1.9
Aluminium	7.5	Hydrogen	0.9
Iron	4.7	Titanium	0.6
Calcium	3.4	Chlorine	0.2
Sodium	2.6	Phosphorus	0.1
Mn, C, S, Ba, Cr, N, F, Zr, Zn, Ni, Sr, V together			0.45

(D. T. Gibson, *Quarterly Review Chemical Society*, 1949, **3**, 266)

The tendency of such a reaction to occur is known as the **affinity of the reaction**.

The first concise treatment of this subject came from Geoffroy who published a table of affinities in 1718. This was extended by Bergman in 1775 but the relative nature of affinities and other factors affecting chemical reactions were clearly expounded for the first time by Berthollet in 1801-3.

While the tendency of a reaction to proceed in the direction indicated may be determined by the study of energies (Thermodynamics), other reasons may prevent it from happening. There can be an Energy Barrier. For a reaction to occur this barrier or 'energy hill' must be crossed just as the summit of a hill has to be reached before an object can roll down the other side. The rate of a reaction depends on the states of reactants, their active masses, i.e. partial pressures or concentrations (strictly, 'activities'), temperature, pressure and the presence of catalysts and light. Secondary chemical reactions may lead to the rejection of an otherwise promising method of extraction. Finally, chemical reduction may prove inconvenient or impossible for the highly reactive metals. Their cations are reduced at the cathode during electrolysis,

$$M^{n+} + n\,e^- \rightarrow M$$

Consider the chemical reduction of oxides, which may be expressed:

$$MO + Y \rightleftharpoons M + YO$$

where the reversible arrows indicate that relative affinities may change with conditions. Whichever element, M or Y, has the greatest affinity for oxygen will be united with it at the end, or with a slight difference in emphasis, the less stable oxide will be reduced. For effective comparison the relative chemical avidity of the element for oxygen, or the affinities of the reaction, must be measured in some standard way.

At first sight it might seem that the heats of formation of corresponding compounds formed by a sequence of elements would yield the required comparison. J. Thomsen and M. Berthelot, who did much experimental work in Thermochemistry, considered this to be so. Berthelot in 1875 listed three principles of Thermochemistry, the third of which was the Principle of Maximum Work: 'all chemical changes occurring without the intervention of outside energy tend toward the production of bodies, or of a system of bodies, which liberate more heat.' This attractive notion is false.

Table 8.2 Enthalpy changes of formation at 298 K (25°C) under constant pressure conditions: $\Delta H^{\ominus}_{298}/\text{kJ mol}^{-1}$

KCl (s)	−436.3	CaO (s)	−636.0
NaCl (s)	−411	Al_2O_3 (s)	−1671
AgCl (s)	−127	Fe_2O_3 (s)	−823.0
PCl_5 (g)	−399	PbO_2 (s)	−277
PCl_3 (g)	−306	CO_2 (g)	−394
HCl (g)	−92.6	CO (g)	−111
HBr (g)	−36	NO_2 (g)	+34
HI (g)	+26	NO (g)	+90.4
KF (s)	−563.0	SO_2 (g)	−297
NaF (s)	−569.4	SO_3 (g)	−396

Note the physical states quoted

Thermochemical data for gases will differ according to whether constant volume or constant pressure conditions appertain, but only constant pressure can be used for liquids and solids. Berthelot applied the terms exothermic and endothermic to describe reactions which proceed with the respective evolution and absorption of heat by the systems. Data are listed in Table 8.2 for the heats of formation (or more usually, enthalpy changes of formation) of 1 mole of each of certain compounds from the constituent elements in their standard states at 298 K (25°C) under constant pressure conditions.

ΔH symbolises the change in heat content of the system, or heat of reaction (enthalpy change of reaction) under standard constant pressure conditions; it is expressed in the units kJ mol^{-1}, a positive sign indicating heat absorbed by the system (the acquisitive convention), and a negative sign, heat evolved. Temperature change alters the value of ΔH according to the different specific heat capacities of reactants and products at constant pressure.

$\Delta H^{\ominus}_{298}$, called the standard enthalpy change for a reaction, refers to the amounts shown in the equation, which should be given, at a pressure of 101 kPa (1 atmosphere, 101325 N m^{-2}), a temperature of 298 K (298.15 K) and with the substances in their normal physical states under these conditions. Solutions must have a concentration of 1 mol dm^{-3}. The sign convention is given above. Most of the reactions occur with evolution of heat but some incur the absorption of heat. The existence of stable compounds formed endothermically, and the fact that some processes are reversible, so that heat is evolved in one direction and absorbed in the other, show immediately that the simple comparison of heats of reaction is insufficient to illustrate the principle of reaction affinities.

The affinity of a reaction, which is the measure of the tendency of the reaction to occur spontaneously, is now defined by a thermodynamical function called the Gibbs' Free Energy, G. While the detailed treatment of the thermodynamics of smelting is undesirable in this book, and done sketchily might prove a hindrance to a later treatment, some qualitative discussion may be developed.

Not all of the energy in a reactant is available for chemical change, some energy, called Bound Energy, being fixed in the substance under the given conditions. Energy is required for the rupture and formation of bonds and the assembly of the new ionic or molecular lattice of solids, or of the liquid state. The Available Energy is called the Gibbs' Free Energy. When a reaction occurs the energy change revealed by the heat evolved or absorbed is the result of the algebraic sum of the changes in Gibbs' Free Energy, available for reactions, and the Bound Energy, which also changes. Only the change in Gibbs' Free Energy is a measure of the spontaneous tendency of the reaction to occur, or affinity.

A loss in **Gibbs' Free Energy** is a thermodynamical measure of the tendency of a reaction to occur by itself ($=$ the affinity of the reaction) and it is measured by the maximum amount of work (in excess of pressure-volume work) which could be extracted by using the system reversibly at constant temperature and pressure, the change continuing until the Gibbs' Free Energy reaches a minimum.

As mentioned before, a reaction which is potentially possible from Energy studies will not necessarily occur.

$\Delta G = \Delta H - T\Delta S$

Some readers may need an introduction to the above thermodynamical equation. G, H, and S are known as extensive properties: they are proportional to the amount of the phase. G is the Gibbs' function (Gibbs' Free Energy) and a process occurs if $\Delta G < 0$ i.e. it is characteristic of G that at constant pressure and temperature a process can only occur in the direction in which G decreases. H is the enthalpy and ignoring any work done on the system by the surroundings, the energy the system gains in any process conducted at constant pressure is ΔH. As in the foregoing discussion these are

related mathematically by the Gibbs-Helmholtz equation (at constant pressure)

$$H = G - T\frac{\partial G}{\partial T}$$

Thus H and G differ except when either $T = 0$ or $\partial G/\partial T = 0$, i.e. at absolute zero or when G does not alter with temperature change.

The term entropy, S, may be defined as

$$S = \frac{\partial G}{\partial T}$$

so that $H = G + TS$

or $G = H - TS$

It is easier to study changes in H and S (i.e. $-\partial G/\partial T$) than changes in G directly

$$\Delta G = \Delta H - T\Delta S$$

This is the beginning of a treatment which leads to the prediction of reactions which *could* happen, other factors being favourable.

Entropy is regarded as some sort of measure of the disorder of a system which is a colloquial way of saying that while the total quantity of energy remains constant, it tends to be re-distributed in increasingly random ways. There are several ways of expressing the term entropy (and entropy change). Boltzmann first defined it in a statistical form relating to gases

$$\Delta S = S_2 - S_1 = k\ln\frac{w_2}{w_1}$$

in which w_1 and w_2 represent the number of ways of achieving the states between which the change occurs, k is the Boltzmann constant, the gas constant (R) for one molecule ($R = Lk$, where L is the Avogadro constant) and $\ln$ ($=\log$) the natural logarithm function.

Gibbs' Free Energy may be briefly considered in relation to different types of reaction. For a reaction to happen spontaneously, $\Delta G < 0$

i.e. $\Delta H - T\Delta S < 0$

Now, ΔH and ΔS may in theory be either positive or negative, giving four different combinations.

(a)   If the reaction is exothermic, ΔH (negative) < 0 and if ΔS is positive, making $-T\Delta S < 0$, the condition $\Delta G < 0$ is fulfilled. Such reactions occur widely.

(b)   If on the other hand ΔH is positive and ΔS negative, the reverse of the conditions above, $\Delta G > 0$ and no reaction occurs. In any case such a reaction would be the reverse of that above.

(c)   If ΔH and ΔS are both positive, the reaction being endothermic and if we assume little variation in the values of these terms with change in temperature, the criterion for the reaction to occur will only be fulfilled when the product $T\Delta S$ exceeds the value of ΔH. Warming a saturated solution (i.e. increasing T) to dissolve more of a salt provides a simple illustration of the change from the order of the crystalline state to the disorder in the solution.

(d)   For exothermic reactions, ΔH negative, in which ΔS is also negative an interesting conflict arises, because the positive value assumed by the product $-T\Delta S$ (ΔS being negative) could prevent the negative ΔH from ensuring $\Delta G < 0$. This happens at high temperatures, if we assume little variation in the values of ΔH and ΔS. Such a reaction will proceed when the temperature has been lowered, thereby reducing $-T\Delta S$, but another problem then arises. A reaction may be favoured thermodynamically but it may prove too slow for exploitation! A not uncommon problem encountered in the chemical industry.

Briefly the change in Gibbs' Free Energy may be computed direct from reaction kinetics, or when a reaction can be arranged to occur in an electric cell by measuring the reversible electrical work ($\Delta G = -nFE$, where E V = potential difference and nF = number of Faradays of change transferred per mole) or indirectly from thermochemical calculations based on experimental measurements.

The smelting of oxides

Standard Gibbs' Free Energies may be assigned under arbitrarily standard conditions where active masses equal unity and the temperature is fixed. For gases, it is usual to take the standard state as that of ideal gases at 101kPa (1 atmosphere) pressure. Gibbs' Free Energies of Formation at 298K (25°C) for some compounds are listed in Table 8.3. Syntheses with the greatest affinity show the greatest changes. These energy changes may be added algebraically in thermochemical calculations. An element with potentially the greater change in Gibbs' Free Energy on formation of an

Table 8.3   Gibbs' Free Energy of Formation at 298 K (25°C) under constant pressure conditions: ΔG/kJ mol^{-1}

KCl (s)	−409	CaO (s)	−604.7
NaCl (s)	−384	Al$_2$O$_3$ (s)	−1577
AgCl (s)	−110	Fe$_2$O$_3$ (s)	−741.5
PCl$_5$ (g)	−325	PbO$_2$ (s)	−219
PCl$_3$ (g)	−286	CO$_2$ (g)	−395
HCl (g)	−95.5	CO (g)	−137
HBr (g)	−53.2	NO$_2$ (g)	+51.9
HI (g)	+1.3	NO (g)	+86.7
KF (s)	−533.5	SO$_2$ (g)	−301
NaF (s)	−541.4	SO$_3$ (g)	−371

Note the physical states quoted

oxide will reduce the oxide of an element with a lesser change.

To sum up, the relative affinities of reactions may be compared using the thermodynamical Gibbs' Free Energies at stated temperatures. Table 8.4 shows the synthesis of various oxides listed so that the reaction with the greatest tendency to occur at the stated temperature is highest in the appropriate column. The reaction is indicated by the symbol of the element combining with oxygen. The first column shows the order of the same elements in the electrochemical series. Metallurgical operations occur over a range of 0–2000°C roughly, and a comparison has been made at 500° intervals, based on the work of H. J. T. Ellingham. A more lucid and rigorous treatment has appeared in D. J. G. Ives, *Principles of the Extraction of Metals* (Royal Society of Chemistry Monographs, No. 3).

In Table 8.4, the order is roughly that of the electrochemical series, based on standard electrode potentials at 298K (25°C), where metal ions leave the metal lattice to become hydrated ions in solution or return by the reverse process. It would be altogether surprising if this order was identical to that obtained by study of the thermodynamical reaction affinities. Carbon needs special mention because of its two oxides, the process with most affinity being shown in rounded brackets, that with least in square brackets. At 710°C there is a changeover from excess of one oxide to excess of the other in the gaseous product from the combustion of carbon; below 710°C, the equilibrium mixture has excess of the dioxide and above this temperature, excess monoxide. With a sufficiently elevated temperature, all but calcium and aluminium should be reduced by carbon but various other factors make the extractions difficult. The equilibrium mixture must be cooled rapidly to

Table 8.4   Affinities of oxide formation compared

E.C.S. order		0°C	500°C	1000°C	1500°C	2000°C
Ca		$O_2 + 2Ca \rightarrow 2CaO$	Ca	Ca	Ca	Ca
Na		$O_2 + 2Mg \rightarrow 2MgO$	Mg	Mg	Al	Al
Mg		$O_2 + \frac{4}{3}Al \rightarrow \frac{2}{3}Al_2O_3$	Al	Al	Mg	(C → CO)
Al		$O_2 + 4Na \rightarrow 2Na_2O$	Mn	Mn	(C → CO)	Mn
Mn		$O_2 + 2Mn \rightarrow 2MnO$	Na	Cr	Mn	[C → CO$_2$]
Zn		$O_2 + \frac{4}{3}Cr \rightarrow \frac{2}{3}Cr_2O_3$	Cr	{Na*	Cr	Cr
Fe	decreasing affinities (all reactions require 1 mole of oxygen as stated)	$O_2 + 2Zn \rightarrow 2ZnO$	Zn	{(C → CO)	[C → CO$_2$]	Mg
Cr		$O_2 + 2CO \rightarrow 2CO_2$	CO → CO$_2$	{[C → CO$_2$]	Fe	Fe
Ni		$O_2 + Sn \rightarrow SnO_2$	Fe	{Zn	H$_2$	H$_2$
Sn		$O_2 + Fe \rightarrow 2FeO$	H$_2$	Fe	CO → CO$_2$	Pb
Pb		$O_2 + 2H_2 \rightarrow 2H_2O$	(C → CO$_2$)	H$_2$	Zn*	CO → CO$_2$
H		$O_2 + 2Ni \rightarrow 2NiO$	[C → CO]	CO → CO$_2$	Pb	Ni
Hg		$(O_2 + C \rightarrow CO_2)$	Ni	Ni	Ni	—
Ag		$O_2 + 2Pb \rightarrow 2PbO$	Sn*	Pb	—	—
		$[O_2 + 2C \rightarrow 2CO]$	Pb	—	—	—
		$O_2 + 2Hg \rightarrow 2HgO$	dec	—	—	—
		$O_2 + 4Ag \rightarrow 2Ag_2O$	dec	—	—	—

(to the corresponding oxidation states)

*Data incomplete over whole temperature range

avoid re-oxidation. Carbide formation may occur and there is the problem of finding suitable furnace linings. Aluminium may be used in the thermite process for the production of a number of metals.

For sulphides and chlorides the direct recovery of metals may also be studied from the thermo-dynamic standpoint.

Oxidation as a means of isolating some elements

The isolation of metals involves reduction. There are far fewer non-metallic elements and their extraction methods vary widely. Some require reduction, some physical separation and others, oxidation. Molten sulphur is pumped up from the depths of the earth. Oxygen, nitrogen and inert gases are separated from the air, except for helium which is obtained from natural gas. Phosphorus is produced by reduction of the pentoxide, displaced from calcium phosphate by silica, using carbon in the electric furnace. The halogens are recovered from halides by oxidation, chemical for bromine but electrolytic for fluorine and chlorine, using molten and aqueous electrolyte respectively:

fluorine: $2F^- \rightarrow F_2 + 2e^-$

chlorine: $2Cl^- \rightarrow Cl_2 + 2e^-$

bromine: $Cl_2 + 2Br^- \rightarrow Br_2 + 2Cl^-$

The application of electrolysis in extraction processes

Very reactive metals, such as sodium, calcium and aluminium, and the most reactive non-metallic element, fluorine, are produced commercially by the electrolysis of fused compounds. The principal method for the production of chlorine is the electrolysis of aqueous salt (sodium chloride) while about half of the world's production of zinc is by electrolysis of an aqueous solution of the sulphate under carefully controlled conditions. In the previous chapter, attention was drawn to the contributions of standard electrode potentials, ionic concentrations, overpotentials and polarization to the decomposition potential. The extra electrical power required to overcome internal resistance and polarization is dissipated as heat, which helps to maintain the electrolyte in the molten state. The electrical energy of the source therefore supplies heat energy to the electrolyte.

Work done or energy supplied/J

= electric charge passed/C
× potential difference/V

= electric current/A × time/s
× potential difference/V

The power (i.e. rate at which work is done)/
W ($=Js^{-1}$)

$$= \text{electric current/A}$$
$$\times \text{potential difference/V}$$

The quantities of products liberated at electrodes are proportional to the charge passed (Faraday's First Law).

The ideal electrolyte will have a low melting-point and a high electrical conductance. Usually, the melting-point is too high and to avoid the liberation of metal at too high a temperature, perhaps very near to the boiling-point, salts are added to lower the melting-point. A knowledge of the discharge potentials under the cell conditions is required in choosing the compound to be added. Lowering the temperature of the electrolyte reduces heat losses and lowers operating costs. To achieve separation of the metal from the electrolyte by tapping the furnace or ladling, there must be a difference in densities and immiscibility, to give the separate layers. If a metal disperses as a fog in the electrolyte, secondary reactions may occur with lowered efficiency. Suitable additions may bring the desired density change. Where the products of electrolysis react together, cell design must achieve the separation of the products. The metal may be protected by molten or solidified electrolyte, or by physical barriers built into the cell design.

Electrolysis may be used in the refining of crude metal obtained by smelting. In copper refining, this enables the valuable, more noble metals, present in traces, to be recovered.

Other problems to be studied

Assuming that no effective kinetic reasons preclude extraction, the systematic study of chemical affinities for chemical reduction and electrode processes for electrolysis provide the principles of the extraction method. But the final decision depends on the cost, which turns on a number of variables which it will be of interest to mention. The quality of the ore is obviously important and that portion required may be but a few parts per hundred. The geographical location of the ore deposits brings into the reckoning transportation problems and costs, the proximity of fuel, coal or oil, electrical power and labour. Low-grade ores may not be economically worthwhile exploiting for the present or the economic viability may fluctuate with the prevailing world price: witness the experience of the companies mining for tin in Cornwall, UK. The demand for the product determines the scale of operation. Large-scale production requires a plentiful, and therefore, cheap reducing agent, such as coke or, in some regions, electricity. Expensive reducing agents may be used for relatively small amounts of high-purity product. It must be decided whether a multistage process followed by a separate refining of the metal or a single operation leaving metal of the required purity is the best investment. An ore consists of more than one element in combination so that by-products may prove important or bring special problems on account of their poisonous or pervading nature. Arsenic trioxide, volatilized during the roasting of zinc sulphide, must be precipitated (using electrostatic precipitation). Arsenic renders metals brittle. Zinc smelting using the sulphide ore also illustrates the other point: sulphur dioxide is liberated during roasting and converted into sulphuric acid. The zinc smelter is also the manufacturer of sulphuric acid. This has advantages: during the post-war decades when a world surplus of zinc appeared and prices fell to uneconomical levels, the extraction was continued because of the shortage of sulphuric acid. Special properties may be used: nickel has been largely extracted from low-grade ore by the formation of the volatile nickel carbonyl, subsequently thermally decomposed. In times of national crisis, the commercial profit motive recedes and special techniques may have to be evolved as in the production of uranium. This point may also be illustrated by the war-time chemical reduction of magnesium oxide by coke at 1900°C with 'crash-cooling' to avoid regeneration of magnesium oxide from the equilibrium shift.

An outline of extraction methods

Preliminary treatment

This will depend on the nature of the ore and may include physical processing, called mineral (or ore) dressing, or chemical treatment, or both.

Ore is usually associated with siliceous matter—rock, clay, earth. Physical separation of this *gangue* will avoid wasteful smelting of worthless material. The study of physical processing has become the science of **Mineral Dressing**. With improved techniques very low-grade ores may be profitably worked.

At the simplest level, preliminary treatment requires hand sorting of roughly crushed material moving on a conveyor belt through a group of hand operatives. If labour is expensive, a higher initial outlay is made and mechanical sorting used. The material is crushed and ground wet, avoiding the formation of slimes and with sieving to avoid over-pulverization. This requires care because different portions of the ore may have different resistances to crushing and grinding. Material susceptible to magnetic fields may be separated by application of electromagnets. Density differences may be enough to allow separation by washing techniques in which crushed material is agitated on a sloping corrugated table on which jets of water play across the line of inclination, so that the dense particles accumulate at the lower end. All of the techniques mentioned are applied in the metallurgy of tin.

Froth Flotation is used extensively in the preliminary treatment of minerals. Taking care to avoid the formation of colloidal slimes, crushing and wet grinding are used to separate the components of the ore. Separation may be effected by using their different wetting characteristics with oil and water. The slurry is treated with oil and suitable additives while compressed air is introduced into the base of each flotation cell so that the required minerals are carried up in the water–air interface of the foam. The stream of air also serves to agitate the mixture. The froth spills over and the mineral is separated and collected. A series of flotation cells is required to remove the gangue effectively. The concentration of copper sulphide ore requires the use of hundreds of cells. Various additives are used: 'frothers' to stabilize the froth, 'collectors' to modify the characteristics of grains and the activation or depression of flotation properties by other reagents. Pine oil is frequently used as a frother, the active constituent being terpineol which has both hydrophilic (water-loving) and hydrophobic (water-hating) parts to the molecule. Ethyl xanthate and potassium ethyl xanthate are examples of collectors which attach themselves by polar groups to grains of mineral which then become water repellent, ready for removal in the froth at the water–air interfaces. By suitable activation or suppression of flotation, an operator may achieve any degree of separation he desires. An example will illustrate this point. Separation of a mixture of galena (lead sulphide), sphalerite (zinc sulphide) and pyrite (iron sulphide) may be achieved in three stages. Galena may be floated by potassium ethyl xanthate with sodium cyanide and alkali acting as depressant, followed by flotation of sphalerite on addition of copper(II) sulphate to activate the grains by rendering them more insoluble in water and finally, acidification leads to the floating of pyrite. Constant supervision of froth flotation is necessary. Quantities of reagents added may vary between a fraction of a kilogram to several kilograms per tonne. The process is very vulnerable to the presence of traces of impurity. Control is maintained through frequent sampling.

Ion exchange and solvent extraction methods have been developed in the USA for the production of uranium, using tributylphosphate as extractant, and for the concentration and separation of the lanthanide and actinide elements. Amines are especially useful for the extraction of uranium compounds from sulphate leach liquors.

Chemical treatment may involve roasting in air to free the oxide from sulphide ore before smelting or the careful purification of a mineral prior to electrolysis so that deleterious impurities are not found in the product.

In spite of unfavourable reaction kinetics under normal conditions, several commercially economic ways of converting sulphide ores to aqueous solutions of sulphate have been developed recently. These are likely to be particularly important in the extraction of copper, zinc, lead, nickel and cobalt. One process involves the oxidation of sulphide with oxygen under pressure, as in the extraction of copper and nickel from concentrates of complex sulphides. In another, fluidized bed reactor techniques enable insoluble sulphide to be converted into water-soluble sulphate, as in the preliminary treatment for concentration of copper and cobalt as sulphates during the roasting of pyrite mineral in the production of sulphuric acid. Finally, sulphur-oxidizing bacteria have been discovered which promote the direct formation of sulphate from sulphide, a field of study which is attracting much research effort. *Thiobacillus oxidans* is active in the oxidation of copper sulphide mineral to copper(II) sulphate, and of zinc sulphide to zinc sulphate.

The main stage

The various methods of extraction will be outlined in a general way.

1 *Chemical reduction: smelting*

Chemical reduction with carbon (coke) is used in the smelting of the less reactive metals although in war-time it has also been used for the production of magnesium. The non-metallic element phosphorus is isolated by carbon reduction in the electric furnace. In smelting practice, fluxes may be added as a help in fusing the mass and impurities may be removed as slag by the judicious addition of basic (lime) and acidic (silica, acid silicates, acid borates) materials. Slag must be separated cleanly from the metal: it must be fluid, easily tapped, without action on the metal and the furnace lining, and have a melting-point below the boiling-point of the metal. Reduction may be brought about by a metal, as by aluminium in the thermite process, where, for example, manganese is liberated from an oxide:

$$3Mn_3O_4 + 8Al \rightarrow 9Mn + 4Al_2O_3$$

and the reduction of titanium(IV) chloride by molten magnesium (or sodium) under an inert atmosphere:

$$TiCl_4 + 2Mg \rightarrow Ti + 2MgCl_2$$

2   Autoreduction: smelting (air reduction)

The sulphide is partly converted to metal oxide by roasting in air, the supply of which is then cut off and heating continued. Oxide and sulphide interact to form metal. Lead was formerly produced in a reverberatory furnace this way:

$$2PbS + 3O_2 \rightarrow 2PbO + 2SO_2$$

$$2PbO + PbS \rightarrow 3Pb + SO_2$$

Copper from copper(I) sulphide, Cu_2S, antimony from its sulphide, Sb_2S_3 and mercury from mercury(II) sulphide ore, HgS, are similarly produced. Autoreduction is also called air reduction.

3   Electrolysis

This method may be used for the isolation of the most reactive metals from molten electrolytes, for fluorine and (sometimes) chlorine, and because of hydrogen overpotential, for the extraction of zinc from aqueous solution under suitable conditions. It is used in the refining of metals.

4   Displacement

Electrochemical displacement of one element by another is sometimes used in extraction. Bromine and iodine may be displaced by the action of chlorine. Bromine is liberated from the mother liquors after crystallization of the chlorides of potassium and magnesium:

$$Cl_2 + 2Br^- \rightarrow Br_2 + 2Cl^-$$

Zinc shavings precipitate silver from the cyanide complex,

$$Ag(CN)_2^- \rightleftharpoons Ag^+ + 2CN^-$$

$$2Ag^+ + Zn(s) \rightarrow 2Ag(s) + Zn^{2+}$$

In electrolytic refining of zinc, the liquid accumulating in the vats is rich in cadmium, which may be displaced by addition of zinc,

$$Zn(s) + Cd^{2+} \rightarrow Zn^{2+} + Cd(s)$$

One of the most important recent developments uses hydrogen under pressure for the reduction of nickel, copper and cobalt sulphates in aqueous solution to the metal in controlled grain size.

5   Special methods: miscellaneous

By exploiting special characteristics, individual extraction methods may be evolved. The extraction of silver and gold by complex cyanide formation followed by displacement with zinc is a good example of the application of complex ion study, the method also illustrating displacement. Nickel carbonyl formation and immediate decomposition is another example of a special method. Chlorination may be employed, as in the direct chlorination of antimony ore to produce the volatile antimony trichloride, which is condensed.

6   Finding deposits of the element

Very noble, unreactive metals may be found in native deposits. Gold prospecting has proved attractive. For non-metallic elements, diamond mining and sulphur extraction provide examples.

Refining the crude metal

Electrolysis has already been mentioned. Fractional distillation may be used to separate zinc and cadmium. Cupellation will be described in silver refining, where traces of lead are oxidized away in a hot air blast. Powder metallurgy and zone melting, in which an ingot is subjected to heating so that a zone of melting travels along it, carrying impurities away, are among the newer techniques.

Scrap metal as a source

The importance of scrap is often overlooked. Iron and steel scrap is used in abundance in steelworks. Tin may be recovered from scrap tinplate by chlorination when volatile tin(IV) chloride is swept away and condensed as the first stage:

$$Sn + 2Cl_2 \rightarrow SnCl_4$$

As people become increasingly conscious of the need to safeguard the environment and natural resources (and concerned, to say the least, at accumulating rubbish around our cities) the recovery, or re-cycling, of materials of all kinds will become increasingly important.

9

The extraction and uses of typical elements

The Presentation

In this chapter we study the production of the commercially more important elements and their uses, Group by Group. The transition metals, and those elements associated with them, appear in the next chapter. The production of nitrogen and oxygen (from liquid air) appears in Chapter 12, and chlorine (from salt) in Chapter 14. The methods adopted generally illustrate the principles set out in Chapter 8, and earlier. They reflect the changes in character from metallic to non-metallic, in electropositive to electronegative character across the Periodic Table, the clarity of this picture being particularly striking in Groups IV and V.

Group I: Lithium, sodium and potassium

Lithium

Occurrence

Although lithium is roughly twice as abundant as lead, economically worth while deposits of lithium compounds are comparatively rare.

The ores are complex. The principal sources are spodumene, lithium aluminium silicate,

$$LiAl(SiO_3)_2$$

lepidolite, a 'lithium mica'

$$K(Li,Al)_3(Al,Si)_4O_{10}(OH,F)_2,$$

petalite, another lithium aluminium silicate,

$$(Li,Na)AlSi_4O_{10},$$

and amblygonite, a fluorophosphate,

$$(Li,Na)AlPO_4(F,OH).$$

All are mined in Zimbabwe and Namibia. The first mineral mentioned also occurs in the USA.

Production

Lithium sulphate is made by roasting petalite, lithium aluminium silicate, with sulphuric acid. The carbonate is then precipitated under carefully controlled conditions and converted into the hydroxide in the Gossage manner, using slaked lime:

$$Li_2CO_3 + Ca(OH)_2 \rightleftharpoons CaCO_3 + 2LiOH$$

The least soluble compound, calcium carbonate, is precipitated and removed. On evaporation, the monohydrate of lithium hydroxide separates on cooling to 40°C. Carbon dioxide must obviously be excluded. The product is converted into lithium chloride by the action of hydrochloric acid.

Electrolysis of a fused mixture of lithium chloride (60%) and potassium chloride (40%) at 475–500°C with a steel cathode, separated from graphite anodes by a wire screen, yields lithium at the cathode. Lithium is the lightest solid known. It is greased before storage. Lithium is extremely reactive.

Uses

In addition to the uses of lithium compounds in organic synthesis, lithium metal is employed as a de-gasifier in metallurgy and by the German railways as a minor constituent (0.04% Li) with lead in Bahnmetall for anti-friction bearing alloy. Missile fuels have been developed from the lithium oxosalts, nitrate and perchlorate. Lithium hydride, because of its reactivity and low relative molecular mass (molar mass), is an extremely convenient portable source of hydrogen, generated by reaction with water, used for air–sea rescue work and in observation balloons. The hydrogen bomb and thermonuclear processes probably depend on lithium deuteride in conjunction with plutonium and uranium.

Sodium and potassium

A detailed treatment of sodium and an outline of potassium appears in Chapter 14, Chemicals from salt: the alkali industry.

Group II: Beryllium, magnesium and calcium

Beryllium

Occurrence

Beryllium is listed as 32nd in the order of estimated abundance of elements in the Earth's crust. The principal mineral is beryl, beryllium aluminium silicate, $2BeO.Al_2O_3.6SiO_2$. This is bluish-green in colour. The deep green translucent form is the precious stone, emerald. The ore is worked in Brazil chiefly but also in Argentina, Zimbabwe, South Africa and India.

Production

Beryl crystals, having about the same density as the accompanying silica, are hand-sorted. The beryllium content is roughly $2\frac{1}{2}\%$ with iron as the main impurity.

The first stage is the isolation of beryllium hydroxide. The ground ore is heated with lime to 1500°C to react with the alumina and silica present. Addition of concentrated sulphuric acid precipitates silica and calcium sulphate, giving a liquor from which ammonia alum is crystallized on addition of ammonium sulphate. Iron is precipitated as hydrated iron(III) oxide after oxidation of the iron (II) to the iron(III) state by sodium peroxide and adjusting the pH (=4.0) with chalk. Beryllium hydroxide is precipitated by addition of ammonia, removed and dried at 100°C.

The hydroxide is bricketted with charcoal using wood tar as a binder, baked to the oxide and chlorinated at 700–800°C in an electrically heated furnace when beryllium chloride volatilizes away with the impurities: chlorides of aluminium, silicon and iron(III);

$$BeO + C + Cl_2 \rightarrow BeCl_2 + CO$$

Purification is effected by fractional distillation in an atmosphere of hydrogen, which reduces any residual iron(III) chloride to the non-volatile iron(II) chloride.

Beryllium chloride, mixed with an equal weight of salt, is electrolysed in nickel cells heated externally to 350°C. Chlorine is evolved at the graphite anode and beryllium collects on the walls and bottom of the cell, which acts as the cathode. The metal is flaked off and sintered at 1400°C in hydrogen to give metal of 99.0–99.5% purity.

Uses

Beryllium copper (up to 4% Be), an alloy of tensile strength greater than that of other non-ferrous alloys and comparable to that of alloy steel, was the chief application until recently. The search for metals with the special characteristics needed for atomic reactor equipment has stimulated research into beryllium production, but it is difficult to work and when finely divided, is toxic. It has been used as a sheathing metal for atomic fuel cans to contain the radioactive metal and fission products, and in advanced gas-cooled reactors designed to work up to 600°C, as it absorbs relatively few neutrons, is chemically inert to both fuel and coolant gas and conducts heat well. Beryllium is used in X-ray 'windows', and in high-speed aircraft, having superior hardness, elasticity and strength to magnesium and aluminium but with a density slightly more than that of the former and only two-thirds that of the latter. Beryllium has been used in the construction of man-made Earth satellites and in rocket guidance equipment.

Magnesium

Occurrence

Magnesium forms an estimated 2.2% by weight of the Earth's crust. It is the eighth most abundant element and the sixth most abundant metal. Solid deposits are widespread, especially dolomite, $MgCO_3.CaCO_3$. Magnesite, $MgCO_3$, occurs in North America, Europe, India and the USSR. Brucite, $MgO.H_2O$, is found in Canada. Mineral deposits formed by the evaporation of inland seas, such as those at Stassfurt in Germany (the former DDR), contain double salts, e.g. carnallite, $MgCl_2.KCl.6H_2O$ and kainite, $MgSO_4.KCl.3H_2O$, and kieserite, $MgSO_4.H_2O$. However, sea water is the main commercial source of magnesium. It contains 0.13% Mg^{2+} by weight. Evaporation of sea water to roughly one-sixth of the original volume results in the separation of much common salt and a liquid remains which is rich in magnesium salts. These impart a bitter taste. The solution is called a 'bittern'.

Because of their strength, lightness and ease of working, magnesium alloys are of increasing importance, but comparatively few countries can produce magnesium economically. In the UK, magnesium production has fluctuated: it was economical during the period 1864–90 before German competition proved too strong, in the First World War (1914–18) and again during the period 1935–47. In 1951, production started again. The cheapest magnesium is formed by electrolysis of the fused chloride, in Norway and at Freeport, Texas, USA. Low cost of raw materials, including sea water and an abundant electricity supply, favours Norway, while at Freeport, sea water and the purest lime in inexhaustible supply with natural salt deposits, natural gas as a source of heat and electrical energy, provide ideal conditions. Natural advantages of mineral wealth and electric power make Canada and Italy important producers of magnesium by chemical reduction.

Production by electrolysis

The electrolysis of fused magnesium chloride yields magnesium and chlorine. Originally (1886) carnallite was used but the process was inefficient due to the hydroxide ions present in the basic salt. Hydrogen was liberated, thereby lowering the yield of magnesium, and wear occurred at the anodes due to the liberation of oxygen. Carnallite was replaced by synthetic magnesium chloride. Two methods of obtaining magnesium chloride will be described but the principal source is sea water.

In the Elektron process, magnesite is calcined at 500°C to give magnesium oxide,

$$MgCO_3 \rightarrow MgO + CO_2$$

The product is bricketted with fine coal using magnesium chloride as binder, coked in rotary kilns and chlorinated above the melting-point of magnesium chloride:

$$MgO + Cl_2 + C \rightarrow MgCl_2 + CO$$

$$2MgO + 2Cl_2 + C \rightarrow 2MgCl_2 + CO_2$$

The anhydrous fused magnesium chloride is transferred to the electrolysis pots as required.

In the Dow process, magnesium is extracted from sea water and from the residual liquors remaining after the recovery of chlorine, bromine, sodium and calcium from the brine wells in Michigan, USA. Sea water is run into settling tanks and mixed with milk of lime. This is formed by slaking the quicklime produced when oyster shells, composed of calcium carbonate and dredged from the ocean floor, are roasted. Magnesium hydroxide

is precipitated. Hydrochloric acid, formed by interaction of chlorine and natural gas, is added. Magnesium chloride is formed. The solution is evaporated to dry flakes, hydrolysis being avoided by leaving the salt partially hydrated.

Electrolysis is done in steel cells lined with fire brick and heated externally. The temperature is 670–730°C and the melt contains sodium chloride (50–60%) and calcium chloride (15%) as fluxes and to increase the density. Magnesium, released at steel cathodes, floats to the surface where it is protected from oxidation by an atmosphere of coal gas. Chlorine is liberated at graphite anodes and ducted away by a pipe surrounding the electrodes to avoid the interaction of the products.

The electrode processes are:

$$Mg^{2+} + 2e^- \rightarrow Mg \qquad 2Cl^- - 2e^- \rightarrow Cl_2$$

The current is 30 000–50 000 A at 6.6 V potential difference and the product, 99.5–99.9% magnesium, is ladled into moulds. Chlorine is re-cycled.

Production by chemical reduction

Magnesium oxide is very stable but processes have been developed to effect reduction. These necessarily require very high temperatures and, unlike electrolysis, have the disadvantage of being non-continuous.

The Hansgirg process requires reduction of calcined magnesite with coke,

$$MgCO_3 \rightarrow MgO + CO_2$$

$$MgO + C \rightleftharpoons Mg + CO$$

The reaction is brought about at 2000°C in an electric arc furnace with the quenching of the vaporized products to 200°C by hydrogen gas, or natural gas (where available), which is subsequently burnt. This avoids the reduction of carbon monoxide by magnesium. The metal, in powder form, is re-distilled at 800°C in a vacuum and the crystalline solid product collected.

Uses

Contrary to what might be at first thought, magnesium is not dangerously flammable. The pure metal is not strong enough for structural use by itself so a range of light alloys has been developed. An alloy is a substance with metallic physical properties formed from a metal by the addition of other elements, usually metals but sometimes metalloids or non-metals. Desirable modifications of magnesium can be brought about by the addition of aluminium, manganese, zinc and zirconium. The alloys are described in Chapter 11, Industrial alloys.

Magnesium is very reactive, and in addition to its being placed high in the electrochemical series there is no hydrogen overpotential. It is used in cathodic protection of other metals, including iron and steel structures: buried pipe-lines, buried structures of all sorts, ships' hulls and the insides of catalytic crackers and distillation plants. In certain circumstances, magnesium alloys require protection. Thus, at joints with other metallic surfaces where moisture cannot be excluded, special jointing compounds are used. Special chemical treatment with a chromate solution induces surface passivity, which can also be produced by electrolytic anodizing. However, epoxy resin paints provide the best protection.

Magnesium alloys are welded in an atmosphere of argon by the tungsten arc process. Alloys are used in manufacturing atomic fuel cans for gas-cooled reactors. Magnesium is used in the reduction of titanium(IV) chloride to the metal, in pyrotechnics and photographic flash work.

Calcium

After aluminium and iron, calcium is the commonest metal found combined in the Earth's crust. It is essential to both animals and plants. Calcium phosphate is the chief constituent of bones. Calcium carbonate occurs widely as limestone, chalk, marble, and also as calcite, Iceland spar, aragonite, etc., or with magnesium carbonate as dolomite, $CaCO_3.MgCO_3$. Calcium sulphate is mined as anhydrite, $CaSO_4$, and as gypsum or alabaster, $CaSO_4.2H_2O$. Calcium silicates are found in rocks.

Calcium is produced by the electrolysis of fused calcium chloride. It is used in the laboratory as a reducing agent, for removing traces of hydrogen and nitrogen from gases and as a deoxidizer in some metallurgical operations.

Group III: Boron and aluminium

Boron

The borax industry

Comparatively little elemental boron is produced, although boron compounds are used widely. The element's principal source lies in southern California, USA, where deposits of sodium borates,

kernite, $Na_2B_4O_7.4H_2O$, and (tincal) borax, $Na_2B_4O_7.10H_2O$, are mined from the Mojave Desert. The borax industry is described later. A black or brownish-black amorphous boron is produced by reduction of the oxide with sodium or magnesium, or fluoroborate with potassium; it is very difficult to purify:

$$B_2O_3 + 6Na \rightarrow 2B + 3Na_2O$$

$$KBF_4 + 3K \rightarrow 4KF + B$$

Almost pure crystalline boron is made by the reduction of boron bromide vapour with hydrogen at 1000–1300°C on a tantalum filament just below atmospheric pressure:

$$2BBr_3 + 3H_2 \rightarrow 6HBr + 2B$$

Uses

Among the elements which are readily available, boron is second only to cadmium as a neutron absorber. The element and boron carbide are used in the shielding of atomic piles, screening operators, and in the control rods used for regulating the chain reaction. Having a low density (2.3 g cm^{-3}), boron has strong claims for use in mobile nuclear reactor units e.g. in submarines and surface ships.

High-purity boron is used as a crystalline semiconductor material and has the advantage of functioning at higher temperatures than germanium and silicon, the limits for which are approximately 100°C and 300°C, respectively.

Boron is an essential minor element for the healthy growth of plants.

Aluminium

Occurrence

Aluminium forms an estimated 7.5% by weight of the Earth's crust. It is the most abundant metal, occurring as silicates in clay, granite, shale and slate, and as the hydrated oxide in the bauxites. The latter are formed by the weathering of clays in warm regions of abundant, intermittent rain falling on land of low relief so that the rain soaks into the deposits, washing out silica and leaving aluminium oxide. In 1821, Berthier discovered the first mineral of high aluminium content (52%) at Les Baux in France, which has led to the use of the name bauxite as a collective term for aluminium oxide deposits. The hydrated aluminium oxide of bauxite may occur in a number of forms: gibbsite, $Al_2O_3.3H_2O$, as böhmite and diaspore, $Al_2O_3.H_2O$.

They range in colour from creamy-white to reddish-brown, having an aluminium content of 50–65%. The main impurities are oxides of titanium, silicon and iron. White bauxite has 1–4% silica and no iron while red bauxites contain up to 27% red iron(III) oxide. Bauxite occurs in widespread but localized deposits which are 3–30 m deep. At present, the extraction of aluminium from clay is uneconomic and it is made solely from bauxite. The degree of hydration varies from $Al_2O_3 . H_2O$ in the French deposits to $Al_2O_3 . 3H_2O$ in Guyana. Deposits are exploited in south-eastern Europe, in the USSR, China, India and south-eastern Asia, Australia, the West Indies, Brazil and southern USA.

Bauxite is quarried by drilling and blasting. Coarse impurities are removed and before transportation it is washed, dried and graded.

The only known source of the other aluminium mineral, cryolite, is at Invigtut in Greenland from where it is exported to the UK. Much sodium aluminium fluoride, or synthetic cryolite, is made industrially for the extraction process.

Preliminary treatment of bauxite

The method used today for nearly all alumina processing was patented in 1888 by J. K. Bayer.

Bauxite is crushed and digested with caustic soda solution under pressure. The resulting liquor is diluted and an insoluble residue containing titania, titanium(IV) oxide, sodium aluminium silicate and iron(III) oxide settles. The supernatant liquid containing sodium aluminate is collected and filtered. A large excess of crystalline aluminium oxide trihydrate, $Al_2O_3 . 3H_2O$, is stirred into the solution. This acts as a seeding agent and crystalline trihydrate is deposited instead of the more usual gelatinous precipitate. It is filtered, washed and calcined at about 1300°C to give anhydrous alumina.

Although the chemical changes are no doubt complex, the reactions may be represented:

$$\underset{\text{decomposition}}{\overset{\text{extraction}}{Al_2O_3 . 3H_2O + 2NaOH \rightleftharpoons 2NaAlO_2 + 4H_2O}}$$

$$\underset{\text{aluminate}}{\text{sodium}}$$

in which sodium aluminate solution is only stable with an excess of free alkali present. The left to right reaction is extraction, the reverse, decomposition.

An alternative formulation of the aluminate anion as a hydroxo-aluminium complex:

$$\underset{\text{decomposition}}{\overset{\text{extraction}}{Al_2O_3 . 3H_2O + 2NaOH \rightleftharpoons 2NaAl(OH)_4}}$$

or

$$Al_2O_3 . 3H_2O + 2OH^- \rightleftharpoons 2Al(OH)_4^-$$

On calcination,

$$Al_2O_3 . 3H_2O \rightarrow Al_2O_3 + 3H_2O$$

The smelting process

Aluminium is formed by electrolysis of alumina dissolved in molten cryolite, between carbon electrodes. Calcium fluoride (fluorspar) is added to lower the melting-point. An outline theory gives the following dissociations:

$$2AlF_3 \rightleftharpoons Al^{3+} + AlF_6^{3-}$$

$$Al_2O_3 \rightleftharpoons AlO_2^- + AlO^+ \quad AlO^+ \rightleftharpoons Al^{3+} + O^{2-}$$

Aluminium is preferentially discharged at the cathode and oxygen at the anodes:

$$Al^{3+} + 3e^- \rightarrow Al$$

$$2AlO_2^- \rightarrow Al_2O_3 + \tfrac{1}{2}O_2 + 2e^-$$

The carbon anodes burn to carbon dioxide,

$$C + O_2 \rightarrow CO_2$$

This is the primary by-product but some aluminium is dispersed as fog in the electrolyte, reducing dioxide to carbon monoxide. Exit gas contains about 30% carbon monoxide.

Undoubtedly such complexes as $Al_2OF_6^{2-}$ and $Al_2O_2F_4^{2-}$ are present, depending on the aluminium concentrations. The overall anode reaction seems to be

$$Al_2O_2F_4^{2-} + 2AlF_6^{3-} + C \rightarrow$$
$$4AlF_4^- + CO_2 + 4e^-$$

and at the cathode,

$$AlF_6^{3-} + 3e^- \rightarrow Al + 6F^-$$

the hexafluoroaluminate ion being in equilibrium with the tetrafluoroaluminate ion

$$AlF_6^{3-} \rightleftharpoons AlF_4^- + 2F^-$$

The anode process may be seen as a smelting process superimposed on an electrolysis. Any calcium (from the flux) deposited is volatilized away.

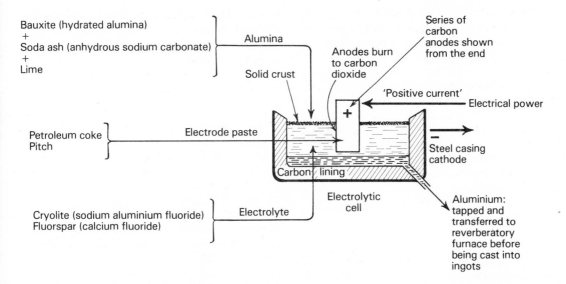

Fig. 9.1 A schematic representation of the production of aluminium by the electrolysis of a mixture of purified bauxite and cryolite

An overall cell reaction may be computed:

$$3Al_2O_2F_4^{2-} + 10AlF_6^{3-} + 3C \rightarrow 12AlF_4^- + 3CO_2 + 4Al + 24F^-$$

Diagrammatic representation of the furnace is shown in Fig. 9.1. Molten cryolite is contained to a depth of *c.* 300 mm in an insulated steel box lined with carbon blocks cemented with carbon paste or lined with a mixture of finely ground carbon and pitch rammed into position. The electrolyte is extremely corrosive. The lining forms the cathode of the cell. About 3–5% alumina is dissolved in the cryolite, lowering the melting-point from 1000°C to 970°C. The temperature is maintained by dissipation of electric power. Carbon anodes dip into the electrolyte which is covered by a crust of frozen electrolyte and alumina, protecting the carbon blocks from oxidation. Large furnaces have about 20 anodes, each *c.* 400 mm square.

A current of about 40 000 A passes in series through about 100 cells. The voltage drop across each cell is about 5 V. Aluminium accumulates in a pool *c.* 100 mm deep on the floor of the cell, while the anodes, suspended *c.* 50–75 mm above the surface of the aluminium, burn to gaseous products, which bubble away. More alumina is added to replace that used and the crust is plugged. As the concentration of alumina falls, fluorine is evolved and the voltage required for electrolysis rises to 30–50 V. This happens every 8 hours. About 275 kg of aluminium is formed daily in each cell and every two days it is either siphoned out by suction or tapped, depending on the design of cell used. The fluid metal is transferred to an oil or gas-fired reverberatory holding furnace at 900°C. From this furnace aluminium is cast into ingots or rolling blocks.

Because the density of the metal (2.3 g cm^{-3}) is not very different from that of its electrolyte (2.1 g cm^{-3}), some aluminium is dispersed as a fog and is oxidized, causing a loss in efficiency. Although the metal is liberated in accordance with Faraday's Laws, the process is only about 85% efficient. In practice about 6.8 kg, instead of 8.0 kg aluminium, is liberated by a current of 1000 A in one (24 h) day. This charge at 5 V potential corresponds to an energy expenditure of 120 kW h, so that every tonne (1000 kg) of aluminium requires about 18 000 kW h of electrical energy†. Cheap hydro-electric power is essential. Aluminium is manufactured in mountainous regions with a generous rainfall, including the western Highlands of Scotland, Norway, Switzerland, parts of France and Italy, and the Canadian Rocky Mountains. Of the widely used metals, aluminium is by far the most expensive to produce in energy terms, which has given an impetus to the recycling of the metal, the collection, melting and reworking of scrap metal. Other methods of extraction may be developed, that involving the chloride as intermediary seeming attractive.

† A = *basic S.I. unit*, W = J s^{-1} = AV, V = JA^{-1}s^{-1}, h = 3.6 × 10^3 s, C = As, kWh = 3.6 × 10^6 J

The formation of 1000 kg of aluminium requires about 2000 kg of alumina and results in the combustion of about 650 kg of carbon anode. To reduce costs, Soderberg electrodes were introduced in which a paste of coke and pitch-tar binder is confined in a casing and bakes to a hard mass as it nears the electrolyte. Ash from the anodes introduces impurities into the aluminium. Anthracite is used in the preparation of cell linings. The aluminium is up to 99.8% pure, with traces of iron and silica. Commercial aluminium is never less than 99% pure. Further purification need not concern us here.

Uses

Pure aluminium is a soft white metal of low tensile strength, but alloying and heat treatment brings the strength near to that of mild steel. A full description of aluminium alloys is given in Chapter 11. Aluminium alloys are used for aircraft frames, car engines and components, bridges, building structures, superstructures on ships, scaffolding and ladders, where lightness is important. Aluminium has a density of only a third that of steel. High electrical conductance and low density give aluminium a great advantage over copper when used for overhead electrical transmission lines. Steel-cored aluminium cables are used in the National Grid. Although aluminium has only two-thirds the electrical conductance of copper, the density of one-third that of copper gives it twice the electrical conductance on a weight-for-weight comparison. Aluminium is characterized by high resistance to corrosion, due to a thin film of oxide, which may be increased in thickness by anodizing. High thermal conductance and resistance to corrosion lead to the application of aluminium in kitchen utensils and the food industry. The reflecting power of the metal, coupled with the lightness and resistance to corrosion of aluminium, bring the application of aluminium alloys as a roofing material, giving thermal insulation by reflecting the heat of the sun. Similarly it is used in the linings of cold-weather clothing (e.g. anoraks) and in survival bags for emergency use on mountains. The appearance of the metal is an asset and the surface may be left unpainted in many of its applications, thereby giving an economic advantage over other materials, which need protection, and even an extra economy of lower weight (no paint!) for London Underground trains.

Group IV: Carbon, silicon, germanium, tin and lead

Graphite (carbon)

Occurrence and production

Deposits of graphite or plumbago are found in Ceylon and Central America. It was formerly mined in Cumberland, being used to make 'black lead' for pencils. Another allotrope of the element is mined in the form of diamonds in South Africa and the USSR. Carbonates and organic compounds are widespread.

Graphite is manufactured by the Acheson process, which requires abundant cheap electrical power. Petroleum coke and sand, with coal tar as a binding agent, are maintained at a very high temperature for 24–30 hours in an electric furnace. The alternating current is led through carbon rods which pass through the mass, supported on a refractory hearth. It is thought that silicon carbide is formed and then dissociates to graphite and silicon, which volatilizes away:

$$SiO_2 + 3C \rightarrow SiC + 2CO$$

$$SiC \rightarrow Si + C$$

Uses

Black lead is produced by mixing graphite with moist plastic clay, extruding into threads and baking. Pure graphite is used for electrodes, brushes for electric motors and carbons for dry cells, because of its good electrical conductance. It is an excellent lubricant and may be used as a colloidal suspension in water or in oil. The suspension in water is stabilized by the presence of tannin. The commercial name 'Dag' is short for deflocculated Acheson Graphite. Carbon fibre has been developed for use in engineering. Rolls Royce aircraft engines are a magnificent example of the product of research and (costly) development of this material, and many of us prize our carbon fibre fishing rods.

Silicon

Occurrence

Silicon is our second most abundant element and is always found combined with the most abundant element, oxygen. It is estimated that silicon forms

27.6% by weight of the Earth's crust. Silicon dioxide, silica, SiO_2, is found comparatively pure as quartz, flint, agate and sand, and is formed by weathering of igneous rocks, which contain silicates. An example of such rock is orthoclase or potash felspar, $K_2O.Al_2O_3.6SiO_2$. Under the action of rain water containing carbon dioxide, acid water from peat bogs, etc., and those regions where volcanic gases occur, the complex silicates disintegrate to form clays and by further changes, silica. China clay has the composition $Al_2O_3.2SiO_2.2H_2O$.

Uses

Pure silicon has poor mechanical properties and finds application in alloy form only.

Ferrosilicon, usually containing iron and silica in equal amounts, is used for the de-oxidation and de-gasification of steel formed by the Open Hearth and Bessemer processes. It is produced by heating high-grade silica with coke and iron in an electric furnace. It is common practice to add lumps of ferrosilicon to acid open hearth steel just prior to tapping. To avoid reduction of phosphates to phosphorus in the basic process, the addition of crushed ferrosilicon is made to the steel in the ladle after tapping. Structural steels contain less than 0.3% silicon but other steels contain as much as 0.6% to increase hardness and generally to improve their properties. Electrical steels and certain special steels use much more.

Silicon of metallic appearance is made by the charcoal reduction of pure silica and is used in alloying. Silumin, a product of the reduction of china clay, containing up to 12% silicon with aluminium, combines strength with hardness and a low coefficient of thermal expansion, which makes it a useful material for pistons and cylinders in car engines.

The oxide, silica, and silicon carbide are used as refractory materials and the latter under the trade name of Carborundum is also used as an abrasive.

Silicon 'chips' are used in micro-electronic circuits. The name has become a household word. The devices are developed from the original invention, based on discoveries using pure germanium, at the Bell Telephone Laboratories (1947–8) in New York. The development of zone-refining and allied techniques enabled very pure silicon to be formed and certain unique aspects of its chemistry made the manufacture of integrated circuits possible, so that silicon has displaced germanium (see below) in this area. The initial impetus came from the need for maximum reliability and minimum weight for the intercontinental ballistic missiles developed for the US Defence Programme. Miniature (or integrated) circuits combining a range of electrical functions may be made all on the surface of a single piece of semiconductor material. The dimensions of such small components are expressed in micrometres† and the development research is intensely competitive. Other materials, e.g. gallium arsenide, have been investigated for use. The Band Theory for semiconductors has been described on page 90.

Germanium

Occurrence

Germanium is found widespread but in small concentration. Minute quantities occur in silicate rocks and it is associated with certain lead and zinc ores. The dioxide is found in small amounts in certain coal ashes. It is extracted commmercially from the flues of gas producer plants, the dust coming from coke ash, and is a by-product after the extraction of zinc. Germanium is used in the electronics industry although the use of silicon is becoming paramount.

Germanium minerals include argyrodite, a silver germanium(IV) sulphide, $4Ag_2S.GeS_2$, found in Bolivia, and germanite, the sulphide associated with copper and iron(II) sulphide, $7Cu.FeS.GeS_2$, in Namibia.

Production

By various methods, germanium tetrachloride, germanium(IV) chloride, is formed from compounds associated with zinc and gas producer plants. It is purified by distillation and hydrolysed with water to the dioxide which is washed, dried and reduced in a stream of hydrogen at 600–650°C to yield germanium powder:

$$GeCl_4 + 2H_2O \rightarrow GeO_2 + 4HCl$$

$$GeO_2 + 2H_2 \rightarrow Ge + 2H_2O$$

The powder is melted at 1100°C under a chemically inert gas, such as nitrogen, and cast into ingots which may be further purified by heating to 1000°C under a vacuum with controlled cooling to separate the impurities.

† micrometre, μm

Uses

Germanium crystal diodes and triodes were originally manufactured to replace thermionic valves. They are small and have the advantages of reliability over a long life, with a much greater resistance to mechanical shock. The development of the germanium crystal diode followed the need to detect very high frequencies in early radar work during World War II. The germanium crystal triode was developed by the Bell Telephone Co., USA, in 1948 and called the 'transistor'. In transistors, crystals of germanium with properties modified by the controlled addition of certain other elements are used to achieve rectification and amplification. Group III elements—boron, aluminium and gallium—give the Positive (P) Type, and Group V—nitrogen, phosphorus and arsenic—the Negative (N) Type; one type is sandwiched as the 'grid' between two layers of the other type.

Tin

Occurrence

The only important ore of tin is tin(IV) oxide, cassiterite or tin-stone, SnO_2. The principal sources are Malaysia, Indonesia and Bolivia, while important deposits are worked in Nigeria, Zaire, Thailand and China. Cornwall, a primary source of tin a century ago, now yields comparatively little. Tin-stone is found in alluvial gravel in the Far East and is recovered by dredging, hydraulic sluicing of gravel banks and pumping from gravel mines. Other deposits yield tin in association with the sulphides of iron, copper, lead, zinc, arsenic, and tungsten as wolframates (tungstates). Tin is also recovered from scrap tin-plate.

Preliminary treatment of cassiterite

The ore is crushed and concentrated by agitation in water, lighter material being washed away. Further concentration is accomplished chemically with the aim of converting the impurities into a soluble or volatile form leaving the tin(IV) oxide unchanged. Roasting enables sulphur, arsenic and antimony to volatilize as oxides, leaving metallic oxides which may be leached out with acids. Salt added during roasting assists volatilization of impurities as chlorides and subsequent leaching out of what remains. Remaining sulphides may be recovered by froth flotation and magnetic minerals of iron and tungsten by electromagnets. The method adopted in the preliminary treatment depends on the impurities.

The smelting process

The tin concentrate is smelted in a reverberatory furnace with charcoal or anthracite and a limestone flux:

$$SnO_2 + 2C \rightarrow Sn + 2CO$$

If much iron remains, a silica flux is used. Because of the amphoteric nature of tin(IV) oxide, calcium stannate(IV) is formed to some extent with a basic flux and tin silicates with an acidic flux. The slag is resmelted at a higher temperature to achieve further separation of tin.

Liquid tin is run off and cast into slabs for refining.

Refining

Tin smelted from Far Eastern ore requires little refining, the principal impurity being iron. Impure tin is also associated with copper, lead, arsenic, antimony and bismuth when smelted from other sources. The procedure depends on the impurities.

Tin has a relatively low melting-point (232°C) and refining by liquefaction exploits this. Impure slabs of metal are heated to melting on the sloping hearth of a small reverberatory furnace: molten tin runs away. The residue is smelted with slag for tin recovery, arsenic and much iron being removed.

Preferential oxidation removes other impurities. Air or steam is blown through strongly heated molten metal when iron and other impurities are oxidized and coagulate to be skimmed away. Similarly, chlorine is used to oxidize lead and sulphur, to remove copper as copper(I) sulphide, which is also skimmed away.

Uses

Tin is a soft, silvery-white, lustrous metal, plastic and malleable, but of low tensile strength and not at all ductile. It is notable for its resistance to chemical action. Tin is subject to an enantiotropic change at 13.2°C with the formation of a grey, powdery and brittle allotrope. Fortunately, the change is subject to a good deal of supercooling. Where the change does occur, the growths of grey tin appear fungoid in character on the white tin surface, justifying the description, tin plague.

Tin is used as foil for the wrapping of cheese, chocolate and tobacco, for electrical condensers and cables and in extrusions for collapsible tubes used for cosmetics, pharmaceutical preparations, toothpastes, adhesives and paints, etc., and for pipes conveying beer and carbonated drinks, chemicals and water. Up to 40% of tin produced, appears as coatings on other materials, mainly steel and copper. Tinning is achieved by hot dipping in molten tin, spraying or wiping molten metal on to the hot article or by electrodeposition. Steel tin-plate is used for food cans. Domestic, brewery and dairy utensils may be tinned copper or iron depend-ing on the quality. Tinned copper, tinned brass and tinned cast iron have other uses as well. About 30% of all tin produced goes into alloys, described in Chapter 11.

Lead

Occurrence

The proportion of lead by weight in the Earth's crust is estimated as 0.002%. Galena, lead(II) sulphide, PbS, is the chief ore and is found asso-ciated with many other sulphides. Principal sources occur in Australia at Broken Hill (NSW) and Canada which are the two largest exporters, Mexico and the USA. Cerrusite, lead(II) carbonate, $PbCO_3$, and anglesite, lead(II) sulphate, $PbSO_4$, are other ores.

Preliminary treatment of galena

Ores range generally from 3 to 10% in galena content, although some ores are richer and at the other extreme, some at under 1% have been econ-omically worked. The mined ore is crushed and concentrated by froth flotation in which galena rises in the froth formed by air with water contain-ing pine oil and other agents, and is skimmed off. Concentrates may contain 60–80% galena.

Before blast-furnace smelting, the ore is spread on a continuous chain grate with limestone and roasted in a sintering machine with the formation of lead(II) oxide and sulphur dioxide:

$$2PbS + 3O_2 \rightarrow 2PbO + 2SO_2$$

Waste gases are trapped as a source of sulphuric acid. A limestone flux is added to obtain the coarse lumps necessary to withstand blast-furnace pressures.

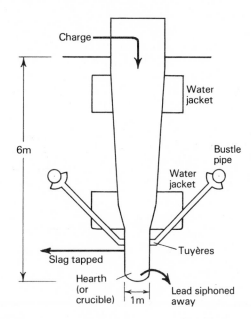

Fig. 9.2 The blast furnace production of lead (not to scale)

The smelting process

The roasted, sintered concentrate is mixed with coke, limestone to remove siliceous matter as calcium silicate slag, scrap iron or iron oxide ore to give iron for reducing unchanged lead(II) sulphide, and fluxes before heating in a blast furnace, shown in Fig. 9.2. This may be 3.5–6 m in height, of square section, tapering from about 2 m at the top to 1 m at the base. The temperature reaches 900°C and in the larger furnaces, the shaft is water-cooled. Lead is formed in the hearth and may be siphoned away. It is called 'base bullion'. From above it, matte (sulphides) and slag (silicates) may be tapped. The reactions are:

$$2C + O_2 \rightleftharpoons 2CO$$

$$PbO + CO \rightarrow Pb + CO_2$$

$$PbS + Fe \rightarrow Pb + FeS$$

$$PbSiO_3 + Fe \rightarrow Pb + FeSiO_3$$

Impurities include copper, arsenic, antimony, tin, zinc, silver, gold and bismuth. These are collected and worked up for marketing.

Refining Stage I: copper removal Sulphur is stirred into molten impure lead. Immiscible copper and copper(I) sulphide rise and are skimmed off with the oxide, other sulphides (matte), arsenides and antimonides (speise).

Refining Stage II: arsenic, antimony and tin removal, the Harris process Molten lead is pumped through a fused mixture of caustic soda and sodium nitrate or other oxidizing agent for some hours when sodium arsenate, antimonate(v) and stannate(iv) are formed. The elements may be recovered and the alkali regenerated.

Refining Stage III: the Parke's process for recovering silver and gold Molten lead is agitated with molten zinc when precious metals are distributed between the almost immiscible metals according to the Partition Law:

$$\frac{\text{Concentration in zinc}}{\text{Concentration in lead}} = constant$$

(at a given temperature for the same molecular species)

This is called the Partition Coefficient for the system. At 800°C for silver, the Partition Coefficient is 300. The melt is cooled and the zinc content, which solidifies first, is skimmed off. Traces of zinc remaining in the lead are removed by oxidation in air using a reverberatory furnace, by distillation under reduced pressure or by preferential oxidation by chlorine. The silver–zinc alloy contains up to 10% silver by weight. The precious metals are recovered after lead has been squeezed away in presses. Zinc is distilled off and residual traces of lead are removed by cupellation for about 12 hours in a small reverberatory furnace in which an air blast oxidizes base metals which are tapped as oxides. Silver remains, with any gold and platinum present. Using suitable oxidizing fluxes up to 99.8% silver may be prepared. The hearth is of bone ash, cement or dolomite refractory and the furnaces are gas or oil-fired. Cupellation is followed by electrolytic refining for the silver.

Refining Stage IV: removal of bismuth by the Betterton process This is essentially similar in principal to the Parkes' process but a calcium–magnesium alloy is used to recover bismuth.

Impurities may be reduced to such a low level that lead is marketed in bulk at 99.9% purity.

Uses

Lead is an extremely dense (11.3 g cm^{-3}), grey metal of low tensile strength, easily worked, being soft, malleable and with a low melting-point (327°C). It is only superficially oxidized at ordinary temperature in air to hydrated oxide (hydroxide) and carbonate. Lead is used for sheathing electrical cables, as sheets for roofing and pipes for plumbing (Latin, *plumbum* = lead). Lead compounds are poisonous. In hard water districts, lead water pipes soon become coated with lead(ii) carbonate and sulphate, preventing uptake of lead compounds. In areas served by soft, acidic, moorland water, silicates are added to water to give an insoluble coating of lead(ii) silicate. Traces of lead in drinking water have a deleterious effect on health, retarding mental development. For the same reason the use of tetraethyl lead, $Pb(C_2H_5)_4$, as an anti-knock agent in petrol is being phased out in countries of the EEC and elsewhere, particularly in the USA. Lead paints must be avoided where a child might chew the surface, as with cots, toys etc. Lead is highly resistant to sulphuric acid and is used in the Lead Chamber process and as an acid-resisting surface in laboratories. Lead is used to absorb harmful radiation from nuclear reactors and X-ray sources.

Alloying improves the tensile strength of lead, as described in Chapter 11. Many solders and fusible alloys contain lead.

Group V: Nitrogen, phosphorus, arsenic, antimony and bismuth

Nitrogen

The production of nitrogen from air is described in Chapter 12. There are many important compounds of nitrogen, finding uses as fertilizers, explosives, dyes and plastics.

Phosphorus

Occurrence

Originally extracted from calcined bone ash, containing about 80% calcium phosphate, by atreatment with sulphuric acid followed by charcoal reduction of the phosphoric acids produced, phosphorus is now produced from natural deposits of calcium phosphate, usually occurring with calcium fluoride in the fluorapatites, $3Ca_3(PO_4)_2 . CaF_2$. The impure mineral is known simply as phosphate rock. Impurities include carbonates of calcium and magnesium, aluminium oxide and iron(iii) oxide. Deposits are worked in the USSR, southern USA, north Africa and on certain Pacific islands.

The phosphate content varies widely. Clay and sand are removed by hosing. Fine ore is sintered by heating with fine coke and a little clay and then broken into lumps. About 30–50% of the fluorine escapes as silicon tetrafluoride.

Production

Calcium phosphate (phosphate rock), silica (sand) and carbon (coke) in calculated proportions are heated to about 1500°C in an electric furnace. This process, electrothermal as distinct from electrolytic, has been in use since 1890. Phosphorus pentoxide, displaced from calcium phosphate by the non-volatile silica, is reduced by carbon to phosphorus vapour, which is led off and condensed under water as the white allotrope:

$$2Ca_3(PO_4)_2 + 6SiO_2 \rightarrow 6CaSiO_3 + P_4O_{10}$$

$$P_4O_{10} + 10C \rightarrow P_4\,(g) + 10CO\,(g)$$
tetratomic molecules
in the vapour state

The white phosphorus rapidly becomes opaque and brownish-yellow on storage.

The electric power consumed by furnaces varies from 1500 to 30 000 kW. Every kg of phosphorus requires 13–15 kW of power to be produced. There are usually three electrodes with a 3-phase AC voltage of 150–300 V and a current of 5000–30 000 ampères. One UK phosphorus works is situated alongside an electricity generating station, near to a port. The furnace is a mild steel shell lined with fire bricks, with the hearth and lower parts of

the walls protected by carbon blocks (Fig. 9.3). The three suspended electrodes may be up to 1 m diameter and are maintained at a constant height above the slag, being replaced as they are worn away. The temperature of the melt is maintained by the dissipation of electrical power. The furnace runs until the hearth is worn out, which may be a matter of nine months or several years. The steel casing is cooled by either a water spray or air.

Phosphorus in the emerging gases passes through an electrostatic precipitator and condenses in sprays of warm water. Molten phosphorus, collected under water, is pumped along steam-jacketed pipes for immediate use or to storage vessels, where it is kept under water. The yield is about 90% of the phosphorus available. From the base of the furnace is tapped calcium silicate slag, used for road construction, and the denser ferro-phosphorus, FeP or Fe₂P, used in the steel industry to prevent steel-plates from sticking together during pack-annealing.

By maintaining the metastable white allotrope of phosphorus at about 270°C for 4–5 days, the stable red modification is produced. The conversion is, of course, exothermic. The element, covered with water, is heated in a steel pot with a firmly secured lid through which passes a safety tube fitted with a short reflux condenser to return escaping phosphorus vapour. Much excess white phosphorus is distilled off at 400°C and the product ground under water. Boiling with sodium carbonate solution removes the remaining white

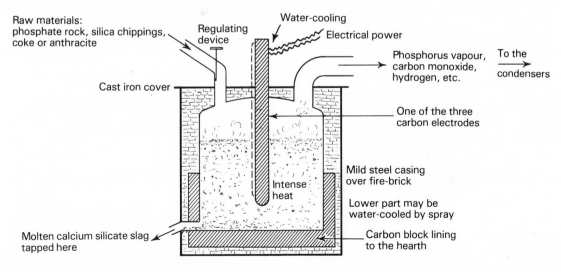

Fig. 9.3 The electrothermal production of phosphorus

phosphorus as phosphine, sodium hypophosphite and phosphate. Filtration and vacuum drying yields red phosphorus.

Uses

Much of the phosphorus is converted into ortho-phosphoric acid and phosphates, for de-rusting and fertilizers.

There is much concern about the excessive use of phosphates in farming because of the adverse effect on the natural environment (e.g. algal blooms on lakes) when the excess is leached out of the ground by rain water. Similarly the use of phosphates in washing powders and detergents has been discontinued in some countries. Phosphorus compounds are used in insecticides, plasticizers, detergents and flotation agents. Phosphorus sesquisulphide, P_4S_3, is produced for 'strike anywhere' watches and red phosphorus for the side of safety matchboxes, the other reactant, potassium chlorate, being placed in the match-head of both types of match. At one time, yellow phosphorus was used in the match industry but constant inhalation of phosphorus vapour by workers produces phosphorus necrosis, 'phossy jaw', and the use of this form of the element has been discontinued.

Arsenic

Occurrence

Thioarsenides are frequently associated with metal ores. Principal sources of arsenic and its compounds are the by-products of smelting these ores. Arsenopyrite or mispickel, $FeAsS$, cobaltite, $CoAsS$, and nickel glance, $NiAsS$, are examples of thioarsenides. More than enough arsenic trioxide is volatilized during the smelting of copper and lead to satisfy world demand. The sulphides of arsenic have been mined for use as pigments: realgar, As_4S_4, is red, and orpiment, As_4S_6, yellow.

Most arsenic is marketed as arsenites and arsenates, formed from sublimed arsenic trioxide. Arsenical compounds are sometimes used as insecticides and for killing potato tops but are giving way to less toxic compounds.

Production

Arsenic trioxide, collected in the course of metallurgical operations, is purified by sublimation in a small reverberatory furnace and reduced by charcoal at red heat in iron or steel retorts:

$$As_4O_6 + 6C \rightarrow As_4(g) + 6CO$$

tetratomic
molecules in the
vapour state

The element distils over and condenses to the crystalline, grey allotrope.

Uses

A little arsenic (0.2%) is used to increase the hardness of lead sheet and is said to cause sphericity in lead shot. Speculum metals, alloys of copper and tin (55:45 by weight approximately) and arsenic, have a high polish resembling cleaned silver, resist tarnishing and corrosion, and are used as reflectors.

Antimony

Occurrence

The principal ore is antimony trisulphide, stibnite, Sb_2S_3, found in Mexico, Bolivia, South Africa and southern China.

Production

In the preliminary treatment, high-grade ores are heated to about 550°C in a reverberatory furnace when molten stibnite may be run off. Low-grade ore is concentrated by flotation methods.

Two methods of antimony smelting may be outlined. In the first, stibnite and scrap iron are heated in a reverberatory furnace, with fluxes to aid the separation of antimony metal and fused iron(II) sulphide matte:

$$Sb_2S_3 + 3Fe \rightarrow 2Sb + 3FeS$$

Otherwise, stibnite is oxidized at 350–450°C with a controlled supply of air and the oxide volatilized. After collection, this is reduced with charcoal or coke for about 10–12 hours using fluxes to hinder the loss of oxide by volatilization:

$$2Sb_2S_3 + 9O_2 \rightarrow Sb_4O_6 + 6SO_2$$

$$Sb_4O_6 + 6C \rightarrow 4Sb + 6CO$$

In both cases, molten antimony is run into moulds. The chief impurities tend to be iron, copper, arsenic and sulphur. The crude metal is heated with the calculated amount of stibnite, liberating more antimony and producing a fused matte of the sulphides of iron and copper, which are skimmed off. Oxidation of arsenic and sulphur is promoted by blowing air through the molten metal.

Uses

Antimony is brittle and is not used by itself. (Antimony alloys will be described in Chapter 11.) It confers hardness to lead, being used in the production of type metals and bearings.

Bismuth

Occurrence

The principal forms in which bismuth is found are the native element, sulphide ore called bismuth glance or bismuthinite, Bi_2S_3, and associated with these, bismuth ochre, or bismite, Bi_2O_3. The richest sulphide ores occur in Bolivia and the native metal was worked in Saxony. However, most bismuth produced is derived as a by-product from the smelting of lead and copper in Peru, Mexico, Canada and the USA.

Production

From the sulphide ore, bismuthinite, the extraction closely follows that of antimony, either an oxidation roast in air followed by carbon reduction in a reverberatory furnace or heating with scrap iron and a lime flux:

$$Bi_2S_3 + 3Fe \rightarrow 2Bi + 3FeS$$

$$2Bi_2S_3 + 9O_2 \rightarrow 2Bi_2O_3 + 6SO_2$$

$$Bi_2O_3 + 3C \rightarrow 2Bi + 3CO$$

Various processes are used to recover bismuth from the impurities separated in the smelting of lead and copper.

Bismuth of high purity is required for pharmaceutical preparations. Electrolysis is the simplest process in principle with impure bismuth anodes, a bismuth trichloride and hydrochloric acid electrolyte and silver cathodes. Bismuth is deposited at the cathodes from which it is stripped, melted and cast into moulds.

Uses

Like the other elements of this family, bismuth is associated with brittleness and this characteristic limits its use in alloys. Bismuth has a low melting-point (269°C) and it is used to prepare fusible alloys for automatic fire-alarms and safety plugs in boilers. These applications and the use in master pattern casting are discussed in Chapter 11.

Group VI: Oxygen and sulphur

Oxygen

Oxygen production from air is described in Chapter 12.

Sulphur

Occurrence

Elemental sulphur is found in the states of Texas and Louisiana in the USA, and in Sicily. It is an important by-product in refineries which process oil of high sulphur content, notably those refining Sahara oil in France. There are many forms of combined sulphur from which sulphuric acid can be made. The sulphides: galena, PbS, zinc blende, ZnS, and iron pyrites, FeS_2, and the sulphates: gypsum, $CaSO_4.2H_2O$, anhydrite, $CaSO_4$, provide sulphur compounds. In Sicily, the element is removed from sulphur-bearing rock by heating. The ore is stacked so as to admit only a controlled supply of air and some of the sulphur ignited. The heat of combustion melts the remainder which is run off. It is later refined by distillation. The USA is the largest producer of sulphur, covering over 90% of the world demand.

Much sulphur is wasted during the combustion of coal in Britain. Fortunately the air of our cities is much cleaner nowadays and we no longer endure the occasional sulphurous smog of the 1950s, and the 'Hollywood film fogs' are no longer a regular feature of London life. However, sulphur dioxide in the waste gases of coal-burning electricity stations is a serious environmental hazard being one of the causes of acid rain. The provision of electrostatic precipitators etc. is planned but these are very expensive and it may be cheaper to use lower grade imported coal.

Production by the Frasch process   In the USA, sulphur occurs at a depth of about 150 m under soft sand and water. Conventional mining is impossible. In the Frasch process, a hole is bored down to the sulphur bed and lined with a steel casing. The Frasch Sulphur Pump consisting of four concentric steel tubes is lowered into this. The diameter of the bore-hole is about 300 mm. Super-heated water at about 180°C is forced down the two outer annular spaces and hot compressed air down the axial tube. Sulphur melts under the influence of the super-heated water and an emulsion of sulphur, water and air is forced up the remaining annular space. This is shown diagrammatically in Fig. 9.4. The emulsion is collected in vats where the sulphur solidifying is about 99.8% pure. Alternatively, the element is pumped away molten for immediate use.

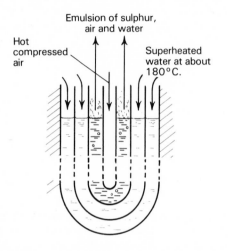

Fig. 9.4 The Frasch Sulphur Pump (diagrammatic). Four concentric pipes fitting into a borehole of approximately 300 mm diameter. Longitudinal section and half-section are shown

Uses

Elemental sulphur is used as a source of sulphur dioxide for the production of sulphuric acid and for making calcium hydrogensulphite for the wood-pulp industry, and directly as the element in vulcanizing rubber and the manufacture of matches (see page 154) and pyrotechnics.

Group VII: Fluorine, chlorine, bromine and iodine

Fluorine

Fluorine is prepared commercially on a relatively small scale by the electrolysis of potassium fluoride in anhydrous hydrofluoric acid.† About 60% by weight of the salt is used. The anodes are of un-graphitized carbon, i.e. without the sheet structure of graphite, because graphite swells and disintegrates

under the action of fluorine. The cathode is of steel and the tank, Monel metal. Hydrogen is discharged at the cathode and fluorine at the anode. Volatilized hydrogen fluoride is condensed at −80°C (b.p. +19°C, m.p. −84°C) and the residual 1–2% by absorption in sodium fluoride:

$$HF + NaF \rightarrow NaHF_2$$

Fluorine is used in the concentration of fissionable uranium by the fractional diffusion of uranium hexafluoride and for the production of sulphur hexafluoride, used in the electrical industry, and various fluorides which are used as fluorinating agents.

Chlorine

The production and uses of chlorine are described with the alkali industry in Chapter 14.

Bromine

Occurrence

Bromides are distributed widely but in relatively small amounts. There are no important mineral sources, the element being recovered from sea water, inland seas and salt lakes, brines and salt deposits left by the evaporation of seas. The richest inland sea is the Dead Sea, containing 0.56% bromine, from which the element is produced in Israel. Other inland salt waters are found in California, USA, and in the Crimea, USSR. The separation of salt by evaporation of sea water leaves a residual liquor, bitter in taste, called a bittern. Natural brines containing up to 0.25% bromine occur in Michigan, Ohio and west Virginia, USA. Recovery of bromine from the European saline deposits of Alsace and Stassfurt commences with the recrystallization of salts less soluble than bromides. The mother liquors contain up to 0.5% bromine. Bromine is also recovered from sea water, although it contains only 6.7×10^{-2} g dm^{-3} of the element.

Bromine will be extracted in plant located near to the source, where the sea is relatively warm and not diluted by fresh water rivers or sewage. Power and transportation costs are kept to a minimum.

Production

The extraction of bromine from sea water means processing a large bulk of water: about 20 000 kg of sea water are piped for the producion of 1 kg of the

† *strictly*, hydrogen fluoride

element. Due to dissolved hydrogencarbonate, carbonates and borates, sea water has a pH = 8.1 at 15°C.

Sea water is acidified to pH = 3.5 with sulphuric acid to avoid interaction of halogen in quantity with water and chlorine is introduced. Bromine is liberated by oxidation:

$$2Br^- + Cl_2 \rightarrow Br_2 + 2Cl^-$$

Using the countercurrent principle, air is blown through the liquid and the bromine-laden air is mixed with sulphur dioxide. The resultant hydrobromic and sulphuric acids, together with hydrochloric acid produced in a similar manner from traces of chlorine, are absorbed in water:

$$Br_2 + SO_2 + 2H_2O \rightarrow 2HBr + H_2SO_4$$

Treatment with a countercurrent of chlorine and steam displaces bromine, which distils in the steam and is collected under a saturated solution of bromine in water. Pure bromine is obtained on fractional distillation. Spent water is discharged away from the intake area and the acid generated above used for the initial acidulation of further incoming water. The yield is about 80%.

With water containing a relatively high bromide content (more than 0.1%), the brine is preheated to 90°C and oxidation with chlorine and steam is applied direct without the preliminary concentration effected above. The yield may reach 95% of the bromine content.

As an alternative to sulphur dioxide, aqueous sodium carbonate may be used for absorption, sodium bromide and bromate resulting.

$$3CO_3^{2-} + 3Br_2 \rightarrow 5Br^- + BrO_3^- + 3CO_2$$

Bromine is liberated on acidification with sulphuric acid,

$$5Br^- + BrO_3^- + 6H^+ \rightarrow 3Br_2 + 3H_2O$$

Uses

The production of tetraethyl lead from 1,2-dibromoethane (ethylene dibromide) accounted until recently for the larger part (90%) of bromine produced. Tetraethyl lead is an anti-knock agent used to aid combustion in petrol engines but its use is being discontinued because of the harmful effects of traces of lead compounds in the air of towns and cities on health, particularly of the young. There is evidence for the retardation of mental growth. A wide range of inorganic and organic bromine compounds including drugs, dyes and chemical intermediates are used, but the element is only of laboratory importance.

Iodine

Occurrence

Iodine is placed 47th in the order of abundance of elements in the Earth's crust, taking the lanthanides as a single element. Iodine compounds are widely distributed.

The principal source of iodine is that dispersed in deposits of sodium nitrate, nitre† or caliche, found in northern Chile. It is present chiefly as the mineral lautavite, calcium iodate, $Ca(IO_3)_2$. Caliche nitrate ore stretches in a belt about 500 km long and 25–35 km wide in a desert tableland about 900 m above sea level. The ore occurs at a depth of 1–15 m and has an iodine content of 0–0.3%.

Sea water contains about 0.05 p.p. million of iodine but certain seaweeds (the Laminariaceae) have the power of iodine absorption. Once of major importance, the extraction of iodine from seaweed ash ('kelp') is now of secondary significance. Brines taken from oil wells in Java, Russia and California have dissolved iodides in quantities worth commercial exploitation as do certain mineral spring waters in northern Italy. About 75% of iodine originates as caliche, about 18% is produced from brines and less than 7% from seaweed.

Production from caliche

The nitrate-bearing rock is blasted, collected, crushed and treated with hot water to leach out various soluble salts. The residue is rock, sand and clay. On cooling, sodium nitrate crystallizes out and the mother liquor is recycled for leaching out again. In this way the iodate concentration rises to 6–12 g dm^{-3} iodine. The solution is filtered and an excess of sodium hydrogensulphite solution added. The liquor is almost neutralized with sodium carbonate solution and an amount of mother liquor, determined by analysis, is added just to complete the generation of iodine. Excess would return some of this into solution as iodide ion, insufficient would leave iodate in solution:

$$3SO_3^{2-} + IO_3^- \rightarrow I^- + 3SO_4^{2-}$$

$$IO_3^- + 5I^- + 6H^+ \rightarrow 3I_2 + 3H_2O$$

† Chile nitre

When iodine has settled, the supernatant liquor is decanted. Iodine is washed, filtered into canvas bags and pressed free of much of the water. It is sublimed from cement-lined retorts, condensing as crystals in earthenware pipes, each $c.\,0.5$ m in diameter and $c.\,1$ m long in sets of 6–10. Water condenses and escapes; mineral salts remain behind. The purity is not less than 99%.

Production from iodiferous brine

In California, USA, brine which approximates in composition to sea water is associated with the oil-drilling regions. The iodide ion concentration is 0.003–0.007% by weight, but enough to permit commercial exploitation. The brine is filtered free of oil and mud, treated with sulphuric acid which causes further precipitation, and then sodium nitrite solution is added. The precipitated iodine is absorbed in activated charcoal:

$$2I^- + 2NO_2^- + 4H^+ \rightarrow I_2 + 2NO + 2H_2O$$

Iodine is extracted from the saturated charcoal with caustic soda solution and re-precipitated by the action of acidified sodium dichromate solution:

$$Cr_2O_7^{2-} + 14H^+ + 6I^- \rightarrow 2Cr^{3+} + 3I_2 + 7H_2O$$

It may be purified by melting under sulphuric acid or by steam sublimation.

Uses

Compounds of iodine have very many uses, of which some are the preparation of dye-stuffs, light- and heat-sensitive papers, insecticides and vermicides, polaroid film and general organic chemical production.

Tincture of iodine (1 g each of potassium iodide, iodine and water in 40 cm^3 of rectified alcohol) is still used as an antiseptic for minor skin cuts and abrasions but it has a strong irritant action. Iodoform is used as an antiseptic.

One mouth antiseptic contains an iodine complex, poly (1-vinyl-2-pyrrolidone)-iodine, shortened to PVP-iodine.

10

The extraction and uses of transition and associated metals

The metals, their physical properties and value

The transition metals are physically very metallic, having high melting-and boiling-points, and with the notable exception of titanium, high densities. Copper, silver and gold melt at about 1000°C and the last group, zinc, cadmium and mercury, at much lower temperatures. Strictly, these are not transition metals, because their penultimate electron shells are complete. Data are given in Table 10.1.

The physical properties of elements in relation to their structures and positions in the Periodic Table was discussed fully in Chapter 5, p. 88. With our selection of elements from the three transition series in mind, we can arrange the subject matter to underline what has been said in relating properties to respective electronic configurations. A fuller discussion appears in Chapters 24–26 inclusive.

Comparatively difficult to produce in useful commercial form, titanium and zirconium are important in nuclear engineering and the former is also used in the construction of ultra-high-speed aircraft and spacecraft.

Iron is of outstanding metallurgical importance owing to its mechanical properties and the easily worked high-grade ores which are readily available. The remaining transition metals find widespread use in ferrous and non-ferrous alloys. Ferrous alloys containing up to about 2% carbon, combined or in solid solution, are called *steels*.

Table 10.1 Physical data of metals the extractions of which are described in this chapter

	22 Ti	23 V	24 Cr	25 Mn	26 Fe	27 Co	28 Ni	29 Cu	30 Zn
m.p./°C	1660	1720	1830	1242	1527	1490	1455	1083	419
b.p./°C	3287	3380	2260	1900	2750	2870	2730	2573	907
density/g cm⁻³	4.5	6.0	7.1	7.4	7.9	8.6	8.9	8.9	7.1
	40 Zr							47 Ag	48 Cd
m.p./°C	1852							961	321
b.p./°C	4377							2152	764
density/g cm⁻³	6.5							10.5	8.6
								79 Au	80 Hg
m.p./°C								1063	−38.9
b.p./°C								2360	357
density/g cm⁻³								19.3	13.6

↑
2nd metal
of each
transition
series

↑
End of
transition
series
proper

Incorporation of certain other elements leads to the emergence of special characteristics.

Copper, silver and gold have been known since history was recorded, being worked by the ancient civilizations of Mesopotamia 6000 years ago. With their characteristic lustres, resistance to corrosion and wear, and ease of working, these metals were used for coinage. Only copper is of any importance for this purpose now. Copper is the principal metal of metallurgy after iron.

Zinc is used widely in alloys and by itself. Both zinc and cadmium are important as protective coatings on iron and steel to prevent rusting. Mercury has industrial uses because it is a liquid.

Elements 22-28 inclusive (Ti – Ni) of electronic configuration (Ar), $(3d)^{2-8}$ $(4s)^{2(1)}$

These metals are of the 1st transition series, the first member (scandium) having been omitted. Titanium seems apart from the others which are of very similar general behaviour as might be expected from the expanding penultimate electron shell.

Titanium

Occurrence

The principal ores of titanium are the dioxide, rutile, TiO_2, and ilmenite, iron(II) titanium(IV) oxide†, $FeO.TiO_2$. The minerals are widespread and often occur in the beach sands formed by the breakdown of rocks. Titanium ores are located in Australia, India, Brazil, Canada, the USA, and Norway. Titanium is the seventh most abundant metal and about the tenth most abundant element in the Earth's crust.

Molten titanium attacks furnace linings and absorbs gases, especially oxygen, hydrogen and nitrogen, so that conventional smelting methods cannot be used. Interest in the metal was stimulated by its very good resistance to corrosion and heat, while having only 60% of the density of stainless steel. In 1949, the Kroll process, which appeared in the USA, provided the first economic large-scale production.

Production

The ore-dressing methods used will depend on the quality of the sands. Concentration is effected by screening or crushing and grinding, washing, flotation techniques and separation by electromagnetic methods.

Titanium(IV) oxide(dioxide) concentrate, mixed with carbon, is chlorinated at 700–800°C when titanium(IV) chloride, a colourless liquid, may be condensed. This is similar to the preparation of the chlorides of beryllium and magnesium prior to the extraction of those metals:

$$TiO_2 + C + 2Cl_2 \rightarrow TiCl_4 + CO_2$$
$$TiO_2 + 2C + 2Cl_2 \rightarrow TiCl_4 + 2CO$$

In the Kroll reduction (1949), a highly exothermic reaction with magnesium at 800–900°C in an atmosphere of helium or argon is used:

$$TiCl_4 + 2Mg \rightarrow Ti + 2MgCl_2$$

Molten magnesium chloride is mostly drained away while any residue is distilled under reduced pressure. Titanium remains as a spongy mass adhering to the reaction chamber. Magnesium and chlorine are regenerated by electrolysis and used again.

The ICI reduction (1955) uses sodium under an atmosphere of argon in a steel bomb. After removal of sodium chloride, dull grey granules of titanium remain:

$$TiCl_4 + 4Na \rightarrow 4NaCl + Ti$$

Titanium melts at 1800°C. Ingots are made by melting in an electric arc.

Uses

Titanium and alloys show marked general resistance to chemicals and corrosion. Their strength is retained at high temperatures and they have a comparatively low density. Accordingly, applications are in aircraft and spacecraft, chemical and electro-chemical engineering. Titanium and some alloys are used in nuclear engineering. Spectacle frames made with titanium (alloys) are astonishingly light and flexible.

Vanadium

Occurrence

Most vanadium is a by-product from the treatment of lead, zinc and uranium ores. In the western USA, production is linked with that of uranium from carnotite, potassium uranyl vanadate,

† Also called iron(II) titanate

$K_2O.2U_2O_3.V_2O_5.3H_2O$. In Peru, a sulphide ore, patronite, VS_4, is mined by open-cast methods. Vanadium is associated with lead and zinc ores as lead vanadate and zinc vanadate in Zimbabwe and Namibia.

Production

Carnotite is hand-picked, crushed, roasted with salt and agitated with alkali to yield sodium vanadate from which vanadium pentoxide, vanadium(v) oxide, is precipitated by sulphuric acid. Simple in principle, the reactions are complex but, using metavanadate, may be expressed as

$$2VO_3^- + 2H^+ \rightarrow V_2O_5 + H_2O$$

Vanadium is required as an alloy with iron, ferrovanadium containing 30–40% vanadium, for the steel industry. The production of ferro-vanadium may be accomplished by the thermite process, made more efficient by external heating in an electric furnace or by reduction with carbon or silicon, also in an electric furnace:

$$3V_2O_5 + 10Al \rightarrow 6V + 5Al_2O_3$$

$$V_2O_5 + 5C \rightarrow 2V + 5CO$$

$$2V_2O_5 + 5Si \rightarrow 4V + 5SiO_2$$

In the thermite (alumino-thermic) process, vanadium pentoxide with powdered aluminium and steel turnings is charged into a steel reaction vessel lined with magnesite. An ignition mixture of barium peroxide and aluminium powder to trigger the main reaction is placed in a small heap on top and fired with a piece of magnesium ribbon. The reaction is violently exothermic, and molten ferro-vanadium with aluminium oxide as slag, later removed, is poured into moulds.

Pure vanadium is not in demand because even a small quantity of oxygen renders the metal brittle and difficult to work.

Uses

Vanadium is added as an alloying element to steels. It is a silvery-white metal, very ductile when pure.

Chromium

Occurrence

The main ore of chromium is chromite, $FeO.Cr_2O_3$, and the principal centres of production are in the USSR, Turkey, Zimbabwe, the Transvaal (South Africa) and the Philippines. Chromium is usually extracted as an alloy with iron, ferro-chrome, for use in the steel industry, and it is interesting to note that, except for the USSR, the chief steel-producing countries are short of chromium ore deposits.

Production

Initially, high-grade ores are hand-picked while those of lower quality are concentrated by washing methods.

Chromite is smelted with powdered anthracite in an electric furnace to yield ferro-chrome with a high carbon content (Cr 60, Fe 35, C 5 per cent by weight approximately), while alloys of low carbon content are produced with silicon as reductant:

$$FeO.Cr_2O_3 + 4C \rightarrow (Fe + 2Cr) + 4CO$$
alloy
high carbon content

$$FeO.Cr_2O_3 + 2Si \rightarrow (Fe + 2Cr) + 2SiO_2$$
alloy
low carbon content

Chromium is too brittle to be of commercial use as the pure metal, but may be extracted by reduction of chromium(III) oxide by thermite reduction using preheated reactants or by silicon reduction in an electric furnace:

$$Cr_2O_3 + 2Al \rightarrow 2Cr + Al_2O_3$$

$$2Cr_2O_3 + 3Si \rightarrow 4Cr + 3SiO_2$$

Chromium(III) oxide is formed by reduction of sodium dichromate with sulphur in a furnace. The production of sodium dichromate from chromite is described with the reactions of dichromates later.

Uses

Chromium is used widely in alloy steels and certain non-ferrous alloys. It is a bluish white, shiny, hard and brittle metal, notable for hardness and toughness, with resistance to wear, heat and extreme conditions of oxidation and corrosion. The special qualities of chromium may be utilized as chromium-plating on iron and steel. Articles to be electroplated form cathodes with anodes of sheet lead in an electrolyte containing 25% chromic acid and 0.25% sulphuric acid. Chromium-plating is used extensively in the motor-car industry. The quality of its durability depends on the under-coating of nickel or copper previously applied by electrolysis.

Manganese

Occurrence

Some 125 minerals containing manganese have been recorded but few are of economic importance. Pyrolusite, manganese dioxide, manganese(IV) oxide, MnO_2, is the principal source, and is also used directly as an oxidant in uranium production and in dry electric cells as a depolarizer. Ores of manganese(III) oxide, manganite, Mn_2O_3, braunite, $3Mn_2O_3.MnSiO_3$, and a hydrated dioxide, psilomelane, are also important. The USSR supplies over one half of the world requirements, the remainder coming principally from Ghana, South Africa, India and Brazil. Except for the USSR, manganese ores occur far from the major steel-producing countries, although certain poor-grade sources are worked in the USA.

Production

High-grade material is mined, crushed and screened while low-grade ores, especially siliceous minerals, require fine grinding, froth flotation and complex processing. Main impurities are oxides of aluminium, silicon and iron.

Manganese–iron alloys are produced direct from the ores for the steel industry. Ferro-manganese (Mn 80, Fe 13, C 6, Si 1 per cent by weight approximately) and spiegeleisen (Mn 18, Fe 76, C 5, Si 1 per cent by weight approximately) are formed by the reduction of manganese ore and iron ore, or scrap iron and steel, with coke in an electric furnace using limestone as flux. Ferro-manganese is used for alloying steels and spiegeleisen to remove combined oxygen and sulphur from molten iron and steel. Addition of silicon yields a silico-manganese iron alloy from the above reduction. This alloy is used to de-oxidize steels and to bring certain desirable mechanical properties.

Manganese metal may be obtained by carbon reduction in an electric furnace, the alumino-thermic process on a suitably prepared ore or the electrolysis of manganese(II) sulphate obtained by leaching ore with sulphuric acid and ammonium sulphate under carefully controlled conditions.

Uses

Manganese is a silvery-white metal, of the greatest importance in steel-making. As sheet and wire, the metal has been used in the manufacture of electric lamps. Manganese steels are notable for hardness, resistance to wear and strength. Non-ferrous alloys with, principally, copper, aluminium and magnesium are important.

Iron

Occurrence

Iron is the fourth most abundant element and second most abundant metal in the Earth's crust, of which it forms roughly 5% by weight. Commercially worthwhile deposits of iron are widespread so that iron is produced in very many countries. The principal ores are of iron(III) oxide, found as haematite, Fe_2O_3, and limonite, $2Fe_2O_3.3H_2O$, of tri-iron tetroxide, magnetite, Fe_3O_4, which is richest in proportion of iron, and iron(II) carbonate, siderite, $FeCO_3$. Economic ores range from 20 to 70% in iron content. In the UK, deposits are, or have been mined in Cumberland, Oxfordshire, Leicestershire, Lincolnshire and Northamptonshire. Scrap metal is also collected and rich iron ore is imported on a large scale. Furnaces are sited to reduce transport costs; coal and fluxes being carried to the ore for smelting, either in ore-bearing regions or near ports in south Wales, the north-east of England or Scotland.

Production

Except for calcining certain ores to expel carbon dioxide, moisture and volatile impurities, such as sulphur and arsenic as oxides, concentration methods are rarely needed. However, in the future, when increasing use will be made of low-grade ores, concentration may become necessary. The product is iron(III) oxide which is usually smelted in a blast furnace.

In addition to iron(III) oxide the materials used are the air-blast, possibly oxygen-enriched, metallurgical coke to provide the heat of combustion and the reducing agent, carbon monoxide, and limestone as a flux, although the exact flux requirement will depend on the nature of the ore blend used. The coke must be capable of withstanding high pressures without crumbling and contain a low proportion of sulphur and phosphorus, leaving little ash. Where hydroelectric power is cheap, electric furnaces are used in which coke is retained to supply the reductant but not the heat. Furnaces with electric arc heating are used in Norway, Sweden and

Italy. In addition, rotary furnaces, various gas reduction methods and low shaft furnaces are used to some extent.

The blast furnace is illustrated in Fig. 10.1. It is a shaft furnace, of characteristic shape, about 30 m high with a hearth 5–8 m diameter which may be lined with carbon blocks for durability or with fire-clay, as is the rest of the shaft. The outside is constructed of steel plates. The solid burden is introduced as required through a hopper and double bell system at the top and meets an upward moving air blast, introduced under pressure through 8–20 nozzles, called tuyères, served by a bustle pipe which encircles the furnace at about 2.5 m from the base.

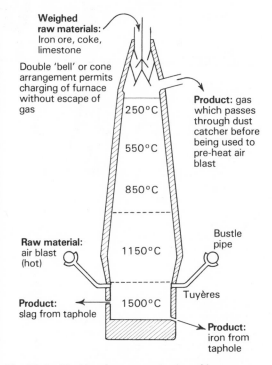

Weighed raw materials: Iron ore, coke, limestone

Double 'bell' or cone arrangement permits charging of furnace without escape of gas

250°C

550°C

850°C

Raw material: air blast (hot)

1150°C

Product: slag from taphole

1500°C

Bustle pipe

Tuyères

Product: gas which passes through dust catcher before being used to pre-heat air blast

Product: iron from taphole

Fig. 10.1   The blast furnace production of iron

The air blast is preheated to 550–850°C by passage through chequered brickwork, previously heated by combustion of part of the waste gases, in Cowper Stoves. Addition of oxygen to the air blast and the application of higher gas pressures throughout to slow down the blast, gives increased reaction time, an improved yield and more economical use of fuel.

The products leave the furnace in three places. Molten iron is tapped from the base of the hearth every 6 hours or so, the less dense molten slag floating on it is tapped more frequently and the waste gases, collected at the top of the furnace, pass along the 'downcomer' pipe and are purified prior to combustion.

The temperature range in the furnace drops from approximately 1500°C at the tuyères to 200–300°C at the top. The flame temperature is about 2200°C. The burden descends into zones of increasing heat. The reactions are complex and may with advantage be simplified here.

Coke burns in the air blast to carbon monoxide,

$$2C + O_2 \rightarrow 2CO$$

The ore takes about 8 hours to pass through the furnace. Near the top, reduction to spongy iron occurs,

$$Fe_2O_3 + 3CO \rightleftharpoons 2Fe + 3CO_2$$

Lower, at about 600°C, limestone dissociates,

$$CaCO_3 \rightleftharpoons CaO + CO_2$$

The products now undergo further changes: carbon dioxide is reduced by coke to the monoxide, bringing further reduction of ore and maintaining the active mass of reductant, while quicklime reacts with silica and alumina, forming a liquid slag, which is considered to be calcium silicate and calcium aluminate,

$$CO_2 + C \rightarrow 2CO$$

$$CaO + SiO_2 \rightarrow CaSiO_3$$

$$CaO + Al_2O_3 \rightarrow Ca(AlO_2)_2$$

Finely divided carbon, of which some is subsequently carried down with molten iron, is also formed by reduction of carbon monoxide by iron with the formation of oxides of iron below red heat, and by the breakdown of carbon monoxide under action of hot spongy iron:

$$2CO \rightarrow C + CO_2$$

In the hottest regions, reduction of remaining iron(III) oxide is completed by carbon,

$$Fe_2O_3 + 3C \rightleftharpoons 2Fe + 3CO$$

Spongy iron, containing carbon, melts at about 1200°C, which is below the melting-point of the pure metal. As the impurities in the iron are removed, its melting-point rises. However, the exothermic nature of the reduction ensures that the temperature of the melt also rises. After tapping, it is run into moulds on an endless belt or into special ladles, lined with fire brick, for the mixer furnace of the adjacent steel-works. The solid product is called cast iron or pig-iron, from the arrangement of channels when sand moulds were used. Carbon is retained as iron carbide, cementite, Fe_3C, and as flakes of uncombined graphite with certain other elements so that cast iron is an alloy (Fe 88, C 4.5, P 2, Si 3.5, Mn 2, (S)† per cent by weight approximately). Slag is used for road-making and on analysis one sample, as an example, contained percentages by weight:

SiO₂	Al₂O₃	CaO	MgO	MnO	S	Fe
32.0	21.7	35.6	7.2	1.26	1.42	0.90

$$SiO_2 : CaO$$
$$0.97$$

Waste gas, used for pre-heating the air blast, steam-raising and heating coke ovens, on analysis after cleaning, may contain roughly the following percentage by volume:

CO	CO₂	N₂	H₂
24	12	60	4

where hydrogen is formed by the water gas reaction from moisture in the air blast and burden.

Virtually pure iron is made by fire-refining pig iron with haematite on the hearth of a reverberatory furnace. Impurities are oxidized away and the mass becomes pasty as the melting-point rises. Slag is squeezed out, leaving wrought iron.

Direct reduction refers to the reduction of iron ore before going through the molten stage. The operating temperature is below 1000°C and the reductant may be a solid or a gas, usually natural gas. This is 'reformed' to a mixture of carbon monoxide and hydrogen

$$CH_4 + CO_2 \rightarrow 2CO + 2H_2$$
$$\text{natural gas}$$

The products reduce iron oxide ore:

$$Fe_2O_3 + 3H_2 \rightarrow 2Fe + 3H_2O$$

$$Fe_2O_3 + 3CO \rightarrow 2Fe + 3CO_2$$

The gaseous product at the top of the shaft contains approximately 30% carbon dioxide at 400°C. It is washed and cooled, thus removing water vapour and most of the dust, and then catalytically reacted with fresh natural gas to produce carbon monoxide and hydrogen in about 93% yield. The temperature (about 950°C) of the reducing mixture is controlled

by passage through a water cooler. The ore passes through the reduction zone in 5–6 hours, descends through a cooling zone against a current of nitrogen (with some reducing gas) for another 5 hours and is discharged at 25–30°C. Pellets containing 93% metallic iron are produced, and are easily handled for steel-making. Figure 10.2 illustrates the process.

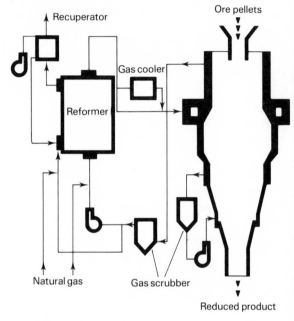

Fig. 10.2 The Midrex process

Uses

Before rusting occurs, iron is a silvery-grey metal. Wrought iron has high tensile strength and may be welded. It contains a small proportion of carbon and may be classified as mild steel.

Alloys of iron are pig, or cast, iron which contains more than 2% carbon and steel with less than this. Less than 0.25% carbon gives mild steel. Incorporation of vanadium, chromium, cobalt, nickel, etc., or extra silicon and manganese yields special alloy cast irons and alloy steels. Notable for excellent engineering properties is nickel cast iron.

Iron and steel are widely used in all aspects of engineering. However, the modern industry endures fluctuating fortunes and in the EEC some rationalization has occurred in the face of competition from countries in the Far East. Old plant has

† (S) indicates a trace of sulphur

been taken out of production and national production quotas have been established. Prolific users of steel, the formerly dominant heavy shipbuilding industry has largely disappeared. The social and economic effect of all of this has been devastating to whole regions in the UK and the rest of Europe. The UK now has one of the most profitable and effective slimmed-down steel industries in the Western World.

Cobalt

Occurrence

Cobalt is found associated with the ores of copper, lead and nickel, and as the arsenical ores, smaltite, $CoAs_2$, and cobaltite, or cobalt glance, $CoAsS$. As cobalt is a by-product from various sources, the metallurgy of the extraction is correspondingly complex. The most abundant source is cobaltiferous copper ore, mined in Zaire and Zimbabwe. Arsenical ores are found associated with gold in Morocco and with silver in Canada. Cobalt is also produced from copper-nickel ores mined in Canada and processed there and in the UK.

Production

Oxidized cobaltiferous copper ore is treated with sulphuric acid containing iron(II) sulphate when cobalt passes into solution with copper. Electrolysis of the acidic solution results in the deposition of copper while cobalt remains in solution. Electrodeposition of cobalt occurs from neutral, not acidic, solution so lime is added and a pulp of suspended cobalt(II) hydroxide is electrolysed to give 91–95% cobalt, later refined in an electric arc furnace.

Cobalt may be produced during the direct electro-refining of nickel from nickel sulphide matte anodes. A *matte* is the product of the first smelting of ores. Selective oxidation by chlorine and hydrolysis leads to precipitation of cobalt(III) hydroxide. Impurities having been removed, the hydroxide is calcined to oxide and smelted with petroleum coke in oil-fired reverberatory furnaces to yield metal which is cast as anodes for electrolytic refining. Nickel, iron and copper are removed from the electrolyte and cobalt is deposited on stainless steel cathodes, stripped off, washed and dried.

Uses

The principal use is in alloy steels and alloys with tungsten where hardness, resistance to wear, heat and corrosion are important. Magnetic materials are also prepared from alloys of cobalt.

Nickel

Occurrence

Three types of nickel ore are mined but the greatest part by far of the world production comes from the sulphide occurring in complex mixtures of pentlandite $(Ni,Fe)_9S_8$, chalcopyrite, $CuFeS_2$, and pyrrhotite, Fe_7S_8, mined in the Sudbury district of Ontario and in Manitoba, Canada. Silicate ores come from the USA, Venezuela, Brazil and New Caledonia, near Australia in the Pacific. Nickel-bearing iron ores are worked in Cuba and nickel sulphide ores are found in Finland and the USSR. Nickel is produced also as a by-product during other metallurgical operations.

Production

High-grade ore is smelted in a blast furnace with coke and a flux, limestone. Fine ore of good quality is previously sintered, sulphur being oxidized. Coke burns in the air blast and the heat produced causes fusion. Quicklime, from decomposition of limestone, forms a silicate slag with the rocky material. The whole is run into a settler, the lighter slag removed and the remaining molten sulphide matte, containing sulphides of nickel and copper with iron, is tapped.

However, most nickel is produced from disseminated ores and direct smelting is too costly. Concentration is effected by grinding, flotation and magnetic methods. The concentrate of nickel-iron sulphide with copper sulphide and iron sulphide is roasted. Iron(II) sulphide burns readily in air and about half of the sulphur is oxidized, the remainder being left for matte formation,

$$2FeS + 3O_2 \rightarrow 2FeO + 2SO_2$$

Sand has already been added and when the calcine is fused on the hearth of a reverberatory furnace, much iron forms the silicate:

$$FeO + SiO_2 \rightarrow FeSiO_3$$

A matte of sulphides of nickel and copper, with iron, cobalt and other elements in combination results.

Direct electro-refining is now practised with the recovery of high-purity sulphur, selenium and cobalt as by-products. Molten sulphide matte is cast into anodes and placed with thin nickel sheet or sheet steel cathodes in an electrolyte containing sodium, nickel, sulphate and chloride ions at pH = 4.0. Iron, cobalt, arsenic, lead and copper compounds are removed by various methods. During electrolysis the sulphide anodes corrode away. The sulphide of nickel, Ni_3S_2, is essentially ionic and the anode processes are

$$Ni_3S_2 - 6e^- \rightarrow 3Ni^{2+} + 2S \ (95\%)$$

and

$$4OH^- - 4e^- \rightarrow 2H_2O + O_2 \ (5\%)$$

The cathode process is mainly (95.5%) discharge of nickel with some hydrogen ion discharge,

$$Ni^{2+} + 2e^- \rightarrow Ni \qquad (95.5\%)$$

$$H^+ + \ e^- \rightarrow H \quad H + H \rightarrow H_2 \ (4.5\%)$$

The nickel content, which would be reduced by the extra cathode efficiency relative to that of the anode, is restored and the pH maintained, by addition of nickel hydroxide.

Uses

While nickel has important uses itself, it is used principally in numerous alloys. Nickel has excellent mechanical properties and marked resistance to corrosion, qualities retained at elevated temperatures. It is one of the toughest metals known, has good thermal conductance and is non-toxic. Nickel vessels and fittings are used in the catalytic oxidation of ammonia, production of very pure caustic soda and soap, and in chlorinations. Nickel is used as a thick undercoating prior to chromium-plating and as a constituent of catalysts for the hydrogenation of unsaturated hydrocarbons and the steam cracking of other hydrocarbons. Nickel is used in electrical accumulators containing alkaline electrolyte: the NiFe cells.

There are many alloys of nickel, with a full range of nickel content from 1 to 98%. Coinage is a well-known application. Resistance to corrosion is developed in alloys for chemical plant, food and pharmaceutical industries. Heat resistance is important for gas turbine and jet engine parts, electrical resistances and heating elements. Magnetic and non-magnetic alloys have been developed. The alloying elements are chromium, manganese, iron, copper and zinc, tin and molybdenum.

Element 40 (Zr) of electronic configuration (Kr), $(4d)^2 (5s)^2$

This element, abundant yet difficult to isolate, is of importance in nuclear engineering as a canning material. There are some similarities to titanium, the element placed directly above it in the Periodic Classification.

Zirconium

Occurrence

Zirconium is found as the silicate, zircon, $ZrO_2.SiO_2$, in the beach sands of Queensland, Australia, and the monazite sands of India. The dioxide, baddeleyite, ZrO_2, is found in gravels in Brazil.

Like titanium, zirconium is an abundant metal which is difficult to isolate because of the reactivity shown to refractories and gases at high temperatures. But extremely good characteristics for use in nuclear reactors stimulated research into the metallurgical extraction.

Production

The Kroll reduction process (1958), similar to that used for titanium, is applied. Chlorination at 500°C of zircon and carbon, previously bricketted at 800–1000°C in an arc furnace, yields zirconium(IV) chloride, which condenses as a solid. This compound is reduced with magnesium under an atmosphere of helium and an ingot is formed by consumable electrode melting as for titanium. Zirconium melts at 1852°C.

Uses

Marked resistance to chemicals and corrosion generally lead to applications in chemical engineering. Already mentioned are uses in nuclear engineering. Photographic flash bulbs filled with zirconium foil or fine wire give a brighter flash than magnesium and are used in indoor colour photography.

Elements 29, 47 and 79 (Cu, Ag and Au, respectively) of outer electronic configurations $(3d)^{10} (4s)^1$, $(4d)^{10} (5s)^1$ and $(5d)^{10} (6s)^1$

Of successive periods these elements hover on the edge of their respective transition series. As soon as more than one electron is lost, there is an incomplete penultimate electron shell and chemical properties akin to those exhibited by proper members (i.e. full members) of the transition series appear. For us, this is especially true of copper but the $(4d)^{10}$ state (oxidation number = 1) remains the principal one for silver. Gold differs in preferring the $(5d)^8$ state (oxidation number = 3) in forming compounds.

Copper

Occurrence

The principal ores of copper are sulphides: chalcopyrite, $CuFeS_2$ ($Cu_2S.Fe_2S_3$), containing much sulphur, chalcocite or copper glance, Cu_2S, and bornite, Cu_5FeS_4 ($2Cu_2S.CuS.FeS$).

There is usually only a small proportion of copper. Oxidized ores, produced by weathering of primary sulphide ores, which were probably ejected from the hot core of the Earth, include malachite, $CuCO_3.Cu(OH)_2$, azurite, $2CuCO_3.Cu(OH)_2$, both basic carbonates, and cuprite, copper(I) oxide, Cu_2O. In addition, native copper is found, notably at Lake Superior in the USA.

Copper ores are widespread. Principal producers include Zaire, Zimbabwe, South Africa, Canada, the USA, Chile and Peru. Finland and the USSR have major deposits, while minor producers include Australia, India, Japan, Cyprus and some European countries. Because copper is virtually indestructible, scrap presents an important source.

Production

The ore is crushed and ground wet. Sulphides are concentrated by flotation. Low-grade ore is treated differently: after leaching out with acidified water, copper is precipitated by the addition of scrap iron. Other grades are smelted after concentration.

If necessary, sulphur in excess of matte requirements is removed by a preliminary roasting in air at 1200–1500°C. Sulphur dioxide and oxides of volatile impurities, such as arsenic, are evolved.

$$2CuFeS_2 + O_2 \rightarrow Cu_2S + 2FeS + SO_2$$

Hot calcine, or wet concentrate, if no preliminary roasting is done, is smelted with a flux of limestone or silica, or both, in a reverberatory furnace. Earthy material forms a slag, which floats and may be removed. Air for the combustion is preheated by hot waste gases. A fusion of copper(I) and iron(II) sulphides, called a matte, with a copper content of 40–45% is formed.

Air under pressure is passed through the molten matte for about 10 hours in converters lined with magnesite bricks. Iron(II) sulphide is oxidized preferentially, iron passing to the slag and sulphur being evolved as dioxide,

$$2FeS + 3O_2 \rightarrow 2FeO + 2SO_2$$

$$FeO + SiO_2 \rightarrow FeSiO_3$$

A silicate slag is formed by use of siliceous copper ore as flux. The reaction is exothermic and is maintained at about 1300°C by addition of cold scrap copper. Copper(I) sulphide is partly oxidized to the oxide, when copper is formed by autoreduction:

$$2Cu_2S + 3O_2 \rightarrow 2Cu_2O + 2SO_2$$

$$Cu_2S + 2Cu_2O \rightarrow 6Cu + SO_2$$

Molten copper is cast into water-cooled moulds. Expelled sulphur dioxide leaves the surface roughened, so the product, 98–99.5% pure, is called 'blister' copper.

High quality blister copper is furnace refined. The process is called 'poling' because of the use of wooden poles. Air is blown through the molten copper, to which a silica flux has been added. Impurities are oxidized; sulphur forms sulphur dioxide which escapes with other volatile matter while other impurities pass into the slag. Partial oxidation of copper to copper(I) oxide also occurs. To counter this, the surface of the molten copper is sealed with coke of low sulphur content and the ends of green hardwood trees submerged below the surface. Hydrocarbon decomposition products of wood reduce any copper(I) oxide to copper. The metal is cast into moulds and quenched with water.

Where the precious metal content is high, electrolysis follows a preliminary furnace refining. The electrolyte contains copper(II) sulphate and dilute sulphuric acid. Anodes of impure copper are hung interleaved with the cathodes of pure copper. Copper passes from anode to the electrolyte and so to the cathodes while noble impurities drop into the anode sludge, which is periodically worked up for gold and silver. Deposited copper is melted, furnace refined and cast. The purity of the electrolyte must be maintained to avoid contamination of copper.

Uses

Copper is malleable and ductile, of excellent thermal and electrical conductance, with marked resistance to corrosion. The properties of copper are altered appreciably by impurities. Copper, of a purity not less than 99.9%, is used for electrical purposes: about 40% of world production going as wire to the electrical industries for use in the windings of electric motors, power lines, submarine cables, telephone lines and, with cadmium added for strength, for overhead wires on electrified railways and for extra long spans of high-voltage transmission lines of the National Grid. Copper is electroplated for two main purposes. A comparatively thin coating is applied to articles before nickel- and chromium-plating while thicknesses of copper may be built-up, or electro-formed, by electrolysis for printing rollers, wave-guides and moulds for gramophone records.

Copper alloys are widely used and being the most important group of non-ferrous alloys, are considered separately in the next chapter. Besides the general engineering purposes to which copper and its alloys are put, further uses include heat exchange units and boilers where good thermal properties are paramount, and in the chemical, brewing and jam-making industries where corrosion resistance is important and building, plumbing and marine engineering. Alloys of copper are also used in coinage.

Silver

Occurrence

Silver sulphide, Ag_2S, is the principal source. It occurs as argentite or silver glance, and as complex sulphides with arsenic and antimony. These may occur alone or with base metal sulphides. About three-quarters of the silver produced comes as a by-product of the extraction of base metals, chiefly lead and copper, but also zinc, nickel and tin. It is also found with gold. Silver chloride, $AgCl$, occurs naturally as the mineral cerargyrite. The American continent supplies most of the world's silver: since the 16th century, Mexico has been the main supplier.

Production

About 20% of silver produced is extracted by cyanidation of sulphide ore in Mexico, the silver content being about 0.5% by weight. After crushing and wet grinding in sodium cyanide or calcium cyanide solution, the resulting pulp is poured into large tanks, the cyanide concentration increased to about 0.25% and the whole agitated by jets of compressed air for about 3 days. Essentially, the insoluble silver sulphide passes into solution as the complex dicyanoargentate(I) ion,

$$Ag_2S + 4CN^- \rightleftharpoons 2Ag(CN)_2^- + S^{2-}$$

The removal of sulphide ion by complex reactions results in more and more silver sulphide passing into solution,

$$2CN^- + 2S^{2-} + 2H_2O + O_2 \rightarrow 2CNS^- + 4OH^-$$

the resulting solution containing the hydroxide and thiocyanate of sodium (or calcium). After filtration on a continuous rotary vacuum filter, air being excluded, silver is displaced by the addition of an emulsion of zinc dust,

$$2Ag(CN)_2^- + Zn \rightarrow Zn(CN)_4^{2-} + 2Ag$$

The residue contains 75–90% silver, excess zinc which is washed out with acid, and other base metals, removed by melting with an oxidizing flux.

About 45% of silver produced comes from lead-zinc ores while 70% of the lead produced comes from argentiferous ores.

Silver from various sources, including scrap, is refined in a two-stage process. Silver bullion is melted on a shallow hearth of basic material in a current of air for about 12 hours. This process is called cupellation. Base metals form oxides which dissolve in fused lead oxide formed by oxidation of lead present and are run off. By using oxidizing fluxes, a purity of 99.8% silver, with gold and platinum, may be obtained. Usually ingots of 98–98.5% purity are cast as anodes for electrolytic refining. The electrolyte contains silver nitrate (40–50 g dm^{-3}) and nitric acid (10 g dm^{-3}). The impure anodes corrode away and silver is deposited on thin cathodes of silver or stainless steel, being removed by mechanically operated wooden scrapers to avoid short-circuiting. The current density is about 400–500 A m^{-2}. The concentration of silver nitrate is maintained by addition of silver nitrate until the copper concentration becomes too high, when the electrolyte is renewed. Every 24 hours, silver is collected, dried and melted down for casting into ingots.

Uses

Silver is noted for excellent qualities of electrical and thermal conductance, resistance to corrosion and reflectivity. About 90–95% of visible light may be reflected from highly polished silver but the surface is adversely affected by tarnishing due to sulphur compounds. Both mechanical strength and hardness of silver are improved by alloying. In chemical engineering, silver linings are used in reaction vessels, where the high initial cost is offset by the recovery value when plant is scrapped. Silver-lined moulds are used for casting very pure caustic soda and caustic potash. Silver is used in electrical equipment as contact material in domestic switches, circuit-breakers and telephone relays. In electronic engineering, the 'skin effect', whereby current is carried by the outer skin of a conductor, is utilized in radio-frequency circuits: here, silver-clad copper or silver electrodeposited on brass, aluminium or steel gives the effective performance of pure silver.

Silver is used in electroplating. Articles of German silver ($Cu\,50$, $Zn\,30$, $Ni\,20$ per cent by weight) are cathodes in an electrolyte of potassium dicyanoargentate(I), $K[Ag(CN)_2]$, with silver anodes. A fine, coherent, strongly adhering deposit results whereas with an electrolyte of silver nitrate, the relatively huge silver ion concentration gives a loose, spongy layer,

$$Ag(CN)_2^- \rightleftharpoons Ag^+ + 2CN^-$$

The concentration of silver ions is relatively minute but ions are replenished as required; the electrode processes are:

at the cathode, $Ag^+ + e^- \rightarrow Ag$

at the anode, $Ag - e^- \rightarrow Ag^+$

$$Ag(CN)_2^- \rightleftharpoons Ag^+ + 2CN^-$$

The concentration of the electrolyte is therefore maintained.

Pure silver is called 'fine silver' to distinguish it from an alloy called sterling silver or standard silver ($Ag\,92.5$, $Cu\,7.5$, percentages by weight). Except for a period in the 17th century, this alloy has been used for plate since 1300 and until 1921, for coinage. The alloy is of excellent appearance and durability, harder and more lustrous than silver itself but still liable to tarnishing.

Gold

Occurrence

That nuggets of gold have been discovered is well known, but most native gold is widely dispersed. The deposits of the Witwatersrand in the Transvaal and other rich deposits in Orange Free State make South Africa the foremost producer. Second comes the USSR. Among other important sources are Canada (the Klondike fields), the USA (Alaska and California) and Australia (Victoria and Western Australia). There are reports of substantial deposits found in Czechoslovakia.

Gold is disseminated in veins of quartz and is liberated by weathering. With other dense minerals, gold is separated naturally by running water, accumulating in river beds and dried-up watercourses. It is frequently associated with sulphides of iron, copper, lead, antimony and arsenic, and sometimes alloyed with silver.

In passing it may be mentioned that gold can be separated by mechanical agitation with water. This may be done by hand: prospectors used shallow dishes, about 0.5 m in diameter, and 'panned' for gold by swirling with water. On the larger scale, vessels called 'rockers' are used and processes require sluicing with water under pressure and dredging. However, the principal process now used is cyanidation, similar to that used for silver.

Production

The ore is ground to a fine slime in sodium cyanide solution and aerated by agitation for 2 days:

$$4Au + 8CN^- + O_2 + 2H_2O \rightarrow$$
$$4Au(CN)_2^- + 4OH^-$$

Gold is precipitated by addition of zinc dust:

$$2Au(CN)_2^- + Zn \rightarrow Zn(CN)_4^{2-} + 2Au$$

Zinc in the precipitate is oxidized by calcining and removed as slag with other impurities by melting with borax and sand. Gold is finally cast into ingots. Further refining and separation of silver may be achieved by controlled selective chlorination.

Uses

Gold is the basis of many currencies. The metal is very malleable and ductile; a 10 mm cube may be beaten out to the size of a football field and 1 g drawn into 2000 metres of fine wire. It is not tarnished in air and is used in jewellery and ornamentation, usually alloyed with copper, or sometimes silver and zinc, to increase hardness. The purity is reckoned in terms of 24 units or carats; pure gold is 24 carat, and 22 carat, called British Standard gold, has two parts in twenty-four of copper. Electrodeposited gold and alloys with silver and platinum are used in electrical and electronic engineering where a constant contact resistance is desired. Gold is sometimes used as a lining to reaction vessels for extreme conditions but platinum, which has better mechanical properties, is preferred.

Elements 30, 48 and 80 (Zn, Cd and Hg, respectively) of outer electronic configurations $(3d)^{10}(4s)^2$, $(4d)^{10}(5s)^2$ and $(5d)^{10}(6s)^2$

With completed inner electron shells, these elements form a group which in many ways is alien to the other elements of the transition series and more akin to the heavier elements of the boron group (Group III).

Zinc

Occurrence

The chief ore of zinc is the sulphide, zinc blende or sphalerite, ZnS, found associated with the sulphides of copper, cadmium, iron, lead and silver. Marmatite, a zinc-iron sulphide ore, is also important. Sulphide ore is mined in the USA, Canada, Mexico and South America, Australia, Africa, Germany, Spain and Poland. In addition, ore is also processed in Belgium and the UK. Carbonate ores, calamine in the UK and smithsonite in the USA, are now unimportant.

Production

The ore is concentrated by froth flotation. Many lead-zinc sulphide ores, however, have minerals too well intergrown for effective separation. This last type of ore and low-grade ores of a simpler nature may now be worked by the blast-furnace smelting process introduced by the Imperial Smelting Corporation at Avonmouth. Of the several chemical methods only this will be described.

The sulphide ore is sintered by passage through an ignitor on a continuous chain grate through which air is drawn. Combustion, which is exothermic, spreads through the mass, yielding a porous clinker:

$$2ZnS + 3O_2 \rightarrow 2ZnO + 2SO_2$$

Sinter containing zinc oxide, with coke and limestone flux, is charged into a blast furnace and subjected to a preheated air blast at 550°C. Coke burns to carbon monoxide and some dioxide. Reduction of zinc oxide occurs and zinc (b.p. 907°C) distils out of the top of the furnace which is maintained at 1000°C:

$$ZnO + CO \rightleftharpoons Zn + CO_2$$
$$ZnO + C \rightarrow Zn + CO$$

To avoid oxidation of zinc as the temperature falls, the products are quenched in a spray of molten lead. Slag containing iron, with any lead and silver, is run off from the base of the furnace while waste furnace gas is burnt either for preheating air blasts or for steam raising. Zinc and lead, only partially miscible in the molten state, may be run off and 98.8% zinc separated.

About half of the zinc produced is electrolytic in origin. It is of very high quality but special problems arise. For effective competition with the smelting process, cheap electrical power is required, and for efficient deposition of zinc, very pure electrolyte. Because of the high hydrogen overpotential, zinc may be deposited with high current efficiency but deposited impurities form areas at which hydrogen is liberated. 93% of the charge passing liberates zinc, the remainder, hydrogen and impurities.

After concentration, ore is calcined to oxide as before and then treated with spent electrolyte containing 10% sulphuric acid. Undissolved solids are filtered and an extensive purification carried out. The solution is electrolysed at 35–40°C with aluminium cathodes and lead anodes. Zinc is stripped from the cathodes every 72 hours, melted in oil-fired reverberatory furnaces and cast into ingots. The purity is 99.986–99.990% zinc, with 0.013% maximum lead, by weight. While zinc ions are discharged at the cathodes, oxygen is evolved by discharge of hydroxide ions at the anodes:

$$4OH^- - 4e^- \rightarrow O_2 + 2H_2O$$

Hydroxide ions are generated from water molecules, so that hydrogen ions accumulate, the acidity rising for the next leaching out process.

The by-products are many. Sulphur dioxide is converted into sulphuric acid from which superphosphate may be produced using phosphate rock. If the ammonia synthesis is worked, ammonium sulphate can be manufactured. During the electrolytic production of zinc, copper and cadmium metals and cobalt(II) oxide are separated as valuable by-products.

Uses

Pure zinc has high resistance to atmospheric corrosion. Its chief application is as a coating on iron and steel to prevent rusting. The motor car industry now uses it extensively. After surface damage, the more electropositive zinc is corroded away so that rust does not appear. Zinc is applied chiefly by hot-dip galvanizing, in which articles are immersed in molten metal at 450–500°C, and by electro-galvanizing with the articles as cathodes, using zinc anodes in an electrolyte of either zinc sulphate and sulphuric acid or alkaline potassium cyanide solution. Small articles may be sherardized, which involves treatment with zinc dust at 350–400°C in a revolving container. Nuts, bolts, springs, small intricate castings and window frames may be sherardized. For bridges, pylons, storage tanks and ships' hulls, spraying 'atomized' molten zinc is more convenient, and for minor applications, the use of paint containing metallic zinc.

Pure zinc is of moderate tensile strength but easily worked. 98.5% zinc is rolled into sheets, for roofing, dry-battery cells, engraving plates and linings.

The chief use of zinc in alloys is in the brasses, the most widely used non-ferrous alloys, where up to 45% zinc is added to copper. For castings, alloys with aluminium, magnesium and copper have been developed as well as alloys in which zinc is a very minor constituent.

Cadmium

Occurrence

The chief mineral is greenockite, cadmium sulphide, CdS. Up to 0.4% cadmium may be found associated with zinc and lead ores. Because of its volatility, cadmium accumulates in flues, electrostatic precipitators and other purification apparatus in lead, copper and zinc works. There is some alarm about ground pollution by cadmium in certain nearby areas.

Production

Cadmium-bearing material from zinc and lead plant is treated with dilute sulphuric acid and the liquid is brought to the boil by steam heating. Lead(II) sulphate is left as a residue after decantation, zinc and cadmium passing into solution. After careful purification, spongy cadmium is precipitated by zinc dust and refined by distillation from graphite retorts at about 1000°C into cast-iron condensers,

$$Cd^{2+} + Zn \rightarrow Zn^{2+} + Cd$$

Uses

Cadmium is a silvery-white, soft, ductile and malleable metal. Electroplating is the chief outlet of cadmium: small steel articles are protected from rusting in a way similar to the protection afforded by zinc. Cadmium-plating is applied to radio and telephone components made of copper and brass. Cadmium may form the negative plate of alkaline electrical storage cells with an electrolyte of caustic potash. In nuclear engineering, cadmium rods have been used in the regulators of atomic piles for the conversion of uranium into plutonium.

Cadmium is a constituent of certain bearing alloys, solders and fusible alloys. It is used to strengthen copper for long spans of the high-voltage overhead electric cables of the National Grid.

Mercury

Occurrence

The principal ore of mercury is cinnabar, mercury(II) sulphide, HgS. The mercury content by weight is usually less than 1% but Spanish ore may contain 5–7%. Mercury is produced from ore mined in Spain, Italy, Yugoslavia, Peru and the USA. The principal countries in which mercury is extracted are Spain, the USA and Yugoslavia.

Production

Beyond hand-picking, crushing and mechanical sifting there is no preliminary treatment of the ore. Mercury(II) sulphide is roasted in air and mercury vapour escapes with sulphur dioxide and other gases,

$$HgS + O_2 \rightarrow Hg + SO_2$$

The concentration of mercury in the gaseous product is exceedingly small and presents special problems. Further, mercury vapour is very toxic. The most efficient furnaces are the rotary type used in the USA. Ore passes down the rotating kiln, 15–16 m long and 1–1.5 m diameter, inclined at a gradient of about 1 in 12, or less. Flames from an oil burner play directly on to the ore reaching the bottom, and hot gases pass up the kiln, decomposing the mineral. Mercury vapour passes through water sprays or electrostatic dust precipitators and so to the condensing system of many connected vertical pipes of glazed stoneware, alloy steel or painted sheet iron. Condensed mercury drips through the open bottoms of the pipes into water in a trough. Filtration through canvas separates mercury from solid 'soot' which is worked up for further yields of mercury.

Uses

The metal is used in scientific instruments, mercury vapour lamps, arc rectifiers, electrolytic cells for the production of pure caustic alkali and in amalgam alloys.

11

Industrial alloys

The need for alloys

Most pure metals are not used as such in engineering because of poor mechanical properties. However, the properties of a metal may be modified by the addition of other elements, usually metals but sometimes metalloids or non-metals, and the resulting product, if it has metallic physical properties, is called an *alloy*. The physico-chemical equilibria of alloys and their crystal structures will not be discussed in more than outline but it may be noted that compounds may be formed: cementite, Fe_3C, in steel is a simple example.

Alloys are generally prepared molten but this is not always possible. For very high-melting-point metals, of which molybdenum (m.p. 2895°C) and tungsten (m.p. 3400°C) are examples, the technique of powder metallurgy is employed: fine powder is compressed into suitable shapes using pressures of about $750 \, N \, mm^{-2}$ followed by heating. Also, by the careful regulation of current density, electrolyte composition and temperature, direct electrodeposition of certain alloys may be accomplished. Brass may be deposited electrolytically under suitable conditions.

In describing the characteristics of metals, including alloys, for the purposes of this chapter, certain terms need to be defined. The general behaviour of a metal, described as its mechanical properties, depends on its strength, malleability, ductility, hardness, toughness and resistance, or otherwise, to corrosion. Strength refers to resistance to applied stresses and tensile strength to the load-supporting characteristics. Hardness describes resistance to cutting, abrasion and indentation, while toughness is the impact strength or resistance to fracture by impact. Rolling-out into sheets, which involves deformation by compression, demands high malleability whereas wiredrawing, deformation by tensile stress, requires

high ductility. Whatever the mechanical properties, however, breakdown of a metal after much use may come about through fatigue, in which cracks start at points of slight structural defect, or points subjected to corrosion or which have been scratched. Metal fatigue became headline news after the crashes of the first jet airliner, the DH *Comet I*.

Some alloys are suitable for casting. This may be done in sand or metal moulds. Die-casting is the production of small and medium-sized articles in quantity by injecting molten metal under pressure into a steel mould or die, when rapid solidification leads to fine grain and good surface texture, truer reproduction and good mechanical properties. Other alloys are cast into ingots, which are subjected to mechanical working to provide stock from which shapes are machined or to yield the required shape direct. Such alloys, whose desired properties develop with working, are termed wrought alloys. The processes may be listed: forging, rolling, pressing, drawing and extruding. The first and last are essentially hot-working processes, pressing and drawing are cold-working processes starting from wrought stock, while rolling may be done hot or cold depending on the characteristics of the metal.

Alloys are usually divided into ferrous and nonferrous, which recognizes the dominance of iron and steel. Steels are alloys of iron containing up to 2% carbon, either combined or in solid solution with other elements incorporated for special properties. With more carbon, the brittle cast iron, or pig iron, results. The most widely used nonferrous alloys are based on copper.

The nature of solid phases in alloy production

Usually homogeneous liquid solutions result when pairs of metals are mixed in various proportions and melted. However, on solidification there may be more than one phase present, each of these being homogeneous until atomic dimensions are met, but distinct and perhaps of very small grain size. There are three principal types of solid phase in alloys: *pure metals, solid solutions* and *intermediate phases (intermetallic compounds)*.

When the atoms of the metals are closely similar in size, a *substitutional solid solution* may be formed in which the minority atoms are distributed randomly in the crystal lattice favoured by the majority atoms: copper and zinc form such a solid solution. However, these metals have different lattice structures and while zinc dissolves in copper in an amount up to *c.* 40 mol%, copper will only dissolve in zinc up to 2–3 mol%. On the other hand copper and nickel have the same crystal structure and similar radii, leading to a continuous range (0–100%) of solid solutions.

Where the size of the atoms differ, *interstitial solid solutions* may be formed in which the smaller element is accommodated in the regular spaces found in the three types of metallic lattice, which are described on pp. 101–5. This type of solution is particularly important when certain non-metallic elements are dissolved (e.g. carbon-iron: martensite) in a metal.

Intermediate phases, or *intermetallic compounds*, are quite different. The two (or more) types of atom may be arranged in a regular repeating pattern in the lattice or apparently at random. Thus, 'CuZn' has the CsCl structure for its lattice, each metal forming its own simple cubic lattice, eight atoms of one surrounding one atom of the other, the two different sub-lattices being components of a so-called *superlattice*. Above 733 K the mixture forms a body-centred cubic lattice with random distribution as distinct from the face-centred cubic structure favoured by copper and the hexagonal close-packed lattice of zinc. The phase may have a range of composition.

A discussion of the phase diagrams for the various types of alloy equilibria would take up too much space here. An advanced text on Physical Chemistry should be consulted and the various types of diagram interpreted with the attendant cooling-curves in mind. The lattices adopted by the metals in this discussion are shown on pp. 98 and 101–5.

Alloys based on copper

The brasses are the principal alloys and consist of copper with zinc. One group, containing up to about 37% zinc, is suitable for cold-working processes, such as press-work. *Cartridge Brass* [Cu 70 ± 2, Zn 30 ± 2]† which is outstanding in this respect is used in the production of electric light bulb caps, door furniture, headlamp reflectors, radiator casings, as well as cartridge and shell cases. These alloys resemble copper in being tough, malleable and ductile. *Admiralty Brass* [Cu 70, Zn 29, Sn 1] was used for marine condenser tubes on account of the extra resistance to corrosion afforded by tin, but has been replaced by *Aluminium Brass* [Cu 76, Zn 22, Al 2, (As)††] which combines the working qualities of Cartridge Brass with extra resistance to corrosion. *Gilding Metals* [Cu 85–95, Zn 15–5] show a range of shades from the red of copper to brassy yellow as the proportion of zinc rises. With 37–45% zinc, brasses are much less plastic and are worked hot. Notable is *Muntz Metal* or *Yellow Metal* [Cu 60, Zn 40] used for a range of engineering purposes; addition of lead forms *Leaded Brass* which has improved machining quality. For structural purposes at sea, *Naval Brass*, Muntz Metal to which has been added about 1% tin, shows greater resistance to corrosion by sea water.

Alloys of copper and tin are called bronzes. With zinc added too, gunmetals result. Bronzes are tough, strong and corrosion-resistant. *Wrought Bronzes*, containing up to 8% tin, are used as wire and strip for springs; addition of phosphorus (0.4%) yields *Phosphor-Bronzes* of improved quality. For casting quality, more tin is added. *Bell Metal* [Cu 80, Sn 20] which is brittle but sonorous, *Speculum Metal* [Cu 60–70, Sn 40–30], which is also brittle but has a white, lustrous surface formerly used for mirrors, and *Leaded Bronzes* [Cu 70–85, Sn 5–10, Pb 25–5] for bearings, are examples of bronzes for casting. *Admiralty Gunmetal* [Cu 88, Sn 10, Zn 2] is cast for marine purposes, valves and fittings for steam plant and ornamental bronzework. *Aluminium Bronzes*, with 4–7% aluminium giving the extra corrosion resistance of an aluminium oxide film, are golden in colour and used in jewellery, cigarette-cases and other accessories. The aluminium bronzes are also used in the form of plate and tube for heat-exchanges, etc., in chemical plant, a use which applies the excellent heat conductance of copper and its good corrosion-resisting properties with the extra protection of aluminium. With 9–10%

† parts per hundred by weight

†† (As) indicates a trace of arsenic

aluminium and 2% iron, strong, tough and cor-rosion-resistant aluminium bronzes result which are used for pump components and valve fittings, propellers, wheels and sluice-gate fittings.

Alloys of copper and nickel are called *Cupro-Nickels* and with zinc added, *Nickel Silvers* on account of their appearance although no silver is present. Up to 3% nickel in cupro-nickel improves mechanical properties, 5–10% nickel gives resist-ance to sea water for marine engineering, and 20–30% nickel, an alloy for marine condenser tubes. *Nickel Silvers* [Cu 55–63, Ni 30–10, Zn 15–27] are used in decoration, for cutlery and in articles to be silver-plated.

Traces of other elements, notably silver, cad-mium, beryllium, tellurium and chromium, are added to copper for specific purposes. *Beryllium Copper* [Cu 98, Be 2] shows an exceptional increase in toughness.

The coinage alloys

Copper, silver and gold are often still called 'the coinage metals'. Silver and gold have now been withdrawn in the UK. Before 1921, silver coinage was mostly silver [Ag 92.5, Cu 7.5] and after 1927 until World War II, half silver [Ag 50, Cu 40, Ni 5, Zn 5]. In the interim period, alloys of slightly dif-ferent composition, [Ag 50, Cu 40, Ni 10] and [Ag 50, Cu 50] were minted. A cupro-nickel alloy, also used for the five-cent piece or 'nickel' in the USA, has replaced silver. 'Silver coins' [Cu 75, Ni 25], and 'coppers' [Cu 95.5, Sn 3, Zn 1.5] are the coinage alloys of the UK. They prove strong and resistant to wear and corrosion.

Light alloys based on aluminium, magnesium, titanium and beryllium

Table 11.1   The light alloy metals

	Atomic number	Density/g cm^{-3}	Melting-point/°C
Beryllium	4	1.85	1283
Magnesium	12	1.74	651
Aluminium	13	2.70	660
Titanium	22	4.51	1800

Alloys of aluminium have been developed which are light, strong and very resistant to atmospheric corrosion. Principally marketed in rolled sheets and extruded form, much is also used in casting,

forging and drawing. Very pure aluminium is soft and ductile, of high electrical and thermal con-ductance, of only a third the density of copper and iron, and able to resist corrosion through a tenaciously adherent film of oxide, spontaneously formed by the action of air. Except for the attack of alkalis, halogen acids and sulphuric acid, it re-sists corrosion by chemicals. However, because of its low strength, the pure metal is of little engineer-ing use. Strength and harness may be increased by addition of other elements, notably copper, mag-nesium, silicon, manganese, zinc and nickel, some-times with sacrifice of corrosion resistance, especially with copper. Commercial aluminium (99.3–99.6% pure) contains iron and silicon, is usefully stronger and may be rolled into sheets. Wrought alloys generally contain up to 6% and casting alloys up to 15% of alloying elements. Alloys with up to 10% magnesium are excep-tionally resistant to corrosion by sea and coastal conditions, besides being of lower density than aluminium, and are used in ship and boat construc-tion. For hollow-ware, packaging and building, 1–1.5% manganese is added. The *Duralumin* type of alloy [Al with Cu 1.5–5, Mn 0.75, Mg 0.5–0.75, Si 0.5–0.75], widely used for stressed structures as in aircraft, has added strength due to copper but less resistance to corrosion. Sheets of duralumin are cladded with a layer of pure aluminium on each side. The omission of copper gives alloys [Al with Mg 0.5–1, Si 0.5–1, Mn 0.5–1] with more resistance to corrosion. Alloys of exceptionally good strength–weight ratio, for the aircraft industry, containing zinc [Al with Zn 5–7, Mg 2–2.5, Cu 1–2] have been produced but they are difficult to fabricate. For casting alloys, as distinct from the wrought alloys above, 5–12% silicon gives high fluidity, good reproduction and reduced thermal expansion. Silicon alloys are used for cylinder heads and with nickel added, pistons, where good heat transfer is also important, crank-cases and gearboxes. Casting alloys also incorporate 1.5–10% magnesium and 4–10% copper. To sum up: aluminium alloys are used for lightness, strength and resistance to corrosion for air-frames, ship superstructures and vehicle bodies, for engine components, in the chemical, brewing and food industries, and in building construction. Finally, although aluminium has but 60% of the elec-trical conductance of copper on a volume com-parison, on a weight basis it is superior. Aluminium wire, stranded on a steel core and, a later develop-ment, special aluminium alloys with 54% of the conductance of annealed copper, are used for overhead electrical transmission cables.

Magnesium is the lightest constructional metal, with a density only two-thirds that of aluminium. It is not highly flammable as might have been supposed. The pure metal, while stronger and harder than aluminium, is unsuitable by itself for engineering purposes and alloys are preferred. Until the end of World War II, the principal alloying elements were aluminium, zinc and manganese: the first two for hardness and strength, the last to improve corrosion resistance. Alloys of this type, suitable for casting or wrought purposes, with up to 9.5% aluminium, 1.5% zinc and 0.3% manganese, are used in aircraft and general engineering, for textile machinery, portable tools, vacuum cleaners, cameras and binoculars, etc. In World War II, aircraft landing wheels, of which nearly one million were manufactured for British aviation, were made from this type of alloy [Mg with Al 8, Zn 0.5, Mn 0.3]. Light alloys of improved strength at high temperatures result from the addition of 0.6–0.7% zirconium, up to 5.5% zinc and, in certain alloys, 1.2–3.0% rare earth elements or 1.8–3.0% thorium. These alloys are used extensively in the aircraft industry, for parts of jet and turboprop engines, landing wheels and undercarriage forks, and aircraft skins, etc., to name a random selection. Magnesium alloys are used also in car engines and tractor transmission cases. Normally, magnesium alloys resist corrosion by the atmosphere but may be further protected by epoxy resin paints or anodizing; they lend themselves to all the usual methods of fabrication, but are usually welded in an inert gas atmosphere by an electric arc between tungsten poles.

Titanium and beryllium are more difficult to produce in a useful industrial form and have specialized applications. Titanium has a very high melting-point and a density which is about half that of copper and iron, the traditional engineering metals. Titanium alloys will withstand fatigue at high temperatures and are used as light alloys, in place of the much heavier steels, for applications in ultra-high-speed flight, compressor blades and by-pass ducts in aircraft engines. These alloys retain their strength at temperatures higher than those considered safe for other light alloys and prove resistant to corrosion by hydraulic fluids and fuels. They are also used in steam turbines and turbo alternators. Beryllium metallurgy was developed for nuclear engineering but alloys for high-speed flight and missiles have been made.

Alloys developed for resistance to chemical attack

The first group is based on *nickel*. Combining good resistance to corrosion, except under oxidizing conditions, with excellent mechanical properties, nickel is one of the toughest metals and retains much strength and ductility at high temperatures. The element may be used with all types of fresh water, sea water, alkalis and various salts but not with mineral acids or where oxidizing conditions prevail. A passive film of oxide affords protection. Copper has high ductility, moderate strength and good resistance to wear and chemical corrosion, except against the strong mineral acids in air, and against ammonia. Combining the optimum properties of strength and resistance to corrosion of the chief constituents, the *Monel* range of nickel–copper alloys contain roughly two-thirds nickel and one-third copper, with manganese and iron in small proportion and silicon added for casting quality [Ni 63–68, Cu balance with Fe up to 3.0, Mn 0.5–1.5, Si 0.5–4.0, C 0.30–0.12]. Monel is used for valve and pump components in steam power plant on land and sea, turbine blading, marine equipment, in chemical engineering for processing alkalis, dry gases and non-oxidizing acids generally, and in equipment for the dyeing, food, laundry and dry-cleaning industries. Superior to Monel under acidic oxidizing conditions is the nickel–chromium range of alloys. Chromium is hard, brittle and extremely resistant to corrosion. *Inconel* [Ni 76, Cr 15, Fe 8, Mn 1] is used for the construction of food-processing equipment, reaction vessels and evaporators in chemical plant, for photographic developing equipment and dyeing equipment in the textile industry. For the best resistance, among industrial alloys widely used, to boiling hydrochloric acid, excellent resistance to boiling phosphoric acid and useful resistance to sulphuric acid, molybdenum is incorporated to give *Corronel B* [Ni 66, Mo 28, Fe 6], which is used in chemical plant where very corrosive conditions are encountered. For resistance to hot, strong acids and mixtures of nitric acid and sulphuric, phosphoric and hydrofluoric acids both chromium and molybdenum are used with iron in nickel alloys of the *Ni–O–Nel* range [Ni 40, Fe 31, Cr 21, Mo 3, Cu 1.75 (Mn, Si, C)]†.

The *stainless steels* are a range of alloys showing marked resistance to chemical attack, wear and

† (Mn, Si, C) indicates a trace of manganese, silicon and carbon

high temperatures. The rate of stainless steel production, half of which goes into consumer durable goods and the rest into capital goods industries, is a fairly good indicator of economic activity generally. It is one of the pointers used by economists to judge the overall health of metal-using industries and the long term outlook for the prices of their raw materials. Stainless steel production slumped in the 1981–83 world recession, the worst period for base metals producers since the Great Depression (1929) but recovered with demand far outstripping economic growth in 1987–88. Nickel producers could not keep pace with the increased demand so that in March 1988 the price reached $23,900 a tonne on the London Metal Exchange. Steel containing 12–20% chromium is used for cutlery and surgical instruments. Nickel is usually added to improve working. Stainless steels are used in the construction of the outer fuselage of ultra-high-speed aircraft.

Resistance to atmospheric corrosion, brine and such powerful chemicals as nitric acid, aqua regia, chlorine and metallic chlorides make *titanium and its alloys* of importance to the chemical industry for the construction of reaction-vessel linings and complete components like agitators and thermostat sheaths. Titanium alloys are not attacked by body fluids and are replacing stainless steel and cobalt alloys for making surgical implants of which internal splints and bone fixation screws provide examples. Zirconium is also highly resistant to corrosion, being superior to titanium in concentrated hydrochloric acid and fused caustic soda but inferior in chlorine, iron(III) chloride solution and aqua regia. *Zirconium alloys* have similar chemical uses to alloys of titanium.

Lead is suitable for certain types of equipment in chemical plant, notably for the production of sulphuric acid. Addition of up to 12% antimony brings hardness to lead so that sheet lead, pipes and other fittings may be machined. Hardened lead is also used in the lead-plate accumulator.

Precious metals are used in chemical engineering and in laboratory apparatus to combat corrosion. Platinum and its alloys are used to withstand molten glasses which require protection from contamination. They have the strongest resistance of all readily worked metals to corrosion. Platinum and platinum–iridium alloys are used as catalytic gauzes in the production of nitric acid by oxidation of ammonia. In this and other catalytic processes, the metal withstands highly corrosive conditions. An alloy of gold with platinum [Au 70, Pt 30] is

used to construct the spinning jets through which rayon filaments are formed by extruding a strong, alkaline viscose solution into a coagulating bath containing acid: high corrosive extremes. Linings of gold and silver on copper-base metals are used in chemical plant, gold for reaction vessels exposed to highly corrosive contents and silver for components and linings in the application of organic chemicals, chlorine and caustic alkalis.

Heat-resisting alloys

Alloys based on the composition, Ni 80 : Cr 20, have been developed for use in gas turbine engines. This range, called *Nimonic Alloys*, have small additions of cobalt, titanium and aluminium and will retain strength and resist oxidation at high temperatures. *Inconel*, already mentioned above, may be used in a sulphur-free atmosphere for furnace components, enamelling racks, protective sheaths for the electrical elements in cookers, and for aircraft exhaust manifolds.

Cobalt–chromium–tungsten alloys of the *Stellite* range [Co 40–60, Cr 30, W 0–20], originally developed for stainless cutlery, will retain hardness at high temperatures and are used in high-speed tools. Other *cobalt-based alloys* [Co 60–65, Cr 20–30, Mo 5] are used in jet and gas turbine engines where tremendous wear and stress at high temperatures are encountered.

Titanium alloys which stand up to stress at high temperatures are used for ultra-high-speed flight, for fireproof bulkheads and exhaust shrouds.

Of steels, *chromium and cobalt steels* withstand high temperatures. With 20–35% chromium, resistance to conditions of oxidation and corrosion at high temperatures is developed and steels of this chromium content have been used for jet engines and gas turbine engines, nose-cones of earth satellites, and chemical plant. Up to 13% cobalt is incorporated in high-speed tool steels, because of retention of hardness, corrosion and oxidation resistance at high temperatures. *Invar*, a nickel steel [Fe 64, Ni 35, (Mn, C)]†, has an extremely low temperature coefficient of expansion and is used for pendulum rods.

Alloys developed for electrical resistance and electrical heating

Certain ductile alloys, with high specific resistance and negligibly low temperature coefficient of

† (Mn, C) indicates a trace of manganese and carbon)

resistance, prove ideal for the manufacture of fixed resistances in electrical instruments. These alloys are usually copper based. With nickel, *Constantan* [Cu 60, Ni 40] is well known and the addition of manganese yields alloys of the *Manganin* type [Cu 84, Mn 12, Ni 4] while there is a similar range of *copper–manganese–aluminium* alloys [Cu 85, Mn 13, Al 2]. *Nickel silver* [Cu 55–63, Ni 10–30, with Zn] is used and also the alloy, *Rheotan* [Cu 52, Ni 25, Zn 18, Fe 5]. Precious metals may be used where corrosion resistance is very important; *palladium–silver alloys* [Pd 60, Ag 40] are endowed with the resistance to corrosion characteristic of palladium, and have maximum hardness and electrical resistivity with an extremely low temperature coefficient. They find application in the potentiometer windings of special instruments.

Alloys used for the electrical heating elements of fires, stoves and furnaces must give long service under conditions of intermittent heating and cooling, involving an absence of sagging, brittleness and tendency to oxidation. Nickel-chromic alloys [Ni 80, Cr 20] form the basis of one important group, of which *Nichrome* [Ni 60, Fe 20, Cr 20] is frequently used.

Alloys for nuclear engineering

Metals used in reactors must absorb relatively few neutrons, upon which the chain reaction depends, have good mechanical strength up to the temperature (about 500°C) reached and be very resistant to corrosion. *Zirconium and its alloys* have been used extensively for the fabrication of sheathing cans for the fuel elements and core components of boiling-water reactors and the pressurized water reactors of the nuclear submarines, the earliest of which were USN *Nautilus* and RN *Dreadnought*. The difficulties encountered in reactor design and control are underlined by the news, which reaches the West from time to time, of the breakdown (or worse) of reactors in Soviet submarines. Alloys of zirconium, which are resistant to hot carbon dioxide coolant gas up to 500°C, have been developed for use in the Berkeley gas-cooled type of nuclear power station, for fuel element supports, flexible hose and packings. Other applications of zirconium alloys include the cladding of graphite moderators in liquid sodium-cooled reactors to avoid interaction of the materials.

Sheathing metal for reactor fuel must absorb few neutrons, contain the radioactive metal and fission products safely without corrosion, be chemically inert to coolant and conduct heat well. *Beryllium* is replacing magnesium in alloys for use in gas-cooled reactors and equipment designed to operate up to 600°C instead of 450–470°C. Beryllium brings superior strength and elasticity but not ductility to magnesium–aluminium alloys. *Magnox* (Mg with Al 0.75, Be 0.005) has been used for fuel cans. The early generation of Magnox nuclear power stations is soon to be decommissioned with all the attendant problems of what to do with the reactor pile and other parts of the plant which may have residual radioactivity. Titanium and certain alloys are used in control rod sheaths and handling gear.

Special steels

Alloys of iron, containing up to 2% carbon in the combined state or as solid solution, are called *steels*. Mild steel contains relatively little carbon and is soft and ductile. Increase in carbon content increases the tensile strength with consequent loss in ductility. Heat treatment modifies the properties of steel. Heated to redness and plunged into cold water, or quenched, steels develop hardness. Tempered steel has been heated to a carefully predetermined temperature to achieve softening. Addition of certain ferro-alloying elements bring special properties.

Vanadium steels are for general engineering purposes. Excellent welding characteristics arise from small quantities, 0.1–0.25% V, while a large proportion, 1–5% V with chromium and tungsten, is used to make high-speed tools noted for hardness retention over a moderate temperature range. Vanadium steels are used in high-pressure boilers, locomotive parts, gears and turbines.

Chromium steels have increased strength, resistance to wear, oxidation and corrosion. Armour plate, ball-bearings and tool steels contain 1–2% Cr, high-speed tool steels 4% Cr, grinding machinery 3–12% Cr, stainless steel for cutlery and surgical instruments 12–20% Cr, and with 20–25% Cr, resistance to extremes of temperature, oxidation and corrosion previously mentioned appear.

Manganese steels are very hard and have high tensile strength with marked resistance to wear. Railway points and cross-overs in the UK contain 0.9–1.2% Mn while steel for rock-crushing equipment contains 14% Mn.

Cobalt steels, containing up to 13% Co with tungsten, are hard, stainless and retain good mechanical properties at high temperatures. They are used in high-speed tool steels.

Nickel steels have strength, resistance to wear and withstand mechanical shock and strain. Up to 5% Ni gives steels for gears, steering components, transmission parts, drills and armour plate.

Type metals and master pattern alloys

Metals usually contract on cooling but some expand. Tin and lead of Group IV contract while antimony and bismuth of Group V expand. The former are metals of low melting-point and are notable for ease of casting while the latter, when added to alloys, counter shrinkage. *Type metal* [Pb 60–80, Sb 13–30, Sn 3–10], *Linotype metal* [Pb 83, Sb 12, Sn 5] and *Stereotype metal* [Pb 52–67, Sb 12–23, Sn 3–17] have little or no contraction on solidification following casting: antimony counters shrinkage and hardens lead while tin improves casting quality. Sharply defined castings, necessary for master pattern making, may be formed by alloys of bismuth and tin or bismuth and lead: a eutectic mixture of bismuth and lead [Bi 55.5, Pb 44.5], melting-point 124°C, may be used for this purpose.

Fusible alloys and solders

Fusible alloys are usually, or nearly, eutectic mixtures and therefore have sharp melting-points, instead of gradually melting over a range of temperature. A simple eutectic, a binary mixture of lead and tin (Pb : Sn = 1 : 2), melts sharply at 180°C. Fusible alloys are based on metals of low melting-point and are ternary (i.e. with three metals as components) or quaternary (i.e. with four metals) eutectic systems. The usual metals for these alloys include tin (m.p. 232°C), bismuth (m.p. 273°C), cadmium (m.p. 321°C) and lead (m.p. 328°C). Alloys associated with the names *Rose* [Bi 50.0, Sn 22.0, Pb 28.0] and *Darcet* [Bi 50.0, Sn 25.0, Pb 25.0] are based on the eutectic mixture [Bi 52.5, Sn 15.5, Pb 32.0] and melt at approximately 96°C. Metals associated with *Lipowitz* [Bi 50.0, Sn 13.3, Pb 26.7, Cd 10.0] and *Wood* [Bi 50.0, Sn 12.5, Pb 25.0, Cd 12.5] are based on the eutectic mixture [Bi 49.5, Sn 13.13, Pb 27.27, Cd 10.10] and melt at about 70°C. Boilers are fitted with fusible plugs at the lowest safe water-level, which, when no longer water-cooled, melt and are ejected by pressure of steam. Wood's metal is used in automatic fire alarm and sprinkler systems where the valve, holding back the water supply, is mounted in a support constructed with solder of the fusible alloy.

Solders are alloys used for joining metals. They should be of low melting-point, with good wetting and adhesive power. The metal surfaces are cleaned with a flux, often incorporated in the core of solder sticks during manufacture. Various combinations of tin, lead and antimony are employed. Resin is an example of a flux. *Examples of solders* include alloys for electrical apparatus [Sn 95, Pb 4.5, Sb 0.5], tinsmiths [Sn 60, Pb 39.5, Sb 0.5], general purposes [Sn 50, Pb 47.5, Sb 2.5] and for wiping joints in plumbing [Sn 32, Pb 66.4, Sb 1.6].

Bearing metal

One important class of bearing metals, the white bearing metals, contains hard crystals imbedded in a relatively soft matrix. These alloys must have good mechanical and antifrictional properties. *Babbit metals*, named after the discoverer and also called white metals, used for bearings in cars and railway rolling stock, are alloys of lead–antimony–tin with 12–18% Sb and either 0–5% Sn or 10–12% Sn, the remainder being lead except for 0.5–2.0 Cu added for hardening. Heavy-duty bearing metal for diesel and aircraft engines may use up to 30% lead with copper and some tin. Other bearing metals are *Frary metal* [Pb 97, Ba 2, Cu 1], *leaded bronze* [Cu 70–85, Sn 5–10, Pb 25–5] already mentioned under copper-based alloys, and *antimonial bronze* [Cu 79, Pb 10, Ni 8, Sn 2, Sb 1].

Magnetic alloys

Because of their ease or working, alloys of copper with iron, cobalt and nickel which happen to be strongly magnetic are manufactured for use in instruments. *Cunife* [Cu 60, Ni 20, Fe 20] and *Cunico* [Cu 50, Ni 21, Co 29] are of importance. Other examples include the *Heusler Alloys* [Cu 65, Mn 25, Al 10].

Miscellaneous alloys of interest

1. (*Old*) *Pewter* [Sn 95, Sb 4, Cu 1] and (*New*) *Pewter* [Sn 80, Pb 20] are used for household utensils and ornaments.
2. *Dental alloys* (amalgams) are formed by grinding mercury with an alloy of tin and silver (Ag_3Sn). Copper amalgam is also used presumably as a cheap substitute for the gold fillings of yester-year.

3. *Cast iron* contains up to 4.5% carbon with Mn, P, Si, S.

4. *Devarda's alloy* [Cu 50, Al 45, Zn 5] is used for reduction of oxonitrogen salts to ammonia using alkali.

The production of steel

Commercial iron—pig iron—may contain up to 10% by weight of impurities, accounting for up to 25% of the volume. The carbon content is 3–4%. Generally, the carbon content of steels lies in the 0.15–0.25% range. The main difference between iron and steel lies in this carbon content. About 85% of all pig iron is converted into steel while the remainder is refined and the composition adjusted for use as cast iron for the manufacture of pipes and guttering, drain and manhole covers, stoves, etc. On solidification, expansion occurs so that a sharp casting is obtained. Relatively pure iron is available in the form of wrought iron produced by refining pig iron with haematite when the impurities are oxidized in reducing the added ore to iron. Slag is squeezed out under steam hammers and as the impurities are removed, the melting-point rises and the mass becomes pasty. Before the principles of steelmaking were introduced by Bessemer in 1856, wrought iron was in large demand for industrial and engineering processes but it is not as strong as steel.

A typical mild steel may have the following percentage by weight of elements alloyed with iron:

$$0.15\,C \quad 0.03\,Si \quad 0.05\,S \quad 0.05\,P \quad 0.50\,Mn$$

The refining process involves the selective oxidation of impurities which form products collectively known as slag. These are immiscible with steel and are removed while the gaseous products escape. This is followed by addition of alloying elements to bring the composition to the desired level. The original pneumatic process, developed by Bessemer for the production of steel, used converters lined with silica brick and sand which, being acidic, did not permit the re-removal of phosphorus and sulphur although the content of carbon, silicon and manganese could be controlled. The elements come from the original ore, the coke and fluxes used in the blast furnace. About 20 years after the Bessemer process came into use, Gilchrist and Thomas introduced the basic lining technique, by which the phosphorus content can be removed and sulphur reduced in addition to the control of other elements. This led the way for the development of

important steel centres in Europe where iron ore of high phosphorus content is available. In 1863, the Siemens brothers designed the open hearth furnace, overcoming the problems of heat conservation by a preheating system. Nowadays, roughly 9 out of every 10 tonnes of steel produced in the UK is open hearth steel. These processes were developed in the UK. More recently the application of 'tonnage' oxygen (see Chapter 12) in Austria has brought a further revolution to steelmaking: the L–D (Linz–Donawitz) process.

Hitherto, about 87% of steel was made in the UK by the Open Hearth process, some 7% by the Bessemer process and 6% using electric furnaces. With the appearance of cheap oxygen in large quantities and the increased supplies of cheaper electric power, the L–D process and electric furnace methods are likely to account for about 12% each of the UK production. The Open Hearth proportion will probably drop to under 70% of the total production. Such figures must be treated with caution in the light of economic difficulties of the steel industry in the face of competition from Far Eastern countries with cheaper labour costs.

In steelmaking, the phosphorus content of the ore has been seen to determine the type of lining and refining used. 'Acid' and 'Basic' steelmaking processes are in operation where the terms refer to the type of refractory lining, i.e. silica or calcined dolomite respectively. By far the most important in the UK is the Basic process. Steelmaking is carried out in areas where high-grade haematite was originally available. Basic processes will be described here.

The Open Hearth process

Molten pig iron of known composition is heated with the calculated quantities of haematite, any available scrap steel and calcined dolomite, $MgO.CaO$, or limestone, in a shallow hearth furnace. The temperatures attained are in the 900–1650°C range. The basic lining is in the form of refractory brick with an outer steel casing. When the charge has melted, most of the silicon has been oxidized and passed into the slag. Carbon forms the monoxide which escapes. Oxygen comes partly from the air and partly from the haematite, iron(III) oxide. Further heating removes the remaining silicon and carbon, and also phosphorus and sulphur, although most of the last named is usually removed by the addition of soda ash (anhydrous sodium carbonate) during a pre-refining process. Silicates and phosphates pass into the slag:

$$Fe_2O_3 + 3C \rightarrow 2Fe + 3CO$$

$$10Fe_2O_3 + 12P \rightarrow 20Fe + 3P_4O_{10}$$

$$P_4O_{10} + 6CaO \rightarrow 2Ca_3(PO_4)_2$$

$$2Fe_2O_3 + 3Si \rightarrow 4Fe + 3SiO_2$$

$$SiO_2 + CaO \rightarrow CaSiO_3$$

A furnace making 80–100 tonnes of steel takes 10–14 hours for refining. However, the process enables a very high-quality product to be formed.

The mass may be heated by burning oil, natural or coal-produced gas. Burning oil under pressure is squirted into preheated air in the region between the surface of the melt and the heat-resisting low roof of the furnace. The waste gases pass through a set of chambers filled with open chequered brickwork. Periodically the direction of the air and waste-gas flow is reversed and the conserved heat taken up by the incoming air. In gas furnaces, both gas and air are heated separately and mix above the hearth.

The direct application of oxygen has been found to shorten the working life of the furnaces due to the intensity of oxidation, leading to violent 'slopping' and the excessive escape of brown fumes of iron(III) oxide. Some pre-refining with an oxygen jet is practised in the ladle before furnace-refining with the addition of 1–2% limestone.

During refining, addition of scrap steel, etc., may be made through sliding doors in the side of the furnace. On completion of refining, after analysis of samples, the furnace is tapped. Larger furnaces are equipped for tilting . The steel is collected in ladles into which the alloying elements, in the form of ferromanganese, ferrosilicon, etc., are added. Open hearth furnaces may have a capacity of up to about 360 tonnes or be as small as 60 tonnes capacity.

The Bessemer process

Steelmaking by the pneumatic method is rapid. The air blast at 140–170 kPa ($kN\ m^{-2}$), often enriched by oxygen, is forced through 10–60 tonnes of molten pig iron to which $1–1\frac{1}{2}$ tonnes of quicklime has been added in a cylindrical converter about 8 m high. The Basic process is sometimes referred to as the Thomas process and the basic lining consists of calcined dolomite bonded with tar and baked by a coke fuel. The lining lasts for 100–120 'blows'. The contents are introduced into the converter while horizontal. On tilting to the vertical position, the air blast is forced up through holes (tuyères) in the base. In a spectacular display of fireworks, a violent exothermic oxidation of impurities occurs and is complete in about 30 minutes. No external heat is necessary. A white-hot plume of brilliant flame appears and violent agitation of the contents occurs. Silicon and manganese form oxides which pass into the slag. Carbon forms gaseous carbon monoxide which burns and after about 20 minutes, when the flame has died down, during the 'after-blow', phosphorus is oxidized to the pentoxide. On completion, the calculated quantity of ferromanganese, etc., is added. Manganese 'de-oxidizes' any iron(III) oxide formed, and the residue remains as an alloying element. The floating slag is skimmed off and the steel poured.

Oxygen and superheated steam are used in some steelworks. This avoids the absorption of nitrogen by steel, which is undesirable. With 20% oxygen, the refining period is cut to 15 minutes during which the temperature is 300°C above normal, while with 30% oxygen the time is only 8 minutes and the temperature raised by 400°C above normal.

The electric arc furnace method

Very special high-quality steels are made by the electric furnace methods. The use of electricity avoids the introduction of oxidizing gases and the refining may pass through both oxidizing and reducing phases, not possible with the other processes. It is especially useful for the economical use of high-grade scrap steel.

Ten to thirty tonnes of steel scrap is melted by means of an electric arc on a shallow circular tank using three carbon electrodes. At one works, this requires 360 A at 34 000 V. The temperature reaches 3400°C. The composition of the original scrap is accurately known and impurities are oxidized by addition of ore, mill-scale (scrap from ingots) or increasingly, oxygen, with lime as flux. The slag is skimmed off. Carbon is added and refining continued under reducing conditions with lime and fluorspar as fluxes. Powdered ferrosilicon is added to 'de-oxidize' and 'de-sulphurize'. The requisite proportions of alloying elements are added and the furnace tapped.

The L–D (Linz–Donawitz) process (*Austria*)

Molten pig iron and scrap steel, with the required proportion of lime, forms the charge in vessels of about 30 tonnes capacity, lined with a basic refractory of tar-dolomite or magnesite. A jet of oxygen at about 1 MPa pressure (i.e. about 10 atmospheres) is delivered at about 1 m above the metal surface for about 20 minutes from a water-cooled tube of diameter 40 mm. Carbon is oxidized away to a final level of about 0.03% by weight while oxidation products of the other impurities pass into the slag. Phosphorus is removed as calcium phosphate, leaving less than 0.015% behind, and the steel poured into the ladles. Final additions are made as before. Quantities on the open hearth scale and of a similar quality are now refined by the L–D process in the time associated with steel production in Bessemer converters. In the search for better steelmaking techniques, the advent of cheap 'tonnage' oxygen has brought a revolution, as shown in Table 11.2.

Table 11.2 World steel production/megatonne (Tg*)

Year	Open hearth	Bessemer	Oxygen	Others
1960	233	42.1	12.0	52.6
1964	272	40.4	52.2	62.1
1965	265	37.5	72.4	73.8
1966	254	33.3	99.2	78.0

*10^{12} tera, T

The Kaldo process (*Sweden*)

This differs principally from the L–D process in the geometry of the containing vessel which, in the Kaldo process, is rotating and inclined.

Spray steelmaking (*United Kingdom*)

A falling stream of molten iron from a blast furnace is atomized to a fine spray by a blast of oxygen, thereby undergoing oxidation to steel in seconds. The spray cone temperature is about 2100°C. Lime is injected into the stream before oxidation and (with added fluxes) forms a molten slag with oxidation products and unwanted elements. Waste gases escape.

Not only is the process rapid (10^5 times faster than the Bessemer technique in theory) and the plant relatively simple, but it is a prime method for inclusion in modern plants designed to produce steel continuously from iron ore, such integrated plants bringing sharply reduced costs after the initially heavy investment.

Finishing process

A description of the various processes to which steel ingots or continuous cast steel are subjected may be found in the literature made available by the steel producers. However, a sense of the magnitude of the steelmaking operations is captured only by a visit to a large iron and steel works or, perhaps, at second hand, by film.

Industrial pollution

In common with much of heavy industry, steelmaking can cause air and soil pollution, not to mention the spoiling of rivers. The pollution will continue until the seriousness of the problem is recognised, the technicalities of reducing it to a minimum are devised and the necessary devices installed. As an illustration, the use of an oxygen jet can be accompanied by a plume of brown smoke issuing from the chimney of the works. The brown coloration is due to iron (III) oxide dust and signals the end of oxidation of impurities and the onset of metal oxidation.

The full horror of industrial pollution at its worst, however, can be seen in eastern Europe. Brandenburg, on the river Havel, in (east) Germany is nowadays one of the grimiest and most dilapidated of towns. Pollution statistics reveal that the chimneys of its steelworks annually belch out about 9000 tonnes of dust (containing heavy metals) into the atmosphere. Over the 40 years or so of the plant's production life, it is claimed that 36 000 tonnes of zinc, 12 000 tonnes of lead, 5000 tonnes of copper, 2000 tonnes of chromium, 100 tonnes of cadmium and 60 tonnes of nickel have been deposited on what was one of Prussia's most attractive cities. Such reports, culled from newspaper stories, bristle with semi-detached statistics but do give some notion of the extent of the tragedy.

Corrosion of iron and steel

See pp. 128 to 130.

12

Commercial processes based on air, water, petroleum and coal

The liquefaction and fractionation of air

The volume composition of dry air is roughly: nitrogen 78%, oxygen 21%, argon 1% and carbon dioxide 0.03%. The detailed composition was given in Table 3.2, p. 29. In addition there are traces of other inert gases in the atmosphere, substances thrown up by factories in industrial regions, including more carbon dioxide, and variable amounts of water vapour. The purification of air prior to liquefaction must include the filtration of dust and, to avoid blockage of liquefaction plant by frozen solids at the low temperatures reached, removal of moisture and carbon dioxide. The accumulation of carbon dioxide and methane, and other gases and vapours, including the freons (organic compounds of fluorine and chlorine, CFCs, seemingly inert and used in refrigerators and aerosols until recently) are a cause of great worry to those concerned with the environment. The former gases are believed to cause the so-called 'greenhouse effect' (which is expected to lead to a general warming of the atmosphere and the surface of the Earth) and the latter halogen compounds are believed to be directly, or indirectly, responsible for a thinning of the protective layer of ozone (keeping out harmful ultraviolet rays) in the upper atmosphere, the so-called 'hole' in the ozone layer, detected over Antarctica.

Oxygen was first liquefied in 1877 by Cailletet and Pictet working independently. The first industrial processes for gas liquefaction on the small scale appeared in 1895, when Linde in Germany and Hampson in the UK demonstrated continuous liquefaction of air. In 1903, Linde produced oxygen, though in relatively low yield, by the fractionation of liquid air and by 1910 he was able to effect the continuous production of relatively pure oxygen and nitrogen by fractional distillation using a 'double rectification column'. Liquefaction was induced by the Joule–Thomson cooling effect observed when compressed gases expand adiabatically below their inversion temperatures. Cumulative cooling was achieved by using the outgoing gas to cool that being forced in. The efficiency of liquefaction was improved by Claude in France by arranging for the compressed air to do work adiabatically in an engine and still further improved by Heylandt in Germany who introduced high pressures of 14–17 MPa (140–170 atm) or more to this technique. Claude expansion machines take in air at about −140°C so that no conventional lubricants may be used, but Heylandt machines, taking in air at about −20°C, use an oil.

Oxygen obtained by the rectification of liquid air produced by these methods is up to 99.5% pure and is used for cutting and welding. Advances in methods of bulk refrigeration and transport came with increased demand. But the production of oxygen was increased still further by the introduction of the Linde–Fränkl liquefaction process in 1930. Medium-purity, 85–95% gaseous oxygen could be obtained from this process with a considerable price reduction where a large continuous demand justified the initial capital expenditure on oxygen plant adjacent to where the demand lay. This 'tonnage oxygen' was produced for the chemical industry and the iron and steel industry. The quantities of oxygen used since the war have risen enormously and in the USA automatic and semi-automatic oxygen plants have been developed. For relatively small users of oxygen, it is cheapest to liquefy and transport in bulk. Thus one volume of liquid oxygen at its boiling-point (−183°C) and atmospheric pressure (101 kPa) expands to 860 volumes at atmospheric temperature and the same pressure. In contrast, the same gaseous bulk at 13 MPa (130 atmosphere) pressure is reduced considerably but is still 6 volumes. Oxygen is the most important gas, as far as demand goes, separated from air. The production of nitrogen exceeds demand but the gas is used extensively in the synthesis of ammonia. Argon is also separated.

All processes in commercial use for separating oxygen from the air involve liquefaction, although the product finally piped away may be gaseous. Various qualities are available: high-quality gas, high-quality liquid and medium-purity tonnage-scale oxygen gas.

Theory of the distillation of miscible liquids

The theoretical treatment of the fractional distillation of mixtures of miscible liquids, based on the application of Raoult's Law, will be found in textbooks of Physical Chemistry.

The production of gaseous oxygen

The simplest plant, shown diagrammatically in Fig. 12.1, has a single rectification column. This may be filled with a special gauze packing or trays, of the bubble cap or sieve tray type, used in the columns of larger plant.

The lower boiling-point gas nitrogen passes from the top of the column, through the heat exchanger in which it cools the incoming gas. Cooling is also helped by the outgoing oxygen and by liquid oxygen in the base of the column. Oxygen gas is removed from just above the liquid oxygen. In larger plant, a Claude expansion engine is included to raise the efficiency.

The yield is about 58% of oxygen available in air, at a purity of 99.5%.

The production of liquid oxygen or liquid nitrogen

Increased refrigeration is required for the production of liquid oxygen from air. High operating pressures of the order of 14–15 MPa (140–150 atm) are used with expansion in an engine on the Claude Principle. In such a Heylandt Cycle, about 90% of the air from the first pre-cooling is expanded in this

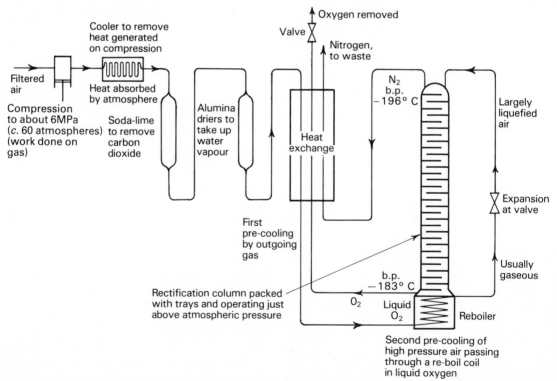

Fig. 12.1 The smallest and simplest plant for gaseous oxygen (adapted from 'Liquefaction and Fractionation of Gases', J. B. Gardener, *Chemical Engineering Practice*, vol 6, by permission of the publishers Butterworth-Heinemann Ltd ©)
N.B. In a large plant, 90% of the compressed air would be made to work adiabatically after 1st pre-cooling, and after cooling in this way would join the 10% which had been liquefied after expansion. For mechanical reasons, appreciable liquefaction is desirable in the exhaust. A double rectification column would also be used

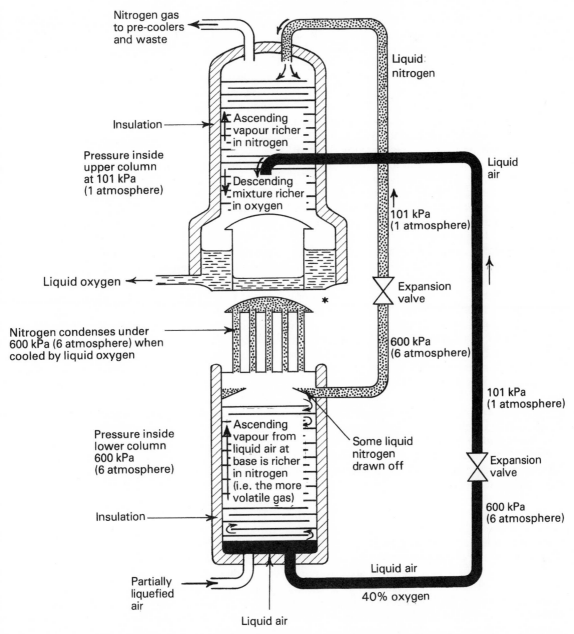

Nitrogen gas to pre-coolers and waste

Liquid nitrogen

Insulation

Ascending vapour richer in nitrogen

Pressure inside upper column at 101 kPa (1 atmosphere)

Descending mixture richer in oxygen

Liquid air

101 kPa (1 atmosphere)

Liquid oxygen

Expansion valve

600 kPa (6 atmosphere)

Nitrogen condenses under 600 kPa (6 atmosphere) when cooled by liquid oxygen

101 kPa (1 atmosphere)

Pressure inside lower column 600 kPa (6 atmosphere)

Ascending vapour from liquid air at base is richer in nitrogen (i.e. the more volatile gas)

Some liquid nitrogen drawn off

Expansion valve

600 kPa (6 atmosphere)

Insulation

Partially liquefied air

Liquid air

40% oxygen

Liquid air

Fig. 12.2 Double rectification column used with Heylandt cycle. For clearness the upper and lower halves of the column have been separated—the liquid oxygen of the upper part cools the nitrogen which condenses in the lower part (*The two parts fit together)

way from c. 15 MPa (150 atm) and −25°C to c. 600 kPa (6 atm) and −170°C, joining the residual 10% which has undergone a second pre-cooling to −170°C through an expansion valve to equalize pressures. A double rectification column shown in Fig. 12.2 is used. In contrast to the upper column which is at atmospheric pressure, the lower part of the column operates at about 600 kPa (roughly 6 atm) pressure which is enough to liquefy nitrogen at the temperature of liquid oxygen. Both gases cannot be liquefied simultaneously because of the presence of argon, which boils at a temperature intermediate ($-186°C$) between the boiling-point of liquid oxygen ($-183°C$) and liquid nitrogen ($-196°C$), and because of the limitations of separation by fractional distillation.

Applications of gases separated from air

Medium-grade 'tonnage' oxygen is used extensively in the iron and steel industry. It is also used for the production of nitric acid by catalytic oxidation of ammonia. In addition to its use in medicine and dentistry, oxygen of high purity is used for joining and cutting metals. An oxygen lance used with acetylene will cut heavy sections of steel and ingots, and may be used for tapping furnaces. Other uses include the gasification of coal, oxidation of ethanal (acetaldehyde) to ethanoic acid (acetic acid) and the oxidation of hydrocarbons to methanol.

Nitrogen is required for the synthesis of ammonia, annealing of steel sheets, the production of calcium cyanamide and for the deep freezing of foodstuffs and biological material, such as frozen fish spawn for restocking waters. Liquid nitrogen has been used as a scrubber to remove carbon monoxide from gases prior to ammonia synthesis.

Liquefied oxygen and nitrogen are stored in double-walled tanks, the space within the walls being filled with a non-flammable efficient insulating material such as basic magnesium carbonate, slag wool and certain commercial preparations. Some loss by evaporation occurs. Small vessels have low-pressure insulation, the insulating region being exhausted to about 15 Pa. In the laboratory, Dewar vessels are used.

Argon is potentially available in plenty from the fractionation of liquid air. It is taken from the upper column in the double column rectification. The principal use of argon is in the arc welding of aluminium, titanium and stainless steel where it provides an inert atmosphere (shield gas) avoiding a reaction between the hot metals and the oxygen and nitrogen of the air. 80–85% of argon produced is used for this purpose. Another 10–11% is used in the refining of titanium and zirconium, where it serves the same purpose. Three to four percent of argon produced goes into incandescent electric light bulbs, providing an inert atmosphere for the filament. It is also used in the metallurgical flushing of nitrogen and hydrogen from metals, this use accounting for 1% of the total.

The production of neon, krypton and xenon necessarily requires the processing of large quantities of air. Helium production from air cannot compete commercially with that from natural gas sources in North America. Krypton is used in special electric lamps, such as those used by miners where high illumination is required. Neon is used in electric signs and fluorescent lamps. Xenon replaces mercury vapour in discharge tubes and lamps for high-intensity illumination.

Chemicals from coal: coal gas

Although natural gas from the extensive North Sea oil and gas 'fields' has replaced coal gas in England, Scotland and Wales, the coal gas industry remains of interest in Northern Ireland and Eire, and in other countries without ready access to such rich natural resources. It is to be hoped that the gas pipelines will be led across the Irish Sea. Natural gas is largely methane and hydrogen, produced no doubt from matter of vegetable origin. The gas piped to our homes is better called 'town gas': the source may be natural, coal or petroleum feedstock depending on the economic circumstances of the region.

Coal is a hard, opaque, black or blackish mineral formed by fossilization of vegetable matter during the long geological history of the earth, when plants and trees decayed incompletely under the conditions of humidity, temperature and pressure associated with the changes in the great swamps. The mineral is found in seams and strata below the Earth's surface. Fossilized remains of plant-life point to the vegetable origin and careful microscopic examination has shown the remains of incompletely decayed vegetable matter.

Coal is a mixture of complex organic substances which will burn with the liberation of much heat energy. Combustion in the open grate is a most inefficient way of utilizing this energy. There are three main classes of coal: *lignite* or *brown coal*, mined in Germany and the USA, *bituminous coal* which provides much gas, burning with a bright flame, and *anthracite* which is very hard, with little volatile matter and therefore burns without flame, leaving little ash. Peat is another product of vegetable decay and may also be burnt.

When coal is heated in the absence of air, or dry distilled, a wide range of products may be separated. This process is called carbonization and yields coal gas, coke, ammonia, tar, benzole and sulphur. The relative quantities depend on the quality of the coal and the temperature of carbonization. One tonne of coal may give 340–370 m^3 gas, 45 dm^3 tarry matter, 3 kg ammonia, 4.5 kg sulphur and

500 kg coke. The composition of coal gas varies: in rounded figures the volume percentage composition is hydrogen 50%, methane 30%, carbon monoxide 7%, ethene 3%, and carbon dioxide, oxygen, nitrogen together 10%. The gas industry was pioneered by William Murdoch, who lit his home at Redruth, Cornwall, with gas in about 1780 and the Soho works of his firm, Boulton and Watt of Birmingham, in 1801. Pall Mall was lit by gas in 1807, the first public gas street lighting. The first gas cooker appeared in 1840 and gas fires by 1855.

Carbonization is effected at about 1000°C although localized temperatures may reach 1400°C. The operation is done in retorts, either continuous or intermittent, and in coking ovens. Dimensions vary. Continuous vertical retorts may be 7.5 m high, 1–2 m long and tapering from 250 mm wide at the top to 500 mm at the bottom. Coal passes through in about 8 hours so that each retort uses 3000†–10 000 kg every day. Horizontal retorts are about 6–7 m long, and carbonize 650–900 kg coal in 10–12 hours. The intermittent vertical retort is 5 m high, 3 m long and 250 mm wide, carbonizing 3000–4000 kg in 10–12 hours. Coking ovens, used primarily for the production of hard metallurgical coke which is required to withstand great pressures in the blast furnaces at ironworks without crumbling, are about 12 m long, 4–5 m high and 500 mm wide, treat 17 000 kg of coal in 18–24 hours, and supply coal gas as a by-product to nearby towns. A retort house may contain 20–300 horizontal retorts or 100 continuous vertical retorts. The retorts are heated by burning producer gas or sometimes, water gas. They are constructed of fire brick, or silica which is lime bonded, or siliceous clay which fuses to a calcium silicate glass. Coking ovens are lined with fire brick and run continuously for many years, because cooling causes crumbling of the brick. Coke is ejected from coking ovens by a rammer falling into special rail wagons which are immediately shunted under a cold water douche. Ordinarily retorts last for about 6 years. Coke contains 85–90% carbon, with a little volatile matter and about 10% ash. Gas coke, as distinct from metallurgical coke, is graded and sold as a smokeless fuel for domestic and industrial use, the production of water gas and producer gas, and as a fuel for electrical power stations.

The hot vapours leaving the retorts or ovens are collected into one main stream and cooled. A number of methods are in use. Atmospheric condensers are merely zig-zag pipes cooled by water spray and with an arrangement to reverse the flow

to remove less volatile products which accumulate. Water-cooled condensers may be used with washing on the countercurrent principle, where gas passes up a packed tower down which weak ammoniacal liquor falls. Impure gas, after cooling, passes to the exhauster, maintaining the flow of gas through the works. To minimize leakage, retorts operate at about atmospheric pressure. The gaseous products leave the retort at about 66°C. A water seal is provided to enable each retort to be emptied for recharging, or where appropriate, repaired. Of the tar, 80–85% condenses in the retort house and 7–14% in the condensers. Residual tar, about 3.5 dm³ per 1000 kg of coal, is removed in Livesey washers, gas being dispersed by bubbles in an ammoniacal liquor, and also by electrostatic precipitation. Tar is used on roads and runways, and among other uses, as a covering for underwater telephone cables. A constituent, creosote, is used as a wood preservative, in sheep dips, winter washes for fruit trees and as a fuel oil in preparing carbon black for printing. But more important, from tar there may be separated a vast range of organic compounds for use in making plastics, medicines, dyes, perfumes and synthetic fibres, explosives and solvents.

Coal gas is associated with a watery ammoniacal liquor. Ammonia is now removed by washing in water or weak liquor recovered earlier until the liquor is saturated, thereby reducing later transport costs. Ammonia is present as the free compound and united with carbon dioxide, hydrogen sulphide and hydrogen cyanide as carbonate, hydrogencarbonate, carbamate, hydrogensulphide and cyanide respectively. Hydrogen cyanide may be removed chemically after ammonia recovery, being converted into potassium hexacyanoferrate(II) by the action of alkali and an iron(II) salt:

$$FeSO_4 + H_2S \rightarrow FeS + H_2SO_4$$

$$FeS + 6HCN + 2K_2CO_3 \rightarrow$$
$$K_4[Fe(CN)_6] + H_2S + 2CO_2 + 2H_2O$$

although hydrogen cyanide is increasingly made during the processing of natural gas. Ammonia is used in refrigerators, explosives manufacture, in photographic chemicals, as household ammonia, as a source of ammonium compounds and, in an extremely important application, as ammonium sulphate fertilizer. Liquor containing cyanides is used for the production of hexacyanoferrates(II),

† 1000 kg = 1 Mg = 1 tonne

hexacyanoferrates(III), Prussian Blue and sodium hexanitrocobaltate(III).

The gas may contain 0.7–2.0% hydrogen sulphide which must by law be reduced to less than 0.7 parts per million. Coal gas, as supplied, must not discolour lead acetate paper in 3 minutes. To meet the requirements of legislation, catalytic removal of hydrogen sulphide by bog iron ore is used. Holland and Denmark are sources of bog iron ore, which consists (as would be expected) of iron(III) oxide in peat. The approximate composition is water 50%, iron(III) oxide 25% and peat 17%. The process occurs in two stages, the second either in the converter or in air when the 'spent' oxide is crushed and watered:

$$Fe_2O_3.H_2O + 3H_2S \rightarrow Fe_2S_3.H_2O + 3H_2O$$

$$2Fe_2S_3.H_2O + 3O_2 \rightarrow 2Fe_2O_3.H_2O + 6S$$

The net reaction is simply:

$$2H_2S + O_2 \rightarrow 2H_2O + 2S$$

$$\Delta H = -2 \times 223 \text{ kJ mol}^{-1}$$

Air is admitted to bring about sulphide oxidation giving about 0.7% by volume as the final proportion of oxygen present. Spent oxide provides a valuable source of sulphur in the UK. When the sulphur content has risen to about 50%, the spent oxide is sold to manufacturers of sulphuric acid. Potassium hexacyanoferrate(II) may be prepared from spent oxide. It is heated with potassium chloride and milk of lime to form calcium potassium hexacyanoferrate(II), which may be decomposed with potassium carbonate:

$$CaK_2[Fe(CN)_6] + K_2CO_3 \rightarrow$$
$$CaCO_3(s) + K_4[Fe(CN)_6]$$

Finally, naphthalene, benzole and compounds of sulphur are removed. Naphthalene would be deposited in mains and meters leading to blockage. It is washed out by gas oil made from benzole. Washing with gas oil is also used for removal of benzole, recoverable by steam distillation. Sulphur compounds are removed at the same time although catalytic decomposition may be induced by a nickel subsulphide catalyst. Benzole is the source of numerous organic compounds, finding application in the production of paints, varnishing lacquers, dyestuffs, drugs, perfumes, explosives, plastics, printing inks and photographic chemicals, etc. Blended with petrol, much benzole was marketed as Benzole Mixture fuels for cars and aircraft.

Water vapour can cause blockage of gas mains unless it is reduced to such a level that the dewpoint of the gas is below the lowest temperature likely to be reached. Calcium chloride solution, being hygroscopic, is used for the absorption of water. It is concentrated for further use by steam heating.

At this stage, coal gas is supplemented with carburetted water gas or petroleum gas to improve the calorific value. It passes through meters into gas holders, where it is stored either over water or in dry-holders with a fabric seal containing a special tar oil. When a gas holder is being brought into use, air is removed by using a purge gas of nitrogen and carbon dioxide. This is also used when the holder is being taken out of use, avoiding the formation of explosive mixtures of gas and air.

Carburetted water gas: supplementing coal gas

The calorific value of coal gas may be increased by the addition of carburetted water gas. This is water gas to which the products of the thermal decomposition of gas oil have been added. Gas oil is petroleum oil boiling in the range 160–350°C and is substantially the same as diesel oil. Heavier, i.e. higher boiling-point, oils are also being used. The plant consists of four units: generator, carburettor, superheater and waste heat boiler.

Water gas is made by forcing steam through incandescent coke, when the following endothermic reactions occur:

$$C + H_2O \rightleftharpoons CO + H_2; \qquad \Delta H = 132 \text{ kJ mol}^{-1}$$

$$C + 2H_2O \rightleftharpoons CO_2 + 2H_2; \qquad \Delta H = 90.0 \text{ kJ mol}^{-1}$$

During this 'steaming' period the proportion of carbon monoxide falls as the temperature drops. Heat must be supplied to the coke if the quality of the gas for heating purposes is not to suffer. This is done during the 'blow' period.

Steam is discontinued and air is blown through the fuel bed, the temperature rising because of the exothermic nature of the reactions:

$$2C + (O_2 + 4N_2) \rightarrow 2CO + 4N_2$$
$$\Delta H = -2 \times 111 \text{ kJ mol}^{-1}$$

$$C + (O_2 + 4N_2) \rightarrow CO_2 + 4N_2$$
$$\Delta H = -394 \text{ kJ mol}^{-1}$$

This gas passes through the carburettor, superheater and waste heat boiler to extract surplus heat. Air is admitted into the first two vessels in sufficient quantity to permit the carbon monoxide to burn, generating more heat.

Water gas, which consists of hydrogen and carbon monoxide with traces of nitrogen and carbon dioxide, has a relatively low calorific value.

There would be no advantage acquired by mixing it with coal gas. It is, therefore, enriched by carburetting. Gas oil is sprayed into the carburettor during the steaming period. Oil vapours and gas pass into the superheater, a vessel containing checkered brickwork. Pyrolysis (cracking, thermal decomposition) of the vapours occurs with the formation of methane, ethene (ethylene), etc. Some hydrogenation due to the hydrogen present also occurs. The product is carburetted water gas.

The use of carburetted water gas makes output more flexible because plant can be started up from cold quickly and it requires little labour. In addition, it makes the problem of controlling the calorific value of the mixture of gases, conveniently called town gas, easier and increased quantities of gas may be quickly generated. The cycle of alternate air 'blows' and steam 'runs' is controlled automatically and takes $2\frac{1}{2}$ minutes. A short period of cycle is preferred to maintain constancy of composition.

Town gas from petroleum

Surplus gas from oil refineries, from the catalytic gasification of heavy oils by steam cracking (the *Shell* and *Texaco* processes) and from the pressure steam reforming of naphtha, a convenient light petroleum feedstock (the *ICI* process), have all found wide application in making town gas, accelerated by the rising costs of coal. The gas is rich in hydrogen and lower saturated hydrocarbons, especially methane.

The discovery and commercial exploitation through a national grid network of pipelines of North Sea, and adjacent mainland, gas in the UK, Norway and Holland has led to the rapid replacement of coal as a source of domestic and industrial fuel gas.

Further along the North Sea coast, a gas pipeline from the North Sea reserves is projected to serve Jutland in southern Denmark. However, in the adjoining German *Land* of Schleswig-Holstein, piped gas (as distinct from having a container tank of liquid propane in the garden to serve the house) seems to have been phased out.

In the border town of Flensburg there is a unique district heating system, started in 1968 and nearing completion. It uses energy from the town's coal-fired electricity power station (connected to the German and Danish National Grids) to supply an underground network with fast flowing water at a temperature of 73 to 130°C, under a nominal working pressure of 25 bar ($2.5 \times 10^6 \, \mathrm{Nm^{-2}}$) and an actual one of 16 bar ($1.6 \times 10^6 \, \mathrm{Nm^{-2}}$) due to staging pumps every 2.5 km. Secondary pipelines operating at 65–100°C and 5 bar ($5 \times 10^5 \, \mathrm{Nm^{-2}}$) pressure serve domestic premises, using heat exchangers to breakdown network pressure. The price of space heating and hot water provision for domestic (and other) premises is certainly lower than that for other methods of heating. It is claimed that replacing many domestic heating boilers with centrally located boilers reduces air pollution dramatically. In addition, the technique that is used for coal combustion (fluidized bed technology) causes less pollution than natural gas, especially in terms of emissions of oxides of nitrogen. Not only has an integrated energy policy been established for this town of 100000 inhabitants but the assumption that gas combusion is the cleanest form of heating has been challenged.

The commercial production of ammonia and nitrogen compounds

Fertilizers and crop nutrition

The assimilation of nitrogen compounds by plants and animals is essential for life. Nitrogen occurs in all proteins, which are the structural units of living matter. With few exceptions, atmospheric nitrogen cannot be used directly by plants but must be 'fixed' as suitable compounds in the soil for absorption by the root system. The exceptions are pod-bearing plants or leguminous plants such as peas, beans, lupins and the agricultural crops, clover and lucerne, which bear nodules in their root systems with which nitrogen-fixing bacteria are associated. Mosses, algae and fungi have this power, too. Nitrogen is taken up from air spaces in soil by leguminous plants. Animals acquire combined nitrogen by eating plants or other animals, restoring it to the soil by excretion. After death, saprophytic bacteria (Greek, *sapros* = rotten, putrid) break down the remains of plants and animals to ammonium salts, which are oxidized to nitrates and then built up into nitrogenous living matter. Some nitrogen is lost by the action of denitrifying bacteria on nitrates, some by the leaching out of salts by rain water, but much more important is the loss through intensive farming, the production of crops and meat and the disposal of sewage. The circulation of nitrogen between soil and living matter is thrown out of balance by the requirements of civilization. The nitrogen cycle is shown in Fig. 12.3.

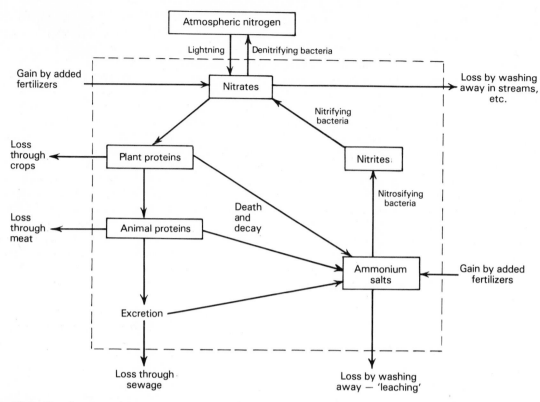

Fig. 12.3 The nitrogen cycle

Ironically, the countries of Western Europe are literally having to pay the price for excessive use of nitrogeneous fertilizers as precious supplies of drinking water show signs of contamination by nitrates. This is likely to be a long term problem because it takes many years for the salts in solution to percolate to the depths of the underground reservoirs. Water purification, the subject of EEC directives, to remove nitrates (and nitrites, suspected of carcinogenic activity) is both difficult and expensive, but of vital importance to us all. Yet without the application of nitrogenous (and other) fertilizers to the land it is difficult to see how great populations could be fed, however attractive the principles of organic farming may seem. Besides nitrogen, the major components of fertilizers are compounds of phosphorus and potassium. While the use of these has remained fairly static in the years 1960–86, that of nitrogen has increased substantially in the UK. This is shown in Fig 12.4.

The rise in application of nitrogenous fertilizers reflects not only increased agricultural production but also changes in husbandry: it is associated with

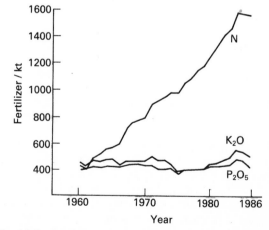

Fig. 12.4 Fertilizer consumption 1960–86 (source: Fertilizer Manufacturers Association)

the introduction of stiff-strawed varieties of barley and wheat, and the larger applications of nitrogen compounds increase the quantity and quality of feed available to animals, which in turn increases the output of milk and meat.

Biological nitrogen fixation

The industrial production of nitrogenous fertilizers consumes a lot of energy and raw materials. A major effort is being made to understand the mechanisms by which bacteria fix nitrogen in the soil with a view to exploiting such knowledge commercially. There are two types of bacteria involved: the symbiotic (which live in association with roots, e.g. rhizobia in the roots of legumes) and those which are independent, e.g. the azotobacter. Nitrogen is reduced to ammonia by nitrogenase enzymes, two oxygen sensitive proteins of relative molecular mass 60 000 (i.e. 60 kDa†: containing 4 Fe and 4 S per mole) and 220 000 (220 kDa: *ca* 30 Fe, 30 S and 2 Mo or 2 V per mole). The two proteins can be extracted from bacterial cells and in the presence of magnesium adenosine triphosphate and an electron source, will reduce nitrogen to ammonia, and substrates such as C_2H_2 (to C_2H_4). These are shown in Table 12.1.

Table 12.1   Reduction of substrates by nitrogenase enzymes (R. L. Richards, *Chem. Br.*, 1988, **24**, 133)

Substrate	Product
N_2	$2NH_3$
H^+	$\frac{1}{2}H_2$
C_2H_2	C_2H_4
	$(C_2H_6)*$
CH_3NC	$CH_3NH_2 + CH_4$
HCN	$CH_4 + NH_3$

*Vanadium nitrogenase only

Biochemical and chemical studies concentrate on postulating a sophisticated model for the way in which nitrogen fixation may work, and involving complexes of molybdenum (or analogues such as tungsten or rhenium) or vanadium.

The aim of the research is to create a low technology system for ammonia synthesis coupled with the use of an alternative energy source, e.g. wind, tide or sun.

Fixing atmospheric nitrogen commercially

The need for sources of nitrogen to be added to the soil was realized in the 19th century. Besides the incorporation of farmyard manure into the soil, material was imported, notably sodium nitrate from Chile and guano, formed from hardened bird droppings, from islands off Peru. Further, with the stimulus to coal gas production, supplies of ammonium sulphate were available. But other sources were required.

In Norway, the Birkeland and Eyde process (1905), now obsolete, utilized the enormously high local temperatures in an electric arc to synthesize nitrogen oxide (nitric oxide) from air,

$$N_2 + O_2 \rightleftharpoons 2NO$$

The yield was very small but electric power was readily available.

The Cyanamide process (1905) developed in Germany and Italy fixed nitrogen by a reaction with crushed calcium carbide in an electric furnace at 1100°C, calcium cyanamide and graphite being formed,

$$CaC_2 + N_2 \rightarrow CaNCN + C$$

The product, black because of the graphite, is marketed as nitrolim. The original process for ammonia production, now obsolete, required hydrolysis of this product,

$$CaNCN + 3H_2O \rightarrow CaCO_3 + 2NH_3$$

Calcium cyanamide is used for the manufacture of melamine, synthetic resins and as a defoliant for cotton plants before harvesting and as a fertilizer. About 4% of fixed nitrogen is obtained this way.

The Haber Bosch process (1905–13) was developed in Germany and accounts for about 75% of the total world production of ammonia by synthesis from the elements,

$$N_2 + 3H_2 \rightleftharpoons 2NH_3$$

Ammonia is required mainly for fertilizer production, for explosives, general chemical production of such compounds as nitric acid and washing soda, plastics and nylon, dyestuffs and drugs.

The principles of ammonia synthesis

$$N_2 + 3H_2 \rightleftharpoons 2NH_3$$

$$\Delta H = -2 \times 46.0 \text{ kJ mol}^{-1}$$

The reaction is balanced but under normal conditions the equilibrium position lies far to the left:

$$K_p = \frac{p_{NH_3}^2}{p_{N_2} \cdot p_{H_2}^3} \text{ at } T$$

in which p_{NH_3}, p_{N_2}, p_{H_2} are the partial pressures of the respective gases in the mixture and K_p is the equilibrium constant at the temperature T. A quantitative treatment may be sought in texts on

† Proteins' molecular masses are sometimes expressed in terms of the unit Dalton or kiloDalton which is self explanatory

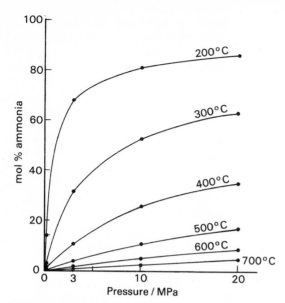

Fig. 12.5 The curves show the percentage of ammonia at equilibrium at the pressure and temperature indicated, starting with a mixture of $3H_2 : 1N_2$

Mol % ammonia for plotting curves

Pressure/MPa	Temperature/°C					
	200	300	400	500	600	700
0.1	15.3	2.18	0.44	0.13	0.05	0.02
3	67.6	31.8	10.7	3.62	1.43	0.66
10	80.6	51.2	25.1	10.4	4.5	2.1
20	85.8	62.8	36.3	17.6	8.2	4.1

Physical Chemistry but Fig. 12.5 provides graphs of the equilibrium percentage of ammonia from a 3 : 1 molar ratio of hydrogen and nitrogen at varying pressures for certain temperatures taken from some of Haber's results.

Qualitatively, increasing the pressure increases the reaction rate, and proportion of ammonia (Le Châtelier's Principle p.106). The equilibrium proportion of ammonia is also increased by lowering the temperature, the reaction being exothermic, but the reaction rates fall. Therefore, a catalyst is required.

The reaction occurs in a large pressure vessel known as a converter. Initially the catalyst is of tri-iron tetroxide fused with promoters of alumina, calcium and potash, but the hydrogen reduces the catalytic material to finely divided iron which is prevented from coalescing by the promoters. Conditions of synthesis vary but a temperature range of 350–600°C and a pressure range of

15–35 MPa (*c.* 150–350 atm) may be used. Without necessarily waiting for equilibrium to be attained, the maximum conversion to ammonia in the minimum time is sought and excess reactants are recycled.

The ICI (1962) integrated ammonia process

There are three successive reactions, each requiring different temperature and pressure conditions, and contact with a different catalyst to select and speed up the desired change. These are now described.

The pressure steam reforming of naphtha (or natural gas)

A gaseous mixture of hydrogen, nitrogen and carbon monoxide is prepared by reacting vaporized hydrocarbon feedstock with steam and air at a furnace temperature of 950°C and 3 MPa (30 atm) pressure over a catalyst, much heat being absorbed. Naphtha is a mixture of hydrocarbons boiling below about 220°C. Using methane to illustrate the reaction:

$$CH_4 + H_2O \rightarrow CO + 3H_2$$

The feedstock is previously freed from sulphur compounds in a vapour phase reaction at 350–400°C over a catalyst based on zinc oxide, to avoid poisoning the catalysts used later.

The process was introduced by ICI in 1962 to meet the needs of countries without natural gas because the steadily increasing cost of coal threatened to make the production of hydrogen, ammonia and methanol commercially uneconomical. Energy costs are the largest item in the final balance sheet of ammonia production economics. The original coke-based plants

$$C + H_2O \rightleftharpoons CO + H_2$$

consumed not less than $88 \, MJt^{-1}$ ($88 \, Jg^{-1}$) and while those using natural gas reduced this requirement to $38–40 \, MJt^{-1}$, the figures remain a long way from a calculated value of $18.4 \, MJt^{-1}$. Coal is no longer used in the UK. Naphtha is a convenient feedstock but the process can easily be adapted to use natural gas.

The water–gas–shift reaction (Bosch reaction)

The water–gas–shift reaction (Bosch reaction)

Carbon dioxide and more hydrogen are formed by reaction of the carbon monoxide and the steam, first at about 500°C with an iron-chromium catalyst and secondly, at 240°C over a catalyst containing copper, the pressure being 3 MPa (30 atm):

$$CO + H_2O \rightleftharpoons CO_2 + H_2; \quad \Delta H = -41 \text{ kJ mol}^{-1}$$

Equilibrium conditions at the lower temperature favour a very high conversion of carbon monoxide. Carbon dioxide is now removed. Residual carbon monoxide and dioxide (poisons for the ammonia catalyst) are removed by a methanation step over a nickel catalyst.

The heat liberated in this stage and the next is used to raise steam for driving compressors to give a fivefold increase in pressure for the synthesis.

The conditions for synthesis

The conditions for synthesis

In the current process, hydrogen and nitrogen, in the required volume and molecular ratio 3 : 1, react at 450°C and 15 MPa (150 atm) pressure over the catalyst based on tri-iron tetroxide described earlier.

The exit gases contain 10–20% ammonia at a temperature of 100–200°C. Water-cooling to atmospheric temperature brings about liquefaction of some ammonia and the residual gas, containing 6–7% ammonia, is re-cycled. More ammonia may be trapped by refrigeration to −20°C when the residual gas contains only 1–2% ammonia. Alternatively, ammonia may be dissolved in water. In a recirculation system, a purge of gas must be made to avoid accumulation of impurities.

The pressure is allowed to drop from the synthesis pressure to a few atmospheres, some vaporization bringing cooling to the boiling-point under these conditions by absorption of latent heat of evaporation. Ammonia is stored in steel tanks or vessels of cast iron, aluminium, natural or synthetic rubber, polythene or neoprene. It is transported in steel cylinders or in rail tank-wagons designed to withstand the vapour pressure of liquid ammonia.

The ICI (1985) AMV process for ammonia

Two surges in world-wide energy costs in the 1970s greatly slowed economic activity and depressed the demand for ammonia. Producers without the immediate benefit of long-term natural gas contracts

were caught in the pincer grip of rising costs and falling prices, at a time when many were commissioning new and expensive plants. Every conceivable way of reducing energy costs was explored in the various stages of ammonia synthesis.

In 1985 the AMV process was introduced at the 1 120 tonne per day plant at ICI's Canadian subsidiary, C-I-L Ltd of Lambton, Ontario. This process produces a superstoichiometric proportion of nitrogen at the secondary reform stage, the surplus gas (and inert substances) being removed in a conventional cryogenic system. About half of the steam requirement is obtained by evaporating condensate from the water gas shift reaction directly into the reformer feed gas in a saturator, thus eliminating liquid effluent. The synthesis itself is run under an unusually low pressure for this process, 85 bar ($8.5 \times 10^6 \text{Nm}^{-2}$), which requires the use of only a single-stage compressor. The ICI catalyst is still iron based, but with special promoters and a three-bed design to give maximum exposure at the comparatively low operating pressure. The complexity of this modern, large scale integrated plant is revealed in Fig. 12.6. It is the basic large scale plant depending on a sophisticated energy recovery system and huge throughput to be profitable.

The ICI (1988) LCA (Leading Concept Ammonia) process

In 1984, ICI saw the need to replace its two ageing ammonia plants at Severnside. The design team decided to go back to basic principles to ensure optimum process economics and minimal environmental impact. Both the production and use of steam have been separated almost entirely from the process into special service areas, to serve both 450 tonne per day synthesis units. The result is a breakthrough in ammonia production economics. The plant is cheaper to build and cheaper to run, quieter and more flexible than conventional plants. The flow sheet for the ICI LCA process is shown in Fig. 12.7.

"Natural gas feed is mixed with recycled hydrogen, heated and desulphurized. It is then cooled by preheating the feed to the desulphurizer before passing to a feed gas saturator where it is contacted with circulating hot process condensate. The feed gas from the saturator is mixed with a further quantity of steam to give a steam to carbon ratio of about 2.5 and preheated in the reformed gas stream. The reactants enter the primary

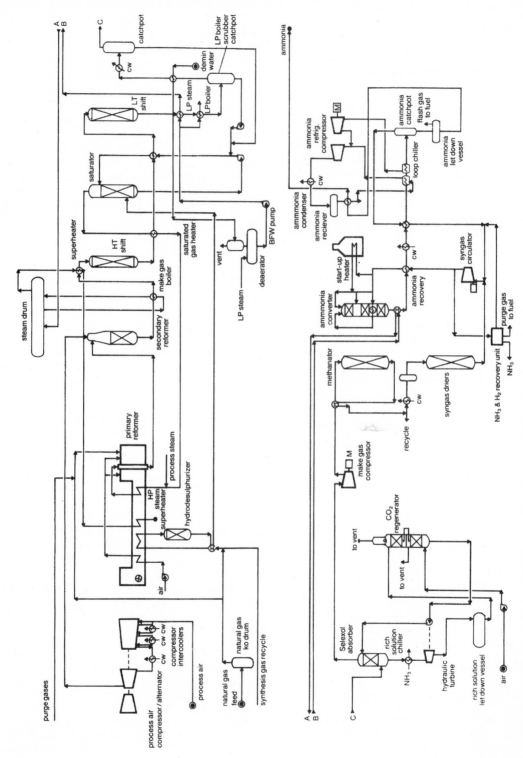

Fig. 12.6   Process Flowsheet for the AMV Process (from *Nitrogen*, 1989, British Sulphur Corporation Ltd)

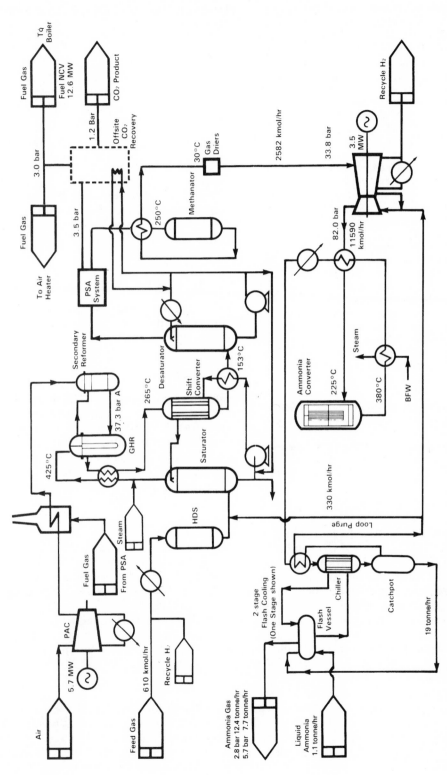

Fig. 12.7   LCA Process Diagram (from *Design and Operating Experience of the ICI LCA Ammonia Process*, J. M. Halstead, A. M. Haslett, A. Pinto and T. Hicks, © ICI)

reformer which operates with an exit temperature of 700–750°C and a pressure of 30–45 bar $(3.0–4.5 \times 10^6 Nm^{-2})$. The gas mixture is then fed to a secondary reformer for further reforming with an excess of process air. The reformed gas is cooled by providing the heat for the primary reforming reaction and preheating the reactants. The cooled reformed gas is shifted in an isothermal water cooled low temperature shift converter.

After shift conversion, the gas is cooled by direct contact with circulating process condensate and then fed to a pressure swing adsorption unit to remove excess nitrogen, carbon dioxide and parts of the inerts. Part of the carbon dioxide is recovered from the pressure swing adsorption waste gas using an aqueous solution of a tertiary amine. At Severnside the carbon dioxide is sold by ICI as an industrial gas.

The gas leaving the pressure swing adsorption unit is methanated and then cooled and dried. The dried gas enters the ammonia synthesis loop at the circulator suction. In the synthesis loop, gas from the circulator is heated and passed over low pressure ammonia synthesis catalysts to produce ammonia. The hot gas leaving the ammonia converter is cooled by generating 60 bar $(6.0 \times 10^6 Nm^{-2})$ steam and by heating the feed gas to the converter. Ammonia is separated from the partially cooled gas by vaporizing product ammonia. The unreacted gas is returned to the circulator. Argon and methane are removed from the synthesis loop by taking a purge and recycling back to the synthesis gas generation section as feed.''

(Halstead *et al*, *Design and Operating Experience of the ICI LCA Ammonia Process,* © ICI)

The low pressure synthesis and use of feed gas separator are borrowed from the AMV process. Fundamental differences from other modern plants include the design of the primary reformer (all heat being taken at 970°C from the process gas leaving the secondary reformer) and the complete absence of high pressure steam formation beyond this point, thus avoiding the expense of furnaces, gas treatment and heat recovery. A feature is a refractory-lined heat exchanger with catalyst packed into tubes. There are many other features. In the synthesis loop the gas is heated to about 225°C against returning process gas and passes through the converter which it then leaves at about 380°C. This is the only point in the synthesis section where steam is raised, providing about one third of the total 60 bar $(6.0 \times 10^6 Nm^{-2})$ steam generated.

ICI states that the energy consumption is $29.4 \, GJt^{-1}$ $(29.4 \, kJg^{-1})$.

Other products associated with the ammonia synthesis

Nitric acid

Ammonia is oxidized with air or oxygen by passing the mixture over a platinum catalyst at up to 1 MPa (1–10 atm) pressure and 850–950°C when the gases unite with the liberation of heat:

$$4NH_3 + 5O_2 \rightarrow 4NO + 6H_2O$$

About 95% of the ammonia forms nitrogen oxide (nitric oxide), the remainder being combusted to nitrogen and steam. Cooling results in further oxidation to nitrogen dioxide which forms nitric acid (finally) with water during absorption in a counter-current of nitric acid and water:

$$2NO + O_2 \rightarrow 2NO_2$$

$$3NO_2 + H_2O \rightarrow 2HNO_3 + NO$$

Nitric acid is widely used industrially for making explosives, dyes and fertilizers and as a fuel in rocket propulsion.

Carbon dioxide

Usually removed prior to the ammonia synthesis by solution in water under a pressure of 1–5 MPa (10–50 atm), carbon dioxide is used for producing urea, ammonium sulphate from calcium sulphate rock (see below), and itself is sold commercially in the solid form for refrigeration and in the liquid form for use in fire-extinguishers.

Fertilizers

Ammonia itself may be directly applied to the soil as a fertilizer and its controlled use in this way is increasing in the USA. Ammonium sulphate is widely used and may be produced either directly from ammonia and sulphuric acid, as a coal gas by-product, or by the action of ammonia and carbon dioxide in solution on a suspension of calcium sulphate, mined as anhydrite or gypsum,

$$2NH_3 + H_2O + CO_2 + CaSO_4 (s) \rightarrow$$
$$CaCO_3 (s) + (NH_4)_2SO_4$$

Calcium carbonate is less soluble than the sparingly soluble calcium sulphate and is precipitated. Made direct from ammonia and the corresponding acids, are the ammonium phosphates and ammonium nitrate, the latter being mixed with chalk (calcium carbonate) to minimize risks of an explosion. Chalk precipitated in ammonium sulphate production may be converted into calcium nitrate, also used as a fertilizer. Another compound which may be favoured as a fertilizer especially in tropical countries is urea, which has a high nitrogen content but is rather too soluble for widespread use.

Urea

Carbon dioxide, recovered at a previous stage of the ammonia synthesis, reacts with ammonia under pressure,

$$CO_2 + 2NH_3 \rightarrow CO(NH_2)_2 + H_2O$$

The integrated manufacture of urea is most effectively accomplished by coupling the process to the ICI LCA process which produces a surplus of carbon dioxide at the reforming stage, where only about 90% recovery is necessary for the reaction. This should be compared with the idealized natural balance of the processes based on methane (natural gas) which are one mole of carbon dioxide short.

$$7CH_4 + 10H_2O + 8N_2 + 2O_2 \rightarrow 7CO_2 + 16NH_3$$

Urea is used in the production of urea–formalde-hyde resins, textiles, adhesives, in paper-making and as a fertilizer.

Nitrogen for metallurgy

For many metallurgical operations an inert atmosphere is essential. Anhydrous ammonia, stored at pressure in cylinders, may conveniently be catalytically cracked when heated at atmospheric pressure in a transportable 'ammonia cracker'. One 240 kg cylinder of ammonia yields more than 675 m³ of gas mixture. If desired, the hydrogen may be combusted with a calculated quantity of air, which introduces more nitrogen to give in all almost 1200 m³ of the gas. Besides applications in metallurgy, cracked ammonia is utilized for the manufacture and spinning of nylon, the manufacture of expanded rubber and in atomic hydrogen welding.

Methanol

This may be associated with the ammonia synthesis; water gas and hydrogen at 450°C and 20 MPa (200 atm) pressure unite over a basic zinc chromate catalyst:

$$(CO + H_2) + H_2 \rightarrow CH_3OH$$

13

The industrial production of sulphuric acid

The paramount importance of sulphuric acid

In 1843, Liebig wrote, 'we may fairly judge of the commercial prosperity of a country from the amount of sulphuric acid it consumes.' Probably the demand for industrial electrical power is a better indication of commercial prosperity today but sulphuric acid remains an industrial chemical of fundamental importance, used by itself and in the production of many other chemicals, and finding its way into a great many manufacturing processes.

Sulphuric acid was originally made in a fairly small way in the 16th and 17th centuries by the distillation of green vitriol, itself the product of the atmospheric oxidation of iron pyrites, FeS_2. The overall distillation of hydrated iron(II) sulphate, or green vitriol, may be expressed:

$$2FeSO_4 . 7H_2O \rightarrow Fe_2O_3 + SO_3 + SO_2 + 14H_2O$$

The distillate contains sulphuric acid, H_2SO_4, or oil of vitriol. Green vitriol was in use as a mordant and sulphuric acid became associated with the finishing processes of the textile industry from the start. Demand stimulated other methods of production. Early in the 18th century, sulphuric acid was produced industrially in England by burning sulphur with nitre, potassium nitrate, and absorbing the product in water. Large glass vessels, or 'bells' of 175–225 dm^3 capacity were used. By manufacturing sulphuric acid 'by the bell', Joshua Ward in 1736 reduced the price† of sulphuric acid from 1s 6d–2s 6d per oz to that price per lb from his Twickenham works. In 1740, he moved to Richmond after complaints about the fumes, the recovery problem being solved by Gay-Lussac and Glover later. Sulphuric acid was required not only in textiles, but in many engineering processes, for pickling metal and cleaning, and in the refining of precious metals. Because of the inadequate road surfaces, sulphuric acid could not be transported safely and was produced near to where it was required. From 1746, Roebuck and Garbett manufactured sulphuric acid at Birmingham, where lead-lined chambers replaced glass vessels, the Chamber process. In 1749, they built a similar works at Prestonpans near Edinburgh. Clément and Désommes showed in 1793 that an increased yield resulted from providing a stream of air and that oxides of nitrogen catalysed the oxidation. Another catalytic process, the Contact process, was devised in 1831 by Peregrine Phillips, a Bristol vinegar manufacturer. It was later developed to give sulphuric acid of higher quality than that obtained from the Chamber process.

Until 1838, Sicily provided the sulphur from which sulphuric acid was made. Following the purchase of some mines by British manufacturers, the King of the Two Sicilies gave a monopoly to a French company and imposed an export duty. The price of sulphur rose from £5.10s to £15 per tonne with disastrous results for the Sicilian economy. The search for fresh supplies began and in 1839 and during the next decade, pyrites burning displaced sulphur as the source of sulphur dioxide. From 1894, the sulphur deposits of Louisiana and Texas have been mined by the Frasch process and the USA has become the principal exporter of sulphur, producing some 90% of world supplies. In 1950–51, a world sulphur shortage occurred due to the restriction of exports from the USA, anxious about its reserves on account of increased home demand. Cutting the export of sulphur to the UK by one-third not only stimulated production of that mineral in Sicily, Japan and South America, but led to a much greater investment in the UK in the

† Money: £1 = 20s (shillings) = 240d (pence)
  Weight: 1 lb (pound) = 16 oz (ounces)

production of sulphur dioxide from calcium sulphate, an indigenous mineral found as anhydrite and in the hydrated form, as gypsum. Sulphur is also produced during petroleum refining. Even minute traces of sulphur compounds would be oxidized ultimately to sulphuric acid in internal combustion engines, causing corrosion. Sahara oil has a relatively high sulphur content which makes France increasingly important as a sulphur producer. A country relying on imported sulphur is very vulnerable in times of emergency.

The manufacture of sulphuric acid requires the formation of sulphur dioxide, which is oxidized to sulphur trioxide, itself the anhydride of sulphuric acid:

$$2SO_2 + O_2 \rightleftharpoons 2SO_3 \quad SO_3 + H_2O \rightarrow H_2SO_4$$

This is accomplished by either the Contact process or the Lead Chamber process. Annual statistics show that the Contact process accounts for about two-thirds of the acid produced in the UK.

Figures from the National Sulphuric Acid Association show the consumption of sulphuric acid in the UK has fallen. In 1989, out of the 2.32 m. tonnes consumed, 2.15 m. tonnes were home produced with 249 000 tonnes imported and 44 000 tonnes recovered. In addition 127 000 tonnes were exported. About 21% of the acid came from indigenous and other sulphur-bearing materials with the remainder made from imported elemental sulphur.

Sulphur dioxide

The production of sulphur dioxide

1 *By combustion of molten sulphur in air*

$$S + O_2 \rightarrow SO_2$$

a very pure supply of sulphur dioxide is obtained. The paramount importance of the USA as a sulphur producer has been mentioned. On a much smaller scale in Europe, Italy and Sicily are main producers, followed by France. The spent oxide resulting from the purification of coal gas by hydrated iron(III) oxide contains free sulphur and iron(II) sulphide. Roasting in air regenerates iron(III) oxide with liberation of sulphur dioxide.

2 *By the combustion of sulphide ores in air*

$$4FeS_2 + 11O_2 \rightarrow 2Fe_2O_3 + 8SO_2$$
$$2ZnS + 3O_2 \rightarrow 2ZnO + 2SO_2$$

Iron pyrites contains about 45% sulphur by weight and is obtained principally from Cyprus, Norway and Spain. Zinc sulphide, or zinc blende, is roasted in air to form the oxides of zinc and sulphur as the first stage in zinc smelting. The demand for sulphuric acid now exceeds that for zinc, and production is augmented by supplying extra sulphur dioxide from pyrites burning or sulphur combustion.

3 *From calcium sulphate*

This occurs widely in SE, Central, NE and NW England as anhydrite, $CaSO_4$, and gypsum, $CaSO_4.2H_2O$. Anhydrite is reduced by coke in the presence of acidic non-volatile oxides. Silica, iron(III) oxide and alumina are used under carefully defined conditions in rotary kilns, over 60 m long, fired by pulverized coal and air so that hot gases pass up through the charge. The gaseous products contain about 9% sulphur dioxide and the solid is valuable as cement clinker. In the firing zone, the temperature reaches 1400–1450°C. A number of reactions occur and may be considered separately. Anhydrite may be reduced to calcium sulphide by coke rapidly at 800–900°C:

$$CaSO_4 + 2C \rightarrow CaS + 2CO_2$$

At 1200°C a reversible reaction occurs:

$$CaS + 3CaSO_4 \rightleftharpoons 4CaO + 4SO_2$$

Production of sulphur dioxide may be assisted by removal of calcium oxide at 1400–1450°C by combination with silica and alumina, added as shale, with ashes and sand in carefully controlled proportions to give Portland cement clinker:

$$CaO + SiO_2 \rightarrow CaSiO_3$$
$$CaO + Al_2O_3 \rightarrow Ca(AlO_2)_2$$

The net reduction becomes

$$2CaSO_4 + C \rightarrow 2CaO + CO_2 + 2SO_2$$

The purification of sulphur dioxide

Much of the dust from anhydrite reduction and sulphide combustion is removed and re-cycled. Sulphur dioxide is dissolved in water, to be later expelled by a counter current of air. The gas is now cool and damp. Residual dust is removed in electrostatic precipitators, operating at a potential of 30 000–50 000 V, and the gas is dried by scrubbing with a counter current of sulphuric acid. Gas from the anhydrite process contains roughly 6.5% sulphur dioxide at this stage.

Sulphuric acid

The production of sulphuric acid

The Chamber process

The Chamber process was the first process to be commercially successful, becoming established in many countries during the 19th century. Sulphur dioxide, oxygen from air, and water react in the presence of oxides of nitrogen to form sulphuric acid, expressed simply by the equations:

$$H_2O + SO_2 + NO_2 \rightarrow H_2SO_4 + NO$$

$$2NO + O_2 \rightarrow 2NO_2$$

or the overall change by:

$$2H_2O + 2SO_2 + O_2 \rightarrow 2H_2SO_4$$

The yield is 95–99% theoretical and the process still provides an economical method of manufacturing moderately concentrated sulphuric acid from a range of sulphur-bearing minerals, where high purity is not essential.

The mechanism of the catalytic change is very complex and involves the formation of sulpho-nitronic acid, SO_5NH_2, and nitrosyl hydrogen-sulphate (nitrosylsulphuric acid), $NOHSO_4$, as intermediates:

(a) $2NO + O_2 \rightarrow 2NO_2$

$$H_2O + SO_2 \rightarrow H_2SO_3$$

(b) $H_2SO_3 + NO_2 \rightarrow SO_5NH_2$

$$2SO_5NH_2 + NO_2 \rightarrow 2NOHSO_4 + H_2O + NO$$

and also

(c) $SO_5NH_2 \rightleftharpoons NO + H_2SO_4$

$$4NOHSO_4 + 2H_2O \rightleftharpoons 4NO + O_2 + 4H_2SO_4$$

$$NOHSO_4 + HNO_3 \rightleftharpoons 2NO_2 + H_2SO_4$$

From the original burning of sulphur with potassium nitrate, the process has undergone much development. Oxides of nitrogen are now introduced from the catalytic oxidation of ammonia,

$$4NH_3 + 5O_2 \rightarrow 4NO + 6H_2O$$

and elaborate arrangements are made for the recycling of the gaseous catalyst. The process is continuous.

Since 1880, the basic units of the Chamber process have been three.

1   The lead chambers from which sulphuric acid is drawn off.
2   The towers for the absorption of the catalytic oxides of nitrogen, developed by Gay-Lussac in 1827, originally to prevent atmospheric pollution.
3   The tower for the recovery and re-cycling of the oxides of nitrogen developed by Glover in 1859.

Inevitably catalyst is lost or reduced and is replenished by catalytic oxidation of ammonia. The plant is shown diagrammatically in Fig. 13.1.

A dust-free mixture of sulphur dioxide and air, at about 600°C, is introduced into the base of the Glover Tower, down which flows weak chamber acid (roughly 66% sulphuric acid) and acid containing oxides of nitrogen (originally 78% sulphuric acid) from the Gay-Lussac Towers. Oxides of nitrogen are expelled and swept away with the gaseous reaction mixture, the temperature of which has dropped to about 90°C for the next stage. Acid flowing from the base of the Glover Tower is roughly 78% sulphuric acid. The water is condensed for use in the chambers. The Glover Tower is lined with lead and is packed with a chequered pattern of acid-resistant brickwork. Catalytic oxidation begins immediately.

The mixture of sulphur dioxide, air and oxides of nitrogen meets a carefully controlled fine spray of water inside the chambers which are lined with lead. The oxidation is exothermic and the overall change may be summarized:

$$2SO_2 + O_2 + 2H_2O \rightarrow 2H_2SO_4$$

The concentration of acid produced in the chambers is in the range 62–70% sulphuric acid at a temperature of 20–30°C to avoid corrosion problems. This concentration of acid is important: a higher strength acid would form nitrosyl hydrogen-sulphate (nitrosylsulphuric acid), $NOHSO_4$, while that of lower strength would form nitric acid by interaction of water and oxides of nitrogen. The acid is concentrated to about 78% in the Glover Tower and the water recovered for re-use.

The mixture of oxides of nitrogen and nitrogen leaving the chambers passes into the bottom of the first of a series of Gay-Lussac Towers down which 78% sulphuric acid from the Glover Tower flows. Oxides of nitrogen are absorbed to leave a waste gas consisting of atmospheric nitrogen (and inert gases) and the excess of oxygen, approximately 5%, required to ensure maximum conversion of sulphur dioxide.

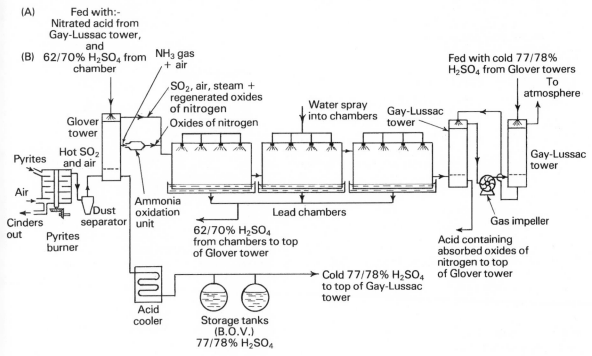

Fig. 13.1   The Chamber Process (From *Sulphuric Acid*, ICI, London)

The Contact process

The principle of the Contact process was discovered in 1831 but its development into the principal method for the production of sulphuric acid did not start until there was a demand for higher grade sulphuric acid and fuming sulphuric acid, oleum, required in the dyestuff and other chemical industries. The physical principles governing the equilibrium mixtures were not elucidated until 1901.

Sulphur dioxide and oxygen, from the air, react reversibly over a solid catalyst of platinum, vanadium(v) oxide (vanadium pentoxide) or iron(III) oxide with a carefully regulated temperature control in a series of cylindrical steel vessels called converters,

$$2SO_2 + O_2 \rightleftharpoons 2SO_3$$

Platinum is expensive and highly susceptible to poisoning, and iron(III) oxide requires the highest temperature with a necessarily lower equilibrium yield, so vanadium(v) oxide, compressed with silica into pellets, is widely used as the contact catalyst. A catalyst increases the rates of reaction but does not increase the equilibrium yield. However, an industrial process is worked to give the maximum rate of production, without waiting necessarily for equilibrium to be established. Working at constant pressure for a given temperature, the equilibrium constant may be expressed in terms of partial pressures:

$$K_p = \frac{p_{SO_3}^2}{p_{SO_2}^2 \cdot p_{O_2}} \text{ at } T$$

A slight excess of oxygen is used to increase the proportion of sulphur dioxide converted but a large excess, though increasing the equilibrium conversion, results in too much dilution of sulphur dioxide. Increased pressure, although favouring the equilibrium yield of sulphur trioxide, is not used because of the difficulty of operation with the corrosive (when moist) gases and the yield is high anyway. The reaction is exothermic:

$$SO_2 + \tfrac{1}{2}O_2 \rightarrow SO_3: \quad \Delta H = -94.6\,\text{kJmol}^{-1}$$

With a rise in temperature, both forward and reverse reaction rates increase but the equilibrium yield falls. Therefore, an optimum temperature must be discovered in conjunction with a suitable catalyst for obtaining a maximum yield. Using vanadium(v) oxide, the temperature range inside the converter is in the range 450–580°C, the gas

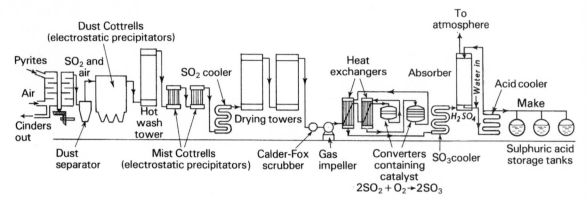

Fig. 13.2   The Contact Process (From *Sulphuric Acid*, ICI, London)

being introduced at 400–450°C and heat being generated as the conversion proceeds. Before passing into the second converter, the gaseous mixture passes through a heat exchanger to reduce the temperature to 400–450°C. Finally, 95–98% of the original sulphur dioxide has been converted to sulphur trioxide, which is obtained at a temperature not exceeding 480°C.

After cooling, the sulphur trioxide is absorbed in concentrated sulphuric acid (about 98%) on the counter-current principle to form 99–100% sulphuric acid. As the acid passes from the tower, it is diluted. Heat is evolved in both processes and the acid is cooled before storage or re-cycling.

A scheme for the Contact process is shown in Fig. 13.2. Comparison with that for the Chamber process shows clearly the more stringent purification programme required in the Contact process involving separation of dust, scrubbing, electrostatic precipitation at a potential difference of 30 000–50 000 V and drying.

The industrial applications of sulphuric acid

Pure concentrated sulphuric acid is a colourless, dense, oily liquid (density 1.84 g cm^{-3}), which has a great affinity for water, with which it reacts violently. It is a strong dehydrating agent, removing water or the elements of water from many substances. It is dibasic and forms a series of important normal salts called sulphates. On boiling, at about 338°C, sulphuric acid begins to dissociate into steam and sulphur trioxide. It forms a maximum constant boiling-point mixture with water, containing 98.7% acid at atmospheric pressure.

Sulphuric acid is used in innumerable industrial processes. However, certain industries absorb the greater part of the production in the UK.

1   *Superphosphate fertilizer*   Approximately one-quarter of the annual production is used to convert imported calcium phosphate rock into the more soluble superphosphate:

$$Ca_3(PO_4)_2 + 2H_2SO_4 \rightarrow Ca(H_2PO_4)_2 + 2CaSO_4$$

Using a larger proportion of acid, phosphoric acid may be produced, and from it a wide range of phosphorus compounds.

2   *Ammonium sulphate*   Ammonia, formed by the destructive distillation of coal in gas works and coking ovens, is converted into ammonium sulphate fertilizer by reaction with sulphuric acid,

$$2NH_3 + H_2SO_4 \rightarrow (NH_4)_2SO_4$$

This accounts for about one-seventh of the acid produced. Ammonium sulphate is also manufactured direct from anhydrite.

3   *Viscose rayon*   On treatment with acid, the viscose solution is re-converted into cellulose: rayon is used in yarn and fibres, finding many industrial uses, and for transparent wrapping paper for foodstuffs. Roughly one-tenth of the sulphuric acid is taken by the rayon industry.

4   *Paint pigments*   Approximately the same proportion (one-tenth) of sulphuric acid is manufactured for the paint industry, for the production of pigments based on barium sulphate, titanium(IV) oxide (titanium dioxide) and lithopone (barium sulphate and zinc sulphide). They are not blackened by sulphuretted hydrogen, which darkens lead-based paints exposed to town air.

Miscellaneous Among industries taking 5% or less, the following are important. Sulphuric acid is used in the iron and steel industry for metal-pickling, involving the removal of oxide film by acid treatment before the application of finishing processes, e.g. galvanizing, tinning, enamelling and plating. Oil refineries use sulphuric acid for the removal of sulphur derivatives and unsaturated hydrocarbons from refinery gases. It is used in the industries which produce textiles, dyestuffs, explosives, soap and synthetic detergents. Sulphuric acid is used in lead-plate accumulators, in acid-carbonate fire-extinguishers and as a weed-killer. There are many other applications.

14

Chemicals from salt: the alkali industry

A brief history of alkali production

The term alkali is derived from the Arabic, *al-qaliy*, ashes, applied about 2000 years ago. But alkali, either as natron, which crystallizes from natural lakes containing soda, or as plant ash, had been used as a flux and detergent much earlier. The formation of a glass bead found in Egypt and estimated roughly as 5000 years old required the use of alkali in its production. The use of soap, made from oil and vegetable ash, is recorded on a Babylonian tablet dating from 2800 B.C. Hippocrates recorded the uses of alkalis known to the Greeks before 400 B.C. but the fullest account is given in *Natural History*, written by the Roman, Pliny the Elder (A.D. 23–79). Here, the alkalis present in the ashes of seashore plants and seaweed, from which are produced hard soaps characteristic of soda, were distinguished from those present in the ashes of inland vegetation, from which are produced the soft soaps formed by potash. The distinction was confirmed in 1735 by du Monceau who demonstrated that the alkali of seashore plants was the same as that of common salt.

Naturally occurring alkali is found crystallizing from the waters of the lakes of the Rift Valley System, a series of depressions formed a few million years ago on the eastern side of the African continent and associated with volcanic activity. The lakes are fed by very hot spring water containing small amounts of the carbonate, hydrogencarbonate and chloride of sodium. The heat is acquired from subterranean lavas. Evaporation is favoured by the hot, dry climate and sodium sesquicarbonate, trona, $Na_2CO_3.NaHCO_3.2H_2O$, crystallizes as a solid mass so that the lake surface may be walked upon except near to the entry of spring waters. Lake Magadi in Kenya is an important commercial source of sodium sesquicarbonate, its associated products and common salt.

Until the Industrial Revolution enough soda could be produced from ashes, but a sharp increase in demand from about 1760 for the manufacture of soaps, mordants and bleaches stimulated the search for other sources, resulting in the founding of the heavy chemical industry. In 1775, when France was at war and the British blockade prevented the importation of barilla (plant ash) from Spain, the French Academy of Sciences offered a prize of 100 000 francs, then equivalent to £40 000, to the inventor of a process for the manufacture of alkali from non-vegetable sources. In 1790, a process which was to be worked for about 130 years was devised by Leblanc. It is tragic to record that he received no prize money, had his patent rights and factory confiscated by the revolutionaries in 1791, and when the works was returned to him later he had no capital to develop it. In 1806, he committed suicide.

In the Leblanc process, sea water or natural brine was evaporated to give salt, which on heating strongly with sulphuric acid generated hydrogen chloride, which escaped, leaving salt-cake, sodium sulphate:

$$2NaCl + H_2SO_4 \rightarrow Na_2SO_4 + 2HCl$$

The product was mixed with finely ground limestone and coke or coal slack. On heating, a mixture of sodium carbonate and calcium sulphide, called black ash, remained:

$$Na_2SO_4 + CaCO_3 + 2C \rightarrow Na_2CO_3 + CaS + 2CO_2$$

On lixiviation with cold water, sodium carbonate dissolved, leaving calcium sulphide, and was crystallized. The raw materials were salt, limestone, sulphuric acid and coke. In addition to sodium carbonate, calcium sulphide and solid ash from the coke and carbon dioxide were formed. Because of the preliminary evaporation of brine and soda solution, and the roasting, the process used a lot of fuel. The waste products were troublesome; gaseous hydrogen chloride killed local vegetation, and corroded ironwork and brickwork, while calcium sulphide, of which two tonnes were obtained for every tonne of sodium carbonate, formed stinking waste heaps, from which hydrogen sulphide escaped, a constant reminder of the loss of sulphur. Gossage applied the countercurrent principle to the absorption of hydrogen chloride by water in 1830 but it was not until 1863 that the first Alkali Act required that not less than 95% of the hydrogen chloride should be absorbed. The economics of the process were improved by recovery and use of the co-products.

Chlorine could be recovered for bleaching purposes by the action of manganese(IV) oxide (manganese dioxide) on hydrogen chloride,

$$4HCl + MnO_2 \rightarrow MnCl_2 + Cl_2 + 2H_2O$$

but the relatively expensive manganese was lost. Chlorine was absorbed in lime as bleaching powder. Weldon had developed a recovery process for the manganese compound by 1870 and immediately the price of paper and calico throughout the world was reduced. In Weldon's process, the liquor was treated with milk of lime and air was blown through the resulting mixture for several hours, the reaction being exothermic. The product, called Weldon mud, was allowed to settle and re-used:

$$MnCl_2 + Ca(OH)_2 \rightarrow Mn(OH)_2 + CaCl_2$$

$$2Mn(OH)_2 + 2Ca(OH)_2 + O_2 \rightarrow$$
$$2CaO.MnO_2 + 4H_2O$$

It will be observed from the two sets of equations that only half of the chloride was converted.

In Deacon's process, hydrogen chloride was oxidized by air over a copper chloride catalyst,

$$4HCl + O_2 \rightleftharpoons 2H_2O + 2Cl_2$$

to give very dilute chlorine but suitable for absorption by lime to form bleaching powder.

In 1882, the most successful sulphur-recovery process appeared, the Chance process. Alkali waste was made into a paste with water and treated with carbon dioxide formed by burning limestone,

$$2CaS + CO_2 + H_2O \rightarrow CaCO_3 + Ca(HS)_2$$

$$Ca(HS)_2 + CO_2 + H_2O \rightarrow CaCO_3 + 2H_2S$$

The hydrogen sulphide was collected in a gas holder and burnt in a Claus kiln using a calculated volume of air in the presence of an iron oxide catalyst,

$$2H_2S + O_2 \rightarrow 2H_2O + 2S$$

The Leblanc process was introduced by Muspratt into the UK in 1822 and was the only source of non-vegetable alkali until 1873 when the introduction of the Ammonia-Soda process started its decline. At first, recovery of the by-products, as outlined above, bolstered up the Leblanc process, but the introduction of the electrolysis of brine for the simultaneous production of chlorine and caustic soda, together with the fact that the sodium carbonate made by the Ammonia-Soda process was of a quality hitherto unknown, provided severe competition and by 1920 the Leblanc process had disappeared. Electrolysis of brine by the mercury cathode cell was patented by Castner in 1892.

The Ammonia-Soda process (see p. 206) is based on the reaction discovered by Fresnel in 1811 but this was not used immediately, since it was economically unattractive at that time in comparison with the established Leblanc process, and required chemical plant of a more advanced design. A scheme for the recovery and re-cycling of the ammonia was the subject of a patent by Dyar and Hemming in 1838. However, while the potential worth of the process was recognized, all attempts to make it commercially successful failed until 1863, when Solvay succeeded in Belgium. It was in 1872 that Brunner, Mond and Company began production under licence in the UK at Winnington in Cheshire. In the face of competition from the high-quality soda, 45 firms working the Leblanc process amalgamated in 1890 to form the United Alkali Company. But by 1902, 92% of all soda here was made by the Ammonia-Soda process. In 1926, the Brunner-Mond Group, United Alkali, Nobel Industries and the British Dyestuffs Corporation united to form Imperial Chemical Industries Limited. ICI nowadays is an international company of world-wide standing, manufacturing a diverse range of products.

There are now six major producers of soda ash, sodium carbonate, in the EEC, producing 8 Mt of the product each year, worth £1000 million, sold in high volume at low prices. The energy-intensive manufacturing process has high fixed capital costs and low margins. Soda ash sells for about 12p per kilogram. At a time of static, though huge demand, the industry faces cheap imports from the US, Turkey, Tanzania and South Africa, produced from naturally occurring trona, mentioned earlier, and selling at roughly half the European prices.

As an alternative to soda ash, users have also turned to caustic soda, a principal *by-product of chlorine production* by electrolysis of brine.

Notice the change in emphasis. Almost 40 million tonnes of chlorine and about 45 million tonnes of caustic soda (sodium hydroxide) are produced throughout the world every year.

The occurrence of brine and rock salt

Sodium chloride occurs widely in nature not only in the oceans but in salt lakes and in extensive deposits where the seas of bygone eras have evaporated. Salt lakes of note include the Dead Sea, an important source of chemicals to Israel, and the Great Salt Lake of Utah, in the USA. In hot, dry regions, evaporation of brine in shallow pools by utilizing solar energy is commercially successful. Solid sodium chloride may be mined as rock salt, and is usually contaminated by compounds of calcium and magnesium which have also come from sea water. Salt deposits are usually up to about 90 m thick although those at Stassfurt reach 300 m.

In the UK, salt deposits of importance occur in Cheshire, Lancashire and South Durham, near to which are located factories for the electrolytic production of chemicals from salt, and also in Staffordshire, where non-electrolytic processes are worked, and in Somerset. Generally, alkali plant is situated near to the raw materials. About 88% of the salt recovered in the UK comes from Cheshire. Less than 1% of the total is mined as solid rock salt, recovery as brine being preferred. There are many examples of subsidence in the salt-producing areas, where underground cavities have been left. Nowadays, fresh water or waste liquor containing calcium chloride is pumped into the ground as the salt is removed. In this country deposits occur at depths ranging from about one hundred to several hundred metres.

The degree to which brine is purified depends on the purpose for which it is required. Addition of sodium carbonate precipitates calcium ions as calcium carbonate, and caustic soda is used to precipitate magnesium hydroxide. For electrolysis in diaphragm cells, sulphate ions, which promote corrosion of the steel net cathodes, are precipitated as barium sulphate, using suitable barium compounds. Iron is precipitated as iron(III) hydroxide during purification.

The applications of solid sodium chloride are numerous. Essential to life, 5–6 kg of common salt is required by each of us every year. It is used in the seasoning and preserving of food, in the production of many chemicals, as a glaze in ceramics, as a cheap means of refrigeration with ice, in tanning hides, and in stock feeds. Products are indicated in Fig. 14.1.

Other products associated with brines

Other compounds occur in the waters of oceans and salt lakes and may be extracted. Substances which are produced on a large scale in this way include magnesite for refractory bricks, magnesium and bromine.

Soda (sodium carbonate): the Ammonia-Soda process

Based on the principle discovered by Fresnel and engineered by Solvay, the overall reaction for the production of sodium carbonate requires the interaction of sodium chloride (brine) and calcium carbonate (limestone):

$$2NaCl + CaCO_3 \rightarrow CaCl_2 + Na_2CO_3$$

The other raw materials required are coal, coke and water. The latter is used for cooling, steam-raising and quicklime-slaking. To replace production losses, ammonia is obtained by distillation of ammoniacal liquor from gas works and coking ovens.

Purified brine (about 25.5% w/v at 20°C) is pumped to the top of a tower and descends, meeting an upward stream of gaseous ammonia, intimate mixing being achieved by means of a series of mushroom-shaped baffles designed to impede the liquid and disperse the gas. Saturation with ammonia causes evolution of heat and the liquid is cooled. The ammoniated brine is pumped in succession to the top of a number of tall Solvay towers and descends through a baffle-system, meeting an upward-moving stream of carbon dioxide gas under about 250 kPa (2.5 atm) pressure. The towers are water-cooled to remove the heat evolved. Sodium hydrogencarbonate, which is sparingly soluble under these conditions, separates out:

$$NH_3 + H_2O \rightleftharpoons NH_4^+ + OH^-$$

$$Na^+ + Cl^- + NH_4^+ + OH^- + CO_2 \rightarrow$$
$$Na^+ + HCO_3^- + NH_4^+ + Cl^-$$

$$Na^+ + HCO_3^- \rightarrow Na^+HCO_3^-(s)$$

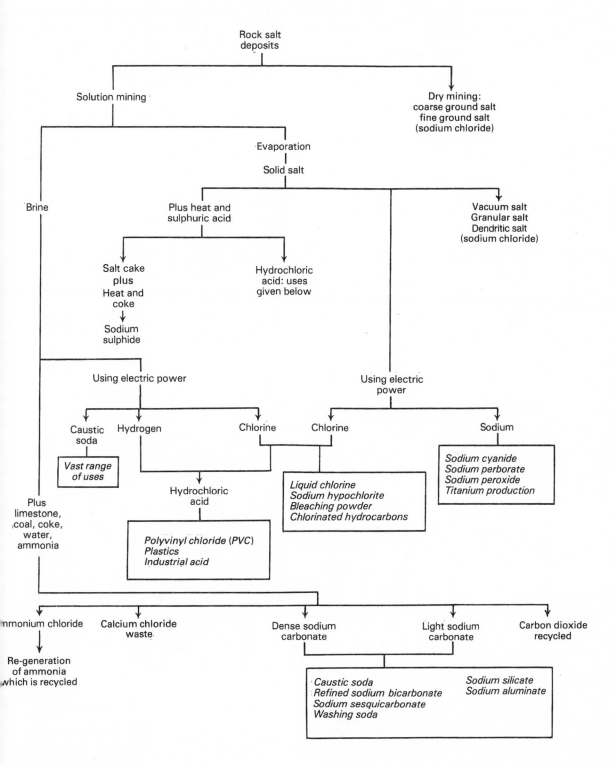

Fig. 14.1   The salt industry

The sodium hydrogencarbonate is collected on rotary filters and washed free of ammonium chloride solution.

It is calcined on rotary driers to give a light bulky form of anhydrous sodium carbonate called light soda ash:

$$2NaHCO_3 \rightarrow Na_2CO_3 + H_2O + CO_2$$

The carbon dioxide is re-cycled. It is obtained in the first place by 'burning' limestone with coke in kilns, a reaction used to augment production losses:

$$CaCO_3 \rightarrow CaO + CO_2$$

Quicklime is also formed and slaked:

$$CaO + H_2O \rightarrow Ca(OH)_2$$

Ammonia is re-generated by treating ammonium chloride filtrate with milk of lime together with some ammoniacal liquor, mentioned previously, and the gas expelled by distillation in steam:

$$2NH_4Cl + Ca(OH)_2 \rightarrow 2NH_3 + CaCl_2 + 2H_2O$$

Calcium chloride is waste.

The process is continuous and depends fundamentally on the relatively low solubility of sodium hydrogencarbonate ($8.2 \text{ g } 100 \text{ g}^{-1}$ water at 10°C) and cannot be used for the production of potassium carbonate because the hydrogencarbonate has a much higher solubility ($28 \text{ g } 100 \text{ g}^{-1}$ at 10°C). The product, anhydrous light soda ash, is dusty and bulky. By moistening and drying it is processed to give a denser, granular form better suited for transportation. Sodium carbonate is used in the production of other sodium compounds of major economic utility, as well as being of great importance itself.

The Ammonia-Soda process may be set out diagrammatically as in Fig. 14.2.

Caustic soda (sodium hydroxide): the Lime-Soda (Gossage) process

Although caustic soda is made electrolytically and the mercury cathode cell gives a very pure product, the older Lime-Soda process has been the principal source of caustic soda in the UK, while there was

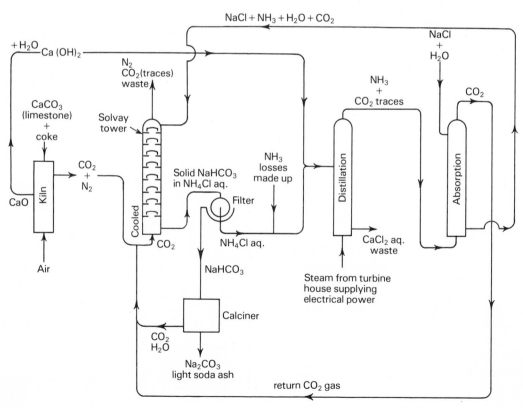

Fig. 14.2 The Ammonia-Soda process

not quite such a demand for the co-product of electrolysis of brine, chlorine, which would have to be wasted. In the USA the position is reversed and enough chlorine is required to leave a surplus of caustic soda, which may be converted into sodium carbonate by carbon dioxide. Recent developments in electrolytic cell design are considered later in the chapter.

Milk of lime, prepared from freshly slaked lime, reacts with aqueous sodium carbonate in large tanks with mechanical stirring and steam heating. Calcium carbonate, being less soluble than calcium hydroxide, is precipitated†:

$$Na_2CO_3 + Ca(OH)_2 \rightleftharpoons 2NaOH + CaCO_3$$

The causticization is best done in dilute solution so the product has to be concentrated by evaporation, usually by steam initially and over a fire finally. Much caustic soda (sodium hydroxide) is sold as liquor (50%) while some is supplied as solid (95%) in flake, stick or pellet form.

Refined sodium hydrogencarbonate (sodium bicarbonate)

Solid sodium carbonate is moistened and treated with carbon dioxide to form sodium hydrogencarbonate which is washed with cold water and dried:

$$Na_2CO_3 + H_2O + CO_2 \rightarrow 2NaHCO_3$$

Sodium silicates

Solid sodium carbonate is fused with sand (silica) in a furnace to produce sodium silicate(s), the mass being extracted with water:

$$Na_2CO_3 + SiO_2 \rightarrow Na_2SiO_3 + CO_2$$
$$\text{(etc.)}$$

Uses of the major alkali products

Dense sodium carbonate, especially suitable for furnace work as it is less dusty than the light form, is used principally in the glass industry for window glass, bottles, jars and other containers. Soda glass is manufactured by the fusion of sand (silica), calcium carbonate and sodium carbonate. Dense soda ash is also used in steelworks, foundries, for enamelling processes and the manufacture of ultramarine, a deep blue synthetic sodium aluminium silicate containing combined sulphur. The lighter form goes to manufacturers of soap and soap

powders, the textile industry for scouring wool and to the refining of gas, tar and oil. It is used in the chemical industry generally. The decahydrate, which contains about two-thirds by weight of water, is readily soluble and is used domestically, softening water by precipitation of calcium and magnesium carbonates. In hot, dry climates where efflorescence of the decahydrate would occur or where it would melt, the monohydrate is supplied for use in toilet and household preparations. The more expensive sesquicarbonate,

$$Na_2CO_3 . NaHCO_3 . 2H_2O$$

is used in certain toilet preparations e.g. bath salts.

Refined sodium hydrogencarbonate is of high purity and supplied to the food industry for self-raising flour and baking powder, as a mild antacid in the preservation of cream in butter-making and in health salts, where the dry powder is mixed with a solid, weak acid, tartaric acid or an acid salt, e.g. an acid phosphate.

Sodium silicates (Na_2SiO_3, $Na_2Si_3O_7$) are manufactured for use in special detergents, soaps and soap powders to avoid re-deposition of dirt. Other principal applications depend on their natural adhesive qualities and include the manufacture of cardboard generally and the sealing of cardboard boxes, and for making concrete building materials and pottery.

Caustic soda is the strongest alkali commonly manufactured. It is produced in various grades. Other methods of manufacture are described later. Much of the total production of very pure alkali goes into the rayon industry and it is supplied to textile manufacturers for the mercerization of cotton, bleaching and dyeing processes and in paper-making. The hydrolysis (saponification) of natural oils and fats, by boiling with caustic soda solution, produces soap, as a mixture of sodium salts of certain organic acids (e.g. sodium stearate) and the trihydric alcohol, glycerol. Caustic soda is used in the removal of acidic compounds, such as phenol and the cresols, during the refining of coal tar. At least 400 products of the chemical industry require the use of caustic soda at some stage of their manufacture. Much caustic soda is now transported as 50% liquor by weight in road and rail tankers.

† In this book, physical states are introduced only to draw attention to special points in the chemistry. From the text it is seen that both Na_2CO_3 and $NaOH$ are in solution and the process depends on the relative 'insolubilities' of the other compounds

Electrolysis of sodium chloride

During the electrolysis of fused sodium chloride between inert electrodes, sodium is liberated at the cathode and chlorine at the anode in accordance with Faraday's Laws of Electrolysis:

$$Na^+ + e^- \rightarrow Na \qquad 2Cl^- - 2e^- \rightarrow Cl_2$$

$$2Na^+Cl^- \rightarrow 2Na + Cl_2$$

To avoid re-combination, the products must be kept apart.

The electrolysis of a saturated aqueous solution of sodium chloride follows a number of routes, depending on the conditions. Using carbon electrodes, the products of electrolysis are hydrogen at the cathode, discharged in preference to sodium, and chlorine at the anode:

$$2H^+ + 2e^- \rightarrow H_2 \qquad 2Cl^- - 2e^- \rightarrow Cl_2$$

Sodium ions remain in solution with the excess of hydroxide ions, formed near the cathode when hydrogen is discharged. The initial products are hydrogen, chlorine and sodium hydroxide solution, and with the use of an ion exchange membrane or a porous diaphragm separating cathode and anode solutions, they may be recovered. Inert metal cathodes, except mercury, will permit discharge of hydrogen. If the products are mixed together as electrolysis proceeds a series of reactions occur of which the more important are given below. With efficient mixing, nearly all of the chlorine forms sodium hypochlorite, otherwise known as sodium chlorate(I):

$$Cl_2 + 2OH^- \rightarrow Cl^- + ClO^- + H_2O$$

Further reactions may occur. Discharge of hypochlorite ions at the anode leads to the formation of sodium chlorate, otherwise known as sodium chlorate(v),

$$ClO^- - e^- \rightarrow ClO$$

$$12ClO + 6H_2O \rightarrow 4ClO_3^- + 12H^+ + 8Cl^- + 3O_2$$

Continued electrolysis with mixing yields more sodium chlorate, while the hypochlorite concentration remains at a steady level. At an optimum temperature of 70°C, interaction of hypochlorous acid and hypochlorite ions forms chlorate,

$$2HClO + ClO^- \rightarrow ClO_3^- + 2H^+ + 2Cl^-$$

Some hypochlorite may be lost by reduction to chloride by hydrogen discharged at the cathode.

The course of the electrolysis of fairly concentrated aqueous sodium chloride is radically changed by the use of a mercury cathode, at the surface of which there is a hydrogen overpotential (see p. 127). Preferential discharge of sodium occurs. It dissolves in the mercury, forming an amalgam,

$$Na^+ + e^- \rightarrow Na$$

Sodium amalgam may be decomposed by water, with liberation of hydrogen, the evolution of which is facilitated by having iron, or some other metal with a low hydrogen overvoltage, in contact with the mercury thereby creating a short-circuited electric cell:

$$2Na + 2H_2O \rightarrow 2(Na^+ + OH^-) + H_2$$

The overall change, whether an ion exchange membrane, a porous diaphragm or mercury is used for separation, is the same:

$$2NaCl + 2H_2O \rightarrow 2NaOH + H_2 + Cl_2$$

During electrolysis, fixed ratios of the various products are formed; a fact of considerable commercial importance in selecting manufacturing methods.

Environmental and economic pressures are changing the industry. Existing plants in Europe and the USA are large while new plants are smaller, with individual capacities of less than 50 000 tonnes of chlorine per annum. Traditional mercury cells are now regarded in many countries as being suspect from health and environmental points of view, adding problems of safety to operating costs. The use of asbestos in diaphragm cells is similarly out of favour. Membrane cells, utilizing sophisticated fluoropolymers similar to PTFE (polytetrafluoroethylene) with sulphonated or carboxylated active sites and about 250 μm thick are coming into use. Outstanding in this respect is the FM21 design by ICI (1977) which has been used under licence world-wide since 1981, maintaining the company's pre-eminence in the field of chlorine production technology.

The greater part of the cost of manufacturing chemicals by electrolysis is, of course, that for the electrical energy, which is the product of the quantity of electricity (i.e. current × time) and the voltage required to pass this quantity of electricity under the given conditions. The former, by Faraday's Laws of Electrolysis, determines the quantities of products liberated but the efficiency, when considered from this viewpoint, is limited by the electrolysis of impurities and by side-reactions. Electrodes can be electrically connected in parallel or in series in the industrial cell. In the first arrangement (monopolar) the current required is proportional to the total membrane area at a voltage which relates to that of one anode-cathode pair, while in the second (bipolar) arrangement the current is limited to that required by a single sheet of membrane but the voltage has to be multiplied by the number of electrode pairs. Bipolar electrodes tend to be large in area (5 m^2, or more) while the monopolar design uses smaller electrodes (less than 2 m^2) in area. There are pros and cons for each system. The FM21 cells mentioned earlier have electrodes in parallel, the design bringing major benefits of simplicity of construction and operational reliability so that damaged units can be replaced as convenient without the need to close down the plant.

The voltage required to overcome the electrical resistance of the electrolyte, resistance at the electrode surfaces and overvoltage effects is calculated to give a desirable commercial rate of production. The decomposition voltage, which is the minimum required to start electrolysis, may be calculated by equating the electrical energy to (roughly) the chemical energy, in the same units, calculated for the overall reaction by application of Hess' Law. In the electrolysis of brine, a mercury cell may work at 4.3 V with 30 000 A passing while a diaphragm cell may use 1000 A at 3.2 V. The respective decomposition voltages are 3.2 V and 2.3 V. The difference is 'recovered' as the EMF of the iron-mercury cell during reaction of sodium amalgam with water, so that the overall reaction has the same energy requirements whichever path it takes. In the electrolysis of fused salt, energy dissipated in overcoming the resistance of the electrolyte helps to maintain it in the molten state.

To provide the huge quantities of electrical power for the large-scale electrolysis of salt, generating stations for electrical power are usually built near to the plant. In addition to energy costs there are other factors which need to be considered before selecting replacement electrolytic plants. These include manpower costs, space availability and the existing technology which needs to be utilized to keep production costs down. Sodium chloride electrolysis is the most important application of electrolysis in the heavy chemical industry and supplies caustic soda, hydrogen and chlorine from aqueous salt, sodium and chlorine from fused salt and by mixing the products of the electrolysis of brine, chlorates and hypochlorites. Note that there is a fixed ratio of product and co-product: with every tonne of chlorine, 1.13 tonne of caustic soda and 0.028 tonne of hydrogen are produced from aqueous salt while from the fused compound, 0.65 tonne sodium is liberated.

The electrolysis of potassium chloride solution may be used for the production of caustic potash and co-products but for the isolation of the metal, the electrolysis of the fused hydroxide is preferred. The metal disperses in potassium chloride and if calcium chloride is added to reduce the melting-point, a calcium–potassium alloy results.

Electrolysis of brine (i) by the Mercury Cell process

Until recently this was the principal method used in the UK and was developed from the original patent of Castner.

Purified saturated brine (25.5% w/v sodium chloride at 20°C) flows through a long steel trough lined with ebonite or graphite, 7–21 m long and 0.5–2 m wide. A very shallow stream of mercury, about 3 mm deep, flows in the same direction and acts as the cathode. Graphite blocks (between 20 and 150) acting as anodes are suspended in the electrolyte and only just above (2 mm) the moving surface of mercury. They are held in position by a gas-tight cover for collecting the chlorine evolved at the anodes. Sodium is liberated and dissolves in the mercury to form an amalgam containing up to 0.2% sodium by weight.

The sodium amalgam passes into another cell where it reacts with a counter current of water. Decomposition occurs and is assisted by the presence of iron grids floating on the mercury. A primary cell is formed and short-circuited. Hydrogen is evolved at the surface of the iron, sodium ions pass into solution and the electrical energy is dissipated as heat:

$$2Na + 2H_2O \rightarrow 2(Na^+ + OH^-) + H_2$$

This soda cell is placed either alongside or below the mercury cell.

This process accounted for most of the chlorine manufactured. The gas is washed, cooled and dried, prior to liquefaction. Caustic soda is concentrated by evaporation and marketed as 50% (by weight) liquor or as solid, in which case the final evaporation is carried out in cast-iron pots at 500–600°C. Mercury is re-circulated by use of a mechanical pump. Unused brine, containing about 21% sodium chloride, is freed from chlorine by an

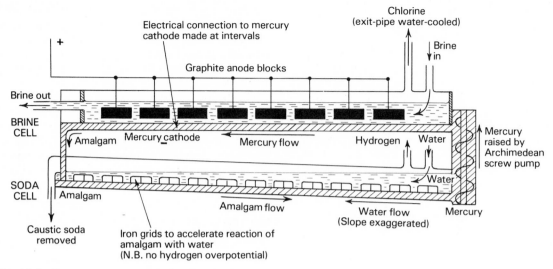

Fig. 14.3   The electrolysis of brine using the Mercury Cell process

air current and run to waste. Hydrogen is collected as a by-product.

Mercury cathode cells have the disadvantage of high capital cost and the continuous loss of small quantities of mercury. Health and environmental hazards have already been mentioned. However, they yield a very pure product. The cells may also be used for the electrolysis of potassium chloride in the production of caustic potash. The plant is shown diagrammatically in Fig. 14.3.

Electrolysis of brine (ii) by the Gibbs diaphragm cell

The cell is shown in Fig. 14.4. Two concentric cylindrical cathodes of steel gauze or perforated sheet steel, separated by a distance of 50–75 mm, are lined with several layers of asbestos paper on their opposing faces. This arrangement rests in a large cast-iron pot and is held in position by a concrete ring at the base and a groove in the concrete cover of the cell at the top. Within the cathode space, which occupies a relatively small part of the pot, a series of graphite anodes, rectangular in section, are suspended from copper bars and reach almost to the concrete base of the compartment. The anodes are partly immersed in brine, the level of which is regulated by a constant flow device. The concrete cover is gas-tight and adapted to keep the gaseous products, hydrogen and chlorine, separate and to duct them away.

About 30 dm^3 of brine at 90–95°C is introduced into the cathode space through a pipe occupying

a position in the ring of anodes. As electrolysis proceeds, chlorine is evolved at the anodes and collected in the circular space above the cathode compartment and led away. Hydrogen is liberated outside the asbestos paper diaphragm at the cathode surface and removed. Residual hydroxide ions and sodium ions remain and constitute caustic soda which collects with some sodium chloride in solution in the base of the pot under an atmosphere of hydrogen. Operational drawbacks of such asbestos diaphragm cells include maintenance hazards to health so that membrane electrolyser designs are to be preferred.

The cell itself is electrically insulated and arrangements are made to break up the stream of incoming brine into a spray at one stage in its delivery and to allow the caustic liquor to drip away, thus obtaining the maximum electrical isolation.

The caustic liquor is concentrated fivefold by using multiple vacuum evaporators to give about 50% strength when most of the sodium chloride separates on cooling, its solubility being suppressed by common ion action. Traces of iron compounds are removed and the caustic soda supplied commercially as 50% liquor (by weight) containing up to 1% sodium chloride and 1% sodium chlorate.

Electrolysis of brine (iii) by the ICI FM21-SP series membrane electrolyser

All of the traditional environmental problems associated with mercury and diaphragm cell

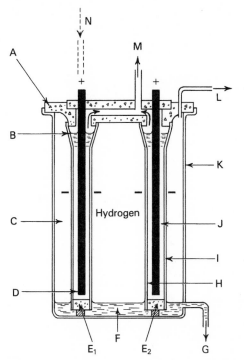

Fig. 14.4 The electrolysis of brine using the Gibbs diaphragm cell

A   Concrete cover to cell
B   Brine level
C   Hydrogen
D   One of 11 graphite anodes, rectangular in section, suspended in a ring, also carrying the brine feed pipe which occupies the 12th space
$E_1 + E_2$ Concrete ring forming base of cathode compartment
F   Caustic liquor
G   Caustic liquor (outflow)
H   Inner cathode cylinder with diaphragm of asbestos paper covering outer surface

I   Outer cathode cylinder of steel (electrical connections not shown) with diaphragm of asbestos paper covering inner surface
J   Cathode compartment: brine enclosed in annular space between diaphragms of asbestos paper supported outside by steel gauze cathodes
K   Cast iron circular pot
L   Hydrogen
M   Chlorine
N   Brine feed pipe occupies one of the anode positions

586 mm high (outside dimensions). The comparatively short height confers a short circuit path, the small electrodes being characterized by a uniform current distribution and a good energy consumption performance. Each anode is a panel of titanium, pressed from a 2 mm thick sheet of metal and is held between compression moulded gaskets. Similarly, each cathode is a nickel panel, also held between compression moulded gaskets. The electrodes are ultra smooth (to avoid damage to the membranes) and are coated suitably to keep the required electrolysis voltage to a minimum. Naturally the success of the cell depends on the quality and longevity of the ion exchange membrane, which is made of a fluoropolymer akin to PTFE (polytetrafluoroethylene) which incorporates sulphonated or carboxylated sites. The membrane must also be handled carefully and fitted precisely. Membranes are about 250 μm thick. The low height of the electrolyser ensures minimal pressure fluctuations. The electrolyser also incorporates a balance header which allows the product streams to be effectively separated on their exit to downstream plant. The exploded view of the cell is shown in Fig. 14.5 which locates the different electrodes the membrane and exits of products and depleted brine.

Sodium chlorate† and sodium hypochlorite produced by the electrolysis of brine

Hot purified saturated brine is electrolysed in a concrete-lined iron tank with graphite anodes and iron cathodes under conditions designed to ensure efficient mixing of chlorine and sodium hydroxide solution formed. To promote chlorate formation, the hydrogen ion concentration is controlled by addition of a little sodium dichromate. Solid salt is added as electrolysis proceeds. On cooling, sodium chlorate separates out:

$$3Cl_2 + 6OH^- \rightarrow 5Cl^- + ClO_3^- + 3H_2O$$

Sodium hypochlorite solution may be produced by the electrolysis of brine which passes between a series of graphite bipolar electrodes, the anode of one pair of electrodes acting as the cathode of the next. However, sodium hypochlorite is usually made by the direct action of chlorine on caustic soda solution:

$$Cl_2 + 2OH^- \rightarrow Cl^- + ClO^- + H_2O$$

† Otherwise known as sodium chlorate(v) and sodium chlorate(ı), respectively

technologies have been eliminated by this design. The result of careful development, the electrolyser exhibits good long term performance and elegant simplicity of operation and maintenance. Further, because the electrodes are linked electrically in parallel (the monopolar design), existing power supplies and the lay out of cell rooms need only minor alterations.

The FM21-SP membrane electrolyser is a nominal 2 040 mm long, 1 354 mm wide and

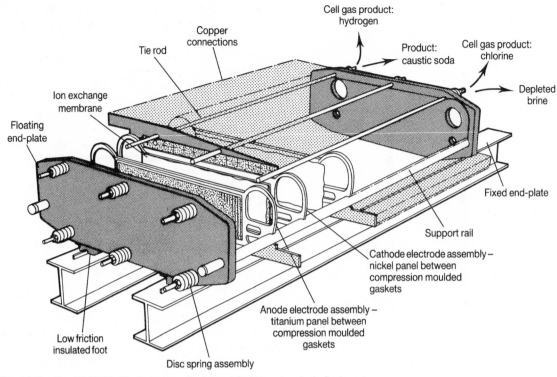

Fig. 14.5 The ICI FM21-SP electrolyser for brine (ICI Chlor-chemicals Business)

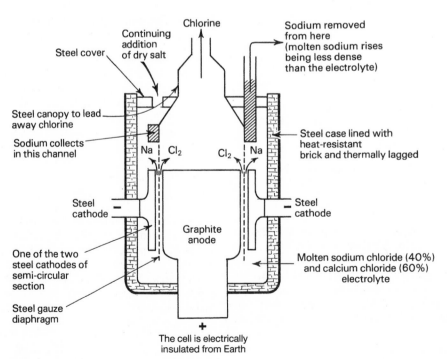

Fig. 14.6 The electrolysis of fused salt using the Downs cell.

Electrolysis of fused salt using the Downs cell

A fused mixture of sodium chloride and calcium chloride (2:3) is electrolysed at 600°C. The products are chlorine at the anode and sodium at the cathode. In spite of the high proportion of calcium chloride, very little calcium is discharged, preference being given to sodium. Calcium chloride is added to lower the melting-point: otherwise, a temperature of about 800°C would be required and the products would be very much more corrosive.

The electrolyte, fused by short-circuiting the electrodes with graphite blocks, later removed, is contained in a thermally and electrically insulated circular steel shell. This is lined with refractory brick through the bottom of which emerges a massive circular graphite anode, surrounded closely by two semicircular steel cathodes. Between the anode and cathodes, a steel gauze cylinder is suspended from a steel cone, within which the chlorine collects as it is evolved from the anode. Molten sodium rises through the electrolyte and being of much lower density than the electrolyte, rises in the collecting pipe. As electrolysis proceeds, solid salt is added to replenish the electrolyte. The cell is shown diagrammatically in Fig. 14.6.

Potassium compounds produced electrolytically

Caustic potash may be produced by electrolysis of saturated aqueous potassium chloride in mercury cells, while potassium chlorate is formed by electrolysis in a similar way to the sodium salt.

Potassium hydroxide is sold as liquor and used mainly as an intermediate in the manufacture of potassium salts in the chemical and agricultural industries. In 1989, when the 1926 'New Factory' at Runcorn was closed by ICI, the mercury cell production was replaced by the latest membrane cell technology.

Potassium chloride 'brine' is fed into the anode compartments of the cell and potassium hydroxide solution into the cathode compartments, and an electric current is passed between the electrodes through the liquors. Chloride ions are discharged at the anode and liberated as chlorine gas. Potassium ions are transported through the chemically inert and selective membrane towards the negatively charged cathode, but hydrogen ions (formed by the dissociation of water) are preferentially discharged as would be expected, leaving the equivalent amount of hydroxide ions. Thus,

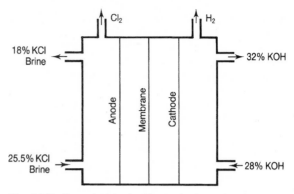

Fig. 14.7 Potassium hydroxide production using the membrane cell (by courtesy of ICI)

potassium hydroxide accumulates, the solution flowing through the cathode compartments becoming more concentrated in the alkali. The membrane serves to separate the gaseous products, hydrogen and chlorine, and the solutions, although some back migration of potassium hydroxide solution into the potassium chloride solution occurs and cannot be prevented. This is measured as a loss in current efficiency. The membrane cell is depicted diagrammatically in Fig. 14.7.

Applications of the elements produced by the electrolysis of sodium chloride

Chlorine is used in the organic chemical industry for the production of intermediates in the manufacture of tetraethyl lead anti-knock (now going out of favour because of the effect of lead residues on health) and ethylene glycol antifreeze, for making polyvinyl chloride plastic and various solvents (chloroethenes) for degreasing and dry-cleaning. Chlorine is used direct or combined as a bleach. It is used in water sterilization and sewage treatment, and for many other processes. About one-quarter of the hydrochloric acid produced in the UK is synthesized from chlorine and hydrogen, the co-products of electrolysis. Chlorine is used in the recovery of tin, titanium and magnesium from scrap and for the manufacture of, chlorinated rubbers, insecticides, dyes and drugs. The refrigerant and aerosol propellant compounds, the freons (CFCs), which are organic compounds of fluorine and chlorine and apparently ideally inert for their purposes, are prime suspects for the decrease in thickness of the ozone layer in the upper atmosphere. By international agreement their manufacture is being discontinued and a search has begun for 'environmentally friendly' substitutes. Further, it seems eminently reasonable to suspect chlorinated organic solvent residues (and petroleum vapours) for the damage to conifer trees seen along the autobahns of Europe. Such trees are noted for their special oil constituents which would dissolve away in these residues. Hydrogen is also used in the synthesis of ammonia and the reduction of organic compounds, as in the manufacture of margarine and nylon.

Metallic sodium is used in the extraction of titanium, for the production of sodium cyanide and sodium peroxide and, used in conjunction with an alcohol, for the industrial reduction of fatty acids. Sodium is utilized as a heat transfer medium in some atomic power stations. Other applications of sodium and its alloys include the familiar yellow street lights, photoelectric cells, with potassium in high-temperature thermometers and with lead as an intermediary in the manufacture of tetraethyl lead (now discredited for health reasons).

15

Hydrogen, the hydrides and water

Hydrogen

Hydrogen is a colourless, odourless, tasteless, gaseous element. It does not support respiration or combusion. Hydrogen burns in air with an almost invisible blue flame to form water. Physical data are listed in Table 15.1.

Table 15.1 Some physical data for hydrogen

Atomic number	1
Electronic configuration	1
Relative atomic mass	1.00797
Relative molecular mass	2.01594
Atomic radius/nm	0.029
van der Waals' radius/nm	0.12
Electronegativity	2.1
Ionization energy/kJ mol^{-1}	1312
Boiling-point (101 kPa)	$-253°C$ (20 K)
Melting-point (101 kPa)	$-259°C$ (14 K)
Density/g dm^{-3} (standard conditions*)	0.0899
Solubility in water/cm^3 dm^{-3}	
(standard conditions)	21.5
Critical temperature	$-241°C$ (32 K)
Critical pressure/MPa	2.0

*exactly: 273.15 K, 101325 Pa (Nm^{-2})

About 0.87% by weight of the Earth's crust (the outer 10 miles) consists of hydrogen. Combined hydrogen occurs in all animal and vegetable tissue, and as water. It is estimated that hydrogen accounts for 15.4% of all atoms in the Earth's crust. On this basis it is the second most abundant element, after oxygen. There are more compounds of hydrogen than any other element, including carbon. The proportion by volume of hydrogen in the atmosphere is less than 1×10^{-6} although it is evolved as 10–30% of natural gas. Molecules of hydrogen are so light that they travel fast enough to escape from the Earth's gravitational field. Spectroscopic studies led to the estimate that 30% by weight of the sun consists of hydrogen. The

element is believed to have the key role in the formation of stars in one theory about the creation of the universe.

Although hydrogen had been collected by other workers previously, it was first systematically investigated by Henry Cavendish in 1766. He observed that the same quantity of hydrogen was evolved when acids (dilute hydrochloric and sulphuric acids) reacted with a fixed weight of a particular metal (zinc, iron, tin) and concluded that the gas came from the metal, calling it 'the inflammable air from metals'. In 1781–4, he demonstrated that water was the product obtained on exploding the gas with air or oxygen. Lavoisier named the gas hydrogen (Greek, *hydor genon* = water former). Hydrogen was liquefied in 1899 by Travers. Dewar was believed to have liquefied hydrogen during the previous year but no account of this was published. At ordinary temperatures, sudden expansion from high to low pressure results in a heating effect, but below $-80°C$ the Joule–Thomson cooling effect is obtained. Before commencing liquefaction, hydrogen is cooled in liquid air ($-185°C$).

Element No. 1 of the Periodic Classification

Hydrogen is the first element in the Periodic Table. With atomic number $= 1$, the nucleus of the most abundant isotope is one proton with one valency electron outside it; the mass number is therefore 1.

There is a naturally occurring isotope of mass number $= 2$. This has a neutron as well as one proton in the nucleus. Of course, the atomic number remains 1 and this isotope occupies the same position (*isos topos*) as ordinary hydrogen in the Periodic Table. However, the doubled mass accentuates the physical difference between the isotopes, which become of great interest and are given special names. The heavier isotope is called *deuterium*, ^2_1H = D, and the lighter, sometimes, *protium*, ^1_1H = H. On Earth, the mass ratio H/D is roughly constant (about 6000/1) in all sources of hydrogen, so that roughly 1 in 12 000 atoms of hydrogen is deuterium. Hydrogen, therefore, consists mostly of H_2 molecules with some D_2 and HD molecules. A third isotope, artificial and radioactive (by β decay), has a mass number = 3. Designated ^3_1H, this isotope is known as *tritium* and has 2 neutrons and 1 proton in the nucleus. While diffusion and thermal diffusion methods may be used to separate the isotopes, the technical production of water rich in deuterium, heavy water, D_2O and HOD, is accomplished by electrolysis in Norway where there is abundant cheap hydroelectric power. The H/D ratio in the gas discharged is less than in water, causing enrichment of the electrolyte liquor. In 1933, using about 2250 dm³ of water from commercial electrolysis cells (already enriched), Lewis obtained 83 cm³ of 99% pure D_2O.

Deuterium and hydrogen (protium) will differ only in properties dependent on mass, enabling diffusion methods to be used in their separation. Chemically, the relatively large difference in mass (D = 2, H = 1) leads to a noticeable difference in reaction rates, a heavier atom being slower than a lighter atom with the same kinetic energy. Farkas found that chlorine in the light at 30°C combines with hydrogen at three times the rate of its reaction with deuterium. The melting-point of deuterium, D_2, is 4.70°C higher and the boiling-point, 3.2°C higher than for hydrogen, H_2. The maximum density for heavy water, D_2O, deuterium oxide, occurs at 11.22°C compared with 4.08°C for H_2O; furthermore, D_2O has a freezing-point 3.8°C higher and a boiling-point 1.4°C higher than water.

Hydrogen in combination

Period 1 of the Periodic Classification starts with hydrogen and is completed by the inert gas, helium, mass number = 4 and atomic number = 2, having a nucleus composed of two protons and two neutrons, with two external valency electrons. An atom of hydrogen may attain the inert gas electronic configuration of helium in two ways: either by sharing a pair of electrons to form a covalent compound or by acquiring another electron to form an ionic compound. This is rather like the behaviour of the halogens but hydrogen is much less electronegative, being assigned to no one group of the Table. Compounds in which hydrogen is united with one other element, binary compounds, are called *hydrides*. Consider examples: note that the valency of hydrogen cannot exceed unity because the valency shell cannot hold more than two electrons.

The hydrogen molecule, H_2

A covalent structure is in agreement with its properties. In passing, it may be mentioned that

$$H \overset{\times}{\underset{\circ}{\circ}} H$$

One shared pair of electrons

H—H

the separate nuclei (of single protons) are spinning, either in the same sense (ortho-hydrogen) or opposed (para-hydrogen). The two forms are in equilibrium, a fact not simply explained. At low temperatures para-hydrogen predominates but forms ortho-hydrogen as the temperature is raised until the limit ortho/para = 3/1 has been reached. The subject is too advanced for treatment here.

The hydrides

Covalent hydrides

These are formed by sharing electrons. Simple hydrides of carbon, nitrogen, oxygen and fluorine provide examples. Although electrons are indistinguishable it helps to indicate their different origins in simple bond diagrams.

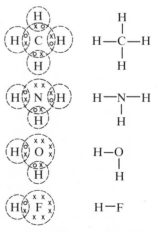

Consisting of molecules, covalent hydrides are colourless, electrically non-conducting compounds of low boiling-point. They are formed by non-metals, except inert gases, and by some weak metals in which the tendency to form cations is not strong. Where chains of atoms are formed by an element, more than one hydride is possible:

$$CH_4, C_2H_6, C_3H_8, \text{etc.}\quad NH_3, N_2H_4, HN_3$$

$$H_2O, H_2O_2$$

Atomic number	Electronic configuration	
	Before	After
1	H 1	2 (He)
6	C 2, 4	2, 8 ⎫
7	N 2, 5	2, 8 ⎪ (Ne)
8	O 2, 6	2, 8 ⎬
9	F 2, 7	2, 8 ⎭

Ionic (salt-like) hydrides

In reactions with certain strong metals, which readily form cations, a hydrogen atom receives an electron to form the negative hydrogen ion, the hydride ion, H^-. Lithium and calcium react:

$$[Li]^+[H_O^x]^-\quad LiH$$

$$[Ca]^{2+}[H_O^x]_2^-\quad CaH_2$$

Atomic number	Electronic configuration	
	Before	After
1	H 1	2 (He)
3	Li 2, 1	2 (He)
20	Ca 2, 8, 8, 2	2, 8, 8 (Ar)

These hydrides are colourless solids, of low volatility and with ionic lattices. They are known as the saline, or salt-like, hydrides. Although fusion brings slow decomposition, rapid electrolysis yields hydrogen at the anode in accordance with Faraday's Laws. The metal appears at the cathode. A few reactions may be mentioned, using lithium hydride and writing ionic equations. With water, lithium hydroxide and hydrogen are formed:

$$H^- + H_2O \rightarrow H_2 + OH^-$$

Anhydrous ammonia yields the amide:

$$H^- + NH_3 \rightarrow H_2 + NH_2^-$$

Carbon dioxide gives the methanoate (formate), $HCOO^-Li^+$; carbon monoxide, the methanoate

and carbon:

$$H^- + 2CO \rightarrow H.COO^- + C$$

and sulphur dioxide, the dithionite,

$$2H^- + 2SO_2 \rightarrow S_2O_4^{2-} + H_2$$

Interstitial (metal-like) hydrides

Hydrogen is occluded in non-stoichiometric proportions by certain transition metals. The compounds are examples of *berthollides*. Atoms of the gas penetrate into the layers of metal atoms (or ions) in the lattice. The strength of the bonding forces indicates a part chemical and part physical nature. Being metal-like, they are also called *the solid alloy type of hydride*. The final composition depends on the physical conditions and the state of division of the metal.

Other hydrides

Certain non-transition metals form non-volatile hydrides which are not ionic and are relatively unstable. Examples are BeH_2, MgH_2, AlH_3, ZnH_2, CdH_2, HgH_2, CuH.

Complex hydrides

Aluminium and boron form complex hydrides of some importance in which the Group III element is in the anion.

Lithium borohydride
Lithium tetrahydrido-borate(III), (page 305)

Lithium aluminium hydride
Lithium tetrahydrido-aluminate(III), (page 309)

Hydride type and the Periodic Table

The classification of elements into metals and non-metals has already been discussed in Chapter 4. A metal is an element which forms simple positive ions (cations) by the loss of electrons. The metals are described in a sort of verbal shorthand as strong or weak according to whether this tendency to form cations is also strong or weak. However, it is difficult to decide whether some elements are best classed as metals or non-metals. The strong metals form ionic hydrides. They are metal elements with low electronegativity (highly electropositive) found in Groups I and II. Beryllium and magnesium, the less electropositive members of Group II, form hydrides which are not of this simple type.

	I	II
Metals forming ionic hydrides	Li	
	Na	
	K	Ca
	Rb	Sr
	Cs	Ba

Calcium hydride is the least ionic. Hydrides of metals below sodium in Group I are highly reactive.

Non-metals form covalent hydrides. But so do certain metals and metalloids. The stability of the hydrides falls off with increase in atomic number. In Group V, BiH_3 was first detected as a radioactive gas in small yield from the action of acid on an alloy of magnesium and a radioactive bismuth isotope. This hydride must be placed in the progression

$$NH_3, \ PH_3, \ AsH_3, \ SbH_3 \ and \ BiH_3,$$

where the hydrides prove progressively less stable with the development of metallic characteristics. The hydride type, and particularly the stability of the covalent hydride, is a good criterion for grading an element.

	III	IV	V	VI	VII
Elements forming covalent hydrides	B	C	N	O	F
		Si	P	S	Cl
		Ge	As	Se	Br
		Sn	Sb	Te	I
		Pb	Bi	Po	At†

The hydrogen cation, H⁺

The bare hydrogen nucleus, a proton of diameter about 10^{-15}m (1fm††), as compared with 10^{-10}m (100 pm††) for other atoms, is much too small a point charge to exist freely in the presence of other atoms, except in a suitable discharge tube. In aqueous solution, it attaches itself to a water molecule to form the oxonium, hydroxonium or hydronium ion.

$$\overset{+}{\underset{\underset{H}{|}}{H-O-H}}$$

Acidity is ascribed to their presence. Pure water contains only 1×10^{-7} mole dm^{-3} of these ions (and the same amount of hydroxide ions) but on dissolving an acid the concentration increases (while the concentration of hydroxide ions decreases in accordance with the Equilibrium Law, p. 106).

$$\underset{\text{strong acid}}{HCl + H_2O} \rightleftharpoons \underset{\substack{\text{ionization almost}\\\text{complete or complete}}}{H_3O^+ + Cl^-}$$

$$\underset{\text{weak acid}}{HCN + H_2O} \rightleftharpoons \underset{\text{ionization poor}}{H_3O^+ + CN^-}$$

These equations are usually written simply:

$$HCl \rightleftharpoons H^+ + Cl^-$$

$$HCN \rightleftharpoons H^+ + CN^-$$

where it is understood that the simple hydrogen ion is actually hydrated as $H^+(H_2O)$ or possibly, as $H^+(H_2O)_2$. Usually the hydrated ion is taken as $H^+(H_2O)$ or H_3O^+ and strictly, the name oxonium refers to this structure alone.

The hydrogen bond

The melting-points and boiling-points of the hydrides of nitrogen, oxygen and fluorine are much higher than would reasonably be expected by comparison with corresponding data for hydrides of other elements in the respective groups. This is best illustrated diagrammatically, in Fig. 15.1. The melting-point and boiling-point of methane, CH_4, follows the general trend of values for Group IV hydrides while abnormally high values appear for the first members of the other groups. This is explained by hydrogen bond formation.

Because of its comparatively minute size and the inadequate screening of the nucleus by the shared pair of electrons, the combined hydrogen atom exerts a strong polarizing effect on other atoms. Hydrogen bonding is formed by compounds of the most electronegative elements, fluorine, oxygen and nitrogen. The bond formed is electrostatic in origin and involves the hydrogen atom and an unshared electron pair. Because of the relatively large size of the atoms bridged and their mutual electrostatic repulsion, only two atoms are linked by a hydrogen atom in this way; one covalently, the other electrostatically. This type of linkage can neither be described nor depicted simply. The aggregates of molecules have reduced volatility and therefore higher boiling-points, stronger bonding forces

† $^{210}_{85}At$ has a half-life of 8 h (one of the synthetic elements)
†† f (fermto) = 10^{-15}; p (pico) = 10^{-12}

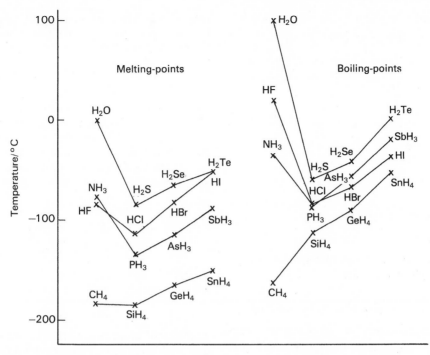

Fig. 15.1   The melting-points and boiling-points of isoelectronic sequences of hydride molecules. (After Pauling (1945) *Nature of the Chemical Bond*, Cornell)

and higher melting-points. Hydrogen fluoride tends to form polymers with the angle, $H-\widehat{F}\cdots H = 140°$. $(HF)_6$ is a very stable molecule of zig-zag structure.

Reactions in which hydrogen is liberated

CAUTION: *Because of explosive mixtures formed with air, hydrogen is an unusually dangerous substance. This danger is often underrated. The reaction between potassium and sodium with water may cause explosions, so use small quantities. On no account heat preparative flasks in which hydrogen is being liberated. Before collection or burning at a jet, test samples of hydrogen collected in small tubes with a flame at a distance from the apparatus, until the gas burns quietly without a sharp explosion. For hydrogen in quantity, use a cylinder.*

1   The action of metals on water

Very electropositive metals liberate hydrogen from cold water, others require hot water or steam. In general, the vigour of action is greater the higher a metal is placed in the electrochemical series.

The alkali metals (Group I) react vigorously with cold water. Less dense than water, globules of metal rush across the surface. The reactions are exothermic, the products being metal hydroxide solution and hydrogen. The reactions increase in violence with increasing atomic number of the metal. Potassium is extremely violent: the gas ignites and burns with a flame coloured lilac by potassium vapour. Sodium does not ignite the hydrogen unless motion is restricted by placing the metal on a floating filter paper when the hydrogen burns with a flame of yellow due to the sodium vapour. Lithium is brisk in action. Taking this as the example: lithium hydroxide solution remains,

$$2Li + 2H_2O \rightarrow 2LiOH + H_2$$

Various devices, none completely safe, have been devised for collecting hydrogen from the sodium reaction. For lithium, submerging the metal in a box made from gauze and holding this with tongs under a gas jar full of water, inverted in a trough of water, is probably the simplest way of collecting the hydrogen evolved.

Metals forming sparingly soluble hydroxides or with a tenacious film of adhering oxide, may react slowly. Calcium sinks and reacts quietly with cold water, the slightly soluble hydroxide accumulating,

$$Ca + 2H_2O \rightarrow Ca(OH)_2 + H_2$$

Strontium and barium are increasingly vigorous, in that order. Magnesium decomposes boiling water slowly.

2   The action of metals on steam

Magnesium burns brilliantly when heated in steam, and leaves the oxide:

$$Mg + H_2O \rightarrow MgO + H_2$$

Zinc behaves similarly, but iron forms tri-iron tetroxide by a reversible process:

$$3Fe + 4H_2O \rightleftharpoons Fe_3O_4 + 4H_2$$

Manganese, cobalt, nickel and tin liberate hydrogen at very high temperatures.

3   The action of metals on acids in solution

Metals above hydrogen in the electrochemical series liberate it from acids. The rate and extent of the reaction is determined by the state of purity of the metal, the solubility or insolubility of the product and the nature of the acid. An oxidizing acid will liberate hydrogen but rarely, as, for example, very dilute (1–2%) nitric acid with magnesium. The acids must be dissolved in water, because the process depends on the presence of hydrogen (oxonium) ions:

$$2H_3O^+ + 2e^- \rightarrow H_2 + 2H_2O$$

Except for aluminium, metals above tin in the electrochemical series liberate hydrogen from dilute acids. Aluminium and tin react with fairly concentrated hydrochloric acid. Very electropositive metals react violently. Magnesium and zinc are often used in the laboratory, the latter for gas in quantity. Neglecting hydration of ions,

$$Mg + 2H^+ \rightarrow Mg^{2+} + H_2$$

e.g.
$$Mg + H_2SO_4 \rightarrow MgSO_4 + H_2$$
$$\text{magnesium}$$
$$\text{sulphate}$$

Zinc reacts very slowly when pure but rapidly when impure, owing to galvanic action:

$$Zn + 2H^+ \rightarrow Zn^{2+} + H_2$$

Transfer of electrons is effected: zinc is oxidized and hydrogen ions reduced. Hydrogen may be dried over granular calcium chloride and collected over mercury. It reacts very slightly with concentrated sulphuric acid,

$$H_2SO_4 + H_2 \rightarrow SO_2 + 2H_2O$$

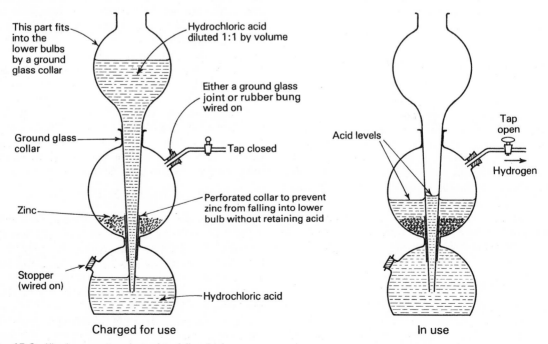

Fig. 15.2   Kipp's apparatus charged to deliver hydrogen

Hydrogen is usually obtained from a cylinder when required in quantity. However, a Kipp's apparatus may be used. Such an apparatus, charged prior to use and during use, is shown in Fig. 15.2.

4   The action of certain metals on alkalis

Hot concentrated solutions of sodium hydroxide will react with aluminium and zinc to form respectively solutions of sodium aluminate and sodium zincate, with evolution of hydrogen. The aluminate and zincate ions may be shown in either the complex or the simplest form:

$$2\,Al + 2OH^- + 6H_2O \rightarrow 2Al(OH)_4^- + 3H_2$$
aluminate ion

or

$$2Al + 2OH^- + 2H_2O \rightarrow 2AlO_2^- + 3H_2$$
aluminate ion

$$Zn + 2OH^- + 2H_2O \rightarrow Zn(OH)_4^{2-} + H_2$$
zincate ion

or

$$Zn + 2OH^- \rightarrow ZnO_2^{2-} + H_2$$
zincate ion

Silicon gives sodium silicate while tin gives a mixture of sodium stannate(II) and sodium stannate(IV). Other alkalis behave similarly.

5   The action of water on salt-like hydrides

Hydrolysis of ionic hydrides yields hydrogen and the alkali; essentially,

$$H^- + H_2O \rightarrow OH^- + H_2$$

Calcium hydride is produced commercially as a portable source of hydrogen, where the use of cylinders is not convenient, under the name hydrolith:

$$CaH_2 + 2H_2O \rightarrow Ca(OH)_2 + 2H_2$$

6   Electrolysis

Pure water is a poor conductor of electricity. However, aqueous solutions of acids, alkalis and certain salts (usually of Group I metals) may be used to prepare hydrogen by electrolysis. For example, electrolysis of separate aqueous solutions of sulphuric acid, sodium hydroxide and sodium sulphate yields two volumes of hydrogen at the cathode for every

one volume of oxygen at the anode, when the electrolyte is saturated with gas. At the cathode, the net process is:

$$2H_3O^+ + 2e^- \rightarrow 2H_2O + H_2$$

For very accurate quantitative work with hydrogen, the purest gas is obtained by the electrolysis of warm baryta (barium hydroxide) solution between nickel electrodes. Contamination of the hydrogen by carbon dioxide is avoided because the latter is precipitated as barium carbonate. At the anode, oxygen is evolved:

$$OH^- \rightarrow OH + e^-   2OH \rightarrow H_2O + O$$

$$O + O \rightarrow O_2$$

Some diffusion of oxygen through the electrolyte occurs but contamination of the desired product, pure hydrogen, is minimized by the use of a U-tube for electrolysis. Hydrogen is purified by passage through a red-hot platinum gauze, which catalyses the combustion of traces of oxygen to water, absorbed in tubes containing solid caustic potash and phosphorus pentoxide. The gas is stored by occlusion in palladium and may be regenerated on warming.

Commercially, hydrogen is a product of the electrolysis of brine (Chapter 14). In Tasmania, hydrogen is produced by the electrolysis of 28% aqueous caustic potash in mild steel cells, with mild steel sheet electrodes and nickel-plated anodes separated by asbestos cloth diaphragms. The cells are water-cooled. Distilled water is added to replace that decomposed.

The industrial production of hydrogen

Electrolytic methods have been described. Until 1939, about 90% of all hydrogen made came from the water gas reaction. However, the pressure steam reforming of naphtha, the use of natural gas, mostly hydrogen and methane, and oil-refinery gas have come into prominence (Chapter 12). In the USA, the 1957 breakdown for hydrogen production was natural gas (72%), refinery gas (11%), water gas (9%), electrolysis (6%) and coke oven gas (2%). Italy, France, Yugoslavia and Pakistan as well as the UK, also use natural gas. Middle East countries and the USA utilize oil-refinery gas. In the steel-producing regions of the UK, Germany, France and Belgium, coke oven gas, which contains 50–60% hydrogen, is important.

The reactions of hydrogen

Although comparatively inert at ordinary temperatures, large quantities of energy are liberated when hydrogen reacts with very electronegative elements. It reacts with fluorine in the dark and with chlorine in sunlight. Catalysis plays an important part in the reactions of hydrogen.

1 With elements; the synthesis of hydrides

Groups I and II

Except for beryllium and magnesium, the non-ionic hydrides of which are made indirectly, ionic hydrides are made by heating the metal in hydrogen. Unusual as it may seem, the metal undergoes oxidation by loss of electrons while hydrogen is reduced:

$$2Na + H_2 \rightarrow 2NaH$$

$$Ba + H_2 \rightarrow BaH_2$$

Group III

No direct action.

Group IV

When an electric arc is struck between carbon poles in an atmosphere of hydrogen, some hydrides of carbon are produced. The temperature is not less than 3000°C. Ethyne (acetylene) is the main product:

$$2C + H_2 \rightarrow C_2H_2$$

Group V

With a suitable catalyst, such as finely divided iron in the Haber process, hydrogen combines reversibly with nitrogen to form ammonia. The synthesis is exothermic. The physical conditions chosen vary but a temperature within the range 350–600°C and a pressure of 15–35 MPa (*c.* 150–350 atmosphere) may be used. The yield is 10–20%:

$$N_2 + 3H_2 \rightleftharpoons 2NH_3$$

Group VI

Hydrogen burns quietly in air with a very pale blue flame but in suitable proportions they form highly explosive mixtures. The product is water with traces of hydrogen peroxide. A more vigorous reaction is obtained with pure oxygen. Finely divided platinum catalyses the synthesis:

$$2H_2 + O_2 \rightarrow 2H_2O$$

In the oxy-hydrogen blow-pipe, the gases are delivered in concentric tubes, hydrogen burning in an atmosphere of oxygen. The temperature is about 2000°C.

When passed through boiling sulphur, hydrogen reacts to form some hydrogen sulphide:

$$H_2 + S \rightleftharpoons H_2S$$

Selenium and tellurium react similarly.

Group VII

The vigour of reaction decreases with increase in atomic number of the halogen. Hydrogen inflames with fluorine, even at −250°C in the dark, while chlorine explodes in sunlight or ultra-violet light from a lamp or in the light of burning magnesium. However, chlorine and hydrogen react quietly in diffuse daylight or in the presence of activated charcoal or platinum as catalyst. Bromine and hydrogen react over a heated platinum catalyst. Hydrogen and iodine react reversibly on heating. Each halogen undergoes covalent reduction:

$$H_2 + F_2 \rightarrow 2HF$$
hydrogen fluoride

$$H_2 + Cl_2 \rightarrow 2HCl$$
hydrogen chloride

$$H_2 + Br_2 \rightarrow 2HBr$$
hydrogen bromide

$$H_2 + I_2 \rightleftharpoons 2HI$$
hydrogen iodide

Hydrogen chloride is synthesized commercially by burning chlorine in hydrogen, both gases being products of the electrolysis of brine.

Transition metals

Interstitial hydrides have been mentioned. In accurate quantitative work with hydrogen, the gas may be stored in palladium.

2 Addition to unsaturated compounds

This is described in Organic texts. Compounds containing double or triple bonds are said to be *unsaturated* and *react by addition*. Hydrogen usually reacts in the presence of heated finely divided nickel or colloidal palladium at room temperature. Examples include the reduction of ethene (an olefine) to ethane (a paraffin hydrocarbon), cyanomethane (a nitrile) to ethylamine (a primary amine), ethanal (an aldehyde) to ethanol (a primary alcohol) and propanone (a ketone) to propan-2-ol (a secondary alcohol). The reactions are general. Industrial examples include the reduction of carbon monoxide to methanol (Chapter 12) and the manufacture of margarine and cooking fat.

3   Reduction of oxides

Metals from iron downwards in the activity series may be obtained by reduction of their heated oxides in hydrogen:

$$PbO + H_2 \rightarrow Pb + H_2O$$
lead monoxide

Monatomic hydrogen

Langmuir discovered that hydrogen, exposed to very low pressures of the order of 1 Pa (i.e. less than 0.01 mmHg) and heated to high temperatures by a tungsten, platinum or palladium wire carrying a heavy electric current, dissociated into atoms. Passage of molecular hydrogen at atmospheric pressure through an electric arc struck between tungsten electrodes also brings about dissociation:

$$H_2 \rightleftharpoons 2H;   \Delta H = + 432\,kJ\,mol^{-1}$$

Experimental data are listed in Table 15.2. Atomic hydrogen is very reactive, forming hydrides with metals and non-metals at room temperature. It is, of course, a powerful reducing agent. At first sight it would seem that immediate re-combination of the atoms in atomic hydrogen would be likely to occur. However, a molecule formed by the collision of two atoms has enough energy to dissociate immediately. The surplus energy must be removed by 'three body' collisions, either with a third molecule, which will be rare, or with the surface of the containing vessel. Low pressures favour the

survival of atomic hydrogen. At 67 Pa (0.5 mmHg) pressure, the 'half-life' of atomic hydrogen is $\frac{1}{3}$ second.

The great heat of re-combination may be used in welding. In the Atomic Hydrogen Blow-pipe or Welding Torch, a jet of hydrogen passes through an electric arc struck between tungsten poles and impinges upon the surface to be heated. The temperature is above 3000°C. Although the hydrogen burns in the atmosphere, this is a secondary effect. Melting and welding is accomplished without surface oxidation in the reducing atmosphere of atomic hydrogen. Tungsten (m.p. 3400°C), tantalum (m.p. 2996°C) and thorium(IV) oxide (thorium dioxide) (m.p. 1825°C) readily melt with this technique.

A summary of the uses of hydrogen

1   In meteorological and cosmic research balloons because of its extremely low density.
2   Producing margarine, cooking fats and soaps.
3   The atomic hydrogen blow-pipe and oxy-hydrogen blow-pipe used for welding.
4   As a reducing atmosphere in certain metallurgical operations.
5   As a coolant for steam-driven turbogenerators because it has the highest heat conductance properties of any gas.
6   In the production of ammonia and nitrogen compounds, used for agricultural fertilizers, explosives and in the plastics industry.
7   Producing organic chemicals, such as methanol.
8   Deuterium oxide (heavy water) is used in the atomic energy industry.

Hydrogen oxide: water

Ice, water and steam are the solid, liquid and gaseous forms of hydrogen oxide, H_2O. Water and ice, which are normally colourless, appear to have a bluish colour when viewed in depth. The melting-point of ice and the boiling-point of water under normal atmospheric pressure are taken as fixed points in thermometry, as 0° and 100° respectively, on the Celsius (Centigrade) scale. The critical temperature of water vapour is 374°C and the critical pressure, 22 MPa (218 atmosphere). At 4°C and normal atmospheric pressure, the density of water is at a maximum. By definition (1901), the volume of 1 kilogram of water became the standard litre under these conditions, so that the density of water, by definition, was 1 g ml^{-1} at 4°C. Scientists

Table 15.2   The thermal dissociation of hydrogen: $H_2 \rightleftharpoons 2H$

Temperature/K	2000	2500	3100	3500	4000
Percentage decomposition					
at 101 kPa	0.33	3.1	17	34	61
at 133 Pa	8.7	57.5	96	99.3	99.9

use internationally agreed units: those used in this book are recommended by the International Union of Pure and Applied Chemistry. By a resolution of the 12th Conference Générale des Poids et Mesures (CGPM) in 1964, the term litre is now recognized as a special name for the cubic decimetre, but is not used to express high precision measurements. The density of water at 4°C is now 1 g cm^{-3}. The 1901 definition still applies in the UK for the purposes of the 1963 Weights and Measures Act. When the Act came into force on 31 January 1964, the yard was redefined as 0.9144 m exactly and the pound as 0.45359237 kg exactly.

Water contracts when heated from 0°C to 4°C and expands on heating above 4°C. When lake water is cooled from ordinary temperatures, by the cold air above, the surface layer becomes denser and sinks. A circulation of water is maintained until 4°C is reached. Further surface cooling produces water which is of a lower density and this remains in the surface until it freezes. If this did not occur, water would freeze from the bottom upwards, as happens in fast-moving steams where the water is thoroughly mixed. Sea water, with a relatively high proportion of dissolved solids, contracts continuously as the temperature falls. The density of ice is 0.917 g cm^{-3}. Freezing is accompanied by expansion; water pipes which are split during a 'freeze-up' appear to 'burst' during the thaw. The disintegration of rocks is brought about by the pressure of ice forming in the cracks and fissures. Plant cells are burst by the freezing of their contents. The triple-point of water is 0.0073°C at 624 Pa (4.68 mm Hg).

Water is a very weak electrolyte and is almost electrically non-conducting at 25°C, the electric conductance† is 4.0 μS m^{-1}††. It is a highly associated liquid, with a high dielectric constant (= 81). The thermal dissociation of steam, imperceptible at 1000°C, is only 4% at 2300°C and 101 kPa pressure, hydrogen and oxygen showing a marked affinity for each other:

$$2H_2 + O_2 \rightarrow 2H_2O; \quad \Delta H = -2 \times 243 \text{ kJ mol}^{-1}$$

The liberation of hydrogen from water may be achieved by the most electropositive metals (e.g. sodium) and from steam by a wider range of metals at higher temperatures.

The formula of steam

Early experiments on the synthesis of water were performed by Macquer (1776) and Priestley and Warltire (1781). Cavendish determined the combining ratio of hydrogen to oxygen by volume, which he found to be 2.01 : 1. Lavoisier realized the significance of these experiments in relation to his ideas on combustion. Nicholson and Carlisle (1800) showed that water could be decomposed by electrolysis into hydrogen and oxygen. Davy (1806) proved that no other substance was formed and that the volume relationship, hydrogen : oxygen is 2 : 1. Gay-Lussac and von Humboldt (1805) determined the volume combining ratio of hydrogen and oxygen in twenty-four experiments; the average value of the ratio was 199.89 : 100 which they assumed equalled 200 : 100, allowing for experimental error. Further experiments by Bunsen confirmed this.

In the laboratory demonstration, the combining ratio of hydrogen and oxygen by volume is determined by explosion in a eudiometer tube in which the gases are confined over dry mercury. In some determinations an excess of hydrogen may be used, the excess being measured afterwards. But the electrolysis of dilute sodium hydroxide solution between nickel electrodes yields a pure gas with hydrogen and oxygen, from cathode and anode respectively, in the volume ratio of 2 : 1. This is dried by calcium chloride and introduced into the eudiometer of which several patterns are available. That of A. W. Hofmann (1865) is illustrated in Fig. 15.3.

With the heating jacket removed, mercury is introduced until the whole of the eudiometer is full. By connecting a reservoir of dry electrolytic gas at the upper tap which is opened and running mercury away at the lower tap, the required quantity of gas is introduced into the graduated portion of the explosion tube. About 30 cm^3 is usually taken. The gas is now heated in the assembled jacket and the levels of mercury equalized. When conditions are steady the volume of gas is noted or alternatively, the level marked on the jacket with a rubber band or gummed paper. Then mercury is run out of the lower tap until it occupies only about 150 mm of the tube above the tap and the open tube is corked to prevent loss of mercury during the explosion. The cork is held and with the apparatus screened, a spark is passed. After levelling the mercury, the volume of residual gas is noted. The steam occupies two-thirds of the original volume. On cooling, this

† Formerly called specific conductance, but specific is now restricted to the meaning *divided by mass*, not the case here

†† 1 S = 1 mho (reciprocal ohm)

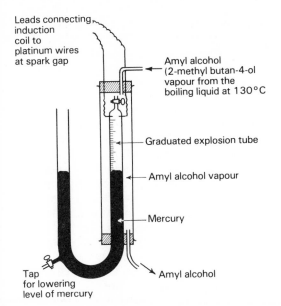

Fig. 15.3 Eudiometer for demonstrating the volume composition of steam (Hofmann)

contracts to a negligible volume of water. Thus,

2 volumes of
hydrogen (H₂)
$+$ under fixed   2 volumes of
conditions   steam (H_xO_y)
1 volume of   of T and P
oxygen (O₂)
(i.e. 3 volumes
in all)

Hence, by Avogadro's Hypothesis,

$$2H_2 + O_2 \rightarrow 2H_xO_y$$

when it is seen,

$$x = 2, \quad y = 1$$

$$\therefore \textit{formula of steam is } H_2O$$

Burt and Edgar (1916) have shown, as the result of 59 determinations, that the volume combining ratio for hydrogen and oxygen is $200.288 : 100$. The ratio is not $200.000 : 100$ because neither gas obeys the ideal gas equation exactly. Oxygen and hydrogen were made by electrolysis of warm baryta solution, purified very carefully and dried. To avoid temperature and pressure corrections, all measurements were made at $0°C$ and $101\ kPa$ ($760\ mmHg$) pressure.

The gravimetric composition of water

Dumas determined the combining ratio of hydrogen and oxygen by mass in 1842. An excess of purified dry hydrogen was passed over a weighed quantity of heated dry copper(II) oxide and the water collected by absorption in calcium chloride, solid potassium hydroxide and sulphuric acid or phosphorus pentoxide. The loss in mass of the copper(II) oxide gave the mass of oxygen which forms water, the mass of which was also measured. Thus, the ratio in which hydrogen and oxygen combine by mass could be calculated:

Mass of hydrogen : mass of oxygen $= 1.002 : 8$

Morley in 1895 employed direct synthesis; his apparatus is shown in Fig. 15.4. The apparatus is evacuated and weighed. Oxygen and hydrogen, purified and collected in weighed globes, are introduced separately as shown and burned at platinum jets, action being started by electric sparking across the spark gap. In later experiments, hydrogen was stored in palladium from which it could be expelled by heating. After the experiment, the water is frozen and the residual gas (oxygen) pumped out, any water vapour being trapped by phosphorus pentoxide. The gas is weighed and analysed. The apparatus is re-weighed. In this method the separate masses of oxygen and hydrogen are measured and also the mass of water formed. This provides a check: the sum of the masses of oxygen

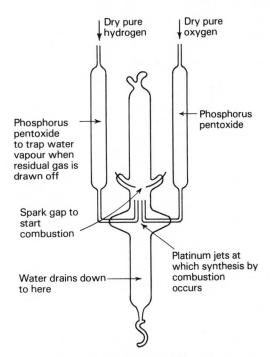

The apparatus is about 150 mm long

Fig. 15.4 The gravimetric composition of water (Morley)

and hydrogen must equal the mass of water. About 30 g of water were synthesized in each experiment. Morley found, as the average of a series of 9 experiments, that the combining ratio of hydrogen to oxygen is 1.00768 : 8 (exactly).

Using Morley's values of the normal densities of oxygen and hydrogen, Burt and Edgar (1916) calculated, from their volumetric composition results, the gravimetric composition of water. The combining ratio of hydrogen and oxygen by mass came to 1.00773 : 8.

Water supplies

Water occurs abundantly in the oceans, lakes and rivers, the atmosphere and as water of hydration in rocks. Sea water contains about 3.6% of dissolved solids by weight, including about 2.7% sodium chloride, common salt. Evaporation of sea water yields sodium chloride, magnesium chloride and sulphate, calcium sulphate and carbonate (from hydrogencarbonate), potassium chloride and bromide; the following ions are present: Na^+, K^+, Mg^{2+}, Ca^{2+}, Cl^-, Br^-, HCO_3^-, SO_4^{2-} together with smaller quantities of other substances. Of natural waters, rain water is the purest form. It is precipitated from water vapour which has evaporated from the seas and lakes. Away from towns, the impurities are the dissolved gases oxygen, nitrogen and carbon dioxide from the atmosphere, and nitrogen and ammonium salts from electrical action on air. Near industrial areas, sulphur dioxide, which forms sulphurous acid and then sulphuric acid, and soot are present. The impurities gathered by stream, river and lake water depend very much on the rock formations over and through which the water has passed. They include chlorides, sulphates, nitrates, hydrogencarbonates and carbonates of sodium, magnesium, calcium and iron to which may be added animal and vegetable refuse, sewage and industrial effluents.

A highly industrialized society like ours depends to a marked degree on a piped supply of clean, safe water for domestic and commercial purposes. Water for consumption must be purified if disease and epidemics are to be avoided. Where water comes from a river like the Thames, it must be stored for weeks or months to allow suspended matter to settle and many harmful bacteria to die. Other sources, such as moorland catchment areas and limestone regions, yield water which can be processed immediately. Coarse debris is removed by filtration through sand and gravel and pressure filtration through sand, coupled with the action of chemical coagulants. These include aluminium sulphate, sodium aluminate and silicate, which remove algae, bacteria and other fine suspended matter by absorption on precipitated aluminium hydroxide and hydrated silica.

There is increasing concern over the pollution of water for drinking by the residues of fertilizers, insecticides and herbicides, used in intensive farming and from industrial effluent. The EEC has produced regulations requiring governments to adopt stringent purity specifications for water, especially over the permitted concentration of nitrate ion, difficult and expensive to remove. The nitrite ion, NO_2^-, is suspected of being capable of causing cancer. Water catchment areas are signposted as such in certain countries e.g. western Germany, a move which has much to commend it when backed up by legal requirements. There is no instant solution to the problem, or series of solutions, because the chemicals which cause the pollution take very many years to reach the levels of the underground water reservoirs. Water for drinking is sterilized by chlorine in a concentration of 1–2 parts per million. While the water is now fit for drinking and domestic use, it may need further treatment for industrial purposes.

Finally, let it be remembered that the fame of not a few beers rests firmly on the very high quality of the water from private wells or reservoirs associated with the brewery and this is used as a selling point e.g. Burton (UK), Bitburger (Germany). Paradoxically, purification measures enforced by law may result in some very pure waters becoming less so. The habit of buying bottled water has spread across Europe from France. It is quite instructive to read the declared analysis of contents on the labels of such bottles, as much for what is stated as for what is not. It should also be remembered that while some brands are no doubt excellent, others have been sharply criticised by consumer associations. Whatever the merits or otherwise of what is written here, it is abundantly clear that EEC concern over the quality of water supplies is not misplaced.

Water pollution by fertilizers

Not all the crop nutrients applied as fertilizers are used by the crops. The unused portions represent a financial loss to the farmer and an environmental hazard to us all. The entry of nitrate into drinking water will cause acute commercial and scientific

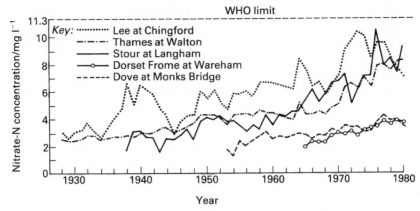

Fig. 15.5   The mean annual concentrations of nitrate in some English rivers – arable and grassland areas (J. K. R. Gasser, *Chem. Br.*, 1988, **24**, 129)

problems for the immediate future. The EEC nitrate target of less than 50 parts per million in drinking water will be extremely difficult to reach in areas of intensive farming such as East Anglia and Lincolnshire and there is informed debate about whether this target is too stringent. The increasingly serious nature of the problem is indicated by the graph in Fig. 15.5, showing the rise in nitrogen concentration/mg l^{-1} over a 50 year period.

The hardness of water

A hard water will not readily form a lather with soap.

Soap solution forms stable bubbles of lather by lowering the surface tension. Soaps are sodium and potassium salts of long-chain fatty acids; e.g. sodium octadecanoate, sodium stearate, $C_{17}H_{35}COONa$. Hard water contains calcium or magnesium ions or both, as hydrogencarbonates, sulphates or chlorides. The stearates of these metals are insoluble. On adding soap to hard water, a precipitate of calcium octadecanoate or magnesium octadecanoate appears as a scum. Extra soap is required before the water has been softened:

$$Ca^{2+} + 2C_{17}H_{35}.COO^- \rightarrow (C_{17}H_{35}.COO)_2Ca(s)$$

The presence of scum will spoil the finish on materials which have been laundered and on fabrics in the textile industry. The use of synthetic detergents avoids this. They have either soluble calcium and magnesium salts or are non-ionic, being described as synthetic to distinguish them

from soaps which are made from vegetable and animal oils. In the context of the domestic washing of clothes the concept of hardness seems distinctly old-fashioned.

Sometimes hardness can be removed by boiling and this leads to the distinction between two types of hardness.

Temporary (or alkaline) hardness of water may be removed by boiling.

Permanent (or non-alkaline) hardness of water cannot be removed by boiling.

Temporary hardness is due to the hydrogen-carbonates of calcium and magnesium. They are formed when rain water containing dissolved carbon dioxide, some of it as carbonic acid, meets limestone, magnesite, dolomite or similar carbonate rocks of calcium and magnesium. The hydrogencarbonate ion is decomposed by heat and the metal carbonate is re-precipitated. For calcium hydrogencarbonate,

$$CaCO_3 + H_2O + CO_2 \underset{\text{heat}}{\overset{\text{formation}}{\rightleftarrows}} Ca(HCO_3)_2$$

Not only is hardness wasteful of soap but it is a source of expense and some danger in industrial plant using boilers. The removal of temporary hardness by boiling leaves behind a boiler 'scale' of calcium and magnesium carbonates. Extra heat must now be supplied to heat the water within the barrier of scale and extra pumping power must be supplied to drive water through congested pipes, which may become blocked.

Permanent hardness is caused by the chlorides and sulphates of magnesium and calcium. They accumulate in steam-raising plant so that eventually the water becomes saturated. Crystallization occurs and deposition on the inside of boilers begins. Problems of hardness affect the running of power stations, ships and hot-water systems.

Minerals dissolved in natural waters are responsible for the distinctive 'Spa' waters, and probably for distinctive types of beer and drinks associated with particular regions. Hardness in water will seriously affect the processes in the dyestuff industry and in tanning leather.

Removal of hardness in water is called, appropriately, softening.

The softening of water

Temporary hardness can be removed by boiling but this is expensive. The hydrogencarbonate ion is decomposed on heating,

$$2HCO_3^- \rightarrow H_2O + CO_2 + CO_3^{2-}$$

and precipitation of carbonate occurs,

$$Mg^{2+} + CO_3^{2-} \rightarrow MgCO_3(s)$$
$$Ca^{2+} + CO_3^{2-} \rightarrow CaCO_3(s)$$

Solutions of hydrogencarbonates are alkaline by salt hydrolysis, which explains the term 'alkaline hardness', although they are, in fact, acid salts. Removal of the hydrogen, the acid part, by alkali leaves carbonate ions which are precipitated as before. Addition of the calculated quantity of slaked lime will destroy temporary hardness,

$$HCO_3^- + OH^- \rightarrow H_2O + CO_3^{2-}$$
$$Ca^{2+} + CO_3^{2-} \rightarrow CaCO_3(s)$$

Extra lime must be avoided because a solution of slaked lime, calcium hydroxide, shows permanent hardness.

Both types of hardness usually occur together. The degree of hardness may be determined by titration against standard soap solution, before and after boiling, to give the relative amounts of temporary and permanent hardness. Alkalinity and the concentration of other ions such as chloride and phosphate (used in water treatment) are also determined titrimetrically.

1   Softening by precipitation:
(a) The lime-soda method

Calculated quantities of lime (calcium hydroxide) and sodium carbonate are added; the former removes temporary hardness and the latter, perma-

nent. In fact, sodium carbonate would remove both kinds of hardness but the use of lime is cheaper. The processes may be summarized by equations; for calcium:

$$HCO_3^- + OH^- \rightarrow CO_3^{2-} + H_2O$$
$$Ca^{2+} + CO_3^{2-} \rightarrow CaCO_3(s)$$

or

$$Ca(HCO_3)_2 + Ca(OH)_2 \rightarrow 2CaCO_3(s) + 2H_2O$$

permanent:

$$Ca^{2+} + CO_3^{2-} \rightarrow CaCO_3(s)$$

or

$$CaSO_4 + Na_2CO_3 \rightarrow CaCO_3(s) + Na_2SO_4$$
$$CaCl_2 + Na_2CO_3 \rightarrow CaCO_3(s) + 2NaCl$$

2   Softening by precipitation:
(b) The use of caustic soda (and sodium carbonate)

Carefully controlled quantities of caustic soda are used to precipitate magnesium hydroxide and calcium carbonate from temporary hard water. Magnesium hydroxide is less soluble than the carbonate and is therefore precipitated. The full equations show that sodium carbonate solution is a product,

$$Ca(HCO_3)_2 + 2NaOH \rightarrow$$
$$CaCO_3(s) + Na_2CO_3 + 2H_2O$$
$$Mg(HCO_3)_2 + 4NaOH \rightarrow$$
$$Mg(OH)_2(s) + 2Na_2CO_3 + 2H_2O$$

Any permanent hardness may be destroyed by two methods. That due to magnesium by caustic soda,

$$Mg^{2+} + 2OH^- \rightarrow Mg(OH)_2(s)$$

and that due to calcium either by the sodium carbonate generated above or by this plus extra sodium carbonate added for the purpose,

$$Ca^{2+} + CO_3^{2-} \rightarrow CaCO_3(s)$$

If the softening treatment results in too much sodium carbonate being left in the water, the lime-soda process is preferred. The coagulation of magnesium hydroxide is helped by the addition of sodium aluminate, the aluminium hydroxide formed aiding flocculation.

3   Exchange methods for softening

The zeolite softening process is frequently used in the home, in the textile industry and by some water authorities. Zeolite mineral is a complex sodium aluminium silicate and hard water flowing through it leaves behind calcium and magnesium ions, acquiring the equivalent quantity of sodium ions. This is also called the Base-Exchange method. Regeneration of the zeolite is achieved by addition of strong salt solution, when the trapped magnesium and calcium ions pass into solution. While the process is expensive, it is useful where hardness is slight.

The Hydrogen Ion Exchange process utilizes the carbonaceous product of the action of sulphuric acid on coal. Metal ions are replaced by hydrogen ions so the product is acidic and is neutralized by passage through another type of exchange material. By the use of special Ion-Exchange Resins both anions and cations can be removed from water and the product is comparable with distilled water.

4 Phosphate treatment

A highly polymerized phosphate, sodium metaphosphate, $(NaPO_3)_n$, with added alkali, is sold under the name 'Calgon' for water treatment. Its applications are many. Small quantities will *sequester* calcium and magnesium ions, removing the ions as complex phosphates. It is used as a domestic softening agent, to prevent scale formation in hot-water supply systems, to prevent corrosion of iron and steel water mains by soft moorland water and to hold in solution compounds of iron which would otherwise be precipitated by oxidation of iron(II) hydrogencarbonate in well waters. In boiler water treatment, Calgon may be used to ensure controlled precipitation of phosphate and carbonate sludge without scale formation. It is used either by itself with soaps in textile scouring and laundry processing or in detergents marketed for dish washing and bottle washing. There are many other industrial applications.

Corrosion by soft waters

Iron and lead pipes in hard-water regions are rapidly coated by compounds formed by minerals in the water carried. Very soft water from moorland catchment areas will corrode pipes made from iron or lead especially if it is relatively acidic. Water is coloured red by iron compounds and lead causes poisoning. The water is just alkaline to methyl orange. Increasing the pH (see p. 232) by carefully controlled additions of lime, sodium carbonate and sodium hydroxide will make the water alkaline enough to prevent corrosion. Sodium silicate is particularly effective in preventing both types of corrosion. It provides, with the deposition of hydrated silica, a silicaceous lining to the pipes and unlike lime, does not introduce hardness.

Stalactites and stalagmites

When rain water, with its dissolved carbon dioxide, falls on to limestone rocks, calcium hydrogencarbonate solution is formed:

$$CaCO_3 + H_2O + CO_2 \rightleftharpoons Ca(HCO_3)_2$$

On reaching a cave, it drips from the roof. Decomposition occurs and calcium carbonate is precipitated. Hanging icicle-like deposits called stalactites are formed, growing at the rate of a fraction of an inch every year. On the floor of the cave, the continuous dripping leads to the upward growth of stalagmites. Wherever a stalactite and stalagmite meet, a pillar is formed. In the Mendip caves and elsewhere, stalactites, stalagmites and pillars of incredible beauty abound.

The reactions of water

1 With elements

The liberation of hydrogen by certain metals acting on water and steam has been described. Of the non-metals, fluorine, chlorine, carbon and silica decompose water.

Fluorine decomposes water with the formation of hydrofluoric acid and ozonized oxygen, while chlorine in sunlight produces hydrochloric acid and oxygen,

$$2F_2 + 2H_2O \rightarrow 4HF + O_2 \text{ (and } O_3)$$

$$2Cl_2 + 2H_2O \rightarrow 4HCl + O_2$$

Steam, passed over carbon at red to white heat, yields water gas,

$$C + H_2O \rightleftharpoons CO + H_2$$

while silicon very slowly forms silica and hydrogen,

$$Si + 2H_2O \rightarrow SiO_2 + 2H_2$$

2 With oxides

The oxides of certain strong metals i.e. those with a powerful tendency to form cations, react with water to form solutions of the hydroxide:

e.g. $BaO + H_2O \rightarrow Ba(OH)_2$

while oxides of many non-metals form acids:

e.g. $SO_3 + H_2O \rightarrow H_2SO_4$

3   Decomposition by water: hydrolysis

Hydrolysis is the decomposition of compounds by water. Where the balance between the hydrogen and hydroxide ion concentrations is disturbed with a resulting change in the pH, but without an observed decomposition, the term *salt hydrolysis* is applied. pH is a measure of the acidity:

$$pH = -\log_{10}[H^+]/mol\,dm^{-3}$$

where $[H^+]$, is the hydrogen ion concentration, $mol\,dm^{-3}$, and the temperature is usually standardized at 25°C or 18°C. Salt hydrolysis appears at the end of this chapter (p. 241).

Binary compounds of metals and non-metals may be hydrolysed to hydrides of the non-metal:

$$Al_4C_3 + 12H_2O \rightarrow 4Al(OH)_3 + 3CH_4(g)$$
aluminium   methane
carbide

$$CaC_2 + 2H_2O \rightarrow Ca(OH)_2 + C_2H_2(g)$$
calcium   ethyne
carbide   (acetylene)

$$Mg_2Si + 4H_2O \rightarrow 2Mg(OH)_2 + SiH_4(g)\ etc.$$
magnesium   silane
silicide

$$Mg_3N_2 + 6H_2O \rightarrow 3Mg(OH)_2 + 2NH_3(g)$$
magnesium   ammonia
nitride

$$Ca_3P_2 + 6H_2O \rightarrow 3Ca(OH)_2 + 2PH_3(g)$$
calcium   phosphine
phosphide

$$MgS + 2H_2O \rightarrow Mg(OH)_2 + H_2S(g)$$
magnesium   hydrogen
sulphide   sulphide

Consider the hydrolysis of chlorides, taking the chlorides XCl_3 of the Group V elements nitrogen and phosphorus as examples. Nitrogen trichloride, a yellow oily liquid, is slowly hydrolysed and the products, ammonia and hypochlorous acid, otherwise chloric(ɪ) acid (shown in the equation), react to form nitrogen, nitric acid and hydrochloric acid:

$$NCl_3 + 3H_2O \rightarrow NH_3 + 3HOCl$$

Phosphorus trichloride, a liquid, is rapidly hydrolysed to phosphorous acid and hydrogen chloride:

$$PCl_3 + 3H_2O \rightarrow H_3PO_3 + 3HCl$$

Hydrolysis usually gives the hydroxy compound of the element and not a hydride as occurred with nitrogen trichloride. Thus, boron trichloride forms boric and hydrochloric acids,

$$BCl_3 + 3H_2O \rightarrow H_3BO_3 + 3HCl$$

The extent of hydrolysis is indicative of the nature of an element: metal or non-metal, weak or strong metal. Salts with oxoacids are formed only by metals, but those of weaker metals are readily hydrolysed or only exist as basic salts. Bismuth nitrate is hydrolysed to an oxonitrate or sub-nitrate, called bismuth oxide nitrate:

$$Bi(NO_3)_3 + H_2O \rightarrow BiONO_3 + 2HNO_3$$

Lead(ɪv) sulphate and acetate are instantly decomposed by water,

$$Pb(SO_4)_2 + 2H_2O \rightarrow PbO_2 + 2H_2SO_4$$

$$CH_3(COO)_4Pb + 2H_2O \rightarrow$$
$$PbO_2 + 4CH_3COOH$$

lead(ɪv) oxide (lead dioxide) being precipitated.

The structure of water and hydration

The bond diagram for water may be written,

104.5°

in which hydrogen and oxygen acquire the electronic configurations of helium and neon respectively. The structure of water has been investigated by infra-red spectroscopy. The $H-O$ internuclear distance is 0.095 nm and the $H\hat{O}H = 104.5°$. That the water molecule is not linear is supported by the fact that the dipole moment is $6.0 \times 10^{-28}Cm$ ($= 1.8D$) whereas for a linear molecule it would

$\delta+$
H
       $O^{2}\delta-$
H$\delta+$

be zero. Oxygen with a high affinity for electrons, has a residual negative charge while the hydrogen atoms are positive. This accounts for the association of water molecules with both anions and cations in solution and other water molecules. The positive hydrogen is electrostatically attracted to anions and the negative oxygen to cations, although oxygen may enter into co-ionic bond formation as in the oxonium ion:

$$H—O \overset{+}{\underset{\backslash H}{\overset{/H}{}}}$$

Without association of water molecules, the melting-point of water might reasonably be expected to be in the region of $-100°C$ and the boiling-point about $-80°C$ (Fig. 15.1).

The hydrated cations of transition metals and copper(II) have characteristic colours: Mn^{2+} (pink), Fe^{2+} (pale green), Ni^{2+} (deep green), Co^{2+} (deep pink) and Cu^{2+} (blue).

Water of crystallization

Cooling saturated aqueous solutions of many salts and certain acids and bases, yields crystals containing the compound associated with water in definite molecular proportions. The isolated crystals are perfectly dry. These additive molecular compounds, containing water of crystallization, are termed *hydrates* and may be easily decomposed by heat to give the anhydrous compound, although some salts undergo hydrolysis. Blue vitriol, copper(II) sulphate pentahydrate, loses four-fifths of the water of hydration at about $100°C$ and the remainder above $200°C$, leaving a white (microcrystalline) powder:

$$CuSO_4.5H_2O \rightarrow CuSO_4.H_2O + 4H_2O(g)$$

$$CuSO_4.H_2O \rightarrow CuSO_4 + H_2O(g)$$

Four molecules of water are associated with the copper(II) ion and one with the anion, holding the crystal structure together. Magnesium chloride, $MgCl_2.6H_2O$, and iron(III) chloride, $FeCl_3.6H_2O$, yield the oxide and lose hydrogen chloride and water on heating. Hydrates of acids and bases include ethanedioic acid (oxalic acid), $(COOH)_2.2H_2O$, and lithium hydroxide, $LiOH.H_2O$.

The zeolite minerals (complex aluminium silicates), used in water-softening, lose water without change of crystal structure. Sometimes, this water is called *zeolytic water* to distinguish it from water of crystallization.

Efflorescence and deliquescence

On exposure to the atmosphere certain hydrates lose water, or effloresce. Translucent crystals of sodium carbonate decahydrate (washing soda), $Na_2CO_3.10H_2O$, become coated with a powdery deposit of monohydrate, appearing white,

$$Na_2CO_3.10H_2O \rightleftharpoons Na_2CO_3.H_2O + 9H_2O$$

Crystals of sodium sulphate decahydrate form the anhydrous salt,

$$Na_2SO_4.10H_2O \rightleftharpoons Na_2SO_4 + 10H_2O$$

For this to happen at a given temperature, the vapour pressure due to water produced by dissociation must exceed the partial pressure of water vapour in the atmosphere. This is usually about two-thirds of the saturated vapour pressure of water at ordinary temperatures, between 1.3–2.0 kPa (10–15 mmHg), while for sodium carbonate decahydrate and sodium sulphate decahydrate the dissociation pressures of water vapour are 3.23 and 3.71 kPa (24.2 and 27.8 mmHg) at $18°C$ respectively.

Deliquescence occurs where a substance absorbs moisture from the atmosphere to form a solution. The saturated vapour pressure of water over the solution formed (saturated initially) is less than the vapour pressure due to water in the atmosphere at that temperature. Absorption of water continues until they are equal. The vapour pressures at saturation due to water over saturated solutions of calcium chloride and iron(III) chloride at $18°C$ are 427 and 800 Pa (3.2 and 6.0 mmHg) respectively. The hexahydrates of calcium and iron(III) chloride are deliquescent.

Where the water vapour pressure is less than that in the atmosphere and the water vapour pressure of the solution exceeds that of the atmosphere, hydrated compounds show neither efflorescence nor deliquescence.

Hygroscopic substances

Solids which absorb moisture from the atmosphere without appearing wet, and liquids which absorb moisture, are termed *hygroscopic*. In quantitative work, copper(II) oxide must be heated and cooled in a desiccator before weighing because it absorbs water. Sulphuric acid on the other hand is used for drying, gases being bubbled through the liquid.

A desiccator is a closed vessel for storing a substance in the presence of a compound which absorbs moisture. Calcium chloride, sulphuric acid and silica gel are among the compounds chosen for their water absorption properties. Two patterns of desiccator are shown in Fig. 6.2 (p. 94).

The detection and estimation of water

Water may be characterized by the determination of its melting- and boiling-points. A qualitative test for traces of water in liquids employs anhydrous copper(II) sulphate which turns from white to the blue hydrate. In quantitative experiments, water may be absorbed in previously weighed tubes of phosphorus pentoxide or calcium chloride, suitably protected from water vapour in the atmosphere. Solids are heated to constant mass with cooling in a desiccator in the determination of water, if the compound is otherwise stable. One volumetric method for estimating water in liquids requires the addition of calcium carbide, the volume of ethyne (acetylene) produced with any water present being measured,

$$CaC_2 + 2H_2O \rightarrow Ca(OH)_2 + C_2H_2(g)$$

Deuterium compounds

Deuterium, the naturally occurring isotope of hydrogen, with mass number = 2, due to the neutron acquired by the proton of the hydrogen nucleus, has been described briefly in the introduction to the chapter (p. 218).

Deuterium oxide, 'heavy water'

By prolonged electrolysis of water, in which the lighter hydrogen (protium) isotope is preferentially discharged, deuterium oxide accumulates. This was first accomplished by Washburn and Urey in 1932. Pure deuterium oxide was prepared by Lewis and Macdonald in the following year. They obtained 83 cm^3 of 99% purity, using 2250 dm^3 of residual water from commercial electrolysis cells. Heavy water contains both D_2O and HOD.

D_2O is available commercially and has been studied in detail. The melting-point is 3.8°C and the boiling-point, 101.4°C.

The preparation of deuterium compounds

Deuterium oxide may be used to prepare deuterium compounds. These have slightly different physical constants from the corresponding compounds containing the lighter isotope, but undergo the same chemical reactions. Although the chemical reactions are the same as with ordinary water, it is instructive to outline the synthesis of deuterium compounds starting from deuterium oxide. Compounds are named by introducing the prefix deutero- before the more usual name or by displacing hydrogen by deuterium where appropriate.

ND$_3$, deuteroammonia:

$$Mg_3N_2 + 6D_2O \rightarrow 3Mg(OD)_2(s) + 2ND_3(g)$$

and $ND_3 + D_2O \rightleftharpoons ND_4^+ + OD^-$

NaOD, sodium deuteroxide:

$$2Na + 2D_2O \rightarrow 2NaOD + D_2(g)$$

D$_2$, deuterium:

Deuterium may be prepared by the action of sodium on D_2O or by the electrolysis of D_2O containing alkali (e.g. NaOD, etc.) or acid (e.g. D_3PO_4, D_2SO_4, etc.) or an oxosalt (e.g. Na_2SO_4).

Deuterophosphoric acid may be prepared by the action of phosphorus pentoxide on deuterium oxide:

$$2D_2O + P_4O_{10} \rightarrow 4DPO_3$$

$$6D_2O + P_4O_{10} \rightarrow 4D_3PO_4$$

D$_2$SO$_4$, deuterosulphuric acid:

$$SO_3 + D_2O \rightarrow D_2SO_4$$

and similarly, DNO$_3$ and DClO$_4$.

Salts:

$$D_2SO_4 + 2ND_3 \rightarrow (ND_4)_2SO_4$$

$$D_2SO_4 + 2NaOD \rightarrow Na_2SO_4 + 2D_2O$$

$$Na_2SO_4 + 10D_2O \rightarrow Na_2SO_4.10D_2O$$

$$CuO + D_2SO_4 \rightarrow CuSO_4 + D_2O$$

$$CuSO_4 + 5D_2O \rightarrow CuSO_4.5D_2O$$

Insoluble deuteroxides:

$$Al^{3+} + 3OD^- \rightarrow Al(OD)_3(s)$$

Halogen deuterides:

$$D_2(g) + Cl_2(g) \rightarrow 2DCl(g)$$

$$NaCl + D_2SO_4 \rightarrow NaDSO_4 + DCl(g)$$

$$P_4 + 10I_2 + 16D_2O \rightarrow 20DI(g) + 4D_3PO_4$$

Deuterocarbons: deuteroethyne (deuteroacetylene):

$$CaC_2 + 2D_2O \rightarrow C_2D_2(g) + Ca(OD)_2$$

C_2D_2 is the source of many organic compounds.

The general properties of acids, bases and salts in relation to water

Acids

Acids belong to a group of compounds which have certain similar characteristics when dissolved in water. They taste sour, deriving their name from the Latin, *acidus* = sour. Acids corrode the more reactive metals liberating hydrogen, produce an effervescence of carbon dioxide with carbonates and change the colour of certain dye solutions, called, appropriately, indicators. The vegetable dye, litmus, is turned from purple to red by acids. Acids were investigated by Boyle in 1663.

The characteristic properties of acids are *neutralized* by reaction with compounds having another group of characteristics in common. These compounds were named *bases* by Rouelle in 1774 and include oxides and hydroxides of metals. In a way, acids and bases are chemical opposites.

When an acid in solution reacts with a metal, hydrogen is evolved. Where does it come from? The acid, the water or both? All acids may be shown to contain hydrogen, replaceable by a metal with the formation of a salt, but this does not distinguish acids alone. For example, caustic soda solution attacks aluminium with the liberation of hydrogen. Solutions of different acids are electrically conducting to varying degrees for corresponding concentrations and this is due to the dissociation of the acid molecule into charged ions. The appearance of 'hydrogen ions' accounts for the closely similar behaviour of acids in aqueous solution. An example will illustrate the role of the water: freshly prepared nitric acid, a liquid, is without action on marble chippings (calcium carbonate) but addition of a little water results in a vigorous reaction. The ionization of nitric acid should be written:

$$HNO_3 + H_2O \rightarrow H_3O^+ + NO_3^-$$

yet it is common practice to use the simpler equation of the Arrhenius Ionic Theory which may be more familiar:

$$HNO_3 \rightleftharpoons H^+ + NO_3^-$$

but it must be remembered that the presence of water is essential for the display of acidic behaviour and that the H^+ ion does not exist, as such, in aqueous solution. From the ionic standpoint for aqueous solutions an acid may be simply defined:

An **acid** is a compound which ionizes to produce hydrogen ions (strictly, oxonium ions) when dissolved in water. The oxonium ion is sometimes called the hydronium or hydroxonium ion. The term is analogous to ammonium.

Examples of acids include the following:

$HClO_4$	perchloric acid	HNO_3	nitric acid
H_2SO_4	sulphuric acid	H_2CO_3	carbonic acid
H_3PO_4	orthophosphoric acid	H_3BO_3	boric acid
HCl	hydrochloric acid	H_2S	hydrogen sulphide
HCN	hydrocyanic acid	HF	hydrofluoric acid

When there is potentially one hydrogen ion in each molecule the acid is monobasic (e.g. nitric acid), two ions, dibasic (e.g. sulphuric acid) and three ions, tribasic (e.g. orthophosphoric acid). The prefixes *per-*, *ortho-* and others will be explained under the appropriate element. Alternatively, and this is increasingly favoured by some, the *Stock Notation* (p. 125) may be used where possible ambiguities arise over name and structure:

$$HNO_3 \quad \text{nitric(v) acid}$$
$$H_3PO_4 \quad \text{phosphoric(v) acid}$$
$$H_2SO_4 \quad \text{sulphuric(vi) acid}$$
$$HClO_4 \quad \text{chloric(vii) acid}$$

The acidity of a solution depends solely upon the hydrogen ion concentration, which will be decided by the total concentration of acid dissolved and the extent to which this is ionized. Acids are broadly classified as *strong* or *weak*. This does not refer to the concentration of the acid but to the *degree (i.e. fraction) of ionization (or dissociation)* of the acid. Strong acids are virtually completely ionized in moderate concentrations (up to about 4M) while weak acids are feebly ionized. The degree of ionization is increased by dilution of the solution. Examples will illustrate this point: hydrochloric acid is 91% ionized in decimolar (0.1 M) solution and 96.0% ionized in centimolar (0.01 M) solution, while ethanoic acid (acetic acid) is 1.3% and 4.2% ionized at these molar concentrations. Hydrochloric acid is strong and acetic acid weak. A quantitative treatment will be found in texts on Physical Chemistry.

Acidity then is characteristic of hydrogen ions. Ethanoic acid (acetic acid) will show acidic properties to a less marked degree than hydrochloric acid. The reaction between an acid and a metal will depend on the nature of the metal too. This will involve the electrode potential (p. 116) and overvoltage under the conditions of the experiment. In addition, the possible insolubility of the products and the formation of complexes influence the expected course of reactions. To these must be added the effect of temperature change, always increasing the rate of reaction for a rise in temperature and influencing each of the other factors. Further complications occur when the acid has oxidizing powers, leading to secondary reactions. Magnesium reacts with very dilute nitric acid to give the nitrate and hydrogen but in more concentrated acid the gaseous product consists of oxides of nitrogen.

Most acids contain oxygen as well as hydrogen, and Lavoisier advanced a theory, based on knowledge in the year 1775, that all acids do contain it. Oxygen had just been prepared by Priestley and the name, which means *acid-former*, was given by the French chemist. When oxides of non-metals react with water, acids are formed. Sulphur trioxide yields sulphuric acid and phosphorus pentoxide, orthophosphoric acid:

$$SO_3 + H_2O \rightarrow H_2SO_4$$

$$P_4O_{10} + 6H_2O \rightarrow 4H_3PO_4$$

These oxides are called *acidic oxides* and belong to a class of compounds called *acid anhydrides*.

An **acidic oxide** reacts with bases to form salts in which the element of the oxide is in the anion,

e.g. $Na_2O + SO_3 \rightarrow (Na^+)_2SO_4^{2-}$

The Lavoisier concept was disproved about 30 years later by Davy. Hydrogen chloride had been discovered by Priestley in 1772 by the action of concentrated sulphuric acid on common salt. The solution, which had typically acidic properties, was called muriatic acid (Latin, *muria* = brine). In 1779, Lavoisier, working by analogy, concluded that the acidic behaviour was due to combined oxygen in muriatic acid. By heating the acid with manganese(IV) oxide (manganese dioxide), Scheele, in 1774, obtained a yellow poisonous gas which he regarded as muriatic acid minus hydrogen. In sunlight, a solution of the gas evolved oxygen leaving muriatic acid in solution, leading Berthollet to name the gas oxymuriatic acid in 1785, although

the compound was not an acid. By 1810, Davy had heated metals and non-metals in the gas but had not obtained oxygen or any recognized compound of oxygen and he concluded: 'there may be oxygen in oxymuriatic gas, but I can find none.' He declared oxymuriatic acid to be an element and named it chlorine (Greek, *chloros* = greenish-yellow). Scheele's preparation may be summarized by the equation:

$$MnO_2 + 4HCl \rightarrow MnCl_2 + Cl_2 + 2H_2O$$
manganese(IV)   manganese(II)
oxide   chloride

while Berthollet's oxygen came from the water:

$$2Cl_2 + 2H_2O \rightarrow 4HCl + O_2 \qquad \text{(net change)}$$
chlorine =   hydrochloric
oxymuriatic   acid =
acid   muriatic acid

As we have seen, the key element in acids is the hydrogen which may be released as hydrogen ions. This is not necessarily all of the hydrogen. Only a quarter of the hydrogen in ethanoic acid, for example, is potentially acidic.

$$CH_3.COOH \rightleftharpoons CH_3.COO^- + H^+$$
ethanoate ion

Bases and alkalis

Acids are neutralized by bases and an *early, but still useful*, definition of the term propounded by Rouelle in 1774 referred directly to this.

A **base** is any substance which will react with an acid to produce a salt.

A base may be soluble or insoluble in water, or react with water to produce a basic solution. Insoluble bases include copper(II) oxide and magnesium hydroxide, which react with sulphuric acid:

$$CuO + H_2SO_4 \rightarrow CuSO_4 + H_2O$$
copper(II)   sulphuric   copper(II)   water
oxide   acid   sulphate

$$Mg(OH)_2 + H_2SO_4 \rightarrow MgSO_4 + 2H_2O$$
magnesium   sulphuric   magnesium   water
hydroxide   acid   sulphate

The products remain in solution.

Potassium hydroxide is a soluble base, reacting with nitric acid to form potassium nitrate solution:

$$KOH + HNO_3 \rightarrow KNO_3 + H_2O$$

Barium oxide reacts violently with water to form the soluble base, barium hydroxide, which may be used to neutralize hydrochloric acid:

$$BaO + H_2O \rightarrow Ba(OH)_2$$

$$Ba(OH)_2 + 2HCl \rightarrow BaCl_2 + 2H_2O$$
barium   hydrochloric   barium
hydroxide   acid   chloride

A solution of barium chloride is formed in this example.

Metal oxides which neutralize acids are called **basic oxides**. A soluble base is commonly called an **alkali**. A basic oxide reacts with acids to form salts in which the element of the oxide is in the cation.

Alkaline solutions, in addition to neutralizing acids, feel slimy to the touch, taste soapy and reverse the colour changes brought about by the action of acids on indicators. Common alkalis include the hydroxides of the metals of Groups Ia and IIa of the Periodic Table (Li to Cs, Be to Ra). They are strong electrolytes because of their ionic nature and the alkaline character is due to the hydroxide, formerly hydroxyl, ion. The ionic structures dissociate in water:

$$Na^+OH^- = Na^+ + OH^-$$

$$Ba^{2+}(OH^-)_2 = Ba^{2+} + 2OH^-$$

An **alkali** is a compound which dissociates to produce hydroxide ions when dissolved in water.

Aqueous solutions of ammonia and its organic derivatives, such as triethylamine, are alkaline by interaction with the solvent,

$$NH_3 + H_2O \rightleftharpoons NH_4^+ + OH^-$$
ammonia   ammonium
             ion

$$N(C_2H_5)_3 + H_2O \rightleftharpoons N(C_2H_5)_3H^+ + OH^-$$
triethyl-   triethyl-
amine   ammonium ion

The presence of hydroxide ions may easily be demonstrated. However, ammonia and the amines will react directly with acids without going into solution first:

$$NH_3 + HBr \rightarrow NH_4^+Br^-$$
hydrogen   ammonium
bromide   bromide

$$N(C_2H_5)_3 + HBr \rightarrow [N(C_2H_5)_3H^+]Br^-$$
triethylammonium
bromine

Therefore:

A **base** is a compound which will unite with hydrogen ions.

The Brønsted–Lowry Theory

Independently in 1923, J. N. Brønsted and T. M. Lowry developed a general theory of acids and bases. So far, we have regarded acids as compounds giving hydrogen ions and soluble bases as yielding hydroxide ions in aqueous solution. Such a view is simple and straightforward but with non-aqueous media, of very great importance nowadays, a more general approach is needed. A modified concept of what is understood by the terms acid and base may be developed.

An **acid** is any compound having a tendency to lose a proton.

A **base** is any compound having a tendency to accept a proton.

An acid and base are said to be *conjugate with one another* when related by the transfer of a proton:

$$Acid \rightleftharpoons H^+ + Base$$

In aqueous solution the base proton forms the oxonium ion; using hydrochloric acid:

$$HCl + H_2O \rightleftharpoons H_3O^+ + Cl^-$$
acid   base   acid   base
1   2   2   1

Both hydrogen chloride and oxonium ions tend to throw off protons, so for convenience they may be labelled acid 1 and acid 2 respectively. The conjugate bases are the chloride ion, base 1, and the solvent, water, base 2, which has accepted protons. The equilibrium lies far to the right, so that hydrochloric acid is a strong acid. The chloride ion must be a weak base, having little tendency to revert to the acid by accepting a proton. The anion of a feebly ionized acid clearly has a strong tendency to form the conjugate acid with protons and is therefore a strong base. Hydrocyanic acid is such an acid and the cyanide ion a strong base:

$$HCN + H_2O \rightleftharpoons H_3O^+ + CN^-$$
hydrogen
cyanide
acid   base   acid   base
1   2   2   1

The strength (N.B. not the concentration) of an acid depends on the strength of the solvent as a base—on its ability to accept protons from the acid. Solvents may be classified as proton-acceptors (*protophilic*), proton-donors (*protogenic*) or as neither of these (*aprotic*). By definition, protogenic solvents are acids and protophilic, bases. Ethanoic acid, weak in water, is strongly ionized in liquid ammonia, a base. Bases are stronger when dissolved in ethanoic acid than in a less protogenic solvent such as water. Pure water itself undergoes self-ionization in which one water molecule *protonates* another:

$$H_2O + H_2O \rightleftharpoons H_3O^+ + OH^-$$

for which the ionic product of water

$$K_w = [H_3O^+][OH^-] = 1 \times 10^{-14} \text{ mol}^2 \text{ dm}^{-6} \text{ at 298 K}$$

Since $[H_3O^+] = [OH^-] = 1 \times 10^{-7}$ mol dm^{-3}

Liquid ammonia has a high dielectric constant, is a poor conductor of electricity and is an ionizing solvent:

$$2NH_3 \rightleftharpoons NH_4^+ + NH_2^-$$

acidic   basic
ion   ion

The proton is carried by an ammonia molecule, which becomes the ammonium ion. Ammono-acids and ammono-bases neutralize each other:

$$CH_3.CO.NHH + Na^+NH_2^- \rightarrow$$

ethanamide   sodamide
(acetamide)

$$CH_3.CO.NH^-Na^+ + NH_3$$

$$NH_4Cl + NaNH_2 \rightarrow 2NH_3 + NaCl$$

ammonium   sodamide
chloride

which corresponds to

$$HCl + NaOH \rightarrow H_2O + NaCl$$

or more accurately

$$OH_3Cl + NaOH \rightarrow 2H_2O + NaCl$$

Acid-base theory has been extended into non-protonic systems, i.e. where protons are not encountered. Many interhalogen compounds ionize slightly and are ionizing solvents:

$$2ICl_3 \rightleftharpoons ICl_2^+ + ICl_4^-$$

$$2BrF_3 \rightleftharpoons BrF_2^+ + BrF_4^-$$

$$2IF_5 \rightleftharpoons IF_4^+ + IF_6^-$$

↑   ↑
acids bases

From this brief survey, final definitions may be stated:

An **acid** is a compound which ionizes to give the cation of the parent solvent.

A **base** is a compound which ionizes to give the anion of the parent solvent.

The Lewis Theory

The transference of a proton from acid to base may be regarded as the reverse transference of a pair of electrons.

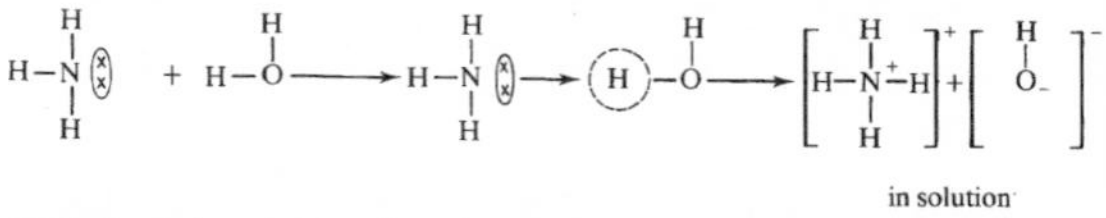

For clarity, the lone electron pairs on the oxygen atom have been omitted

G. N. Lewis (1923) used the electron pair as the basis of his definitions.

A Lewis acid accepts an electron pair donated by a Lewis base.

In this theory, neutralization corresponds to the formation of a co-ionic, or dative bond.

Thus boron trifluoride behaves as a Lewis acid:

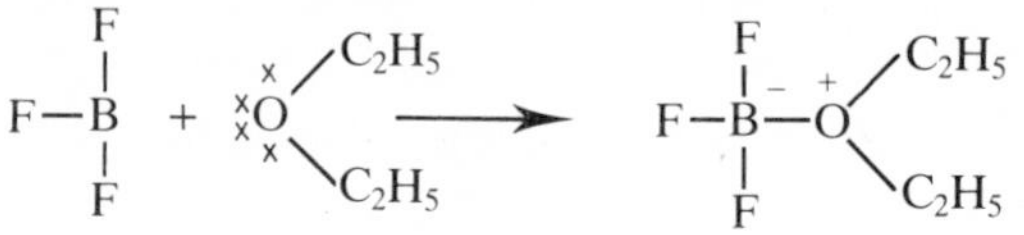

diethyl ether ethoxyethane
(showing lone electron-pairs)

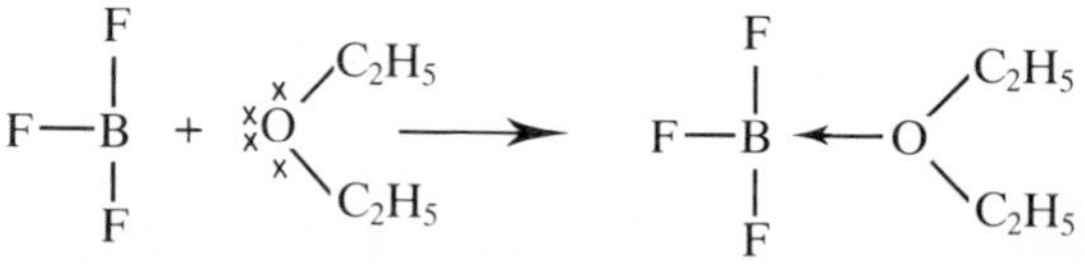

alternative formulation

Similarly

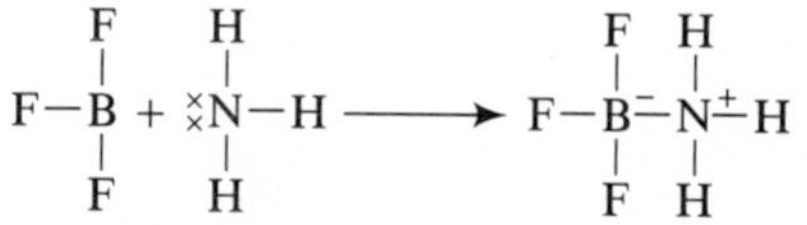

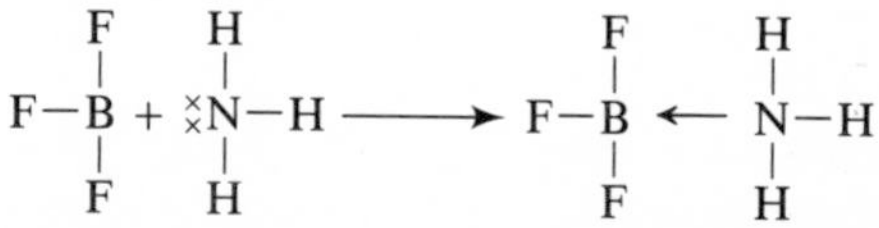

In these examples the boron atom attains an octet of electrons by accepting a lone-pair into its energy levels, and attains an enhanced stability in its molecular environment (see page 62).

Highly charged cations are known to be hydrated in solution. Electron pairs are donated by molecules of water which form ligands around the central cation. The metal ions act as Lewis acids and water as a Lewis base. Thus acidity in the Lewis sense will be greatest for cations with high charge and small radii, e.g. Al^{3+}, Fe^{3+}, and lowest for the large cations of Groups I and II, e.g. K^+, Ca^{2+}.

hexa-aquo iron(III) cation
octahedral structure (shape)

Thus Fe^{3+} is characterized by high Lewis acidity, although normally this is expressed traditionally as low basicity. This will be seen shortly in a discussion of the hydrolysis of salts of insoluble bases.

Normal, acid and basic salts

A salt is formed by replacement of acidic hydrogen in an acid by a metal or a basic group (such as $NH_4{}^+$). Very few of these compounds are not ionic and the term salt is sometimes restricted to those with an essentially ionic structure. For example, sodium sulphate, $(Na^+)_2SO_4{}^{2-}$, is composed of sodium ions and sulphate ions in the ratio $2:1$. The sulphate ion itself is covalent internally. Ion-pair formation and various strong electrolyte effects occur in concentrated solutions so that determinations of the apparent degree of dissociation by colligative methods will indicate that ionization is apparently incomplete. Colligative properties (relative lowering of the vapour pressure, depression of the freezing-point, elevation of the boiling-point and osmotic pressure) depend on the number of particles (molecules, ions) and not on their size or chemical nature. In some rare cases, e.g. CdI_2, complex ion formation occurs in solution. The positive ion is sometimes called the basic radical and the negative ion, the acidic radical. Ammonium nitrate, $NH_4{}^+NO_3{}^-$, contains no metal, having the ammonium radical as the cation and the nitrate radical as the anion. However, such usage

of the term radical is best avoided. The term is best regarded as describing a group of atoms which occurs repeatedly in a number of different compounds. Examples are $COCl_2$, *carbonyl* chloride, $PSCl_3$, *thiophosphoryl* chloride, $POCl$, *phosphoryl*(III) chloride, CrO_2Cl_2, *chromyl* chloride.

When all of the acidic hydrogen has been replaced the product is called a *normal salt*; it contains neither H^+ from the acid, nor OH^- from the base:

$$2NaOH + H_2SO_4 \rightarrow Na_2SO_4 + 2H_2O$$
sodium
sulphate

$$3NaOH + H_3PO_4 \rightarrow Na_3PO_4 + 3H_2O$$
trisodium
orthophosphate

Both sodium sulphate and trisodium orthophosphate are normal salts. Polybasic acids may form *acid salts* in which some acidic hydrogen remains in each anion. Such salts release hydrogen ions in water being also acids. Examples include, sodium hydrogensulphate, monosodium dihydrogenorthophosphate and disodium hydrogenorthophosphate:

$$NaOH + H_2SO_4 \rightarrow NaHSO_4 + H_2O$$
sodium
hydrogen
sulphate

$$H_3PO_4 + NaOH \rightarrow NaH_2PO_4 + H_2O$$
monosodium
dihydrogen
orthophosphate

$$NaH_2PO_4 + NaOH \rightarrow Na_2HPO_4 + H_2O$$
disodium
hydrogen
orthophosphate

Ionization of the anions occurs:

$$HSO_4{}^- \rightleftharpoons H^+ + SO_4{}^{2-}$$

$$H_2PO_4{}^- \rightleftharpoons H^+ + HPO_4{}^{2-}$$

$$HPO_4{}^{2-} \rightleftharpoons H^+ + PO_4{}^{3-}$$

Sodium hydrogensulphate is sometimes known as sodium bisulphate. Sodium hydrogencarbonate (sodium bicarbonate) and sodium hydrogensulphite (sodium bisulphite) are other examples of acid salts.

As the name suggests, *basic salts* retain a portion of the base, either as oxygen or as a hydroxide (hydroxyl) group. The normal salt may usually be formed with strong acids. The formation of basic salts is a sign of weakness in metallic nature of the participating cation of the salt. Bismuth oxide chloride, formerly oxy-chloride, BiOCl, yields the normal salt, bismuth(III) chloride (bismuth trichloride), with hydrochloric acid,

$$BiOCl + 2HCl \rightarrow BiCl_3 + H_2O$$

and tin(II) hydroxide chloride, $[Sn^+(OH)]Cl^-$, reacts similarly

$$Sn(OH)Cl + HCl \rightarrow SnCl_2 + H_2O$$

Basic copper(II) ethanoate (acetate), $Cu(OH)_2 . Cu(CH_3COO)_2$, or verdigris, white lead, basic lead(II) carbonate, $2Pb(OH)_2 . PbCO_3$, and malachite, basic copper(II) carbonate, $Cu(OH)_2 . CuCO_3$, are other illustrations.

Nomenclature for acids and salts

The prefixes and suffixes marked in Table 15.3 should be noted. Sodium chloride and hydrochloric acid each consist of two elements only; the suffix -ide denotes this for the salt while -ic is derived from the way Lavoisier named acids in connection with his oxygen theory of acids. However, note in the other cases that an acid with a name ending in -ic forms salts which have names ending in -ate. With one oxygen atom per molecule less than chlorous acid, HClO is called hypochlorous acid, hypo (Greek = below or lower) indicating the lower oxygen content; the salts are called hypochlorites. The prefix per (Latin = completely) is used to show a maximum of combined oxygen so that $HClO_4$, with one more oxygen atom per molecule than chloric acid, is called perchloric acid and the salts, perchlorates.

Table 15.3   The naming of acids and salts

Formula of acid	Name of acid	Formula of sodium salt	Name of sodium salt
HCl	Hydrochloric	NaCl	Chloride
HClO	Hypochlorous	NaClO	Hypochlorite
HClO₂	Chlorous	NaClO₂	Chlorite
HClO₃	Chloric	NaClO₃	Chlorate
HClO₄	Perchloric	NaClO₄	Perchlorate

When an element or radical forms salts in which it exhibits more than one valency, until recently the suffix -ous denoted the lower valency state and -ic the higher. This now applies to certain oxo-acids only (Table 15.4). Ferrous sulphate is now known as iron(II) sulphate and ferric sulphate as iron(III)

Table 15.4   Naming acids and salts where more than one valency or oxidation state occurs

Formula of acid or base with element of variable valency	Name	Formula of a salt	Name of salt
H₂SO₃	Sulphurous acid	Na₂SO₃	Sodium sulphite
H₂SO₄	Sulphuric acid	Na₂SO₄	Sodium sulphate
FeO	Ferrous oxide	FeSO₄	Ferrous sulphate
Fe₂O₃	Ferric oxide	Fe₂(SO₄)₃	Ferric sulphate

sulphate where the roman numerals indicate the oxidation state.

If the *Stock Notation* is applied to acids and their anions, H_2SO_3 becomes sulphuric(IV) acid and H_2SO_4, sulphuric(VI) acid, and the respective salts sodium sulphate(IV) and sodium sulphate(VI). Alternatively, the proportion of oxygen might be indicated as in sodium trioxosulphate and sodium tetraoxosulphate, to use the salts Na_2SO_3 and Na_2SO_4 as illustrations. Note that the terms sulphuric and sulphate, used originally to specify a particular acid and its anion (salt), should now designate quite generally an acid or a negative group containing sulphur as the central atom (linked to other atoms) irrespective of its state of oxidation. While the oxidation number is stated in the first set of examples, it is implied in the second.

Table 15.5 shows the Stock Notation extended to those acids and their salts already named in Table 15.3. The alternative names should be compared.

Table 15.5   The Stock Notation for acids and salts

Formula of acid	Oxidation number of chlorine	Name of acid	Name of anion (i.e. salt)
HCl	− 1	Hydrochloric	Chloride
HClO	+ 1	Chloric(I)	Chlorate(I)
HClO₂	+ 3	Chloric(III)	Chlorate(III)
HClO₃	+ 5	Chloric(V)	Chlorate(V)
HClO₄	+ 7	Chloric(VII)	Chlorate(VII)

While the terms hydrochloric and chloride remain unchanged, chloric and chlorate have become general family terms, requiring the addition of oxidation numbers to be specific.

The use of the prefix -ortho, as in orthophosphoric acid, H_3PO_4, will have been noticed. *Ortho*-(Greek = correct) is used to indicate the presence of the maximum quantity of the elements of water uniting. Loss of water from each molecule yields metaphosphoric acid, HPO_3, where *meta* (Greek = after, i.e. H_3PO_4) shows the relationship. An intermediate acid formed by the action of heat on orthophosphoric acid is pyrophosphoric acid, $H_4P_2O_7$, where *pyros* (Greek = fire) indicates the mode of formation.

Salt hydrolysis

A normal salt is the product of the neutralization of an acid by a base, all of the acidic hydrogen of the acid being replaced. If the salt is soluble, tests with indicators may show that the pH $(= -\log_{10}[H^+]/\text{mol dm}^{-3})$ is not 7 at 25°C; neutralization has not produced a neutral solution! The pH of a salt solution depends on the relative strengths of the combining acids and bases. By interaction with the solvent, the equilibrium,

$$2H_2O \rightleftharpoons H_3O^+ + OH^-$$

is disturbed, with the result that the concentrations of hydrogen (oxonium) ion and hydroxide ion are no longer equal and the solution no longer neutral. This phenomenon is called salt hydrolysis and will be examined under two headings.

Hydrolysis of salts of soluble bases

Salts of strong acids with strong bases are not hydrolysed and neutral solutions are formed. However, hydrolysis occurs where weak acids and weak bases are used.

1 Salts of weak acids and strong bases, e.g. potassium cyanide solution:

$$(K^+ +) CN^- + H_2O \rightleftharpoons HCN + OH^- (+ K^+)$$

Hydrocyanic acid is feebly ionized. The solution contains an excess of hydroxide ions over hydrogen ions. It is alkaline to litmus (pH > 7).

2 Salts of strong acids and weak bases, e.g. ammonium chloride solution:

$$(Cl^- +) NH_4^+ + H_2O \rightleftharpoons NH_3 + H_3O^+ (+ Cl^-)$$

Aqueous ammonia is a weak base with only a small tendency to compete for the proton. There is an excess of oxonium (hydrogen) ions and the solution is acidic to litmus (pH < 7).

3 Salts of weak acids and weak bases, e.g. ammonium ethanoate (acetate) solution:

$$(NH_4^+ +) CH_3COO^- + 2H_2O \rightleftharpoons$$
$$NH_3 + H_3O^+ + CH_3COOH + OH^-$$

which becomes

$$NH_4^+ + CH_3COO^- \rightleftharpoons NH_3 + CH_3COOH$$

Although each ion is involved in hydrolysis, the final acidity of the resulting solution will depend on the relative extent of the two processes. Actually, in this case the processes of salt hydrolysis cancel each other out and the solution is neutral.

Hydrolysis of salts of insoluble bases

Metal hydroxides formed by elements other than those of the metals of Group I (alkali metals) and II (alkaline earths) of the Periodic Classification are weak bases, being virtually insoluble in water. Salts of these bases are often acidic in solution. Copper(II) sulphate, aluminium sulphate and iron(III) chloride are examples.

Consider aluminium sulphate solution. The aluminium ions are hydrated as $Al(OH_2)_6^{3+}$, the hexaquoaluminium ion, in which the water molecules are probably linked by co-ionic bonds from oxygen to metal in octahedral formation (i.e. producing an octahedral shape to the cation) although the bonding might be electrostatic in nature. Competition between hydrated aluminium ions and solvent molecules leads to the loss of protons:

$$Al(OH_2)_6^{3+} + H_2O \rightleftharpoons$$
$$Al(OH_2)_5(OH)^{2+} + H_3O^+$$

$$Al(OH_2)_5(OH)^{2+} + H_2O \rightleftharpoons$$
$$Al(OH_2)_4(OH)_2^+ + H_3O^+$$

$$Al(OH_2)_4(OH)_2^+ + H_2O \rightleftharpoons$$
$$Al(OH_2)_3(OH)_3 + H_3O^+$$
$$\text{hydrated aluminium}$$
$$\text{oxide or hydroxide}$$

Unless alkali is added, the hydrolysis will not go as far as the third stage. Iron(III) chloride solution may become reddish-brown on storage indicating the presence of hydrated iron(III) oxide in colloidal solution. The addition of acid reduces hydrolysis and gives a solution with a yellow tinge. The products of hydrolysis of the hexaquoiron(III) ion, $Fe(OH_2)_6^{3+}$ will be

$$Fe(OH_2)_5(OH)^{2+}, \quad Fe(OH_2)_4(OH)_2^+$$

and $Fe_2O_3.xH_2O$ or hydrated $Fe(OH)_3$

This type of hydrolysis is pronounced with hydrated ions of high charge (Al^{3+}, Sb^{3+}, Bi^{3+}, Fe^{3+}), much less so with bivalent ions (Zn^{2+}, Cu^{2+}

are hydrolysed while Fe^{2+} is not) and not at all with ions of single charge (e.g. Ag^+).

The key to the discussion lies in the polarizing power (Fajans' Rules) in determining the acidity of $[M(H_2O)_6]^{n+}$ ions from the Lowry-Brønsted viewpoint. This polarizing power may be expressed as charge/radius, the strong bond formed between the metal atom (aluminium) and oxygen weakening the $O-H$ bond of the attached water molecules leading to dissociation and an acidic solution.

16

Oxygen, oxides and peroxides

Oxygen

Oxygen is a colourless, odourless, tasteless, gaseous element. The molecules are diatomic. Oxygen supports the related phenomena, combustion and respiration. Although the solubility of oxygen in cold water is only slight, there is enough present to support the life of fish and other aquatic life. Some physical data for oxygen are listed in Table 16.1.

Table 16.1 Oxygen (*p* block element): physical data

Atomic number	8
Electronic configuration	2, 6
[Core] outer electrons	$[He](2s)^2(2p)^4$
Relative atomic mass	15.9994 (16.00)
Relative molecular mass	31.9988 (32.00)
Atomic radius/nm	0.066
Ionic radius (O^{2-})/nm	0.140
van der Waals' radius/nm	0.140
Electronegativity	3.5
1st ionization energy/kJ mol^{-1} (273 K)	1312
2nd ionization energy/kJ mol^{-1} (273 K)	3387
Boiling-point (101 kPa)	$-183°C$ (90 K)
Melting-point (101 kPa)	$-219°C$ (54K)
Density/g dm^{-3} (standard conditions*)	1.429
Solubility in water/cm^3 dm^{-3} (standard conditions*)	48.9
Critical temperature	$-119°C$ (154 K)
Critical pressure/(MPa)	5.20

exactly: 273.15 K, 101325 Pa/Nm^{-2}

Oxygen occurs freely in the atmosphere of which it comprises 20.8% by volume, roughly one-fifth, and 23.0% by mass, roughly one-quarter. It is the most abundant element in the Earth's crust and with carbon, nitrogen and hydrogen is essential to living matter. Water contains 89% by mass, or eight-ninths, of combined oxygen and in all it has been estimated that 49.4% of the Earth's crust consists of the combined element, rocks containing 40–50% by weight. Oxides are extremely important.

Ozone, trioxygen, is an allotropic modification of oxygen. It is triatomic, of formula, O_3. It occurs in the higher levels of the atmosphere and reduces the amount of ultraviolet light which penetrates to the surface of the Earth and which would otherwise cause problems of health, notably melanosis, cancer of the skin.

Scheele discovered oxygen some time during the period 1771–3, but did not publish his results until 1777. He heated nitre and other compounds, including mercury(II) oxide, collecting the oxygen evolved in bladders. He observed the brilliant combustion of charcoal in what he called 'fire air'. In the meantime, Priestley prepared oxygen on 1st August 1774. Using a 'burning lens', he focused the rays of the sun on to mercury(II) oxide (mercurius calcinatus per se), contained in an inverted vessel over quicksilver (mercury). He also produced oxygen by heating minium (red lead) similarly. He named the gas dephlogisticated air, observed that a candle burned with a vivid flame and recorded that the gas supported the life of a mouse better than did 'common air'. Two months later, in Paris, he related his discovery to Lavoisier who immediately grasped its significance in relation to his own work. Subsequently, combustion, respiration and the calcination of metals were proved by Lavoisier to involve a common process. He recognized the gas as an element, finally naming it oxygen (Greek, *oxys*, sour or acid; *genon*, to form) because products of combustion appeared to him to be acidic. This is only partially true. Some acids were proved to contain no oxygen while some oxides formed alkalis, not acids, with water. Hydrogen cyanide, HCN, was shown by C. L. Berthollet in 1787 to contain no oxygen. The work of Davy was mentioned in the previous chapter. During this period Lavoisier did much to establish chemistry as a science.

Oxygen was liquefied by Pictet in 1877, using the 'cascade process', and independently in the same year by Cailletet who used the principle of adiabatic expansion, the Joule–Thomson effect.

Element No. 8 of the Periodic Classification

Oxygen is the first member of Group VI of the Periodic Family, which also contains sulphur, selenium, tellurium and polonium. There is a greater difference between oxygen and sulphur than between later successive pairs. With atomic number 8, the electronic configuration is 2,6. The most abundant isotope by far has a mass number 16, the nucleus having 8 neutrons as well as 8 protons. For isotopes of mass number 17 and 18, there are 9 and 10 neutrons respectively in addition to 8 protons in the nucleus. Two atomic mass scales were in use until recently. The physical scale was based on the $^{16}_{8}O$ isotope = 16 exactly and the chemical on the usual isotopic mixture being assigned this value. The interconversion factor is 1.00028.

Physical Atomic mass =
Chemical Atomic mass × 1.00028

Elements have slightly larger atomic masses on the Physical Scale. The $^{12}_{6}C$ = 12 (exactly) scale is now preferred.

Oxygen is second only to fluorine in electronegativity. This leads to its somewhat anomalous behaviour as the first member of Group VI and, in particular, involves it in hydrogen bonding.

Oxygen in combination

Oxygen is in the second period of the Periodic Table, two places from neon. An atom of oxygen may acquire the electronic configuration of neon in two ways: either by sharing electrons to form covalent or co-ionic bonds, depending on the original distribution of electrons, or by taking two electrons to form an ion. The simple valency of oxygen in covalent and ionic compounds is two. A marked polarity occurs in covalent compounds due to the high electronegativity of oxygen and this is manifested in a dipole moment in unsymmetrical molecules. Binary compounds of oxygen are called *oxides*.

Oxygen, O_2

A covalent structure could be written in which the octet of electrons corresponding to neon is acquired. Do not think of electrons as localized particles:

   O=O

However, the structure is not straightforward. Oxygen is paramagnetic, as a gas, liquid and solid, and its addition reactions and other characteristics lead to the structure being written:

   O—O

There are two unpaired electrons in the molecule: each atom has an odd electron configuration. This cannot be explained by the basic ideas of the octet theory, but requires the theory of molecular orbitals which may be developed from the atomic orbital theory outlined in the earlier chapters (or fundamentally, developed in parallel to atomic orbital theory).

In this theory the combination of atomic orbitals (or rather the $+/-$ mathematical functions from which they are conceived) results in a re-enforcement of electron density between the approaching atomic nuclei (to form a bonding molecular orbital), or the electron density disappears (an anti-bonding molecular orbital). *Sigma* (σ_s, σ_p) bonds are formed along the internuclear axis (by s, p_x orbitals), while *pi* (π_p) bonds are formed by overlapping of the p_y and p_z orbitals ('sideways on') above and below the internuclear axis but *not* along it. Bonding orbitals are filled in order of energy, minimum first and bonding orbitals (σ, π) are filled before anti-bonding orbitals (σ^*, π^*), in accordance with Hund's Rule.

Compare oxygen, which has two unpaired electrons in the π_p^* orbitals with fluorine which has no electrons unpaired.

$$O(2s)^2(2p)^4 + O(2s)^2(2p)^4 \rightarrow$$
$$O(\sigma_s)^2(\sigma_s^*)^2(\sigma_p)^2(\pi_p)^4(\pi_p^*)^2$$

$$F(2s)^2(2p)^5 + F(2s)^2(2p)^5 \rightarrow$$
$$F(\sigma_s)^2(\sigma_s^*)^2(\sigma_p)^2(\pi_p)^4(\pi_p^*)^4$$

The net counts of bonding electrons are, respectively, 4(= double bond) and 2(= single bond). If a hypothetical Ne_2 molecule is postulated the next orbital to be filled is the anti-bonding σ_p^*. The resulting net electron count is zero, and there is no bonding, and no molecule.

The two unpaired electrons in the $\pi_p{}^*$ orbital are responsible for the paramagnetism mentioned earlier. (The treatment may be extended to comparisons of CO and NO, CN⁻ and NO⁺ etc.)

Oxides

Covalent (and co-ionic) oxides

These structures are formed by sharing electrons, oxygen usually forming single bonds, but double bonds are not unknown. Some examples are listed below in which the electrons have been distinguished to show the atoms from which they originated.

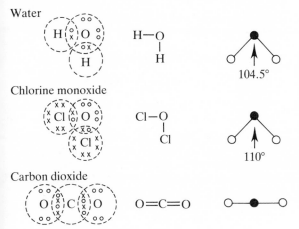

Water

Chlorine monoxide

Carbon dioxide

Atomic number	Electronic configuration	
	Before	After
1	H 1	2, (He)
6	C 2, 4	2, 8 (Ne)
8	O 2, 6	2, 8 (Ne)
17	Cl 2, 8, 7	2, 8, 8 (Ar)

Some covalent oxides also contain co-ionic bonds. Consider the two chief oxides of sulphur:

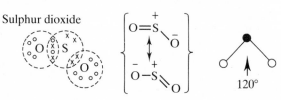

Sulphur dioxide

Sulphur trioxide

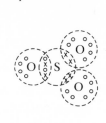 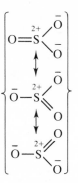 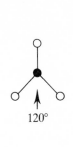

Atomic number	Electronic configuration	
	Before	After
8	O 2, 6	2, 8 (Ne)
16	S 2, 8, 6	2, 8, 8 (Ar)

The sulphur–oxygen links are hybrids as indicated by the bracketed structures. See the Theory of Resonance in Chapter 5 (p. 81).

Covalent oxides belong to the non-metals, to many transition metals in their highest oxidation states and to the metalloids and metals of high atomic number in the middle of the Periodic Table. Most are acidic; either reacting with water to form acids or, if insoluble, reacting with aqueous or fused alkali:

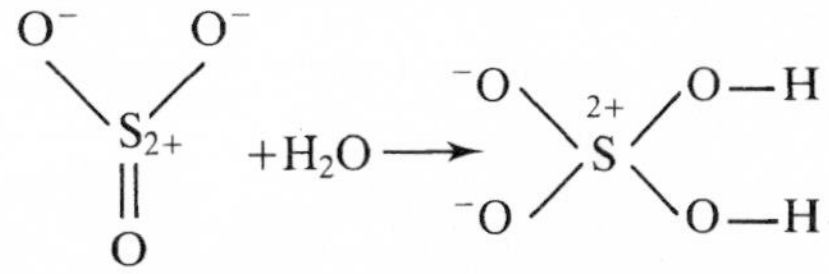

sulphur trioxide   sulphuric acid

$$SiO_2 + 2NaOH \longrightarrow Na_2SiO_3 + H_2O$$

silica
macromolecule
(insoluble
in water)

sodium
silicate

Acidic oxides that react with water are also sometimes called *acid anhydrides*. Neutral oxides also occur in this structural group. Besides water, these include carbon monoxide, dinitrogen oxide (nitrous oxide) and the odd electron molecule of nitrogen oxide (nitric oxide):

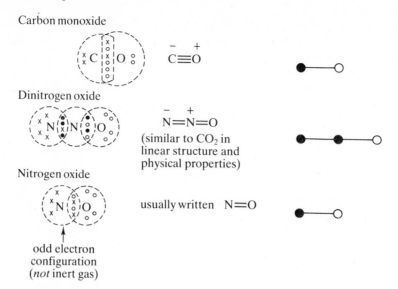

Carbon monoxide

$$\overset{-}{C}\equiv\overset{+}{O}$$

Dinitrogen oxide

$$\overset{-}{N}=\overset{+}{N}=O$$

(similar to CO_2 in linear structure and physical properties)

Nitrogen oxide

usually written $N=O$

odd electron configuration
(*not* inert gas)

Atomic number	Electronic configuration	
	Before	After
6	C 2,4	2,8 ⎫
7	N 2,5	2,8* ⎬ (Ne)
8	O 2,6	2,8 ⎭

*but 2, 7 in nitrogen oxide

Ionic oxides

By the acceptance of two electrons, an oxygen atom forms the oxide ion. This type of oxide is formed by the strongly electropositive metals of Groups I and II except beryllium. Taking an example from each group, lithium monoxide and calcium oxide:

$$\left[Li\right]_2^+ \left[O\right]^{2-} \qquad (Li^+)_2 O^{2-}$$

$$\left[Ca\right]^{2+} \left[O\right]^{2-} \qquad Ca^{2+}O^{2-}$$

Atomic number	Electronic configuration	
	Before	After
1	H 1	2 ⎫ (He)
3	Li 2,1	2 ⎭
8	O 2,6	2,8 (Ne)
20	Ca 2,8,8,2	2,8,8 (Ar)

These oxides are basic, uniting directly with water to produce the corresponding hydroxides, also basic and ionic: Li^+OH^-, $Ca^{2+}(OH^-)_2$. In each group the vigour of the reaction of the metal oxide with water to form the hydroxide increases sharply with rise in atomic number of the metal:

$$Li_2O + H_2O \rightarrow 2LiOH$$

$$BaO + H_2O \rightarrow Ba(OH)_2$$

Lithium oxide reacts quietly. Sodium monoxide reacts vigorously. Magnesium oxide is very slow to react and the product is only slightly soluble. Calcium oxide gives out much heat on slaking while barium oxide may become incandescent. Acids usually react with great vigour to form salts,

$$Na_2O + 2HCl \rightarrow 2NaCl + H_2O$$
$$\text{sodium chloride}$$

although magnesium oxide reacts quietly:

$$MgO + 2HCl \rightarrow MgCl_2 + H_2O$$
$$\text{magnesium chloride}$$

Oxides of intermediate bond type

Some oxides cannot conveniently be classified as ionic or covalent; they are formed by metals other than those of Groups I and II (except beryllium) already mentioned and include some oxides of transition metals. Their hydroxides are made indirectly. Both oxide and hydroxide may display properties which are both basic and acidic. They are *amphoteric*. Zinc oxide forms the chloride with hydrochloric acid and the zincate with potassium hydroxide, both being salts.

$$ZnO + 2HCl \rightarrow ZnCl_2 + H_2O$$

$$ZnO + 2KOH \rightarrow K_2ZnO_2 + H_2O$$
$$(\text{or } K_2[Zn(OH)_4])$$
hydrated form as complex

The hydroxide behaves similarly.

Peroxides and hyperoxides (superoxides)

Peroxides are best regarded as salts of the weak acid hydrogen peroxide, liberated on reacting with acids. They are formed by the metals of Groups I and II, except beryllium. The peroxides of sodium and barium provide examples:

$$\left[Na \right]_2^+ \left[\overset{\circ\circ}{\underset{\circ\circ}{O}} \overset{\bullet\bullet}{\underset{\bullet\bullet}{O}} \right]^{2-} \quad (Na^+)_2 O_2^{2-}$$

$$\left[Ba \right]^{2+} \left[\overset{\circ\circ}{\underset{\circ\circ}{O}} \overset{\bullet\bullet}{\underset{\bullet\bullet}{O}} \right]^{2-} \quad Ba^{2+} O_2^{2-}$$

(electrons are distinguished for convenience only)

Atomic number	Electronic configuration	
	Before	After
8	O 2,6	2,8
11	Na 2,8,1	2,8 (Ne)
56	Ba 2,8,18,18,8,2	2,8,18,18,8 (Xe)

Hyperoxides or superoxides (e.g. KO_2) contain the O_2^- ion.

Using the molecular orbital ideas touched on briefly earlier, we may compare the electronic configurations of O_2, O_2^- and O_2^{2-}:

oxygen molecule O_2 $(\sigma_s)^2(\sigma_s*)^2(\sigma_p)^2(\pi_p)^4(\pi_p*)^2$ **4**
hyperoxide ion O_2^- $(\sigma_s)^2(\sigma_s*)^2(\sigma_p)^2(\pi_p)^4(\pi_p*)^3$ **3**
peroxide ion $O_2^{2-}(\sigma_s)^2(\sigma_s*)^2(\sigma_p)^2(\pi_p)^4(\pi_p*)^4$ **2**

The net counts of bonding electrons are shown on the right and correspond, respectively, to a double bond, a 'one-and-a-half' bond and a single bond (compare with F_2). The hyperoxide with three electrons in the π_p orbitals (which can accommodate four electrons) has one of these unpaired and shows paramagnetism.

The classification of oxides

The oxides are among the most important of the compounds formed by elements. An attempt has been made to classify them along valency lines and this proves useful. The types of oxide already described include: acidic, neutral, basic, amphoteric and peroxides. In addition, *polyoxides or higher oxides* are classed separately. Sometimes, *dioxides* are considered on their own. However, certain oxides may be regarded as salts of a lower basic oxide of a metal with the acidic oxide of the higher oxidation state. These may be called *salt oxides*, *compound oxides* or *mixed oxides*.

Red lead behaves as a combination of the monoxide and dioxide, *i.e.* oxides of Pb^{II} and Pb^{IV} respectively:

$$Pb_3O_4 = 2PbO \cdot PbO_2$$

With nitric acid, lead(II) nitrate and the dioxide, lead(IV) oxide, are formed:

$$2PbO \cdot PbO_2(s) + 4HNO_3 \rightarrow$$
$$2Pb(NO_3)_2(s) + PbO_2(s) + 2H_2O$$

Tri-iron tetroxide reacts with hydrochloric acid to produce chlorides of iron(II) and iron(III):

$$Fe_3O_4 = FeO \cdot Fe_2O_3$$

$$FeO \cdot Fe_2O_3 + 8HCl \rightarrow FeCl_2 + 2FeCl_3 + 4H_2O$$

Ceramic oxides in superconductor research

In press releases about superconductor research in 1987 it was announced that certain unspecified ceramic oxides have the property of carrying electric current without resistance at temperatures (liquid nitrogen was used) far higher than that which superconductors had required hitherto (that of liquid helium). By early 1990 at least 12 compounds in four main groups had been discovered. The US Navy hopes to use the technology in satellite communications in 1991 and it is anticipated that instruments to detect very tiny changes in electromagnetic signals will be in use in oil field exploration before 1995. Such materials are discussed further in the chemistry of copper (p. 482).

Oxide type, oxidation state and the Periodic Table

In Chapter 7, it was seen how the early attempt to classify certain reactions, which involved the addition of oxygen as oxidation reactions led inexorably to the more sophisticated notion of oxidation as the loss of electrons by ions, radicals, atoms or molecules (in redox processes) and thence to the use of the term oxidation state (or oxidation number) in developing a systematic nomenclature for inorganic compounds. The oxidation number of an element is a precise (and limited) term fixed by application of a set of rules. Its use avoids the controversies which might arise over use of the term valency, which attempts to portray by means of theoretical concepts what distinguishes one state of chemical combination of an element from another. This notion will be explained in more detail at appropriate points in the text which follows.

Metals form simple cations by the loss of electrons; those with the stongest tendency to do this are the strongest metals. Hybrid elements, between metals and non-metals in their characteristics, are called metalloids.

Consider the oxides in a period of typical elements. Taking Period 3 and writing down the formulae and nature of oxides and the corresponding hydroxides we see that the oxides become progressively less basic and more acidic as the oxidation number increases in steps from 1 to 7. The oxide bringing out the group oxidation number or valency is chosen in each case:

I	II	III
Na_2O	MgO	Al_2O_3
$NaOH$	$Mg(OH)_2$	$Al(OH)_3$
strong base	strong base	amphoteric

IV	V	VI	VII
SiO_2	P_4O_{10}	SO_3	Cl_2O_7
H_2SiO_3	HPO_3	H_2SO_4	$HClO_4$
weak acid	moderate acid	strong acid	strong acid

In a periodic group or family, the acidity of the corresponding oxide of each element decreases with increasing atomic number:

$$N_2O_5 \quad P_4O_{10} \quad As_4O_{10} \quad Sb_2O_5 \quad Bi_2O_5$$

Bismuth 'pentoxide' which has not been isolated pure is only feebly acidic. With groups of strong metals, the oxides are all basic but the basic nature increases in intensity with increasing atomic number. The change in character of oxides in a group emphasizes the transition from non-metallic to metallic nature, or through lessening degrees of non-metallic character to increasing degrees of metallic nature.

When an element forms a series of oxides, the acidity increases with the proportion of oxygen and the valency type becomes more covalent. Manganese affords a good example. The oxidation state of manganese is shown above each formula:

I	II	III
(Mn_2O)	MnO	Mn_2O_3
not isolated	basic	basic

IV	V	VI	VII
MnO_2	—	(MnO_3)	Mn_2O_7
amphoteric		not isolated	acidic

Compounds of MnO_3 are known which indicate that its nature would be acidic. Many other transition metals have similarly graded oxides while others, such as nickel, form only basic oxides.

An interesting division of the elements of the Periodic Table, omitting transition and associated elements, may be made using oxide classification as in Table 16.2. For completeness, various elements not studied in depth here are included but bracketed. Fluorine, which is more electronegative than oxygen, forms oxides (it is thereby reduced), but these are not acidic, basic or amphoteric: therefore, its symbol is ringed.

The lines of demarcation are in some cases a matter of opinion. Arsenic trioxide, As_4O_6, and arsenious acid, H_3AsO_3, are apparently amphoteric

Table 16.2　The Periodic Table used to show the classification of oxides

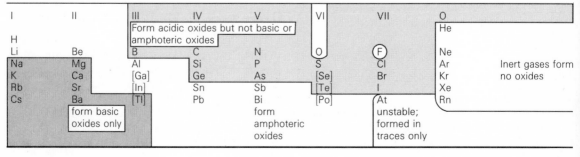

but no cations, As^{3+}, $As(OH)^{2+}$ or $As(OH)_2^+$ are known. The formation of $AsCl_3$ with hydrochloric acid is best classed as alcoholic (by analogy to the covalent alcohols which form covalent alkyl halides) and arsenic excluded from the amphoteric region. The next element of the group, antimony, forms one acidic oxide and one amphoteric oxide, while bismuth trioxide is basic and the higher oxide, acidic. The vertical transition from acidic to basic through amphoteric is well shown. The formation of covalent chlorides by the action of hydrochloric acid on oxides and acids is also indicative of the approach of metallic nature. The position of Indium (In) is also arguable.

Reactions in which oxygen is liberated

1   The thermal decomposition of metallic oxides

The affinity of oxygen for elements is generally high and only fluorine will displace it immediately from compounds. The affinity table showing the metal most reactive to oxygen at the head and listing metals in order of decreasing activity is roughly the same as that showing the tendency to produce cations in solution: the electrochemical series. Metals low in this series (e.g. mercury, silver, platinum and gold) form oxides easily decomposed by heat alone. Mercury(II) oxide and silver oxide are decomposed to metal and oxygen, mercury volatilizing away:

$$2HgO \rightarrow 2Hg(g) + O_2(g)$$

$$2Ag_2O \rightarrow 4Ag + O_2(g)$$

Various higher oxides, dioxides and mixed oxides, and peroxides are thermally decomposed to oxygen and a lower oxide. On gentle heating, both red lead and lead dioxide, lead(IV) oxide, break down to yield lead monoxide, lead(II) oxide, as the solid product:

$$2Pb_3O_4 \rightarrow 6PbO + O_2$$

$$2PbO_2 \rightarrow 2PbO + O_2$$

Trimanganese tetroxide is formed from manganese dioxide, manganese(IV) oxide, at red heat:

$$3MnO_2 \rightarrow Mn_3O_4 + O_2$$

2   The thermal decomposition of salts with anions rich in oxygen

Nitrates, permanganates, dichromates, chlorates, bromates, iodates and peroxodisulphates (persulphates) are decomposed on heating.

At red heat, potassium nitrate yields the nitrite and oxygen:

$$2KNO_3 \rightarrow 2KNO_2 + O_2$$

On warming to 240°C, potassium permanganate readily gives the manganate(VI), manganese(IV) oxide (dioxide) and oxygen:

$$2KMnO_4 \rightarrow K_2MnO_4 + MnO_2 + O_2$$

Potassium dichromate decomposes at white heat to chromate(VI), chromium(III) oxide and oxygen:

$$4K_2Cr_2O_7 \rightarrow 4K_2CrO_4 + 2Cr_2O_3 + 3O_2$$

Potassium chlorate forms the perchlorate when heated to just above the melting-point, 347°C, and then this decomposes to the chloride and oxygen. The optimum temperature for the first, a disproportionation reaction, is 395°C:

$$4KClO_3 \rightarrow 3KClO_4 + KCl$$

$$KClO_4 \rightarrow KCl + 2O_2$$

The net change is

$$2KClO_3 \rightarrow 2KCl + 3O_2$$

The thermal decomposition of chlorates is hastened by catalysis; various metal oxides may be used, including those of manganese(IV) (manganese dioxide), iron(III), cobalt(II), nickel(II) and copper(II). The first is well known although the mechanism of the breakdown is not clear. Decomposition is endothermic, starting just above 200°C and being rapid below 300°C, both temperatures being below the melting-point of the chlorate. In this temperature range, manganese(IV) oxide is chemically unchanged. Usually potassium chlorate and granular manganese(IV) oxide are mixed in the proportion of 3 : 1 by weight, the latter being dried beforehand by heating on an iron tray with a luminous bunsen flame. Care must be taken: manganese(IV) oxide containing carbon or antimony sulphide may cause a dangerous deflagration or explosion on heating. A trial portion of the mixture should be heated in an ignition tube before the main experiment. The *danger of confusing powdered charcoal with manganese(IV) oxide* is obvious but accidents have occurred.

3   The action of hot concentrated sulphuric acid on certain higher oxides and their anions in salts

Manganese(IV) oxide and chromium(VI) oxide, when separately heated with concentrated sulphuric acid, produce the sulphate of a lower oxidation state, water and oxygen:

$$2MnO_2 + 2H_2SO_4 \rightarrow 2MnSO_4 + 2H_2O + O_2$$

$$4CrO_3 + 6H_2SO_4 \rightarrow$$
$$2Cr_2(SO_4)_3 + 6H_2O + 3O_2$$

Potassium dichromate yields oxygen readily on heating with concentrated sulphuric acid,

$$2K_2Cr_2O_7 + 10H_2SO_4 \rightarrow$$
$$4KHSO_4 + 2Cr_2(SO_4)_3 + 8H_2O + 3O_2$$

The decomposition of potassium permanganate *must not be attempted*: the heptoxide of Mn^{VII}, Mn_2O_7, breaks down with explosive violence.

4   The catalytic decomposition of hypochlorites

Addition of a solution of a cobalt(II) salt (sulphate, chloride or nitrate, etc.) to a hypochlorite, otherwise chlorate(I), solution results in a brisk evolution of oxygen:

$$2NaClO \rightarrow 2NaCl + O_2$$

It has been suggested that cobalt(II) oxide or hydroxide formed by the alkali present is converted into the cobalt(III) compound which decomposes to regenerate the cobalt(II) compound:

$$Co^{2+} + 2OH^- \rightarrow Co(OH)_2$$

$$OCl^- + H_2O + 2Co(OH)_2 \rightarrow 2Co(OH)_3 + Cl^-$$
$$4Co(OH)_3 \rightarrow 4Co(OH)_2 + O_2\uparrow + 2H_2O$$

Cobalt(III) hydroxide has not been isolated.

5   By the decomposition of peroxides

Hydrogen peroxide readily decomposes on heating the solution to boiling:

$$2H_2O_2 \rightarrow 2H_2O + O_2$$

The decomposition of hydrogen peroxide is influenced by many catalysts; including finely divided metals, gold, silver, platinum especially in the colloidal state, and various oxides.

Potassium permanganate in acid solution is used in the titrimetric (volumetric) estimation of hydrogen peroxide, the gaseous product being oxygen,

$$5H_2O_2 + 2KMnO_4 + 4H_2SO_4 \rightarrow$$
$$2KHSO_4 + 2MnSO_4 + 8H_2O + 5O_2$$

Water dripped on to sodium peroxide yields a steady stream of oxygen,

$$2Na_2O_2 + 2H_2O \rightarrow 4NaOH + O_2$$

This can be dangerous because any free metallic sodium present may cause an explosion.

The peroxides of Group II metals decompose on heating. The alternate decomposition and regeneration of barium peroxide was the basis of the now obsolete Brin's process (1881):

$$2BaO_2 \rightarrow 2BaO + O_2$$

6   By electrolysis

Oxygen appears at the anode during the electrolysis of alkalis, oxoacids and certain of their salts. Pure oxygen is prepared by electrolysis of warm barium hydroxide solution. Any hydrogen which has diffused through the electrolyte into the anode gas may be removed by catalytic oxidation over a red-hot platinum gauze, followed by drying with phosphorus pentoxide.

The reactions of oxygen

Except for the noble (inert) gases and certain very noble metals, such as gold and platinum, oxygen reacts with all elements, although rather special conditions are required, as with the halogens, for some of the reactions. While a reactive element like yellow phosphorus inflames immediately, action is usually slow unless the element is heated, as for copper, or ignited, as for sulphur and hydrogen. Frequently the reaction is vigorous with the evolution of heat and light. The reaction rate is increased by fine division: heated iron and lead powder inflame, and sulphur vapour from the boiling element burns vigorously.

In Chapter 7, the concept of oxidation and reduction was developed.

Covalent oxidation is the addition of a more electronegative element or removal of hydrogen involving covalent (or co-ionic) bonds.

Ionic oxidation is the loss of electrons by atoms, groups of atoms, ions or molecules.

Reduction is the converse of oxidation and takes place at the same time as oxidation.

It is interesting to note that the faster reactions of oxygen are those in which the O–O bond remains unbroken. Examples include the synthesis of sodium peroxide, Na_2O_2, and barium peroxide, BaO_2.

1   With elements; the synthesis of oxides and peroxides

Hydrogen burns with a pale blue flame to form water,

$$2H_2 + O_2 \rightarrow 2H_2O$$

Groups I and II

While excess of metal in Group I produces the monoxide (e.g. Li_2O),

$$4Li + O_2 \rightarrow 2Li_2O$$

there is an increasing tendency to form a peroxide (e.g. Na_2O_2) or hyperoxides (KO_2, RbO_2, CsO_2). The use of the prefix *hyper* (Greek = over, excessive) draws attention to the unusually high molar proportion of oxygen. Heated sodium burns in oxygen with a yellow flame to leave yellowish sodium peroxide,

$$2Na + O_2 \rightarrow Na_2O_2$$

Potassium burns with a lilac flame to produce the hyperoxide,

$$K + O_2 \rightarrow KO_2$$

For lithium, action stops with the monoxide; sodium goes further with the formation of the peroxide while potassium, rubidium and caesium form the hyperoxide. All are ionic compounds.

Group II metals form the oxide when heated in oxygen. Magnesium and calcium burn with an intense white light,

$$2Mg + O_2 \rightarrow 2MgO$$

$$2Ca + O_2 \rightarrow 2CaO$$

Excess oxygen yields a mixture of oxide and peroxide with heated strontium ($SrO + SrO_2$) and the peroxide with barium (BaO_2).

Beryllium oxide is covalent; the others, ionic.

Except for beryllium, metals of both groups undergo suface oxidation in the cold when exposed to air.

Group III

Finely divided boron burns to the trioxide when heated in air or oxygen,

$$4B + 3O_2 \rightarrow 2B_2O_3$$

Aluminium undergoes surface oxidation in the air, the oxide layer protecting the metal from further action.

$$4Al + 3O_2 \rightarrow 2Al_2O_3$$

Groups IV, V and VI

In Period 2, heated charcoal burns with a shower of sparks, but nitrogen will only react reversibly with oxygen at the temperature of an electric arc. Carbon dioxide and nitrogen oxide (nitric oxide) are the respective products:

$$C + O_2 \rightarrow CO_2$$

$$N_2 + O_2 \rightleftharpoons 2NO$$

The yield of nitrogen oxide is only 1.2% by volume at 2000°C and 5.3% at 3000°C.

In Period 3, silicon burns brilliantly when heated in oxygen to form the dioxide. Phosphorus burns, either spontaneously as the white allotrope or when heated as the red, with a brilliant white flame to give trioxide and progressively pentoxide as the proportion of oxygen is increased, and heated sulphur burns with a blue flame to the dioxide,

$$Si + O_2 \rightarrow SiO_2$$

$$4P + 3O_2 \rightarrow P_4O_6$$

$$4P + 5O_2 \rightarrow P_4O_{10}$$

$$S + O_2 \rightarrow SO_2$$

All other elements are oxidized on heating. For example, tin forms the dioxide, tin(IV) oxide, when fused in air or oxygen, and lead, the monoxide, lead(II) oxide. The valency drop indicates increased metallic character for the metal of higher atomic number, in the same group,

$$Sn + O_2 \rightarrow SnO_2$$

$$2Pb + O_2 \rightarrow 2PbO$$

The halogens (Group VII) require special conditions and as the products are not important to us, the reactions are not included.

Transition and associated metals

Steel wool or iron wire, previously heated or dipped into burning sulphur, burns brilliantly with sparks to form tri-iron tetroxide,

$$3Fe + 2O_2 \rightarrow Fe_3O_4$$

On heating in air or oxygen, copper becomes

coated with a film of black copper(II) oxide. The metal shows little affinity for oxygen:

$$2Cu + O_2 \rightarrow 2CuO$$

Mercury reacts reversibly according to the temperature and the product, mercury(II) oxide, is of considerable historical interest in the researches of Scheele, Priestley and Lavoisier. The oxide is decomposed on further heating to higher temperature:

$$2Hg + O_2 \rightleftharpoons 2HgO$$

2   Combustion of compounds in oxygen

Very many substances burn in oxygen. Consider hydrides. Hydrocarbons form carbon dioxide and water;

$$C_3H_8 + 5O_2 \rightarrow 3CO_2 + 4H_2O$$

Petrol and kerosene, which consist of hydrocarbons, burn similarly. In air, some soot (carbon) may be formed due to incomplete combustion. The exhaust systems of new cars are fitted with catalytic converters to ensure maximum oxidation of the products of petrol combusion and so avoid the awful smog and other harmful effects of excessive use of motor cars in our cities. The USA has led the way with the most stringent measures, now adopted by the EEC, to combat the problem. Lean-burn engines are being developed in the UK as a fundamental approach to the problem.

Hydrogen sulphide burns in air and oxygen, with a blue flame, and ammonia in oxygen with a peach-coloured flame:

$$2H_2S + 3O_2 \rightarrow 2H_2O + 2SO_2$$

$$4NH_3 + 3O_2 \rightarrow 6H_2O + 2N_2$$

3   Addition of oxygen to oxides and hydroxides

Barium oxide forms the peroxide on heating in air or oxygen:

$$2BaO + O_2 \rightarrow 2BaO_2$$

Oxides of Group I metals, except lithium, also take up oxygen.

Carbon monoxide burns with a blue flame to form the dioxide, nitrogen oxide (nitric oxide) forms the brown dioxide on exposure to air and phosphorus trioxide takes up oxygen on heating to form the pentoxide:

$$2CO + O_2 \rightarrow 2CO_2$$

$$2NO + O_2 \rightarrow 2NO_2$$

$$P_4O_6 + 2O_2 \rightarrow P_4O_{10}$$

Sulphur dioxide reacts reversibly with oxygen over a heated catalyst of vanadium(V) oxide (pentoxide) or platinized asbestos:

$$2SO_2 + O_2 \rightleftharpoons 2SO_3$$

The hydroxides of certain transition metals in the bivalent state are oxidized in air. Manganese(II) hydroxide, covalent $Mn(OH)_2$, forms Mn_2O_3 (Mn^{III}) and MnO_2 (Mn^{IV}), the sesqui- and di-oxides respectively. Iron(II) hydroxide, $Fe(OH)_2$, forms hydrated iron(III) oxide, iron(III) hydroxide, $Fe(OH)_3$. Vanadium(II) and chromium(II) ions, V^{2+} and Cr^{2+}, are very readily oxidized as are iron(II) ions, Fe^{2+}, in neutral solution. The manganese(II) ion, Mn^{2+}, is the most stable form of combined manganese.

4   Organic oxidations

Giving one example: ethanal (acetaldehyde) vapour, mixed with air and passed over vanadium(V) oxide (pentoxide) catalyst, is oxidized to ethanoic acid (acetic acid),

$$2CH_3.C\!\!\begin{array}{c}\nearrow H \\ \searrow O\end{array} + O_2 \rightarrow 2CH_3.C\!\!\begin{array}{c}\nearrow OH \\ \searrow O\end{array}$$

A summary of methods used to prepare oxides

1   Synthesis

Direct combination of an element with oxygen requires heating or the previous ignition of the element, which may be metallic or non-metallic. Thus, lithium forms the monoxide,

$$4Li + O_2 \rightarrow 2Li_2O$$

2   The action of steam at red heat

Heated in steam, iron forms tri-iron tetroxide. Carbon monoxide is made industrially (as water gas) by the action of steam on white-hot coke. Hydrogen is also formed:

$$3Fe + 4H_2O \rightleftharpoons Fe_3O_4 + 4H_2$$

$$C + H_2O \rightleftharpoons CO + H_2$$

3 The thermal decomposition of hydroxides, carbonates and nitrates

These compounds of all but the strongest metals may be decomposed on heating. For calcium:

$$Ca(OH)_2 \rightleftharpoons CaO + H_2O$$

$$CaCO_3 \rightleftharpoons CaO + CO_2$$

$$2Ca(NO_3)_2 \rightarrow 2CaO + 4NO_2 + O_2$$

4 From other oxides

Higher oxides may decompose on heating, while some lower oxides may take up further oxygen under similar conditions. Lead(IV) oxide (dioxide) and tin(II) oxide from Group IV provide examples. The former gives litharge, the oxide of bivalent lead, while the latter forms tin(IV) oxide, the oxidation number increasing.

$$2PbO_2 \rightarrow 2PbO + O_2$$

$$2SnO + O_2 \rightarrow 2SnO_2$$

Lead is more metallic in chemical character than tin.

5 The action of nitric acid

Concentrated nitric acid acts on non-metals and weak metals to form either an oxide, or if this is soluble, the corresponding acid. Iodine forms iodic acid when boiled with very concentrated nitric acid. Iodine pentoxide is formed on heating:

$$3I_2 + 10HNO_3 \rightarrow 6HIO_3 + 10NO + 2H_2O$$

$$2HIO_3 \rightarrow I_2O_5 + H_2O$$

Photosynthesis

In the light, green plants utilize carbon dioxide, water and light energy to form plant sugars (carbohydrates) with liberation of oxygen. Chlorophyll, the green plant pigment, a complex organic compound of magnesium, is essential to this process which occurs in chloroplasts. These are corpuscles containing chlorophyll united to protoplasm. They are usually in the cell layers near the surfaces of leaves. Carbon dioxide is taken up from the air, or for water plants, from aqueous solution:

$$CO_2 + H_2O + \underset{\text{energy}}{\overset{\text{light}}{\longrightarrow}} (CH_2O)_n + O_2$$
$$\underset{(unbalanced)}{} \qquad\qquad \underset{\text{plant sugars}}{}$$

The liberated oxygen has been shown by isotopic ratio studies to come entirely from water and not from the carbon dioxide. Historically, Priestley studied the role of plants in replenishing the air. The work was extended by Ingenhousz (1730–99), who showed that plants exhale carbon dioxide, and by Jean Senebier (1742–1809).

Respiration

Plants and animals require energy for work and for chemical processes connected with the building up of tissue. The energy comes from the oxidation of suitable substances by atmospheric oxygen: this is respiration. Plants use oxygen during respiration. Fish take up oxygen in solution via their gills. In man and higher animals, oxygen is taken in during breathing, absorbed by the red blood corpuscles of the blood in the lungs and transported as an easily dissociated substance, oxyhaemoglobin. Haemoglobin is the red blood pigment and is a complex organic substance, molecular mass about 68 000. It contains iron. Air, which has been filtered, warmed and moistened in the air passages of the nose and wind-pipe, passes into the lungs, the expansion and contraction of which is controlled by the diaphragm and intercostal muscles of the chest. In man, the maximum capacity which can be breathed in is 4.0–4.5 dm³ and is closely related to the surface area of the body, about 1.5 m². This is about 2.5 dm³ in men and 2.0 dm³ in women per square metre of body surface. The lung surface area, consisting of small sacs called alveoli made up of sheets of cells (Fig. 16.1), is about fifty times the area of the body surface. The term *alveolus* refers to the individual cavity of a honeycomb structure. About 0.5 dm³ air is breathed in at rest in every breath, totalling about 5 dm³ in every minute.

Under the relatively high partial pressure of oxygen in the lungs, oxyhaemoglobin is formed.

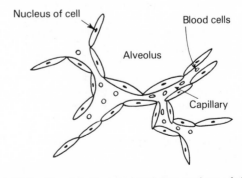

Fig. 16.1 Alveoli: in the lung, small sacs made up of sheets of cells showing blood capillaries

The blood is pumped by the heart to the tissues which are served by a fine, intricate mesh of capillaries into which this bright red blood passes. Under the low partial pressure of oxygen found in the tissues, oxyhaemoglobin dissociates into haemoglobin and oxygen. The maximum quantity required fully to saturate blood is about 19 cm^3 per $100\,cm^3$ of blood. There is about 14.4 g haemoglobin in this volume and 5.5×10^6 red cells per cubic mm in the male and 4.8×10^6 cells per cubic mm in the female. By a complex cycle of processes, oxygen is used in the generation of energy, and waste products, water and carbon dioxide accumulate. Carbon dioxide dissolves in the blood (corpuscles and plasma) and returns in the veins as carbonic acid and hydrogencarbonate ions, being subsequently discharged from the lungs via the alveoli. The acidity of the blood is controlled by a chemical buffer system.

The uses of oxygen

1   As an aid to respiration, oxygen is used in medical and dental practice. It is supplied where lungs are weak or injury makes artificial respiration impracticable and in the resuscitation of victims of coal gas poisoning and other accidents. It may be used, where appropriate, mixed with anaesthetics. It is essential for high-altitude flying.
2   Oxygen is used in the steel industry for the Bessemer, Open Hearth, Spray and L–D processes in addition to the enrichment of gas used for blast furnaces. An oxygen lance is used to tap a furnace.
3   With combustible gases, ethyne (acetylene), hydrogen, natural gas, it is used for the welding and cutting of ferrous and non-ferrous metals.
4   Chemical processes requiring oxygen include the underground gasification of coal by combustion, the oxidation of hydrocarbons to produce hydrogen/carbon monoxide mixtures for methanol production, the oxidation of ammonia to nitric acid and the oxidation of ethanal (acetaldehyde) to ethanoic acid (acetic acid).
5   Liquid oxygen is used as part of the fuel for the rocket propulsion units required for the various space programmes, and for sustaining life aboard the space stations which orbit the earth.

The allotropy of oxygen: ozone (trioxygen)

Ozone has the formula O_3. In 1785, van Marum noticed the distinctive smell of ozone associated with electrical machinery but it was 1840 before ozone was recognized as a separate chemical substance by Schönbein. It is called ozone from the Greek, *ozein* = to smell.

Ozone is formed by the action of ultra-violet light on oxygen. It has been estimated that there is a belt of ozone at 25 km height in the atmosphere. The overall quantity of ozone in the atmosphere is estimated as 2 parts per million by volume. There is international concern about the apparent depletion of the ozone layer, after the discovery of a 'hole' in the layer over Antartica. The layer protects life from the harmful effects of excessive ultraviolet radiation, which causes skin cancers. The loss of ozone is attributed to the action of (among other agents) organic compounds containing fluorine and chlorine (CFCs) used hitherto in aerosols, refrigerators etc., and manufactured because of their apparent chemical inertness. The problem is intensified over the Antarctic subcontinent where the polar vortex reduces mixing of the atmosphere and produces a critical concentration of chlorine atoms. This enables the compounds to penetrate to the stratosphere where subsequent photolytic decomposition produces chlorine atoms, which are regenerated in a chemical cycle that destroys ozone molecules.

In the ordinary atmosphere at ground level there is not more than 1.1 parts per million of ozone. Sea air contains no extra ozone whatever the brochures and publicity posters say. While minute traces destroy bacteria, ozone is otherwise poisonous.

Ozone melts at $-250°C$ and boils at $-112°C$. It forms a blue liquid which may explode spontaneously. With organic matter it explodes. Ozone is the fourth most powerful oxidizing agent, being behind fluorine, oxygen difluoride and atomic oxygen, listed in decreasing power of oxidation.

The formation of ozone from oxygen is endothermic:

$$3O_2 \rightleftharpoons 2O_3; \quad \Delta H = +285\,kJ\,mol^{-1}$$

It may be shown thermodynamically that ozone is always unstable and breaks down to oxygen. In forming ozone, oxygen exhibits monotropic allotropy or monotropy. The structure is a resonance hybrid:

The preparation of ozone

1 From oxygen using a silent electrical discharge

Oxygen is passed between electrical conductors but separated from them by the glass walls of the tube acting as an insulator (see Fig. 16.2). The conductors are connected to the secondary current of an induction coil and maintained at a high potential difference. Sparking must be avoided because ozone is decomposed rapidly as the temperature is raised. Some ozone is formed,

$$3O_2 \rightleftharpoons 2O_3$$

Two ozonizers are in common use, named after Siemens (1858) and Brodie (1872), and shown in Fig. 16.2. In Siemens' ozonizer, a slow stream of oxygen passes through the annular space between two glass tubes on the outer sides of which are tinfoil coatings between which a silent electrical discharge is maintained. In Brodie's ozonizer, the conductor is dilute sulphuric acid which is connected to the potential source by copper wires; the apparatus is necessarily mounted vertically. The mixture of ozone and oxygen, ozonized oxygen, is led away. But ozone attacks rubber, so that an inert connection must be made to the delivery tube. These days it is common practice to use ground glass jointed apparatus. The yield is said to be about 10%.

2 The electrolysis of cold dilute sulphuric acid

Ozonized oxygen is evolved at the anode when ice-cold dilute sulphuric acid is electrolysed. A high current density is arranged by using a smooth platinum wire electrode, just protruding from the end of a glass tube, into which it is sealed. A 14% yield is said to be obtained.

The reactions of ozone

Ozone is an extremely powerful oxidizing agent. It decomposes to oxygen slowly at room temperature but at an increasing rate as the temperature is raised until at 300°C the decomposition is immediate. Decomposition is catalysed by finely divided platinum, silver(I) oxide and manganese(IV) oxide (dioxide). Except in its reaction with tin(II) chloride and sulphur dioxide, which are oxidized by oxygen under suitable conditions to form tin(IV) chloride and sulphur trioxide respectively, ozone donates only one oxygen atom per molecule to the substance oxidized:

$$O_3 + X \rightarrow XO + O_2$$

but

$$3SO_2 + O_3 \rightarrow 3SO_3$$

$$3SnCl_2 + 6HCl + O_3 \rightarrow 3SnCl_4 + 3H_2O$$

Iron(II) chloride is oxidized to iron(III) chloride in the presence of hydrochloric acid,

$$2Fe^{2+} + 2H^+ + O_3 \rightarrow 2Fe^{3+} + H_2O + O_2$$

and potassium hexacyanoferrate(II) to hexacyanoferrate(III),

$$2Fe(CN)_6^{4-} + H_2O + O_3 \rightarrow 2Fe(CN)_6^{3-} + 2OH^- + O_2$$

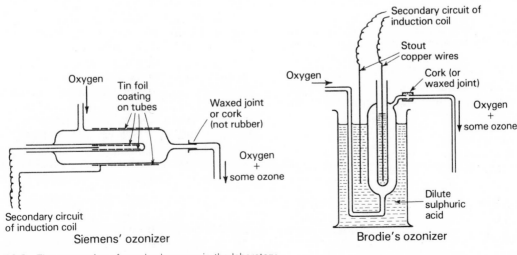

Fig. 16.2 The preparation of ozonized oxygen in the laboratory

Dark brown lead(II) sulphide is oxidized to white lead(II) sulphate, and hydrogen sulphide to sulphuric acid:

$$PbS + 4O_3 \rightarrow PbSO_4 + 4O_2$$

$$H_2S + 4O_3 \rightarrow H_2SO_4 + 4O_2$$

Potassium iodide in neutral solution liberates iodine, but this reaction is not specific for ozone, being general for oxidizing agents including the halogens, acids and oxides of nitrogen and hydrogen peroxide:

$$2I^- + H_2O + O_3 \rightarrow 2OH^- + O_2 + I_2$$

Except for fluorine, the halogens are displaced from the corresponding hydrides, HCl, HBr, HI,

$$2HBr + O_3 \rightarrow H_2O + Br_2 + O_2$$

Acidified dilute potassium permanganate and dichromate are unchanged by ozone unlike the reactions which occur with hydrogen peroxide.

Mercury loses its mobility when treated with ozone; it 'tails' and smears the glass, the globules and meniscus no longer being visible. This phenomenon is probably due to superficial oxidation affecting the surface tension. It may be demonstrated by shaking globules of mercury with ozonized oxygen in a corked flask. On adding water, the film of mercury gives way to globules again. Silver foil is attacked by ozone with formation of a brown oxide layer. The tarnishing action of ozone on these two metals distinguishes the gas from pure oxygen.

Of great value to the organic chemist is the oxidation of unsaturated compounds by ozone. Unstable ozonides are formed in anhydrous solvents. Hydrolysis with acid yields aldehydes and ketones. If these are identified, an attempt can be made to picture the original molecule. Consider the normal butenes:

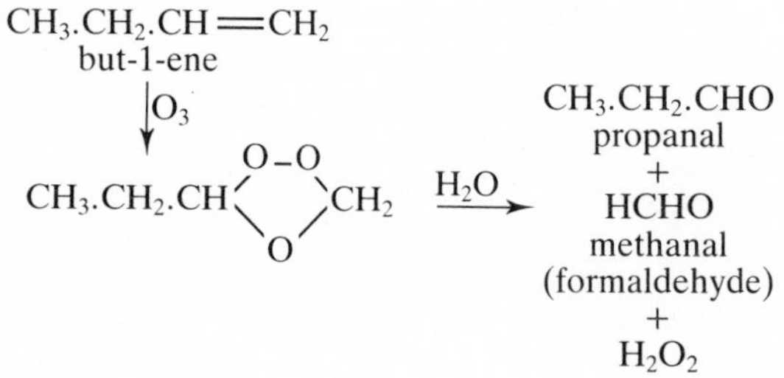

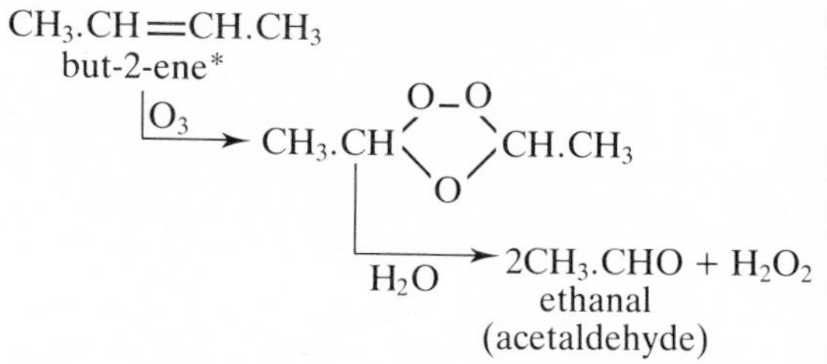

The $\rangle C = C \langle$ group is replaced by $\rangle C = O \; O = C \langle$. Oil of turpentine contains two unsaturated hydrocarbons, α-pinene and β-pinene, and may be used to absorb ozone. Cinnamon oil may also be used.

The formula of ozone

Ozone is prepared from oxygen alone and on thermal decomposition, it yields oxygen only. Its formula may be written O_n.

On passing ozonized oxygen through a tube cooled in liquid air, liquefaction occurs. Liquid ozone has been separated by fractional distillation. It is dark blue. Careful evaporation gives gaseous ozone and a Dumas determination on the small scale shows a relative density corresponding to O_3. This was done by Riesenfeld and Schwab in 1922. But the explosive nature of liquid ozone poses special difficulties.

During the period 1866-8, Soret determined the constitution of ozone. In the first series of experiments, he confined equal volumes of the same sample of ozonized oxygen over water. Oil of turpentine was introduced into one sample whereupon ozone reacted and was absorbed, dense fumes being produced initially. The contraction was measured. The other sample was heated, ozone forming oxygen with an increase in volume. This was measured under the same conditions as the first contraction. It was one-half of the contraction. Average values obtained by Soret for $250 \, cm^3$ of ozonized oxygen were a contraction of $6.26 \, cm^3$ and an expansion of $3.26 \, cm^3$. Thus, 1 volume of ozone yields $1\frac{1}{2}$ volumes of oxygen measured under the same conditions of temperature and pressure. By Avogadro's Hypothesis,

$$O_n \rightarrow \tfrac{3}{2}O_2$$

and by inspection ozone is O_3.

In a second series of experiments, Soret measured the relative rates of diffusion of ozone from ozonized oxygen into pure oxygen and chlorine into oxygen from a mixture of chlorine and oxygen. The volume of gas which remained and the volume which had diffused in a given time were determined chemically. For ozone:

Relative rate of diffusion

$$= \frac{\text{Volume of ozone which diffused}}{\text{Total volume of ozone}}$$

Similar determinations for chlorine and carbon dioxide were made, the former having a molecular weight larger than ozone and the latter, a little smaller. The relative rates were:

Ozone : chlorine : carbon dioxide
$$= 0.271 : 0.227 : 0.290$$

By Graham's Law of Diffusion,

$$\frac{\text{Rate of diffusion of chlorine}}{\text{Rate of diffusion of ozone}}$$

$$= \sqrt{\frac{\text{Relative density of ozone, } O_n}{\text{Relative density of chlorine, } Cl_2}}$$

$$= \sqrt{\frac{\text{Molecular mass of ozone}}{\text{Molecular mass of chlorine}}}$$

$$\therefore \frac{0.227}{0.271} = \sqrt{\frac{16n}{71}}$$

Whence $n = 3.11$ or 3 to the nearest integer.

$$\therefore \text{ ozone is } O_3$$

For carbon dioxide,

$$\frac{0.290}{0.271} = \sqrt{\frac{16n}{44}}$$

Whence $n = 3.15$ or 3 to the nearest integer and ozone is O_3.

The detection and estimation of ozone

The distinctive odour of ozone is not enough to identify it in small quantities. In any case the sense of smell alone can be misleading. The tailing of mercury and the blackening of clean silver together with the liberation of iodine from potassium iodide are characteristic of ozone. However, the sensitive starch–iodide paper test is also positive for other oxidizing agents, halogens, oxides of nitrogen (and

nitrous acid) and hydrogen peroxide. Of these, only ozone and hydrogen peroxide vapour are destroyed by passage along a heated tube. The problem is resolved into the distinction between hydrogen peroxide and ozone in small quantities if the mercury and silver tests are not suitable. Test results are indicated in Table 16.3. Two organic spot reagents are included; they are alcoholic solutions of tetramethyl-parapara'-diaminodiphenylmethane ('tetramethyl base') and benzidine. The detail of other tests is given in practical texts.

Table 16.3 A comparison of ozone and hydrogen peroxide

	Ozone	Hydrogen peroxide
Dilute acidified potassium permanganate	Unchanged	Decolorized
Dilute acidified potassium dichromate and ether	Unchanged	Blue ethereal layer
Tetramethyl base	Violet	Unchanged
Benzidine	Brown	Unchanged
Titanium(IV) oxide (dioxide) in concentrated sulphuric acid	Unchanged	Yellow

Quantitative analysis of ozonized oxygen may be done by absorbing ozone in oil of turpentine. Otherwise, iodine may be liberated from potassium iodide in neutral or somewhat alkaline solution,

$$2KI + O_3 + H_2O \rightarrow 2KOH + O_2 + I_2$$

$$2KOH + I_2 \rightarrow KI + KIO + H_2O$$

Acidification, not done at first because direct oxidation by oxygen must be avoided, liberates iodine which may be titrated against standard sodium thiosulphate solution with starch as indicator,

$$KI + KIO + 2HCl \rightarrow 2KCl + H_2O + I_2$$

Hydrogen peroxide

Hydrogen peroxide was discovered in 1818 by Thénard. It has the formula H_2O_2. Traces are said to be found associated with water and snow, and in the atmosphere. Hydrogen peroxide is used as part of a defence mechanism in plants and animals. Within minutes injured plant cells attack all sorts of chemical intruders (microbes, insects, etc.) by use of oxygen from hydrogen peroxide, liberated by the action of enzymes. This 'nascent' oxygen destroys many organic chemicals, as is seen anyway in the use of hydrogen peroxide in bleaching textiles and wood pulp, and in proprietory lavatory cleaners, marketed as an alternative to hypochlorite solution. In animals white blood cells destroy the protective membranes of invading cells with hydrogen peroxide. Thus it occurs widely in the natural world.

When a hydrogen flame impinges on ice, traces of hydrogen peroxide are readily detected by the yellow coloration given with titanium(IV) oxide (dioxide) reagent. It is also formed in traces by the action of ultraviolet light or strong sunlight on water containing dissolved oxygen and directly from the elements in the presence of palladium black as catalyst. Hydrogen peroxide is prepared in the laboratory by the action of acids on peroxides and commercially by the hydrolysis of peroxodisulphuric acid, produced from fairly concentrated sulphuric acid by electrolysis, and by auto-oxidation.

Hydrogen peroxide is a fairly viscous liquid, density $1.46\,g\,cm^{-3}$, with an odour reminiscent of nitric acid, clear and colourless in small quantities but bluish in bulk. It blisters the skin. Hydrogen peroxide is miscible with water and is usually met as a colourless, odourless solution with a sharp metallic taste. The pure liquid freezes at $-1.7°C$ and explodes when heated to boiling at atmospheric pressure. At a pressure of $9.1\,kPa$ (68 mmHg), hydrogen peroxide boils at $85°C$, and at $3.5\,kPa$ (26 mmHg), $69°C$. It has a higher dielectric constant than water, 89 at $0°C$, is a good ionizing solvent and is associated. It dissolves in alcohols and ethers, both oxygen-bearing organic solvents, and in basic liquids, such as aminobenzene (aniline). Hydrogen peroxide is a very weak acid.

The thermal decomposition of hydrogen peroxide is exothermic:

$$2H_2O_2 \rightarrow 2H_2O + O_2; \quad \Delta H = -2 \times 98\,kJ\,mol^{-1}$$

The pure liquid is fairly stable if stored in a smooth, clean, dark bottle in a cool place. Stabilized 27.5% and 35% solutions should not lose more than 1% of their strength in a year. Storage vessels for concentrated solutions, pipes, etc., should be made from 99.5% aluminium. Bleaching vessels for alkaline hydrogen peroxide are made of stainless steel. Polythene and polyvinyl chloride gaskets and flexible hose are used in handling the liquid. Decomposition occurs readily on rough edges, scratch marks and on dust particles. Pure hydrogen peroxide decomposes explosively when heterogeneous catalysts such as finely divided, or colloidal platinum, silver and gold are introduced. The ions of iron, copper and manganese catalyse decomposition. Rapid decomposition may be brought about by as little as 1 part per million of iron or 0.05 parts per million of copper. Stabilizing agents may be added. These include alcohol, glycerine and N-phenylethanamide (acetanilide). Concentrations greater than 60% will often inflame with organic matter.

Solutions of hydrogen peroxide are graded and sold by their 'volume strength'. The volume strength of a solution of hydrogen peroxide is measured by the number of volumes of oxygen liberated under standard conditions (273 K, 101 kPa) by one volume of solution according to the equation:

$$2H_2O_2 \rightarrow 2H_2O + O_2$$

$2 \times 34\,g$ hydrogen peroxide yields $22.4\,dm^3$ of oxygen at STP. If 68 g hydrogen peroxide is made up to $1\,dm^3$ of aqueous solution, the volume strength is 22.4. Typical strengths are '10 vol' and '20 vol'.

10 volume hydrogen peroxide contains

$$\frac{68}{22.4} \times \frac{10}{1}\,g\,dm^{-3}\ \text{hydrogen peroxide, or 30 g dm}^{-3}.$$

10 volume hydrogen peroxide is a 3% solution and 20 volume, 6%, by volume, if the density of each solution is assumed to be unity. Table 16.4 shows the relationship between the various ways of expressing the strength of solutions of hydrogen peroxide. It includes the 'volumetric strength,' now rarely used, but derived from the analytical method used for the determination of weak solutions: the volume/cm^3 of $0.02\,M$ potassium permanganate solution equivalent to $2\,cm^3$ of hydrogen peroxide.

Table 16.4 The relationship between the various ways in which the strengths of hydrogen peroxide solutions are expressed

% H$_2$O$_2$ by mass	% H$_2$O$_2$ by volume	Volume strength	Volumetric strength, /cm^{-3}	% Active oxygen by mass
3.0	3.0	10	36	1.41
5.9	6.0	20*	72	2.8
27.5*	30.3	100	356	12.9
30*	33.3	110	392	14.1
35*	39.7	130	466	16.5
50*	59.8	197	703	23.6
90*	125.4	413	1474	42.3

*Usual trade description (Laporte Chemicals Ltd)

The formula and structure of hydrogen peroxide

The determination of relative molecular mass by the freezing-point method gives the value 34.

Two moles of pure hydrogen peroxide on decomposition yield two moles of water and one mole of oxygen. This supports the molecular formula H_2O_2. The decomposition of very nearly pure hydrogen peroxide over mercury by heat and the discovery of the catalytic action of manganese(IV) oxide (dioxide) were achieved by Thénard.

Investigations based on spectroscopy and crystal structure in addition to knowledge of organic peroxide reactions leads to the formulation given overleaf. The H—O bonds are at an angle of 94° to each other and at 97° to the O—O bond which forms the axis of the molecule. The molecule is thus askew with H—O bonds roughly at right angles to each other.

The structure of hydrogen peroxide is probably

$$
\begin{array}{c}
H-O \\
| \\
O-H
\end{array}
$$

There is some evidence for the existence of another form, but the matter is still in dispute:

$$
\begin{array}{c}
H \\ \diagdown \\[-4pt] \overset{+}{} \quad \overset{-}{} \\[-6pt] O-O \\[-4pt] \diagup \\ H
\end{array}
$$

Interconversion could occur through ionisation:

$$
H-O-O-H \rightleftharpoons H^+ + H_2O^- \rightleftharpoons \begin{array}{c} H \\ \diagdown \\ \overset{+}{O}-\overset{-}{O} \\ \diagup \\ H \end{array}
$$

The preparation of hydrogen peroxide

Hydrogen peroxide is formed by the action of dilute acids on peroxides, usually either barium or sodium peroxide, essentially:

$$2H^+ + O_2^{2-} \rightarrow H_2O_2$$

A dilute solution is obtained.

1   From barium peroxide

Barium peroxide reacts with sulphuric acid, precipitating barium sulphate. However, commercial barium peroxide is soon coated with insoluble barium sulphate and action ceases. A porous, hydrated form is required.

Commercial barium peroxide is gradually added to a slight excess of ice-cold dilute hydrochloric acid with stirring until it has dissolved. Addition of the equivalent quantity of dilute sodium hydroxide solution, enough to have neutralized the acid taken, results in the precipitation of hydrated barium peroxide, $BaO_2.8H_2O$, which is collected by suction filtration using a Buchner filter funnel:

$$BaO_2 + 2H^+ \rightarrow Ba^{2+} + H_2O_2$$

$$H_2O_2 + Ba^{2+} + 2OH^- + 6H_2O \rightarrow BaO_2.8H_2O$$

The freshly prepared hydrated barium peroxide is added to ice-cold dilute sulphuric acid. If a slight excess of acid is used, it may be removed by addition of precipitated chalk, or barium carbonate. A solution of hydrogen peroxide remains after filtration:

$$BaO_2.8H_2O + H_2SO_4 \rightarrow BaSO_4 + H_2O_2 + 8H_2O$$

2   From sodium peroxide

Sodium peroxide is added to a slight excess of ice-cold dilute sulphuric acid, a little at a time with constant stirring. The decahydrate of sodium sulphate, $Na_2SO_4.10H_2O$, crystallizes out, accounting for some of the water,

$$Na_2O_2 + H_2SO_4 \rightarrow Na_2SO_4 + H_2O_2$$

Alternatively, the calculated quantity of either phosphoric acid or sodium dihydrogenphosphate may be used and partial concentration effected by the crystallization of disodium hydrogenphosphate with twelve molecules of water of crystallization, $Na_2HPO_4.12H_2O$,

$$Na_2O_2 + H_3PO_4 \rightarrow Na_2HPO_4 + H_2O_2$$

$$Na_2O_2 + 2NaH_2PO_4 \rightarrow 2Na_2HPO_4 + H_2O_2$$

The concentration of solutions of hydrogen peroxide

Hydrogen peroxide as ordinarily prepared is readily decomposed on heating, the process being catalysed by rough edges, dust and other surface catalysts. Concentration may be started by careful evaporation in a smooth porcelain or platinum dish over a water bath to about 60–70°C when effervescence will be observed. Pure hydrogen peroxide has an estimated boiling-point of 144°C at atmospheric pressure, being less volatile than water. A

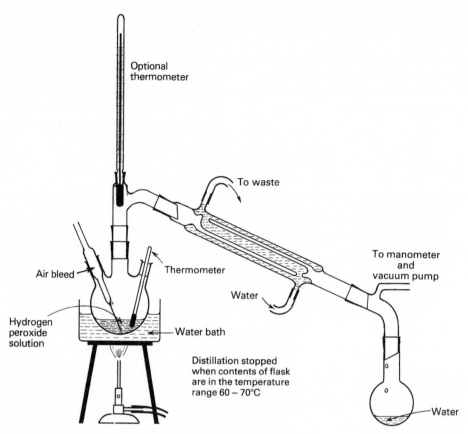

Fig. 16.3 Concentration of hydrogen peroxide solution under reduced pressure: vacuum distillation of water

solution of about 45% concentration is obtained by evaporation in the way described. Unfortunately this method leaves impurities in correspondingly increased concentration, so that the concentrate is more liable to catalytic decomposition. Under reduced pressure, using the apparatus shown in Fig. 16.3, water is distilled off at 35–40°C under a pressure of about 2 kPa (15 mmHg). The boiling-point of the contents of the flask increases as concentration proceeds. Distillation is stopped in the range 60–70°C.

Further concentration may be achieved by evaporation *in vacuo* over concentrated sulphuric acid in a vacuum desiccator as shown in Fig. 6.2. A dish containing the already fairly concentrated solution of hydrogen peroxide rests over concentrated sulphuric acid on a perforated support. The lid and tap are well greased and the desiccator exhausted. Water vapour is absorbed by the sulphuric acid.

Concentrations of more than 91% may be achieved by recrystallization using a freezing mixture of solid carbon dioxide and methanol for cooling.

The commercial production of hydrogen peroxide

Until recently most of the hydrogen peroxide made commercially was electrolytic in origin.

An intermediate, peroxodisulphuric acid, may be prepared by the electrolysis of cold 50% sulphuric acid with a high current density. The anode is bright platinum to keep the oxygen overpotential high and thereby avoid discharge of that gas. The cathode is of graphite or lead. While there is some doubt about the mechanism, peroxodisulphuric acid is formed at the anode. The simplest explanation involves hydrogensulphate ions,

$$HSO_4^- \rightarrow HSO_4 + e^- \quad 2HSO_4 \rightarrow H_2S_2O_8$$

The product is hydrolysed by distillation with sulphuric acid under reduced pressure to give about 30% hydrogen peroxide which may be further concentrated by fractionation:

$$\begin{array}{l} \text{O–SO}_2\text{OH} \\ | \\ \text{O–SO}_2\text{OH} \end{array} + \begin{array}{l} \text{H}_2\text{O} \\ \\ \text{H}_2\text{O} \end{array} \longrightarrow \begin{array}{l} \text{O–H} \\ | \\ \text{O–H} \end{array} + 2\text{H}_2\text{SO}_4$$

In one process, ammonium peroxodisulphate (persulphate), $(\text{NH}_4)_2\text{S}_2\text{O}_8$, is formed by the electrolysis of ammonium sulphate and sulphuric acid and then hydrolysed to ammonium sulphate, sulphuric acid and hydrogen peroxide. The electrolytic method has now largely been superseded by the 'AO' or auto-oxidation process.

The AO process produces hydrogen peroxide from hydrogen and oxygen in the presence of a quinone as catalyst. This yellowish organic solid is converted into a quinol in the course of the reactions. A mixture of solvents is required. The solution is relatively expensive. The plant is constructed out of 99.5% aluminium and certain aluminium alloys, with stainless steel valves and pumps, etc. The first stage involves the hydrogenation of the quinone to the quinol in the presence of a special catalyst, which is then removed by a highly efficient filtration process, to avoid later catalytic decomposition of hydrogen peroxide. The second stage involves auto-oxidation of the quinol in the presence of molecular oxygen dissolved in the solution, achieved by passing the solution and air up a packed column. The reactions, using ordinary quinone as the example, because the name of the quinone used has not been disclosed, may be written:

quinone   quinol

The hydrogen peroxide is extracted by passing demineralized water down the extraction columns in a current counter to the flow of solution. After purification by treatment with various organic solutions, hydrogen peroxide of 18% strength is drawn off. Much of the product is concentrated to high-test product, HTP hydrogen peroxide.

The reactions of hydrogen peroxide solution

The stability of hydrogen peroxide solutions has been discussed already. High-purity hydrogen peroxide has an excellent stability especially if small amounts (less than 0.035%) of organic and/or inorganic stabilizers are added. In addition the pH is adjusted to the range 2–3. The pH of pure aqueous 35% hydrogen peroxide by weight is 5. If the pH is increased, i.e. in alkaline solution, hydrogen peroxide becomes much less stable, decomposition being rapid at pH = 10. Unstabilized hydrogen peroxide, especially in the presence of traces of certain metals, decomposes to water and oxygen:

$$2\text{H}_2\text{O}_2 \rightarrow 2\text{H}_2\text{O} + \text{O}_2$$

In its reactions hydrogen peroxide may be considered to assume three rôles:
(a)   oxidizing agent
(b)   reducing agent, and
(c)   acid, forming salts called peroxides.

1   Hydrogen peroxide as an oxidizing agent

In oxidations, the ion-electron equation of hydrogen peroxide as the oxidant may be expressed:

$$\text{H}_2\text{O}_2 + 2\text{H}^+ + 2\text{e}^- \rightleftharpoons 2\text{H}_2\text{O}$$

Oxidation by hydrogen peroxide is therefore promoted by the presence of acids.

Potassium hexacyanoferrate(II) (ferrocyanide) solution in the presence of hydrochloric acid is oxidized to the hexacyanoferrate(III) (ferricyanide)

$$2\text{Fe(CN)}_6{}^{4-} + \text{H}_2\text{O}_2 + 2\text{H}^+ \rightarrow 2\text{Fe(CN)}_6{}^{3-} + 2\text{H}_2\text{O}$$

Iron(II) sulphate solution containing dilute sulphuric acid is oxidized to iron(III) sulphate solution,

$$2\text{Fe}^{2+} + \text{H}_2\text{O}_2 + 2\text{H}^+ \rightarrow 2\text{Fe}^{3+} + 2\text{H}_2\text{O}$$

Potassium iodide solution, acidified with hydrochloric acid, is oxidized to iodine and potassium chloride,

$$2\text{I}^- + \text{H}_2\text{O}_2 + 2\text{H}^+ \rightarrow \text{I}_2 + 2\text{H}_2\text{O}$$

Black lead(II) sulphide is oxidized to the white lead(II) sulphate. Both are insoluble. This process may be used in cleaning oil paintings where the white pigment, white lead, $2\text{PbCO}_3 . \text{Pb(OH)}_2$, has been darkened by hydrogen sulphide in the air to form lead(II) sulphide,

$$PbS + 4H_2O_2 \rightarrow PbSO_4 + 4H_2O$$

Sulphur dioxide, bubbled into hydrogen peroxide, is oxidized to sulphuric acid, sulphites are oxidized to sulphates, nitrites to nitrates, and arsenites to arsenates,

$$SO_2 + H_2O_2 \rightarrow H_2SO_4$$

$$SO_3^{2-} + H_2O_2 \rightarrow H_2O + SO_4^{2-}$$

$$NO_2^- + H_2O_2 \rightarrow H_2O + NO_3^-$$

$$AsO_3^{3-} + H_2O_2 \rightarrow H_2O + AsO_4^{3-}$$

Hydrogen peroxide is used in its oxidizing capacity as a mild bleaching agent for wool, hair, teeth, straw, silk, feathers and increasingly, for wood-pulp, etc., in the paper industry. It is the standard bleaching agent for both woollen and union blankets. Medicinally, hydrogen peroxide is used as a mild antiseptic.

2   Hydrogen peroxide as a reducing agent

The ion-electron equation for hydrogen peroxide as a reductant is

$$H_2O_2 + 2OH^- - 2e^- \rightleftharpoons 2H_2O + O_2$$

Alkaline conditions favour this action. In the presence of potassium hydroxide, aqueous potassium hexacyanoferrate(III) (ferricyanide) is reduced to a solution of the hexacyanoferrate(II) (ferrocyanide):

$$2Fe(CN)_6^{3-} + 2OH^- + H_2O_2 \rightarrow$$
$$2Fe(CN)_6^{4-} + 2H_2O + O_2$$

Iron(III) salts are reduced when alkali and hydrogen peroxide are added:

$$2Fe^{3+} + H_2O_2 + 2OH^- \rightarrow 2Fe^{2+} + 2H_2O + O_2$$

Compare this with the reaction under acidic conditions. Potassium permanganate (tetraoxomanganate(VII)), a very powerful oxidant, oxidizes hydrogen peroxide in acid solution: using sulphuric acid, potassium and manganese(II) sulphates are formed in solution, with oxygen evolution,

$$5H_2O_2 + 2MnO_4^- + 6H^+ \rightarrow$$
$$2Mn^{2+} + 8H_2O + 5O_2$$

Manganese(IV) oxide (dioxide) in the presence of sulphuric acid reacts with hydrogen peroxide to give manganese(II) sulphate and oxygen.

$$MnO_2 + 2H^+ + H_2O_2 \rightarrow Mn^{2+} + 2H_2O + O_2$$

However, under alkaline conditions the reverse happens,

$$Mn^{2+} + H_2O_2 + 2OH^- \rightarrow MnO_2(s) + 2H_2O$$

Chlorine is reduced to hydrochloric acid, ozone to oxygen and the oxides of silver and gold to the metals by hydrogen peroxide in alkaline solution:

$$H_2O_2 + Cl_2 \rightarrow 2HCl + O_2$$

$$H_2O_2 + O_3 \rightarrow H_2O + 2O_2$$

$$H_2O_2 + Ag_2O \rightarrow 2Ag + H_2O + O_2$$

$$3H_2O_2 + Au_2O_3 \rightarrow 2Au + 3H_2O + 3O_2$$

3   Hydrogen peroxide as an acid

Addition of hydrogen peroxide to a saturated solution of barium hydroxide precipitates hydrated barium peroxide, a salt,

$$Ba(OH)_2 + H_2O_2 + 6H_2O \rightarrow BaO_2 . 8H_2O$$

Pure hydrogen peroxide turns litmus solution red but as an acid it is so weak that dilute solutions do not act on indicators,

$$H_2O + H_2O_2 \rightleftharpoons H_3O^+ + HO_2^-$$

Dropwise addition of sodium carbonate solution to hydrogen peroxide solution leads to an evolution of carbon dioxide and the formation of sodium peroxide, Na_2O_2:

$$CO_3^{2-} + H_2O_2 \rightarrow CO_2 + O_2^{2-} + H_2O$$

Addition of hydrogen peroxide to sodium carbonate leads to the evolution of oxygen, because the decomposition of hydrogen peroxide is catalysed by the alkaline conditions.

The detection and estimation of hydrogen peroxide

Hydrogen peroxide reacts with potassium dichromate (μ-oxobistrioxochromate(VI)) solution acidified with sulphuric acid, and in the presence of ether imparts a blue coloration to the ether. This test is very sensitive and may be used to detect chromates (which on acidification produce dichromates) and dichromates. Covalent chromium peroxide, CrO_5, is formed. It is relatively stable in ether. Otherwise, it is rapidly decomposed:

$$4H_2O_2 + Cr_2O_7^{2-} + 2H^+ \rightarrow 2CrO_5 + 5H_2O$$

$$4CrO_5 + 12H^+ \rightarrow$$
$$4Cr^{3+} + 6H_2O + 7O_2$$

A colourless solution of titanium(IV) oxide (dioxide) in fairly concentrated sulphuric acid is converted into the peroxide, TiO_3, by hydrogen peroxide. Even 1 part of hydrogen peroxide in 180 000 parts of solution gives a light yellow coloration while with increasing concentration the solution deepens in colour and is finally orange-yellow. (This is the most sensitive test.)

Hydrogen peroxide may be estimated volumetrically by direct titration against standard potassium permanganate solution or cerium(IV) sulphate, both in the presence of sulphuric acid:

$$MnO_4^- + 8H^+ + 5e^- \rightleftharpoons Mn^{2+} + 4H_2O$$

$$Ce^{4+} + e^- \rightleftharpoons Ce^{3+}$$

$$H_2O_2 + 2OH^- - 2e^- \rightleftharpoons 2H_2O + O_2$$

(This equation could be written:

$$H_2O_2 - 2e^- \rightleftharpoons 2H^+ + O_2)$$

$$\therefore \quad \frac{KMnO_4}{5} \equiv \frac{H_2O_2}{2} \equiv \frac{Ce(SO_4)_2}{1}$$

Hydrogen peroxide is diluted accurately to give reasonable titration readings. Permanganate is run into acidified hydrogen peroxide until a slight permanent pink coloration is obtained:

$$5H_2O_2 + 2MnO_4^- + 6H^+ \rightarrow$$
$$2Mn^{2+} + 8H_2O + 5O_2$$

Hydrogen peroxide is run into yellow cerium(IV) sulphate solution. Phenylanthranilic acid is added near the end-point, imparting a mauve coloration, discharged at equivalence:

$$2Ce^{4+} + H_2O_2 \rightarrow 2Ce^{3+} + 2H^+ + O_2$$

In another method, hydrogen peroxide liberates iodine from warm acidified potassium iodide solution,

$$2I^- + H_2O_2 + 2H^+ \rightarrow I_2 + 2H_2O$$

To increase the speed of precipitation, a catalyst of molybdic acid or ammonium molybdate is added, and iodine titrated immediately against standard sodium thiosulphate (trioxothiosulphate) solution,

$$I_2 + 2S_2O_3^{2-} \rightarrow 2I^- + S_4O_6^{2-}$$

The concentration/g dm^{-3} of hydrogen peroxide solution may be converted into the *volume strength*, which is the volume of oxygen generated under standard conditions (STP: 273 K, 101 kPa)

by one volume of hydrogen peroxide solution when decomposition occurs according to the equation:

$$2H_2O_2 \rightarrow 2H_2O + O_2$$

Where (2×34) g hydrogen peroxide liberates 22.4 dm^3 of oxygen under STP which is a step from the answer required.

Further, 2 mol hydrogen peroxide liberates 22.4 dm^3 of oxygen under STP, and 1 mol hydrogen peroxide liberates 11.2 dm^3 of oxygen.

Hence volume strength $= 11.2 \times$ Molarity.

The uses of hydrogen peroxide

1   As a mild oxidising agent, hydrogen peroxide has been used to restore oil paintings and other works of art, where lead pigments have been darkened to the sulphide by atmospheric hydrogen sulphide. Lead white pigment (basic lead carbonate $2PbCO_3 . Pb(OH)_2$) often used in Italian renaissance drawings, which has gone black may be restored by conversion to the white lead sulphate using ether containing a little hydrogen peroxide, the two immiscible liquids having been well shaken. Similarly, red lead pigments, which have gone brown or blotchy, are restored, the resulting film of lead being so thin that its whiteness is not seen. Its use as a mild and safe bleach for natural products has been described.

2   Reducing properties make hydrogen peroxide suitable as an antichlor to remove traces of chlorine or hypochlorite (chlorate(I)) remaining after strong bleaching.

3   Hydrogen peroxide is used as an oxidizing agent in the chemical and allied industries. It is especially valuable because the products of decomposition are water and oxygen only. The oxidation of iron(II) compounds, phosphorous acid, arsenious acid, and the oxidation of selenium to selenic acid and tellurium to telluric acid are accomplished with hydrogen peroxide.

4   Hydrogen peroxide is used to prepare the peroxides of calcium, strontium, magnesium and zinc on the industrial scale and in the manufacture of peroxo compounds, including sodium peroxoborate and sodium peroxocarbonate, which find widespread application in household detergents as mild and safe bleaching agents.

5   Hydrogen peroxide is used in large-scale synthesis of organic compounds usually through the intermediary of organic peracids, particularly perethanoic (peracetic) acid, made *in situ* by the action of hydrogen peroxide on an ethanoic acid solution of the compound to be oxidized.

6 The ease of decomposition of hydrogen peroxide has led to its application in the production of sponge rubber and lightweight concrete where it acts as an aerating agent.

7 Hydrogen peroxide is a useful source of oxygen and has been used with concentrated aqueous sodium permanganate as a propellant for submarine torpedoes and rockets, the mixture generating oxygen and steam.

8 Liquid hydrogen peroxide has been used with liquid hydrazine as a rocket fuel. It has been used to fuel turbines in submarines, notably U-boats in World War II.

9 In the laboratory, hydrogen peroxide is used in qualitative analysis for the oxidation of chromium hydroxide and for taking manganese(IV) oxide (dioxide) and higher oxides of lead into solution.

17

Group I: the alkali metals

Lithium, sodium, potassium, rubidium and caesium

3
Li
2, 1
11
Na
2, 8, 1
19
K
2, 8, 8, 1
37
Rb
2, 8, 18, 8, 1
55
Cs
2, 8, 18, 18, 8, 1
87
Fr
2, 8, 18, 32, 18, 8, 1

The alkali metals are a distinctive family of the chemically most reactive metals, showing a progressive increase in electropositive character with increasing atomic number.

The overall picture shows the steady gradation of properties of very similar elements and their compounds. Sodium and potassium compounds in general use are described in detail. The radioactive element, francium, will not be mentioned further but rubidium and caesium are included in a general way to emphasize the close family similarity. The elements of the second period of the Periodic Classification comprise the first members, except for the noble (inert) gases, of the periodic groups and they show, in varying degrees, anomalies in behaviour although clearly belonging to their assigned family groups. Compounds of lithium are included to illustrate this point.

The alkali metals are all soft, white and lustrous but rapidly tarnish in air. They are usually stored under oil or solvent naphtha. With increasing atomic number, the metals become softer, easier to fuse and more volatile. Rubidium and caesium are denser than water while the others are less dense. Lithium has the lowest density and highest specific heat capacity of any solid element at room temperature. With increase in atomic number, the general reactivity increases markedly, lithium being somewhat apart from the others in this respect. This is illustrated by the action on water which ranges in intensity from the quiet evolution of hydrogen with lithium to the violence of caesium. In the original Mendeléeff Periodic Table, lithium and sodium were the typical elements and the family branched into subgroups: Ia comprised potassium, rubidium and caesium, and Ib, copper, silver and gold. The only resemblance between the subgroups lies in the numerical value of the group valency, namely, one.

In 1807, Sir Humphrey Davy announced the isolation of potassium and sodium by the electrolysis of fused (caustic) potash and fused (caustic) soda respectively. As a source of direct current electricity he used a Voltaic Pile, the principle of which had been discovered by Volta in 1800. Solid potash becomes moist in air and gives a conducting surface. The solid was fused by passing an electric current, a platinum spoon acting both as a container and negative pole with a platinum wire as the positive pole. Vivid action occurred and small globules, resembling quicksilver (mercury), appeared on the spoon, some of which burnt with a flash and explosion while others were protected by fresh tarnishing. Davy isolated sodium and other metals using the same principle. From a study of their characteristics he concluded that the newly discovered products were metallic elements and accordingly named them after the parent alkalis with the accepted Latin suffix to denote the metallic condition. Lithium was detected by Arfvedsen in 1817 combined in certain minerals (Greek, *litheous* = stony). It was isolated in 1855 by Bunsen and Mathiessen using the electrolysis of lithium bromide containing lithium chloride. Rubidium and caesium were detected in 1860 by the spectroscopic examination of mineral water at Dürkheim by Bunsen and Kirchhoff; two lines in the dark red region of the spectrum were

Table 17.1 Group I (alkali metals): physical data   s block elements* (ns¹) $n \neq 1$

	Li	Na	K	Rb	Cs
Atomic number	3	11	19	37	55
Electronic configuration	2, 1	2, 8, 1	2, 8, 8, 1	2, 8, 18, 8, 1	2, 8, 18, 18, 8, 1
[Core]outer electrons	$[He](2s)^1$	$[Ne](3s)^1$	$[Ar](4s)^1$	$[Kr](5s)^1$	$[Xe](6s)^1$
Relative atomic mass	6.939	22.9898	39.102	85.47	132.905
Metallic radius/nm	0.155	0.191	0.235	0.248	0.267
Atomic radius/nm	0.134	0.154	0.196	0.216	0.235
Ionic radius/nm	0.060	0.095	0.133	0.148	0.169
Atomic volume/cm³ mol⁻¹**	12.97	23.68	45.36	55.80	69.95
Standard electrode potential (H Scale: 298 K)/V	−3.05	−2.71	−2.93	−2.99	−3.02
Boiling-point/°C***	1336	882.9	757.5	700	670
Melting-point/°C***	179	97.9	63.5	39.0	28.45
Electronegativity	1.0	0.9	0.8	0.8	0.7
Density/g cm⁻³ (at 293 K)	0.535	0.971	0.862	1.532	1.90
Co-ordination number	4, 6	6	6	6	6, 8
Abundance (% by mass)	7×10^3	2.8	2.6	3.1×10^{-2}	7×10^{-4}

*Because hydrogen $(1s)^1$ is unique, there should be the restriction that $n \neq 1$, where n is the appropriate (principal) quantum number, here and in later tables
**Traditional term. It is the volume of one mole of atoms as a solid element
***°C (not K) The author (at least) finds it easier to relate (say) 39°C, the melting point of rubidium, rather than (273 + 39) K, to everyday experience

Electronic configurations

Inert gases	Ions		
2 **He** 2	3 **Li⁺** 2	Li^+F^- Na^+Cl^- K^+Br^-	lithium fluoride sodium chloride potassium bromide
10 **Ne** 2, 8	11 **Na⁺** 2, 8	Rb^+I^- Cs^+H^- $(Li^+)_2O^{2-}$	rubidium iodide caesium hydride lithium monoxide
18 **Ar** 2, 8, 8	19 **K⁺** 2, 8, 8,	$(Na^+)_2(O_2^{2-})$ $K^+(O_2^-)$ $(Rb^+)_2S^{2-}$	sodium peroxide potassium hyperoxide rubidium sulphide
36 **Kr** 2, 8, 18, 8	37 **Rb⁺** 2, 8, 18, 8	Cs^+OH^- $Li^+ClO_4^-$ $Na^+NO_3^-$	caesium hydroxide *lithium perchlorate sodium nitrate
54 **Xe** 2, 8, 18, 18, 8	55 **Cs⁺** 2, 8, 18, 18, 8	$K^+IO_3^-$ $(Rb^+)_2SO_4^{2-}$ $(Cs^+)_2CO_3^{2-}$	potassium iodate rubidium sulphate caesium carbonate

simple valency   1
oxidation state   + 1

Note: to avoid ambiguity, certain ions have been enclosed in brackets.   * or lithium chlorate VII

Table 17.1(a)   Binary compounds with oxygen

	Li	Na	K	Rb	Cs
Monoxides $(M^+)_2(O^{2-})$ ionic basic	Li₂O white	Na₂O white	K₂O white	Rb₂O pale yellow	Cs₂O orange red
Peroxides $(M^+)_2(O_2^{2-})$	Li₂O₂ white	Na₂O₂ very pale yellow	K₂O₂ orange	Rb₂O₂ dark brown	Cs₂O₂ yellow
Hyperoxides $(M^+)(O_2^-)$			KO₂	RbO₂ deep yellow	CsO₂

attributed to a new element, rubidium (Latin, *rubidus* = dark red) and two lines in the blue, to caesium (Latin, *caesius* = sky blue).

The alkali metals are too reactive to occur uncombined in nature. Sodium and potassium compounds occur abundantly although only sodium is produced commercially on a large scale. Lithium is widely dispersed but in small quantities, yet is of increasing importance. Compounds of rubidium and caesium may be recovered from the mother liquors left after re-crystallization of other naturally occurring salts of the alkali metals.

Potassium is a principal element in plant and animal metabolism. A harmful radioactive isotope of caesium, $^{137}_{55}Cs$, is present in nuclear fall-out.

Physical data for the metals of Group I (alkali metals) are given in Table 17.1.

A general survey of bonding involving the alkali metals of the s block

Atoms of the alkali metals are relatively very large and the single outer valency electron is easily attracted away from the controlling influence of the relatively distant nucleus, leaving an ion of inert gas electronic configuration and unit positive charge. The elements have low first ionization energies, but second ionization energies are high and ionization does not go beyond ejection of the first electron. This is emphasized by the bold typeface in Table 17.2 which also shows that an increase in atomic radius brings a decrease in ionization energy, further illustrated by Figs. 17.1 and 17.2. Values should be compared with those given later for other groups, and to periodic relationships established in Chapter 3.

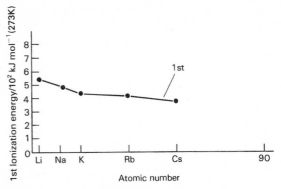

Fig. 17.1 1st ionization energy plotted against atomic number for Group I elements

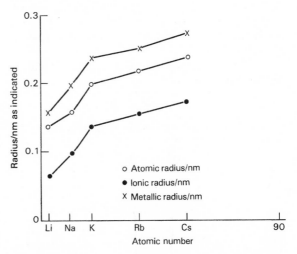

Fig. 17.2 Atomic, ionic and metallic radii plotted against atomic number for Group I elements

Table 17.2 Group I (alkali metals): ionization energies/ kJ mol^{-1} (273 K)

Ionization		Li	Na	K	Rb	Cs
1st	$M(g) \rightarrow M^+(g) + e^-$	**520**	**496**	**419**	**403**	**375**
2nd	$M^+(g) \rightarrow M^{2+}(g) + e^-$	7304	4564	3068	2644	2258

The small size of the Li$^+$ ion leads to the comparatively distinctive behaviour of lithium and its compounds within the alkali metal group. To some degree lithium resembles magnesium, the sizes of atom and ion are comparable for the two, and the changes in properties associated with change of group are matched broadly by those accompanying

change of period. The electronegativity decreases from lithium, showing in a quantitative way the decline in affinity of the elements for electrons as atomic number rises. This trend, illustrated in Fig. 17.3, should be compared with those for other

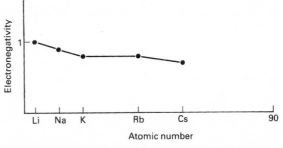

Fig. 17.3 Electronegativity plotted against atomic number for Group I elements

groups (Tables 5.13 and 5.14, and Fig. 5.7). Note the gradation involving lithium and sodium, and the close similarity for the last three members of the group. The electronic configurations of the atoms are the simplest found and this is reflected in the regularity of their properties. The alkali metals are characterized by their unit ionic valency.

Except for certain compounds of lithium, the compounds formed by the alkali metals are ionic and generally have the typically ionic attributes, notably, relatively high melting- and boiling-points with low volatility, good solubility in water and poor solubility in organic solvents. The ionic nature of the compounds is confirmed by X-ray methods. The formation of hydrated ions may be explained by the formation of co-ionic bonds between the central ion and the oxygen atoms of water molecules or simply, by close electrostatic attraction.

Standard electrode potentials (redox, oxidation potentials) for M^+/M

$$M^+ + e^- \rightleftharpoons M$$

by which we mean the potential difference for the cell

$$\text{Pt, H}_2|\text{H}^+|M^+|M$$

implying the reaction

$$\tfrac{1}{2}\text{H}_2 + M^+ \rightarrow \text{H}^+ + M$$

may be plotted against atomic number (Fig. 17.4). The high negative potentials show the tendency of

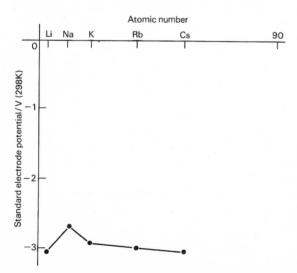

Fig. 17.4   Standard electrode potential plotted against atomic number for Group I elements

the metals, as the strongest reducing agents, to form hydrated cations.

An atom of an alkali metal acquires the electronic configuration of the nearest inert gas by the loss of one electron. The resulting cation retains the nuclear charge of the parent alkali metal. In the illustrations on page 266, the upper number in each case refers to the nuclear charge, the atomic number or proton number, and the lower numbers to the electronic configuration of each atom or ion. Since alkali metal compounds are ionic, by the rules for assigning oxidation numbers (p. 124), the alkali metals exhibit the oxidation state = +1 in their compounds, this being the charge on the cation.

The application of Fajans' Rules brings a deeper understanding of the varying degrees of ionic character encountered in compounds formed by the alkali metals. The lithium cation, being the smallest in the series, has the greatest power of deformation and covalent traits appear in certain lithium compounds. Lithium hydroxide is a weaker base than the other hydroxides. Of the halides, the iodide ion is the largest and most easily polarized. Lithium iodide is the least ionic of the alkali metal halides.

Another trend may be interpreted in this way: the final and most stable products formed when the metals are separately heated in oxygen are, in family order, lithium monoxide, Li_2O, sodium peroxide, Na_2O_2, and for the others, hyperoxides, KO_2, RbO_2 and CsO_2. The first two are colourless, or nearly so, but the others are deep yellow in colour and are odd-electron structures stabilized in the anion by resonance. It is argued that the small cations, by their polarizing power, prevent the dispersal of negative charge which would occur during the progressive increase in proportion of oxygen from monoxide to hyperoxide,

$$\underset{\substack{\text{monoxide}\\ \text{ion}}}{\text{O}^{2-}} \xrightarrow{\frac{1}{2}\text{O}_2} \underset{\substack{\text{peroxide}\\ \text{ion}}}{\text{O}_2^{2-}} \xrightarrow{\text{O}_2} \underset{\substack{\text{hyperoxide}\\ \text{ion}}}{2(\text{O}_2^{\,-})}$$

An alternative explanation based on molecular orbital theory was outlined in the last chapter on p. 247. Note that hyperoxides are also called superoxides.

Again, polyhalide complex ions are not formed by the first two metals, but have been isolated for others. Caesium forms CsI_3 and CsI_4, rubidium forms RbI_3, while for potassium, only a hydrate, $\text{KI}_3.\text{H}_2\text{O}$, has been isolated.

The metals

The metals are soft, white and very good conductors of heat and electricity, a direct consequence of the weak interatomic forces in the metallic state where only one electron is available for bonding. When freshly cut, the surface shows a characteristic lustre which is, however, quickly tarnished. The metals are stored out of contact with air, usually in an inert liquid. They are easily extruded as wire. Comparison of the physical data, in particular, densities, melting-points and boiling-points, shows that lithium is somewhat apart from the others; it is also the hardest metal in the group, although by comparison with a metal regarded as soft by everyday standards, softer than lead.

The metals may be isolated by electrolysis of fused salts or hydroxides. Sodium may be obtained from either the fused chloride or caustic soda while potassium is usually obtained from caustic potash, because the metal diffuses into the fused chloride. Lithium is most readily prepared by the electrolysis of a fused mixture of equal masses of lithium chloride and potassium chloride using a carbon anode and iron cathode. Lithium and chlorine are discharged and should be replenished by addition of lithium chloride. Potassium chloride is added to lower the melting-point. Reduction of lithium ions to the metals occurs (by definition) at the cathode and oxidation of chloride ions to chlorine at the anode:

$$2Li^+Cl^- \rightarrow 2Li + Cl_2$$

Rubidium and caesium catch fire spontaneously on exposure to air. Sodium and potassium burn when heated, while lithium may be melted in air without action but burns when heated further. On combustion, lithium forms the monoxide only, sodium, the monoxide and peroxide, excess air or oxygen giving the latter, while the others† form progressively the hyperoxides:

$$4Li + O_2 \rightarrow 2Li_2O$$

$$2Na + O_2 \rightarrow Na_2O_2$$

$$K + O_2 \rightarrow KO_2$$

A damp atmosphere causes corrosion of any exposed metal. Sodium forms successively the monoxide, hydroxide and by the action of carbon dioxide, a mixture of the efflorescent decahydrate of sodium carbonate and the monohydrate:

$$4Na + O_2 \longrightarrow 2Na_2O$$
$$Na_2O + H_2O \longrightarrow 2NaOH$$
$$2NaOH + CO_2 \longrightarrow Na_2CO_3 + H_2O$$
$$Na_2CO_3.10H_2O \rightleftharpoons Na_2CO_3.H_2O$$

Similar reactions occur with the other metals; for potassium, the carbonate is deliquescent and a solution results.

The intensity of the action with cold water increases with the atomic number of the metal. Hydrogen is displaced and a solution of the hydroxide remains. Lithium floats and reacts quietly. Small pieces of sodium and potassium fuse to shining globules which dart across the surface of the water. The heat of reaction with potassium is enough to ignite hydrogen which burns with a lilac flame due to potassium vapour. Sodium, confined to a wet floating filter paper, will cause ignition of hydrogen, which burns with a yellow flame due to sodium vapour. Rubidium and caesium, both denser than water, react with even greater vigour than potassium, for which the reaction may be expressed:

$$2K + 2H_2O \rightarrow 2KOH + H_2$$

Sodium wire is used for drying organic solvents, such as ether, with which it does not react. With absolute ethanol, the ethoxide and hydrogen are formed: for sodium,

$$2Na + 2C_2H_5OH \rightarrow 2C_2H_5ONa + H_2$$

The metals react with aqueous solutions of strong acids with violence to give the appropriate salt:

$$2Na + 2H^+ \rightarrow 2Na^+ + H_2$$

The metals dissolve in mercury to form alloys, called amalgams, which are liquid or solid depending on the proportion of solid metal incorporated. Amalgams react quietly with water and acids. They are used as reducing agents.

Alkali metals react with hydrides. Heated in ethyne (acetylene), carbides, more properly called ethynides, are formed which will react violently with water to give ethyne and the hydroxide, essentially an action of the ethynide ion:

$$2Na + 2C_2H_2 \rightarrow 2Na(C_2H) + H_2$$
$$2Na + 2Na(C_2H) \rightarrow 2Na_2C_2 + H_2$$

$$\begin{array}{c} C^- \\ \| \\ C^- \end{array} + 2H_2O \rightarrow 2OH^- + \begin{array}{c} CH \\ \| \\ CH \end{array}$$

† To avoid repetition, where two or more elements undergo similar reactions, only one equation will be written. The others follow by substituting the appropriate symbol

In dry ammonia, the heated metals yield amides. Sodium reacts at 350°C and the product, soda-mide, will yield ammonia in a vigorous action with cold water, essentially a reaction of the amide ion,

$$2Na + 2NH_3 \rightarrow 2NaNH_2 + H_2$$

$$NH_2^- + H_2O \rightarrow OH^- + NH_3$$

When heated in hydrogen sulphide, the alkali metals form sulphides which are hydrolysed in aqueous solution to varying extents,

$$2Na + H_2S \rightarrow Na_2S + H_2$$

$$S^{2-} + H_2O \rightleftharpoons OH^- + SH^-$$

$$SH^- + H_2O \rightleftharpoons OH^- + H_2S$$

The heated metals burn in hydrogen chloride giving chlorides which do not react with water, although lithium chloride is slightly hydrolysed,

$$2Na + 2HCl \rightarrow 2NaCl + H_2$$

The alkali metals readily react with a wide range of non-metals. Lithium alone forms a well-defined nitride with nitrogen even at ordinary tempera-tures, while of the other Group V elements, phos-phorus forms phosphides, arsenic and antimony reacting similarly,

$$6Li + N_2 \rightarrow 2Li_3N$$

$$3K + P \rightarrow K_3P$$

Oxide formation has been described. Of the other elements of Group VI, sulphur, selenium and tel-lurium react similarly; sulphur forms sulphides:

$$2K + S \rightarrow K_2S$$

The Group VII elements react to form halides. Sodium burns with a yellow flame when warmed in chlorine,

$$2Na + Cl_2 \rightarrow 2NaCl$$

Sodium reacts slowly with liquid bromine whereas potassium explodes,

$$2K + Br_2 \rightarrow 2KBr$$

Finally, hydrogen reacts with the heated metals to give hydrides, which are ionic in structure, the hydride ion, H^-, having the electronic configura-tion of helium,

$$2Na + H_2 \rightarrow 2NaH$$

The high affinity of the alkali metals for electro-negative elements leads to their use in the reduction of certain compounds which are otherwise difficult to reduce. Oxides of some metals and non-metals are reduced to the element:

$$SiO_2 + 4K \rightarrow Si + 2K_2O$$
silica   silicon

$$Bi_2O_3 + 6K \rightarrow 2Bi + 3K_2O$$
bismuth(III)   bismuth
oxide
bismuth
trioxide

$$Al_2O_3 + 6K \rightarrow 2Al + 3K_2O$$
alumina   aluminium

Chlorides may be reduced similarly to the element, or to hydrocarbons in the case of the alkyl halides:

$$TiCl_4 + 4Na \rightarrow Ti + 4NaCl$$
titanium(IV)   titanium
chloride
titanium
tetrachloride

$$BeCl_2 + 2Na \rightarrow Be + 2NaCl$$
beryllium   beryllium
chloride

$$CH_3.CH_2.I + 2Na + I.CH_2.CH_3 \rightarrow$$
iodoethane
$$CH_3.CH_2.CH_2.CH_3 + 2NaI$$
butane

Tetrachloromethane (carbon tetrachloride) CCl_4, used as a fire-extinguishing agent, will explode with burning sodium. In the Lassaigne sodium test, used in Organic Qualitative Analysis, compounds with a high halogen content, such as trichloromethane (chloroform), $CHCl_3$, will also explode.

The hydrides, oxides and hydroxides

Hydrides, M^+H^-

Hydrides have been synthesized by union of the elements for all of the alkali metals. Lithium reacts at red heat, with incandescence, to form the most stable hydride of the series, m.p. 680°C:

$$2Li + H_2 \rightarrow 2LiH$$

The other metals react in the region of 400°C. Sodium hydride melts with decomposition above 700°C while the others dissociate below their melting-points.

All are colourless, crystalline solids with the sodium chloride lattice in which hydrogen is present as the negative anion, H^-. Structurally, these hydrides are classified as saline, or salt-like.

Lithium hydride merits closer study. It is formed as a transparent, vitreous, deliquescent mass. It melts without decomposition and on electrolysis yields lithium at the cathode and hydrogen at the anode in quantities required by Faraday's Laws,

$$2H^- - 2e^- \rightarrow H_2 \quad 2Li^+ + 2e^- \rightarrow 2Li$$

In cold water a vigorous reaction occurs with the formation of hydrogen and the hydroxide, involving essentially the hydride ion:

$$H^- + H_2O \rightarrow H_2 + OH^-$$

Oxides (see Table 17.1(a))

Strongly exothermic reactions occur on heating alkali metals in oxygen, the final end-products being indicated by the framed formulae. Lithium alone forms the monoxide only, but the other monoxides may be formed from the elements under reduced pressure and the excess metal distilled away. Alternatively, the nitrate may be heated strongly with the metal, when nitrogen is expelled:

$$10K + 2KNO_3 \rightarrow 6K_2O + N_2$$

The monoxides are strongly basic and react violently with water to form the hydroxides and explosively with acids to form salts, processes involving the oxide ion:

$$O^{2-} + H_2O \rightarrow 2OH^- \quad O^{2-} + 2H^+ \rightarrow H_2O$$

Sodium peroxide is the most important of the remaining oxides. Peroxides are salts of the extremely weak acid, hydrogen peroxide.

Sodium peroxide, $(Na^+)_2O_2^{2-}$

The reaction between sodium and oxygen is strongly exothermic. At a temperature maintained at 300°C by drawing sodium in aluminium trays along a pipe against a current of air previously dried and freed from carbon dioxide, sodium peroxide is manufactured as a very pale yellow solid.

Sodium peroxide is hydrolysed in the cold, giving an alkaline reaction, while concentrated solutions break down to oxygen and sodium hydroxide:

$$O_2^{2-} + 2H_2O \rightleftharpoons 2OH^- + H_2O_2$$

$$2O_2^{2-} + 2H_2O \rightarrow 4OH^- + O_2$$

An octahydrate, $Na_2O_2.8H_2O$, may be separated on cooling an aqueous solution.

Well-cooled dilute acid liberates hydrogen peroxide:

$$Na_2O_2 + 2H^+ \rightarrow 2Na^+ + H_2O_2$$

The acidic oxide, carbon dioxide, displaces oxygen from solid sodium peroxide:

$$2Na_2O_2 + 2CO_2 \rightarrow 2Na_2CO_3 + O_2$$

Sodium peroxide is a powerful oxidizing agent and may be used in analysis to convert chromium(III) hydroxide, suspended in water, to sodium chromate(VI) solution,

$$2Cr(OH)_3 + 3Na_2O_2 \rightarrow$$
$$2Na_2CrO_4 + 2NaOH + 2H_2O$$

Hydroxides, M^+OH^-

	Solubility/g(anhydrous compound in 100 g water)		Melting-point
LiOH	13.0	(25°C)	450°C
NaOH	108	(25°C)	318°C
KOH	113	(25°C)	360°C
RbOH	198	(30°C)	301°C
CsOH	386	(15°C)	272°C

Alkalis are soluble bases, caustic in solution when concentrated and slimy to the touch. They neutralize acids. As a group the hydroxides of this family are called the caustic alkalis. Sodium hydroxide is also known as caustic soda, and potassium hydroxide as caustic potash.

The caustic alkalis are colourless solids. Lithium hydroxide is somewhat distinctive in being a weaker base, much less soluble in water and decomposed on strong heating:

$$2LiOH \rightarrow Li_2O + H_2O$$

It is sold as the monohydrate. The others are freely soluble, the solubility rising with atomic number, are deliquescent and absorb carbon dioxide from the atmosphere. They are not decomposed by heat. Caustic soda and caustic potash are widely used. With the exception of potassium they also form monohydrates. On exposure to air the former becomes wet and finally sets to a hydrated sodium carbonate while the latter forms a pool of the deliquescent carbonate. Caustic soda is an exceedingly important industrial chemical. The more expensive caustic potash is prepared similarly, either by electrolysis or the Gossage process.

Neutralizations of strong acids by caustic alkalis may all be reduced to the process:

$$H^+ + OH^- \rightleftharpoons H_2O$$

the salts being fully ionized and not hydrolysed. This is shown by the value of the heat of neutralization (enthalpy change of neutralization) of the acid: the quantity of heat evolved, or enthalpy change, during the neutralization of sufficient acid by a base in dilute solution to form one mole of water in addition to the salt as products. The values are usually determined at 298K (25°C) and for strongly ionized acids are always near 57.4 kJ mol^{-1}, the heat of formation (enthalphy change of formation) of feebly ionized water from its ions:

$$H^+ + OH^- \rightleftharpoons H_2O; \quad \Delta H = -57.4 \text{ kJ mol}^{-1}$$

Weak acids yield solutions of salts, which are fully ionized but somewhat alkaline in reaction due to hydrolysis; examples include sodium borate and sodium carbonate. Alkalis react with acidic oxides: in the absorption of gases, potassium hydroxide solution is generally used for the absorption of carbon dioxide and sulphur dioxide in preference to sodium hydroxide because of the greater solubilities of the products:

$$CO_2 + 2OH^- \rightarrow CO_3{}^{2-} + H_2O$$

$$SO_2 + 2OH^- \rightarrow SO_3{}^{2-} + H_2O$$

Porcelain and glass are attacked slowly but appreciably with the formation of the silicate. Alkalis should not be stored in standard flasks for titrimetric analysis, aspirators with ground-in stopcocks or ground-glass stoppered bottles,

$$SiO_2 + 2OH^- \rightarrow SiO_3{}^{2-} + H_2O$$

Other bases may be displaced from their salts by the alkalis. Volatile bases, ammonia and the amines, are expelled on mixing the solids or warming in solution. The high hydroxide ion concentration produces the weak base:

$$OH^- + NH_4{}^+ \rightarrow NH_3(g) + H_2O$$

In full,

$$(NH_4{}^+)(Cl^-) + NaOH \rightarrow$$
$$\underset{\substack{\text{ammonium} \\ \text{chloride}}}{} NH_3(g) + NaCl + H_2O$$

$$(CH_3\overset{+}{N}H_3)(Cl^-) + NaOH \rightarrow$$
$$\underset{\substack{\text{methyl ammonium} \\ \text{chloride}}}{} CH_3NH_2(g) + NaCl + H_2O$$

Insoluble hydroxides, or oxides where the hydroxide breaks down, are precipitated on mixing

solutions of an alkali and a convenient salt:

$$Al^{3+} + 3OH^- \rightarrow Al(OH)_3$$
$$\underset{\substack{\text{aluminium} \\ \text{hydroxide} \\ \text{colourless}}}{}$$

$$Zn^{2+} + 2OH^- \rightarrow Zn(OH)_2$$
$$\underset{\substack{\text{zinc hydroxide} \\ \text{colourless}}}{}$$

$$Cu^{2+} + 2OH^- \rightarrow Cu(OH)_2$$
$$\underset{\substack{\text{copper(II)} \\ \text{hydroxide, blue}}}{}$$

$$2Ag^+ + 2OH^- \rightarrow Ag_2O + H_2O$$
$$\underset{\substack{\text{silver(I) oxide,} \\ \text{brown}}}{}$$

Certain hydroxides, precipitated at first, re-dissolve in excess alkali, behaving as acids in the presence of the excess alkali, although under other circumstances, in the presence of acids, they behave as bases: salts are formed in each case. Of the hydroxides precipitated above, those of aluminium and zinc do this. They are classed as amphoteric:

$$Al(OH)_3 + OH^- \rightarrow AlO_2{}^- + 2H_2O$$
$$\underset{\text{aluminate ion}}{}$$

or $\quad Al(OH)_3 + OH^- \rightarrow Al(OH)_4{}^-$
$$\underset{\substack{\text{complex aluminate} \\ \text{ion}}}{}$$

$$Zn(OH)_2 + OH^- \rightarrow Zn(OH)O^- + H_2O$$

$$Zn(OH)O^- + OH^- \rightarrow ZnO_2{}^{2-} + H_2O$$
$$\underset{\text{zincate ion}}{}$$

or $\quad Zn(OH)_2 + OH^- \rightarrow Zn(OH)_3{}^-$

$$Zn(OH)_3{}^- + OH^- \rightarrow Zn(OH)_4{}^{2-}$$
$$\underset{\substack{\text{complex zincate} \\ \text{ion}}}{}$$

Potassium hydroxide yields potassium aluminate and potassium zincate solutions respectively.

Solutions of caustic alkalis are used for the hydrolysis of organic esters, potassium hydroxide being sometimes used in alcoholic solution,

ethyl ethanoate   ethanoate ion   ethano
ethyl acetate   acetate ion

This process is called saponification, having originated in soap-making (Latin, *sapo, -onis* = soap).

The caustic alkalis, aqueous or fused, react with a wide range of weak metals and non-metals.

Laboratory fusions involving alkali are usually done in nickel or silver basins. Using caustic soda, consider the reactions in which oxosalts are formed.

Group II

Beryllium alone reacts, sodium beryllate being formed:

$$Be + OH^- + H_2O \rightarrow HBeO_2^- + H_2$$

Group III

Boron reacts very slowly to form the metaborate, while aluminium gives sodium aluminate with vigour:

$$2B + 2OH^- + 2H_2O \rightarrow 2BO_2^- + 3H_2$$

$$2Al + 2OH^- + 2H_2O \rightarrow 2AlO_2^- + 3H_2$$

The simple form AlO_2^- for the aluminate ion is usually preferred to the complex form, $Al(OH)_4^-$, although the latter has much to commend it.

Group IV

Silicon and germanium form sodium silicate and sodium germanate while tin forms stannate(II) and some stannate(IV):

$$Si + 2OH^- + H_2O \rightarrow SiO_3^{2-} + 2H_2$$

$$Sn + OH^- + H_2O \rightarrow HSnO_2^- + H_2$$

Carbon reduces fused sodium hydroxide to the metal and carbonate,

$$6NaOH + 2C \rightarrow 2Na + 2Na_2CO_3 + 3H_2$$

Group V

White, but not red, phosphorus reacts with fairly concentrated alkali solution to form phosphine (compare with above) and sodium hypophosphite solution while the other elements of the group do not react:

$$4P + 3OH^- + 3H_2O \rightarrow PH_3(g) + 3H_2PO_2^-$$

Group VI

Only sulphur reacts, forming a mixture of sodium thiosulphate (trioxothiosulphate), sulphide and polysulphides in a complex series of reactions with aqueous alkali:

$$6OH^- + 4S \rightarrow 2S^{2-} + S_2O_3^{2-} + 3H_2O$$
$$\text{sulphide} \qquad \text{thio-}$$
$$\text{sulphate}$$

$$S^{2-} + 4S \rightarrow S_5^{2-}$$
$$\text{penta-}$$
$$\text{sulphide}$$

Group VII

Fluorine yields the monoxide, oxygen difluoride, and the metal fluoride with very dilute solutions, but oxygen and the metal fluoride with more concentrated alkali. Chlorine forms a mixture of sodium chloride and sodium hypochlorite (sodium chlorate(I)) solutions with cold, dilute alkali and a mixture of sodium chlorate(v) and sodium chloride from hot concentrated sodium hydroxide solution. Bromine and iodine react similarly. Equations may be written:

$$2OH^- + 2F_2 \rightarrow OF_2(g) + 2F^- + H_2O$$
$$\text{oxygen difluoride}$$
$$\text{(fluorine monoxide)}$$

$$4OH^- + 2F_2 \rightarrow O_2 + 4F^- + 2H_2O$$
$$\text{fluoride}$$

$$2OH^- + Cl_2 \rightarrow ClO^- + Cl^- + H_2O$$
$$\text{hypo-} \quad \text{chloride}$$
$$\text{chlorite}$$
$$\text{chlorate(I)}$$

$$6OH^- + 3Cl_2 \rightarrow ClO_3^- + 5Cl^- + 3H_2O$$
$$\text{chlorate(v)}$$

Of importance among the elements not included above, zinc reacts readily with sodium hydroxide solution to give sodium zincate and hydrogen:

$$Zn + OH^- + H_2O \rightarrow HZnO_2^- + H_2$$

$$HZnO_2^- + OH^- \rightarrow ZnO_2^{2-} + H_2O$$
$$\text{zincate}$$

Important salts formed by the alkali metals

Unless the anion is coloured, salts of the alkali metals are colourless solids. The industrial production and applications of compounds derived from sodium chloride are described in Chapter 14 on the alkali industry. Where the detailed description of a compound is more appropriately included under the study of the non-metallic radical, a page reference is given. A compound is presented under the periodic group of the principal non-metal, excluding oxygen, which it contains.

Group III: boron

Borax, sodium pyroborate, sodium tetraborate, $Na_2B_4O_7 . 10H_2O$, p. 306.

Group IV: carbon, silicon

Cyanides and derivatives: see nitrogen compounds below. Sodium silicate: see silica and silicates, p. 330.

Carbonates, $(M^+)_2CO_3^{2-}$, and Hydrogencarbonates (bicarbonates), $M^+HCO_3^-$

Carbonic acid is a very weakly ionized dibasic acid and forms both normal and acid salts:

H_2CO_3	$KHCO_3$	K_2CO_3
carbonic acid	potassium hydrogencarbonate, (bicarbonate)	potassium carbonate

The thermal stability of the carbonate and the existence, or not, of a solid hydrogencarbonate are criteria of the degree of electropositive character of the combined metal.

Except for lithium carbonate, the carbonates of the alkali metals are not decomposed by heat. Lithium carbonate yields lithium monoxide and carbon dioxide:

$$Li_2CO_3 \rightarrow Li_2O + CO_2$$

The solubilities of the carbonates increase with atomic number. The carbonates of potassium, rubidium and caesium are extremely soluble but that of lithium is only sparingly soluble (solubility = 1.33 g in 100 g water at 25°C). Lithium carbonate may be precipitated using solutions of ammonium carbonate and lithium sulphate, or other convenient salts of suitable concentration:

$$2Li^+ + CO_3^{2-} \rightarrow Li_2CO_3(s)$$

Lithium carbonate bears some resemblance to carbonates of Group II. The other carbonates are prepared by the action of carbon dioxide on the hydroxide:

$$2OH^- + CO_2 \rightarrow CO_3^{2-} + H_2O$$

Solid hydrogencarbonates are formed by all of the alkali metals except lithium, the hydrogencarbonate of which is known only in solution. Hydrogencarbonates are formed by the action of carbon dioxide and water on the carbonate in the cold, and are decomposed by heat:

$$CO_3^{2-} + CO_2 + H_2O \rightleftharpoons 2HCO_3^-$$

Carbonates and hydrogencarbonates liberate carbon dioxide with acids stronger than carbonic acid:

$$2H^+ + CO_3^{2-} \rightarrow H_2O + CO_2$$

$$H^+ + HCO_3^- \rightarrow H_2O + CO_2$$

Both carbonates and hydrogencarbonates are hydrolysed giving solutions on the alkaline side of neutrality. The carbonates are more alkaline than the hydrogencarbonates, which are classed as acid salts:

$$CO_3^{2-} + H_2O \rightleftharpoons HCO_3^- + OH^-$$

$$HCO_3^- + H_2O \rightleftharpoons H_2CO_3 + OH^-$$

The greater alkalinity of carbonate solutions results in the precipitation of a basic carbonate, while hydrogencarbonates give a normal carbonate, with solutions of the salts of some metals:

$$Zn^{2+} + 2HCO_3^- \rightarrow ZnCO_3(s) + CO_2 + H_2O$$

$$3Zn^{2+} + CO_3^{2-} + 4OH^- + 2H_2O \rightarrow$$
$$ZnCO_3 . 2Zn(OH)_2 . 2H_2O(s)$$

Sodium hydrogencarbonate (sodium bicarbonate), $Na^+HCO_3^-$

For commercial importance, see the alkali industry, Chapter 14.

Sodium hydrogencarbonate may be prepared by passing carbon dioxide into a cold concentrated solution of sodium hydroxide, when white crystals separate:

$$2OH^- + CO_2 \rightarrow CO_3^{2-} + H_2O$$

$$CO_3^{2-} + H_2O + CO_2 \rightarrow 2HCO_3^-$$
$$(as\ NaHCO_3(s))$$

The product is filtered, washed in cold water and dried.

Sodium hydrogencarbonate is sparingly soluble in cold water; solubility = 8.2 g in 100 g water at 10°C.

Boiling a solution of sodium hydrogencarbonate leads to decomposition. Very pure sodium carbonate may be prepared by heating the solid hydrogencarbonate, the other product escaping

$$2NaHCO_3 \rightarrow Na_2CO_3 + CO_2 + H_2O$$

Sodium hydrogencarbonate is used as baking powder and in self-raising flour, to 'aerate' the dough, for which they are mixed with disodium dihydrogenpyrophosphate (diphosphate), $Na_2H_2P_2O_7$, and calcium tetrahydrogendiorthophosphate, $Ca(H_2PO_4)_2$, acid salts strong enough to react with the carbonate formed.

Sodium sesquicarbonate, $(Na^+)_2CO_3^{2-} . Na^+HCO_3^- . 2H_2O$

This is a double salt which may be prepared by mixing solutions of sodium carbonate and hydrogencarbonate in the correct proportions and allowing crystallization to occur. The crystals are not efflorescent, differing in this respect from those of washing soda. Sodium sesquicarbonate is found in nature on the shores of certain lakes in America and Africa, notably at Lake Magadi. It reacts as a mixture of sodium carbonate and hydrogencarbonate.

Sodium carbonate, soda, soda ash, $(Na^+)_2CO_3^{2-}$

The common hydrates are:

$Na_2CO_3 . 10H_2O$ decahydrate
$Na_2CO_3 . H_2O$ monohydrate.

Sodium carbonate is freely soluble in water; solubility = 21.4 g in 100 g water at 25°C. On crystallization, a hydrate is obtained, the composition of which depends on the temperature. The anhydrous compound is formed by heating a hydrate or the hydrogencarbonate to constant mass. It melts at 854°C. The solubility curve of sodium carbonate is shown in Fig. 17.5. Below 32.5°C, the decahydrate separates; above 37.5°C, the monohydrate is formed and, in the interval, a heptahydrate may be obtained. Note the fall in solubility of monohydrate with rise in temperature. Solid sodium carbonate monohydrate is conveniently obtained by evaporation of the solution at the boiling-point. Anhydrous sodium carbonate slowly becomes hydrated in damp air. The decahydrate, which forms large translucent crystals, commonly used domestically as washing soda, effloresces in air, becoming coated with the monohydrate, seen as a white powdery coating.

In titrimetric (volumetric) analysis, sodium carbonate is used for the standardization of strongly ionized acids. If necessary, absorbed moisture should be expelled by heating, with stirring to avoid caking, and the anhydrous sodium carbonate cooled in a desiccator.

In qualitative analysis, sodium carbonate solution is boiled with mixtures containing metals which form insoluble products, before testing for anions. This avoids possible misinterpretation of observations: thus, in the test for sulphate ions,

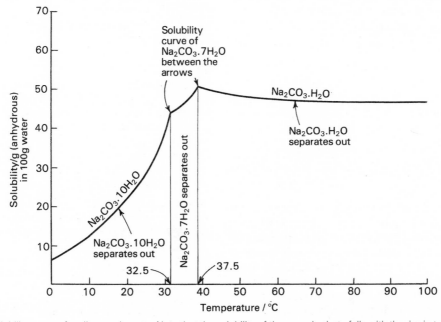

Fig. 17.5   Solubility curve of sodium carbonate. Note that the solubility of the monohydrate falls with the rise in temperature

addition of barium chloride solution to a solution of a lead(II) salt would produce a white precipitate of lead(II) chloride, to be confused possibly with the anticipated precipitate of barium sulphate. Fusion mixture, consisting of equimolar proportions of sodium carbonate and potassium carbonate, is used as a flux in charcoal block reductions. The powdered specimen is mixed with three times its bulk of fusion mixture and heated in the reducing blow-pipe flame. Carbonates of heavy metals are decomposed to the oxide as the residual ions pass into the block. Further action may occur with the oxide being reduced to metal. By using a mixture of carbonates, the fusion point of the flux is lowered as much as possible.

The manufacture of sodium carbonate forms a major part of the alkali industry (Chapter 14).

Potassium carbonate, pearl-ash, potash, $(K^+)_2CO_3^{2-}$

There are three hydrates: $K_2CO_3.2H_2O$, $2K_2CO_3.3H_2O$ and $K_2CO_3.H_2O$.

Potassium carbonate cannot be produced commercially by the Solvay process (see p. 000) because potassium hydrogencarbonate is rather too soluble to be precipitated. It is obtained from potassium chloride, which occurs in salt deposits, by the Precht process. Hydrated magnesium carbonate is added to a concentrated solution of potassium chloride and carbon dioxide is blown in with the resulting precipitation of a double salt:

$$2KCl + 3(MgCO_3.3H_2O) + CO_2 \rightarrow$$
$$2(KHCO_3.MgCO_3.4H_2O)(s) + MgCl_2$$

Treatment with magnesium oxide regenerates magnesium carbonate and potassium carbonate dissolves:

$$2(KHCO_3.MgCO_3.4H_2O)(s) + MgO(s) \rightarrow$$
$$3(MgCO_3.3H_2O)(s) + K_2CO_3$$

After filtration, evaporation yields potassium carbonate. It is freely soluble in water, 113 g dissolving in 100 g water at 25°C. The commercial product is known as pearl-ash.

In the laboratory, potassium carbonate solution can be prepared by boiling a hydrogencarbonate solution, obtained by the action of carbon dioxide on fairly concentrated caustic potash solution. Evaporation and crystallization yields a hydrate. Usually, the di-hydrate separates from solution but others with less water of crystallization may be prepared. Above 130°C, the anhydrous salt is formed. It melts at 900°C.

Potassium carbonate is deliquescent and is used as a desiccating agent for drying organic liquids and to reduce the solubility of ethanol. Moist solid potassium carbonate will absorb carbon dioxide to form the hydrogencarbonate. Alternatively, when carbon dioxide is passed into a concentrated aqueous solution of potassium carbonate, the hydrogencarbonate is precipitated. After filtration, it is left to dry. On heating it decomposes to the carbonate, carbon dioxide and steam.

Potassium carbonate may be extracted from various animal and vegetable products including wood ash. It is a by-product of beet sugar manufacture. Potassium carbonate is used in the production of hard glass and soft soap.

Group V: nitrogen, phosphorus

Transition metal complexes: hexacyanoferrates(II) and (III) (formerly ferro- and ferricyanides, p.470), hexanitrocobaltates(III) (formerly cobaltinitrites, p. 473), pentacyanonitrosylferrates(III) (formerly nitroprussides, p. 471).

Amides, $M^+NH_2^-$, and cyanides, M^+CN^-

The alkali metals react with ammonia to form amides from which cyanides may be prepared with charcoal.

Sodamide, $Na^+NH_2^-$

Sodamide is formed by the action of dry ammonia on sodium at 250°C:

$$2Na + 2NH_3 \rightarrow 2NaNH_2 + H_2$$

It is a white, wax-like solid, melting-point 208°C, instantly hydrolysed by cold water, releasing ammonia and leaving sodium hydroxide solution, essentially a reaction of the amide ion.

$$NH_2^- + H_2O \rightarrow OH^- + NH_3$$

Sodium cyanide, Na^+CN^-

In commercial production, sodamide reacts with red-hot charcoal to form sodium cyanamide and then sodium cyanide, which is run off and cast into moulds:

$$2NaNH_2 + C \rightarrow Na_2N.CN + 2H_2$$

$$Na_2N.CN + C \rightarrow 2NaCN$$

Sodium cyanide is a white solid, melting at 564°C, freely soluble in water and extremely poisonous. The solution is strongly hydrolysed, which is made evident by the bitter almond odour of hydrogen cyanide and the alkaline reaction,

$$CN^- + H_2O \rightleftharpoons HCN + OH^-$$

Sodium cyanide is used in the extraction of silver and gold, and in electroplating.

Potassium cyanide, K^+CN^-

Potassium cyanide may be prepared by the thermal decomposition of potassium hexacyanoferrate(II) (ferrocyanide) alone or with potassium metal,

$$K_4Fe(CN)_6 \rightarrow 4KCN + Fe + 2C + N_2$$

$$K_4Fe(CN)_6 + 2K \rightarrow 6KCN + Fe$$

A mixture of potassium and sodium cyanides is made industrially using sodium in the last reaction,

$$K_4Fe(CN)_6 + 2Na \rightarrow 4KCN + 2NaCN + Fe$$

These reactions involve the breakdown of the complex hexacyanoferrate(II) ion, $Fe(CN)_6^{4-}$. The cyanide product is extracted with water and crystallized. Potassium cyanide is very similar to the sodium salt; it is used as a laboratory agent. The melting-point is $635°C$.

Potassium cyanide is sometimes referred to as 'prussic acid', a name which properly belongs to hydrocyanic acid.

Potassium thiocyanate, K^+SCN^-

Potassium thiocyanate is formed by fusion of sulphur with potassium cyanide, extraction with hot aqueous alcohol, evaporation and cooling,

$$KCN + S \rightarrow KSCN$$

It melts at $161°C$.

Commercially, potassium hexacyanoferrate(II) is fused with sulphur and potassium carbonate.

Potassium thiocyanate is a colourless, deliquescent solid used in the laboratory as a sensitive reagent for the iron(III) ion and with silver nitrate solution for the titrimetric (volumetric) estimation of halides. Alternatively, the cheaper ammonium salt may be used. A blood-red coloration is produced with iron(III) ions: the test is very sensitive and solutions of iron(II) ions may give a pale coloration because of incipient oxidation in air to the iron(III) state; a covalent thiocyanatoiron(III) complex is formed:

$$Fe^{3+} + SCN^- \rightleftharpoons Fe(SCN)^{2+}$$

Because of its deliquescent nature, potassium thiocyanate cannot be made up as a standard solution by weighing. It is standardized against a measured volume of standard silver nitrate (p. 492), to which a little 10% iron(III) alum solution is added as indicator together with some nitric acid to reduce hydrolysis. Thiocyanate is run in until a reddish tinge is observed. The titration may be done in a large white porcelain evaporating dish which gives an admirable background to the colour change. Silver thiocyanate is formed as a white curdy precipitate:

$$SCN^- + Ag^+ \rightarrow AgSCN(s)$$

Nitrates, $M^+NO_3^-$, and nitrites, $M^+NO_2^-$

The solubilities of the nitrates of the alkali metals follow no orderly pattern. On heating strongly, the nitrates decompose to nitrite and oxygen:

$$2CsNO_3 \rightarrow 2CsNO_2 + O_2$$

Sodium nitrate, Chile saltpetre, $Na^+NO_3^-$

Sodium nitrate is mined in a rainless desert region of Chile. Deposits of caliche, containing about 30% of sodium nitrate, are broken up by means of explosives and treated with water. Sodium nitrate is crystallized out. The iodate-bearing mother liquors are the most important world source of iodine. Sodium nitrate is a white solid, somewhat hygroscopic and very soluble in water. At $25°C$, the solubility is 91.8 g in 100 g water. Sodium nitrate melts at $316°C$ and decomposes readily above $550°C$. It forms highly combustible mixtures with organic materials, ethanoates (acetates), cyanides, thiocyanates, hypophosphites and sodamide. The elements carbon, phosphorus and sulphur form carbonate, phosphate and sulphate explosively on heating.

Sodium nitrate may be used as a nitrogenous fertilizer, and in the manufacture of explosives, flares and fireworks, the production of lead glass, for curing meat, and in chemical processes, including nitration and the Lead Chamber production of sulphuric acid.

Potassium nitrate, nitre, saltpetre, $K^+NO_3^-$

Potassium nitrate is obtained on the commercial scale by mixing boiling concentrated solutions of sodium nitrate and potassium chloride from natural sources. Sodium chloride separates as the least soluble pairing of ions and is filtered off. On cooling to about 20°C, potassium nitrate becomes the least soluble salt and crystallizes out. Potassium nitrate is much more soluble in hot water than cold, while the solubility of sodium chloride is little affected by change of temperature. The solubility curves are drawn in Fig. 6.1, p. 92. At 25°C, the solubility of potassium nitrate is 31.6 g in 100 g water.

Potassium nitrate is a white crystalline solid, melting at 340°C and similar in its reactions to sodium nitrate but preferred in the manufacture of gunpowder and pyrotechnics because it does not absorb moisture from the atmosphere. Gunpowder contains potassium nitrate, charcoal and sulphur in the approximate proportions, 6 : 1 : 1 by weight. The mixture is ground to a fine powder when moist and slowly dried out. The release of a great volume of hot expanding gases, nitrogen and oxides of carbon, produces the disruptive power. It is exploded by mechanical shock.

Sodium nitrite, $Na^+NO_2^-$

Sodium nitrite may be prepared by the thermal decomposition of sodium nitrate, but the reduction of nitrate is usually effected by stirring lead parings or copper filings into the molten salt:

$$NaNO_3 + Pb \rightarrow PbO + NaNO_2$$

After cooling, the mass is extracted with hot water, filtered and sodium nitrite crystallized after evaporation to small bulk.

Industrially, sodium nitrite is formed by the action of nitrogen oxide (nitric oxide) and nitrogen dioxide together, obtained by the catalytic oxidation of ammonia, on sodium hydroxide or sodium carbonate solutions:

$$NO + NO_2 + 2OH^- \rightarrow 2NO_2^- + H_2O$$

Sodium nitrite is a white, or very pale yellow, hygroscopic solid. At 15°C, 75.8 g sodium nitrite will saturate 100 g water. In the laboratory, it is used principally as a source of nitrous acid. A cold solution of sodium nitrite on acidification becomes pale blue, an effervescence of nitrogen oxide is seen and brown fumes of nitrogen dioxide appear:

$$H^+ + NO_2^- \rightleftharpoons HNO_2$$

$$3HNO_2 \rightarrow 2NO + H_2O + H^+ + NO_3^-$$

$$2NO + O_2 \rightarrow 2NO_2$$

Sodium nitrite melts at 282°C.

At high temperatures, sodium nitrite will support the combustion of organic materials, like ethanoates (acetates), cyanides and thiocyanates.

Sodium nitrite is a source of nitrous acid for industrial dyeing and the production of azo dyestuffs. It is also used as a rust inhibitor in antifreeze and in the manufacture of vitreous enamel.

In the food industry sodium nitrite is used to cure meat, at the same time improving its colour and preserving it. Frying can convert this nitrite to potentially cancer-causing nitrosamines, the optimum temperature being 185°C. Microwave cooking avoids this problem because the optimum temperature is not reached long enough for significant conversion to occur.

The sodium orthophosphates

Orthophosphoric acid is tribasic and three series of salts may be drived from it:

H_3PO_4	NaH_2PO_4	Na_2HPO_4	Na_3PO_4
ortho phosphoric acid	sodium dihydrogen phosphate	disodium hydrogen phosphate	trisodium phosphate or normal sodium phosphate

Disodium hydrogenphosphate, $(Na^+)_2HPO_4^{2-}$

The common hydrate $Na_2HPO_4.12H_2O$, effloresces to the heptahydrate, $Na_2HPO_4.7H_2O$. It is known just as sodium phosphate. It is formed by titration of phosphoric acid with sodium hydroxide, using phenolphthalein as indicator. The titration curve is shown in Fig. 17.6. The indicator may be adsorbed from the boiling solution by use of animal charcoal, which is filtered off. Evaporation yields efflorescent crystals of the dodecahydrate, which melt at 35°C. On heating to about 250°C, tetra-sodium pyrophosphate (diphosphate) is formed, water being expelled:

$$2Na_2HPO_4 \rightarrow Na_4P_2O_7 + H_2O$$

This is used in the textile industry.

Sodium dihydrogenphosphate, $Na^+H_2PO_4^-$

Titration of phosphoric acid against sodium hydroxide using methyl orange as indicator yields the dibasic phosphate which crystallizes as the monohydrate. The indicator may be removed as above. On heating, sodium metaphosphate results with

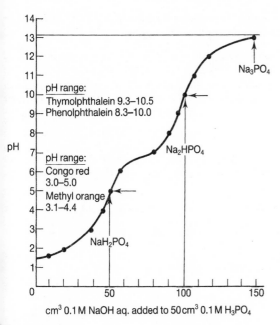

Fig. 17.6 Titration curve for the neutralisation of tribasic phosphoric acid with sodium hydroxide

loss of water:

$$NaH_2PO_4.H_2O \rightarrow NaPO_3 + 2H_2O$$

Normal sodium phosphate, sodium tetraoxophosphate, $(Na^+)_3PO_4{}^{3-}$

This cannot be produced in one stage by titration because of hydrolysis near the stoichiometric point. By a preliminary titration using methyl orange or phenolphthalein, the proportions of acid and base required for the formation of the normal salt may be calculated. A dodecahydrate separates on crystallization. It is used for softening boiler water in industry.

Sodium ammonium hydrogenphosphate, microcosmic salt, $Na^+NH_4{}^+HPO_4{}^{2-}.4H_2O$

Disodium hydrogenphosphate is dissolved in boiling water and the calculated mass of ammonium chloride stirred in. After filtration of any precipitated sodium chloride, microcosmic salt crystallizes on cooling. It is re-crystallized from the minimum quantity of hot water containing a little ammonia and dried between filter papers.

In qualitative analysis, the characteristic colours of the metaphosphates of certain metals serve in their identification. The metaphosphates are formed by fusing a compound of the metal with microcosmic salt in a loop of platinum wire in the bunsen flame, when sodium metaphosphate is formed and may then react with (effectively) the metal oxide:

$$NaNH_4HPO_4.4H_2O \rightarrow$$
$$NaPO_3 + NH_3 + 5H_2O$$

$$NaPO_3 + CoO \rightarrow NaCoPO_4$$
blue sodium cobalt
phosphate as a bead

Group VI (excluding oxygen): sulphur

Potassium peroxosulphate (persulphate), p. 000.
Potassium thiocyanate—see above, under Group V (nitrogen) compounds.

Sulphides, $(M^+)_2S^{2-}$, hydrogensulphides, M^+HS^-, and polysulphides

The sulphides may be prepared by union of the elements, excess metal being distilled away under reduced pressure at about 300°C, or by reduction of the sulphate with carbon, a method used industrially for sodium sulphide. The sulphides are deliquescent and readily hydrolysed. In the laboratory, the sulphides of sodium and potassium are usually made by the action of hydrogen sulphide on aqueous alkali. Hydrogen sulphide is an extremely feebly ionized dibasic acid, forming acid and normal salts.

Sodium hydrogensulphide (hydrosulphide), Na^+HS^-

Aqueous sodium hydroxide absorbs hydrogen sulphide to form sodium sulphide and when saturated, sodium hydrogensulphide:

$$2OH^- + H_2S \rightarrow S^{2-} + 2H_2O$$
$$S^{2-} + H_2S \rightarrow 2HS^-$$

The solution is strongly hydrolysed:

$$HS^- + H_2O \rightleftharpoons H_2S + OH^-$$

The trihydrate, $NaHS.3H_2O$, may be isolated from solution but any attempt to drive off water of crystallization by heating results in hydrolysis.

Acidification expels hydrogen sulphide:

$$HS^- + H^+ \rightarrow H_2S$$

The anhydrous salt may be made by the action of

hydrogen sulphide on sodium ethoxide and is precipitated by the addition of ether:

$$H_2S + C_2H_5ONa \rightarrow NaSH + C_2H_5OH$$

On heating anhydrous sodium hydrogensulphide, anhydrous sodium sulphide is obtained:

$$2NaHS \rightarrow Na_2S + H_2S$$

Sodium sulphide, $(Na^+)_2S^{2-}$

Sodium sulphide solution is prepared by saturating exactly one-half of the available sodium hydroxide solution with hydrogen sulphide, to form the hydrogensulphide, which is then converted into the normal salt with the reserve alkali:

$$HS^- + OH^- \rightarrow S^{2-} + H_2O$$

The solution is strongly hydrolysed:

$$S^{2-} + H_2O \rightleftharpoons HS^- + OH^-$$

Acidification produces hydrogen sulphide:

$$S^{2-} + 2H^+ \rightarrow H_2S$$

From the aqueous solution a nonahydrate, $Na_2S.9H_2O$, buff in colour, may be crystallized. The anhydrous salt, melting-point $1180°C$, may be prepared by the thermal decomposition of the hydrogensulphide as above. It is very soluble in water.

Industrially, sodium sulphide is made by fusing sodium sulphate with powdered coal at about $1000°C$:

$$Na_2SO_4 + 4C \rightarrow NaS + 4CO$$

Sodium sulphide is used in tanning for stripping hair from hides and in the manufacture of dyestuffs.

Polysulphides

The existence of various polysulphides is claimed: Na_2S_2, Na_2S_3, Na_2S_4, Na_2S_5; K_2S_2, K_2S_5, K_2S_6. Sodium pentasulphide, Na_2S_5, is one of the products of boiling sulphur with caustic soda. 'Liver of sulphur', maroon in colour, as might be expected, is formed when sulphur is fused with potassium carbonate and contains potassium sulphate, thiosulphate and polysulphides, principally the pentasulphide. It is used in horticultural sprays as a fungicide.

Sulphite (trioxosulphates), $(M^+)_2SO_3^{2-}$, and hydrogensulphites (bisulphites), $M^+HSO_3^-$

Sulphurous acid is a fairly strongly ionized di-basic acid from which both normal and acid salts may be derived. The saturation of aqueous alkali metal carbonate with sulphur dioxide yields a solution of the hydrogensulphite,

$$CO_3^{2-} + 2SO_2 + H_2O \rightarrow 2HSO_3^- + CO_2$$

Addition of an equivalent quantity of carbonate converts this to a solution of the normal salt, the sulphite,

$$2HSO_3^- + CO_3^{2-} \rightarrow H_2O + CO_2 + 2SO_3^{2-}$$

Alternatively, alkali may be used in a similar manner:

$$SO_2 + H_2O \rightleftharpoons H_2SO_3 \rightleftharpoons$$
$$H^+ + HSO_3^- \rightleftharpoons 2H^+ + SO_3^{2-}$$
$$\downarrow OH^- \qquad \downarrow 2OH^-$$
$$H_2O + HSO_3^- \quad 2H_2O + SO_3^{2-}$$

Sulphite solutions are somewhat alkaline because of hydrolysis while hydrogensulphite solutions are acidic due to ionization of the hydrogensulphite ion:

$$SO_3^{2-} + H_2O \rightleftharpoons HSO_3^- + OH^-$$
$$HSO_3^- \rightleftharpoons H^+ + SO_3^{2-}$$

Sodium sulphite, sodium trioxosulphate $(Na^+)_2SO_3^{2-}$

At $10°C$, $20.0\,g$ sodium sulphite dissolves in $100\,g$ water. The heptahydrate, $Na_2SO_3.7H_2O$, is a colourless, crystalline solid. Sodium sulphite is made industrially by the action of sulphur dioxide on sodium carbonate solution. Sodium sulphite is used as a mild bleaching agent for wool and silk, and as an antichlor after bleaching by chlorine.

Sodium disulphite, sodium pyrosulphite (sodium metabisulphite), $(Na^+)_2S_2O_5^{2-}$

A well-cooled solution of sodium bisulphite deposits the disulphite. It is used in photography.

Sulphates (tetraoxosulphates), $(M^+)_2SO_4^{2-}$, and hydrogensulphates (bisulphates), $M^+HSO_4^-$

Sulphuric acid is a strong dibasic acid forming both normal and acid salts by the complete or half-neutralization with an alkali or carbonate solution.

By titration, using methyl orange as indicator, the proportions by volume of the reactants for complete neutralization may be determined and the sulphate prepared. Doubling the proportion of sulphuric acid yields the acid salt, the hydrogensulphate. The salts may be crystallized after evaporation. The reactions may be summarized:

$$H_2SO_4 \rightleftharpoons H^+ + HSO_4^- \quad \rightleftharpoons \quad 2H^+ + SO_4^{2-}$$

$$\downarrow OH^- \qquad\qquad\qquad \downarrow 2OH^-$$

$$H_2O + HSO_4^- \qquad\quad 2H_2O + SO_4^{2-}$$

$$\updownarrow \qquad\qquad\qquad\qquad \text{normal sulphate}$$

$$H^+ + SO_4^{2-} \qquad\qquad \text{unhydrolysed:}$$

hydrogensulphate ionizes: \qquad\qquad neutral solution
acidic solution

Sodium sulphate, sodium tetraoxosulphate,
$(Na^+)_2SO_4^{2-}$

The industrial production is part of the alkali industry chapter. Anhydrous sodium sulphate melts at 883°C. At 15°C, 13.4 g of the anhydrous salt dissolves in 100 g water. The solubility curve of sodium sulphate, drawn in Fig. 6.1, p. 92, shows a transition-point at 32.38°C. Below this temperature, the decahydrate separates out, while above it, the anhydrous salt is formed. At the transition-point, the solid phase in equilibrium with the solution contains both anhydrous salt and hydrate. Sodium sulphate decahydrate, $Na_2SO_4.10H_2O$, forms colourless translucent crystals which effloresce to the anhydrous salt. It is known as Glauber's salt, and has been used as a purgative.

Sodium hydrogensulphate (sodium bisulphate),
$Na^+HSO_4^-$

Sodium hydrogensulphate results from the action of concentrated sulphuric acid on sodium chloride at room temperature:

$$NaCl + H_2SO_4 \rightarrow NaHSO_4 + HCl$$
$$\text{hydrogen}$$
$$\text{chloride}$$

It is also prepared by adding anhydrous sodium sulphate to concentrated sulphuric acid:

$$Na_2SO_4 + H_2SO_4 \rightarrow 2NaHSO_4$$

Sodium hydrogensulphate melts at 300°C approximately. On gentle heating, the disulphate (pyrosulphate) is formed, which decomposes further at red heat to the sulphate with loss of sulphur

trioxide:

$$2NaHSO_4 \rightarrow H_2O + Na_2S_2O_7$$

$$Na_2S_2O_7 \rightarrow Na_2SO_4 + SO_3$$

From aqueous solutions below 50°C, a monohydrate crystallizes, $NaHSO_4.H_2O$. At 25°C, 28.6 g anhydrous sodium hydrogensulphate saturates 100 g water.

Potassium sulphate, potassium tetraoxosulphate,
$(K^+)_2SO_4^{2-}$

Potassium sulphate occurs in natural salt deposits from which it is extracted with water and crystallized out. Kainite,

$$K_2SO_4.MgSO_4.MgCl_2.6H_2O$$

essentially $K^+Mg^{2+}SO_4^{2-}Cl^-.3H_2O$, is mined at Stassfurt in eastern Germany. Potassium sulphate is widely used as a fertilizer. At 20°C, 11.1 g potassium sulphate dissolves in 100 g water. Potassium sulphate melts at 1050°C.

Sodium thiosulphate, sodium trioxothiosulphate,
$(Na^+)_2S_2O_3^{2-}$

Sodium thiosulphate is prepared by gently boiling a solution of sodium sulphite with finely crushed roll sulphur until the solution is no longer alkaline by hydrolysis,

$$SO_3^{2-} + S \rightarrow S_2O_3^{2-}$$

After filtration, the solution is left to crystallize. Inoculation with a tiny 'seed' crystal of sodium thiosulphate induces prompt crystallization. After filtration, the crystals are dried between filter papers. Large translucent monoclinic crystals of the pentahydrate, $Na_2S_2O_3.5H_2O$, are formed. At 15°C, 65 g of the pentahydrate dissolves in 100 g water. Supersaturated solutions are easily prepared. Alternatively, sulphur may be boiled with sodium hydroxide solution:

$$6OH^- + 12S \rightarrow 2S_5^{2-} + S_2O_3^{2-} + 3H_2O$$

On heating, hydrated sodium thiosulphate melts at 45°C into its own water of crystallization which escapes as steam until above 200°C, sodium sulphate and sodium pentasulphide are formed:

$$4Na_2S_2O_3 \rightarrow Na_2S_5 + 3Na_2SO_4$$

Acids liberate sulphur dioxide and sulphur: acidification of aqueous sodium thiosulphate gives a clear solution which after a short interval becomes cloudy as white, and then yellow, sulphur is seen, largely in the colloidal state. Some sulphur dioxide excapes from solution. It is suggested that the latent period covers the formation and breakdown of (trioxo)thiosulphuric acid, although this compound has never been isolated:

$$S_2O_3{}^{2-} + 2H^+ \rightarrow H_2S_2O_3$$

$$H_2S_2O_3 \rightarrow H_2SO_3 + S(s)$$

Mild oxidation yields sodium tetrathionate, $Na_2S_4O_6$:

$$2S_2O_3{}^{2-} - 2e^- \rightleftharpoons S_4O_6{}^{2-}$$

Sodium thiosulphate is used in conjunction with potassium iodide for the titrimetric (volumetric) determination of oxidizing agents: iodine, liberated by oxidation of iodide ions in the presence of acid, is titrated against a standard solution of sodium thiosulphate:

$$I_2 + 2S_2O_3{}^{2-} \rightarrow 2I^- + S_4O_6{}^{2-}$$

A more detailed examination of the use of iodine in titrimetric analysis appears under that element.

Sodium tetrathionate is also formed by the mild oxidizing action of iron(III) chloride, which is reduced to the iron(II) state:

$$2Fe^{3+} + 2S_2O_3{}^{2-} \rightarrow S_4O_6{}^{2-} + 2Fe^{2+}$$

The powerful oxidizing agents, bromine and chlorine, cause oxidation of thiosulphate to the sulphate:

$$S_2O_3{}^{2-} + 4Cl_2 + 5H_2O \rightarrow$$
$$2SO_4{}^{2-} + 8Cl^- + 10H^+$$

Sodium thiosulphate may be used as an antichlor after bleaching by chlorine.

Sodium thiosulphate, called 'hypo' by photographers, is used as a fixer in photography. Undeveloped silver ions are removed from the developed negative as complex thiosulphates: $Ag(S_2O_3)^-$, $Ag(S_2O_3)_2{}^{3-}$, $Ag_3(S_2O_3)_4{}^{5-}$, by reaction with thiosulphate:

$$Ag^+ + S_2O_3{}^{2-} \rightleftharpoons Ag(S_2O_3)^-, \text{ etc.}$$

Group VII: halogens

Oxosalts formed between the halogens and the alkali metals are described under the appropriate halogen oxoacid, Chapter 23, p. 432. Complex polyhalides of the alkali metals have been mentioned in the introduction, p. 268.

Halides, M^+X^-

The halides may be prepared by the neutralization of the hydrogen halide acid by aqueous alkali or carbonate. However, hydrobromic and hydroiodic acids are not usually available so bromides and iodides are prepared from the free halogen. Fluorides are distinctive in forming acid salts; potassium hydrogenfluoride, acid potassium fluoride (potassium bifluoride), $K^+HF_2{}^-$, is the salt of a dimerized hydrofluoric acid, H_2F_2, which behaves as a completely ionized acid for the first ionization into H^+ and $HF_2{}^-$ ions.

The halides of lithium, sodium and potassium form cubic crystals with the characteristic *sodium chloride lattice*. Lithium forms a fluoride which is sparingly soluble and a hydrated chloride, $LiCl.2H_2O$, which is deliquescent and somewhat hydrolysed, both compounds showing similarities to corresponding compounds of Group II rather than Group I.

Sodium chloride, common salt, Na^+Cl^-

Well known as an article of diet and food preservative, sodium chloride is of supreme importance as the starting-point for the production of many chemicals in the alkali industry (Chapter 14). It occurs in natural brines (e.g. Northwich brine contains 25.79% by mass), in the waters of oceans and seas (e.g. English Channel water contains 27.2% by mass) and as natural deposits of rock salt which occur widespread. Notable salt lakes containing a high proportion of brine include the Dead Sea (about 25% dissolved solids), in the Desert of Judaea between Israel and Jordan, and Salt Lake, Utah, USA. In the UK, important deposits occur in Cheshire, Staffordshire, Yorkshire and Durham, where the salt is extracted with water and pumped up as brine, being replaced by fresh water to avoid surface subsidence. Solid rock salt is obtained from at least one mine in the UK. The salt mines of Poland are extremely well known.

Sodium chloride melts at 801°C. At 20°C, 36.0 g saturates 100 g water.

Potassium chloride, K^+Cl^-

Potassium chloride is a colourless crystalline solid, of which 34.4 g dissolves in 100 g water at 25°C. It is produced by the evaporation of Dead Sea brines by solar energy. During evaporation the concentration rises from 1.2% in the brine to 23% in the liquor which crystallizes in the pans. After refining, the product is 97% potassium chloride. The minerals sylvine, KCl, and carnallite, $KCl.MgCl_2.6H_2O$, occur in natural salt deposits, notably those at Stassfurt in eastern Germany.

Potassium chloride is used as a fertilizer and for the production of potassium hydroxide by electrolysis. It melts at 770°C.

Potassium bromide, K^+Br^-

The addition of bromine to warm, concentrated, aqueous caustic potash yields a mixture of potassium bromide and potassium bromate(v). After evaporation to dryness, the solid is mixed with powdered charcoal and heated to reduce bromate to bromide:

$$3Br_2 + 6OH^- \rightarrow 5Br^- + BrO_3^- + 3H_2O$$

$$2KBrO_3 + 3C \rightarrow 2KBr + 3CO_2$$

Potassium bromide is extracted with water, the solution filtered, evaporated and allowed to crystallize. The solubility is 65.0 g of the salt in 100 g water at 20°C.

Commercially, potassium bromide is formed by the addition of potassium carbonate solution to iron(II) and (III) bromides, formed by the action of bromine on iron borings:

$$3Fe + 4Br_2 \rightarrow Fe^{2+} + 2Fe^{3+} + 8Br^-$$

$$(Fe^{2+} + 2Fe^{3+}) + 4CO_3^{2-} \rightarrow Fe_3O_4 \; hydrated + 4CO_2$$

After filtration, evaporation yields potassium bromide. It is used in medicine as a sedative. It melts at 750°C.

Potassium iodide, K^+I^-

This very soluble salt, of which 145 g dissolves in 100 g water at 25°C, is prepared in a similar way to the bromide. Potassium iodide melts at 705°C. It is a useful source of iodine, with which excess reacts to form a soluble complex iodide:

$$I_2 + I^- \rightleftharpoons I_3^-$$

This is the so-called iodine solution of titrimetric analysis.

Qualitative tests for the ions of lithium, sodium and potassium

The flame test

Alkali metal salts, preferably moistened with hydrochloric acid because of the greater volatility of halides, colour the non-luminous bunsen flame characteristically: lithium is shown by a carmine-red, sodium by a persistent golden yellow and potassium, a lilac coloration. The lilac of potassium is easily masked by the presence of sodium ions, even in traces. However, by viewing through a thick cobalt-blue glass, the yellow is cut out and lilac appears as pink. In qualitative analysis, the most convenient method of detecting potassium is by applying the flame test to the solution obtained by extraction with sodium carbonate, when all other metals except alkali metals have been eliminated.

Positive tests for sodium and potassium

Very few salts of the alkali metals are insoluble in water. In qualitative analysis salts of sodium, potassium and magnesium remain at the end and are obtained by evaporation of the solution. Ammonium salts, which give positive results for many reactions of potassium, are eliminated by sublimation. Solutions of sodium salts react with zinc uranyl ethanoate (acetate) solution to give a yellow precipitate of sodium zinc triuranyl ethanoate hexahydrate,

$$NaZn(UO_2)_3(CH_3COO)_9.6H_2O .$$

Precipitation is aided by the addition of alcohol and the presence of sodium confirmed by the flame test on the precipitate. Fairly concentrated solutions of potassium salts react with sodium perchlorate (sodium chlorate(VII)) solution to give a white precipitate of potassium perchlorate, $KClO_4$. Alternatively, sodium hexanitrocobaltate(III) (cobaltinitrite) solution yields a yellow precipitate of potassium hexanitrocobaltate(III), $K_3[Co(NO_2)_6]$, insoluble in ethanoic acid. The presence of potassium is finally checked by the flame test on the precipitate.

Simple tests illustrating the distinctive behaviour of lithium compounds

The following reactions give precipitates of lithium salts, the corresponding compounds of the other alkali metals being soluble.

1 Disodium hydrogenphosphate solution in the presence of sodium hydroxide, effectively a solution of the normal phosphate, yields a white precipitate of lithium phosphate, Li_3PO_4.

2 Ammonium carbonate solution and ammonia, to reduce the hydrogencarbonate ion concentration present in ammonium carbonate, yields a white precipitate of lithium carbonate, Li_2CO_3, from concentrated solutions.

3 White gelatinous lithium fluoride, LiF, is precipitated by ammonium fluoride solution.

4 Lithium, as the chloride, may be separated from compounds of the other alkali metals by evaporation with concentrated hydrochloric acid and extraction with anhydrous organic solvents, including propanone (acetone), absolute alcohol or dioxan. Lithium chloride may be obtained on evaporation of the solvent.

18

Group II: the alkaline-earth metals

Beryllium, magnesium, calcium, strontium and barium

4	
Be	
2, 2	
12	
Mg	
2, 8, 2	
20	
Ca	
2, 8, 8, 2	
38	
Sr	
2, 8, 18, 8, 2	
56	
Ba	
2, 8, 18, 18, 8, 2	
88	
Ra	
2, 8, 18, 32, 18, 8, 2	

The alkaline-earths are a family of very reactive metals in which the electropositive nature increases sharply with increase in atomic number.

Calcium, strontium, barium and radium form a close series of very similar elements and compounds with a steady gradation of properties. Radium, isolated by the Curies in 1898 from pitchblende, a mineral of uranium which showed unexpectedly high radioactivity, is highly radioactive and while those salts which have been studied show the normal alkaline-earth characteristics, neither the element nor its compounds will be described further here. Magnesium is somewhat apart from metals which follow it, showing some resemblance to zinc, which heads the other subgroup assigned to Group II. Magnesium and calcium compounds in general use are described in detail as are a few compounds of barium. Strontium and its salts are comparatively rare. Some compounds of strontium and barium are mentioned to reveal their overall similarity to the corresponding compounds of calcium. Beryllium is very distinctive; it differs from magnesium more than that element differs from the others, and while it undoubtedly belongs to Group II, it resembles the diagonally placed element, aluminium, so much that at one time beryllium was considered to be tervalent.

The metals magnesium, calcium, strontium and barium are white, lustrous when freshly exposed but quickly tarnished by the atmosphere, and fairly soft. Barium is usually stored in oil and when finely divided is spontaneously flammable. The melting-points and boiling-points do not follow a regular pattern, while the densities at room temperature decrease from beryllium to calcium and then rise to barium. This is accounted for by differences in crystal structure. The reactivity increases with atomic number, well illustrated by the reactions with water: calcium, strontium and barium react with increasing vigour in that order, while magnesium reacts at a reasonable rate with steam only, and beryllium not at all. In the Mendeléeff Periodic Table, calcium, strontium and barium form the IIa subgroup while another subgroup, IIb, is made up of the elements zinc, cadmium and mercury. The elements of both subgroups show a valency of two but are otherwise quite different. Although magnesium shows some resemblance to zinc, it is closer to the calcium subgroup in its properties.

The individual metals are named after minerals known long before their isolation. Impure magnesium, calcium, strontium and barium were all isolated by Davy in 1808 using electrolysis. The term 'earth' was one of the four Aristotelian Elements and came to be applied by early chemists to residual compounds which were apparently unchanged by further heating or by water. Magnesia and lime, because of their alkaline reaction, were called alkaline earths. Following his experiments on oxygen and combustion in 1774, Lavoisier suggested that these alkaline earths were oxides of metals. Accordingly the group has become known as the alkaline-earth metals.

The alkaline-earth metals are too reactive to occur uncombined in nature. After aluminium and iron, calcium is the most abundant metal in the Earth's crust. Strontium and barium occur in

Table 18.1   Group II (alkaline-earth metals): physical data   *s* block elements* $(ns)^2$

	Be	Mg	Ca	Sr	Ba
Atomic number	4	12	20	38	56
Electronic configuration	2, 2	2, 8, 2	2, 8, 8, 2	2, 8, 18, 8, 2	2, 8, 18, 18, 8, 2
[Core]outer electrons	$[He](2s)^2$	$[Ne](3s)^2$	$[Ar](4s)^2$	$[Kr](5s)^2$	$[Xe](6s)^2$
Relative atomic mass	9.0122	24.305	40.08	87.62	137.34
Metallic radium/nm	0.112	0.160	0.197	0.215	0.222
Atomic radius/nm	0.090	0.130	0.174	0.191	0.198
Ionic radius/nm	0.031	0.065	0.099	0.113	0.135
Atomic volume/$cm^3 mol^{-1}$**	4.85	14.00	26.08	34.01	38.26
Standard electrode potential (H Scale: 298 K)/V	−1.70	−2.34	−2.87	−2.89	−2.90
Boiling-point/°C	2970	1107	1187	1366	1537
Melting-point/°C	1283	651	851	800	725
Electronegativity	1.5	1.2	1.0	1.0	0.9
Density/$g cm^{-3}$ (at 293 K)	1.85	1.74	1.55	2.60	3.59
Co-ordination number	2, 4	6	6	6	6
Abundance (% by mass)	0.006	2.1	3.63	0.042	0.039

*excluding helium $(1s)^2$
**Traditional term. It is the volume of one mole of atoms as solid element

Electronic configurations

Inert gases	Ions		
10 **Ne** 2, 8	12 **Mg^{2+}** 2, 8	$Mg^{2+}O^{2-}$	magnesium oxide
		$Mg^{2+}(NO_3^-)_2$	magnesium nitrate
		$Mg^{2+}NH_4^+PO_4^{3-}$	magnesium ammonium phosphate
18 **Ar** 2, 8, 8	20 **Ca^{2+}** 2, 8, 8	$Ca^{2+} \overset{C^-}{\underset{C^-}{\|\|\|}}$	calcium ethynide (carbide)
		$Ca^{2+}(H^-)_2$	calcium hydride
		$(Ca^{2+})_3(PO_4^{3-})_2$	calcium phosphate
36 **Kr** 2, 8, 18, 8	38 **Sr^{2+}** 2, 8, 18, 8	$Sr^{2+}(Cl^-)_2$	strontium chloride
		$Sr^{2+}(OH^-)_2$	strontium hydroxide
		$Sr^{2+}CO_3^{2-}$	strontium carbonate
54 **Xe** 2, 8, 18, 18, 8	56 **Ba^{2+}** 2, 8, 18, 18, 8	$Ba^{2+}(O_2^{2-})$	barium peroxide
		$Ba^{2+}S^{2-}$	barium sulphide
		$Ba^{2+}SO_4^{2-}$	barium sulphate

simple valency   2
oxidation state   + 2

Note: to avoid ambiguity, certain ions have been enclosed in brackets

Table 18.1(a)   Binary compounds with oxygen

	covalent *structure*	ionic *structure*				
Monoxides	BeO	$M^{2+}O^{2-}$ white solids	MgO	CaO	SrO	BaO
Peroxides		$M^{2+}(O_2^{2-})$ white solids	$(MgO + MgO_2)$	CaO_2	SrO_2	BaO_2

localized deposits, each with about 1% of the abundance of calcium. Nuclear fall-out contains a radioactive isotope of strontium, $^{90}_{38}Sr$, which may be absorbed into bones with harmful effects on the formation of blood cells. Magnesium occurs widely in natural salt deposits, mineral spring water, as at Epsom, in sea water and salt lakes. It is the important engineering metal of the group. Beryllium has been produced in the pure state only recently and has specialized applications in nuclear reactors, X-ray windows and high-speed missiles. The McDonnell Douglas F-4 Phantom aircraft was the first aircraft to have a major load-carrying component made of beryllium; namely, the rudder. Special decontamination precautions are necessary during processing to avoid the inhaling of powdered beryllium, which is poisonous.

The green colouring matter of plants, chlorophyll, contains magnesium, the presence of which is essential for healthy growth.

Physical data for Group II (alkaline-earth) metals are given in Table 18.1.

A general survey of bonding involving the alkaline-earth metals of the *s* block

The atoms of alkaline-earth metals, except for beryllium, which, although it is in Group II, is not strictly an alkaline-earth according to the original meaning of the term, are comparatively large and readily lose both valency electrons to give ions of inert gas electronic configuration and double positive charge. No cations of unit positive charge are ever formed. This is shown by comparing the ionization energies shown in Table 18.2 and Fig. 18.1. The bold type facilitates this. Associated with the decrease in ionization energy as atomic number increases is an increase in atomic (and other) radii, shown in Fig. 18.2. The illustrations should be compared with those given for elements of other groups and reference should be made to the relationship between ionization energies and electronic configuration (and periodic patterns) in Chapter 3.

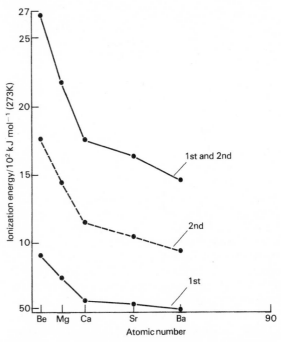

Fig. 18.1 1st, 2nd and 1st + 2nd ionization energies plotted against atomic number for Group II elements

Table 18.2 Group II (alkaline-earth metals): ionization energies/kJ mol^{-1} (273 K)

Ionization		Be	Mg	Ca	Sr	Ba
1st	$M(g) \rightarrow M^+(g) + e^-$	899	**737**	**590**	**549**	**503**
2nd	$M^+(g) \rightarrow M^{2+}(g) + e^-$	1756	**1447**	**1148**	**1061**	**965**
3rd	$M^{2+}(g) \rightarrow M^{3+}(g) + e^-$	14858	7728	4940	4210	—

The electronegativities fall with increase in atomic number, the cations becoming increasingly stable. This trend, illustrated in Fig. 18.3 should be compared with what is shown in Tables 5.13 and 5.14 and Fig. 5.7. Once again the close similarity between calcium, strontium and barium is seen, as well as the sharp change in the beryllium, magnesium and calcium sequence. The alkaline-earth metals form ionic compounds in which they have an ionic valency of two. This is the Group Valency. Since the charge on the cations of Group II metals is +2, this is also the oxidation number

(see p. 124) and in their compounds the metals are in the oxidation state = +2.

Ionic bonding in compounds is usually associated with high melting-points and boiling-points, good solubility in water and poor solubilities in organic solvents, although conclusions based on solubilities alone should be judged with caution.

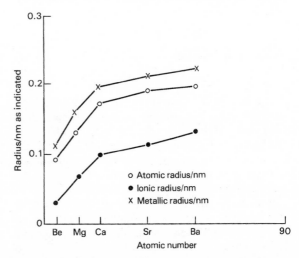

Fig. 18.2 Atomic, ionic and metallic radii plotted against atomic number for Group II elements

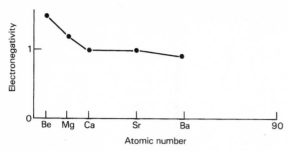

Fig. 18.3 Electronegativity plotted against atomic number for Group II elements

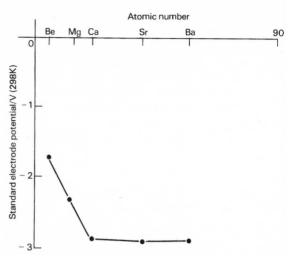

Fig. 18.4 Standard electrode potential plotted against atomic number for Group II elements

Many compounds of the alkaline-earth metals are characterized by these properties and their ionic structure may be shown by X-ray methods. However, it is interesting to note that the alkaline-earth metals form many insoluble compounds, although most are undoubtedly of ionic structure. In examining the connection between lattice energy and solubility in Chapter 5, pp. 85–87, we noted the fourfold effect of increasing the ionic charge (Table 5.18) from M^+X^- to $M^{2+}X^{2-}$. This polarizing aspect offers at least part of the explanation for the generally reduced solubility of compounds in passing from Group I to Group II. Other factors involved in determining the value of lattice energies include the ionic radii and the geometry of the lattices compared.

Standard electrode potentials may be plotted for M^{2+}/M against atomic number

$$M^{2+} + 2e^- \rightleftharpoons M$$

by which we mean the potential difference for the cell

$$\text{Pt, H}_2|\text{H}^+|M^{2+}|M$$

implying the reaction

$$\tfrac{1}{2}\text{H}_2 + \tfrac{1}{2}M^{2+} \rightarrow \text{H}^+ + \tfrac{1}{2}M$$

when it will be seen (Fig. 18.4) that the large negative potentials of the higher members show their very powerful reducing nature and their tendency to form hydrated cations. Again, the close similarity between calcium strontium and barium is seen and the gradation for the earlier members of the family.

Ions of the alkaline-earth metals have twice the net positive charge and are smaller than the corresponding ions of the adjacent alkali metals. According to Fajans' Rules, the cations of the alkaline-earth metals will have a greater polarizing effect on anions than the alkali metals and show traces of latent covalency in certain compounds. The hydroxides are relatively weaker and certain salts are slightly hydrolysed. The salts are more prone to thermal decomposition. The tendency to recapture electrons diminishes with increasing ionic radius, because the electrical attraction between nucleus and electron falls off with distance (inverse square law). Barium forms completely ionic compounds while beryllium, which lies on the diagonal separating metals on the left from non-metals on the right in the Periodic Classification, forms both covalent and ionic compounds, but the latter have a strong tendency always to revert to the covalent condition.

Cations of the alkaline-earth metals, formed by the ejection of two electrons from each atom, have the electronic configuration of the nearest inert gas, but the nuclear charge of the original metal. In the illustrations on p. 286, the upper number in each case refers to the nuclear charge, the atomic number, and the lower numbers to the electronic configuration of each atom or ion.

The strong polarizing effect of the doubly charged beryllium cation, much more marked than that of the singly charged lithium cation, is due to the inadequate shielding of the relatively tiny nucleus by the duplet of electrons. Consider the halides, which are always good for the diagnosis of covalency formation: the anhydrous halides of beryllium have comparatively low melting-points, are non-conducting (or almost so) when fused and readily sublime, dissolve in many organic solvents and dissolve in water with hydrolysis. Even beryllium fluoride is not completely ionized in solution.

The covalent tendencies of beryllium are also shown by the hydrolysis of oxosalts, the ease of formation of basic salts and the amphoteric nature of the hydroxide. Beryllium hydroxide, $Be(OH)_2$, is precipitated from solutions of beryllium salts by addition of alkali, the precipitate dissolving in excess. Because beryllates, e.g. K_2BeO_2 are hydrolysed by water, beryllium hydroxide is reacted with alkali dissolved in absolute alcohol. The high lattice energy of these beryllium salts, due to the metal's high ionization energy and small ionic size, makes dispersal of the lattice in water difficult and causes the compounds to have low solubilities compared with the salts of the other elements in the family.

In covalent compounds, as in ionic, beryllium exerts the Group Valency. Therefore, in covalency the inert gas electronic configuration is not reached. Beryllium chloride may be used in place of aluminium chloride or boron trifluoride in the Friedel-Crafts' reaction (p. 313), there being a tendency to take up a halide ion leaving a positively charged organic ion, a carbonium ion.

The tendency to form complexes decreases with increase in atomic number. Beryllium forms many complexes but these do not concern us here. Magnesium forms complexes with inorganic and organic compounds of oxygen and to a lesser extent, with nitrogen. Increasing atomic number brings greater size and less attraction for molecules with lone pairs of electrons available for donation. Notable among the complexes are hydrated cations and ammines. Magnesium salts may have as many as twelve molecules of water of crystallization. Ammines, which are complexes of ammonia, are usually formed by the chlorides, bromides and iodides, those of magnesium being the best defined. Examples of magnesium chloride hydrates and ammines follow showing the range of co-ordination numbers:

$$MgCl_2 . xH_2O \quad \text{where } x = 12, 8, 6 \text{ or } 4$$

and

$$MgCl_2 . yNH_3 \quad \text{where } y = 6, 4 \text{ or } 2$$

In assigning structures for complexes, co-ionic bonds are drawn between the central ion and the electron-donating atom, oxygen or nitrogen, of the molecules being taken on. Alternatively, water molecules may be regarded as held by electrostatic attraction.

BeCl$_2$
beryllium chloride

BeCl$_2$.4H$_2$O
hydrated
beryllium chloride

Electronic configuration
acquired by
beryllium: 2, 4
(no dipole moment:
linear structure)

tetrahedral

2, 8 (Ne)

hexammine formed by
magnesium chloride
Mg(NH$_3$)$_6$Cl$_2$
Octahedral outline for
cation (mutual repulsion
of NH$_3$)

BeSO$_4$
anhydrous
beryllium sulphate
Be^{2+}SO$_4^{2-}$
Electronic configuration
acquired by
beryllium: 2 (He)

BeSO$_4$.4H$_2$O
hydrated
beryllium sulphate
$[Be(OH_2)_4]^{2+}SO_4^{2-}$

2, 8 (Ne)

The co-ordination number is two in covalent beryllium chloride in the vapour state. It becomes four in the linear chains of the solid state, chlorine atoms acting as bridging ligands and adopting a tetrahedral configuration about beryllium.

Magnesium is intermediate in position between beryllium and the main body of alkaline-earth metals but its compounds are predominantly ionic and are written with ionic structures, as are those of the remaining metals of this family. Magnesium halides are hydrolysed to the basic salts on evaporation of aqueous solutions.

Tetrahedral disposition by $[sp^3]$ hybridization occurs about the beryllium atom here and in the hydrated ion, $[Be(H_2O)_4]^{2+}$ shown earlier.

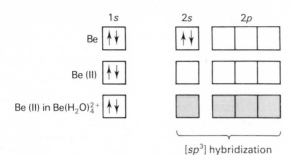

[sp^3] hybridization

While there is little co-ordination chemistry of the alkali metals and the alkaline earths, because of their very low attraction for lone pairs of electrons, the analytical reagent ethylenediaminetetra-acetic acid (ethane-1 : 2-diaminetetraethanoic acid), abbreviated to EDTA, which has powerful chelating tendencies, wraps itself around magnesium and calcium ions, thereby forming the basis of their titrimetric estimation. The complex is 6– co-ordinated. Presumably the extra charge density due to the double charge on the cation is crucial to this slight tendency to form certain complexes when a comparison is made with the alkali metals.

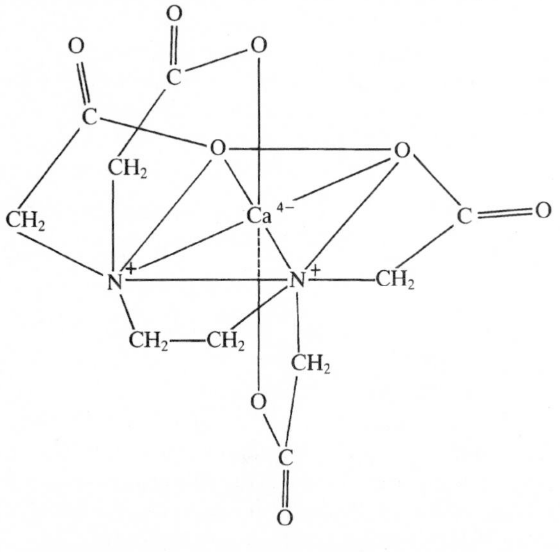

$[Ca((OOCCH_2)_2NCH_2CH_2N(CH_2COO)_2)]^{2-}$

EDTA complex of calcium

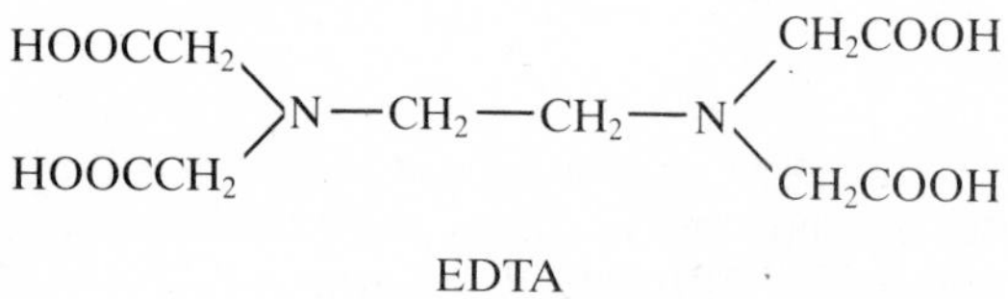

EDTA

If EDTA is represented by H_4Y, and its anion Y^{4-}, the complex shown above is MY^{2-}. It is formed during titration of a solution of a magnesium or calcium salt against a solution made up from the disodium salt, $Na_2H_2Y.2H_2O$

$$Ca^{2+} + H_2Y^{2-} \rightarrow CaY^{2-} + 2H^+$$

EDTA reacts with many cations, all complexes having a 1 : 1 = metal : EDTA anion relationship:

$$M^{3+} + H_2Y^{2-} \rightarrow MY^- + 2H^+$$

The metals, except beryllium

The metals are white, fairly soft when compared with common everyday metals, and good conductors of heat and electricity. Freshly cut surfaces are lustrous, but quickly tarnish. Magnesium and calcium are usually found in the laboratory, the former as ribbon, wire or powder, all grey in colour, and the latter usually as whitish-grey rather brittle turnings, necessarily stored in a tightly sealed container. Beryllium is somewhat different from the others. Calcium, strontium and barium form a closely related family. Calcium metal is by far the most important while strontium metal is very rarely encountered. The reactions of the elements of this group may with advantage be compared with those of the alkali metals.

Magnesium and calcium may be prepared by the electrolysis of the fused chlorides, to which fluxes may be added to lower the melting-points of the electrolytes. Strontium and barium may also be prepared by electrolysis under special conditions. However, they are more conveniently made by reduction of oxides with aluminium *in vacuo*, and distilled away.

In bulk, the metals are protected from further oxidation by an oxide film, except for calcium which largely forms the nitride. Freshly exposed calcium is white and very lustrous, but tarnishes within a few minutes. All will burn when heated in air but finely divided barium catches fire spontaneously on exposure. The product is a mixture of oxide and nitride. The reactions are strongly exothermic. Calcium burns with a reddish flame to give mostly nitride, and magnesium with a brilliant white light, giving off a smoke in which the nitride proportion rises as the air supply is restricted:

$$2Mg + O_2 \rightarrow 2MgO,$$

$$3Mg + N_2 \rightarrow Mg_3N_2$$

Further heating of the oxides in air results in the

formation of peroxides with strontium and barium, although in the former case it is desirable to use oxygen under pressure:

$$2BaO + O_2 \rightarrow 2BaO_2$$

Calcium may be used for removing residual air in the production of high vacuum, for the removal of oxygen from molten casting alloys and for the freeing of inert gases from nitrogen.

The reactions with water increase markedly in vigour with increase in atomic number. Magnesium powder liberates hydrogen extremely slowly from cold water but the metal burns brilliantly in steam to the oxide and hydrogen:

$$Mg + H_2O \rightarrow MgO + H_2$$

Calcium reacts briskly with cold water, liberating hydrogen with evolution of heat and formation of the sparingly soluble hydroxide:

$$Ca + 2H_2O \rightarrow Ca(OH)_2 + H_2$$

Strontium is more vigorous, and barium even more so. Calcium does not react with ethanol and may be used for the final removal of water from that alcohol, just before distillation. Barium does react with ethanol to form the ethoxide with evolution of hydrogen:

$$Ba + 2C_2H_5OH \rightarrow Ba(OC_2H_5)_2 + H_2$$

Vigorous reactions occur with dilute solutions of strongly ionized acids, hydrogen being evolved. This even occurs when very dilute nitric acid acts on magnesium, the other product being magnesium nitrate:

$$Mg + 2H^+ \rightarrow Mg^{2+} + H_2$$

Unlike beryllium, the remaining metals are not attacked by aqueous alkali.

Reactions occur readily with many non-metals. All form nitrides with nitrogen on heating. Calcium reacts at about 400°C:

$$3Ca + N_2 \rightarrow Ca_3N_2$$

These nitrides are hydrolysed to ammonia with water:

$$Ca_3N_2 + 6H_2O \rightarrow 3Ca(OH)_2 + 2NH_3(g)$$

When fused with phosphorus under an inert atmosphere, phosphides are formed which are also hydrolysed with water to the hydride, the phosphine catching fire spontaneously on meeting the atmosphere:

$$3Ca + 2P \rightarrow Ca_3P_2$$

$$Ca_3P_2 + 6H_2O \rightarrow 3Ca(OH)_2 + 2PH_3$$

$$4PH_3(g) + 8O_2 \rightarrow P_4O_{10} + 6H_2O$$

Worth noting here are the compounds formed between magnesium and the remaining elements of Group V. Mg_3As_2, Mg_3Sb_2, Mg_3Bi_2 are decomposed by dilute acids to AsH_3, SbH_3 and the extremely unstable BiH_3 respectively.

From Group VI, sulphur, selenium and tellurium react similarly. On strong heating, magnesium reacts violently with sulphur to form the sulphide, MgS, hydrolysed even in damp air to H_2S (SH_2: compare above). The metals are attacked by the halogens, bromine and iodine being used in the vapour phase. In chlorine, magnesium burns brilliantly to the chloride, $MgCl_2$. Ionic hydrides (e.g. CaH_2) are formed when hydrogen is passed over heated calcium, strontium or barium, with decreasing ease in that order, but not with magnesium or beryllium.

$$Mg + S \rightarrow MgS$$

$$Ca + Br_2 \rightarrow CaBr_2$$

$$Sr + H_2 \rightarrow SrH_2$$

Many compounds may be reduced by the alkaline-earth metals to the elements. At red heat, magnesium reduces carbon monoxide, carbon dioxide, nitrogen oxide (nitric oxide), dinitrogen oxide (nitrous oxide) and sulphur dioxide among the gases. Silica is reduced to silicon and magnesium silicide,

$$SiO_2 + 2Mg \rightarrow Si + 2MgO$$

$$2Mg + Si \rightarrow Mg_2Si$$

Magnesium is used in the metallurgical extraction of certain metals which prove difficult to form by other means. Titanium may be produced by reduction of the tetrachloride with magnesium, beryllium by reduction of the (di)fluoride and in the production of fissionable material, the tetrafluorides of uranium and plutonium are reduced to the metals with magnesium:

$$TiCl_4 + 2Mg \rightarrow Ti + 2MgCl_2$$

$$BeF_2 + Mg \rightarrow Be + MgF_2$$

$$UF_4 + 2Mg \rightarrow U + 2MgF_2$$

$$PuF_4 + 2Mg \rightarrow Pu + 2MgF_2$$

The alkaline-earth metal undergoes oxidation in forming the cation (by loss of electrons). Finally, electrochemical displacement becomes of interest with magnesium, which does not react vigorously with water. Metals below magnesium are displaced from solution of their salts: thus, copper is displaced from copper(II) sulphate, with the simultaneous displacement of some hydrogen:

$$Cu^{2+} + Mg \rightarrow Mg^{2+} + Cu$$

The hydrides, oxides and hydroxides

Hydrides, $M^{2+}(H^-)_2$

Ionic hydrides are formed by calcium, strontium and barium. Calcium hydride is formed when dry hydrogen is passed over calcium turnings heated to a dull redness in an iron boat:

$$Ca + H_2 \rightarrow CaH_2$$

It is a colourless, crystalline solid which reacts with water giving a rapid evolution of hydrogen and forming calcium hydroxide,

$$CaH_2 + 2H_2O \rightarrow Ca(OH)_2 + 2H_2$$

Oxides (see Table 18.1(a))

The alkaline-earth metals react with oxygen on heating to form the monoxides. Further heating in oxygen results in the formation of a peroxide with barium only at atmospheric pressure, but with strontium also when under considerable pressure. The final products obtained on heating the elements in oxygen under atmospheric pressure are indicated by the framed formulae. However, the peroxides are best obtained by neutralization using hydrogen peroxide on the hydroxides or by using sodium peroxide and a solution of the appropriate salt, the octahydrate being the product for calcium, strontium and barium:

$$Ca^{2+} + O_2^{2-} + 8H_2O \rightarrow CaO_2 . 8H_2O(s)$$

Magnesium peroxide has a variable composition.

Thermal decomposition of the carbonates may be used for the preparation of all monoxides, except that of barium:

$$SrCO_3 \rightarrow SrO + CO_2$$

The thermal decomposition of barium carbonate is facilitated by heating with carbon, when a very porous, highly reactive form of the oxide results:

$$BaCO_3 + C \rightarrow BaO + 2CO$$

Alternatively, nitrates or hydroxides may be decomposed by heat.

The oxides are strongly basic white solids, ionic in structure and melting at very high temperatures. Magnesium oxide dissolves only slightly in water but enough for the solution to turn litmus blue. The others are slaked by water, reacting with great vigour when freshly prepared, barium oxide becoming incandescent. Excess moisture escapes as steam.

$$BaO + H_2O \rightarrow Ba(OH)_2$$

Magnesium oxide, magnesia, $Mg^{2+}O^{2-}$

The melting-point of magnesium oxide is 2800°C.

Magnesia is used in the production of magnesite and chrome-magnesite bricks for the steel and other metallurgical industries where a basic furnace lining is required. In granular form, it is used for the basic process in the open hearth furnace in steel-making. The raw materials required in the production of magnesia are dolomite, $CaCO_3 . MgCO_3$, sea water, which contains magnesium chloride and sulphate (more accurately, Mg^{2+}, Cl^- and SO_4^{2-}) and coal as fuel. Dolomite is burnt with coal to yield 'Dolime':

$$CaCO_3 . MgCO_3 \rightarrow CaO . MgO + 2CO_2$$

This is slaked, made into a slurry and residues from the coal removed. The slurry is mixed with sea water from which calcium hydrogencarbonate has been removed as a precipitate of carbonate by an earlier addition of a little 'Dolime'. As the least soluble combination of ions present, magnesium hydroxide is precipitated:

$$CaO . MgO + 2H_2O + Mg^{2+} \rightarrow$$
$$2Mg(OH)_2(s) + Ca^{2+}$$

A thick, creamy slurry settles, is filtered and heated above 1600°C to give magnesium oxide of not less than 90% purity.

$$Mg(OH)_2 \rightarrow MgO + H_2O$$

Calcium oxide, quicklime, $Ca^{2+}O^{2-}$

Calcium oxide melts at 2572°C.

Quicklime is made in great quantity by heating limestone, calcium carbonate, to 900–1200°C in a lime kiln, fired by burning producer-gas. Thermal dissociation of calcium carbonate becomes thermal decomposition because the carbon dioxide is swept away:

$$CaCO_3 \rightleftharpoons CaO + CO_2$$

Quicklime is usually obtained as white lumps. Addition of water yields slaked-lime, calcium hydroxide:

$$CaO + H_2O \rightarrow Ca(OH)_2$$

As the reaction proceeds, the mass swells and cracks, becoming very hot and steaming, finally crumbling to a fine white bulky powder. In dry air, quicklime keeps indefinitely. It is slaked just before use in making mortar and plaster in the building trade. In the cold, quicklime reacts with acids but not with acidic oxides. Industrially, it is used in metallurgical processes to remove acidic oxides as fusible slags:

$$\underset{\text{silica}}{SiO_2} + CaO \rightarrow \underset{\substack{\text{calcium} \\ \text{silicate}}}{CaSiO_3}$$

$$\underset{\substack{\text{phosphorus} \\ \text{pentoxide}}}{P_4O_{10}} + 6CaO \rightarrow \underset{\substack{\text{calcium} \\ \text{phosphate}}}{2Ca_3(PO_4)_2}$$

Quicklime is used in Gossage's process (see p. 208) for making caustic soda and in the production of bleaching powder. Soda-lime is made by slaking quicklime with concentrated aqueous caustic soda. It is used in the laboratory as an absorbent for carbon dioxide, and waste acidic gases, and for decarboxylation of salts of carboxylic acids. Quicklime is used to dry ammonia, which reacts with the usual drying agents, and sometimes, to dry alcohol.

At red heat, calcium oxide reacts with chlorine to yield calcium chloride and oxygen:

$$2CaO + 2Cl_2 \rightarrow 2CaCl_2 + O_2$$

Barium peroxide, $Ba^{2+}O_2^{2-}$

When barium oxide is heated to a dull redness in air or oxygen, barium peroxide is formed in a reversible reaction, accompanied by evolution of heat:

$$2BaO + O_2 \rightleftharpoons 2BaO_2$$

Further heating, or reduction in pressure, results in decomposition. In Brin's oxygen process, now obsolete, barium peroxide was formed by heating barium oxide in the air under increased pressure and the oxygen extracted, drawn off under reduced pressure.

Barium peroxide is a dense off-white powder used for bleaching straw and silk, and in the laboratory preparation of hydrogen peroxide. The commercial form is relatively inactive and a much more reactive, bulky, porous and crystalline octahydrate results from dissolving it in dilute ice-cold hydrochloric acid followed by neutralization with alkali:

$$BaO_2(s) + 2H^+ \rightarrow Ba^{2+} + H_2O_2$$

$$Ba^{2+} + H_2O_2 + 2OH^- + 6H_2O \rightarrow BaO_2.8H_2O(s)$$

Barium peroxide liberates hydrogen peroxide with acids, being a salt of hydrogen peroxide and therefore a true peroxide. Carbon dioxide displaces oxygen from barium peroxide:

$$2BaO_2 + 2CO_2 \rightarrow 2BaCO_3 + O_2$$

With potassium hexacyanoferrate(III), potassium barium hexacyanoferrate(II) is formed and oxygen evolved,

$$BaO_2 + 2K_3Fe(CN)_6 \rightarrow K_6Ba[Fe(CN)_6]_2 + O_2$$

This reaction may be employed for the quantitative estimation of barium peroxide.

Hydroxides, $M^{2+}(OH^-)_2$

	Solubility/g in 100 g water (25°C)
$Mg(OH)_2$	0.01
$Ca(OH)_2$	0.15
$Sr(OH)_2$	0.89
$Ba(OH)_2$	3.32

The hydroxides of the alkaline-earth metals, magnesium to barium, are white, ionic solids and strong bases. They are sparingly soluble in water, the solubility rising with the atomic number as does the difficulty of thermal decomposition to the oxide and steam, barium hydroxide being the most difficult to decompose:

$$Ba(OH)_2 \rightarrow BaO + H_2O$$

The hydroxides neutralize acids, react with acidic oxides and will displace ammonia from salts:

$$Mg(OH)_2 + 2H^+ \rightarrow Mg^{2+} + 2H_2O$$

$$Ca(OH)_2 + SO_2 \rightarrow CaSO_3 + H_2O$$

$$Ca(OH)_2 + 2NH_4Cl \rightarrow CaCl_2 + 2NH_3(g) + 2H_2O$$

The less soluble hydroxides may be precipitated by addition of aqueous caustic alkali to a suitable salt solution,

$$Mg^{2+} + 2OH^- \rightarrow Mg(OH)_2(s)$$

Hydration of the oxide may prove more convenient for the other hydroxides, the oxide being prepared by heating the carbonate, as for calcium oxide. Alternatively, by passing steam over the heated carbonate, the hydroxides of strontium and barium may be formed directly:

$$BaCO_3 + H_2O \rightarrow Ba(OH)_2 + CO_2(g)$$

Calcium hydroxide, slaked lime, hydrated lime, $Ca^{2+}(OH^-)_2$

Calcium hydroxide is produced commercially by slaking quicklime. On heating to red heat, it rapidly evolves steam, leaving quicklime. A suspension of slaked lime in water is called milk of lime and is used as a cheap industrial alkali. The filtered, saturated solution is called lime-water, well known as a test solution for carbon dioxide, which gives initially a white milkiness due to a precipitate of calcium carbonate, re-dissolving with excess carbon dioxide to form a clear solution of calcium hydrogencarbonate, $Ca(HCO_3)_2$:

$$Ca^{2+} + 2OH^- + CO_2 \rightarrow CaCO_3(s) + H_2O$$

$$CaCO_3(s) + H_2O + CO_2 \rightarrow Ca^{2+} + 2HCO_3^-$$

The solubility of calcium hydroxide in water decreases with rise in temperature.

The action of chlorine is important (p. 436). Cold, dry slaked lime yields bleaching powder, a mixture of calcium hypochlorite (calcium chlorate(I)) and basic calcium chloride:

$$3Ca(OH)_2 + 2Cl_2 \rightarrow$$
$$Ca(ClO)_2 + CaCl_2 . Ca(OH)_2 . H_2O + H_2O$$

Cold milk of lime yields calcium hypochlorite while the hot alkali yields calcium chlorate, together with calcium chloride in solution:

$$2Ca(OH)_2 + 2Cl_2 \rightarrow$$
$$CaCl_2 + Ca(ClO)_2 + 2H_2O$$
calcium
hypochlorite
(chlorate(I))

$$6Ca(OH)_2 + 6Cl_2 \rightarrow$$
$$5CaCl_2 + Ca(ClO_3)_2 + 6H_2O$$
calcium
chlorate
(chlorate(v))

Slaked lime is used for making mortar and plaster in the building industry, the production of bleaching powder, the absorption of acidic gases as in the purification of coal gas and town gas, the recovery of ammonia from waste liquors in the Ammonia-Soda process and the softening of water. Mortar is a mixture of slaked lime and sand (about 1 : 3) or powdered clinker ash made into a paste with water. The mortar in a brick wall sets initially by evaporation of water into the air or absorption into the bricks. Afterwards, slow absorption of carbon dioxide occurs with the formation of calcium carbonate. No other chemical reaction occurs.

Barium hydroxide, baryta, $Ba^{2+}(OH^-)_2$

The octahydrate, $Ba(OH)_2 . 8H_2O$, is usually obtained. The aqueous solution, baryta-water, may be used instead of lime-water as a more sensitive test for carbon dioxide. Barium hydroxide is soluble enough to be very useful in titrimetric (volumetric) analysis for the estimation of weak acids using phenolphthalein as indicator. The use of sodium hydroxide is usually made difficult by the absorption of carbon dioxide as carbonate ions, the consequent pH change rendering the end-point indefinite. Using baryta, absorbed carbon dioxide is precipitated as the carbonate, the ions of which cannot therefore interfere with the titration of weakly ionized acids.

Important salts formed by the alkaline-earth metals

Compounds formed by the alkaline-earth metals, except those of beryllium, are essentially ionic. Unless the anion is coloured, they are colourless solids. A compound is given under the Periodic Group of the principal non-metal, excluding oxygen, which it contains.

Group IV: carbon

Carbonates, $M^{2+}CO_3^{2-}$, and hydrogencarbonates (bicarbonates), $M^{2+}(HCO_3^-)_2$

Carbonic acid is an extremely weak dibasic acid which forms both normal salts, carbonates, and acid salts, hydrogencarbonates (bicarbonates), with the stronger, more electropositive metals. Basic carbonates are formed by the less electro-positive metals.

H_2CO_3	$Ca(HCO_3)_2$	$CaCO_3$
carbonic acid	calcium hydrogen-carbonate	calcium carbonate

The carbonates are insoluble in water. Thermal dissociation occurs on heating and by allowing the carbon dioxide formed to escape, the oxide is formed:

$$CaCO_3(s) \rightleftharpoons CaO(s) + CO_2(g)$$

The dissociation pressure of carbon dioxide reaches 101 kPa (atmospheric pressure), signifying complete decomposition, at temperatures which are too high for the thermal decomposition of strontium and barium carbonate using normal laboratory facilities; the approximate temperatures are:

MgCO₃ (540°C)   CaCO₃ (900°C)
SrCO₃ (1290°C)   BaCO₃ (1360°C)

The relative thermal stability of the carbonate indicates the degree of electropositive character of the metal. Beryllium carbonate is very unstable and is prepared in an atmosphere of carbon dioxide. The carbonates may be precipitated by mixing solutions containing the required ions, but the addition of a soluble carbonate to magnesium salt solutions precipitates a basic carbonate of variable composition, magnesia alba, e.g.,

$$3MgCO_3 . Mg(OH)_2 . 3H_2O$$

and $$3MgCO_3 . Mg(OH)_2 . 4H_2O$$

Only magnesium and beryllium form basic carbonates. Their hydroxides are the least soluble in the group and consequently are precipitated by hydroxide ion formed in the hydrolysis of the added soluble carbonate. Carbon dioxide converts the basic carbonate into normal carbonate. Alternatively, addition of sodium hydrogencarbonate results in the slow separation on standing of the normal carbonate.

Calcium carbonate crystallizes in two forms: the stable calcite (hexagonal system) and the meta-stable aragonite (rhombic system). Calcite is the usual form and occurs widely under various names. Iceland spar consists of colourless crystals. Marble, capable of taking a polish, and limestone are masses of interlocking crystals while chalk is microcrystalline, formed from the shells of dead marine creatures sinking into the ocean mud of bygone ages. Aragonite is found in coral and the shells of molluscs. It is mined in Cumberland. Strontium carbonate occurs as strontianite, SrCO₃, found in the lead mine of Strontian in Argyllshire in 1791, after which the metal and its compounds are named. Barium carbonate is mined as witherite, BaCO₃.

Calcium carbonate is used widely as the source of quicklime and is added to the furnace-charge as a flux in many metallurgical operations. Decomposition to the oxide occurs with formation of a fusible slag with acidic oxides. Precipitated chalk and magnesia alba are used in cosmetics and toothpastes.

The hydrogencarbonates are usually met only in solution. They are formed by the action of carbon dioxide on a suspension in cold water of the carbonate to which they revert on heating the solutions:

$$CaCO_3(s) + CO_2 + H_2O \rightleftharpoons Ca^{2+} + 2HCO_3^-$$

Hardness of water is due to the presence of calcium and magnesium compounds which form insoluble salts, seen as a scum, with soap. If hydrogencarbonate ions are also present, boiling produces a precipitate of carbonate, thereby removing the offending metal ions. Hardness which is removed by boiling is called *temporary hardness*.

Hydrogencarbonates and carbonates react with solutions of acids stronger than carbonic acid, with evolution of carbon dioxide:

$$HCO_3^- + H^+ \rightarrow H_2O + CO_2$$

$$MgCO_3(s) + 2H^+ \rightarrow Mg^{2+} + CO_2 + H_2O$$

Portland cement and concrete

Portland cement, so-called because of the resemblance to Portland stone, is a mixture of calcium aluminate and calcium silicate which sets to a hard, strong mass after being mixed with water. Clay, $Al_2O_3 . 2SiO_2 . 2H_2O$, and limestone, $CaCO_3$, or a naturally occurring marl of roughly this composition, is quarried, made into a fine sludge with water and sprayed into the top of a long, inclined, rotating, cylindrical kiln of length 45–110 m and diameter 2–4 m, fired by oil or powdered coal. The highest temperature reached is about 1500°C. After evaporation of water, the fine granules sinter together into pebbles which are ground to a fine powder. On hydration, complex physical and chemical changes occur and hydrated aluminates and silicates are formed in a rock-like mass of interlacing crystals.

Concrete is a mixture of cement and a ballast of gravel, chippings or crushed brick, of a size selected for the type of work. The mixture is made thoroughly wet with water, but of a thick consistency, poured into a mould and allowed to set. Reinforced concrete contains a mesh work of steel rods over which the mixture is poured: it is used in the construction of bridges and large buildings.

Calcium carbide (ethynide, acetylide), $Ca^{2+}\left(\begin{array}{c}C^-\\ \parallel\\ C^-\end{array}\right)$, and calcium cyanamide, $Ca^{2+}(\overset{2-}{N}-C\equiv N)$

Calcium carbide is produced commercially by the reaction between quicklime and coke in an electric furnace. The process is continuous. A large alternating current passes through the coke, between the carbon blocks used to line the furnace and a large carbon electrode. The product fuses and carbon monoxide is evolved. It is white when pure, but the commercial product is obtained as greyish chippings, stored in tightly sealed tins to exclude moisture:

$$CaO + 3C \rightarrow CaC_2 + CO$$

Strontium and barium carbides are formed similarly.

Calcium carbide, proper chemical name calcium ethynide (acetylide) yields the gaseous unsaturated hydrocarbon ethyne (acetylene) with water:

$$CaC_2 + 2H_2O \rightarrow \overset{\textstyle CH}{\underset{\textstyle CH}{\parallel}} + Ca(OH)_2$$

Impurities of phosphine and hydrogen sulphide, produced by hydrolysis of calcium phosphide and sulphide, originating as impurities in the coke, may be removed by passing the gas through acidified copper(II) sulphate when copper phosphide and sulphide are precipitated.

Ethyne is extremely important as the start-point for the synthesis of many organic compounds. It is used in the oxy-acetylene blow-pipe for welding and cutting. It is an endothermic compound and when liquefied is apt to explode spontaneously to the elements. Cylinders of ethyne contain the gas dissolved in propanone (acetone).

Calcium carbide is also used for the production of calcium cyanamide (p. 191), formed by the action of nitrogen at red heat:

$$CaC_2 + N_2 \rightarrow CaN.CN + C$$

Group V: nitrogen, phosphorus

Nitrides and phosphides: see reactions of the metals.

Nitrites, $M^{2+}(NO_2^-)_2$, and nitrates, $M^{2+}(NO_3^-)_2$

Nitrites of the alkaline-earth cannot be prepared by thermal decomposition of the nitrates. When the nitrates are heated, water of crystallization is evolved except with barium nitrate which has none, and the anhydrous salts break down to form the oxides, nitrogen dioxide and oxygen:

$$2Ca(NO_3)_2 \rightarrow 2CaO + 4NO_2 + O_2$$

The nitrates are prepared by neutralizing dilute nitric acid with an excess of the solid hydroxide or carbonate, and for magnesium nitrate, the oxide or metal:

$$Ca(OH)_2(s) + 2H^+ + 2NO_3^- \rightarrow$$
$$Ca^{2+} + 2NO_3^- + 2H_2O$$

$$SrCO_3(s) + 2H^+ + 2NO_3^- \rightarrow$$
$$Sr^{2+} + 2NO_3^- + H_2O + CO_2$$

$$MgO(s) + 2H^+ + 2NO_3^- \rightarrow$$
$$Mg^{2+} + 2NO_3^- + H_2O$$

The nitrates of strontium and barium are produced commercially in the manner used for potassium nitrate by mixing hot concentrated solutions of sodium nitrate and the chloride of the metal.

Barium nitrate is one of the less soluble nitrates. The others are very soluble. Calcium nitrate is the most soluble, followed by magnesium nitrate, both deliquescing. The common hydrates are:

$$Mg(NO_3)_2.6H_2O \quad Ca(NO_3)_2.4H_2O$$

$$Sr(NO_3)_2.4H_2O$$

Barium nitrate imparts a green coloration and strontium nitrate a red coloration to fireworks and signal flares. Thermal decomposition yields oxygen which aids combustion.

The nitrites are much less important even than the nitrates. To illustrate methods of preparation: barium nitrite could be prepared by mixing solutions of sodium nitrite and barium chloride, separating the less soluble sodium chloride first.

Magnesium ammonium phosphate, hexahydrate $Mg^{2+}NH_4^+PO_4^{3-}.6H_2O$

The addition of disodium hydrogenphosphate solution, ammonia to release phosphate ions and ammonium chloride to a solution containing magnesium ions produces a colourless, crystalline precipitate of hydrated magnesium ammonium phosphate:

$$Mg^{2+} + NH_4^+ + PO_4^{3-} + 6H_2O \rightarrow$$
$$MgNH_4PO_4.6H_2O(s)$$

This reaction may be used as a qualitative test for magnesium ions or for their quantitative determination when the precipitate is decomposed by heating to constant mass, magnesium pyrophosphate, dimagnesium diphosphate, remaining:

$$2MgNH_4PO_4.6H_2O \rightarrow$$
$$Mg_2P_2O_7 + 2NH_3 + 13H_2O$$

The calcium orthophosphates

Phosphoric acid is a weak, tribasic acid forming three series of salts which may be named as follows:

H_3PO_4	$CaHPO_4.2H_2O$
orthophosphoric acid	calcium hydrogen orthophosphate dihydrate
$Ca(H_2PO_4)_2.H_2O$	$Ca_3(PO_4)_2$
calcium tetrahydrogen di-orthophosphate monohydrate	tricalcium di-orthophosphate or normal calcium phosphate

The normal salt, commonly called calcium phosphate, is precipitated by mixing a solution containing calcium ions with the usual phosphate laboratory reagent, disodium hydrogenphosphate solution and alkali, to liberate phosphate ions:

$$HPO_4^{2-} + OH^- \rightarrow PO_4^{3-} + H_2O$$

$$3Ca^{2+} + 2PO_4^{3-} \rightarrow Ca_3(PO_4)_2(s)$$

Calcium phosphate is a white powder, insoluble in water but soluble in strongly ionized acids, due to the formation of weakly ionized hydrogenphosphate ions:

$$H^+ + PO_4^{3-} \rightleftharpoons HPO_4^{2-}$$

$$H^+ + HPO_4^{2-} \rightleftharpoons H_2PO_4^-$$

$$H^+ + H_2PO_4^- \rightleftharpoons H_3PO_4$$

The maximum concentrations of calcium and phosphate ions co-existing in the saturated solution at equilibrium with solid are defined by the value of the solubility product at that temperature:

$$[Ca^{2+}]^3[PO_4^{3-}]^2 = \text{solubility product}$$

If the solubility is s mol dm^{-3}, $[Ca^{2+}] = 3s$ and $[PO_4^{3-}] = 2s$ mol dm^{-3}, so that $[Ca^{2+}]^3[PO_4^{3-}]^2 = 108s^5$ mol^5 dm^{-15} = solubility product at this temperature. Removal of PO_4^{3-} ions (as weak acid by addition of hydrogen ions from a strongly

ionized acid) means that $[Ca^{2+}]^3[PO_4^{3-}]^2 < 108s^5$ and more calcium phosphate dissolves until the solubility product is reached. With a suitable quantity of acid, the whole of the calcium phosphate dissolves.

The calcium orthophosphates are important fertilizers. Normal calcium phosphate is only slightly soluble and slowly absorbed. It may be applied as bone meal or basic slag, containing calcium phosphate and calcium silicate, $Ca_3(PO_4)_2$ and $CaSiO_3$, from steel-making. However, impure calcium phosphate, mined as rock phosphate, and fluorapatite, $3Ca_3(PO_4)_2.CaF_2$, are the chief sources of phosphates and phosphorus. To achieve quicker action as a fertilizer, ground normal calcium phosphate is converted into superphosphate of lime by the prolonged action of crude 65–70% sulphuric acid over a period of 10–14 days:

$$Ca_3(PO_4)_2 + 2H_2SO_4 + 5H_2O \rightarrow$$
$$Ca(H_2PO_4)_2.H_2O + 2CaSO_4.2H_2O$$

This is a major application of sulphuric acid. Rather more than one-half of the superphosphate is the inactive calcium sulphate. Freight charges would be considerably lower if this sulphate were absent. To achieve this, triple superphosphate is made by the action of orthophosphoric acid instead of sulphuric acid:

$$Ca_3(PO_4)_2 + 4H_3PO_4 + 3H_2O \rightarrow$$
$$3Ca(H_2PO_4)_2.H_2O$$

Group VI (excluding oxygen): sulphur

Sulphides, $M^{2+}S^{2-}$

The sulphides may be prepared by union of the elements or by reduction of the sulphate. Calcium, strontium and barium sulphates may be reduced by prolonged heating with wood charcoal in a furnace, the reduction of barium sulphate being the most difficult:

$$CaSO_4 + 4C \rightarrow CaS + 4CO$$

Calcium sulphide is sparingly soluble, strontium sulphide dissolves readily in warm water while barium sulphide is readily soluble in cold water, all passing into solution as the hydrogensulphide and hydroxide formed by hydrolysis:

$$CaS + H_2O \rightleftharpoons Ca^{2+} + HS^- + OH^-$$

$$HS^- + H_2O \rightleftharpoons H_2S + OH^-$$

Magnesium sulphide is rapidly decomposed by water according to the same scheme.

Calcium hydrogensulphide (hydrosulphide), $Ca(HS)_2$, is made industrially by acting on milk of lime with hydrogen sulphide:

$$OH^- + H_2S \rightleftharpoons SH^- + H_2O$$

It is used for removing hair from ox-hides and in depilatories generally.

Calcium sulphite, calcium trioxosulphate, $Ca^{2+}S_3^{2-}$, and calcium hydrogensulphite (bisulphite), $Ca^{2+}(HSO_3^-)_2$

Calcium sulphite is conveniently prepared as a white precipitate obtained by mixing solutions of a sulphite and a calcium salt. Calcium hydrogensulphite is made industrially by passing an excess of sulphur dioxide into milk of lime:

$$Ca(OH)_2 + SO_2 \rightarrow CaSO_3(s) + H_2O$$

$$CaSO_3(s) + H_2O + SO_2 \rightarrow Ca(HSO_3)_2$$

Boiling calcium hydrogensulphite solution is used for the extraction of lignin from wood-pulp, leaving cellulose for paper-making. The solution is also used for sterilizing beer casks.

Sulphates (tetraoxosulphates), $M^{2+}SO_4^{2-}$

Magnesium sulphate differs from the others in being freely soluble in water:

	Solubility/ g anhydrous salt in 100 g water	
$MgSO_4$	35.8	(20°C)
$CaSO_4$	0.11	(25°C)
$SrSO_4$	0.011	(18°C)
$BaSO_4$	0.00022	(25°C)

This determines the most convenient methods of preparation. Magnesium sulphate is prepared by the neutralization of dilute sulphuric acid using metal, oxide, hydroxide or carbonate, excess of which may be filtered off. The solution, which is not hydrolysed, deposits the heptahydrate, $MgSO_4.7H_2O$. The other sulphates are prepared by precipitation.

In testing for sulphate ions in solution, the addition of barium chloride solution and hydrochloric acid, to eliminate carbonate ions, etc., which would also cause precipitation, yields a white precipitate of barium sulphate. An interesting application of the difference in solubilities of sparingly soluble sulphates of calcium and strontium appears in qualitative analysis: strontium sulphate

is precipitated by addition of a saturated solution of the more soluble calcium sulphate, barium ions having been already precipitated as the chromate. The presence of calcium and magnesium ions, associated with sulphate and chloride ions, accounts for the permanent hardness of water.

Magnesium sulphate, magnesium tetraoxosulphate, $Mg^{2+}SO_4^{2-}$

Magnesium sulphate was discovered in spring water at Epsom, Surrey, in 1695. The colourless, crystalline heptahydrate, $MgSO_4.7H_2O$, is known as Epsom salts. Magnesium sulphate also occurs in natural salt deposits at Stassfurt in Germany and elsewhere which may contain:

kieserite $MgSO_4.H_2O$

schönite $K_2SO_4.MgSO_4.6H_2O$

kainite $K_2SO_4.MgSO_4.MgCl_2.6H_2O$, (or, more simply, $KCl.MgSO_4.3H_2O$)

Epsom salts is made commercially by reacting magnesite or dolomite with sulphuric acid or by re-crystallization of the naturally occurring salt. It is used as a purgative and in the tanning and dyeing industries.

Calcium sulphate, calcium tetraoxosulphate, $Ca^{2+}SO_4^{2-}$

Calcium sulphate occurs naturally as anhydrite, $CaSO_4$, and as the dihydrate, gypsum, $CaSO_4.2H_2O$, various forms of which are known as selenite (transparent), satin spar (fibrous) and alabaster (opaque).

When gypsum is heated to 97°C, loss of water occurs and the hemihydrate results, a change which is essentially:

$$2(CaSO_4.2H_2O) \rightleftharpoons (CaSO_4)_2.H_2O + 3H_2O$$

The product is plaster of Paris. Addition of water reverses the change and a solid mass of interlocking crystals of gypsum results, accompanied by a slight expansion and generation of heat. Plaster of Paris gives sharply defined castings which are cheap and easy to make. If surface pores in the object are filled by application of paraffin wax dissolved in a volatile solvent, such as petrol, the casting formed from it has a smooth, hard, ivory-like surface. Plaster of Paris is widely used in surgery for keeping parts of the body rigidly in position.

Setting takes up to 15 minutes and may be accelerated by addition of salt or delayed by the use of borax or alum.

Further heating to 200°C yields anhydrous calcium sulphate which also takes up water but stronger heating above 400°C produces a 'dead-burnt' product which is re-hydrated only very slowly. Hard-wearing plaster surfaces (Parian Cement, Estrich's Plaster) use the slow-setting form which sets without expansion, or plaster of Paris and alum (Keene's Cement). Stucco is plaster of Paris made with glue.

Calcium sulphate is used as a filling in the manufacture of glazed paper and for the production of sulphuric acid.

Strongly heated to above 960°C, calcium sulphate begins to dissociate:

$$2CaSO_4 \rightleftharpoons 2CaO + 2SO_3$$
$$\Updownarrow$$
$$2SO_2 + O_2$$

The relatively more volatile sulphur trioxide may be displaced by heating calcium sulphate with silica, the reaction commencing at 870°C:

$$CaSO_4 + SiO_2 \rightarrow CaSiO_3 + SO_3(g)$$
$$\text{calcium silicate}$$

This is similar to the displacement of phosphorus pentoxide from calcium phosphate in the production of phosphorus.

Strontium sulphate, strontium tetraoxosulphate, $Sr^{2+}SO_4^{2-}$

Strontium sulphate melts at 1605°C.

Strontium sulphate is found naturally as celestine or celestite. It is a white anhydrous solid.

Barium sulphate, barium tetraoxosulphate, $Ba^{2+}SO_4^{2-}$

Barium sulphate melts at 1580°C.

Barium sulphate is found naturally as heavy spar or barytes. It is a white solid and very insoluble in water but as it is an important source of barium compounds, methods have been devised to use it in preparations.

On boiling barium sulphate with a concentrated solution of sodium carbonate, some of the less soluble barium carbonate is formed:

$$BaSO_4(s) + CO_3^{2-} \rightleftharpoons BaCO_3(s) + SO_4^{2-}$$

The precipitate is filtered off, washed free of sodium sulphate and boiled with fresh sodium carbonate solution. The cycle is repeated. Alternatively, barium sulphate is fused with 4–6 times its weight of sodium carbonate, cooled and sodium sulphate extracted with boiling water. The barium carbonate resulting from either treatment is dissolved in acid.

Barium sulphate may be reduced to the sulphide by heating to a dull redness in a fireclay crucible with powdered charcoal for two hours:

$$BaSO_4 + 4C \rightarrow BaS + 4CO$$

The solid product, barium sulphide, is extracted with boiling water and decomposed with acid.

Barium sulphate is used in paper-making to improve opacity and weight. Nearly all white and white tinted paints, oil and water, contain the white pigment lithopone, which is a mixture of barium sulphate and zinc sulphide. Barium sulphide is formed by reduction of barytes with coke in a rotary inclined kiln heated by producer gas, extracted with water, purified and mixed with purified zinc sulphate solution:

$$Ba^{2+} + S^{2-} + Zn^{2+} + SO_4^{2-} \rightarrow$$
$$\text{(hydrolysed)} \qquad BaSO_4(s) + ZnS(s)$$

The mixture, which is not blackened in sulphurous town air, is filtered off, dried and processed further for incorporation in paints. As an alternative, titanium(IV) oxide, titanium dioxide is used as a pigment in paints, giving a brilliant white finish which is not discoloured in polluted air.

Barium sulphate begins to dissociate slightly near the melting-point.

Group VII: halogens

Comparison of solubilities for fluorides and chlorides, p. 427. Calcium hypochlorite and bleaching powder, p. 436. Barium iodate, p. 442.

Halides, $M^{2+}(X^-)_2$

The fluorides are all sparingly soluble and prepared by mixing solutions containing the required ions:

$$Ca^{2+} + 2F^- \rightarrow CaF_2(s)$$

Calcium fluoride, formed here as a white gelatinous precipitate, is the most important alkaline-earth fluoride. Generally, the other halides are very soluble, those of barium having the lowest solubility. Chlorides are prepared by the neutralization of hydrochloric acid, on addition of hydroxide or carbonate, excess being filtered off. For magnesium chloride, the oxide or metal may be used. The most stable hydrates formed by the chlorides are:

$$MgCl_2 . 6H_2O \quad CaCl_2 . 6H_2O$$

$$SrCl_2 . 6H_2O \quad BaCl_2 . 2H_2O$$

The hexahydrates are very soluble in water and deliquescent. On heating, the magnesium salt is somewhat hydrolysed to the oxide chloride, formerly oxychloride, while calcium chloride is less hydrolysed and strontium chloride and barium chloride not at all:

$$MgCl_2 . 6H_2O \rightarrow Mg(OH)Cl + HCl(g) + 5H_2O(g)$$

followed by

$$Mg(OH)Cl \rightarrow MgO + HCl(g)$$

The tendency of the halides to form complex ammines (structure p. 289) and the amount of hydration declines with increase in atomic number of the metal.

Calcium fluoride, $Ca^{2+}(F^-)_2$

Calcium fluoride is sparingly soluble (solubility = 0.0025 g in water at 18°C). It melts at 1330°C.

Calcium fluoride occurs naturally as fluorite or fluorspar. The blue variety, used ornamentally, is known as Blue John. Calcium fluoride is used in the production of hydrofluoric acid by treatment with concentrated sulphuric acid:

$$CaF_2 + H_2SO_4 \rightarrow CaSO_4 + 2HF$$

Magnesium chloride, $Mg^{2+}(Cl^-)_2$

The common hydrate is the hexahydrate $MgCl_2 . 6H_2O$, although combination with 12, 8, 6, 4 and 2 molecules of water of crystallization may be encountered.

Magnesium and chloride ions are present in sea water and natural brines. Magnesium chloride is found in salt deposits formed by the evaporation of salt lakes; thus, carnallite has the composition $KCl . MgCl_2 . 6H_2O$.

Anhydrous magnesium chloride cannot be prepared by direct evaporation of the aqueous solution because of hydrolysis. It is formed by passing dry chlorine over a mixture of magnesium oxide and carbon,

$$MgO + C + Cl_2 \rightarrow MgCl_2 + CO$$

It melts at 715°C. At 20°C, 54.1 g anhydrous magnesium chloride dissolves in 100 g water. Other general methods of preparation include the action of dry chlorine or dry hydrogen chloride on the heated metal. Alternatively, the hexahydrate may be heated in a stream of hydrogen chloride, the high concentration of which reduces hydrolysis, or a double salt, formed by evaporation of a solution of magnesium chloride and ammonium chloride in equi-molar quantities, may be heated:

$$MgCl_2 . NH_4Cl . 6H_2O \rightarrow$$
$$MgCl_2 + NH_4Cl + 6H_2O(g)$$

$$MgCl_2 + NH_4Cl \rightarrow$$
$$MgCl_2 + NH_3(g) + HCl(g)$$

Calcium chloride, $Ca^{2+}(Cl^-)_2$

The common, deliquescent hydrate has six molecules of water of crystallization, $CaCl_2 . 6H_2O$, and melts at 30°C. A tetrahydrate and dihydrate may also be prepared.

Calcium chloride is formed in great quantity as a by-product of the Ammonia-Soda process. There are few demands for it. Previously fused calcium chloride is a useful drying agent but may not be used for ammonia or alcohols, with which it forms compounds normally given as $CaCl_2 . 8NH_3$, $CaCl_2 . 4CH_3OH$ and $CaCl_2 . 6C_2H_5OH$.

At 20°C. 81.5 g anhydrous calcium chloride saturates 100 g water. The anhydrous salt melts at 782°C.

Barium chloride, $Ba^{2+}(Cl^-)_2$

Barium chloride may be made in the laboratory from barium sulphate by conversion into the carbonate or sulphide followed by reaction with hydrochloric acid. Commercially, calcium chloride is heated with barytes and coke, the product being extracted from the residual ash and sparingly soluble calcium sulphide with water and purified:

$$BaSO_4 + 4C + CaCl_2 \rightarrow BaCl_2 + CaS + 4CO$$

It crystallizes with two molecules of water of crystallization. Barium chloride dihydrate is neither deliquescent nor hygroscopic and yields the anhydrous salt, without hydrolysis, on heating. Anhydrous barium chloride melts at 955°C. At 20°C, 37.2 g saturates 100 g water.

Qualitative tests for ions of the alkaline-earth metals

The flame test

Salts of calcium, strontium and barium, thoroughly moistened with concentrated hydrochloric acid, to take advantage of the greater volatility of the halides, colour the non-luminous flame respectively yellowish-red, carmine-red and yellowish-green. Traces of sodium compounds will provide a yellow tinge to the flame colorations rendering the distinction between strontium and calcium difficult. However, when seen through a spectroscope, the light due to calcium compounds has red, green and violet as principal lines while the lines of the strontium flame are red, orange, blue and violet. The sodium lines are yellow.

Separation of ions (excluding beryllium) from solution

After the removal of the elements of (Qualitative) Groups I–IV inclusive in qualitative analysis, addition of ammonia and ammonium carbonate solution, ammonium chloride being already present, precipitates the carbonates of calcium, strontium and barium, but not that of magnesium, the solubility product of which is not reached at this pH. Precipitation is usually slow. The filtrate contains any magnesium ions present.

(a)   The mixture of carbonates is dissolved in dilute ethanoic acid, boiled and potassium chromate solution added to give a yellow precipitate of barium chromate:

$$Ba^{2+} + CrO_4^{2-} \rightleftharpoons BaCrO_4(s)$$

$[Ba^{2+}][CrO_4^{2-}] = 2 \times 10^{-10} \, mol^2 \, dm^{-6}$ at room temperature. Calcium and strontium chromates would be precipitated from fairly concentrated solution in the absence of acid, but even weakly ionized ethanoic acid lowers the concentrations of chromate ions (as hydrogenchromate ions, etc.) sufficiently for the solubility products not to be reached. In the presence of a strongly ionized acid, hydrochloric acid, barium chromate would not be precipitated by the potassium chromate solution, which now contains hydrogenchromate ions:

$$H^+ + CrO_4^{2-} \rightleftharpoons HCrO_4^-$$

and the ionic product $[Ba^{2+}][CrO_4^{2-}]$ does not reach $2 \times 10^{-10} \, mol^2 \, dm^{-6}$. Therefore, in qualitative analysis, the carbonates are dissolved in ethanoic acid. The presence of barium in the precipitate is confirmed by the flame test on the precipitate.

(b)   After removal of the precipitate, the orange solution is concentrated by evaporation and divided into two portions. To one portion, an equal volume of saturated calcium sulphate solution is added, the liquid heated and set aside. A fine white precipitate of strontium sulphate appears. The presence of strontium may be confirmed by the flame test, using strontium chloride as a comparison, or by use of a spectroscope. Strontium ions must be removed from the other portion before applying precipitation tests for calcium, because strontium salts are usually less soluble than those of calcium. Addition of a fairly concentrated solution of ammonium sulphate, followed by heating, causes the precipitation of strontium sulphate, which takes about 10 minutes to be completed. The precipitate is discarded. The liquid is heated with ammonia and ammonium ethanedioate (oxalate) solution, when a white precipitate of calcium ethanedioate (oxalate) separates. Recall the use of soluble ethanedioates of sodium and potassium in titrimetric analysis (p. 121). The presence of calcium is confirmed by the flame test, using calcium chloride as a comparison, or spectroscopically.

(c)   The presence of magnesium ions in the filtrate from the initial precipitation of carbonates may be demonstrated in two ways. Addition of sodium hydroxide solution to a solution of a magnesium salt containing a trace of a suitable dye yields a coloured precipitate, or lake, of magnesium hydroxide in a colourless medium. Magneson I (2:4-dihydroxy-4′-nitro-azobenzene) and Magneson II (4-paranitrobenzene-azo-α-naphthol) yield blue lakes while Titan Yellow (also called Clayton Yellow, or Thiazole Yellow) produces a cherry-red precipitate. Alternatively, addition of disodium hydrogenphosphate solution and ammonia to liberate PO_4^{3-} ions in the presence of ammonium chloride to prevent precipitation of magnesium hydroxide, yields a white crystalline precipitate of magnesium ammonium phosphate. Magnesium ions give no flame coloration.

Two reactions showing the distinctive behaviour of beryllium salts

(a)   While magnesium, calcium, strontium and barium have white insoluble ethanedioates (oxalates), the addition of ammonium ethanedioate solution to a solution of a beryllium salt gives no precipitate but an auto-complex beryllium ethanedioate: $Be[Be(C_2O_4)_2]$. Complexes are rarely formed by the other alkaline-earth metals.

(b)   Addition of sodium hydroxide solution gives a white, gelatinous precipitate of beryllium hydroxide, soluble in excess of the reagent but re-precipitated on boiling. This distinguishes the precipitate from aluminium hydroxide:

$$Be(OH)_2(s) + OH^- \rightleftharpoons HBeO_2^- + H_2O$$

beryllate ion

19

Group III: the boron group

Boron and aluminium

5	
B	
2, 3	
13	
Al	
2, 8, 3	
31	
Ga	
2, 8, 18, 3	
49	
In	
2, 8, 18, 18, 3	
81	
Tl	
2, 8, 18, 32, 18, 3	

Boron is the most electronegative element of Group III and has little in common with the other elements. It is non-metallic. This is the first group in which the change from the non-metallic towards the metallic condition occurs with increasing atomic number. The detailed chemistry of aluminium and its compounds will be described, while of the other elements, selected compounds of boron only are included.

Group III is unusual. Reference to the open form of the Periodic Table reveals that the group is placed to the right of the gap opened for the inclusion of transition and associated metals. It is the first group of the *p* block elements. No elements separate beryllium and boron in Period 2, or magnesium and aluminium in Period 3, but in the successive periods scandium (21), yttrium (39), lanthanum (57) and actinium (89) appear as the first members of transition series, formed by successive expansion of penultimate electron shells. In addition, Periods 6 and 7 contain the inner transition series, lanthanides and actinides. Scandium, yttrium, lanthanum and actinium form the A subgroup of the Mendeléeff Table while the B subgroup consisted of gallium, indium and thallium. The subgroups are very much more alike than in the previous groups. However, only boron and aluminium will be considered further. Both have unusual properties. Aluminium is on the 'diagonal borderland' between metals and non-metals in the chemical sense.

The most widely known compound of boron is borax, the name being derived from an Arabic word given to fluxes in general. Impure boron was first isolated in 1808 by Gay-Lussac and Thénard using reduction of boric oxide with potassium. Aluminium was isolated by the reduction of anhydrous aluminium chloride with potassium by Oersted (1825) and Wöhler (1827). The name is derived from a general term for astringents, *alumen*. From 1886, when Hall (USA) and Héroult (France) independently devised the commercial process to produce aluminium by electrolysis, the metal has become of major importance in the light-alloy industry.

Both boron and aluminium are too reactive to occur native. Aluminium is the most abundant metal in the Earth's crust, being present in rocks and clays as alumino-silicates, occurring also as the oxide, and in Greenland, as a complex fluoride, cryolite, Na_3AlF_6. Boron is found as borates of calcium, magnesium and sodium of complex formulae but the most important deposits are kernite, $Na_2B_4O_7 . 4H_2O$, and tincal (borax), $Na_2B_4O_7 . 10H_2O$, occurring in California, USA. Boron is emitted as boric acid in the volcanic vapours from the fumaroles of Tuscany in Italy. Boron is a rare element, essential to plant and animal life and widely, but sparingly, distributed. The scarcity of boron may be due to the ready disruption of the nucleus by sub-atomic particles from natural sources of radioactivity.

Physical data for boron and aluminium are given in Table 19.1.

A general survey of bonding involving boron and aluminium of the *p* block

In Groups I and II, chemical combination is accompanied by the ejection of electrons to form positive ions. The Group III elements have three

Table 19.1 Group III (boron and aluminium): physical data *p* block elements, ns² np¹

	B	Al
Atomic number	5	13
Electronic configuration	2, 3	2, 8, 3
[Core]outer electrons	[He]$(2s)^2(2p)^1$	[Ne]$(3s)^2(3p)^1$
Relative atomic mass	10.811	26.9815
Metallic radius/nm	0.098	0.143
Atomic radius/nm	0.082	0.125
Ionic radius/nm	0.020*	0.050
Atomic volume/cm³ mol⁻¹*** $\begin{cases}4.64** \\ 4.42\end{cases}$		9.99
Standard electrode potential (H Scale: 298 K)/V	−0.73	−1.66
Boiling-point/°C	*c*.2550	2270
Melting-point/°C	2300	660
Electronegativity	2.0	1.5
Density/g cm⁻³ (at 293 K) $\begin{cases}2.33** \\ 2.45\end{cases}$		2.70
Co-ordination number	4	4, 6
Abundance (% by mass)	3×10^{-4}	7.5

*See p. 77, Table 5.9
**There are two forms of boron
***Traditional term. It is the volume of one mole of atoms as a solid element

valency electrons more than the nearest inert gas and their loss does not occur readily and not at all for boron. Boron does not form a simple ion, always being covalent in combination: in salts, it appears united with other more electronegative elements (oxygen, fluorine) in the anion. The hypothetical boron ion of small size and triple charge would immediately cause polarization of the electron-clouds of the anion to the covalent condition.

Therefore, it is hardly surprising that for the next congener aluminium(III) compounds are essentially covalent while hydrated salts, containing $Al(H_2O)_6^{3+}$ are ionic.

Ionization energies for successive ionizations of boron to thallium are compared in Table 19.2 (note the abrupt jump in values immediately below those

Table 19.2 Group III (boron group): ionization energies/ kJ mol⁻¹ (273 K)

	B	Al	Ga	In	Tl
1st $M(g) \rightarrow M^+(g) + e^-$	801	**577**	**579**	**559**	**590**
2nd $M^+(g) \rightarrow M^{2+}(g) + e^-$	2422	1814	1968	1814	1959
3rd $M^{2+}(g) \rightarrow M^{3+}(g) + e^-$	3657	2740	2952	2692	2866
4th $M^{3+}(g) \rightarrow M^{4+}(g) + e^-$	24988	11578	6155	5577	4872

in bold typeface). In passing it may be noted that from gallium onward there is an increasing tendency for singly charged positive ions to be formed, the outer *s* pair of electrons becoming inert so the Tl^+ is usually more stable than Tl^{3+}. This *inert pair effect* will be seen again in later groups where a valency of the *group number less 2* emerges

e.g. Pb^{II} corresponding to oxidation state = +2

Bi^{III} corresponding to oxidation state = +3 .

Of the two elements, only aluminium forms cations. Ionic compounds include salts of the oxoacids, the oxide and hydrated chloride, where the anhydrous covalent compound is stabilized in the ionic state by hydration of the cation. The fluoride is possibly ionic, although its distinctive behaviour could be due to a macromolecular structure.

	Inert gas	*Ion*	
Simple valency 3	10 **Ne**	13 **Al³⁺**	$(Al^{3+})_2(O^{2-})_3$
			$(Al^{3+})_2(SO_4^{2-})_3$
Oxidation state = +3	2, 8	2, 8	$Al^{3+}(NO_3^-)_3$
			$[Al(H_2O)_6^{3+}](Cl^-)_3$
			$Al^{3+}(F^-)_3$

Hydrated salts contain the hydrated aluminium ion, in which the water molecules may be regarded as being attached by co-ionic bonding to the central ion; alternatively, the attraction might be purely electrostatic. Hydrolysis of aluminium salts in solution is explained by the transference of a proton from the hydrated aluminium ion to a solvent molecule, giving an excess of hydrogen ions (H_3O^+):

$$Al(OH_2)_6^{3+} + H_2O \rightleftharpoons Al(OH_2)_5(OH)^{2+} + H_3O^+$$

Both boron and aluminium exert a covalency of three in simple compounds. In exerting this group valency, both elements fail to gain the configuration of an inert gas. Boron acquires the electronic configuration of oxygen but, of course, retains the nuclear charge (atomic number) of boron (proton number = 5). Aluminium acquires the electronic configuration of sulphur, retaining its own nuclear charge (proton number = 13). In their compounds, both elements are in the oxidation state = +3, the oxidation number being calculated according to the rules on page 124. This is irrespective of the type of bonding, covalent or ionic, exhibited.

Electronic configurations

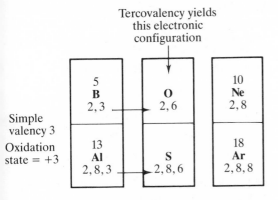

Tercovalency yields
this electronic
configuration

5 **B** 2, 3	**O** 2, 6	10 **Ne** 2, 8
13 **Al** 2, 8, 3	**S** 2, 8, 6	18 **Ar** 2, 8, 8

Simple
valency 3

Oxidation
state = +3

Illustrative compounds and their structures

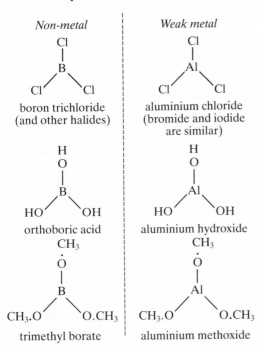

Non-metal

boron trichloride
(and other halides)

orthoboric acid

trimethyl borate

Weak metal

aluminium chloride
(bromide and iodide
are similar)

aluminium hydroxide

aluminium methoxide

The octet of electrons may be achieved by the formation of co-ionic bonds with atoms having lone pairs of electrons:

ammines

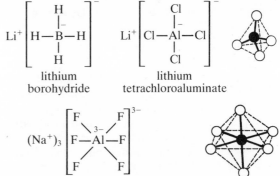

ether complexes

Electron diffraction studies show that aluminium chloride vapour is dimeric:

On the other hand, boron trichloride is not so, the structure being sometimes written with *back-co-ordination* to achieve an octet of electrons:

In the formation of complex ions both boron and aluminium achieve the inert gas electronic configuration, but only aluminium can exceed it:

lithium
borohydride

lithium
tetrachloroaluminate

sodium hexafluoroaluminate; cryolite
(octet exceeded)

Aluminium differs from boron in that $3d$ orbitals are available for this increase in co-ordination number.

Aqueous solutions of aluminium salts contain the hexa-aquo-aluminium(III) ion, $[Al(H_2O)_6]^{3-}$, of octahedral configuration. A co-ionic bond diagram may be constructed and the shape of the complex explained either in terms of mutual repulsion of the ligands around the aluminium ion or of occupancy of a hybridized [sp^3d^2] orbital by the lone-pair electrons of the water molecules. This complex tends to lose hydrogen ions in solution in competition with water molecules, an example of *salt hydrolysis*. Ligand exchange reactions occur in solution, fluoride forming [AlF_6]$^{3-}$, ethanedioate (oxalate) forming [$Al(C_2O_4)_3$]$^{3-}$.

Interestingly, thallium(III) forms rather more stable complexes in solution, e.g. $Tl(C_2O_4)_3^{3-}$ and $TlCl_6^{3-}$

The similarity of aluminium compounds to those of iron(III) may be noted and related to the respective sums of the 1st, 2nd and 3rd ionization energies (p. 304 and p. 447); this is seen particularly with the chlorides and sulphates. Finally, as will be apparent, aluminium behaves anomalously when its position in the electrochemical period is recalled, above those of the transition elements.

Boron, the boric acids and borates

Boron

Boron is a non-metallic element, which may be prepared as a very reactive amorphous black powder or in an inert crystalline form, as hard as boron carbide. Boron does not form positive ions.

Impure amorphous boron is prepared by heating a mixture of boron trioxide and magnesium powder when a vigorous reaction occurs. The product is treated with dilute hydrochloric acid, leaving impure boron:

$$B_2O_3 + 3Mg \rightarrow 2B + 3MgO$$

$$MgO + 2HCl \rightarrow MgCl_2 + H_2O$$

On heating, amorphous boron will react with nitrogen, oxygen, sulphur, fluorine, chlorine and bromine direct, to form respectively the nitride, BN; oxide, B_2O_3; sulphide, B_2S_3; and the halides BF_3, BCl_3, BBr_3. The oxide forms boric acid with

water while the other compounds may be hydrolysed to the acid and corresponding hydride, the nitride requiring steam and the fluoride finally forming a complex fluoroboric acid:

$$B_2O_3 + 3H_2O \rightarrow 2H_3BO_3$$

$$BN + 3H_2O \rightarrow H_3BO_3 + NH_3$$

$$B_2S_3 + 6H_2O \rightarrow 2H_3BO_3 + 3H_2S$$

$$4BF_3 + 3H_2O \rightarrow H_3BO_3 + 3HBF_4$$

$$BCl_3 + 3H_2O \rightarrow H_3BO_3 + 3HCl$$

$$BBr_3 + 3H_2O \rightarrow H_3BO_3 + 3HBr$$

The hydrides of boron are produced indirectly.

At red heat, boron decomposes steam:

$$2B + 3H_2O \rightarrow B_2O_3 + 3H_2$$

Boron reduces many oxides, sulphides and chlorides. Orthophosphoric acid is reduced to phosphorus at 800°C. Silica may be reduced to silicon:

$$3SiO_2 + 4B \rightarrow 2B_2O_3 + 3Si$$

Boron is used in some metallurgical processes as a 'de-oxidizer'. It is attacked by oxidizing acids. Hot concentrated nitric acid is reduced to nitrogen dioxide, leaving boric acid, a typical reaction with a non-metallic element:

$$6HNO_3 + 2B \rightarrow 2H_3BO_3 + 6NO_2$$

Boron is slowly attacked by hot strong alkali:

$$2B + 2OH^- + 2H_2O \rightarrow 2BO_2^- + 3H_2$$

The (meta)borate is formed.

Boron has excellent neutron-absorbing characteristics and is present in the rods used to moderate nuclear reactions in some nuclear reactors, once known as atomic piles. The nuclear chain reaction in the pile is maintained at a suitable rate by controlling the number of neutrons captured by $^{235}_{92}U$. Excess neutrons are absorbed by adjusting the extent to which the neutron-absorbing control rods project into the pile.

The boric acids and their salts: borax, disodium tetraborate, $Na_2B_4O_7.10H_2O$

Orthoboric acid is a very weakly ionized acid which readily assumes *condensed*† forms in salts. Metaboric acid is formed at 100°C, decomposing on

† i.e. elimination of H_2O from the —OH groups of adjacent molecules, giving a doubled structure with a —O—O— bridge

further strong heating to boron trioxide,

$$H_3BO_3 \xrightarrow{100°C} HBO_2 + H_2O$$

$$2HBO_2 \longrightarrow B_2O_3 + H_2O$$

Ortho (Greek = correct) indicates the presence of the maximum amount of the elements of water uniting while *meta* (Greek = after) has less. X-ray and tensimetric studies have shown that other acids, such as $H_2B_4O_7$, are hypothetical. Salts corresponding to this acid are known but other apparently much more complex borates form simple borate ions, BO_3^{3-}, on fusion or dissolving in water. When formed by the thermal decomposition of orthoboric acid, boron trioxide remains as a clear glass. It will react with bases to form borates, many of which are coloured, and displaces more volatile acidic oxides from their salts on heating:

$$CoO + B_2O_3 \rightarrow Co(BO_2)_2$$

$$Fe_2(SO_4)_3 + 3B_2O_3 \rightarrow 2Fe(BO_2)_3 + 3SO_3$$

The two most important naturally occurring compounds of boron are both hydrates of disodium tetraborate, kernite, $Na_2B_4O_7.4H_2O$, and tincal (borax), $Na_2B_4O_7.10H_2O$, which are mined in southern California, USA. Borax, the deca-hydrate, $Na_2B_4O_7.10H_2O$, is obtained by extraction with water and crystallization. Borates find a great many applications, being incorporated into heat-resistant borosilicate glass and fibreglass, porcelain enamels and ceramics, detergents, adhesives and sizes, fluxes and fireproofing preparations.

Borax is a colourless, crystalline solid which is much more soluble in hot water than cold. A solution of borax is hydrolysed in water:

$$B_4O_7^{2-} + 7H_2O \rightleftharpoons 2OH^- + 4H_3BO_3$$

Now, a 0.05 M solution of boric acid has pH $\simeq$ 5 so that borax may be titrated against an acid provided an indicator is used which shows its 'alkaline' colour to boric acid. Hydrochloric acid is frequently standardized against borax using methyl orange (pH range: 2.9–4.0) as indicator:

$$Na_2B_4O_7 + 2HCl + 5H_2O \rightarrow 2NaCl + 4H_3BO_3$$

This is similar to the titration of carbonates, where methyl orange is unaffected by weak carbonic acid.

In qualitative analysis, the borax bead test gives characteristic colours with certain cations. When borax is heated in a small loop in a platinum wire, it swells and forms a colourless, transparent bead,

possibly containing metaborate and boric oxide:

$$Na_2B_4O_7 \rightarrow 2NaBO_2 + B_2O_3$$

The cold bead is moistened and a minute quantity of the unknown substance is taken up, too much giving a dark, opaque bead in the subsequent heating. The bead is heated in the reducing zone of the bunsen flame and finally, in the oxidizing zone. Coloured borates are formed; for copper:

$$CuO + B_2O_3 \rightarrow Cu(BO_2)_2$$

Under oxidizing conditions, copper gives green (hot) and blue (cold), manganese gives amethyst (violet), cobalt, blue, and nickel, reddish-brown (all cold) coloured beads.

Orthoboric acid may be obtained by adding concentrated hydrochloric acid to a hot strong solution of borax. On cooling, it crystallizes out. Orthoboric acid is used as a mild antiseptic and in glazes. Addition of the calculated quantity of sodium hydroxide to borax yields the metaborate:

$$Na_2B_4O_7 + 2NaOH \rightarrow 4NaBO_2 + H_2O$$

Sodium metaborate is more soluble than borax and is also more alkaline by hydrolysis.

Electrolysis of sodium metaborate solution containing sodium carbonate or the addition of hydrogen peroxide to an alkaline solution of sodium metaborate gives sparingly soluble sodium perborate, monosodium trioxoborate, which may be formulated $NaBO_3.4H_2O$. The structural formula, $NaBO_2.H_2O_2.3H_2O$, has been used but this implies more than the experimental observations warrant. Sodium perborate is used as a mild bleaching agent in detergents and toothpastes.

Qualitative tests for borates

The solid is mixed into a paste with concentrated sulphuric acid, calcium fluoride and methanol, which is ignited. A green flame is a specific test for a borate. The green coloration is due to methyl borate, $B(OCH_3)_3$, and boron trifluoride, BF_3. Hydrogen fluoride liberated from calcium fluoride suppresses interference from copper and barium, which are also associated with green colorations.

Aluminium and its important compounds

Aluminium

Aluminium is a white metal of low density, with good electrical and thermal conducting properties, fairly hard except when very pure but malleable, ductile and easily extruded. Aluminium and its alloys are widely used. Industrially, it is prepared by the electrolysis of aluminium oxide (purified bauxite) in molten sodium hexafluoroaluminate, sometimes called sodium aluminium fluoride (cryolite). The metal is protected from attack by air, water and certain acids by a tenaciously adhering film of oxide which may be increased in strength and durability by anodic oxidation: the surface is then said to be anodized. When the oxide film is removed, aluminium is revealed as a very reactive element. Removal may be accomplished by treatment with warm mercury(II) chloride solution when mercury, displaced by the more electropositive aluminium, amalgamates with the surface which may then be wiped clean. The piece of aluminium becomes warm and white, wispy growths of the oxide appear while hydrogen is slowly evolved when it is immersed in water. Brine will also enable penetration of the oxide layer to occur. Aluminium alloy fishing reels can be badly damaged if not cleaned after contact with sea water. At about 800°C, aluminium burns in air with the formation of oxide and nitride. The terrible fires caused by missiles hitting warships constructed with aluminium alloys has already been mentioned.

Aluminium will react with carbon, nitrogen, phosphorus, oxygen, sulphur and the halogens on heating to form respectively: at over 1000°C, the carbide (methide), Al_4C_3; nitride, AlN; phosphide, AlP; oxide, Al_2O_3; sulphide, Al_2S_3; and halides, AlF_3, $AlCl_3$, $AlBr_3$ and AlI_3. Both oxide and fluoride are insoluble in water. The other halides ionize and are somewhat hydrolysed to give acidic solutions. The other binary compounds are completely hydrolysed to hydrated aluminium oxide or hydroxide and the non-metallic gaseous hydride:

$$Al_4C_3 + 12H_2O \rightarrow 4Al(OH)_3 + 3CH_4$$
aluminium   methane
methide

$$AlN + 3H_2O \rightarrow Al(OH)_3 + NH_3$$
aluminium   ammonia
nitride

$$AlP + 3H_2O \rightarrow Al(OH)_3 + PH_3$$
aluminium   phosphine
phosphide

$$Al_2S_3 + 6H_2O \rightarrow 2Al(OH)_3 + 3H_2S$$
aluminium   hydrogen
sulphide   sulphide

In the reactions represented by these equations aluminium resembles boron, but the elements differ in the behaviour of their halides:

$$AlCl_3 + 6H_2O \rightarrow Al(OH_2)_6^{3+} + 3Cl^-$$
aluminium   hydrated
chloride   aluminium
covalent   ion

$$Al(OH_2)_6^{3+} + H_2O \rightleftharpoons Al(OH_2)_5(OH)^{2+} + H_3O^+$$
(*Salt hydrolysis*)   hydrogen
                                          (oxonium)
                                            ions
                                          ∴ acidic

Aluminium hydride is prepared indirectly and is unimportant.

Aluminium reacts with hydrochloric acid under all conditions, but action is slow with very pure metal. Hydrogen is liberated and the chloride remains in solution.

$$2Al + 6H^+ \rightarrow 2Al^{3+} + 3H_2$$

Commercial aluminium dissolves vigorously in warm dilute sulphuric acid, generating hydrogen and leaving a solution of the sulphate. Hot concentrated sulphuric acid also forms aluminium sulphate with aluminium but is reduced to sulphur dioxide:

$$2Al + 6H_2SO_4 \rightarrow Al_2(SO_4)_3 + 3SO_2 + 6H_2O$$

The formation of salts with acids as above is characteristic of metals. However, pure aluminium does not react with nitric acid: the surface is rendered passive. Aluminium vessels may be used to store concentrated nitric acid.

Aqueous solutions of alkalis attack aluminium with the formation of hydrogen and the corresponding aluminate. The action becomes violent with the finely divided metal:

$$2Al + 2OH^- + 2H_2O \rightarrow 2AlO_2^- + 3H_2$$

Alternatively, the aluminate ion may be formulated $Al(OH)_4^-$ by incorporating $2H_2O$, or $Al(H_2O)_2(OH)_4^-$.

Aluminium has a strong affinity for oxygen and may be used for the preparation of chromium, manganese and iron, etc., in small quantities. A mixture of the oxide and the calculated quantity of coarse, grease-free aluminium is placed in a fireclay crucible, embedded in sand in a fire bucket. An ignition mixture of barium peroxide and aluminium powder is placed in a depression in the charge and a piece of magnesium ribbon stuck into it. This fuse is lit and the contents of the crucible undergo a violent exothermic reaction, accompanied by a shower of sparks and a white glare. The temperature may reach 2500–3000°C, the molten metal sinking to the bottom of the crucible, below a mass of fused aluminium oxide (corundum). This is the aluminothermic or Goldschmidt Thermite process. In addition to the extraction of metals, the reaction may be used for the production of steel for welding *in situ*. For manganese and iron:

$$3Mn_3O_4 + 8Al \rightarrow 4Al_2O_3 + 9Mn$$
trimanganese
tetroxide

$$Fe_2O_3 + 2Al \rightarrow Al_2O_3 + 2Fe$$
iron(III)
oxide

The disease *dialysis dementia* has been found to afflict certain patients undergoing renal dialysis on so-called 'kidney machines', who had been given certain aluminium compounds during treatment. The accumulation of combined aluminium in the brain was identified as the probable cause of *dialysis encephalopathy syndrome*. *Alzheimer*'s disease, the onset of premature senility, is also associated with abnormal concentrations of aluminium in certain parts of the brain. This concentration of aluminium compounds in the brain may, of course, be the result of the disease, and not its cause. Such a progressive nervous disease may be due to toxic amino-acids or other biochemical factors. The currently accepted dose of aluminium in the UK is $6\,\mu g$ day^{-1}. The silicic acid content of water may limit the absorption of aluminium compounds from drinking water, for which the EEC directive limit for aluminium is $200\,\mu g$ dm^{-3}. The normal use of aluminium cooking utensils is not considered a hazard, although it is a fairly safe assumption that stainless steel saucepans etc. will be bought by the more affluent. It is probably wise to avoid cooking foods which apparently leave an aluminium saucepan unusually clean, e.g. rhubarb. Accurate analysis of aluminium intake is extremely difficult because the tiny quantities involved are less than the experimental errors liable in the techniques employed.

Lithium aluminium hydride, LiAlH₄

When finely divided lithium hydride reacts with an ethereal solution of aluminium chloride, lithium aluminium hydride is formed in solution. This compound has outstanding reducing powers in

Organic Chemistry. Excess aluminium chloride yields aluminium hydride, $(AlH_3)_n$, of uncertain constitution:

$$4LiH + AlCl_3 \rightarrow LiAlH_4 + 3LiCl$$

$$3LiAlH_4 + AlCl_3 \rightarrow 4AlH_3 + 3LiCl$$

Carboxylic acids may be reduced to the corresponding alcohols with high yields by reduction in ethereal solution and subsequent hydrolysis by 10% sulphuric acid or caustic soda solution. Certain ethenic (ethylenic) linkages are not affected and this simplifies syntheses. Simple examples illustrate these reactions:

$$C_6H_5.CH_2.COOH \rightarrow C_6H_5.CH_2.CH_2OH$$
phenylethanoic acid   2-phenyl ethanol
(phenylacetic acid)

$$C_6H_5.CH=CH.CHO \rightarrow$$
cinnamaldehyde
3-phenylpropenal $C_6H_5.CH=CH.CH_2OH$
     cinnamyl alcohol
     3-phenylpropenol

Lithium aluminium hydride may be used to prepare hydrides: silicon tetrachloride is reduced to silane,

$$LiAlH_4 + SiCl_4 \rightarrow LiCl + AlCl_3 + SiH_4(g)$$

Finally, carbon dioxide is reduced to methanol in 81% yield by lithium aluminium hydride.

Aluminium oxide, alumina, Al₂O₃

Aluminium oxide is a soft, white powder formed usually by gently heating the hydroxide,

$$2Al(OH)_3 \rightarrow Al_2O_3 + 3H_2O$$

It melts at about 2000°C.

In nature, the anhydrous oxide is found as colourless crystals of the very hard corundum or the impure, fine-grained emery. Specimens of corundum, coloured by traces of oxides of certain metals, are valued as the coloured gems, rubies (red), sapphires (blue) and amethysts (violet). Hydrated alumina is mined as bauxite, gibbsite and diaspore. Bauxite, which has been fused at 3000°C in an electric furnace, crushed and fired with clay in a porcelain kiln at 1500°C, is known as alundum, being used in abrasives and as a basic refractory lining. Ciment Fondu, bauxite cement, is obtained by fusing bauxite with lime and resists the action of sea water.

Aluminium oxide is amphoteric, dissolving in mineral acids to form aluminium salts and in caustic alkalis, to form the aluminate:

$$Al_2O_3 + 6H^+ \rightarrow 2Al^{3+} + 3H_2O$$

$$Al_2O_3 + 2OH^- \rightarrow 2AlO_2^- + H_2O$$

$$Al_2O_3 + 2OH^- + 3H_2O \rightarrow 2Al(OH)_4^-$$

When hydrated to give the octahedrally shaped anion, the formula would be, $Al(H_2O)_2(OH)_4^-$, although the reaction can result in the formation of $Al(OH)_6^{3-}$ as might have been expected. However, after heating to red heat a change in properties occurs. Aluminium oxide will then only react with concentrated solutions of alkali or with fused alkali or fused potassium hydrogensulphate.

Alumina will catalyse the dehydration of alcohols: when vapour from boiling ethanol is passed over heated alumina, ethene (ethylene) is formed:

$$\begin{array}{ccc} CH_3 & & CH_2 \\ | & \rightarrow & \| & + H_2O \\ CH_2OH & & CH_2 \end{array}$$

Alumina is very suitable for use as an absorbent in column chromatography, the most elementary method of separating the components of a mixture based on differences in adsorption affinities, originally developed for the separation of natural pigments (Greek, *chroma* = colour). A solution of the mixture is poured on to the top of a column packed with fine, absorbent alumina or another suitable medium. Using pigments, coloured layers appear near the top of the column. Pure solvent is added to the top of the column and as it and successive portions percolate downwards, pigmented zones or bands separate from each other. The pigments are adsorbed to differing degrees. After development with solvent, the column containing adsorbed pigments, called the chromatogram, is pushed out or extruded and the zones cut out. The various fractions are washed with suitable solvents, or eluents, to dissolve the pigments. Colourless compounds may be separated in the same way, some being detected by their fluorescence in ultra-violet light and others being converted into coloured derivatives either before or after development. Other methods of chromatographic analysis include paper and vapour-phase chromatography. In addition to the separation of complex mixtures of organic compounds, mixtures of inorganic anions or cations may be separated. The pH of the column is an important factor in determining the order of adsorption. A solution containing bismuth nitrate and mercury(II) nitrate in dilute nitric acid may be

separated into the two components on alumina. Development with sodium hydroxide solution gives a yellow zone of mercury(II) oxide, and addition of hydrogen sulphide water converts this to a black band and shows at the same time an upper broad brown band of bismuth sulphide.

Aluminium hydroxide, $Al(OH)_3$

Addition of aqueous ammonia to a solution containing aluminium ions results in the formation of a colourless, gelatinous precipitate of aluminium hydroxide:

$$Al^{3+} + 3OH^- \rightarrow Al(OH)_3(s)$$

or to indicate the gelatinous, hydrated nature of the precipitate the salt hydrolysis equation may be used:

$$Al(OH_2)_6^{3+} + H_2O \longrightarrow$$
$$Al(OH_2)_5(OH)^{2+} + \boxed{H_3O^+} \quad OH^-$$
$$\downarrow H_2O$$
$$Al(OH_2)_4(OH)_2^+ + \boxed{H_3O^+} \xrightarrow{OH^-} 6H_2O$$
$$\downarrow H_2O$$
$$Al(OH_2)_3(OH)_3 + \boxed{H_3O^+} \quad OH^-$$

all of which occurs around the octahedral configuration adopted by the central aluminium(III) cation because of mutual repulsion of the incoming ligands.

Aluminium salts are acidic in solution because of the loss of protons by hydrated aluminium ions. Aluminium hydroxide is amphoteric, forming aluminium salts with mineral acids and aluminates with caustic alkalis:

$$Al(OH)_3 \rightleftharpoons H_2O + HAlO_2$$

$$3H^+ \swarrow \qquad \searrow OH^-$$

$$\begin{array}{cc} Al^{3+} + 3H_2O & AlO_2^- + H_2O \\ \text{e.g. } Al(NO_3)_3 & \text{e.g. } KAlO_2 \\ \text{aluminium nitrate} & \text{potassium aluminate} \end{array}$$

Aluminium hydroxide is a much weaker acid than the very weak boric acid and the reaction with alkali is better formulated:

$$Al(OH)_3 + OH^- \rightarrow Al(OH)_4^-$$
$$\text{aluminate anion}$$

and the salt becomes $KAl(OH)_4$.

Solutions of carbonates are sufficiently alkaline by hydrolysis to precipitate aluminium hydroxide from solutions of aluminium salts, carbon dioxide being evolved. Certain fire extinguishers are charged with solutions of sodium hydrogencarbonate and aluminium sulphate: on mixing, a foam of carbon dioxide is ejected, in which the bubbles are stabilized by the presence of aluminium hydroxide. Carbon dioxide will precipitate aluminium hydroxide from solutions of aluminates, carbonic acid being a stronger acid than aluminium hydroxide.

Aluminium hydroxide will readily adsorb colouring matter. When precipitated from solutions containing dyes, alizarin, congo red, litmus, etc., the precipitate is coloured, and is called a lake, while the medium above is rendered colourless. In a similar way, sewage is precipitated by adsorption on to aluminium hydroxide. In the dyeing industry, aluminium hydroxide is used as a mordant (Latin, *mordere* = to bite): it is precipitated inside the fibres of the cloth and takes up the dyestuff. Aluminium hydroxide may be used for waterproofing cloth: the material is soaked in a solution of aluminium ethanoate (acetate), which is then hydrolysed by steam or boiling.

The preparation of oxosalts: aluminium nitrate and sulphate (tetraoxosulphate)

Aluminium carbonate does not exist.

Aluminium nitrate cannot be prepared by the action of nitric acid on aluminium, because the metal is rendered passive, but may be prepared by dissolving freshly precipitated aluminium hydroxide in nitric acid or by arranging a suitable precipitation reaction. Solutions of lead(II) nitrate and aluminium sulphate in calculated proportions interact to precipitate lead(II) sulphate:

$$2Al^{3+} + 3SO_4^{2-} + 3Pb^{2+} + 6NO_3^- \rightarrow$$
$$3PbSO_4(s) + 2Al^{3+} + 6NO_3^-$$

On crystallization, a hydrate, $Al(NO_3)_3.9H_2O$, separates. This may be thermally decomposed to the oxide, nitrogen dioxide, oxygen and steam all escaping:

$$4Al(NO_3)_3.9H_2O \rightarrow$$
$$2Al_2O_3 + 12NO_2 + 3O_2 + 36H_2O$$

Aluminium sulphate is prepared by dissolving the oxide or hydroxide in hot, fairly concentrated sulphuric acid when the hydrate, $Al_2(SO_4)_3.16H_2O$, crystallizes out on cooling. Commercially, purified bauxite is used. Aluminium sulphate is used as a source of aluminium hydroxide in sizing paper, waterproofing and as a mordant.

The alums

The alums are double salts of general formula $M^+M^{3+}(SO_4^{2-})_2.12H_2O$, which form isomorphous crystals and in which M^+ and M^{3+} are respectively univalent and tervalent. In structure, the alums consist of simple ions, being not complexes, but double salts. Potash alum or potassium alum is the common alum, with the formula $KAl(SO_4)_2.12H_2O$ which, for convenience, may be written $K_2SO_4.Al_2(SO_4)_3.24H_2O$. The univalent cation may belong to the alkali metals or be silver, thallium or ammonium. Aluminium may be replaced as the tervalent metal by the ions: Cr^{3+}, Mn^{3+}, Fe^{3+} or Co^{3+}. In solution the simple ions disperse.

The alums may be prepared by mixing the appropriate sulphates in proportions of the formula weight of the anhydrous salt and evaporating to crystallization. But certain alums may be prepared by specially adapted methods. The alums are isomorphous, appearing as octahedral crystals which form overgrowths and solid solutions.

Potassium aluminium sulphate, alum, potash alum, $KAl(SO_4)_2.12H_2O$

Alum may be prepared from aluminium which is reacted in excess with caustic potash solution, and the decanted liquid mixed with four times its original molar equivalent of sulphuric acid:

$$2Al + 2K^+ + 2OH^- + 2H_2O \rightarrow$$
$$2K^+ + 2AlO_2^- + 3H_2$$

$$2K^+ + 2AlO_2^- + 8H^+ + 4SO_4^{2-} \rightarrow$$
$$(2K^+ + SO_4^{2-}) + (2Al^{3+} + 3SO_4^{2-}) + 4H_2O$$

The alum may be readily crystallized being very much more soluble in hot water than cold. Easier to purify by crystallization than the more soluble aluminium sulphate, it is used commercially as a source of aluminium compounds, being manufactured from the component sulphates. On heating, alum melts at 92°C, loses water of crystallization at 200°C to form 'burnt' alum and on further strong heating, evolves sulphur trioxide to form potassium sulphate and alumina. This reaction is essentially

$$Al_2(SO_4)_3 \rightarrow Al_2O_3 + 3SO_3$$

Ammonium aluminium sulphate, ammonia alum, $NH_4Al(SO_4)_2 . 12H_2O$

Formed by mixing solutions of the two constituent sulphates and crystallizing, ammonia alum melts at 95°C and on ignition leaves alumina only:

$$2NH_4Al(SO_4)_2 . 12H_2O \equiv$$
$$(NH_4)_2SO_4 . \qquad Al_2(SO_4)_3 . 24H_2O$$
$$\downarrow \qquad\qquad \downarrow$$
$$2NH_3 + H_2SO_4 + Al_2O_3 + 3H_2SO_4 + 21H_2O$$
$$\text{residue}$$

Other alums given under the transition element

Chrome alum, potassium chromium sulphate,

$$KCr(SO_4)_2 . 12H_2O$$

Iron(III) alum, iron(III) ammonium sulphate,

$$NH_4Fe(SO_4)_2 . 12H_2O$$

The halides

Table 19.3 shows a comparison of some properties of the aluminium halides.

The abrupt change in properties going from the fluoride to the other halides has usually been explained by a change from ionic to covalent structures. Pauling has argued that aluminium fluoride is essentially macromolecular in structure, providing an alternative and additional reason for its lack of volatility. This explanation seems to be more likely.

All may be prepared from the elements. Aluminium fluoride is inert chemically but may be hydrolysed by steam. It is found in Greenland as a complex fluoride, Na_3AlF_6, cryolite or sodium hexafluoroaluminate. Aluminium bromide is prepared by heating the metal in bromine vapour, a very violent reaction occurring with the liquid halogen. Aluminium powder reacts spontaneously with iodine at room temperature if a drop of water is added to the mixture. Much heat is evolved. The compound volatilizes and burns in air. Anhydrous aluminium iodide may be prepared by the action of excess aluminium on iodine, unchanged halogen being volatilized by heating. An inert atmosphere should be used. The chloride, bromide and iodide of aluminium are very similar to each other, having covalent characteristics and forming dimers.

Aluminium chloride, $Al_2Cl_6 \rightleftharpoons 2AlCl_3$

When aluminium metal reacts with concentrated hydrochloric acid, a solution of aluminium chloride is formed, from which a hexahydrate separates on crystallization:

$$2Al + 6H^+ \rightarrow 2Al^{3+} + 3H_2$$

On heating, the hexahydrate is hydrolysed to alumina, the other products escaping:

$$2AlCl_3 . 6H_2O \rightarrow Al_2O_3 + 6HCl + 9H_2O$$

The partial hydrolysis of aluminium salts has been described previously, aluminium chloride solution giving an acidic reaction:

$$Al(OH_2)_6{}^{3+} + H_2O \rightleftharpoons Al(OH_2)_5(OH)^{2+} + H_3O^+$$

In 0.1 M solution, the hydrolysis is 2%.

Anhydrous aluminium chloride must be prepared in the absence of moisture. It is readily purified by sublimation in an inert atmosphere where electron diffraction and vapour density studies indicate that the molecules are dimerized. Thermal dissociation to monomers starts at 400°C and is complete by 750°C. The covalent characteristics of aluminium chloride include ready volatility, solubility in organic solvents and hydrolysis. Anhydrous aluminium chloride gives fumes of hydrogen chloride in moist air, due to hydrolysis.

Table 19.3 A comparison of the aluminium halides

	AlF₃	AlCl₃	AlBr₃	AlI₃
Melting-point (227 kPa)/°C	1290	193	97.5	180
Boiling-point (227 kPa)/°C	Sublimes above m.p.	Sublimes 180	255	381
Conditions for hydrolysis	Steam	Water	Water	Water
Solubility in water	Insoluble	Readily soluble	Readily soluble	Readily soluble
Solubility in organic solvents	Insoluble	Soluble	Soluble	Soluble

Aluminium oxide, mixed with charcoal and heated in dry chlorine, yields aluminium chloride which sublimes; alternatively the metal may be heated in chlorine:

$$Al_2O_3 + 3C + 3Cl_2 \rightarrow Al_2Cl_6(g) + 3CO(g)$$

$$2Al + 3Cl_2 \rightarrow Al_2Cl_6(g)$$

Usually, dry hydrogen chloride is passed over aluminium powder, heated moderately at first to avoid fusion, the combustion being strongly exothermic:

$$2Al + 6HCl \rightarrow Al_2Cl_6(g) + 3H_2(g)$$

The apparatus must be thoroughly dried and all air displaced to avoid the possibility of an explosive mixture of hydrogen and air being formed. The aluminium chloride sublimes as colourless, hexagonal plates. It fumes in air and is very hygroscopic. The apparatus, shown in Fig. 19.1, is used generally for the preparation of halides of non-volatile elements which sublime.

Aluminium chloride readily forms addition products with ammonia and organic compounds of oxygen and nitrogen, acquiring the inert gas configuration through the formation of co-ionic bonding (p. 305).

Aluminium chloride is widely used in the Friedel–Crafts' reaction (1877) in Organic Chemistry, involving a reaction between alkyl or acyl halides and aromatic hydrocarbons:

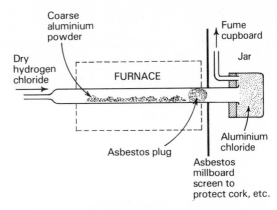

benzene* + CH₃.CO.Cl + AlCl₃ →
ethanoyl chloride
(acetyl chloride)

COCH₃
+ HCl + AlCl₃

phenylethanone
(acetophenone)

Letting R = an alkyl or acyl group and ArH the aromatic hydrocarbon, the mechanism of the reaction is considered to be:

$$RCl + AlCl_3 \rightarrow R^+ + AlCl_4^-$$

$$R^+ + ArH \rightarrow (ArH,R)^+$$

$$(ArH,R)^+ + AlCl_4^- \rightarrow Ar{-}R + HCl + AlCl_3$$

It will be seen that aluminium chloride is an electron pair acceptor in this reaction, i.e. it behaves as a Lewis acid (p. 238). Boron trifluoride may also be used to catalyse this type of reaction.

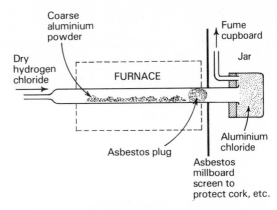

Fig. 19.1 The preparation of aluminium chloride

Qualitative tests for aluminium ions

In qualitative analysis, a mixture of ammonium chloride and aqueous ammonia yields a colourless, gelatinous precipitate of aluminium hydroxide. The precipitate is sometimes extremely difficult to see. On warming, it is coagulated and may be carried upwards by entrapped gas. In Group III of the Qualitative Analysis Scheme, the addition of ammonium chloride reduces the hydroxide ion concentration sufficiently to permit the precipitation of the hydroxides of aluminium, chromium(III) and iron(III) only: $Al(OH)_3$, $Cr(OH)_3$ and $Fe(OH)_3$ respectively. Aluminium hydroxide may be precipitated by addition of aqueous ammonia or caustic soda alone, but re-dissolves in excess of the latter to form sodium aluminate.

The presence of aluminium may be confirmed by precipitating the hydroxide in the presence of a trace of a suitable dye, when the mordant action of the hydroxide is revealed in the formation of a coloured lake. Alizarin S (Alizarin Sulphonate), which gives a yellow coloration in solution with acids and bluish-red with alkalis, imparts a bright red colour to the precipitate.

Solid aluminium compounds give a white infusible mass of oxide when heated in the blow-pipe flame, using fusion mixture on a charcoal block. The residue is just moistened with cobalt(II) nitrate solution and re-heated, when a blue infusible mass of cobalt aluminate is formed. Excess cobalt(II) nitrate would give a black coloration due to tricobalt tetroxide, Co_3O_4:

$$2Al_2O_3 + 2Co(NO_3)_2 \rightarrow$$
$$2Co(AlO_2)_2 + 4NO_2 + O_2$$

* A resonance hybrid, Chapter 5, p. 81

20

Group IV: the carbon group

Carbon, silicon, germanium, tin and lead

6	
C	
2, 4	
14	
Si	
2, 8, 4	
32	
Ge	
2, 8, 18, 4	
50	
Sn	
2, 8, 18, 18, 4	
82	
Pb	
2, 8, 18, 32, 18, 4	

The notable general feature about Group IV is the range and gradation of the properties shown by the elements and their compounds due to the change in character of the elements from weakly electronegative to weakly electropositive as the atomic number increases. All show a valency of 4 but with the progressive increase in metallic character and emergence of metallic properties, there is a reduction in valency by two, a pair of electrons remaining inert. The principal valency of carbon–tin inclusive is four while that of lead is two. Germanium and tin have a subsidiary valency of two, which is well defined, while the less stable valency state of lead is the group valency, four. In the naming of compounds, the higher valency used to be indicated by the suffix -*ic* and the lower by -*ous*. Germanium(II) and tin(II) compounds have a strong tendency to revert to the germanium(IV) and tin(IV) states, thereby undergoing oxidation. Therefore, they are strong reducing agents. Lead(IV) compounds, on the other hand, are oxidizing agents.

Carbon is unique in forming compounds containing chains of carbon atoms, in which the covalent bonding is strong, the octet being complete with no unshared electron pairs and no tendency to form co-ionic bonds. This property is known as *catenation*. The systematic study of these compounds belongs to Organic Chemistry. Certain individual compounds of carbon are included here to illustrate family behaviour. In addition, miscellaneous compounds of carbon which also contain nitrogen or oxygen as well as other elements and having no parallels elsewhere in the family are grouped at the end. Aspects of germanium are touched on very briefly where the gradation of chemical properties is well illustrated. Elaborate comparisons are sometimes drawn between boron and silicon, which are diagonally placed but beyond noting that both are typical non-metals and as such have certain similarities anyway, nothing further need be said. Carbon does not especially resemble phosphorus. In the Mendeléeff Periodic Table, carbon and silicon were recognized as resembling the B subgroup, germanium, tin and lead rather than the A subgroup, titanium, zirconium, hafnium and thorium, which are transition metals. In the open form of the table this is emphasized.

Various forms of carbon have been known since remote times as have the metals tin and lead. Bronze, an alloy of tin and copper, was used in Egypt about 2000 B.C. Silicon was isolated in 1823 by Berzelius, who reduced potassium hexafluorosilicate with potassium:

$$K_2SiF_6 + 4K \rightarrow 6KF + Si$$

Germanium was discovered by Winkler in 1876 as a constituent of argyrodite, $4Ag_2S.GeS_2$.

Carbon occurs widely in nature, as the element in diamond and graphite, combined in vegetable and animal matter, in the carbonate rocks, in the carbon dioxide of the air and dissolved in water, and as petroleum and coal, formed by the decay of organic matter in geological aeons. Coal results from the decay of forest vegetation under the influence of bacteria and conditions of high humidity, temperature and pressure: anthracite contains over 85% carbon; bituminous coal, the common type, about 70%; and lignite, brown coal, about 60% when dried. Silicon occurs as the oxide, silica,

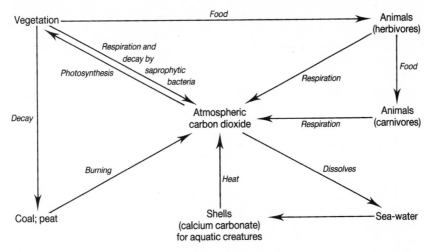

Fig. 20.1 The carbon cycle

in quartz, flint and sand, and as silicates in rocks; it is second only to oxygen in order of abundance in the Earth's crust. Natural deposits of tin and lead have been reported but the former is usually extracted from the oxide and the latter from the sulphide. Other compounds of both metals are also mined. Alluvial deposits of cassiterite (tin-stone, tin(IV) oxide) found in Malaya and Nigeria are main sources of tin. The Cornish tin mines, once very important, are almost worked out.

Carbon is essential to living matter. It has always been assumed that the proportion of carbon dioxide remains constant and so it was made the centre of the carbon cycle, the essentials being shown in Fig. 20.1. However, there is increasing international concern about the accumulation of carbon dioxide and certain other compounds in the atmosphere which are expected to cause the aptly named 'greenhouse effect' with consequent warming of the atmosphere, melting of ice from the polar land masses bringing a rise in sea level, and a change in weather patterns. The origins of the 'greenhouse' gases are many and the problem is aggravated by the destruction (often by burning) of vast tracts of so-called rain forest in countries like Brazil to create agricultural land.

Lead is the final product of three radioactive series resulting from the decay of radium ($\rightarrow ^{206}_{82}Pb$), actinium ($\rightarrow ^{207}_{82}Pb$) and thorium ($\rightarrow ^{208}_{82}Pb$), belonging to the uranium series, actinium series and

Table 20.1 Group IV: physical data p block elements $(ns)^2 (np)^2$ $\qquad n \neq 1$

	C	Si	Ge	Sn	Pb
Atomic number	6	14	32	50	82
Electronic configuration	2, 4	2, 8, 4	2, 8, 18, 4	2, 8, 18, 18, 4	2, 8, 18, 32, 18, 4
[Core]outer electrons	$[He](2s)^2(2p)^2$	$[Ne](3s)^2(3p)^2$	$[Ar,3d^{10}](4s)^2(4p)^2$	$[Kr,4d^{10}](5s)^2(5p)^2$	$[Xe,4f^{14},5d^{10}](6s)^2(6p)^2$
Relative atomic mass	12.01115	28.086	72.59	118.69	207.19
Atomic radius/nm	0.077	0.117	0.122	0.141	0.154
Ionic radius (M^{2+}; M^{4+})*/nm;	$-$; 0.015	$-$; 0.041	$-$; 0.053	0.093; 0.071	0.120; 0.084
Atomic volume/cm³ mol⁻¹**	Diamond 3.42	12.1	13.6	White 16.2	18.3
Standard electrode potential (H Scale: 298 K)/V	$-$	-0.84	-0.3	-0.140	-0.126
Boiling-point/°C	4827 sublimes	2355	2830	2362	1755
Melting-point/°C	3570	1414	959	232	328
Electronegativity	2.5	1.8	1.7	1.7	1.6
Density/g cm⁻³ (at 293 K)	Diamond 3.51 Graphite 2.22	2.33	5.36	Grey 5.75 White 7.31	11.3
Co-ordination number	4, 3, 2	4(6)	4, 6	4, 6	4, 6
Abundance (% by mass)	0.08	27.6	7×10^{-4}	4×10^{-3}	16×10^{-4}

*See Table 5.13, p. 80
**Traditional term. It is the volume of one mole of atoms as a solid element

thorium series respectively (p. 24). The relative atomic mass of any sample will depend on the original source. There are two allotropes of carbon, diamond and graphite, and two allotropes of tin, grey and white tin.

Physical data for the Group IV elements are listed in Table 20.1.

A general survey of bonding involving the carbon group of the *p* block

In passing along the second and third periods of the Periodic Table, the formation of cations becomes more difficult as the nuclear positive charge, the proton or atomic number, rises. We have seen that boron never forms a simple cation while aluminium does so only with difficulty, and that in simple covalent compounds they do not acquire an octet of electrons. Conversely, the elements of Group IV have four electrons each and can acquire an octet of electrons. As might be expected, carbon and silicon form covalent bonds only, being the first elements of their respective periods to acquire an octet of electrons in this way. This agrees with Fajans' Rules: the small, intensely charged, hypothetical C^{4+} and Si^{4+} ions would cause polarization of the electron-clouds of anions with the formation of covalencies.

This preference for covalency in the early congeners may be illustrated by comparing ionization energies, displayed in Table 20.2 and Fig. 20.2, although in the latter only the sum of the 1st and 2nd ionization energies corresponding to $M(g) \rightarrow M^{2+}(g)$ have been shown because the 4th ionization energies lie far beyond the area of the graph (as determined by the scale used). The sums of the first four ionization energies for each of the elements are shown in Fig. 20.3 with a tenfold change of scale on the energy axis. Carbon, silicon and germanium do not form M^{4+} ions but tin and lead do, and this can be understood.

Atomic radii, ionic radii and van der Waals' radii for the elements are plotted against atomic number

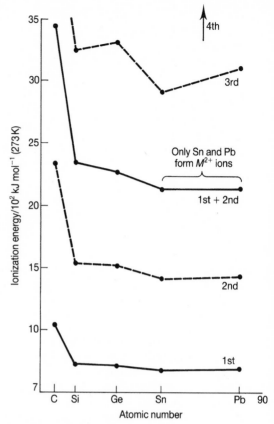

Fig. 20.2 Early ionization potentials plotted against atomic number for Group IV elements

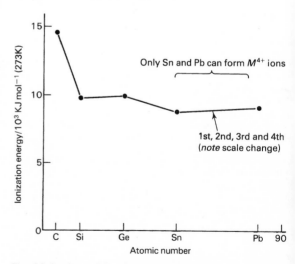

Fig. 20.3 Sum of first four ionization energies for Group IV elements plotted against atomic number. Sn^{4+}, Pb^{4+} are rare: the others are unknown (Note: tenfold change of scale on energy axis)

Table 20.2 Group IV (carbon group): ionization energies/ kJ mol^{-1} (273 K)

	C	Si	Ge	Sn	Pb
1st $M(g) \rightarrow M^+(g) + e^-$	1086	786	760	**707**	**715**
2nd $M^+(g) \rightarrow M^{2+}(g) + e^-$	2354	1573	1534	**1409**	**1447**
3rd $M^{2+}(g) \rightarrow M^{3+}(g) + e^-$	4621	3232	3300	2943	3087
4th $M^{3+}(g) \rightarrow M^{4+}(g) + e^-$	6223	4351	4409	3821	4081
5th $M^{4+}(g) \rightarrow M^{5+}(g) + e^-$	37820	16112	8973	7786	6696

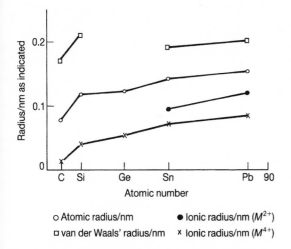

o Atomic radius/nm   ● Ionic radius/nm (M^{2+})

□ van der Waals' radius/nm   × Ionic radius/nm (M^{4+})

Fig. 20.4 Atomic, ionic and van der Waals' radii plotted against atomic number for elements of Group IV

in Fig. 20.4 and electronegativities similarly in Fig. 20.5. Refer to Chapter 5 for essential theory. What is seen here harmonizes with the foregoing discussion. The characteristic distancing in overall similar family behaviour between the first and second elements is seen clearly, and the fairly close similarity between silicon and germanium. Finally, carbon forms the anions C^{4-} (in Be_2C, Al_4C_3) and C_2^{2-} in CaC_2 etc.) as might be expected from the foregoing.

The elements, carbon (diamond), silicon, germanium, grey tin, and silicon carbide, SiC, exist as

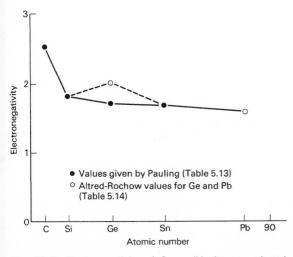

Fig. 20.5 Electronegativity of Group IV elements plotted against atomic number (Pauling values)

macromolecular structures of the zinc blende type. Diamond and graphite illustrate *monotropy* and have been described on pp. 105, 111. Grey and white tin, which exhibit *enantiotropy* appear in the same chapter. With increasing atomic number, a few compounds containing the cation in the quadrivalent state appear: Sn^{4+} and Pb^{4+}, but are very rare. The group valency, four, is the principal valency for all elements of the family except lead, for which it is subsidiary to a valency of two.

Compounds in which carbon and silicon are bivalent do not appear to exist while those of germanium(II) (germanous) have a very strong tendency to revert to the quadrivalent germanium(IV) (germanic) condition, and of tin(II) (stannous) a milder tendency to behave similarly; both act as reducing agents. On the other hand, lead(IV) (plumbic) compounds readily revert to the lead(II) (plumbous) state, and are oxidizing agents. There are a great many examples of macromolecule formation. In saturated compounds the structural configuration adopted is usually tetrahedral.

In addition to the difference between first and second members of a family which may be associated with the size of atoms and seen with the *s* block elements (Li, Na; Be, Mg), *p* block elements of higher atomic number have *d* orbitals available for bonding, and consequent increase of co-ordination number. Thus silicon [Ne]$(3s)^2(3p)^2$ has the 3*d* orbitals available between the 3*p* and 4*s* levels. This accounts for the hydrolysis of $SiCl_4$ (via $Si(OH_2)Cl_4$) in contrast to the inertness of CCl_4.

The oxidation states follow the same pattern as the valencies as shown in the comparison of electronic configurations except that the oxidation number of C in CO is +2, Si in SiO, +2 and Ge in GeO, also +2. This illustrates the earlier statement that oxidation number does not necessarily equal the valency in a particular state of combination. Note that the rules for assigning oxidation numbers (p. 124) can be interpreted as making the oxidation number equal the electric charge on a particular ion, when a covalent structure is replaced by its necessarily hypothetical ionic equivalent, in its way a reversal of the Fajans' Rules (p. 82). Thus $C^-\equiv O^+$ becomes $C^{2+}O^{2-}$ and the oxidation number of C = +2 (see page 320).

Carbon atoms will link together in long chains or in rings with strong covalent bonds: each atom acquires an octet of electrons and has no lone pairs of electrons for donation. Further, the compounds of carbon with hydrogen, oxygen and chlorine have roughly the same stability so that a wide range of

Electronic configurations

Comparison of electronic configurations shows that the relationship between silicon and germanium is not the same as that between sodium and potassium or magnesium and calcium; this is reflected in the properties of the elements and their compounds.

Group IV	Inert gases
6 **C** 2, 4	10 **Ne** 2, 8
14 **Si** 2, 8, 4	18 **Ar** 2, 8, 8
32 **Ge** 2, 8, 18, 4	36 **Kr** 2, 8, 18, 8
50 **Sn** 2, 8, 18, 18, 4	54 **Xe** 2, 8, 18, 18, 8
82 **Pb** 2, 8, 18, 32, 18, 4	86 **Rn** 2, 8, 18, 32, 18, 8

Group IV
Ions (rare)

	Group IV *Ions*
32 **Ge⁴⁺** 2, 8, 18	
50 **Sn⁴⁺** 2, 8, 18, 18	50 **Sn²⁺** 2, 8, 18, 18, 2
82 **Pb⁴⁺** 2, 8, 18, 32, 18	82 **Pb²⁺** 2, 8, 18, 32, 18, 2

← ions — | — covalent molecules →

Simple valency:	4	2	4
Oxidation state:	+4	+2	±4

In covalent molecules, each element acquires the electronic configuration of an inert gas, but retains its own nuclear charge (atomic number). To quote one example, silicon assumes the electronic configuration of argon (2, 8, 8) but still retains 14 protons in the nucleus in SiH_4(O.N. = −4) and $SiCl_4$(O.N. = +4)

Tetrahydrides:

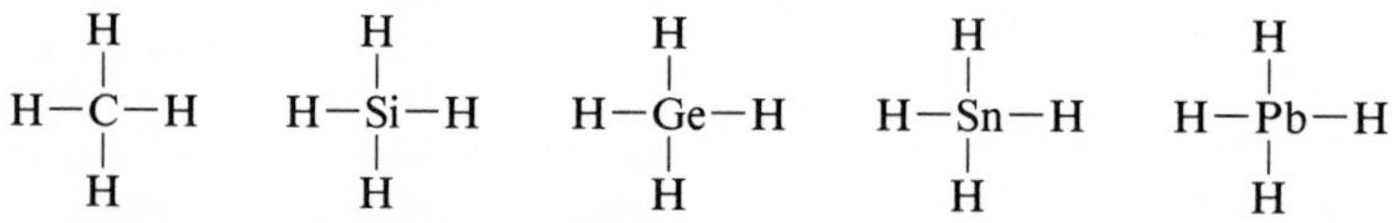

H | H−C−H | H	H | H−Si−H | H	H | H−Ge−H | H	H | H−Sn−H | H	H | H−Pb−H | H
methane	silane	germane	stannane	plumbane

Tetrachlorides:

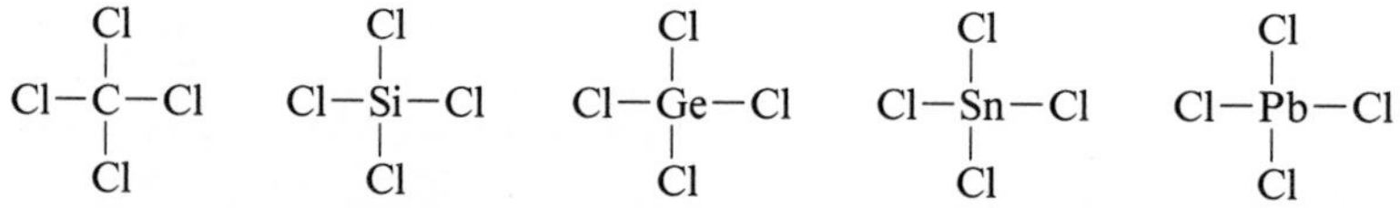

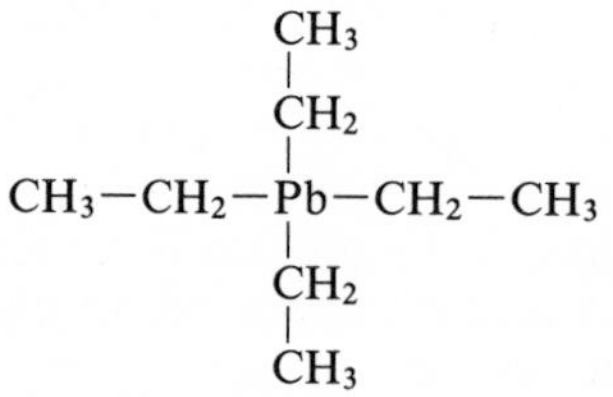

Cl | Cl−C−Cl | Cl	Cl | Cl−Si−Cl | Cl	Cl | Cl−Ge−Cl | Cl	Cl | Cl−Sn−Cl | Cl	Cl | Cl−Pb−Cl | Cl

$$CH_3-CH_2-Pb-CH_2-CH_3$$
with CH₃, CH₂ above and CH₂, CH₃ below the Pb

lead tetraethyl
tetraethyl lead $Pb(C_2H_5)_4$
(similar compounds are formed by the other elements)

Dioxides

O=C=O $(SiO_2)_n$ $(GeO_2)_m$ and $Ge^{4+}(O^{2-})_2$ $Sn^{4+}(O^{2-})_2$ $Pb^{4+}(O^{2-})_2$

carbon   'silicon dioxide'   dimorphic:
dioxide   actually a   one form isomorphic with
macromolecular   quartz, the other with
structure: tin(IV) oxide
silica, quartz

different combinations can be formed. Following the postulate of Kekulé, double and triple bonds are used to retain the quadrivalency of carbon:

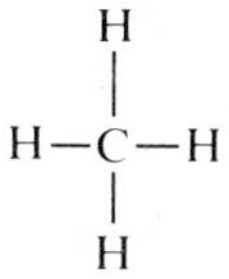

methane

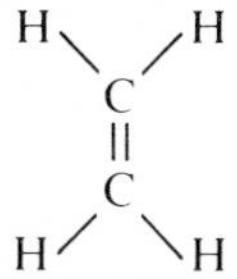

ethene (ethylene)

ethyne (acetylene)

Silicon is characterized by the formation of single covalent bonds with oxygen, multibonding being rare. Silica has a *three-dimensional* macromolecular structure which may be represented diagrammatically:

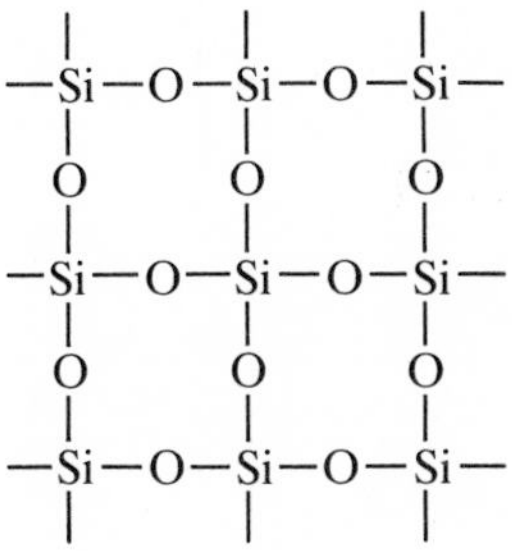

Single bonds between silicon atoms are rare and multibonding unknown. The tetrahedral configuration is adopted. Simple and macromolecule structures, three-dimensional and layered, for mineral silicates exist and are based fundamentally on the SiO_4^{2-} ion but may be considered to be formed by removal of water from the hypothetical $Si(OH)_4$:

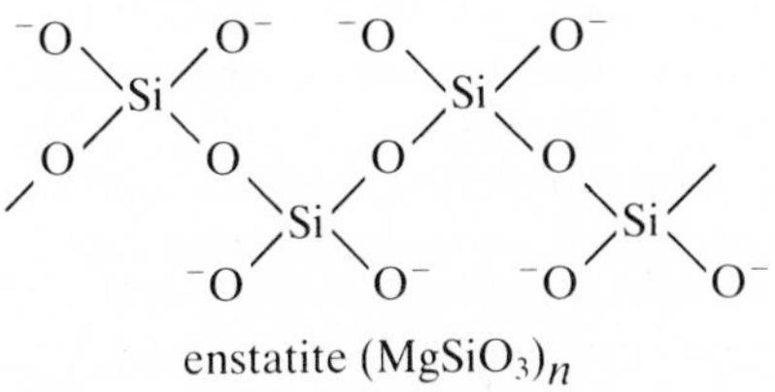

enstatite $(MgSiO_3)_n$

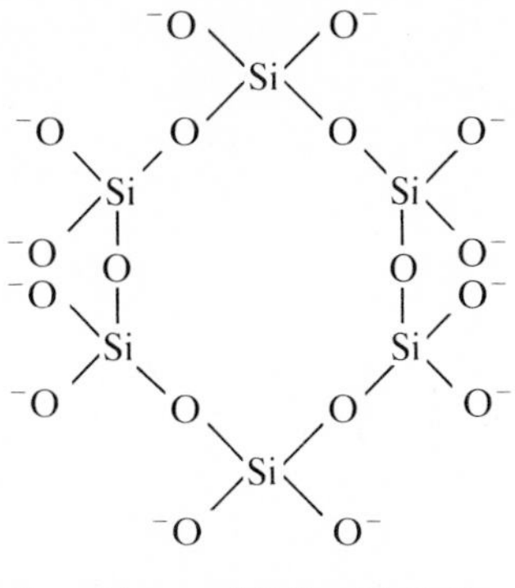

beryl (emerald) $Be_3Al_2Si_6O_{18}$

Carbonates do not form macromolecules. Carbonic acid and its salts may be assigned simple structures:

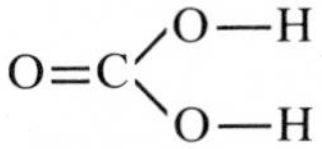

carbonic acid

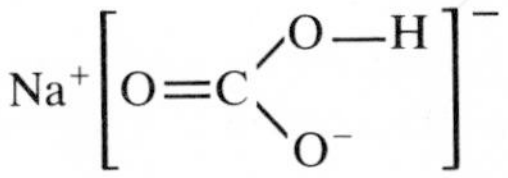

sodium hydrogencarbonate 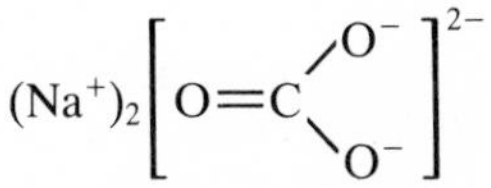   sodium carbonate

Strictly, other structures ought to be added to emphasize the spreading of the negative charge to give a more stable resonance hybrid anion, with equal carbon–oxygen bond lengths; for the carbonate ion:

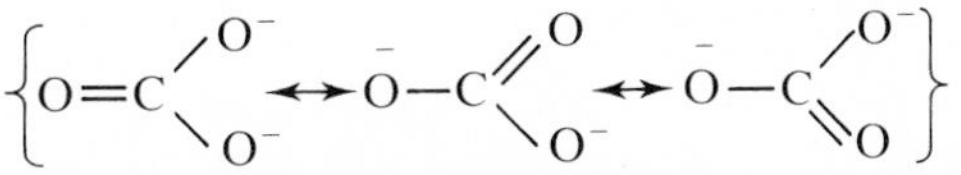

trigonal planar shape

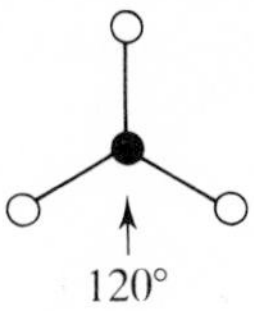

120°

At first sight, carbon monoxide might seem to have carbon in the bivalent state but a satisfactory structure using a co-ionic bond with two covalent bonds may be devised. The structure is isoelectronic with that of nitrogen and it is not altogether surprising that the gases have many physical constants almost the same. Furthermore, the co-ionic bond superimposes a charge which neutralizes the natural polarity of the carbonyl group,

$$\underset{\delta+\quad\delta-}{>C=O,}$$

leaving an otherwise unexpected lack of polarity:

$$\underset{2,5\quad 2,5}{N + N} \longrightarrow \underset{2,8\ 2,8}{N\equiv N}$$

$$\underset{2,4\quad 2,6}{C + O} \xrightarrow[\text{electron}]{\overset{\text{transfer}}{\text{one}}} \underset{2,5\quad 2,5}{\overset{-}{C} + \overset{+}{O}} \rightarrow \underset{2,8\ 2,8}{\overset{-}{C}\equiv\overset{+}{O}}$$

Complex compounds between carbon monoxide and transition metals are called carbonyls, nickel carbonyl being the most important. Co-ionic bonds are formed and nickel achieves the electronic configuration of the nearest inert gas, krypton. The valency of the metal in this type of compound is reckoned as zero as is the oxidation number, since the molecule is uncharged and the oxidation number of O = −2 (by definition) and of C = +2 in carbon monoxide. A molecule (or ion) which attaches itself to a central metal atom (or ion) in this way is called a *ligand*.

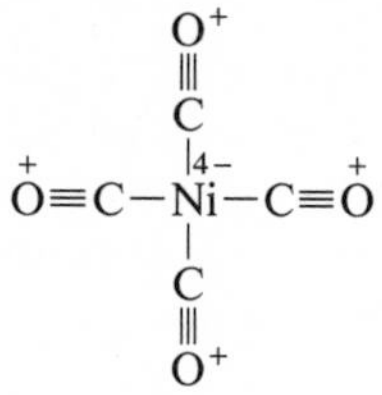

$$\overset{+}{O}\equiv C-\overset{4-}{Ni}-C\equiv\overset{+}{O}$$

(with C≡O⁺ groups above and below the Ni)

Ni(CO)₄
nickel carbonyl

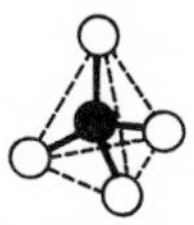

Tetrahedral
shape

In the bivalent state, the only solid compounds which have been isolated are formed by germanium, tin and lead. The oxide, sulphide and four halides of germanium(II), although unstable, have been prepared. The compounds are covalent and are relatively involatile due to co-ionic bonding, or co-ordination:

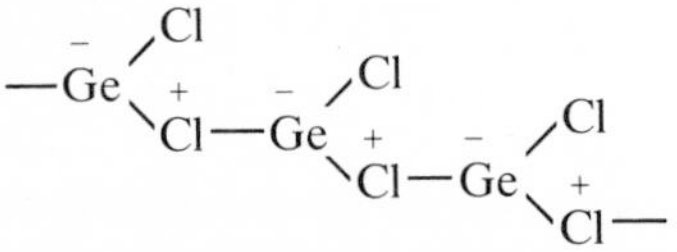

(chain of Ge–Cl groups with co-ionic bonds)

For tin, the simple cation, the tin(II) ion, Sn^{2+}, appears in oxosalts. The oxide is a covalent macro-molecule. The halides are covalent in the vapour phase and in organic solvents, but highly ionized, with complex formation, in water. Germanium(II)

and tin(II) compounds readily pass into the quadrivalent state. Bivalent lead forms more ionic compounds but the oxide is a covalent macro-molecule. The electronic configuration of the M^{2+} cation is (core), 18, 2, called the zinc group electronic configuration:

$$Pb^{2+}SO_4^{2-} \quad Pb^{2+}(NO_3^-)_2$$

Important complex ions, as distinct from the carbonyl complex molecules, formed by carbon include the complex cyanides, which are many and of varying complexity. The hexacyanoferrate(II) (ferrocyanide) and hexacyanoferrate(III) (ferricyanide) ions are important:

$$\underset{2,8\quad 2,8}{\overset{-}{C}\equiv N}$$
cyanide
ion

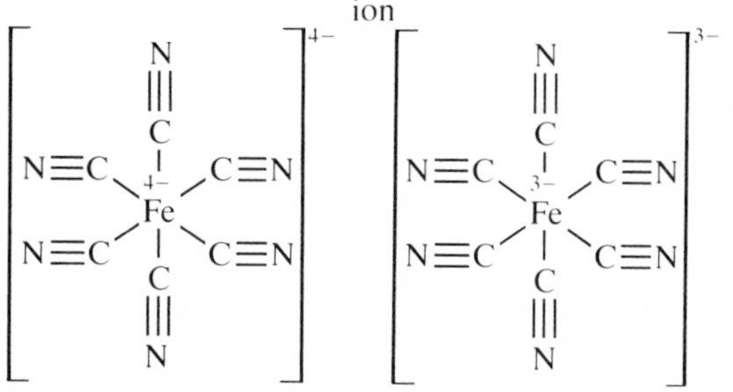

(hexacyanoferrate(II) ion, $[Fe(CN)_6]^{4-}$, and hexacyanoferrate(III) ion, $[Fe(CN)_6]^{3-}$, both octahedral)

hexacyanoferrate (II) ion   hexacyanoferrate (III) ion

octahedral shape

Other notable complexes include the type:

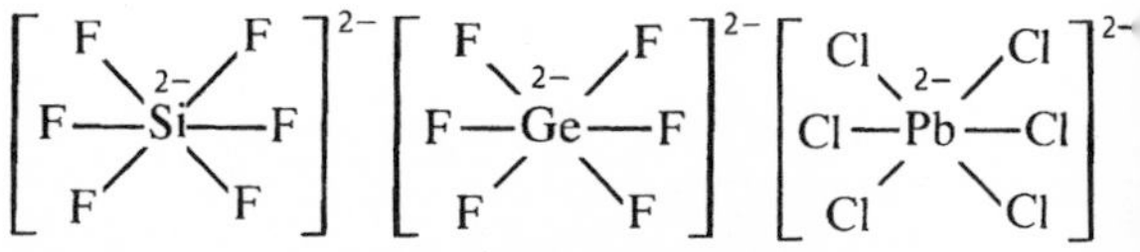

$$\left[SiF_6\right]^{2-} \quad \left[GeF_6\right]^{2-} \quad \left[PbCl_6\right]^{2-}$$

also octahedral arrangements

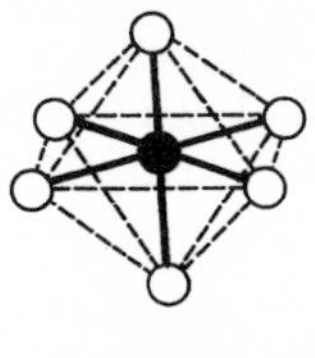

in which the valency octet has been expanded, by use of the *d* orbitals (energy levels). Carbon and other elements of Period 2 cannot do this.

In the bivalent state, tin(II) ions form $SnCl_4^{2-}$ and for lead(II), $PbCl_3^-$, $PbCl_4^{2-}$ and $PbCl_6^{4-}$ have been described. Finally, double, possibly complex, ethanoates (acetates)

$$K[Pb(O.OC.CH_3)_3]$$

and $$K_2[Pb(O.OC.CH_3)_4]$$

have been obtained, providing a simple explanation for the ready solubility of lead(II) sulphate, insoluble in water, in ammonium ethanoate (acetate) solution.

The elements

Carbon and tin form important allotropes which have been described separately in Chapter 6.

Diamond and graphite are the two allotropes of carbon. Diamond is transparent when pure and the hardest naturally occurring substance. Diamonds have a high dispersion and refractive index when cut and are prized as precious stones, the value being maintained by an agreed restriction on the number marketed. Diamond is a non-conductor, while graphite conducts heat and electricity. Impure diamonds, being less valuable, and small synthetic diamonds are used industrially for drill tips in rock-boring, glass-cutting and wire-drawing. Graphite is used on account of its electrical conduction for dynamo and motor brushes, as electrodes and furnace linings. Other forms of graphite include charcoal, soot, coke and lamp-black. Carbon black is manufactured by the incomplete but controlled combustion of heavy oils and is used as a filler in rubber tyres. Charcoal is formed by the thermal destruction of organic matter. Ordinary charcoal, formed from wood, is also highly porous and suitable for adsorption of gases, and is especially useful when cooled in liquid air for removing traces of gases in highly evacuated apparatus. Animal charcoal, formed by calcining bones, is also highly porous and contains calcium phosphate; it is used commercially for decolorizing sugar-cane juice and in the laboratory generally. Concentrated sulphuric acid removes the elements of water from cane sugar, leaving very pure carbon, sugar carbon:

$$C_{12}H_{22}O_{11} - 11H_2O \xrightarrow{H_2SO_4} 12C$$

Silicon is usually prepared in the laboratory as a brown amorphous powder by reduction of silver sand (silica) with magnesium powder, or by a thermite reaction using aluminium powder with sulphur to raise the temperature:

$$SiO_2 + 2Mg \rightarrow 2MgO + Si$$

$$3SiO_2 + 4Al \rightarrow 2Al_2O_3 + 3Si$$

The product is repeatedly treated with hydrochloric acid and evaporated, leaving silicon contaminated with some silica, which may be removed by treatment with hydrofluoric acid in a polythene vessel. The product is filtered, washed and dried. X-ray examination shows that silicon has the diamond structure.

Germanium is a lustrous grey, very brittle metal used in electronics for making transistors.

Zone refining is the technique used to produce silicon and germanium in the very high degrees of purity (better than 1 in 10^9) required for the manufacture of semiconductor devices. The element is formed into a rod. A narrow molten zone is created at one end using a suitable heater. As the heater is moved slowly along the rod, the molten zone moves correspondingly carrying the impurities with it, leaving a rod of extremely high purity behind it.

Tin is a white, very lustrous metal which resists atmospheric corrosion, has a high thermal and electrical conductance and exists as two allotropes, the normal white allotrope and a grey powdery form. It is extracted by smelting tin(IV) oxide with powdered anthracite. Lead is a soft, light-grey metal which is easily cut when pure, exposing a lustrous surface which is quickly tarnished. It is often hardened by the addition of approximately 1% antimony. Tin and lead are used widely as the metals, alloys and in compounds.

Carbon and silicon are non-metals while tin and lead are metals. Germanium is on the borderline between the distinct metals and non-metals. Mendeléeff predicted the properties of germanium and certain of its compounds (Table 2.4), having left a gap for the hitherto undiscovered element in his Periodic Table.

The chemical reactions of the elements may be briefly compared, the different allotropes of each element having about the same reactivity. The so-called amorphous forms of silicon and carbon are more reactive because of their increased surface area. In this connection it may be noted that specially prepared, finely divided lead is spontaneously flammable in air. It is said to be pyrophoric. As normally used, lead is noted for its resistance to chemical attack.

The group valency (4) is the principal valency for all except lead. The greater stability of the bivalent state of lead and the comparative instability of the quadrivalent state, means that the common salts of lead are described as lead(II) salts only when a particular distinction is being drawn in a comparison with the lead(IV) state, the Roman numerals referring to the respective oxidation states, otherwise they are described simply as lead salts. On the other hand, for tin the terms tin(II) and tin(IV) must be used. Germanium forms germanium(II) and germanium(IV) compounds but they hardly concern us here. Note the relative stability of bivalent lead in the following reactions.

All react with air or oxygen on heating, forming the dioxide except for lead, which yields the monoxide and red lead, minium, Pb_3O_4:

$$Sn + O_2 \longrightarrow SnO_2$$
tin(IV) oxide

$$2Pb + O_2 \longrightarrow 2PbO$$
lead monoxide
lead(II) oxide,
litharge

$$6PbO + O_2 \xrightleftharpoons[470°C]{400°C} 2Pb_3O_4$$
red lead
minium

Charcoal smoulders and can be fanned into a bright glow in an air draught. Diamond will burn at about 700°C. Silicon burns at red heat while molten tin and lead gradually form a scum of oxide.

All react on heating with sulphur to form a disulphide, except for lead which yields lead(II) sulphide, PbS; carbon disulphide is best prepared on the industrial scale when sulphur vapour reacts with white-hot coke, heated in an electric furnace:

$$C + 2S \rightarrow CS_2$$
carbon
disulphide

$$Sn + 2S \rightarrow SnS_2$$
tin(IV)
sulphide

$$Pb + S \rightarrow PbS$$
lead(II)
sulphide

Except for carbon, all react with the halogens. They form the tetrahalide with the exception of lead, which yields the lead(II) salt:

$$Sn + 2Cl_2 \rightarrow SnCl_4$$
tin(IV)
chloride

$$Pb + I_2 \rightarrow PbI_2$$
lead(II)
iodide

When non-metallic elements react with the oxidizing acids, acidic oxides or acids are formed.

The actions are complex. No reactions occur between non-metallic elements and dilute acids. Concentrated nitric acid reacts with all, especially germanium and tin, to give the dioxide, except for lead, which gives lead(II) nitrate. Tin reacts with great vigour to yield β-hydrated tin(IV) oxide, often called β-metastannic acid. All reactions are complex and may be expressed somewhat arbitrarily:

$$3C + 4HNO_3 \rightarrow 3CO_2 + 4NO + 2H_2O$$

$$3Sn + 4HNO_3 \rightarrow$$
$$3SnO_2(s) \; hydrated + 4NO + 2H_2O$$

$$3Pb + 8HNO_3 \rightarrow 3Pb(NO_3)_2 + 2NO + 4H_2O$$

Cold dilute nitric acid reacts with tin to form a mixture of tin(II) nitrate and ammonium nitrate solution, without evolution of gas:

$$4Sn + 10HNO_3 \rightarrow$$
$$4Sn(NO_3)_2 + NH_4NO_3 + 3H_2O$$

The formation of a salt with an oxoacid is typical of a metal. Hot concentrated sulphuric acid yields the dioxide when acting on the non-metallic elements, being reduced to sulphur dioxide. However, tin is sufficiently metallic to form a mixture of tin(IV) oxide and sulphate, while lead forms lead(II) sulphate:

$$C + 2H_2SO_4 \rightarrow CO_2 + 2SO_2 + 2H_2O$$

$$Sn + 4H_2SO_4 \rightarrow Sn(SO_4)_2 + 2SO_2 + 4H_2O$$
(hydrated $SnO_2(s)$ also formed)

$$Pb + 2H_2SO_4 \rightarrow PbSO_4 + SO_2 + 2H_2O$$

Strongly ionized non-oxidizing acids in solution attack tin and lead:

$$Sn + 2H^+ \rightarrow Sn^{2+} + H_2$$

$$Pb + 2H^+ \rightarrow Pb^{2+} + H_2$$

Under reducing conditions, tin gives tin(II) salts. Fairly concentrated hydrochloric acid reacts vigorously with tin to form tin(II) chloride. Boiling hydrochloric acid attacks lead, the lead(II) chloride formed remaining dissolved whereas in the cold it would have precipitated and retarded the action. Sulphuric acid gives an insoluble coating of lead(II) sulphate, hot or cold, and no further action occurs. Lead is used in chemical plant as a chemically resistant lining, a notable example being the Lead Chamber process for sulphuric acid.

White hot carbon (coke) reacts with steam to give a mixture of carbon monoxide and hydrogen, known as water-gas:

$$C + H_2O \rightleftharpoons CO + H_2$$

Steam or boiling water forms silica with silicon:

$$Si + 2H_2O \rightarrow SiO_2 + 2H_2$$

Tin is not attacked by water or steam and has been used as a lining in distilled water plant. Lead is slowly attacked by soft water containing dissolved air, with the formation of hydroxide and carbonate. Lead ions act as a cumulative poison and traces formed in domestic lead pipes could prove dangerous, especially if organic acids from moorland reservoirs are present. Addition of sodium silicate forms a lining of lead silicate in the pipes, thus avoiding the hazard. In hard water districts protection is afforded by a layer of lead(II) carbonate.

Carbon may be used to reduce the heated oxides of the other elements of the group, silicon requiring the heat of an electric furnace; for tin(IV) oxide and lead(II) oxide (monoxide):

$$SnO_2 + 2C \rightarrow Sn + 2CO$$

$$PbO + C \rightarrow Pb + CO$$

Tin and lead may be displaced from aqueous solutions of their salts by the addition of a metal higher in the electrochemical series:

$$Zn(s) + Pb^{2+} \rightarrow Zn^{2+} + Pb(s)$$

Molten caustic soda or a concentrated solution reacts: silicon and germanium form the silicate and germanate(IV) while tin and lead form the stannate(II) (mainly) and plumbate(II), oxosalts of the bivalent condition:

$$Si + 2OH^- + H_2O \rightarrow SiO_3^{2-} + 2H_2$$
silicate

$$Sn + 2OH^- \rightarrow SnO_2^{2-} + H_2$$
stannate(II)

$$Pb + 2OH^- \rightarrow PbO_2^{2-} + H_2$$
plumbate(II)

Carbon reduces sodium hydroxide to sodium metal and sodium carbonate on intense heating:

$$2C + 6NaOH \rightarrow 2Na + 2Na_2CO_3 + 3H_2$$

Carbon reacts with a range of metals to form various types of carbide. Silicon forms silicides to some extent, and germanium behaves similarly.

Carbides are binary compounds containing carbon. In combination with more electronegative elements, compounds of carbon are described as oxides, halides, etc., not as the carbides of the elements concerned.

Carbides and silicides

Only carbides are of any importance although magnesium silicide is met in connection with the silicon hydrides and is prepared by heating magnesium powder with silver sand in suitable proportions:

$$4Mg + SiO_2 \rightarrow Mg_2Si + 2MgO$$

Carbides may be classed as ionic, covalent or interstitial (metallic).

Ionic carbides

Ionic carbides are formed by the metals of Groups I, II and III of the Periodic Classification. They are colourless, crystalline solids, non-conductors of electricity in the solid state and hydrolysed to hydrocarbons by water or dilute acids. They may be sub-classified according to the hydrocarbon produced.

(a)   Methanides or methides

Beryllium carbide, Be_2C, and aluminium carbide, Al_4C_3, are hydrolysed to methane and some hydrogen:

$$Al_4C_3 + 12H_2O \rightarrow 4Al(OH)_3 + 3CH_4$$

Each is formed by union of the elements at high temperatures and the lattice dimensions are such as to allow the presence of what may be considered to be C^{4-} ions. The hydrolysis is essentially:

$$C^{4-}(s) + 4H_2O \rightarrow 4OH^- + CH_4(g)$$

(b)   Ethynides (Acetylides)

Compounds of Group I (Li_2C_2, etc.), Group II (CaC_2, etc., including BeC_2) and Group III (Al_2C_6) elements are formed by heating the metals in ethyne (acetylene), or for calcium, strontium and barium, heating the alkaline-earth oxide with coke in an electric furnace. Hydrolysis yields ethyne (acetylene)

$$CaC_2 + 2H_2O \rightarrow Ca(OH)_2 + C_2H_2$$

Ethynides contain the ion, $\overset{C^-}{\underset{C}{\vphantom{|}\|}}$, the hydrolysis being essentially:

$$\overset{C^-}{\underset{C^-}{\vphantom{|}\|}} \text{(s)} + 2H_2O \rightarrow 2OH^- + \overset{CH}{\underset{CH}{\vphantom{|}\|}} \text{(g)}$$

ethyne
acetylene

Covalent carbides

Covalent carbides include the hydrocarbons. Two other important carbides are silicon carbide, sold under the name *carborundum*, SiC, in which silicon and carbon atoms alternate in a diamond lattice giving inertness, extreme hardness and a high dissociation temperature, and boron carbide, B_4C, which has similar properties and a highly complex lattice. Both are formed by reduction of the oxides with carbon in the electric furnace.

Interstitial carbides

Interstitial carbides (metallic carbides), with varible compositions, are formed by the transition metals. They possess a lustrous appearance, high electrical conductance, high melting-points, hardness and chemical inertness. The relatively small carbon atoms are accommodated inside the metal lattice. They are examples of compounds called Berthollides (p. 3).

Hydrides

All form volatile covalent hydrides of formula XH_4, though with increasing difficulty as the atomic number increases. The hydrides of carbon belong to Organic Chemistry but a number of hydrides corresponding to the saturated paraffin hydrocarbons are formed by the other elements:

alkanes	silanes	germanes	stannane	plumbane
CH_4	SiH_4	GeH_4	SnH_4	PbH_4
C_2H_6	Si_2H_6	Ge_2H_6		
C_3H_8	Si_3H_8	Ge_3H_8		
C_4H_{10}				
etc.				

Silicon hydrides are obtained by the action of hydrochloric acid on magnesium silicide, followed by liquefaction at the temperature of liquid air (SiH_4, b.p. $-112°C$) and fractionation. They are spontaneously flammable in air. For silane, silicomethane, the reactions may be summarized:

$$Mg_2Si + 4HCl \rightarrow 2MgCl_2 + SiH_4$$

$$SiH_4 + 2O_2 \rightarrow SiO_2 + 2H_2O$$

Hydrides of germanium, prepared similarly, are less reactive. Tin(IV) hydride, stannane, is prepared in small yield by electrolytic reduction of tin(IV) sulphate or better, by the reduction of tin(IV) chloride by lithium aluminium hydride. Lead(IV) hydride, plumbane, has been prepared and was first detected as a radioactive gas by the action of hydrochloric acid on a magnesium–lead alloy, Mg_2Pb, containing radioactive lead (thorium-B). It is very unstable.

The formation of covalent, volatile hydrides is favoured by non-metallic elements. The temperatures at which thermal dissociation of the tetrahydrides occur are as follows:

CH_4	SiH_4	GeH_4	SnH_4	PbH_4
800°C	450°C	285°C	150°C	0°C

The decline in stability and ease of formation of the covalent hydrides illustrates the change in character of the elements in Group IV the non-metallic character giving way to the metallic, the electronegative to the weakly electropositive.

Simple organic compounds

All of the elements form a range of organic compounds with carbon groups structurally evolved from the simple hydrides. The use of tetraethyl lead in petrol as an anti-knock to prevent pre-ignition of petrol vapour and air in engines of high compression ratio is being discontinued because of the health hazards posed by lead residues in the air (and elsewhere). There is evidence of mental retardation in the young caused by the metal. Intensively used motorway junctions, as well as main routes in our cities have been shown to have far too much residual lead compounds in the immediate environment. To prevent the accumulation of lead on the sparking plugs, an *additive*, such as 1:2 dibromoethane, is used. The alkyls provide simple examples of organic compounds:

$$C(CH_3)_4 \quad Si(CH_3)_4 \quad Ge(CH_3)_4$$
$$C(C_2H_5)_4 \quad Si(C_2H_5)_4 \quad Ge(C_2H_5)_4$$
$$Sn(CH_3)_4 \quad Pb(CH_3)_4$$
$$Sn(C_2H_5)_4 \quad Pb(C_2H_5)_4$$

The elements silicon to lead have an extensive

organometallic chemistry of interest to pharmaceutical and other chemical manufacturing industries.

Dioxides, monoxides and oxoacids/hydroxides

All form both dioxides and monoxides. The dioxides are the most stable oxides for all except lead. Those of carbon and silicon are completely acidic but some amphoteric properties begin to emerge with germanium while for lead, the plumbate(IV) state is more stable than the plumbate(II). The dioxides remain predominantly acidic. Carbon monoxide is $C^- \equiv O^+$ while the other monoxides are macromolecular, little being known about that reported for silicon. Carbon monoxide is neutral, the monoxide of tin is amphoteric while lead monoxide, lead(II) oxide, is also amphoteric but predominantly basic.

Carbon dioxide, carbonic acid, hydrogen-carbonates (bicarbonates) and carbonates

Carbon dioxide is a colourless gas, with a faint, pleasant odour. It is about 1.5 times denser than air; density relative to hydrogen $= \frac{1}{2}$ (relative molecular mass) = 22, while that of air = 14.4, and of hydrogen = 1 (as standard). Carbon dioxide is relatively inert. The gas is moderately soluble in water, one volume dissolving one volume of gas at 15°C. Soda-water contains carbon dioxide under pressure, with a little added sodium hydrogencarbonate; various soft mineral waters are manufactured by the further addition of sweetening, flavouring and possibly, colouring substances. Carbon dioxide reacts with water to form the weakly ionized carbonic acid. The gas is, therefore, an acidic oxide and an acid anhydride, carbonic anhydride:

$$CO_2 + H_2O \rightleftharpoons \overset{HO}{\underset{HO}{>}}C=O \rightleftharpoons$$

$$H^+ + HCO_3^- \rightleftharpoons 2H^+ + CO_3^{2-}$$

| hydrogen carbonate ion salts: hydrogencarbonates (bicarbonates) e.g. $NaHCO_3$, $Mg(HCO_3)_2$ | carbonate ion salts: carbonates e.g. K_2CO_3, $CaCO_3$ |

The ortho-acid, $C(OH)_4$, H_4CO_4, has not been isolated. Solutions of carbonates and hydrogencarbonates are hydrolysed in solution, hydrogen ions from the solvent, water, being mopped up by carbonate or hydrogencarbonate ions to form carbonic acid, leaving an excess of residual hydroxide ions and an alkaline solution.

Carbon dioxide is formed by the thermal decomposition of carbonates, except those of sodium-caesium of Group I and with just the normal laboratory facilities, strontium and barium of Group II:

$$CaCO_3 \rightleftharpoons CaO + CO_2$$

It is usually prepared by the action of dilute acids on carbonates:

$$CO_3^{2-} + 2H^+ \rightarrow H_2O + CO_2$$

The few solid hydrogencarbonates are easily dissociated on heating and readily react with acids:

$$2NaHCO_3 \rightarrow Na_2CO_3 + H_2O + CO_2$$

$$HCO_3^- + H^+ \rightarrow H_2O + CO_2$$

The complete combustion of carbonaceous material produces carbon dioxide.

Industrially, carbon dioxide is an important by-product of brewing in which ethanol is produced by the fermentation of sugar using yeast. In the laboratory, glucose may be used. The enzyme, zymase, in yeast brings about decomposition:

$$\underset{\text{glucose}}{C_6H_{12}O_6} \rightarrow \underset{\text{ethanol}}{2C_2H_5OH} + 2CO_2$$

Carbon dioxide is a by-product in the commercial production of hydrogen from water gas, the carbon monoxide of which reacts with steam in the presence of an iron–chromium catalyst at 500°C:

$$H_2 + CO + H_2O \rightleftharpoons 2H_2 + CO_2$$

Carbon dioxide is absorbed in water under 1–5 MPa (10–50 atmospheres) pressure.

Carbon dioxide is readily liquefied. It was used by Andrews in his classical work on critical phenomena in 1869. The critical temperature of carbon dioxide is 31.1°C, so that liquefied gas may be stored under pressure at room temperature. By rapid expansion of highly compressed gas through a nozzle into a bag, solid carbon dioxide may be collected. This sublimes at −78°C at 101 kPa (atmospheric) pressure. It is used in bulk refrigeration for the transport of perishable food-stuffs and ice-cream, and is known as Dry Ice. Mixed with alcohol, it provides a useful low-temperature bath (approximately $-80°C$) in the laboratory. Carbon dioxide under pressure is used as a heat-transference medium in some nuclear power stations, transferring heat from reactor fuel cans to the steam turbines.

Carbon dioxide is present to about 0.03% by volume in the air. It is used by plants, with energy from sunlight and water in the presence of the green colouring matter, chlorophyll, to synthesize starch (photosynthesis, see p. 253). There are fears that the natural equilibrium governing the concentration of carbon dioxide in the atmosphere has been disturbed. The more than gradual accumulation of carbon dioxide contributes to the aptly named greenhouse effect which may have grave environmental consequences.

During respiration, both plants and animals convert carbohydrates into carbon dioxide and water, at the same time releasing energy. In animals, the concentration of carbon dioxide in the blood acts on certain organs called respiratory centres, and affects the rate of breathing. When the carbon dioxide level rises markedly, as after a sprint, the rate of respiration rises. This causes the carbon dioxide level to fall and breathing slows until the level is raised again. The result is periodic breathing, which continues until conditions in the blood are normal again. In cases of near-drowning or electric shock, breathing may be stimulated by administering a mixture of oxygen and carbon dioxide. Expired air contains about 4% carbon dioxide by volume.

Carbon dioxide may be detected in the laboratory by the formation of a precipitate with lime-water (calcium hydroxide solution). For a more sensitive test, baryta-water (barium hydroxide solution) may be used. The turbidity due to precipitated carbonate disappears with excess carbon dioxide due to the formation of the soluble hydrogen carbonate,

$$CO_2 + Ba^{2+} + 2OH^- \rightarrow BaCO_3(s) + H_2O$$

$$BaCO_3(s) + CO_2 + H_2O \rightarrow Ba^{2+} + 2HCO_3^-$$

By absorption in aqueous caustic potash or baryta, carbon dioxide may be quantitatively determined.

In order to deduce the formula of carbon dioxide, two experiments are done. Carbon is completely combusted in an excess of oxygen using an apparatus such as that shown in Fig. 20.6. After an initial expansion of gas due to the heat of combustion, the mercury level returns to its original position on cooling.

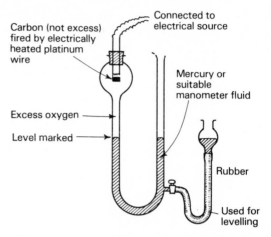

Fig. 20.6   The volume composition of carbon dioxide: when carbon is combusted in excess oxygen and the apparatus allowed to cool to the original temperature without change of pressure, no volume change is noted

$\therefore$ 1 volume of oxygen + solid carbon $\rightarrow$
$$\text{1 volume of carbon dioxide}$$

In the second experiment, the relative density is determined: this is 22 relative to hydrogen ($= 1$).

By Avogadro's Hypothesis,

n molecules of oxygen + solid carbon $\rightarrow$
$$n \text{ molecules of carbon dioxide}$$

$\therefore$ 1 molecule of oxygen + solid carbon $\rightarrow$
$$\text{1 molecule of carbon dioxide}$$

Assuming that the molecular formula of oxygen is O_2 (based on experiment and Avogadro's Hypothesis), the molecular formula of carbon dioxide is C_xO_2. Note that Avogadro's Hypothesis gives information about the gaseous state only. But the relative density of carbon dioxide is 22, and since this is half of the relative molecular mass (based on Avogadro's Hypothesis and the atomicity of hydrogen, determined as for oxygen):

$$C_xO_2 \equiv 44$$

(the relative atomic mass standard chosen does not affect the value at this level of accuracy)

$$\therefore 12x + (2 \times 16) = 44$$

$$\therefore x = 1$$

Carbon dioxide is CO_2

Important carbonates and hydrogencarbonates are described fully under their metals. The carbonates of the Group I metals, except lithium, and ammonium carbonate are the only soluble carbonates met here. The only solid hydrogencarbonates encountered are those of sodium and potassium. Soluble carbonates may be prepared by the action of carbon dioxide on the alkali:

$$2OH^- + CO_2 \rightarrow H_2O + CO_3^{2-}$$

Excess of carbon dioxide yields the hydrogencarbonate:

$$CO_3^{2-} + H_2O + CO_2 \rightarrow 2HCO_3^-$$

For the removal of carbon dioxide from gases, caustic potash is used because the products are freely soluble in water. One very important use of carbon dioxide is in the Ammonia–Soda process for the manufacture of sodium carbonate (see Chapter 14, p. 206). Hydrogencarbonates of calcium and magnesium cause temporary hardness of water (p. 229). Insoluble carbonates may be precipitated by mixing solutions containing the required ions:

$$Ba^{2+} + CO_3^{2-} \rightarrow BaCO_3(s)$$

Solutions of both carbonates and hydrogencarbonates are hydrolysed, the former being the more alkaline:

$$CO_3^{2-} + H_2O \rightleftharpoons HCO_3^- + OH^-$$

$$HCO_3^- + H_2O \rightleftharpoons H_2CO_3 + OH^-$$

Where both normal carbonates and mixtures of carbonate and hydroxide, called basic carbonates, exist, the former are generally precipitated by addition of sodium hydrogencarbonate solution and the latter by sodium carbonate solution:

$$Zn^{2+} + 2HCO_3^- \rightarrow ZnCO_3 + CO_2 + H_2O$$

$$3Zn^{2+} + CO_3^{2-} + 4OH^- + 2H_2O \rightarrow$$
$$ZnCO_3.2Zn(OH)_2.2H_2O(s)$$

Addition of sodium carbonate solution to solutions containing the ions of aluminium, chromium(III) and iron(III) precipitates very insoluble hydroxides with evolution of carbon dioxide and not the carbonates, which have so far not been prepared:

$$3CO_3^{2-} + 3H_2O + 2Al^{3+} \rightarrow 2Al(OH)_3(s) + 3CO_2$$

The thermal decomposition of carbonates and hydrogencarbonates and the action of acids have been described. Weak metals form carbonates which are readily decomposed by heat while the strongest metals of Group I and II form carbonates which cannot be decomposed by heat. Only the metals of Group I, except lithium, form hydrogen-

carbonates which are readily isolated as solids. In the laboratory, carbonates and hydrogencarbonates are recognized by the action of heat and of dilute acids. The only soluble carbonates normally encountered are those of ammonium, sodium and potassium. The only hydrogencarbonates used in the laboratory are those of sodium and potassium. A mixture of a carbonate and hydrogencarbonate in solution may be identified by addition of magnesium sulphate solution. An immediate precipitation of white magnesium carbonate reveals the presence of carbonate ions while further precipitation on boiling the filtrate is due to the decomposition of magnesium hydrogencarbonate to magnesium carbonate, similar to the removal of temporary hardness and confirms the presence of hydrogencarbonate. Alternatively, calcium chloride solution is used, the initial precipitate of calcium carbonate removed and hydrogencarbonate ions neutralized by addition of ammonium carbonate with a resultant precipitation of calcium carbonate.

Soluble carbonates and hydrogencarbonates may be estimated in titrimetric (volumetric) analysis by titration against standard hydrochloric acid using (screened) methyl orange as indicator. 0.05 M sodium carbonate solution has pH = 11.5 and 0.10 M sodium hydrogencarbonate, pH = 8.4, while carbonic acid has pH = 4.4–5.4 under titration conditions. Phenolphthalein is deep pink at pH 11.5. At pH 8.4 it assumes a faint pink hue, carbonate (CO_3^{2-}) having been converted to hydrogencarbonate (HCO_3^-) by the acid. Methyl orange (pH range 2.9–4.0) reacts alkaline (yellow, pH $\geqslant$ 4.0) to carbonate, hydrogencarbonate and carbonic acid but assumes the acidic (red) colour in excess hydrochloric acid at the end-point. This affords a way of estimating mixtures of carbonate and hydrogencarbonate using two indicators. The titration of 25 cm³ of solution with standard acid is done using phenolphthalein as indicator and the volume recorded [$=x$ cm³] as soon as the colour is discharged. Methyl orange is added and the titration is continued until at the end-point the total volume of acid required for the whole reaction is known = [$(x + y)$ cm³].

Stage 1:
$$Na_2CO_3 + HCl \rightarrow NaCl + NaHCO_3$$
$$\text{requires } x \text{ cm}^3 \text{ acid}$$

Stage 2:
$$NaHCO_3 + HCl \rightarrow NaCl + H_2O + CO_2$$
$$\text{requires another } x \text{ cm}^3$$

for the hydrogencarbonate formed in Stage 1. An additional $(x + y) - 2x = (y - x)$ cm^3 acid is required for the original hydrogencarbonate.

$$25 \text{ cm}^3 \text{ original sodium carbonate} \equiv 2x \text{ cm}^3 \text{ standard acid}$$

and

$$25 \text{ cm}^3 \text{ original sodium hydrogencarbonate} \equiv (y - x) \text{ cm}^3 \text{ standard acid}$$

Mixtures of alkali and soluble carbonate may be analysed similarly. Insoluble carbonates may be determined by back titration. A weighed quantity is added to an excess of standard acid of known volume and the residual acid is titrated against a standard sodium carbonate solution using (screened) methyl orange as indicator.

Carbon monoxide

The structure of carbon monoxide is $C^- \equiv O^+$.

Carbon monoxide is an extremely dangerous gas (b.p. $-190°C$). It is colourless, of about the same density as air, odourless and poisonous, reacting with the haemoglobin of the red blood corpuscles to form very stable carboxyhaemoglobin, the haemoglobin of which is no longer available to transport oxygen from the lungs to the tissues. It has been stated that 0.05% by volume in air produces giddiness, 0.2% loss of consciousness and 1.0%, rapid death. Victims show the characteristic cherry-pink of carboxyhaemoglobin. Sources of carbon monoxide include coal gas, car exhaust gases, and the incomplete combustion of paraffin in stoves where ventilation is inadequate. Accidental deaths can also occur when gas heaters have been incompetently installed, in spite of regulations covering such work in the UK and abroad. Faulty car exhausts have been known to cause death in cars that are waiting in traffic queues with their engines running. Carbon monoxide burns with a characteristic blue flame to form carbon dioxide, and will form explosive mixtures with air:

$$2CO + O_2 \rightarrow 2CO_2$$

Carbon monoxide may be prepared in the laboratory by the dehydration of methanoic acid (formic acid), ethanedioic acid (oxalic acid) or their salts by warm concentrated sulphuric acid:

methanoic acid
(formic acid)

ethanedioic acid
(oxalic acid)

Where ethanedioic acid and its salts are used, carbon dioxide is removed by bubbling the gas through caustic potash solution. Concentrated sulphuric acid will also act on potassium hexacyanoferrate(II) to form carbon monoxide:

$$K_4Fe(CN)_6 . 3H_2O + 6H_2SO_4 + 3H_2O \rightarrow$$
$$3(NH_4)_2SO_4 + FeSO_4 + 2K_2SO_4 + 6CO$$

More simply, pure dry carbon dioxide may be passed over red-hot charcoal:

$$C + CO_2 \rightarrow 2CO$$

Residual carbon dioxide is removed by aqueous caustic potash or granular soda lime. The characteristic blue flame on a coke fire is due to the combustion of carbon monoxide formed by this reaction. Being only slightly soluble, carbon monoxide is collected over water. Great care must be exercised to ensure that none escapes into the laboratory.

The unsaturated nature of carbon monoxide is shown by the addition of chlorine in the presence of activated charcoal to form phosgene, carbonyl chloride, another poisonous gas and an industrial intermediary,

$$CO + Cl_2 \rightarrow COCl_2$$

With sulphur vapour, carbonyl sulphide results:

$$CO + S \rightarrow COS$$

Carbonyls are formed by the action of carbon monoxide on many metals, the best known being nickel carbonyl, used industrially in one extraction of nickel:

$$Ni + 4CO \rightleftharpoons Ni(CO)_4$$

Carbon monoxide is a strong reducing agent, converting many oxides on heating to the metals:

$$PbO + CO \rightarrow Pb + CO_2$$

Fehling's solution is reduced to copper(I) oxide and ammoniacal silver nitrate to silver. A solution of palladium(II) chloride is reduced to palladium; a white filter paper soaked in a solution of palladium(II) chloride assumes a yellowish hue, which darkens and becomes jet black in carbon monoxide:

$$PdCl_2 + CO + H_2O \rightarrow Pd + CO_2 + 2HCl$$

Carbon monoxide is a neutral oxide. It will not give methanoic acid with water although formed by its dehydration. However, sodium methanoate is manufactured by the action of carbon monoxide on sodium hydroxide at 160°C under pressure:

$$NaOH + CO \rightarrow H.COONa$$

Carbon monoxide is present in the industrial gaseous fuels, water gas and producer gas, and in coal gas.

In the laboratory, carbon monoxide may be detected by the blue flame given on combusion resulting in the formation of carbon dioxide and by the dark stain produced with palladium(II) chloride paper. Gaseous mixtures containing carbon monoxide may be analysed by absorbing the gas in ammoniacal copper(I) chloride solution to form $CuCl.CO.2H_2O$, any oxygen having been removed previously by means of pyrogallol in sodium hydroxide to avoid oxidation of the copper(I) chloride. Alternatively, carbon monoxide may be quantitatively oxidized by iodine pentoxide and the carbon dioxide formed precipitated as barium carbonate by passage through baryta-water, the precipitate being filtered, washed and weighed:

$$I_2O_5 + 5CO \rightarrow 5CO_2 + I_2$$

The formula of carbon monoxide may be deduced from the results of eudiometric explosion of carbon monoxide and oxygen to form carbon dioxide, of previously determined formula.

Roughly equal measured volumes of carbon monoxide and oxygen are introduced into the eudiometer, shown in Fig. 20.7, at atmospheric pressure and sparked. After the slight explosion, the mercury is levelled after cooling and the volume of carbon dioxide and excess oxygen measured. Fairly concentrated caustic soda solution is introduced through the tap and the carbon dioxide absorbed. After levelling, the volume of excess oxygen is measured. It is found that

2 volumes carbon monoxide +
1 volume oxygen →
2 volumes carbon dioxide under the same
conditions of temperature and pressure

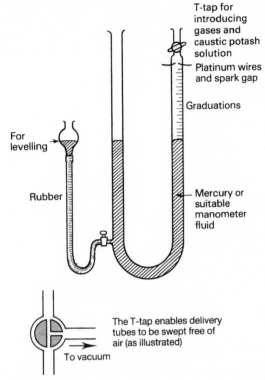

Fig. 20.7 The volume composition of carbon monoxide: the explosion eudimeter

By Avogadro's Hypothesis,

$2n$ molecules carbon monoxide +
n molecules oxygen →
$2n$ molecules carbon dioxide

∴ 2 molecules carbon monoxide +
1 molecule oxygen →
2 molecules carbon dioxide

But carbon dioxide has been proved to be CO_2 (see above) and oxygen, O_2 (by experiment and the application of Avogadro's Hypothesis by Cannizzaro's Argument),

∴ 2 molecules carbon monoxide + $O_2 \rightarrow 2CO_2$

∴ carbon monoxide is CO

This is confirmed by the determination of the relative density, which is 14 (H = 1).

∴ relative molecular mass =
28, corresponding to CO.

Silica and the silicates

Silica (silicon dioxide) is found widely in nature and assumes several crystalline forms, cristobalite, tridymite and quartz, of which the last is the commonly known crystalline form, and a hydrated amorphous form, consisting of the siliceous remains of extinct diatoms, called kieselguhr, which has remarkable powers of absorption. On disintegration, quartz forms sand which is white when pure but often tinged yellow by the presence of iron(III) oxide. Quartz, or rock-crystal, may be found as colourless transparent hexagonal prisms or rendered opaque or coloured by impurities. Silica has a macromolecular structure, already described (see p. 319), so that there are no discrete SiO_2 molecules.

Silica may be fused in the oxy-hydrogen blowpipe flame, softening at 1500–1600°C and fusing above 1700°C. The softening plastic mass may be worked like glass or drawn into threads. The amorphous vitreous product is a supercooled liquid, silica glass, quartz glass, fused quartz or just silica. It has a very low thermal linear expansivity $(5 \times 10^{-7} \, K^{-1})$ which is less than one-tenth that of soda-glass. Apparatus made of silica will withstand thermal shock, even being quenched with cold water when red hot without fracture. Silica pyrometer tubes are used in measuring the temperature of furnaces. Silica plates, lenses and prisms are used in optical work, especially for the transmission of ultra-violet light. Vitreosil is a translucent form of silica made from sand.

Oxygen and silicon are respectively the most abundant and next most abundant elements in the Earth's crust. Both occur in silica and the silicate and alumino-silicate rocks. Sand is used in enormous quantities in civil engineering for cement, concrete and mortar, and for the manufacture of all types of glass. Kieselguhr is used as an absorbent in pharmaceutical dressings and for the packing of dangerous chemicals, where the containing vessels might be broken during transport.

Silica is an acidic oxide, insoluble in water and because of its macromolecular structure, nonvolatile. It reacts with hot concentrated or fused alkali to form the (meta) silicate:

$$SiO_2 + 2OH^- \rightarrow SiO_3^{2-} + H_2O$$

Silica will displace other more volatile acidic oxides from their salts on strong heating: sulphur trioxide being displaced from sodium sulphate, carbon dioxide from sodium carbonate and phosphorus pentoxide from calcium phosphate, leaving the corresponding silicate:

$$Na_2SO_4 + SiO_2 \rightarrow Na_2SiO_3 + SO_3$$

$$Na_2CO_3 + SiO_2 \rightarrow Na_2SiO_3 + CO_2$$

$$2Ca_3(PO_4)_2 + 6SiO_2 \rightarrow 6CaSiO_3 + P_4O_{10}$$

Each of these acidic oxides forms a stronger acid than silica and in solution will cause precipitation of silicic acid, or hydrated silica:

$$SiO_3^{2-} + 2H^+ \rightarrow H_2SiO_3(s) \quad \text{or} \quad SiO_2 . nH_2O(s)$$

Silica will react with hydrofluoric acid only amongst acids, forming silicon tetrafluoride, a reaction important in the etching and frosting of glass:

$$SiO_2 + 4HF = SiF_4(g) + 2H_2O$$

Silicon carbide is made industrially as carborundum by reduction of silica with excess coke in an electric furnace:

$$SiO_2 + 3C \rightarrow SiC + 2CO$$

It has the diamond structure and is an extremely hard abrasive. Heated with magnesium powder, silica is reduced to silicon while excess magnesium forms magnesium silicide:

$$SiO_2 + 2Mg \rightarrow 2MgO + Si$$

$$SiO_2 + 4Mg \rightarrow Mg_2Si + 2MgO$$

Within the limits set by the reactions given above, silica apparatus and furnace linings may be used for high-temperature reactions involving a high degree of chemical corrosion. The obvious general limitation is susceptibility to alkali.

Pure amorphous silica is a very fine, white powder, easily dispersed and the most reactive form. The reactivity is due to its large surface area. It may be produced from impure silica (sand) which is in abundant supply by forming silicon tetrafluoride, hydrolysing it, and heating the hydrated silica formed to drive off water. This is described in detail under silicon tetrafluoride. Alternatively, fusion with sodium carbonate yields sodium silicate, from a solution of which hydrated silica is precipitated by acid, collected and heated. The relative proportions by volume of sodium silicate solution and hydrochloric acid may be determined by use of methyl orange as indicator in a rough titration, delivering the acid from a measuring cylinder. Hydrochloric acid is then added to an excess (about 5%) of sodium silicate solution and the mixture boiled. The precipitate is filtered off, washed and dried in an oven at 120°C:

$$SiO_3^{2-} + 2H^+ \rightarrow H_2O + SiO_2 \text{ hydrated}$$

A colloidal solution of hydrated silica is formed when sodium silicate solution is slowly added with stirring to an excess (about 5%) of dilute hydrochloric acid. A clear solution is produced which may be dialysed to remove much of the electrolyte. An excess of silicate ions causes coagulation. On standing, a stiff jelly, or 'gel' may result. On careful heating, most of the loosely held water is expelled from the gel and the resulting solid, 'silica gel', may be used as a desiccant. Applications include drying blast-furnace gas on the large scale down to keeping the interior of chemical balances dry and for keeping electrical components free from deterioration through dampness during storage.

There is some doubt about the definite existence of the silicic acids. It has been claimed that the vapour pressure curve of hydrated silica shows breaks corresponding to orthosilicic acid, H_4SiO_4, and metasilicic acid, H_2SiO_3. Neither has been crystallized and the gelatinous precipitate obtained on acidifying a silicate solution may be called silicic acid or hydrated silica. However, a wide range of silicates based on these (hypothetical) acids and their condensed forms exist. The hypothetical acids are formulated:

HO $\diagdown$ $\diagup$ OH
      Si   or $SiO_2.2H_2O$
HO $\diagup$ $\diagdown$ OH

orthosilicic   hydrated
acid   silica

HO $\diagdown$
      Si $=$ O   or $SiO_2.H_2O$
HO $\diagup$

metasilicic   hydrated
acid   silica

Only the alkali metal silicates are soluble in water. Sodium silicate is made by fusing sand with soda ash and the glassy product boiled with water under pressure to give the commercial product, water glass. This is used as an egg preservative, sealing the pores of egg shells with calcium silicate, in china adhesives and in waterproofing. Solutions are alkaline by hydrolysis, as are carbonates. Two sodium silicates, Na_2SiO_3 (meta) and Na_4SiO_4 (ortho), have been prepared. Other silicates may be prepared by fusing the oxide of the metal with silica,

$$PbO + SiO_2 \rightarrow PbSiO_3$$

or by mixing solutions containing the required ions,

$$Pb^{2+} + SiO_3^{2-} \rightarrow PbSiO_3(s)$$

The study of igneous rocks and the production of pottery and glass requires a knowledge of complex silicates and alumino-silicates, much of which is beyond the scope of this book. Slow cooling produced crystalline rocks (e.g. granite and basalt) while rapid cooling gave glassy rocks (e.g. sanidine and obsidian). The composition of minerals is usually expressed in terms of the proportions of metallic oxides, silica and water contained in the macromolecules. Felspar is $K_2O.Al_2O_3.6SiO_2$. Compositions vary slightly because cations of roughly the same size may replace each other, if electrical neutrality is maintained. Olivine is written $(Mg,Fe)SiO_4$ to show iron(II) replacing magnesium. The weathering of rocks of the felspar type yields clay. China clay, or kaolin, has the approximate composition $Al_2O_3.2SiO_2.2H_2O$.

Pottery and china are manufactured from clay which is moulded, heated to dehydrate aluminium silicate and a glaze applied. This is fused by further heating to give a mixture of silicate as a surface coating. The three main raw materials in fine bone china are stone flux, china clay and ox-bone. It is the bone, reduced to a fine ash, which imparts to china its desired whiteness, translucency and comparatively great strength. Approximately half of the weight of a Wedgwood bone china plate consists of calcined ox-bone, accounting for its surprising weight and strength. Alumino-silicate minerals, called the zeolites, which have macromolecular open structures, are used to soften water.

Glass is a supercooled liquid containing silica dissolved in a mixture of fused silicates. It has taken on the appearance of the solid state as the viscosity increased and is amorphous, that is, without crystalline structure. Soda-glass, of approximate composition $Na_2O.CaO.6SiO_2$, is manufactured by fusing together silica (pure sand), an alkali (interpreted widely as sodium carbonate or sodium sulphate) and lime (chalk, limestone or lime). It has no definite melting-point and gradually softens on heating. Harder potash-glass is obtained using potassium carbonate as alkali. For heat-resistant glass, alumina and boric oxide, with up to about 80% silica, are used to give the borosilicate glasses. These have a low thermal linear expansivity and may be used to make thicker and stronger apparatus, with complicated joints and seals which will survive heating, and for domestic ovenware. Borosilicate glass is worked in the oxy–coal gas blow-pipe flame. Soda-glass is worked in the air–coal gas flame but requires careful annealing (slow cooling) and is apt to devitrify (crystallize) with age. Coloured glasses are made by the addition of metal oxides giving coloured silicates; blue glass requires cobalt oxide, red glass, copper(I) oxide, and green glass, iron(II) salts. Armoured glass is cooled very rapidly during production and the surface forms a very hard skin and if cut or worked, the inside crumbles to a powder.

Silicon monoxide

Not much is known about this oxide, but it is included for completeness. It is said to be obtained by heating silicon with silica under special conditions. On heating, it reverts to these substances:

$$2SiO \rightleftharpoons SiO_2 + Si$$

Sn^{II} and Sn^{IV}: the oxides of tin

Tin forms oxides in which it exerts the principal valency of 4 and the subsidiary valency of 2. The respective oxidation numbers are incorporated into the names. Tin(II) oxide (stannous oxide) smoulders when exposed to air and may become incandescent, forming tin(IV) oxide, (stannic oxide)

$$2SnO + O_2 \rightarrow 2SnO_2$$

Tin(IV) oxide is found as the ore, cassiterite or tinstone. Hydrated forms of both oxides exist. Tin(II) hydroxide has not been isolated as such and is probably best regarded as hydrated tin(II) oxide. The same is true of tin(IV) hydroxide, known more frequently as stannic acid, but better regarded as hydrated tin(IV) oxide of which two forms, α and β, will be described. These probably differ in degree of hydration and size of particle. The properties of tin(IV) oxide depend very much on whether it is hydrated or dried.

Tin(IV) oxide may be prepared by heating molten tin in oxygen or air or by igniting the β-hydrated tin(IV) oxide. This is formed in a way reminiscent of non-metallic elements, by the action of concentrated nitric acid on tin, the complex exothermic reaction proceeding with increasing vigour to leave a bulky white residue:

$$Sn + 4HNO_3 \rightarrow$$
$$SnO_2(s) \ hydrated + 4NO_2 + 2H_2O$$

Tin(IV) oxide is a white powder which is reduced to the metal on heating with carbon,

$$SnO_2 + 2C \rightarrow Sn + 2CO$$

and reacts with fused alkalis to form the stannate(IV):

$$SnO_2 + 2OH^- \rightarrow SnO_3^{2-} + H_2O$$
$$\text{stannate(IV)}$$

The reaction expressed in this way is similar to those which occur with carbon dioxide or silica and alkali. But sodium stannate(IV), $Na_2SnO_3 \cdot 3H_2O$, used commercially as a mordant in dyeing, cannot be dehydrated and is thought to be $Na_2Sn(OH)_6$, the hexahydroxostannate(IV). The simpler ion is retained usually in equations. Tin(IV) oxide, as prepared above, does not react with dilute acids. It dissolves in hot concentrated sulphuric acid from which it is precipitated on addition of water. Tin(IV) oxide is used to manufacture white glass, white tiles and glazes and as a polishing medium.

Addition of aqueous ammonia to a solution of tin(IV) chloride, or of dilute hydrochloric acid to sodium stannate(IV) solution precipitates α-hydrated tin(IV) oxide:

$$SnCl_4 + 4NH_3 + 2H_2O \rightarrow$$
$$4(NH_4^+ + Cl^-) + SnO_2(s) \ hydrated$$

$$SnO_3^{2-} + 2H^+ \rightarrow$$
$$\underset{\substack{complex \\ as \ above}}{} \qquad SnO_2(s) \ hydrated + H_2O$$

This form of tin(IV) oxide is much more amphoteric. Hydrochloric acid forms tin(IV) chloride. As this compound contains some (hydrated) tin(IV) ions, it is considered to be a salt:

$$SnO_2 \ hydrated + 4HCl \rightarrow SnCl_4 + 2H_2O$$

At red heat, α-hydrated tin(IV) oxide loses water.

Unlike tin(IV) oxide, which is assigned the ionic structure, $Sn^{4+}(O_2^-)_2$, tin(II) oxide has a covalent macromolecular structure. It is a dark powder, variously described as brown, black and olive-green, formed when tin(II) ethanedioate (oxalate) is heated in the absence of air,

$$\begin{matrix} COO \\ | \\ COO \end{matrix} Sn \rightarrow SnO + CO + CO_2$$

Alternatively, the hydrated oxide ('tin(II) hydroxide') may be precipitated by addition of aqueous ammonia to tin(II) chloride solution, collected, dried and heated, all operations being done in an inert atmosphere. Tin(II) oxide is amphoteric: tin(II) salts are formed with acids, and stannate(II) with alkalis:

$$SnO + 2H^+ \rightarrow H_2O + Sn^{2+}$$

$$SnO + OH^- \rightarrow HSnO_2^-$$
$$\text{stannate(II)}$$

With concentrated alkali, disproportionation occurs followed by formation of sodium stannate(IV):

$$2SnO \rightarrow Sn + SnO_2$$

$$SnO_2 + 2OH^- \rightarrow SnO_3^{2-} + H_2O$$

The action of alkali, and sometimes acids, on so-called amphoteric oxides is not as straight-forward as it might seem. The oxide may pass into colloidal solution, the micelles being stabilized by adsorption of ions of electrolyte. This is called *peptization*.

Pb^{II} and Pb^{IV}: the oxides of lead: lead monoxide, PbO; red lead, Pb$_3$O$_4$; lead dioxide, PbO$_2$; (lead sesquioxide, Pb$_2$O$_3$)

The principal valency state of lead is two, while the group valency, four, has become subsidiary; these are also the oxidation states indicated in the names below. Lead(II) oxide (monoxide), PbO, is the most stable oxide, the others reverting to it on heating. Red lead, minium, trilead tetroxide, Pb$_3$O$_4$, may be regarded as a mixed oxide, 2PbO.PbO$_2$, dilead(II) lead(IV) oxide, but is better assigned the constitution, lead(II) orthoplumbate(IV), (Pb^{2+})$_2$ PbO$_4^{4-}$. The plumbate(IV) state is much more stable than that of lead(IV).

Lead(II) oxide is prepared in the laboratory by heating the nitrate, carbonate, 'hydroxide' (hydrated oxide) or any other oxide to above 600°C. Commercially, a yellow form, massicot, is obtained by gently heating lead in air while fusion produces litharge, a reddish-yellow form, which is then ground to a powder:

$$2Pb + O_2 \rightarrow 2PbO$$

Red lead is a scarlet powder obtained by heating massicot in air to 400°C; on further heating above 470°C, litharge results:

$$6PbO + O_2 \xrightarrow[470°C]{400°C} 2Pb_3O_4$$

Lead(IV) oxide (dioxide) cannot be prepared from the monoxide and oxygen because it decomposes when heated to 300°C:

$$2PbO_2 \rightarrow 2PbO + O_2$$

Dilead trioxide, lead sesquioxide, Pb$_2$O$_3$, is much less important. It is an orange powder. There is some uncertainty about its exact composition but there is experimental evidence for regarding it as lead(II) metaplumbate(IV), Pb^{2+}PbO$_3^{2-}$.

Lead(II) oxide is an amphoteric oxide. It is insoluble in water, but reacts with both acids and alkalis. Because most salts of lead are insoluble, the nature of the acid used is important. Dilute nitric acid gives lead(II) nitrate and boiling dilute hydrochloric acid, lead(II) chloride, which though sparingly soluble in the cold is moderately soluble in hot solution:

$$PbO + 2H^+ \rightarrow H_2O + Pb^{2+}$$

Plumbate(II) is formed in reaction with alkali:

$$PbO + OH^- \rightarrow HPbO_2^-$$

$$HPbO_2^- + OH^- \rightarrow PbO_2^{2-} + H_2O$$

Lead(II) hydroxide, Pb(OH)$_2$, has not been isolated: addition of aqueous ammonia to a solution of lead(II) nitrate yielding a hydrated oxide, (PbO)$_2$.H$_2$O, as a white precipitate. It is amphoteric, but much less acidic than the corresponding tin(II) compound. Lead(II) oxide is easily reduced to the metal on heating with carbon, carbon monoxide and hydrogen, the charcoal block reduction being very convenient:

$$PbO + C \rightarrow Pb + CO$$

$$PbO + CO \rightarrow Pb + CO_2$$

$$PbO + H_2 \rightarrow Pb + H_2O$$

Lead(II) oxide is used in the manufacture of lead plate accumulators, as a drier in varnishes, for the production of lead compounds and in the manufacture of optical glass.

Lead(IV) oxide is a puce-coloured solid, insoluble in water and nitric acid. It is precipitated by boiling a solution of lead(II) ethanoate (acetate) with sodium hypochlorite (chlorate(I)) solution, the colour of the mixture changing from orange to dark brown. The precipitate is filtered at the pump, washed with hot water and dried at 120°C in an oven:

$$(CH_3COO)_2Pb + ClO^- + H_2O \rightarrow$$
$$PbO_2(s) + 2CH_3COOH + Cl^-$$

Lead(IV) oxide is a strong oxidizing agent. On heating, it evolves oxygen, leaving litharge. At 0°C, with concentrated hydrochloric acid, some lead(IV) chloride, lead tetrachloride, is formed but this soon loses chlorine on warming. At ordinary temperatures, chlorine is evolved immediately leaving lead(II) chloride (compare MnO_2):

$$PbO_2 + 4HCl \rightarrow PbCl_4 + 2H_2O$$

$$PbCl_4 \rightarrow PbCl_2 + Cl_2$$

Oxygen is liberated when lead(IV) oxide is heated with concentrated sulphuric acid and lead(II) sulphate remains (compare MnO_2):

$$2PbO_2 + 2H_2SO_4 \rightarrow 2PbSO_4 + 2H_2O + O_2$$

Lead(II) sulphate is formed with incandescence on warming lead(IV) oxide in sulphur dioxide, or with sulphur in air:

$$PbO_2 + SO_2 \rightarrow PbSO_4$$

Lead dioxide and nitric acid will oxidize manganese(II) salts to permanganic acid, $HMnO_4$:

$$2Mn^{2+} + 4H^+ + 5PbO_2 \rightarrow$$
$$2MnO_4^- + 2H_2O + 5Pb^{2+}$$

Lead(IV) oxide reacts with strong alkalis to form the plumbate(IV):

$$PbO_2 + 2OH^- \rightarrow PbO_3^{2-} + H_2O$$

Plumbates of the type Ca_2PbO_4 (calcium(II) ortho-plumbate(IV)) and Na_2PbO_3 (sodium(I) metaplumbate(IV)) corresponding to $Pb(OH)_4$ or H_4PbO_4 and H_2PbO_3, both being hypothetical, are known. Calcium plumbate is used in priming paints for iron and steelwork to resist rusting. Many plumbates(IV) are hydrated and being isomorphous with stannates(IV) may be assigned similar formulae: $Na_2PbO_3 . 3H_2O$ becomes $Na_2[Pb(OH)_6]$ the hexahydroxoplumbate(IV). Some, but not sodium plumbate(IV), decompose on dehydration. They are decomposed by carbon dioxide. The general preparation of the plumbate(IV) requires fusion of lead(IV) oxide and the appropriate base. Because of the formation of certain lead(IV) salts from lead(IV) oxide with acids under carefully regulated conditions, lead(IV) oxide is considered to be weakly basic as well as acidic and is, therefore, classified as an amphoteric oxide. Lead(IV) oxide is used in the grids of lead plate accumulators.

Red lead, trilead tetroxide, Pb_3O_4, is a scarlet powder also called minium. Its commercial production and decomposition has been described. In its reactions it may be regarded as a mixed oxide, $2PbO . PbO_2$, or as lead(II) orthoplumbate(IV), Pb_2PbO_4. With nitric acid, lead(II) nitrate and the puce dioxide are formed:

$$Pb_2PbO_4 + 4HNO_3 \rightarrow$$
$$2Pb(NO_3)_2 + Pb(OH)_4$$

or

$$2PbO . PbO_2(s) + 4HNO_3 \rightarrow$$
$$2Pb(NO_3)_2 + 2H_2O + PbO_2(s)$$

Warm concentrated hydrochloric acid is oxidized to chlorine, lead(II) chloride remaining:

$$2PbO . PbO_2 + 4HCl \rightarrow 2PbCl_2 + 2H_2O + PbO_2$$

and $$PbO_2 + 4HCl \rightarrow PbCl_2 + 2H_2O + Cl_2$$

$$\therefore Pb_3O_4 + 8HCl \rightarrow 3PbCl_2 + 4H_2O + Cl_2$$

Warm concentrated sulphuric acid yields oxygen and white lead(II) sulphate:

$$2[2PbO . PbO_2(s) + 2H_2SO_4 \rightarrow$$
$$2PbSO_4(s) + 2H_2O + PbO_2(s)]$$

$$2PbO_2(s) + 2H_2SO_4 \rightarrow$$
$$2PbSO_4(s) + 2H_2O + O_2$$

$\therefore$ *Adding,*

$$2Pb_3O_4(s) + 6H_2SO_4 \rightarrow$$
$$6PbSO_4(s) + 6H_2O + O_2$$

Red lead may be reduced to lead on heating with carbon, carbon monoxide or hydrogen. Red lead is used as a priming paint for wood and steel work, although other primers (zinc chromate and calcium plumbate(IV)) have been developed for steel, and paints containing a suspension of aluminium metal are used to seal the surface of resinous woods and hardboard. Mixed with linseed oil, it is used by plumbers as a sealing compound for the older type of lead gas pipe connections. As an oxidizing agent, red lead is incorporated in some matchheads and it is also used in the production of optical glass.

Disulphides, monosulphides and thiosalts

Sulphides and thiosalts of the quadrivalent state are formed for all except lead. In the bivalent state, sulphides (and thiosalts) of carbon and silicon do not appear to exist. Thiosalts are analogous to oxosalts, sulphur having replaced some or all of the oxygen. Their existence reveals the acidic nature of the sulphides.

Carbon disulphide and thiocarbonates

Carbon disulphide is a colourless, very volatile liquid (b.p. 46°C) which gives off a highly flammable poisonous vapour with a repellent odour, due to impurities. It is an endothermic compound with a very low ignition temperature, even being ignited in air by a glass rod at just below red heat. Carbon disulphide forms explosive mixtures with air. In all these respects it is more dangerous than ether. Carbon disulphide is typically covalent, is miscible with many organic solvents, immiscible with water and an excellent solvent for sulphur, yellow phosphorus and iodine.

Carbon disulphide is made commercially by the action of sulphur vapour on white-hot carbon. The reaction is endothermic:

$$C + 2S \rightarrow CS_2; \quad \Delta H = +92\,kJ\,mol^{-1}$$

Sulphur and coke are fed into a shaft furnace heated by passage of a large electric current through the mass. Carbon disulphide is evolved, condensed and re-distilled.

In the Courtauld's process, carbon disulphide is made by a continuous process involving methane and hydrogen sulphide:

$$CH_4 + 4S \rightarrow CS_2 + 2H_2S$$

No details are available but the hydrogen sulphide is either oxidized back to sulphur and re-used, or purified and sold. The process requires expertise in handling smelly and toxic offensive gases and liquid wastes. The process can be adapted for the production of fine intermediate chemicals for pharmaceuticals, agrochemicals and oil additives. Carbon disulphide is used in the viscose industry as an industrial solvent, as a vermicide and chemical intermediary.

Complete combustion of carbon disulphide in air yields carbon dioxide and sulphur dioxide:

$$CS_2 + 3O_2 \rightarrow CO_2 + 2SO_2$$

A mixture of carbon disulphide vapour and nitrogen oxide (nitric oxide) reacts when ignited with a spectacular blue flash, leaving sulphur:

$$CS_2 + 2NO \rightarrow CO_2 + 2S + N_2$$

Above 150°C, carbon disulphide is hydrolysed by superheated water to give carbon dioxide and hydrogen sulphide:

$$CS_2 + 2H_2O \rightarrow CO_2 + 2H_2S$$

Chlorine reacts with carbon disulphide in the presence of a catalyst, manganese(II) chloride, iodine or aluminium chloride, to form tetrachloromethane (carbon tetrachloride) and disulphur dichloride, which is finally reduced to sulphur:

$$CS_2 + 3Cl_2 \rightarrow CCl_4 + S_2Cl_2$$

$$2S_2Cl_2 + CS_2 \rightarrow CCl_4 + 6S$$

Carbon disulphide reacts with an alcoholic solution of sodium sulphide to form sodium (tri)-thiocarbonate, Na_2CS_3:

$$S^{2-} + CS_2 \rightarrow CS_3^{2-}$$

Acidification yields thiocarbonic acid as an unstable yellow oil:

$$2H^+ + CS_3^{2-} \rightarrow H_2CS_3$$
$$\text{(compare with carbonic acid)}$$

Carbon disulphide slowly reacts with sodium hydroxide solution to form a mixture of thiocarbonate and carbonate:

$$3CS_2 + 6OH^- \rightarrow CO_3^{2-} + 2CS_3^{2-} + 3H_2O$$

Sn^{II} and Sn^{IV}: sulphides of tin

Both tin(IV) sulphide (stannic sulphide), SnS_2, and tin(II) sulphide (stannous sulphide), SnS, may be formed by union of the elements. On heating above 265°C, tin(II) sulphide slowly disproportionates into a mixture of tin and tin(IV) sulphide, the sulphide of the principal valency state:

$$2SnS \rightarrow Sn + SnS_2$$

The sulphides are more usually prepared by precipitation using hydrogen sulphide and solutions of the chlorides, slightly acidified to reduce hydrolysis,

$$SnCl_4 + 2H_2S \rightarrow SnS_2 + 4HCl$$

$$SnCl_2 + H_2S \rightarrow SnS + 2HCl$$

Tin(II) sulphide is brown and the tin(IV) compound, yellow.

Tin(IV) sulphide dissolves in warm (colourless) ammonium sulphide solution to form ammonium thiostannate(IV), $(NH_4)_2SnS_3$:

$$SnS_2 + S^{2-} \rightarrow SnS_3^{2-}$$

It is usually more convenient to use yellow ammonium (poly)sulphide, $(NH_4)_2S \cdot S_x$. Tin(II) sulphide will not dissolve in colourless ammonium sulphide, but will react with warm yellow ammonium sulphide, being oxidized to the tin(IV) state and forming ammonium thiostannate(IV):

$$SnS + S^{2-} . S \rightarrow SnS_3^{2-}$$

Addition of acid precipitates the yellow tin(IV) sulphide:

$$SnS_3^{2-} + 2H^+ \rightarrow SnS_2(s) + H_2S$$

(compare carbonates)

When yellow ammonium sulphide is used for the separation of Group II into the Groups IIa and IIb of the Qualitative Analysis Tables, tin(II) sulphide is converted into the tin(IV) compound. Alternatively, separation of the sulphides which form thiosalts may be effected by boiling with lithium hydroxide reagent, which dissolves the sulphides of arsenic, antimony and tin, although with the last-named some hydrolysis occurs. The reagent contains potassium nitrate to prevent dispersal of sulphides into the colloidal condition:

$$2SnS + 2OH^- \rightarrow \underset{\text{stannate(II)}}{HSnO_2^-} + \underset{\substack{\text{thio-}\\\text{stannate(II)}}}{HSnS_2^-}$$

$$3SnS_2 + 6OH^- \rightarrow \underset{\text{stannate(IV)}}{SnO_3^{2-}} + \underset{\substack{\text{thio-}\\\text{stannate(IV)}}}{2SnS_3^{2-}} + 3H_2O$$

Acidification precipitates sulphides in the original valency state.

Both tin(II) sulphide and tin(IV) sulphide dissolve rapidly in concentrated hydrochloric acid with liberation of hydrogen sulphide,

$$SnS + 2HCl \rightarrow SnCl_2 + H_2S$$

$$SnS_2 + 4HCl \rightarrow SnCl_4 + 2H_2S$$

Mosaic gold for gilding is manufactured by heating a mixture of tin filings, sulphur and ammonium chloride to give a mass of tin(IV) sulphide.

Pb^{II}: lead(II) sulphide, PbS

Lead(IV) sulphide (PbS$_2$) has not been prepared but a lead(II) disulphide (Pb^{2+}S.S^{2-}) has been made. However, lead(II) sulphide, PbS, is the important sulphide. It is found naturally as galena.

Lead(II) sulphide may be produced by union of the elements but more usually as a black precipitate obtained on passing hydrogen sulphide into a solution of a lead salt:

$$Pb^{2+} + H_2S \rightarrow PbS(s) + 2H^+$$

In qualitative analysis, some lead passes through from the initial precipitation of lead(II) chloride in Analysis Group I to Group II. It is precipitated as the sulphide in the presence of hydrochloric acid, the precipitate appearing red or yellow at first, due to PbS.PbCl$_2$. and then black as PbS is formed. Paint containing lead compounds will darken in the sulphurous atmosphere of our towns due to the formation of lead(II) sulphide.

Lead(II) sulphide is insoluble in water, colourless and yellow ammonium sulphide, and in alkali. In these respects it differs from sulphides of tin. Concentrated nitric acid oxidizes it to lead(II) sulphate, but at lower concentrations, lead(II) nitrate and sulphur result. Hydrogen peroxide oxidizes lead(II) sulphide to the sulphate, which is white:

$$PbS + 4H_2O_2 \rightarrow PbSO_4 + 4H_2O$$

This reaction has been applied to the restoration of darkened oil paintings.

Halides

Carbon, silicon, germanium and tin form the complete range of halides in the principal quadrivalent state while lead, the principal valency of which is two, forms only an unstable tetrachloride. The tetrafluorides of tin and lead are involatile solids, unlike the others which are volatile liquids (of which the boiling-points rise with increasing atomic number and relative atomic mass). The solid structures are akin to that of AlF$_3$ and the properties of the fluorides show some similarity to those formed by Group III elements (the boron group). In addition to the simple tetrahalides, various mixed halides are well known, especially fluorine products. The hitherto commercially successful freons (substituted chloro- and fluoro- hydrocarbons of apparent inertness) were used widely as refrigerants and aerosol propellants, and are prime suspects in the destruction of atmospheric ozone.

Only germanium, tin and lead form halides in the bivalent state, those of lead being the most stable for that element while those of germanium and tin are readily oxidized. Tin(II) chloride is used regularly as a powerful reducing agent although it is less powerful than germanium(II) chloride, which has the greater tendency to become quadricovalent:

$$Sn^{2+} - 2e^- \rightleftharpoons Sn^{4+} \quad \text{or} \quad Sn^{IV} \text{ (covalent)}$$

Sn^{IV} refers to quadrivalent tin i.e. in the covalent oxidation state of four, Sn(IV). The halides of the bivalent state are of the type formed generally by metals (ionic, except for germanium(II) chloride) while those of the quadrivalent state are typical of those formed by non-metals (covalent).

Silicon tetrafluoride, SiF₄

When hydrofluoric acid attacks silica or a silicate, silicon tetrafluoride is formed:

$$SiO_2 + 4HF \rightarrow SiF_4(g) + 2H_2O$$

This reaction is used in etching and frosting glass. In the laboratory, silicon tetrafluoride is prepared by warming dried silver sand, calcium fluoride and concentrated sulphuric acid, hydrogen fluoride being formed *in situ*. Silicon tetrafluoride is a colourless gas which fumes in moist air. It is rapidly hydrolysed by water to form hydrated silica and fluorosilicic acid, and should be collected over mercury:

$$SiF_4 + 2H_2O \rightarrow SiO_2(s) \; hydrated + 4HF$$

$$SiF_4 + 2HF \rightleftharpoons H_2SiF_6$$

This preparation and hydrolysis affords a method of producing fine amorphous silica starting from impure sand. The apparatus, which must be absolutely dry, is shown in Fig. 20.8. The silica and silicate in the glass of the flask is attacked by the hydrogen fluoride but the flask lasts for several experiments. The gas is led through a mercury seal, as shown, to avoid blockage by gelatinous hydrated silica. The gelatinous pieces are filtered and heated strongly to drive off water of hydration, leaving fine, white, amorphous silica, which is so light that it can be dispersed by the slightest puff of air.

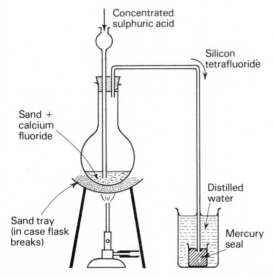

Concentrated sulphuric acid

Silicon tetrafluoride

Sand + calcium fluoride

Distilled water

Sand tray (in case flask breaks)

Mercury seal

Fig. 20.8 The preparation and hydrolysis of silicon tetrafluoride

Because of its exceedingly large surface area, finely divided silica is very much more reactive than the original sand. Careful evaporation of the filtrate concentrates the solution of fluorosilicic acid which breaks down on further heating:

$$H_2SiF_6 \rightarrow 2HF + SiF_4$$

Addition of a potassium salt in solution causes the precipitation of potassium hexafluorosilicate, K_2SiF_6.

Tetrachlorides

The tetrachlorides are covalent liquids, colourless, except for lead(IV) chloride which is yellow, and readily volatile.

The following data

	Boiling-point/°C:
CCl₄	76.4
SiCl₄	57.0
GeCl₄	86.5
SnCl₄	114
PbCl₄	decomposes

show that silicon tetrachloride has the lowest boiling-point of the stable tetrachlorides. Lead(IV) chloride readily decomposes. Carbon tetrachloride, otherwise tetrachloromethane, is used as a solvent for fats and greases in dry-cleaning and, because the vapour does not support combusion, in fire-extinguishers. The formation of tin(IV) chloride is used in the recovery of tin from tin-plate by the action of chlorine. Of special interest is a comparison of the action of water on each tetrachloride.

Carbon tetrachloride, formed by the action of chlorine on carbon disulphide industrially, is inert and not affected by water. The tetrachlorides of silicon, germanium and tin may be formed by union of the elements. When a stream of dry chlorine impinges on warm granulated tin, the metal melts and catches fire, burning with a silvery-grey flame. Tin(IV) chloride may be condensed in a Büchner flask, to which is attached a calcium chloride tube. The apparatus is shown in Fig. 20.9.

Silicon tetrachloride which fumes in damp air, is completely and vigorously hydrolysed by water:

$$SiCl_4 + 2H_2O \rightarrow 4HCl + SiO_2(s) \; hydrated$$

Unlike carbon, silicon (and the others) have d orbitals available at lower energy levels than the s electrons of the next quantum shell (or level). This allows silicon (in SiCl₄) to acquire a water molecule

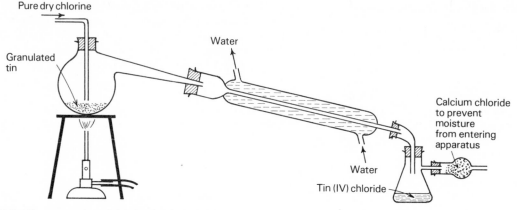

Fig. 20.9   The preparation of tin(IV) chloride (in fume cupboard)

by co-ordination, and hydrolysis occurs by this reaction path of lower energy than any which might be available to carbon (in CCl_4), the reaction intermediate $Si(OH_2)Cl_4$ losing HCl in a step-wise process. Complexes of silicon tetrachloride, $SiCl_4X$ and $SiCl_4X_2$ with organic compounds have been isolated; a discovery which serves to consolidate these ideas.

Germanium(IV) chloride is less hydrolysed. Tin(IV) chloride, a colourless fuming liquid having an unpleasant odour, reacts with a little water to form a solid pentahydrate,

$$SnCl_4 + 5H_2O \rightarrow SnCl_4 . 5H_2O$$

known commercially as *butter of tin*. Excess water causes complete hydrolysis:

$$SnCl_4 + 2H_2O \rightarrow 4HCl + SnO_2(s) \ hydrated$$

Ammonium chloride reacts with tin(IV) chloride to form ammonium hexachlorostannate(IV), known as *pink salt*:

$$2NH_4Cl + SnCl_4 \rightarrow (NH_4)_2SnCl_6$$

Lead tetrachloride, lead(IV) chloride, $PbCl_4$, is a yellow, dense, oily liquid which decomposes on warming to give chlorine and lead(II) chloride:

$$PbCl_4 \rightarrow PbCl_2 + Cl_2$$

Decomposition may become explosive on strong heating. Lead(IV) chloride may be isolated through the formation of ammonium hexachloroplumbate(IV), as a complex intermediate. Lead(II) chloride ($PbCl_2$) is dispersed by grinding in concentrated hydrochloric acid, and chlorine passed into the suspension for several hours, when a yellow solution results. Addition of an ice-cold concentrated solution of ammonium chloride gives a yellow

crystalline precipitate of ammonium hexachloroplumbate(IV):

$$PbCl_2 + Cl_2 + 2Cl^- \rightarrow PbCl_6^{2-}$$

$$2NH_4^+ + PbCl_6^{2-} \rightarrow (NH_4)_2PbCl_6(s)$$

When ammonium hexachloroplumbate(IV) is added in small portions to ice-cold concentrated sulphuric acid, and constantly stirred, lead tetrachloride, lead(IV) chloride, separates:

$$(NH_4)_2PbCl_6 + H_2SO_4 \rightarrow$$
$$2HCl + (NH_4)_2SO_4 + PbCl_4(l)$$

Lead(IV) chloride is rapidly hydrolysed by water with separation of brown lead(IV) oxide:

$$PbCl_4 + 2H_2O \rightarrow 4HCl + PbO_2(s)$$

The formation of silicones

The systematic study of the possible 'organic chemistry' of silicon was conducted by F. S. Kipping early in this century. During the course of this work he investigated the hydrolysis of di-alkyl-dichlorosilanes, e.g. $(CH_3)_2SiCl_2$. An analogue of a ketone might have been expected on hydrolysis (see equation) so that the product was called a silicone:

However, water was lost between separate molecules

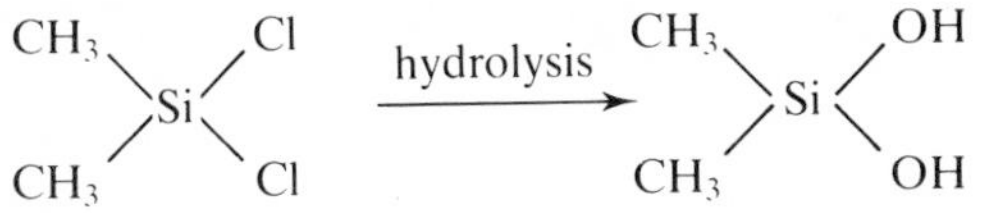

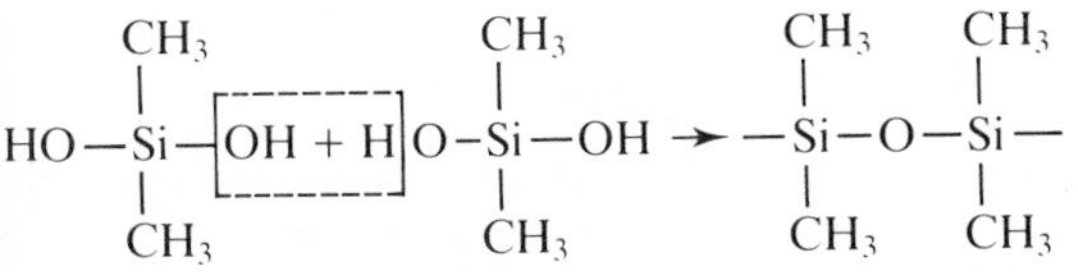

The name silicone was retained for these long-chain polymers although they are more correctly called *polysiloxanes*. The general formula of a linear polymer would be

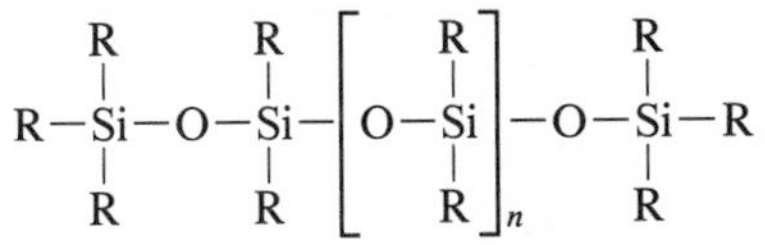

Other structures may be built up by hydrolysis of suitable compounds, forming cross links between chains:

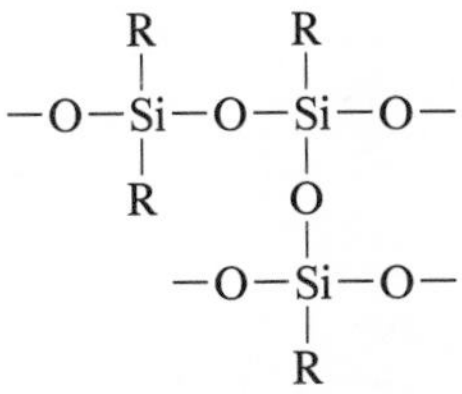

Silicones are noted for their resistance to heat and oxidation, good water repellency and electrical insulation properties. Applications are numerous, including water-repellent finishes on textiles, as a medium for polishes containing hard waxes, in electrical condensers and specialized lubricating rubbers and paint-resins. In physical properties, they range from non-volatile liquids to rubber-like solids.

Complex halide anions formed by tetrahalides

Certain complexes have been mentioned already. The complete range is listed in Table 20.3.

Halides of the bivalent state

Germanium, tin and lead but not carbon and silicon form all four halides of this type. Those of germanium and tin readily pass into the germanium(IV) and tin(IV) state. The lead(II) halides are the stable halides of lead. Certain halides deserve special mention. The few complexes are unimportant.

Sn^{II}: tin (II) chloride, $SnCl_2 . 2H_2O$

Tin(II) chloride (stannous chloride) is formed by the action of fairly concentrated hydrochloric acid on granulated tin, hydrogen being evolved. After evaporation to small bulk, the liquid is stored in a desiccator over sulphuric acid until colourless crystals of the dihydrate separate. Tin(II) chloride dihydrate is used as a mordant in dyeing under the name *tin-salt*.

Tin(II) chloride is very soluble in water and in organic solvents. The aqueous solution is ionized, possibly containing the complex $SnCl_3^-$ ion, and on dilution, becomes milky as a basic salt separates:

$$SnCl_2 + H_2O \rightarrow Sn(OH)Cl(s) + HCl$$

In preparing the laboratory reagent, tin(II) chloride is usually dissolved in hydrochloric acid. The complexes $SnCl_3^-$ and $SnCl_4^{2-}$ are formed. The anhydrous salt may be obtained by heating the dihydrate in a current of hydrogen chloride vapour, hydrolysis resulting from any attempt at dehydration by heating in air:

$$SnCl_2 . 2H_2O \rightarrow Sn(OH)Cl + HCl + H_2O$$

In an alternative preparation of anhydrous tin(II)

Table 20.3 Complex halide anions formed by the tetrahalides of Group IV elements

C	Si	Ge	Sn	Pb
None	SiF_6^{2-} only e.g. K_2SiF_6 potassium hexafluorosilicate	GeF_6^{2-} $GeCl_6^{2-}$	SnF_6^{2-} $SnCl_6^{2-}$ $SnBr_6^{2-}$ SnI_6^{2-} e.g. $(NH_4)_2SnCl_6$ ammonium hexachlorostannate(IV) (sometimes called stannichloride)	$HPbF_8^{3-}$ etc. (irregular) $PbCl_6^{2-}$ e.g. $(NH_4)_2PbCl_6$ ammonium hexachloroplumbate(IV) (sometimes called plumbichloride)

chloride, tin is heated in a stream of dry hydrogen chloride:

$$Sn + 2HCl \rightarrow SnCl_2 + H_2$$

Tin(II) chloride is a powerful reducing agent:

$$Sn^{2+} - 2e^- \rightleftharpoons Sn^{4+}$$

Pieces of tin may be added to tin(II) chloride solution to prevent oxidation. Tin(II) chloride reduces iron(III) chloride to the iron(II) state, the solution changing in colour from yellow to green:

$$2Fe^{3+} + Sn^{2+} \rightarrow 2Fe^{2+} + Sn^{4+}$$

and mercury(II) chloride, which is covalent, is reduced to a white precipitate of dimercury(I) dichloride†, rapidly becoming grey and black as finely divided mercury is formed:

$$2HgCl_2 + Sn^{2+} \rightarrow Hg_2Cl_2(s) + Sn^{4+} + 2Cl^-$$

$$Hg_2Cl_2 + Sn^{2+} \rightarrow 2Hg(s) + Sn^{4+} + 2Cl^-$$

Acidified potassium dichromate changes in colour from orange to green due to reduction of chromium from the sexavalent to the tervalent state:

$$Cr_2O_7^{2-} + 14H^+ + 3Sn^{2+} \rightarrow$$
$$2Cr^{3+} + 7H_2O + 3Sn^{4+}$$

Tin(II) chloride is employed as a reducing agent in Organic Chemistry, tin and hydrochloric acid being used to reduce nitrobenzene to aminobenzene (aniline), and tin(II) chloride reducing benzene diazonium chloride to phenylhydrazine hydrochloride and cyanobenzene (phenylcyanide) to the aldimine, readily hydrolysed to the aldehyde (Stephen Reaction).

Pb^{II}: lead(II) chloride, $PbCl_2$

Lead(II) chloride is formed as a white precipitate on adding a solution of a chloride to a solution of lead(II) nitrate:

$$Pb^{2+} + 2Cl^- \rightarrow PbCl_2(s)$$

It is very much more soluble in hot water than cold. At 25°C, 1.08 g, and at 100°C, 3.34 g, are required to saturate 100 g water. Alternatively, finely ground litharge may be boiled with dilute hydrochloric acid to give colourless, needle-like crystals of lead(II) chloride on cooling:

$$PbO + 2HCl \rightarrow PbCl_2 + H_2O$$

With excess of concentrated hydrochloric acid the complex ion $PbCl_4^{2-}$ is formed. Lead(II) chloride melts at 298°C.

Pb^{II}: lead(II) iodide, PbI_2

Lead(II) iodide may be precipitated by addition of aqueous potassium iodide to a solution of lead(II) nitrate. Lead(II) iodide is much more soluble in hot water than in cold, the 'near-colourless' solution depositing glistening golden scales on cooling.

Salts of the oxoacids

The formation of ionized salts with the oxoacids is confined to elements which have at least some electropositive, or metallic, characteristics. While a sulphate of germanium has been reported, the oxosalts of importance belong to tin and lead, especially bivalent lead. Weaker metals may form nitrates and sulphates, but do not form carbonates. Also, their oxosalts are easily hydrolysed.

Sn^{II} and Sn^{IV}: oxosalts

Tin(IV) nitrate, $Sn(NO_3)_4$, is prepared by the addition of hydrated tin(IV) oxide to nitric acid:

$$SnO_2 \text{ hydrated } + 4HNO_3 \rightarrow Sn(NO_3)_4 + 2H_2O$$

The solution is hydrolysed to hydrated tin(IV) oxide on dilution.

Tin(II) nitrate, $Sn(NO_3)_2$, is formed by the action of very dilute nitric acid on tin, no gas being evolved:

$$4Sn + 10HNO_3 \rightarrow$$
$$4Sn(NO_3)_2 + NH_4NO_3 + 3H_2O$$

The action of concentrated nitric acid on tin yields hydrated tin(IV) oxide.

Tin(IV) sulphate, $Sn(SO_4)_2$, is formed by the addition of hydrated tin(IV) oxide to sulphuric acid and *tin(II) sulphate*, $SnSO_4$, by the reaction between dilute sulphuric acid and tin(II) oxide:

$$SnO_2 \text{ hydrated } + 2H_2SO_4 \rightarrow Sn(SO_4)_2 + 2H_2O$$

$$SnO + H_2SO_4 \rightarrow SnSO_4 + H_2O$$

No carbonates of tin have been isolated.

Pb^{IV}: oxocompounds

Compounds formed between quadrivalent lead and oxoacids are rare and very unstable.

†This illustrates the occasional differences which arise between valency (Hg_2^{II}) and oxidation number (Hg(I)) when describing the same state of chemical combination as in $Hg_2^{II} Cl_2$ and $(Hg(I))_2 Cl_2$

Lead(IV) sulphate, $Pb(SO_4)_2$, is formed as yellow crystals by the electrolysis of fairly concentrated sulphuric acid below 30°C using a lead anode. It is hydrolysed by water to lead(IV) oxide (the dioxide) and acid:

$$Pb(SO_4)_2 + 2H_2O \rightarrow PbO_2(s) + 2H_2SO_4$$

A *lead(IV) pyrophosphate*, lead(IV) diphosphate, PbP_2O_7, has been prepared by the reaction between phosphoric acid and lead(IV) oxide at 300°C. However, the easiest compound to prepare is the ethanoate (acetate). *Lead tetra-ethanoate* or *lead(IV) ethanoate*, $Pb(O.OC.CH_3)_4$ is made by the addition of red lead in successive small portions to warm glacial ethanoic acid. On cooling, colourless crystals separate. Lead(II) ethanoate is also formed and remains in solution:

$$Pb_3O_4 + 8CH_3COOH \rightarrow$$
$$Pb(O.OC.CH_3)_4 + 2Pb(O.OC.CH_3)_2 + 4H_2O$$

Experiments have shown that lead tetra-ethanoate is virtually covalent. Compare this reaction to what happens with oxoacids, the salts of which are ionic (p. 333). It is hydrolysed on warming with water to the dioxide and ethanoic acid:

$$Pb(O.OC.CH_3)_4 + 2H_2O \rightarrow$$
$$PbO_2(s) + 4CH_3.COOH$$

Lead(IV) ethanoate is a strong oxidizing agent, used in Organic Chemistry to oxidize diols (glycols) with adjacent hydroxyl groups smoothly to aldehydes or ketones.

Pb^{II}: oxosalts

The only common soluble salts of lead are the nitrate and the ethanoate (acetate), formed by the addition of litharge, 'hydroxide' or carbonate to a slight excess of hot dilute acid followed by evaporation to yield colourless crystals. Insoluble salts of lead may be precipitated by mixing solutions containing the necessary ions.

Lead(II) nitrate, $Pb(NO_3)_2$, is much more soluble in hot water than in cold, 59.6 g dissolving in 100 g water at 25°C and 134 g at 100°C. On heating, lead(II) nitrate decomposes with decrepitation, forming nitrogen dioxide, oxygen and lead(II) oxide:

$$2Pb(NO_3)_2 \rightarrow 2PbO + 4NO_2 + O_2$$

The crystals are anhydrous, unusual for a heavy metal nitrate.

Lead(II) ethanoate, $Pb(O.OC.CH_3)_2.3H_2O$, is known as sugar of lead because of its sweet taste, although it is poisonous. It is a soluble compound of lead and is largely covalent.

Lead(II) sulphate, $PbSO_4$, occurs naturally as anglesite: it is only very slightly soluble in water (solubility = 0.0045 g in 100 g water at 25°C) but readily soluble in ammonium ethanoate solution due to the removal of lead(II) ions as covalent lead(II) ethanoate and complex ethanoates.

Lead(II) chromate, $PbCrO_4$, is a yellow solid, sold as a pigment, chrome yellow. It is less soluble even than the sulphate and will not dissolve in ammonium ethanoate solution. A basic lead(II) chromate, the pigment chrome red is formed with hot alkali:

$$2PbCrO_4(s) + 2OH^- \rightarrow$$
$$PbCrO_4.PbO(s) + CrO_4^{2-} + H_2O$$

Normal lead(II) carbonate, is formed as a white precipitate on adding sodium hydrogencarbonate solution to lead(II) nitrate solution:

$$Pb^{2+} + 2HCO_3^- \rightarrow PbCO_3(s) + H_2O + CO_2$$

Sodium carbonate solution, which contains a greater concentration of hydroxyl ions due to hydrolysis, forms a basic carbonate:

$$3Pb^{2+} + 2CO_3^{2-} + 2OH^- \rightarrow Pb(OH)_2.2PbCO_3(s)$$

Cerussite, $PbCO_3$, is a naturally occurring form of lead(II) carbonate.

Basic lead carbonate or white lead is used in many white and coloured paints. It is prepared by the action of ethanoic acid (acetic acid) vapour on lead in an atmosphere containing carbon dioxide at about 60°C, a basic ethanoate being initially formed. Various processes have been devised to achieve a fine product of good covering power. The Dutch process uses vinegar as a source of ethanoic acid vapour and fermenting dung to generate carbon dioxide and warmth. White lead darkens in the sulphurous atmosphere of towns, due to the formation of lead(II) sulphide. The contents of the noticeably heavy paint tins need vigorous stirring because the lead constituents settle to a viscous mass. These paints (of excellent covering power) are poisonous and must not be used on items such as cot frames or toys which are likely to be 'chewed' by young children.

Qualitative and quantitative analysis for tin and lead and their compounds

The flame test shows a pale blue coloration with lead compounds but it is inconclusive. The metals must be taken into solution with appropriate acid treatment.

Separation of lead and tin in qualitative analysis

In Group I of the traditional qualitative analysis scheme, addition of hydrochloric acid precipitates white lead(II) chloride which is sparingly soluble in cold water but much more soluble in hot. Any lead(II) ions not precipitated in Group I are precipitated in Group II as black lead(II) sulphide by passage of hydrogen sulphide in the presence of a controlled concentration of hydrochloric acid. In the same group, tin(II) forms brown tin(II) sulphide and tin(IV), yellow tin(IV) sulphide. The sulphides of tin dissolve in warm yellow ammonium sulphide or boiling lithium hydroxide reagent, containing potassium nitrate to coagulate any colloidal particles, and are analysed in Group IIb. The trace of lead remaining after most of the metal has been precipitated in Group I, is analysed as the Group IIa sulphide.

Tests for lead ions in Group I

Lead(II) chloride is separated from silver chloride and dimercury(I) chlorides, also of Group I, by boiling water. On cooling, lead(II) chloride crystals may separate out. To the filtrate is added potassium chromate solution, when a yellow precipitate confirms the presence of lead. Alternatively, potassium iodide solution yields golden iridescent scales of lead(II) iodide which dissolve in hot water to form a colourless solution. Further, a charcoal block reduction of the precipitate yields a bead of lead which will mark paper.

Identification of lead in Group IIa

The precipitate of mixed sulphides is boiled with dilute nitric acid when all but mercury(II) sulphide dissolve. Addition of concentrated sulphuric acid and evaporation causes precipitation of lead(II) sulphate. This is dissolved in ammonium ethanoate (acetate) solution, from which potassium chromate solution precipitates yellow lead(II) chromate.

Identification of tin in Group IIb

Careful addition of dilute hydrochloric acid to the lithium hydroxide extract precipitates tin(II) sulphide or tin(IV) sulphide again. However, acidification of the yellow ammonium sulphide extract precipitates tin(IV) sulphide, whichever form of tin was present originally. On heating with concentrated hydrochloric acid, sulphides of tin and antimony dissolve, leaving arsenic sulphide behind. (There are three elements in this analytical subgroup.) Tin is confirmed by addition of a little 1% aqueous tannin followed by dilute aqueous ammonia to the gently boiling solution until congo red paper is only faintly blue (still acidic) when a cream flocculent precipitate of a tin–tannin complex confirms tin. This is a sensitive test.

To distinguish between tin(II) and tin(IV) in separate solutions

1   Hydrogen sulphide precipitates either brown tin(II) sulphide or yellow tin(IV) sulphide.
2   Mercury(II) chloride solution is reduced to a white precipitate of dimercury(I) dichloride which is further reduced to a grey-black deposit of mercury by a tin(II) solution but no reduction occurs with tin(IV).
3   Excess sodium hydroxide solution causes the gelatinous white precipitate of whichever hydrated oxide is formed initially, to re-dissolve. However, sodium stannate(II) reduces bismuth nitrate solution to elemental bismuth, which appears as a black deposit, while sodium stannate(IV) has no reducing properties.

The titrimetric (volumetric) estimation of tin and lead

Tin in foil, alloys and compounds may be determined by conversion to the tin(II) state and titration against a standard solution of iodine in the presence of acid:

$$SnCl_2 + 2HCl + I_2 \rightarrow SnCl_4 + 2HI$$

Since 1 mole of iodine reacts with 1 mole of tin(II) chloride which has, in turn, been formed from 1 mole of tin, the calculation is straightforward ($I_2 \equiv SnCl_2 \equiv Sn$). The metal is dissolved in fairly concentrated hydrochloric acid. To ensure that the metal is in the tin(II) state when a tin compound is analysed, reduction is effected by means of iron nails in fairly concentrated hydrochloric acid. The nails are removed and pieces of marble are added to provide an atmosphere of carbon dioxide by reaction with acid.

Lead may be estimated by precipitation of the chromate. $25\,cm^3$ of an approximately $0.05\,M$ solution of a lead(II) salt is acidified with dilute ethanoic (acetic) acid and sodium ethanoate (acetate) added. To the boiling mixture, $50\,cm^3$ of standard $0.01667\,M$ (M/6) potassium dichromate is added from a pipette, the contents of the flask cooled, transferred to a 250-ml graduated flask and made up to the mark. Lead(II) chromate settles. The concentration of residual potassium chromate is determined by titration with ammonium iron(II) sulphate and the quantity equivalent to the precipitated lead calculated.

For the precipitation of lead(II) chromate, chromate ions are formed from the dichromate:

$$Cr_2O_7{}^{2-} \xrightarrow[\text{buffer}]{\text{ethanoate}} 2CrO_4{}^{2-}$$

$$CrO_4{}^{2-} + Pb^{2+} \longrightarrow PbCrO_4(s)$$

Since 1 mole of lead(II) ions reacts with 1 mole of chromate ions which in turn is formed from 0.5 mole of dichromate ions, the calculations may be completed. However, note that in the redox titration (see p. 122) 1 mole of dichromate ions reacts with six moles of iron(II) ions, while for the precipitation 1 mole of dichromate ions accounts for 2 moles of lead(II) ions.

Miscellaneous inorganic compounds containing cyano ($-C\equiv N$) and carbonyl ($\rangle C=O$) groups

Many of these compounds are extremely poisonous and their preparation without special precautions could lead to death.

Cyanogen, $(CN)_2$

Cyanogen is evolved on mixing fairly concentrated solutions of potassium cyanide and copper(II) sulphate when copper(II) cyanide, precipitated initially, breaks down to cyanogen and copper(I) cyanide:

$$Cu^{2+} + 2CN^- \rightarrow Cu(CN)_2(s)$$

$$2Cu(CN)_2(s) \rightarrow 2CuCN(s) + \begin{matrix} C\equiv N \\ | \\ C\equiv N \end{matrix}$$

It may also be formed by thermal decomposition of

mercury(II) cyanide both products being volatile:

$$Hg(CN)_2 \rightarrow Hg(g) + (CN)_2(g)$$

Cyanogen is a colourless, extremely poisonous gas (b.p. $-21°C$), which burns with a violet flame to form carbon dioxide and nitrogen:

$$(CN)_2 + 2O_2 \rightarrow 2CO_2 + N_2$$

It is soluble in water and slowly hydrolysed, to give a range of products:

$$\begin{matrix} C\equiv N \\ | \\ C\equiv N \end{matrix} + 4H_2O \rightarrow \begin{matrix} COO^-NH_4{}^+ \\ | \\ COO^-NH_4{}^+ \end{matrix}$$
ammonium
ethanedioate
(oxalate)

$$(CN)_2 + H_2O \rightarrow HCN + HCNO$$
hydro-   cyanic
cyanic acid acid

With caustic alkali, a mixture of the cyanide and cyanate results:

$$2OH^- + (CN)_2 \rightarrow CN^- + CNO^- + H_2O$$

Cyanogen will react with certain heated metals to form cyanides:

$$2K + (CN)_2 \rightarrow 2KCN$$

Because of the superficial resemblance of these reactions to those of the halogens, cyanogen has been called a pseudo-halogen.

Hydrogen cyanide (prussic acid), hydrocyanic acid, HCN and the cyanides

Hydrogen cyanide is formed by the action of dilute sulphuric acid on potassium cyanide:

$$KCN + H_2SO_4 \rightarrow KHSO_4 + HCN$$

It is an extremely poisonous, very volatile liquid (b.p. $26°C$), the vapour of which smells of bitter almonds. Hydrocyanic acid is a very weak acid so that soluble cyanides are hydrolysed in solution, having the characteristic odour of the free acid:

$$CN^- + H_2O \rightleftharpoons HCN + OH^-$$

It is slowly hydrolysed in water to ammonium methanoate (formate):

$$H-C\equiv N + 2H_2O \rightarrow H.C\!\!\begin{matrix} \diagup\!\!\diagup O \\ \diagdown O^-NH_4{}^+ \end{matrix}$$

ammonium methanoate

Hydrogen cyanide burns in air with a purple flame to form steam, carbon dioxide and nitrogen:

$$4HCN + 5O_2 \rightarrow 2H_2O + 4CO_2 + 2N_2$$

Derivatives of aldehydes and ketones are used to characterize these compounds:

$$CH_3.C{\overset{H}{\underset{CN}{\diagdown}}}OH \qquad CH_3{\overset{CH_3}{\underset{CH_3}{\diagup}}}C{\overset{OH}{\underset{CN}{\diagdown}}}$$

ethanal (acetaldehyde) propanone (acetone)
cyanhydrin   cyanhydrin

Hydrocyanic acid is an equilibrium mixture of two forms:

$$H-C\equiv N \rightleftharpoons H-\overset{+}{N}\equiv C^-$$
$$99.5\% \qquad 0.5\%$$

These must not be confused with resonance hybrids (p. 81). Both ionic and covalent cyanides are formed, but the cyanide ion is chiefly notable (as a *ligand*) for the formation of complex cyanides, its unshared electron pairs being available for bonding to a metal ion. Complex and normal cyanides of importance are described under the metal. Cyanides are detected by the formation of Prussian Blue. The test frequently appears in organic analysis. During the Lassaigne Sodium Test, organic compounds containing nitrogen form sodium cyanide on heating with sodium.

The composition of Prussian Blue is described in Chapter 24, p. 471.

Cyanates and cyanic acid, HOCN

Potassium cyanate is formed when red lead is fused with potassium cyanide,

$$4KCN + Pb_3O_4 \rightarrow 4KOCN + 3Pb$$

It is a colourless, crystalline solid. When mixed with ammonium sulphate and heated, urea and potassium sulphate are formed. The transformation of ammonium cyanate into urea was achieved by Wöhler in 1828 and was the first preparation from inorganic compounds of an animal product:

$$2KOCN + (NH_4)_2SO_4 \rightarrow K_2SO_4 + 2NH_4OCN$$
$$\Updownarrow$$
$$2NH_2.CO.NH_2$$
$$\text{urea}$$

This was a breakdown in the supposed barrier between the inorganic and the organic, the latter having been associated with the concept of a vital force. Acidification of a cyanate yields cyanic acid but this is hydrolysed immediately:

$$HOCN + H_2O \rightarrow NH_3 + CO_2$$

Pure cyanic acid is a colourless liquid obtained by heating cyanuric acid, $(HOCN)_3$, which may in turn be prepared from urea. Evidence from Raman and ultra-violet spectra of the acid points to the structure $H-N=C=O$, which corresponds to one series of esters, $R.N=C=O$, and the silver and mercury(II) derivatives. The ionized potassium, tetramethylammonium and lead(II) salts contain the resonance hybrid cyanate ion,

$$\{N^-=C=O \leftrightarrow N\equiv C-O^-\}$$

Thiocyanates and thiocyanic acid, HSCN

Potassium thiocyanate (see p. 277) is obtained when potassium cyanide is fused with sulphur,

$$KCN + S \rightarrow KSCN$$

It is a colourless, deliquescent, crystalline solid. Thiocyanic acid is obtained by distillation of potassium thiocyanate with dilute sulphuric acid at low temperatures, more concentrated acid yielding carbonyl sulphide, etc.

$$2KSCN + H_2SO_4 \rightarrow K_2SO_4 + 2HSCN$$

Thiocyanic acid behaves as a *tautomeric* mixture of two forms, $H-S-C\equiv N$, and $S=C=N-H$, both giving a resonance hybrid thiocyanate ion,

$$\{\bar{S}-C\equiv N \leftrightarrow S=C=\bar{N}\}$$

Suboxides of carbon

Methanedioic anhydride (malonic anhydride), C_3O_2, is formed by the dehydration of methanedioic acid (malonic acid) with phosphorus pentoxide at $140°C$ *in vacuo*:

$$CH_2{\overset{COOH}{\underset{COOH}{\diagup\diagdown}}} - 2H_2O \rightarrow C_3O_2$$

It is a colourless gas (b.p. $6.8°C$), with an objectionable odour. At $200°C$, it decomposes:

$$C_3O_2 \rightarrow CO_2 + 2C$$

The structure is $O=C=C=C=O$. C_5O_2 has been reported. Mellitic anhydride is $C_{12}O_9$. A sulphur analogue, C_3S_2, of methanedioic anhydride has been made in small quantities.

Carbonyl chloride, phosgene, $COCl_2$

When chlorine and carbon monoxide are passed together through animal charcoal at 30–40°C, phosgene is formed:

$$CO + Cl_2 \longrightarrow O{=}C\begin{smallmatrix}Cl\\[4pt]Cl\end{smallmatrix}$$

It is a colourless gas (b.p. 8°C), very poisonous and with a smell of musty hay. Carbonyl chloride is the acid chloride of carbonic acid, which is formed on hydrolysis:

$$O{=}C\begin{smallmatrix}Cl\\[4pt]Cl\end{smallmatrix} + 2H_2O \longrightarrow O{=}C\begin{smallmatrix}OH\\[4pt]OH\end{smallmatrix} + 2HCl$$
$$\downarrow$$
$$H_2O + CO_2$$

Ammonia gives carbamide, or urea:

$$O{=}C\begin{smallmatrix}Cl\\[4pt]Cl\end{smallmatrix} + 4NH_3 \longrightarrow O{=}C\begin{smallmatrix}NH_2\\[4pt]NH_2\end{smallmatrix} + 2NH_4Cl$$

Carbonyl sulphide, carbon oxysulphide, SCO (COS)

Carbon monoxide and sulphur vapour react to form carbonyl sulphide:

$$CO + S \rightarrow SCO$$

It is usually prepared by the hydrolysis of thiocyanic acid, formed by the action of fairly concentrated sulphuric acid on potassium thiocyanate:

$$KSCN + H_2SO_4 \rightarrow KHSO_4 + HSCN$$

$$HSCN + H_2O \rightarrow S{:}C{:}O + NH_3$$
$$\underset{H_2SO_4}{\longrightarrow} (NH_4)_2SO_4$$

Carbonyl sulphide is a colourless, very poisonous gas (b.p. −50°C), which is hydrolysed slowly to carbon dioxide and hydrogen sulphide, and burns in air to carbon dioxide and sulphur dioxide. Normally it has an objectionable odour, but is odourless when pure.

Metal carbonyls

Various methods have been developed for the preparation of carbonyls, including the action of carbon monoxide on the heated metal, its oxide or chloride. Carbonyls are fairly stable but are decomposed into the metal and carbon monoxide on heating. They are insoluble in water and ionizing solvents but soluble in organic solvents, and are decomposed by the halogens to the metal chloride and carbon monoxide. The most important is nickel carbonyl, $Ni(CO)_4$, which is tetrahedral in structure. A range of carbonyls is formed by some metals; iron forms $Fe(CO)_5$, $Fe_2(CO)_9$ and $Fe_3(CO)_{12}$. In these compounds carbon monoxide is referred to as a *ligand*. Because the complexing agent is assumed not to have changed its oxidation number in compounds such as these by convention, the element is assigned a zero oxidation state even though combined[†].

tetrahedral structure

$$\begin{array}{c}O^+\\|||\\C\\|\\{}^+O{\equiv}C{-}\overset{\displaystyle 4-}{Ni}{-}C{\equiv}O^+\\|\\C\\|||\\O^+\end{array}$$

Metal clusters for organometallic catalysis

That many transition metals behave as catalysts for gaseous reactions is well known. The industrial production of plastic materials, synthetic fibres and various pharmaceutical chemicals may be triggered off by the atoms of a metal catalyst, although the metal is not incorporated into the final product. The reactants are often carbon compounds in the gaseous phase. At an early stage in the sequence of chemical reactions it is necessary for a bond to form between the selected metal, which must have available energy levels, and an electron-rich compound, e.g. ethene (ethylene), $CH_2{:}CH_2$, or propene (propylene) $CH_3.CH{:}CH_2$, as in the production of polyethylene (polythene) or polypropylene. Such organometallic reactions are crucial to these manufacturing processes. Many metals (e.g. Pt, W, Fe, Co, Ni, etc.) are suitable and until recently were used in a finely divided form on a trial-and-error basis. Nowadays, molecules are designed in which there are clusters of metal atoms grouped together and protected by simple ligand molecules. This protective barrier may be broken down.

Carbon monoxide is commonly used as a ligand and is of prime importance in this application. On the cluster catalyst surface, bonds between carbon

[†]Which is hardly a satisfactory state of affairs, yet the valency would also be assigned a zero number ($+4 - 4$)!

and oxygen are replaced by links between the separated atoms and the metal, to form what is known as a *carbido cluster*, which signifies that the carbon atoms become incorporated into the framework of metal atoms. Two such structures, taken from the literature, are given below.

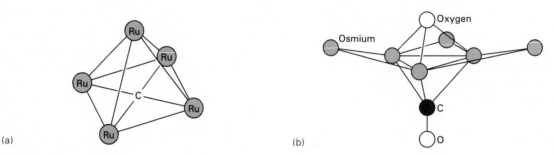

(a)   (b)

Fig. 20.10   (a) A 'carbido' cluster where a carbon atom sits in the base of a framework of ruthenium (Ru) atoms that form a pyramid. The full formula is [Ru$_5$C(CO)$_{15}$]. (b) An oxygen atom sits above an almost planar cluster of osmium metal atoms. Note how a carbon monoxide molecule, CO, is also attached to the same three metal atoms as the oxygen. The diagram shows the skeleton of [Os$_6$O(CO)$_{19}$].

21

Group V: the nitrogen group

Nitrogen, phosphorus, arsenic antimony and bismuth

7	
N	
2, 5	
15	
P	
2, 8, 5	
33	
As	
2, 8, 18, 5	
51	
Sb	
2, 8, 18, 18, 5	
83	
Bi	
2, 8, 18, 32, 18, 5	

The gradation of properties shown by compounds of Group V elements covers a wide range. The elements change in character from non-metallic gaseous nitrogen and solid phosphorus, to arsenic, antimony and the metallic bismuth, all solids. The changeover from non-metal to metal is seen in the middle families of the Periodic Table. This was seen in Group IV but the elements of Group V are more electronegative and the metallic state, when it appears is less pronounced. The borderline between non-metal and metal comes between arsenic and antimony, the latter forming positive ions and having the properties of a weak metal. Nitrogen has the distinctly anomalous properties associated with the first members of Groups V, VI and VII, and also apparent with the metallic families. Nitrogen has a valency of three in simple compounds. Phosphorus, arsenic and antimony show valencies of three and five, the latter being the principal valency state of phosphorus. The valency of bismuth is three, although some poorly defined oxides of a higher valency state exist and 'sodium bismuthate' is assigned the approximate formula $NaBiO_3$. The higher valency state used to be denoted by the suffixes *-ic* and *-ate* and the lower, by *-ous* and *-ite*. Oxidation numbers range from -3 to $+5$ for nitrogen (p. 124) while the others exhibit oxidation states of $-3 (PH_3)$, $+3 (AsCl_3)$ and $+5 (Sb_2O_5)$, the illustrative examples applying generally.

Many of the compounds formed by nitrogen are different enough from those of the other elements of Group V to warrant the consideration that nitrogen ought to be separately classified from the remaining elements, the Phosphorus Group. Nitrogen has a high electronegativity ($= 3.0$) which makes it a very strong non-metal. Unlike the other elements of Group V, it forms multiple bonds but it is remarkable that nitrogen gas, $N{\equiv}N$, is so chemically inert while apparently similar bonding in ethyne, $HC{\equiv}CH$, brings great chemical activity. Nitrogen forms a range of compounds which have no counterparts with the other elements. They include oxides, some of which have odd electron configurations, the explosive halides (except for nitrogen trifluoride), hydrogen azide and other compounds, nominally hydrides and their derivatives. Phosphorus has some resemblance to silicon, which is the adjacent element of Group IV, in linking through oxygen, $-P-O-P-$. The metallic state is uppermost with antimony and bismuth. In the Mendeléeff Periodic Table, nitrogen and phosphorus were placed above the B subgroup, arsenic, antimony and bismuth, rather than with the A subgroup, vanadium, niobium, tantalum and protoactinium.

Nitrogen was probably first recognized as a distinct gas by Rutherford in 1772 and its elemental nature established by Lavoisier in his work on combustion from 1774 onwards. Because it would not support life, Lavoisier called the gas azote (Greek, *a zoe* = no life) but the name nitrogen was preferred when the relationship to nitre was discovered (Greek, *nitron genon* = nitre former). Phosphorus was discovered in 1669 by Brand, who evaporated urine to dryness and heated the residue, which contains microcosmic salt, the decomposition products of which were reduced by carbon formed by charring of the organic matter also present:

$$NaNH_4HPO_4 . 4H_2O \rightarrow NaPO_3 + 5H_2O + NH_3$$

$$8NaPO_3 + 10C \rightarrow 2Na_4P_2O_7 + 4P + 10CO$$

Its elemental nature was not established until a century later by Lavoisier. The other elements have been known since about the 14th century.

Elemental nitrogen accounts for 78% by volume of the atmosphere from which all nitrogen used in commerce is produced. Nitrogen occurs combined in sodium nitrate, which forms about 30% of the mineral caliche, found in a rainless area of Chile. Phosphorus is always combined and occurs as phosphate, the main source being calcium phosphate in rock phosphate and the apatites. Arsenic, antimony and bismuth occur in sulphide ores, although relatively small deposits of native bismuth are mined.

Both nitrogen and phosphorus are essential to living matter. Nitrogen is present in the amino acids from which animal and vegetable proteins are constructed. Phosphorus occurs as calcium phosphate in bones and teeth, and as organic phosphates. In hot dry regions, in Peru and Islands in the Pacific, guano deposits formed from the droppings of countless ocean birds, containing both combined nitrogen and phosphorus, are exploited commercially. The production of nitrogenous and phosphatic fertilizers are major industries. The fixation of nitrogen is described separately in Chapter 12. Unlike nitrogen, phosphorus does not escape from the soil, but the natural cycle is disturbed and phosphorus lost by harvesting crops, sewage disposal and burial in cemeteries. The deficit is replaced by inorganic phosphate chemicals, by animal and fish manure. Excessive use of phosphates and nitrates in intensive farming is a major cause of environmental concern because the chemicals can be leached out of the soil by rainfall contaminating local lakes, rivers and underground water supply reservoirs. Algal blooms were observed in certain Swedish lakes and estuaries in the 1960s. The water was being choked by the bright green organic matter somewhat reminiscent of pea soup. This biological explosion of growth depleted the supply of oxygen in the water and caused aquatic life to be extinguished so that the repulsive sight was accompanied by a repellant smell. This has been the fate (to varying extents) of some lakes and pools in the UK, as fishermen well know. In some countries the use of phosphates in detergents has been discontinued, and has become a selling point, emphasizing the fact that the problem is not one to be faced by the farmers alone.

Physical data for the elements of Group V are given in Table 21.1.

A general survey of bonding involving the nitrogen group of the *p* block

The elements of Group V have electronic configurations with five valency electrons. The formation of a valency octet by loss of five electrons from each atom is considered to be impossible because of the extremely high polarizing effect that the hypothetical cation would then have. Further, cation formation would not be expected from a study of the types of bonding encountered along any period, working successively along from the alkali metals of Group I. Possibly nitrogen and phosphorus acquire the three electrons to form triply charged anions in certain nitrides and phosphides thereby showing the electronic configurations of the inert gases neon and argon

Table 21.1   Group V: Physical Data   *p* block elements $(ns)^2 (np)^3$ $n \neq 1$

	N	P	As	Sb	Bi
Atomic number	7	15	33	51	83
Electronic configuration	2, 5	2, 8, 5	2, 8, 18, 5	2, 8, 18, 18, 5	2, 8, 18, 32, 18, 5
[Core]outer electrons	$[He](2s)^2(2p)^3$	$[Ne](3s)^2(3p)^3$	$[Ar, 3d^{10}](4s)^2(4p)^3$	$[Kr, 4d^{10}](5s)^2(5p)^3$	$[Xe, 4f^{14}5d^{10}](6s)^2(6p)^3$
Relative atomic mass	14.0067	30.9738	74.9216	121.75	208.980
Atomic radius/nm	0.070	0.110	0.121	0.141	0.152
Ionic radius/nm (X^{3-})	0.171	0.212	0.222	0.245	—
Ionic radius/nm (M^{3+})	—	—	—	0.092	0.108
van der Waals' radius/nm	0.15	0.19	0.20	0.22	—
Atomic volume/cm³ mol⁻¹	15.95	17.0 (White)	13.1 (Grey)	18.3 (Grey)	21.3
Boiling-point/°C	−196	416 (sublimes)	633 (sublimes)	1325	1560
Melting-point/°C	−210	589 (4.3 MPa)	818 (3.6 MPa)	631	273
Electronegativity	3.0	2.1	2.0	1.8	1.7
Density/g cm⁻³ (at 293 K)	1.25×10^{-3}	White 1.83 Violet (red) 2.35	Grey 5.73	Grey 6.67	9.80
Co-ordination number	3, 4	3, 4, 5, 6	3, 4, (5), 6	3, 4, (5), 6	3, 6

Electronic configurations

Comparison of electronic configurations shows that the relationship between phosphorus and arsenic is not the same as that between sodium and potassium or magnesium and calcium; this is reflected in the properties of the elements and their compounds.

	Group V	Inert gases
	7 **N** 2, 5	10 **Ne** 2, 8
	15 **P** 2, 8, 5	18 **Ar** 2, 8, 8
	33 **As** 2, 8, 18, 5	36 **Kr** 2, 8, 18, 8

Ions

51 **Sb³⁺** 2, 8, 18, 18, 2	51 **Sb** 2, 8, 18, 18, 5	54 **Xe** 2, 8, 18, 18, 8
83 **Bi³⁺** 2, 8, 18, 32, 18, 2	83 **Bi** 2, 8, 18, 32, 18, 5	86 **Rn** 2, 8, 18, 32, 18, 8

ions ← → covalent
molecules
(or N^{3-}, P^{3-} ions)

Simple valency:	3		3
Oxidation number	+3	0	±3

The elements may be compared. Nitrogen forms discrete molecules containing a triple covalent bond, $:N\equiv N:$, having a lone pair of electrons on each atom. The yellow allotropes of phosphorus, arsenic and possibly, antimony form pyramidal tetratomic molecules (Figure 21.4), each atom radiating three single bonds and having one lone pair of electrons. In covalent molecules of oxidation states $+3$ (PCl_3) and -3 (PH_3), each element acquires the electronic configuration of an inert gas, but retains its own nuclear charge (proton, or atomic number). Thus as an example, phosphorus assumes the electronic configuration of argon (2, 8, 8) but still retains 15 protons in the nucleus

respectively, but this has not been firmly established. The chemistry of the elements of Group V is largely the study of the formation and rupture of covalent bonds. In simple compounds the valency of nitrogen is three, while the others exhibit also the quinquevalent state, this being the principal state for phosphorus but of minor importance for bismuth. The oxidation states exhibited by nitrogen in its compounds are listed in Chapter 7 on p. 124.

Electronegative characteristics become less pronounced and electropositive characteristics appear as the atomic number increases. Metallic properties emerge with antimony and become established in bismuth. As would be expected, the elements of Group V are more electronegative than those occupying the corresponding positions in Group IV. For arsenic the hydrides, principal oxide and halides belong to the tercovalent state. For antimony and bismuth, but apparently not arsenic, positive ions emerge in the principal tervalent state, a pair of electrons remaining inert to give Sb^{3+} and

Bi^{3+}. The ionization energies in bold typeface in Table 21.2 illustrate this. A plot of the sum of the first three ionization energies for elements of this group appears in Fig. 21.1, which has a tenfold change of scale on the energy axis when compared with the graphs shown for Groups I, II and III.

Atomic radii, ionic radii and van der Waals' radii for these elements are plotted against atomic number in Fig. 21.2 and electronegativities in Fig. 21.3. Refer to Chapter 5 for essential theory.

The marked difference of nitrogen from the other elements of the group is shown clearly and is

Table 21.2 Group V (nitrogen group): ionization energies/ $kJ\,mol^{-1}$ (273 K)

		N	P	As	Sb	Bi
1st	$M(g) \rightarrow M^+(g) + e^-$	1399	1061	1013	**834**	**820**
2nd	$M^+(g) \rightarrow M^{2+}(g) + e^-$	2856	1901	1959	**1796**	**1621**
3rd	$M^{2+}(g) \rightarrow M^{3+}(g) + e^-$	4573	2914	2701	**2383**	**2451**
4th	$M^{3+}(g) \rightarrow M^{4+}(g) + e^-$	7477	4959	4814	4245	4390

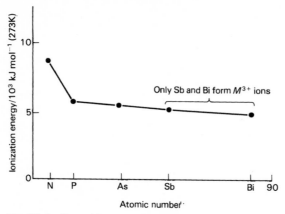

Fig. 21.1   Sum of first three ionization energies for Group V elements plotted against atomic number (note: tenfold change of scale on energy axis when comparing)

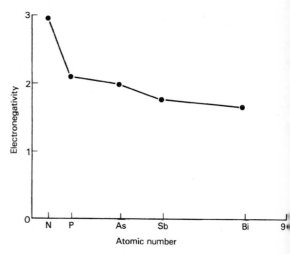

Fig. 21.3   Electronegativity plotted against atomic number for Group V elements

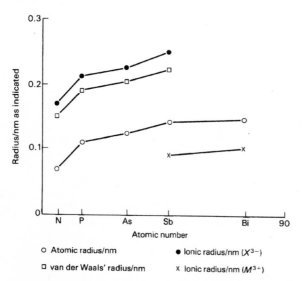

o Atomic radius/nm   • Ionic radius/nm (X^{3-})

□ van der Waals' radius/nm   x Ionic radius/nm (M^{3+})

Fig. 21.2   Atomic, ionic and van der Waals' radii plotted against atomic number for elements of Group V. (Refer to Tables 5.10 and 5.11, and the accompanying text, pp. 74–78)

reflected in the richness and variety of the properties of nitrogen compounds.

Except for nitrogen, and the other elements of Period 2, it is usually accepted that the simple octet rule may be extended to permit the use of more electrons for bonding. Nitrogen has no *d* orbitals available for forming co-ordination compounds while the others do have such energy levels available below those of the *s* electrons of the next quantum shell (level). NCl_3 is hydrolysed in a different manner from the trichlorides of the other elements, PCl_3 (for example) presumably attaching a water molecule through the $3d$ energy level as a preliminary step in the hydrolysis sequence of reactions. Such levels are available for the formation of PF_5, $K[PF_6]$ etc. Odd electron structures are required for certain oxides of nitrogen which have no counterpart with the other elements. The bonding of the group V elements will be dealt with element by element.

Nitrogen

(a)   *Simple covalent molecules*

H H	CH₃ CH₃	Cl Cl	C≡N

ammonia   trimethylamine   nitrogen trichloride   cyanogen

trigonal pyramidal shapes

(b) *Structures excluding oxides and oxoacids, containing co-ionic bonding*

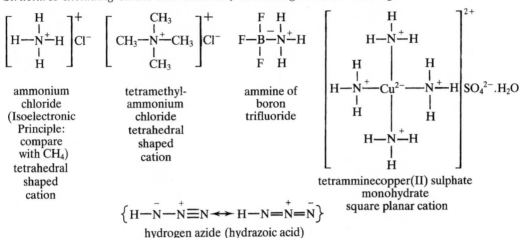

ammonium chloride
(Isoelectronic Principle: compare with CH_4)
tetrahedral shaped cation

tetramethyl-ammonium chloride
tetrahedral shaped cation

ammine of boron trifluoride

tetramminecopper(II) sulphate monohydrate
square planar cation

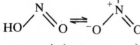

$\{H-\overset{-}{N}-\overset{+}{N}{\equiv}N\longleftrightarrow H-N{=}\overset{+}{N}{=}\overset{-}{N}\}$

hydrogen azide (hydrazoic acid)

(c) *Structures of oxides and oxoacids with odd electrons, resulting in paramagnetic behaviour*

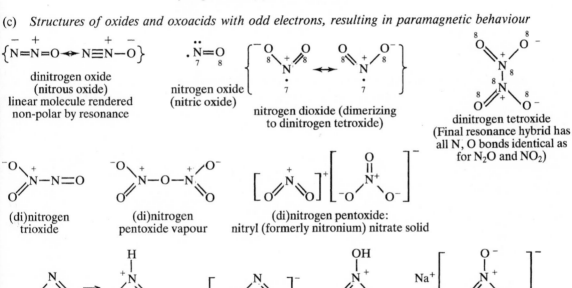

$\{\overset{-}{N}{=}\overset{+}{N}{=}O\longleftrightarrow\overset{+}{N}{\equiv}\overset{-}{N}-O\}$

dinitrogen oxide
(nitrous oxide)
linear molecule rendered non-polar by resonance

nitrogen oxide
(nitric oxide)

nitrogen dioxide (dimerizing to dinitrogen tetroxide)

dinitrogen tetroxide
(Final resonance hybrid has all N, O bonds identical as for N_2O and NO_2)

(di)nitrogen trioxide

(di)nitrogen pentoxide vapour

(di)nitrogen pentoxide:
nitryl (formerly nitronium) nitrate solid

tautomeric (not resonance) forms of nitrous acid

sodium nitrate
bent (V-shaped) anion

nitric acid

sodium nitrate
trigonal planar shaped anion

Phosphorus

(The formulae of certain compounds have been doubled since acquiring their common names: the trioxide is now P_4O_6 and the pentoxide, P_4O_{10}.)

(a) *Simple covalent molecules*

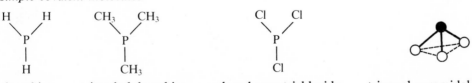

phosphine

trimethylphosphine

phosphorus trichloride

trigonal pyramidal shape

(b)   *Co-ionic bonding, excluding oxoacids*

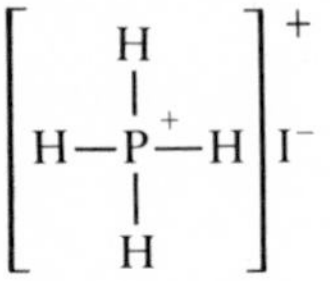

$$\left[\begin{array}{c} H \\ | \\ H-\overset{+}{P}-H \\ | \\ H \end{array} \right]^{+} I^{-}$$

phosphonium iodide
very unstable

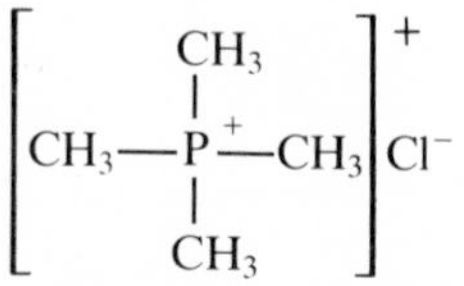

$$\left[\begin{array}{c} CH_3 \\ | \\ CH_3-\overset{+}{P}-CH_3 \\ | \\ CH_3 \end{array} \right]^{+} Cl^{-}$$

tetramethylphosphonium
chloride

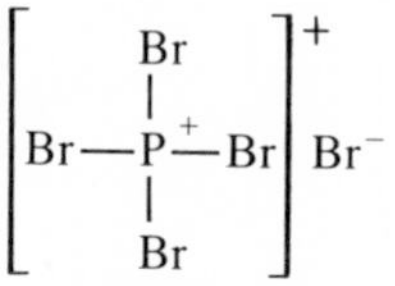

$$\left[\begin{array}{c} Br \\ | \\ Br-\overset{+}{P}-Br \\ | \\ Br \end{array} \right]^{+} Br^{-}$$

solid phosphorus
pentabromide

tetrahedral
shaped cations

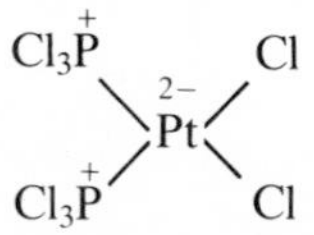

platinum(II) chloride
complex of
phosphorus trichloride

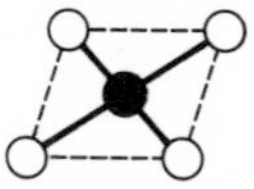

square planar
shaped molecule

(c)   *Extending the octet rule*

While the octet may be retained by judicious use of co-ionic bonding and resonance forms, there are certain compounds for which the octet rule is accepted as inadequate, and the *d* orbitals (energy levels) are brought into use:

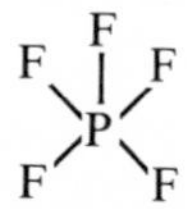

phosphorus
pentafluoride

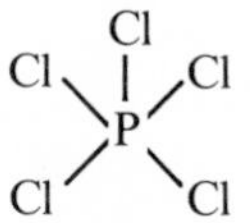

vaporized phosphorus
pentachloride

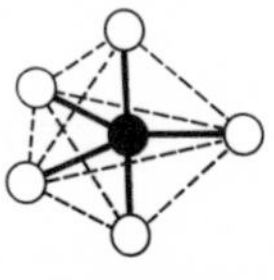

trigonal bipyramidal
shaped molecules

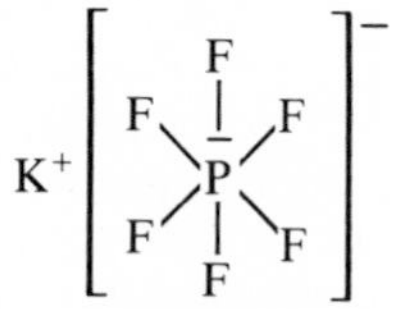

$$K^{+} \left[\begin{array}{c} F \\ F \diagdown | \diagup F \\ P \\ F \diagup | \diagdown F \\ F \end{array} \right]^{-}$$

potassium
hexafluorometaphosphate

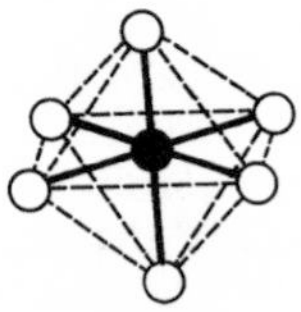

octahedral shaped
anion

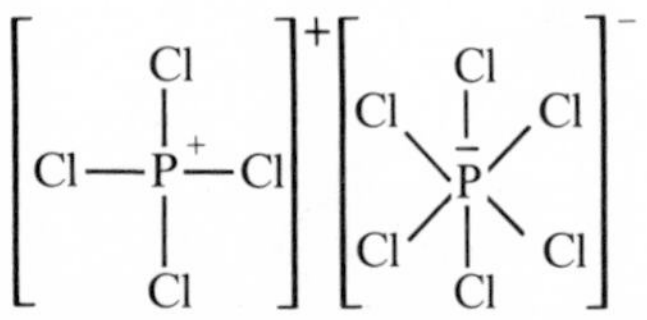

solid phosphorus pentachloride

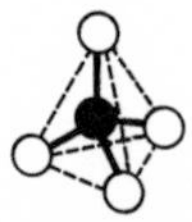

tetrahedral shaped
cation

octahedral shaped
anion

Cl
|
O=P—Cl
|
Cl

phosphoryl(v) chloride
phosphorus oxychloride

irregular tetrahedral
shaped molecule

(d)   The structures of the oxides

The structures of the oxides are related to that of the P_4 molecule, which is tetrahedral.

The P—P single bonds are replaced by oxygen bridges which curve outwards to form the P_4O_6 molecule. Adding four additional oxygen atoms, one at each corner of the P_4O_6 molecule, yields P_4O_{10}, resulting in a larger tetrahedron. The structures of molecules in the vapour state are shown diagrammatically in Fig. 21.4.

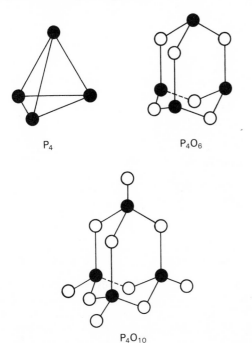

P_4 P_4O_6

P_4O_{10}

Fig. 21.4   The structure of phosphorus and its oxides. P_4 is a regular tetrahedron. P_4O_6 has one O atom separating each pair of P atoms. P_4O_{10} similarly, but O atoms at each corner of the P_4 tetrahedron form a larger tetrahedron.

(e)   Oxoacids and oxosalts containing phosphorus

In writing structures for the oxoacids of phosphorus and their anions, the octet of electrons surrounding the phosphorus atom could be maintained by use of co-ionic bonding. However, the

bond lengths are significantly shorter than would be expected from a single linkage (P—O in PO_4^{3-}: observed = 0.155 nm, theory = 0.176 nm; shortening = 0.021 nm). They clearly have some double bond character. However attractive co-ionic bond diagrams might appear, it seems more accurate to retain classical bond diagrams and where more detailed study is required, to add the various resonance forms; these need not detain us here.

The Greek prefixes in the names of the oxoacids merit comment.

Hypo—means *under, below*, and refers to the oxygen content of the molecule in relation to the acid then named.
Ortho—means *correct*, and relates to the proportion of the elements of water present.
Meta—means *after*, and signifies a lower proportion of the elements of water than the ortho-acid.
Pyro—means *fire*, and tells of the method of preparation.

The suffixes *-ous* and *-ic* refer to the lower and upper oxidation states of the principal element, here, phosphorus.

O
‖
H—O—P—H
|
H

hypophosphorous acid
H_3PO_2,
monobasic

$\begin{bmatrix} & O & \\ & \| & \\ \bar{O}-&P&-H \\ & | & \\ & H & \end{bmatrix}^-$

hypophosphite ion
$H_2PO_2^-$

H
|
HO—P—OH
‖
O

phosphorous acid
H_3PO_3,
dibasic

$\begin{bmatrix} & H & \\ & | & \\ \bar{O}-&P&-\bar{O} \\ & \| & \\ & O & \end{bmatrix}^{2-}$

phosphite ion
HPO_3^{2-}

HO—P=O

metaphosphorous acid
HPO_2,
monobasic

$[\bar{O}—P=O]^-$

metaphosphite ion
PO_2^-

$$\text{HO}\overset{\overset{\displaystyle H}{|}}{\underset{\underset{\displaystyle O}{||}}{P}}\text{—O—}\overset{\overset{\displaystyle H}{|}}{\underset{\underset{\displaystyle O}{||}}{P}}\text{—OH}$$

pyrophosphorous acid
diphosphorous acid
$H_4P_2O_5$,
dibasic

$$\left[\overset{-}{O}\overset{\overset{\displaystyle H}{|}}{\underset{\underset{\displaystyle O}{||}}{P}}\text{—O—}\overset{\overset{\displaystyle H}{|}}{\underset{\underset{\displaystyle O}{||}}{P}}\overset{-}{O}\right]^{2-}$$

pyrophosphite ion
diphosphite ion
$H_2P_2O_5^{2-}$

$$\text{HO}\overset{\overset{\displaystyle OH}{|}}{\underset{\underset{\displaystyle O}{||}}{P}}\text{—OH}$$

orthophosphoric acid
H_3PO_4,
tribasic

$$\left[\overset{-}{O}\overset{\overset{\displaystyle O}{|}}{\underset{\underset{\displaystyle O}{||}}{P}}\overset{-}{O}\right]^{3-}$$

orthophosphate ion
PO_4^{3-}

$$\text{HO}\overset{\overset{\displaystyle OH}{|}}{\underset{\underset{\displaystyle O}{||}}{P}}\text{—O—}\overset{\overset{\displaystyle OH}{|}}{\underset{\underset{\displaystyle O}{||}}{P}}\text{—OH}$$

pyrophosphoric acid
diphosphoric acid
$H_4P_2O_7$,
either di- or tetrabasic

$$\left[\overset{-}{O}\overset{\overset{\displaystyle O}{|}}{\underset{\underset{\displaystyle O}{||}}{P}}\text{—O—}\overset{\overset{\displaystyle O}{|}}{\underset{\underset{\displaystyle O}{||}}{P}}\overset{-}{O}\right]^{4-}$$

pyrophosphate ion
diphosphate ion
$P_2O_7^{4-}$

$$\text{HO}\text{—}\overset{\overset{\displaystyle O}{||}}{\underset{\underset{\displaystyle O}{||}}{P}}$$

metaphosphoric acid
HPO_3,
monobasic

$$\left[\overset{-}{O}\text{—}\overset{\overset{\displaystyle O}{||}}{\underset{\underset{\displaystyle O}{||}}{P}}\right]^{-}$$

metaphosphate ion
PO_3^-

Of these, the phosphoric acids are the most important.

Arsenic, antimony and bismuth

Covalent structures may be drawn to represent all compounds of arsenic. However, weak metallic properties appear with antimony and bismuth. Salts containing the simple ions, Sb^{3+} and Bi^{3+}, are formed. Bismuth is more electropositive than antimony and its salts are less hydrolysed. Both antimony and bismuth also form many covalent compounds.

Hydrides:

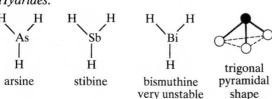

arsine   stibine   bismuthine
                                    very unstable

trigonal
pyramidal
shape

Trimethyl arsine, etc:

$$\overset{\overset{\displaystyle CH_3 \quad CH_3}{\diagdown \quad \diagup}}{\underset{\underset{\displaystyle CH_3}{|}}{As}} \qquad \overset{\overset{\displaystyle CH_3 \quad CH_3}{\diagdown \quad \diagup}}{\underset{\underset{\displaystyle CH_3}{|}}{Sb}} \qquad \overset{\overset{\displaystyle CH_3 \quad CH_3}{\diagdown \quad \diagup}}{\underset{\underset{\displaystyle CH_3}{|}}{Bi}}$$

tri-alkyl derivatives corresponding to the tertiary amines

Trioxides:

(former names retained as for phosphorus)

As_4O_6	Sb_4O_6	$(Bi_2O_3)_n$
similar to P_4O_6 (Fig. 21.4)	similar to P_4O_6 below 570°C	relative molecular mass unknown

Halides:

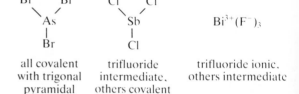

all covalent with trigonal pyramidal arrangement	trifluoride intermediate, others covalent	trifluoride ionic, others intermediate

Oxoacid salts:

$(Sb^{3+})_2(SO_4^{2-})_3$ antimony sulphate	$(Bi^{3+})_2(SO_4^{2-})_3$ bismuth sulphate (various basic salts on hydrolysis)
$Sb^{3+}(NO_3^-)_3$ antimony nitrate	$Bi^{3+}(NO_3^-)_3$ bismuth nitrate

Antimony salts are much more hydrolysed than those of bismuth. It is now considered doubtful whether the groups $(Sb{=}O)^+$ and $(Bi{=}O)^+$ exist and so the use of the terms antimonyl and bismuthyl, for the oxide salts, have been discontinued. The common stable salts of bismuth are oxide (and hydroxide) salts.

$Bi^{3+}O^{2-}NO_3^-$ bismuth oxide nitrate	$(Bi^{3+}O^{2-})_2SO_4^{2-}$ bismuth oxide sulphate
$[(Bi^{3+}O^{2-})_2CO_3^{2-}]_2 . H_2O$ bismuth oxide carbonate hemihydrate†	$Bi^{3+}O^{2-}ClO_4^- . H_2O$ bismuth oxide perchlorate monohydrate††

† i.e. $(BiO)_2CO_3 . \frac{1}{2}H_2O$
†† perchlorate = chlorate(VII)

In BiOCl, bismuth oxide chloride, the solid structure has complex layers consisting of a sheet of coplanar oxygen atoms with a sheet of chlorine atoms on each side, and with the bismuth atoms between the successive sheets.

The elements

A detailed description and comparison of the allotropes of phosphorus appears in Chapter 6. Arsenic and antimony, but not nitrogen and bismuth, also exist in allotropic crystalline modifications, both having yellow, black and grey (the stable) forms.

The fractionation of liquid air to yield nitrogen has been mentioned in Chapter 12.

Nitrogen

Nitrogen is a colourless, odourless, tasteless gas (b.p. $-196°C$) which is only very slightly soluble in water (about 1 volume in 50 volumes of water at room temperature) and chemically inert at room temperature. The atmosphere has 78% by volume nitrogen with 21% oxygen, 1% argon and 0.03% carbon dioxide, plus water vapour, dust, etc. Large quantities of nitrogen for laboratory use may be obtained from a cylinder of compressed gas. However, for the infrequent preparation, a choice is usually made from the following two methods.

1 *Atmospheric nitrogen* which contains residual argon, is obtained by removal of oxygen and carbon dioxide from the atmosphere. Air is forced through a wash-bottle containing caustic potash solution to remove carbon dioxide, through concentrated sulphuric acid to remove water vapour and through a combustion tube containing copper turnings heated in a small furnace:

$$2Cu + O_2 \rightarrow 2CuO$$

Nitrogen, contaminated with argon, is collected over water. Traces of the other gases are present in negligible amount. Alternatively, the gas may be dried with concentrated sulphuric acid or calcium chloride and collected over mercury.

2 *Chemical nitrogen* may be prepared by warming an aqueous solution of ammonium nitrite:

$$NH_4^+NO_2^- \rightarrow N_2 + 2H_2O$$

This is similar to the action of nitrous acid on primary amines ($NH_4NO_2 \equiv HNO_2 + H.NH_2$),

nitrogen being evolved. As ammonium nitrite is very unstable, concentrated equimolar solutions of sodium nitrite and ammonium chloride are usually mixed:

$$Na^+ + NO_2^- + NH_4^+ + Cl^- \rightarrow$$
$$Na^+ + Cl^- + N_2 + 2H_2O$$

Warming starts the decomposition which becomes very vigorous. The gas is collected over water.

The presence of argon in atmospheric nitrogen means that the density of atmospheric nitrogen is slightly greater than that of chemical nitrogen. In 1894, Rayleigh measured the density of nitrogen from various sources to an experimental accuracy of about 0.01%. He found that there was a discrepancy amounting to 0.47% between the values for atmospheric nitrogen and various chemical forms. In a joint paper in 1894, Rayleigh and Ramsay published an account of the discovery of argon. Shortly afterwards the remaining members of the new periodic family, the inert gases, were discovered.

Nitrogen is liberated in other reactions of interest. Aqueous ammonia may be oxidized with sodium hypochlorite† solution, sodium hypobromite solution or an aqueous suspension of bleaching powder, the nitrogen containing some dinitrogen oxide (nitrous oxide) as impurity:

$$NH_3 + OCl^- \rightarrow NH_2Cl + OH^-$$

$$2NH_2Cl + OCl^- + 2OH^- \rightarrow$$
$$N_2 + 3Cl^- + 3H_2O$$

Gaseous ammonia may be oxidized by passage over heated copper(II) oxide, which is reduced to copper:

$$2NH_3 + 3CuO \rightarrow 3Cu + 3H_2O + N_2$$

Oxides of nitrogen may be reduced to nitrogen by passage over a heated metal: dinitrogen oxide reacts with red-hot copper turnings,

$$Cu + N_2O \rightarrow CuO + N_2$$

The thermal decomposition of ammonium dichromate continues by itself when once started, the orange powder burning to a bulky green mass of chromium(III) oxide:

$$(NH_4)_2Cr_2O_7 \rightarrow N_2 + Cr_2O_3 + 4H_2O$$

Nitrogen is inert at ordinary temperatures. On heating, it will form nitrides with certain metals (e.g. Li_3N, Ca_3N_2) and non-metals (e.g. BN,

† hypochlorite = chlorate(I), hypobromite = bromate(I)

Si_3N_4). Binary compounds of nitrogen with more electronegative elements are usually not called nitrides, but oxides, chlorides, etc. Industrial processes based on the reactions of nitrogen with oxygen, hydrogen and calcium carbide are described in Chapter 12.

Phosphorus, arsenic, antimony and bismuth

The remaining four elements of Group V are solids and except for bismuth, exhibit allotropy. Phosphorus appears normally as the white (or yellow) and red (or violet) allotropes and has been prepared in a black form at 200°C and 1.22 GPa† (12 000 atmospheres). The white and red forms of phosphorus have been compared in Chapter 6, p. 109. The allotropes of arsenic and antimony are not important. Arsenic has yellow, black and grey modifications, of which the last is the common form, with a metallic appearance and an electrical conductance only 4.2% that of silver. Antimony exists in four forms: yellow, black, explosive (amorphous) and metallic, the last named being the common form with an electrical conductance 4.4% that of silver. Amorphous antimony is, as the name suggests, not crystalline. The yellow allotrope in each case is covalent, transparent, soluble in carbon disulphide, the least stable form and there is little doubt, consisting of tetrahedral molecules in the solid state: P_4, As_4, Sb_4. Bismuth is metallic, forming large reddish-white crystals. On freezing, it expands by about 4%, is diamagnetic and has an electrical conductance about 1.4% that of silver at 0°C. The conductivity is greatly increased by the application of intense magnetic fields. The melting-point data in Table 21.1 belong to the most stable allotropes of the elements. Densities of the different modifications are compared in Table 21.3. In the following account of chemical properties, a distinction will be made between the white and red forms of phosphorus, because they differ markedly in chemical properties, but for the other elements the common modifications will be assumed. The elements are not usually prepared in the laboratory.

Table 21.3 Comparison of the densities at room temperature for the allotropes of the phosphorus group/g cm^{-3}

Phosphorus		Arsenic		Antimony	
White	1.83	Yellow	1.97	Yellow	—
Red	2.35	Black	4.73	Black	5.3
Black	2.69	Grey	5.73	Grey	6.67

Their industrial production has already been described (pp. 152–55).

Great care must be taken when using white phosphorus. It is usually supplied in sticks which are stored in water, and should be cut under water, gripping the phosphorus in a pair of tongs in a strong vessel. Pieces of phosphorus ignite in air at about 35°C and have been known to catch fire spontaneously on being taken out of water in a warm room. White phosphorus is readily soluble in carbon disulphide, benzene, turpentine and other organic solvents, which on evaporating leave the phosphorus in a finely divided state, in which it is spontaneously flammable. There is a fire hazard from careless use of solutions containing elemental phosphorus. The fumes from white phosphorus are very poisonous. At one time, when white phosphorus was used in strike-anywhere matches, workers were afflicted with *phosphorus necrosis* (phossy jaw), causing the bones of the face to rot. The non-toxic trisulphide, P_4S_3, is now used.

Although there is little doubt that all elements of this group form polyatomic molecules in the solid state, in accordance with general custom for solid elements, and for simplicity in this comparison, simple symbols will be used in equations for phosphorus, arsenic, antimony and bismuth.

The ease of oxidation falls off with increase in atomic number. White phosphorus inflames in air at 35°C while the red allotrope must be heated to about 260°C. In a restricted oxygen or air supply the trioxide is formed but otherwise the pentoxide is the product. This is the most stable oxide of phosphorus:

$$4P + 5O_2 \rightarrow P_4O_{10}$$

Arsenic, antimony and bismuth are not affected by air at ordinary temperatures but burn on strong heating to form the trioxide:

$$4As + 3O_2 \rightarrow As_4O_6$$

$$4Sb + 3O_2 \rightarrow Sb_4O_6$$

$$4Bi + 3O_2 \rightarrow 2Bi_2O_3$$

All react with sulphur on heating. Antimony and bismuth form the trisulphides, although a pentasulphide of antimony may be synthesized:

$$2Sb + 3S \rightarrow Sb_2S_3$$

$$Sb_2S_3 + 2S \rightarrow Sb_2S_5$$

$$2Bi + 3S \rightarrow Bi_2S_3$$

† giga, $G = 10^9$

A wider range of sulphides is formed by arsenic (As_4S_4, As_4S_6, As_2S_5) and phosphorus (P_4S_3, P_4S_7, P_4S_{10}).

All react with the halogens, the action of yellow phosphorus on bromine being explosive. White phosphorus burns spontaneously in chlorine, while red phosphorus sometimes requires heating, forming the trichloride or with excess chlorine, the pentachloride:

$$2P + 3Cl_2 \rightarrow 2PCl_3$$

$$PCl_3 + Cl_2 \rightarrow PCl_5$$

The other elements burn to the trichloride ($AsCl_3$, $SbCl_3$, $BiCl_3$) on warming, although when finely divided, arsenic and antimony take fire spontaneously.

Various compounds with metals are formed, the tendency being greatest with phosphorus and least with bismuth. Dry phosphorus reacts with warm sodium in the absence of air with a flash, to form sodium phosphide:

$$P + 3Na \rightarrow Na_3P$$

With water, sodium phosphide yields phosphine:

$$Na_3P + 3H_2O \rightarrow PH_3 + 3NaOH$$

Calcium forms calcium phosphide which behaves similarly:

$$2P + 3Ca \rightarrow Ca_3P_2$$

$$Ca_3P_2 + 6H_2O \rightarrow 3Ca(OH)_2 + 2PH_3$$

Nitrogen and nitrides behave similarly. Arsenic, antimony and bismuth form compounds or alloys with magnesium: Mg_3As_2, magnesium arsenide; Mg_3Sb_2, magnesium antimonide; and Mg_3Bi_2, the bismuthide. Acids react with these alloys to form the hydrides arsine, AsH_3; stibine, SbH_3; and bismuthine, BiH_3, but the yield of the last is very small indeed.

White phosphorus is distinctive in reacting with hot fairly concentrated aqueous alkali to form phosphine (not hydrogen) and sodium hypophosphite solution as the principal products:

$$4P + 3OH^- + 3H_2O \rightarrow PH_3 + 3H_2PO_2^-$$

Non-metallic elements react with oxidizing acids to form acids or acidic oxides while metals give the salt. Concentrated nitric acid reacts vigorously on warming with red phosphorus in the presence of a trace of iodine as catalyst, and explosively with white phosphorus, to form orthophosphoric acid:

$$P + 5HNO_3 \rightarrow H_3PO_4 + 5NO_2 + H_2O$$

Concentrated nitric acid oxidizes arsenic to arsenic acid:

$$As + 5HNO_3 \rightarrow H_3AsO_4 + 5NO_2 + H_2O$$

while diluted nitric acid yields some arsenic trioxide, As_4O_6. Antimony forms the hydrated pentoxide with nitric acid, while bismuth forms the nitrate:

$$Bi + 6HNO_3 \rightarrow Bi(NO_3)_3 + 3NO_2 + 3H_2O$$

Hot concentrated sulphuric acid in the course of complex reactions yields orthophosphoric acid with phosphorus, arsenic trioxide with arsenic and the sulphates of antimony and bismuth:

$$4P + 8H_2SO_4 \rightarrow$$
$$4H_3PO_4 + S + 7SO_2 + 2H_2O$$

$$4As + 6H_2SO_4 \rightarrow As_4O_6 + 6H_2O + 6SO_2$$

$$2Sb + 6H_2SO_4 \rightarrow Sb_2(SO_4)_3 + 6H_2O + 3SO_2$$

$$2Bi + 6H_2SO_4 \rightarrow Bi_2(SO_4)_3 + 6H_2O + 3SO_2$$

Bismuth and, to a lesser extent, antimony have metallic characteristics, oxidation becoming more difficult and stopping at the lower valency state. Antimony and bismuth may be displaced from solutions of their salts by a metal higher in the electrochemical series. Magnesium (ribbon) rapidly displaces antimony:

$$3Mg(s) + 2Sb^{3+} \rightarrow 2Sb(s) + 3Mg^{2+}$$

Simultaneous evolution of hydrogen will be observed. It is displaced from the acid added to reduce hydrolysis of the antimony salt.

Nitrides

Nitrides are binary compounds containing nitrogen. The term is usually applied to nitrogen compounds containing elements which are electropositive or less electronegative than nitrogen. The class excludes salts of hydrazoic acid, hydrogen azide, e.g. NaN_3, sodium azide. Nitrides may be grouped as ionic, covalent or interstitial. There is some doubt about the exact nature of the bonding in the first and last groups.

Ionic nitrides include those formed by the metals of Groups I and II. Lithium is the only alkali metal to form a well-defined nitride, which may be $Li^+[Li-N^--Li]^-$. All of the alkaline-earth metals react with nitrogen on heating. The ionic nitrides are hydrolysed to ammonia and the metal hydroxide:

$$Li_3N + 3H_2O \rightarrow NH_3 + 3LiOH$$

$$Ca_3N_2 + 6H_2O \rightarrow 2NH_3 + 3Ca(OH)_2$$

Covalent nitrides include ammonia, NH_3, cyanogen, $(CN)_2$, and the oxides and halides of nitrogen. Boron and aluminium react with nitrogen at high temperatures to form nitrides which have macromolecular structures and are hydrolysed to give ammonia:

$$BN + 3H_2O \rightarrow H_3BO_3 + NH_3$$

$$AlN + 3H_2O \rightarrow Al(OH)_3 + NH_3$$

Silicon reacts with nitrogen at about 1300°C to form the nitride, Si_3N_4. Nitrides of germanium, Ge_3N_4, and tin, Sn_3N_4, are formed indirectly. Several nitrides of sulphur, including N_4S_4, are known.

Transition metals of Groups IIIa, IVa, Va absorb nitrogen into their lattices to give interstitial or metallic nitrides, which are very hard refractory materials.

Hydrides and related compounds

All elements of this group form covalent gaseous hydrides of the type XH_3. The stability, ease of formation and basicity decrease rapidly with increase in atomic number. The much greater stability of ammonia and also its greater solubility and basicity may be attributed to the much greater electronegativity of nitrogen, with its lone pair of electrons ready for donation or hydrogen bonding. The molecule of ammonia is pyramidal, undergoing rapid oscillation between two forms:

Boiling-points and melting-points of the hydrides of the Group V elements of the XH_3 type are given in Table 21.4. Ammonia has an unexpectedly high melting-point and boiling-point, attributed to association by hydrogen bonding (Fig. 15.1). Ammonia is useful as a refrigerant.

Hydrazine, N_2H_4, hydroxylamine, NH_2OH, and hydrogen azide, HN_3, will also be described.

Ammonia, NH_3, and ammonium salts

The commercial production of ammonia and certain ammonium salts, with their applications, appear in Chapter 12 (see p. 191).

Ammonia is prepared in the laboratory by the action of any alkali on an ammonium salt. Usually, ammonium chloride and slaked lime are mixed and warmed:

$$Ca(OH)_2 + 2NH_4Cl \rightarrow CaCl_2 + 2NH_3 + 2H_2O$$

The gas is dried over quicklime and collected by downward displacement of air or over mercury. Ammonia is a colourless gas, with a characteristic pungent odour. It is less dense than air, the relative density = 17/2 = 8.5 (H = 1; air = 14.4). Ammonia is very soluble in water, of which one volume takes up 1300 volumes of gas at 0°C and 101 kPa (one atmosphere) pressure. Less (700 volumes) dissolves in one volume of water at 20°C and 101 kPa (one atmosphere) pressure. The usual commercial concentrated solution is 0.880 ammonia, the name referring to the density/gcm^{-3} or kgdm^{-3}. The solution contains 35% ammonia by mass.

Quicklime is used to dry ammonia in place of the more usual drying agents, sulphuric acid reacting to form ammonium sulphate and calcium chloride absorbing ammonia to form the ammine, $CaCl_2 . 8NH_3$. Ammonia is readily liquefied by cooling or at 0°C by a pressure of approximately 700 kPa (c. 7 atm), the critical temperature being 132.5°C.

Ammonia is readily oxidized. A lighted taper is extinguished in a mixture of the gas and air but a transient green flame is observed. A jet of ammonia burns in oxygen with a peach-coloured flame, forming nitrogen and steam, a silica tube being used at the jet to avoid the yellow coloration caused by ordinary glass:

$$4NH_3 + 3O_2 \rightarrow 2N_2 + 6H_2O$$

Explosive mixtures of gaseous ammonia and oxygen may be prepared. Mixed with air and passed through a heated platinum gauze, ammonia forms nitrogen oxide (nitric oxide) and steam:

$$4NH_3 + 5O_2 \rightarrow 4NO + 6H_2O$$

Table 21.4 A comparison of the boiling-points and melting-points of the hydrides XH_3 formed by the Group V elements/°C

	NH$_3$	PH$_3$	AsH$_3$	SbH$_3$	BiH$_3$
Boiling-point/°C	−33.5	−87	−55	−18	+20 (approx.)
Melting-point/°C	−78	−136	−114	−88	—

This reaction is used in nitric acid production (p. 196). Ammonia reduces warm copper(II) oxide to copper and lead(II) oxide to lead:

$$3PbO + 2NH_3 \rightarrow 3Pb + N_2 + 3H_2O$$

An excess of aqueous ammonia is oxidized by chlorine to nitrogen but an excess of chlorine gives the dangerously explosive nitrogen trichloride:

$$8NH_3 + 3Cl_2 \rightarrow N_2 + 6NH_4Cl$$

$$NH_3 + 3Cl_2 \rightarrow NCl_3(l) + 3HCl$$

On heating in dry ammonia the alkali metals react to form amides. Potassium yields potassamide, $K^+NH_2^-$, and hydrogen:

$$2K + 2NH_3 \rightarrow 2KNH_2 + H_2$$

This reacts vigorously with water to regenerate ammonia:

$$KNH_2 + H_2O \rightarrow KOH + NH_3$$

Ammonia is a base, accepting protons to form the ammonium ion:

$$
\begin{array}{c}
H \\
| \\
H-N + H^+ \\
| \\
H
\end{array}
\rightarrow
\left[
\begin{array}{c}
H \\
|+ \\
H-N-H \\
| \\
H
\end{array}
\right]^+
$$

Gaseous ammonia and hydrogen chloride react by transference of a proton to give ammonium chloride:

$$NH_3 + HCl \rightarrow NH_4^+Cl^-$$

Aqueous ammonia is a weak base (apparent dissociation constant = 1.8×10^{-5} mol dm^{-3} at 25°C), acquiring protons from the solvent and leaving hydroxide ions,

$$NH_3 + H_2O \rightleftharpoons NH_4^+OH^-$$

The high solubility of ammonia has been interpreted in terms of hydrogen bonding with the formation of an intermediate:

$$
\begin{array}{c}
H \\
| \\
H-N\cdots H \\
| \\
H
\end{array}
\begin{array}{c}
O \\
\diagdown \\
H
\end{array}
\text{ and }
\begin{array}{cc}
H & \\
| & \\
H-N-H\cdots O \\
| & | \\
H & H
\end{array}
$$

By analogy with the alkalis by comparison of corresponding salts, aqueous ammonia is sometimes called ammonium hydroxide, with the formula NH_4OH. But nitrogen is not quinquevalent. Aqueous ammonia contains uncombined ammonia (46.2%), hydrated ammonia $H_3N\cdots HOH$ (52.4%)

and relatively few (1.4%) ammonium and hydroxide ions, NH_4^+ and OH^- in 0.1 M solution at 25°C. The substitution of alkyl groups for hydrogen, by reactions between alkyl iodides and alcoholic ammonia, gives successively primary, secondary and tertiary amines, which are very soluble, the strength of the base solution rising with the number of substituted groups. Further reaction with the alkyl iodide yields the quaternary ammonium salt. The strongest alkali results from the removal of the possibility of hydrogen bonding altogether by substitution of all four hydrogen atoms of the ammonium ion to form a quaternary ammonium hydroxide, of about the same strength as potassium hydroxide and wholly ionic. Tetraethylammonium hydroxide can be formed by the action of moist silver oxide on the quaternary ammonium salt in aqueous solution, the silver compounds being insoluble:

$$\underset{\text{triethylamine}}{N(C_2H_5)_3} + \underset{\text{iodoethane}}{C_2H_5I} \rightarrow \underset{\substack{\text{tetraethylammonium} \\ \text{iodide}}}{[\overset{+}{N}(C_2H_5)_4]^+I^-}$$

$$\underset{\substack{\text{quaternary} \\ \text{ammonium} \\ \text{salt}}}{N(C_2H_5)_4I} + \text{'AgOH'(s)} \rightarrow AgI(s) + N(C_2H_5)_4OH$$

The filtered solution is evaporated under reduced pressure. The dissociation constants (mol dm^{-3}) at 18°C of ammonia and its ethyl derivatives in aqueous solution show the increase in strength as substitution progesses; the ions are also listed:

$$
\text{Weak}
\begin{cases}
NH_4^+ & OH^- \; 2.94 \times 10^{-5} \text{ mol dm}^{-3} \\
(C_2H_5)NH_3^+ & OH^- \; 6.73 \times 10^{-4} \text{ mol dm}^{-3} \\
(C_2H_5)_2NH_2^+ & OH^- \; 1.06 \times 10^{-3} \text{ mol dm}^{-3} \\
(C_2H_5)_3NH^+ & OH^- \; 7.87 \times 10^{-4} \text{ mol dm}^{-3}
\end{cases}
$$
$$\quad (C_2H_5)_4N^+ \quad OH^- \qquad \text{wholly ionic}$$

Aqueous ammonia may be used to precipitate insoluble hydroxides of metals from solutions of their salts:

$$Al^{3+} + 3OH^- \rightarrow Al(OH)_3(s)$$

With some solutions, basic salts are formed. The concentration of hydroxide ion in a solution of ammonia is susceptible to the common ion effect, being reduced in the presence of a dissolved ammonium salt:

$$NH_3 + H_2O \rightleftharpoons NH_4^+ + OH^-$$

for which

$$\frac{[NH_4^+][OH^-]}{[NH_3]} = 1.8 \times 10^{-5} \text{ mol dm}^{-3} \text{ at } 25°C$$

Addition of ammonium chloride, producing an excess of ammonium ions, leads to a reduction in the concentration of hydroxide ions. This common ion effect is used in the selective precipitation of the hydroxides of aluminium, chromium(III) and iron(III) (more properly, hydrated oxides) in qualitative analysis. Mixtures of aqueous ammonia and ammonium chloride are used as buffer solutions.

The lone pair of electrons held by nitrogen in ammonia may be used for co-ionic bonding. In this way the ammonium ion is formed. But ammonia combines with a wide range of the compounds of metals, as solids and in solution, to form complexes known as ammines. Important examples will be given under the metals involved. Here, one example will suffice: on continued addition of ammonia solution to aqueous copper(II) sulphate, the initial light blue precipitate redissolves to give a deep royal blue solution containing the tetramminecopper(II) ion, $Cu(NH_3)_4^{2+}$, from which alcohol precipitates tetramminecopper(II) sulphate monohydrate, $Cu(NH_3)_4SO_4 \cdot H_2O$ which is a complex salt (p. 478), not to be confused with the pale blue double salt, composed of simple ions, ammonium copper(II) sulphate, $CuSO_4 \cdot (NH_4)_2SO_4 \cdot 6H_2O$. When ammonia is passed over anhydrous copper(II) sulphate a deep blue pentammine is formed, with evolution of heat:

$$CuSO_4 + 5NH_3 \rightarrow CuSO_4 \cdot 5NH_3$$

Liquefied ammonia is an excellent solvent. The alkali metals dissolve without reaction in the absence of the catalysts and light that favour amide formation, to give blue solutions which conduct electricity. Ammonia ionizes slightly:

$$2NH_3 \rightleftharpoons NH_4^+ + NH_2^-$$

The acidic ion is the ammonium ion, bearing a proton, and the basic ion, the amide ion, corresponding to H_3O^+ and OH^- ions respectively. (In ammonia, the imide ion, NH^{2-}, and nitride ion, N^{3-}, are also basic ions.) Ammonium salts, e.g. ammonium chloride, are ammono-acids and metal amides, e.g. potassamide, ammono-bases, neutralization occurring when an ammono-acid reacts with an ammono-base, for which phenolphthalein may be used as indicator:

$$NH_4^+ + NH_2^- \rightleftharpoons 2NH_3$$

It might help to refer to pp. 237–38 at this point. Amphoteric behaviour also occurs: zinc amide dissolving in potassamide solution to form an ammono-zincate ion;

$$Zn(NH_2)_2 + 2NH_2^- \rightarrow Zn(NH_2)_4^{2-}$$

In the Haber process (p. 191) ammonia is synthesized by a reversible reaction between the elements. Ammonia gas, subjected to electrical sparking from an induction coil, is almost, but not entirely, decomposed:

$$2NH_3 \rightleftharpoons N_2 + 3H_2$$

Therefore, direct combination or decomposition of gaseous ammonia will not provide a method for deciding the composition of the gas, unlike, for example, the case of hydrogen chloride. Three quantitative experiments are necessary to deduce the molecular formula of ammonia.

1 Ammonia is oxidized by sodium hypobromite (bromate(I)) solution with the result that the gas yields half of its volume of nitrogen. There is no danger of a bromide analogous to the explosive NCl_3 being formed.

A burette and its jet is filled with dry ammonia by upward displacement and closed with a rubber stopper. It is clamped with the tap immersed in sodium hypobromite solution and the tap opened. Solution enters the burette in a few seconds, especially if air has been displaced from the jet, itself of negligible volume, and an effervescence occurs. As the reaction slows down the tap is closed and the burette and its contents shaken. After cooling to room temperature, the burette is opened with the jet under water in a tall jar and after levelling, the level of the meniscus is recorded at room temperature and pressure. Before results can be calculated, the volume from the zero of the burette to below the rubber stopper and the volume from the 50-ml mark to the tap of the burette must be measured. This is done by determining the volume of water required to fill these spaces.

The reaction may be written:

$$2NH_3 + 3OBr^- \rightarrow N_2 + 3H_2O + 3Br^-$$

2 Aqueous ammonia is decomposed by gaseous chlorine to give nitrogen which occupies one-third of the volume of chlorine.

The Hofmann tube, shown in Fig. 21.5, is filled with chlorine and placed in a water bath at room temperature with a cotton-wool pad at the bottom. Ammonia (0.880 g cm^{-3}) is placed in the funnel

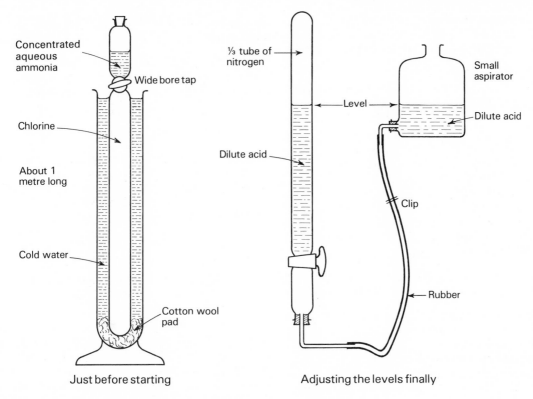

Fig. 21.5   Hofmann: the ratio by volume of nitrogen to hydrogen in ammonia

and the tap turned cautiously to admit a few drops of liquid without escape of gas. An immediate reaction occurs with yellow flashing and the appearance of white fumes. Cautiously, ammonia is admitted in successive portions until flashing has ceased and the green colour of chlorine has disappeared. Unused ammonia is poured away and dilute sulphuric acid added to completely fill the funnel which is now connected to an aspirator of very dilute sulphuric acid and inverted. The tap is opened and the levels adjusted. It is then closed. From the total volume of the Hofmann tube and the volume of the liquid finally contained, the volume of residual nitrogen is obtained by subtraction. The reactions may be written:

$$3Cl_2 + 2NH_3 \rightarrow N_2 + 6HCl$$

$$6(HCl + NH_3 \rightarrow NH_4Cl)$$

We now require the volume ratio in which hydrogen and chlorine combine.

3   In determining the formula of hydrogen chloride, it was found that equal volumes of hydrogen and chlorine unite.

By application of Avogadro's Hypothesis we may deduce:

(a)   1 molecule of nitrogen combines with hydrogen to form 2 molecules of ammonia;
(b)   1 molecule of nitrogen is formed by 3 molecules of chlorine acting on ammonia;
(c)   3 molecules of chlorine react with 3 molecules of hydrogen;
∴   1 molecule of nitrogen + 3 molecules of hydrogen unite to form 2 molecules of ammonia.

It may be deduced by a separate argument based on experiment that hydrogen is H_2 and nitrogen N_2 (Cannizzaro's Argument).

∴ the results obtained may be interpreted by the equation:

$$N_2 + 3H_2 = 2NH_3$$

∴ The molecular formula of ammonia is NH_3.

The commercial production of several ammonium salts and their uses as fertilizers have already been described (Chapter 12, pp. 189–97).

Ammonium chloride (sal-ammoniac) is produced commercially as a sublimate obtained by heating ammonium sulphate with sodium chloride:

$$2NaCl + (NH_4)_2SO_4 \rightarrow$$
$$Na_2SO_4 + 2(NH_3(g) + HCl(g))$$

$$NH_3(g) + HCl(g) \rightarrow NH_4Cl(s)$$

Alternatively, it may be produced by the action of ammonia on hydrogen chloride at 250°C. Ammonium chloride is used in dry cells and as a flux for soldering and in tin-plating.

Ammonium nitrate may be made commercially by neutralization of nitric acid with ammonia or from ammonium sulphate solution by addition of calcium nitrate solution, when sparingly soluble calcium sulphate is precipitated:

$$2NH_4^+ + SO_4^{2-} + Ca^{2+} + 2NO_3^{2-} \rightarrow$$
$$CaSO_4(s) + 2(NH_4^+ + NO_3^-)$$

In addition to being a fertilizer, it is used in explosives: with aluminium powder as ammonal, and TNT as amatol.

'Ammonium carbonate' is formed by a reaction between carbon dioxide, ammonia and steam when a mixture of ammonium carbamate, $NH_2.COONH_4$, and ammonium hydrogencarbonate results. On dissolving in water to which ammonia is added, ammonium carbonate is formed:

$$NH_4^+HCO_3^- \ + \ NH_3 \longrightarrow (NH_4^+)_2CO_3^{2-}$$

Ammonium carbonate is used in smelling-salts (sal-volatile).

Ammonium salts tend to dissociate on heating. When heated, ammonium chloride sublimes. At 340°C, it has been shown that the vapour contains only the mixed gases ammonia and hydrogen chloride and no ammonium chloride:

$$\underset{\text{solid}}{NH_4^+Cl^-} \rightleftharpoons \underset{\text{vapour}}{NH_3(g) + HCl(g)}$$

On cooling, the gases recombine.

Ammonium sulphate melts on heating and gives off ammonia (turning litmus blue) while on further heating residual sulphuric acid distils (changing the litmus to red):

$$(NH_4)_2SO_4 \rightarrow 2NH_3(g) + H_2SO_4(l)$$

Qualitative and quantitative tests for the ammonium ion

In inorganic qualitative analysis, the liberation of ammonia by the action of warm caustic alkali on the solid or by heating it with soda-lime is sufficient to confirm the presence of the ammonium ion:

$$NH_4^+ + OH^- \rightarrow NH_3(g) + H_2O$$

Ammonia has a characteristic pungent smell and turns litmus blue. Further confirmation is afforded by the black stain formed on a piece of filter paper soaked in dimercury(I) dinitrate solution but arsine also blackens this reagent,

$$Hg_2(NO_3)_2 + 2NH_3 \rightarrow \underset{\text{black}}{NH_2Hg_2NO_3} + NH_4NO_3$$

$$NH_2Hg_2NO_3 \rightarrow \underset{\text{also black}}{\underline{Hg + Hg(NH_2)NO_3}}$$

Ammonia may be determined by titration against standard acid. Ammonium salts may be estimated by boiling with an excess of a measured volume of standard alkali until all ammonia has been expelled, the excess alkali being determined by titration against standard acid. Alternatively, ammonium salts may be determined by liberation of ammonia by boiling with alkali, followed by absorption of the gas in a measured volume of excess standard acid, the excess being determined by titration against standard alkali. However, the best method for the determination of ammonia in ammonium salts requires addition of methanal (formaldehyde):

$$4NH_4Cl + 6HCHO \rightarrow$$
$$\underset{\substack{\text{hexamethylene} \\ \text{tetramine}}}{(CH_2)_6N_4} + 4HCl + 6H_2O$$

Phenolphthalein is used as indicator. Hexamethylene tetramine is a very weak base (K = 8.4×10^{-10} mol dm^{-3} at room temperature) and will not affect phenolphthalein. Accordingly 40% methanal solution (traditionally called formalin) is titrated against alkali to a faint rose-pink, to neutralize formic acid and about 10 ml added to 25 ml of the approximately decimolar solution of the ammonium salt. After standing for a minute, the liberated acid is titrated against standard alkali free from carbonate. The final point is a faint rose-pink hue. The molarity of acid calculated equals that of the ammonium ion.

Hydrazine, $H_2N.NH_2$

When sodium hypochlorite (chlorate(I)) solution oxidizes an excess of ammonia in the presence of gelatine or glue, hydrazine is formed. Chloramine is formed initially and without glue would react with excess hypochlorite, liberating nitrogen (see p. 355). Addition of glue retards this reaction, possibly by removal of any heavy metal ions which catalyse the process, allowing chloramine to react with ammonia

$$NH_2Cl + H.NH_2 \rightarrow H_2N-NH_2 + HCl$$

Hydrazine is a colourless, fuming liquid which has been used in the production of rocket fuel. It is a weak diacid base ($K \simeq 1 \times 10^{-6} \, mol^2 \, dm^{-6}$) forming salts:

$$(NH_2.\overset{+}{N}H_3)Cl^- \quad \text{and} \quad (\overset{+}{N}H_3.\overset{+}{N}H_3)(Cl^-)_2 .$$

It is very soluble in water from which a solid hydrate $H_2N.NH_2.H_2O$ may be prepared but cannot be dehydrated. Hydrazine is liable to explode on distillation in the presence of oxygen. Derivatives, phenylhydrazine and $2:4$-dinitrophenylhydrazine, are used to prepare solid derivatives for the identification of aldehydes and ketones. Hydrazine is also used as a reducing agent.

Hydroxylamine, NH_2OH, and hydroxylammonium chloride (hydroxylamine hydrochloride), $(\overset{+}{N}H_3OH)Cl^-$

Hydroxylamine is prepared by the electrolytic reduction of nitric acid at a lead cathode:

$$HNO_3 + 6H^+ + 6e^- \rightarrow NH_2OH + 2H_2O$$

Hydroxylamine is an unstable, weak monoacid base. It is a colourless crystalline solid (m.p. $33°C$), freely soluble in water and liable to decompose spontaneously and explosively. It begins to decompose at $15°C$:

$$3NH_2OH \rightarrow NH_3 + N_2 + 3H_2O$$

$$4NH_2OH \rightarrow 2NH_3 + N_2O + 3H_2O$$

The oxidation number of nitrogen in NH_2OH is -1.

The hydrochloride (hydroxylammonium chloride) is usually obtained in the laboratory. Addition of sulphuric acid to the electrolyte above enables the sulphate, hydroxylammonium sulphate, $(HONH_3)_2SO_4$, to be crystallized. This reacts with barium chloride solution to give a precipitate of barium sulphate and a solution of hydroxylammonium chloride, $(NH_3OH)Cl$.

Hydroxylamine and its salts are powerful reducing agents. Fehling's solution and tetramminecopper(II) sulphate are reduced to copper(I) oxide, ammoniacal silver nitrate reduced to silver and a solution of mercury(II) chloride to dimercury(I) dichloride and mercury, with the liberation of dinitrogen oxide (nitrous oxide):

$$2NH_2OH + 4OH^- - 4e^- \rightleftharpoons N_2O + 5H_2O$$

$$\underset{\text{(as complex)}}{4Cu^{2+}} + 2NH_2OH + 8OH^- \rightarrow \\ 2Cu_2O(s) + N_2O + 7H_2O$$

$$\underset{\text{(as complex)}}{4Ag^+} + 2NH_2OH + 4OH^- \rightarrow \\ 4Ag(s) + N_2O + 5H_2O$$

$$\underset{\text{covalent}}{4HgCl_2} + 2NH_2OH + 4OH^- \rightarrow \\ 2Hg_2Cl_2(s) + N_2O + 5H_2O + 4Cl^-$$

$$2Hg_2Cl_2(s) + 2NH_2OH + 4OH^- \rightarrow \\ 4Hg(s) + N_2O + 5H_2O + 4Cl^-$$

Nitrogen is also formed. Hydroxylamine is used to prepare crystalline derivatives of aldehydes and ketones called oximes, the melting-points of which may serve to identify the organic compound concerned within its homologous series.

Hydrogen azide, hydrazoic acid, HN_3, and azides

Hydrogen azide may be prepared by the oxidation of hydrazine by nitrous acid (acidified sodium nitrite) in the presence of a phosphate buffer, when the yield may reach 60%:

$$H_2N.\overset{+}{N}H_3 + HNO_2 \rightarrow HN_3 + H_3O^+ + H_2O$$

Hydrazoic acid is a colourless, toxic, highly explosive liquid. As an acid, it has about the same strength as ethanoic (acetic) acid.

The structure is a resonance hybrid:

$$\{H-\overset{-}{N}-\overset{+}{N}\equiv N \leftrightarrow H-N=\overset{+}{N}=\overset{-}{N}\}$$

The oxidation number of nitrogen in HN_3 is $-\frac{1}{3}$. Sodium azide is prepared by the action of dinitrogen oxide (nitrous oxide) on sodamide at $190°C$:

$$2NaNH_2 + N_2O \rightarrow NaN_3 + NaOH + NH_3$$

Except for the salts of lead, silver and dimercury(I), azides of the common metals are soluble. Azides are potentially explosive. Those of the alkali and alkaline-earth metals may be melted and decomposed on strong heating but azides of the heavy metals, especially those of lead and silver, explode violently when struck, even gently. Lead(II) azide, $Pb(N_3)_2$, is used as an initiator in the explosives industry.

Phosphine, PH_3, and phosphonium iodide, PH_4I

Phosphine is usually made by the action of hot concentrated alkali solution on white phosphorus. Red phosphorus does not react. When a small piece of white phosphorus (pea-size) is warmed gently with a solution, containing (say) 10 g caustic potash in 20 cm^3 water, phosphine is evolved but burns spontaneously as each bubble of gas reaches the atmosphere. The products of combustion are phosphorus pentoxide and steam. The flammability is due to the hydride, P_2H_4, 'diphosphine', which may be removed by passing the contaminated phosphine through a freezing-mixture. Hydrogen is the main impurity. A solution of potassium hypophosphite is the other main product. The reaction producing phosphine may be expressed:

$$4P + 3OH^- + 3H_2O \rightarrow PH_3 + 3H_2PO_2^-$$

Pure phosphine is not spontaneously flammable. It is a colourless gas (b.p. $-88°C$) associated with the unpleasant smell of rotting fish and very poisonous.

Impure phosphine may be prepared by the action of water or dilute acids on metal phosphides. Calcium phosphide is usually decomposed by water while aluminium phosphide is treated with dilute sulphuric acid:

$$Ca_3P_2 + 6H_2O \rightarrow 2PH_3 + 3Ca(OH)_2$$
$$2AlP + 3H_2SO_4 \rightarrow 2PH_3 + Al_2(SO_4)_3$$

Pure phosphine is prepared by the decomposition of phosphonium iodide with alkali,

$$PH_4I + OH^- \rightarrow PH_3 + I^- + H_2O$$

which is similar to the preparation of ammonia. Alternatively, the reduction of phosphorus trichloride with lithium aluminium hydride in dry ether illustrates an important general method of preparing covalent hydrides:

$$4PCl_3 + 3LiAlH_4 \rightarrow 4PH_3 + 3LiCl + 3AlCl_3$$

Phosphine is very much less basic but a stronger reducing agent than ammonia. It is only slightly soluble in water, the solution being neutral to litmus. The few phosphonium salts readily dissociate and are immediately hydrolysed by water.

Phosphine burns readily in air, spontaneously if it contains P_2H_4 but not below 150°C if pure:

$$4PH_3 + 8O_2 \rightarrow P_4O_{10} + 6H_2O$$

Ammonia burns in oxygen only. A further difference is seen with chlorine: phosphine inflames to produce phosphorus trichloride and hydrogen chloride:

$$PH_3 + 3Cl_2 \rightarrow PCl_3 + 3HCl$$

Solutions of copper, silver and gold salts are reduced to the phosphide and finally to the metal by phosphine:

$$PH_3 + 6Ag^+ + 3H_2O \rightarrow 6Ag(s) + H_3PO_3 + 6H^+$$

Phosphine has a structure similar to that of ammonia, with a lone pair of electrons but very much less attraction for protons. Substitution of hydrogen atoms by alkyl groups, e.g. $P(CH_3)_3$, makes strong bases. Phosphonium salts even of very strong acids are readily dissociated into the hydride and acid. Phosphonium chloride dissociates completely below 0°C. Phosphonium iodide is the only salt usually prepared and this dissociates on gentle warming. It is prepared by the controlled addition of water to an intimate mixture of white phosphorus and iodine in the absence of air. Phosphorus and iodine are dissolved in ice-cold carbon disulphide in an atmosphere of carbon dioxide. The solvent is distilled off. Dropwise addition of water yields phosphine and hydrogen iodide which pass over in a slow stream of carbon dioxide and condense as colourless crystals of phosphonium iodide. The principal reaction is:

$$5I_2 + 18P + 32H_2O \rightarrow 10PH_4I + 8H_3PO_4$$

Phosphonium iodide is hydrolysed by water:

$$PH_4I + H_2O \rightarrow PH_3 + H_3O^+ + I^-$$

in which PH_3 loses control of H^+ to H_2O. Addition of some alkali avoids volatilization of hydrogen iodide in the preparation of pure phosphine.

The hydrides of arsenic, antimony and bismuth

The stability of the hydrides AsH_3, SbH_3 and BiH_3 falls off rapidly with increase in atomic number. They are the only hydrides formed by these elements.

Arsine, AsH_3, and stibine, SbH_3, are colourless extremely poisonous gases, having distinctive odours, the former reminiscent of garlic. Each is generated by the reduction of the element, or any of its compounds, by hydrogen generated *in situ* by zinc and acid, aluminium and alkali or by electrolysis. Alternatively, an arsenide, such as zinc arsenide, may be treated with dilute sulphuric acid or a magnesium alloy of antimony used similarly with acid:

$$Zn_3As_2 + \cdot 3H_2SO_4 \rightarrow 3ZnSO_4 + 2AsH_3$$

$$Mg_3Sb_2 + 6HCl \rightarrow 3MgCl_2 + 2SbH_3$$

These reactions are similar to the hydrolysis of nitrides and phosphides. The hydrides may be separated from excess hydrogen by freezing in liquid air when hydrogen passes on: arsine freezes at $-114°C$ and stibine at $-88°C$. Stibine decomposes into the elements even at room temperature. Alkyl derivatives of both hydrides are much more stable.

Bismuthine, BiH_3, is extremely unstable and has only been prepared in very small yield by the action of acid on a magnesium–bismuth alloy, Mg_3Bi_2.

Arsine and stibine dissociate into the elements when passed through a heated tube, leaving a grey deposit of the solid element: an arsenic mirror is soluble in sodium hypochlorite (chlorate(I)) solution while an antimony mirror is not. The hydrides have no basic properties and there are no salts analogous to ammonium and phosphonium compounds. Both are reducing agents, giving a black stain of silver on a filter paper soaked in silver nitrate solution, a reduction also accomplished by phosphine:

$$AsH_3 + 6Ag^+ + 3H_2O \rightarrow$$
$$H_3AsO_3 + 6Ag(s) + 6H^+$$

$$SbH_3 + 3Ag^+ \rightarrow Ag_3Sb(s) + 3H^+$$

followed by

$$4Ag_3Sb + 12Ag^+ + 6H_2O \rightarrow$$
$$24Ag(s) + Sb_4O_6 + 12H^+$$

The identification of arsine and stibine is important in forensic science and will be dealt with more fully under qualitative tests for the Group V elements.

Oxides and oxoacids/hydroxides

These are given in Table 21.5: compounds not mentioned again are indicated by square brackets and odd-electron molecules are starred. The individual nature of nitrogen is shown in the range and nature of its oxides and the properties of its acids. Simple acids of phosphorus readily lose water to assume condensed forms for which the nomenclature has been stated in the bond diagram section earlier.

Important oxides and oxoacids of nitrogen

Structures have been given on p. 351.

Dinitrogen oxide, nitrous oxide, N_2O

Discovered by Priestley (1772), dinitrogen oxide is a colourless gas (b.p. $-88°C$) with a sweet taste and pleasant odour. It is readily liquefied under pressure, the critical temperature being about $36°C$. Dinitrogen oxide, mixed with oxygen, is still used as an anaesthetic by dentists, and for minor operations, and is sometimes called 'laughing gas' on account of the exhilaration and sometimes, hysteria it produces. Dinitrogen oxide is fairly soluble in water, 1 volume of water dissolving 1 volume of gas at $6°C$. It is a neutral oxide although it may be regarded as the anhydride of hyponitrous acid, from which it is formed on heating, but the reaction cannot be reversed and dinitrogen oxide does not react with alkali to form salts:

$$H_2N_2O_2 \rightarrow H_2O + N_2O$$
$$\text{hyponitrous acid}$$

The dinitrogen oxide molecule is linear and unsymmetrical, with a very small dipole moment, explained by its being a resonance hybrid of two structures each of which would be highly polar in opposite senses:

$$\{ \overset{-}{N} = \overset{+}{N} = O \leftrightarrow N \equiv \overset{+}{N} - \overset{-}{O} \}$$

Dinitrogen oxide may be prepared by thermal decomposition of ammonium nitrate:

$$NH_4NO_3 \rightarrow N_2O(g) + 2H_2O(g)$$

The decomposition is exothermic, starting at $185°C$ and becoming violent at $250°C$. Impurities in the gaseous product include nitrogen oxide, nitrogen dioxide, nitrogen, oxygen and carbon dioxide. To reduce the possibility of an explosion, a mixture of ammonium sulphate and sodium nitrate may be heated, sodium sulphate remaining as the residue. The gas is often collected over hot water on account of its solubility, but rapid collection over cold water is much more convenient. Very pure dinitrogen oxide is prepared by other methods. Solutions of sodium nitrite and hydroxylammonium chloride are mixed and warmed:

$$\overset{+}{N}H_3OH + NO_2^- \rightarrow N_2O + 2H_2O$$

Table 21.5 Oxides and oxoacids/hydroxides of the Group V elements, with oxidation number listed on the left

	N		P		As		Sb		Bi	
+1	N_2O	Dinitrogen (nitrous) oxide			As_4O_6	Arsenic trioxide	$(Sb_2O_3)_n$ or Sb_4O_6	Antimony trioxide	$(Bi_2O_3)_n$	Bismuth trioxide
+2	NO^*	Nitrogen (nitric) oxide								
+3	N_2O_3	Nitrogen sesquioxide / Dinitrogen trioxide	P_4O_6	Diphosphorus trioxide / Phosphorus trioxide						
+4	NO_2^*	Nitrogen dioxide ⇅†								
+4	N_2O_4	Dinitrogen tetroxide	P_4O_8 / $(PO_2)_n$	Phosphorus tetroxide / Phosphorus dioxide	As_4O_8 / $(AsO_2)_n$	Arsenic tetroxide / Arsenic dioxide	Sb_4O_8 or $(SbO_2)_n$	Antimony tetroxide / Antimony dioxide		
+5	N_2O_5	Nitrogen pentoxide / Dinitrogen pentoxide	P_4O_{10}	Diphosphorus pentoxide / Phosphorus pentoxide	As_2O_5	Arsenic pentoxide	Sb_2O_5	Antimony pentoxide	Bi_2O_5†	Bismuth pentoxide
[+6]	$[NO_3^*]$	Nitrogen trioxide								
[+6]	$[N_2O_6]$	Dinitrogen hexoxide	$[P_2O_6]$	Phosphorus peroxide						
+1	$[H_2N_2O_2]$	Hyponitrous acid	H_3PO_2	Hypophosphorous acid						
+3	HNO_2	Nitrous acid	H_3PO_3	Phosphorous acid	H_3AsO_3	Arsenious acid	$Sb(OH)_3$† *Amphoteric*		$Bi(OH)_3$ *basic:*	*Bismuth hydroxide*
+3			$H_4P_2O_5$	Diphosphorous (pyrophosphorous) acid						
+3			HPO_2	Metaphosphorous acid						
+4			$[H_4P_2O_6]$	Hypophosphoric acid						
+5	HNO_3	Nitric acid	H_3PO_4	Phosphoric acid	H_3AsO_4	Arsenic acid	H_3SbO_4	Antimonic acid		
+5			$H_4P_2O_7$	Diphosphoric (pyrophosphoric) acid	$H_4As_2O_7$	Diarsenic (pyro-arsenic) acid				
+5			HPO_3	Metaphosphoric acid	$HAsO_3$	Meta-arsenic acid				
[+6]			[Peroxodiphosphoric acid $H_4P_2O_8$]							
[+7]			[Peroxomonophosphoric acid H_3PO_5]							

*odd-electron structures, see text opposite and p. 351

†not isolated pure

Nitrates may be reduced: nitric acid is reduced by tin(II) chloride in the presence of hydrochloric acid:

$$4Sn^{2+} + 2NO_3^- + 10H^+ \rightarrow$$
$$4Sn^{4+} + N_2O + 5H_2O$$

Dinitrogen oxide is chemically unreactive unless heated when it becomes a strong oxidizing agent, giving up its oxygen:

$$2N_2O \rightarrow 2N_2 + O_2; \quad \Delta H = -2 \times 71 \text{ kJ mol}^{-1}$$

Dinitrogen oxide supports the combustion of substances which are hot enough to decompose it, the proportion of oxygen given being greater than that in air. Fiercely burning sulphur forms sulphur dioxide, and phosphorus gives the pentoxide, leaving nitrogen:

$$2N_2O + S \rightarrow SO_2 + 2N_2$$

$$10N_2O + 4P \rightarrow P_4O_{10} + 10N_2$$

Dinitrogen oxide re-kindles a glowing splint and is decomposed by heated metals such as copper and iron:

$$N_2O + Cu \rightarrow CuO + N_2$$

Dinitrogen oxide could easily be confused with oxygen. Points of distinction are the sweet taste and solubility, combustions in dinitrogen oxide leave residual nitrogen in equal volume and, unlike oxygen, it does not react with an alkaline solution of pyrogallol, nor does it give brown fumes of nitrogen dioxide with nitrogen oxide.

To determine the molecular formula of dinitrogen oxide two determinations are necessary: a measured volume of dinitrogen oxide is decomposed by heated metal when it yields an equal volume of residual nitrogen and, in the other experiment, the relative density of dinitrogen oxide is found to be 22 (hydrogen = 1; air = 14.4). The first experiment may be done conveniently by enclosing the gas over mercury in a stout bent tube (Fig. 21.6) and heating potassium in it. A clean piece of metal, about pea-size, is introduced into the tube and rises to the top of the mercury. The tube is closed with the thumb and with an upward cant, the potassium is jerked into position as shown. The tube is clamped with the open end on the bottom of the trough and heated. A slight explosion occurs as the potassium reacts. The level of mercury returns to its original position on cooling, the residual gas being shown to be nitrogen. Alternatively, an iron wire could be heated electrically in a measured volume of dinitrogen

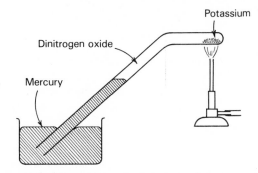

Fig. 21.6 The volume of nitrogen in dinitrogen oxide. (Horizontal arm < 10 cm long, sloping arm < 25 cm long, tube not more than 1.5 cm bore)

oxide confined over mercury:

1 volume of dinitrogen oxide →
1 volume of nitrogen + metal oxide

Applying Avogadro's Hypothesis,

1 molecule of dinitrogen oxide gives
1 molecule of nitrogen

But nitrogen may be shown to be N_2,

∴ dinitrogen oxide is N_2O_x

where x is unknown. However, the relative density is 22 (hydrogen = 1) so that the relative molecular mass is 44,

$$\therefore N_2O_x \equiv 44$$

$$28 + 16x = 44$$

$$\therefore x = 1$$

Dinitrogen oxide is N_2O

Alternatively, the difference in mass between equal volumes of dinitrogen oxide and nitrogen, obtained from their densities, gives the mass of combined oxygen. Using the density of oxygen, the volume which would be occupied by this mass of the free element may be calculated, and it is found that 1 volume dinitrogen oxide gives 1 volume of nitrogen $+\frac{1}{2}$ volume of oxygen from which the formula N_2O may be deduced.

Nitrogen oxide, nitric oxide, NO

First investigated by Priestley (1772) although prepared earlier, nitrogen oxide is a colourless gas (b.p. −152°C) which is difficult to liquefy and combines immediately with oxygen of the air to form brown fumes of nitrogen dioxide:

$$2NO + O_2 \rightarrow 2NO_2$$

Its taste and odour are not known. Nitrogen oxide is only very slightly soluble in water. It is a neutral oxide. Nitrogen oxide has an odd electron structure (p. 351), accounting for its greater reactivity than dinitrogen oxide and paramagnetic behaviour.

In the laboratory, nitrogen oxide is prepared by the reduction of nitric acid, or nitrates in acid solution, or of nitrites. Copper is commonly used: concentrated nitric acid is diluted with an equal volume of water and added to copper turnings. Nitrogen oxide is collected over water. At first, brown fumes of nitrogen dioxide are observed until all oxygen has been used, the fumes dissolving in the water in the trough. The principal reaction is:

$$3Cu + 8HNO_3 \rightarrow 3Cu(NO_3)_2 + 2NO + 4H_2O$$

The gas may be absorbed in iron(II) sulphate solution to form a brown solution containing the complex $Fe^{2+}.NO$ ion, which decomposes on warming to give pure nitrogen oxide. The last method may be used in another way: pure nitrogen oxide is evolved on warming iron(II) sulphate solution and concentrated sulphuric acid with nitric acid or a nitrate:

$$2NO_3^- + 6Fe^{2+} + 8H^+ \rightarrow$$
$$2NO + 6Fe^{3+} + 4H_2O$$

Alternatively, iron(II) sulphate solution and dilute hydrochloric acid or sulphuric acid may be warmed with a fairly concentrated solution of sodium nitrite:

$$2NO_2^- + 2Fe^{2+} + 4H^+ \rightarrow$$
$$2NO + 2Fe^{3+} + 2H_2O$$

Nitrogen oxide may be synthesized from air in very small yield: at atmospheric pressure the yield is 1.2% by volume at 2000°C and 5.3% at 3000°C (the Birkeland-Eyde process). The yield is increased at high temperatures, the forward reaction being endothermic. The mixture of gases was cooled rapidly to 'freeze' the equilibrium.

Thermal dissociation of nitrogen oxide requires strong heating to become appreciable so that the gas will only support the combustion of fiercely burning substances. Phosphorus and magnesium continue to burn in the liberated oxygen to their oxides, leaving residual nitrogen, but sulphur, charcoal and a lighted splint are extinguished. Red-hot copper decomposes the gas to form nitrogen

and copper(II) oxide:

$$2Cu + 2NO \rightarrow N_2 + 2CuO$$

Nitrogen oxide is unsaturated, reacting by addition. With oxygen, brown fumes of nitrogen dioxide are formed while chlorine forms nitrosyl chloride in the presence of activated charcoal as catalyst:

$$2NO + O_2 \rightarrow 2NO_2$$
$$2NO + Cl_2 \rightarrow 2NOCl$$

Remember that carbon monoxide undergoes similar reactions to form carbon dioxide and carbonyl chloride.

Nitrogen oxide gives the brown complex ion, $Fe^{2+}.NO$, with solutions containing ions of iron(II). The iron(II) ion is normally hydrated so that $Fe(OH_2)_6^{2+}$ becomes $Fe(OH_2)_5(NO)^{2+}$. A range of metal nitrosyls similar to carbonyls are known, and nitrogen oxide may be incorporated into other complex ions, a common example being the pentacyanonitrosylferrate(III)† ion (formerly nitroprusside), $Fe(CN)_5(NO)^{2-}$.

Nitrogen oxide is oxidized by acidified potassium permanganate and by iodine, dissolved in potassium iodide:

$$\underset{\text{nitrate}}{NO + 2H_2O - 3e^- \rightleftharpoons NO_3^- + 4H^+}$$

$$3MnO_4^- + 4H^+ + 5NO \rightarrow$$
$$3Mn^{2+} + 5NO_3^- + 2H_2O$$

$$3I_2 + 4H_2O + 2NO \rightarrow 6I^- + 2NO_3^- + 8H^+$$

Nitrogen oxide is reduced to dinitrogen oxide by sulphurous acid:

$$SO_2 + H_2O + 2NO \rightarrow N_2O + H_2SO_4$$

When nitrogen oxide supports combustion it is reduced to nitrogen.

The simplest and quickest method of determining the volume of nitrogen given by a measured volume of nitrogen oxide is by reduction with high-quality reduced iron. The reaction proceeds at 200°C with incandescence, nitrogen and black iron(II) oxide being formed:

$$2NO + 2Fe \rightarrow N_2 + 2FeO$$

1–2 g of freshly reduced iron is introduced into a tube about 350 mm long and 10–15 mm bore drawn

† Oxidation number of NO = 0. The salt may be called *either* sodium pentacyanonitrosylferrate(III) *or* disodium pentacyanonitrosylferrate

out at the ends and fitted with rubber tubing and clips. The air is displaced by coal gas and this in turn by pure nitrogen oxide. After heating, with rotation to allow the reaction to proceed to completion, and cooling, the tube is opened with one end immersed under water which has reached room temperature and the levels adjusted. The volume of nitrogen remaining is half that of the nitrogen oxide taken. Alternatively, an electrically heated iron wire is used to decompose a measured volume of nitrogen oxide confined over mercury, heating being continued until the volume is constant. The reaction is much slower.

> 2 volumes of nitrogen oxide give
> 1 volume of nitrogen + iron(II) oxide

Applying Avogadro's Hypothesis:

> 2 molecules of nitrogen oxide give
> 1 molecule of nitrogen + iron(II) oxide

Nitrogen may be shown to be N_2, so that nitrogen oxide must be NO_x where x is unknown.

But the relative density is 15 (hydrogen = 1), and therefore the relative molecular mass equals 30,

$$\therefore NO_x \equiv 30$$

$$14 + 16x = 30$$

$$\therefore x = 1$$

Nitrogen oxide is NO

Nitrogen sesquioxide, dinitrogen trioxide, N_2O_3

Solid nitrogen sesquioxide forms blue crystals, m.p. $-102°C$. Thermal dissociation commences on melting,

$$2N_2O_3 \rightleftharpoons 2NO + N_2O_4 \rightleftharpoons 2NO + 2NO_2$$

and continues until it reaches 90% at room temperature and is virtually complete at 100°C. Nitrogen sesquioxide is also nitrous anhydride, forming a pale blue solution containing nitrous acid in water and with alkalis being completely converted into nitrites,

$$N_2O_3 + 2OH^- \rightarrow 2NO_2^- + H_2O$$

Oxidation of arsenic trioxide with 60% nitric acid provides nitrogen oxide and nitrogen dioxide in the correct proportions to yield nitrogen sesquioxide on cooling:

$$As_4O_6 + 4HNO_3 \rightarrow As_4O_{10} + 2`N_2O_3` + 2H_2O$$

Dinitrogen tetroxide $\rightleftharpoons$ nitrogen dioxide, $N_2O_4 \rightleftharpoons 2NO_2$

The thermal decomposition of the nitrates of metals other than those of Group I yields nitrogen dioxide, oxygen and the oxide of the metal, which in some cases dissociates further. Lead(II) nitrate is used in the preparation because it is anhydrous, so avoiding a reaction between nitrogen dioxide and water. Even so, lead(II) nitrate should be ground to a fine powder and dried at 150–200°C before use. The reaction proceeds with decrepitation, fragments of crystal being ejected from the mass, and gas is evolved from the fused solid:

$$2Pb(NO_3)_2 \rightarrow 2PbO + 4NO_2 + O_2$$

The gases pass through a vessel cooled in crushed ice when nitrogen dioxide condenses as a pale yellow liquid, oxygen passing on.

The liquefied gas solidifies to a colourless solid at $-11.2°C$. This is dinitrogen tetroxide. On warming, the liquid darkens and is light brown at the boiling-point, 21.1°C. The gas is reddish-brown in colour with a dank acrid odour and darkens on further heating until it is very dark brown at 140°C. The colour change is due to the increasing dissociation of dinitrogen tetroxide into nitrogen dioxide, commencing at the melting-point and being complete at about 140°C. On cooling, the changes are reversed. Dinitrogen tetroxide is colourless and nitrogen dioxide, brown. The increase in number of molecules is accompanied by an increase in volume at constant pressure. Since the mass remains constant, the course of thermal dissociation as the temperature rises may be studied by making measurements of the density of the vapour relative to hydrogen (vapour density):

	$N_2O_4 \rightleftharpoons$	$2NO_2$
Relative molecular mass	92	46
Vapour density	46	23

Mixtures have intermediate value of vapour density between 23 and 46.

Let α be the degree of dissociation of n moles of dinitrogen tetroxide at a given temperature and pressure (changes of which would affect the position of equilibrium). At the given temperature:

> number of moles of $N_2O_4 = n(1 - \alpha)$

> number of moles of $NO_2 = n.2\alpha$

> and the total number of
> moles in the mixture = $n(1 + \alpha)$

Applying the ideal gas law:

if no dissociation had occurred:
$$pV_0 = nRT$$

but with dissociation:
$$pV_d = n(1 + \alpha)RT$$

At constant pressure and the given temperature:

$$\frac{V_{\text{dissociation}}}{V_{\text{no dissociation}}} = \frac{1 + \alpha}{1}$$

and since the total mass is constant,

$$\frac{\text{Calculated vapour density for no dissociation}}{\text{Measured vapour density}}$$

$$= \frac{1 + \alpha}{1}$$

In round figures, 20% of the N_2O_4 molecules have dissociated at 27°C, 50% at 60°C, 90% at 100°C and 100% at 140°C. Additional information about the composition of the gas may be obtained from decomposition with heated metals, as for nitrogen oxide.

Further dissociation occurs above 140°C. The colour lightens appreciably above 500°C when the dioxide has become about 60% dissociated into nitrogen oxide and oxygen:

$$2NO_2 \rightleftharpoons 2NO + O_2$$

Dissociation is complete at 620°C, the gases being colourless. This change is reversed on cooling. Measurements of the density relative to hydrogen may be used to follow the course of the dissociation, the increasing number of molecules leading to an increased volume and decreased density for a given mass at constant pressure. Both dissociations may be summarized

N_2O_4 $\rightleftharpoons$	$2NO_2$ $\rightleftharpoons$	$2NO + O_2$
colourless	dark	colourless
m.p. −11.1°C	brown	
Dissociation commences in the liquid which boils at 21.1°C	complete at 140°C	complete at 620°C

Volumes:	1	2	3
Densities:	46	23	15.3
(H = 1)			

The structures of nitrogen dioxide and dinitrogen tetroxide are shown on page 351. Nitrogen dioxide has an odd number of electrons, causing it to be paramagnetic. The nitrogen–oxygen bonds are equalized by resonance.

Nitrogen dioxide, the name by which the $N_2O_4 \rightleftharpoons 2NO_2$ mixture is known, dissolves in water to form a mixture of nitric and nitrous acids, of which the last undergoes further decomposition, the whole of the gas eventually forming nitric acid:

$$2NO_2 + H_2O \rightarrow HNO_3 + HNO_2$$
$$\text{nitric acid} \quad \text{nitrous acid}$$

$$3HNO_2 \rightarrow HNO_3 + 2NO + H_2O$$

$$2NO + O_2 \rightarrow 2NO_2 \text{ etc. etc.}$$

Solutions of alkalis yield a mixture of the nitrite and nitrate:

$$2OH^- + 2NO_2 \rightarrow NO_2^- + NO_3^- + H_2O$$

Nitrogen dioxide supports the combustion of substances hot enough to cause dissociation. Burning phosphorus, sulphur, charoal and magnesium form the oxides, leaving nitrogen. Red-hot copper and other metals also form oxides and nitrogen:

$$4Cu + 2NO_2 \rightarrow 4CuO + N_2$$

Reduction proceeds as far as nitrogen oxide with carbon monoxide, hydrogen sulphide, sulphurous acid (as the first stage) and iodides:

$$CO + NO_2 \rightarrow CO_2 + NO$$

$$H_2S + NO_2 \rightarrow S + H_2O + NO$$

$$SO_2 + H_2O + NO_2 \rightarrow H_2SO_4 + NO$$

$$2I^- + H_2O + NO_2 \rightarrow I_2 + 2OH^- + NO$$

Moist starch iodide paper is turned blue: a test for an oxidizing agent.

Nitrogen dioxide is oxidized by the more powerful potassium permanganate solution:

$$2MnO_4^- + 10NO_2 + 2H_2O \rightarrow$$
$$2Mn^{2+} + 4H^+ + 10NO_3^-$$

Nitrogen dioxide is used as a catalyst in the production of sulphuric acid by the Lead Chamber process (Chapter 13, p. 200).

Nitrogen pentoxide, dinitrogen pentoxide, N_2O_5

When anhydrous nitric acid is dehydrated by distillation with phosphorus pentoxide, nitrogen pentoxide and metaphosphoric acid are formed:

$$P_4O_{10} + 4HNO_3 \rightarrow 4HPO_3 + 2N_2O_5$$

Anhydrous nitric acid is well cooled and an excess

of phosphorus pentoxide added in small portions. The mass is heated gently. Nitrogen pentoxide is evolved and may be condensed in a well-cooled receiver. It forms colourless crystals, when pure, which start decomposing above 0°C and melt with decomposition at 32.5°C. It rapidly decomposes when heated:

$$2N_2O_5 \rightarrow 2N_2O_4 + O_2$$

Nitrogen pentoxide is nitric anhydride: it is hygroscopic, uniting vigorously with water to form nitric acid:

$$N_2O_5 + H_2O \rightarrow 2HNO_3$$

Nitrogen pentoxide is a powerful oxidizing agent.

Nitrous acid, HNO_2, and nitrites

Nitrous acid is a weakly ionized unstable acid, known only in solution. It is prepared when required by addition of dilute acid to well-cooled sodium nitrite solution:

$$H^+ + NO_2^- \rightleftharpoons HNO_2$$

A pale blue solution is formed and brown nitrogen dioxide is observed as the acid breaks down:

$$3HNO_2 \rightarrow HNO_3 + H_2O + 2NO$$

$$2NO + O_2 \rightarrow 2NO_2$$

The dissociation constant of nitrous acid is 4.5×10^{-5} mol dm^{-3} at room temperature, and a decimolar solution has pH = 2.17 and is 6.7% ionized.

Nitrous acid gives rise to two series of covalent organic compounds. The reaction between sodium nitrite and an alkyl halide yields the alkyl nitrite, an ester, while silver nitrite yields a mixture of the nitrite and an isomeric nitro-compound:

$$CH_3-O-N=O \qquad CH_3-\overset{+}{N}\overset{\displaystyle O}{\underset{\displaystyle O^-}{\diagup}}$$

methyl nitrite   nitromethane

Undissociated nitrous acid is a tautomeric mixture (not a resonance hybrid) of two forms:

Nitrous acid yields diazonium compounds, e.g.

$$(C_6H_5\overset{+}{N}\equiv N)\overset{-}{C}l$$

below 10°C with aromatic primary amines. Otherwise, the $-NH_2$ group is replaced by $-OH$. Both reactions are important in Organic Chemistry.

Nitrous acid may be oxidized to nitric acid, and nitrites to nitrates, by strong oxidizing agents:

$$NO_2^- + H_2O - 2e^- \rightleftharpoons NO_3^- + 2H^+$$

If the Stock Notation were to be applied, nitrous acid would be called nitric(III) acid and a nitrite, nitrate(III). Acidified potassium permanganate, magenta in colour, is slowly decolorized at room temperature but rapidly at 40°C, by sodium nitrite solution as is also yellow cerium(IV) sulphate solution:

$$2MnO_4^- + 6H^+ + 5NO_2^- \rightarrow$$
$$5NO_3^- + 2Mn^{2+} + 3H_2O$$

$$2Ce^{4+} + NO_2^- + H_2O \rightarrow$$
$$2Ce^{3+} + NO_3^- + 2H^+$$

Either reaction may be used in the titrimetric (volumetric) estimation of nitrites, but to avoid the loss of nitrite by decomposition of nitrous acid formed in an acidified solution of a nitrite, the solution of nitrite is placed in the burette and delivered into a warm acidified measured volume of the standard solution of oxidant. Bromine water is decolorized by nitrous acid:

$$Br_2 + NO_2^- + H_2O \rightarrow 2Br^- + NO_3^- + 2H^+$$

Nitrous acid and acidified nitrites may be reduced to nitrogen oxide, or sometimes further, to ammonia, causing oxidation:

$$2HNO_2 + 2H^+ + 2e^- \rightleftharpoons 2NO + 2H_2O$$
weakly ionized

Nitrogen oxide then forms nitrogen dioxide with oxygen of the air. Acidified potassium iodide solution is oxidized to iodine, remaining in solution as the complex I_3^- ion, or falling as a dark grey precipitate:

$$2I^- + 2HNO_2 + 2H^+ \rightarrow I_2 + 2NO + 2H_2O$$

Iron(II) sulphate or chloride solution is oxidized to the iron(III) state, changing in colour from pale green to yellow through brown (see below):

$$2Fe^{2+} + 2HNO_2 + 2H^+ \rightarrow$$
$$2Fe^{3+} + 2NO + 2H_2O$$

Reduction of nitrous acid may be accomplished by

tin(II) chloride in hydrochloric acid, and by sulphurous acid:

$$Sn^{2+} + 2HNO_2 + 2H^+ \rightarrow$$
$$Sn^{4+} + 2NO + 2H_2O$$

$$SO_3^{2-} + 2HNO_2 \rightarrow$$
$$SO_4^{2-} + 2NO + H_2O$$

In qualitative analysis, nitrites are detected by addition of dilute acid, which gives a blue solution from which nitrogen oxide, forming nitrogen dioxide in air, is evolved. When acidified iron(II) sulphate solution and a solution of a nitrite are mixed, a brown addition iron(II)-nitrogen oxide complex is formed:

$$Fe^{2+} + NO \rightarrow Fe^{2+} \cdot NO$$

The complex decomposes on gentle warming. In common with other nitrogenous salts, nitrites are reduced to ammonia by hot alkali and Devarda's Alloy (Cu 50, Al 45, Zn 5). Finally, an aromatic primary amine may be diazotized and coupled with a phenol to produce a characteristic colour. A 1% solution of 4-aminobenzenesulphonic acid (sulphanilic acid) in 5M ethanoic acid and a solution containing a nitrite gives a red coloration when poured on to an excess of naphthalen-1-ol(α-naphthol) in sodium hydroxide solution:

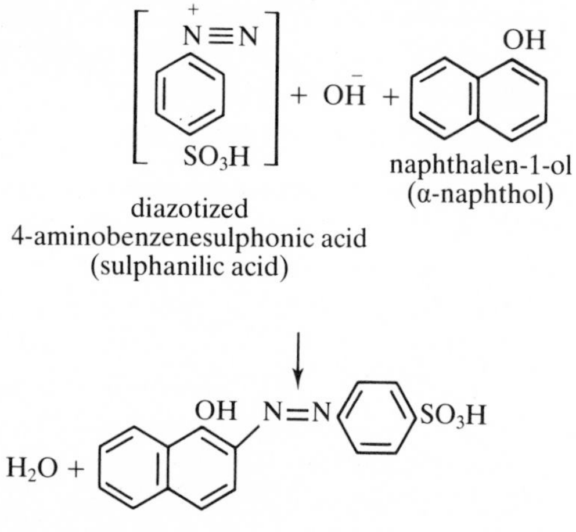

diazotized
4-aminobenzenesulphonic acid
(sulphanilic acid)

naphthalen-1-ol
(α-naphthol)

Nitric acid, HNO_3, and nitrates

Fuming nitric acid is a colourless fuming liquid (m.p. $-41.4°C$, b.p. $86°C$ at atmospheric pressure), with a density of 1.52 g cm^{-3} and containing 98% acid. It decomposes slightly on standing, the liquid becoming tinged with yellow due to dissolved nitrogen dioxide. In sunlight, decomposition goes further until the acid assumes a yellow-brown hue.

The same decomposition occurs on heating:

$$4HNO_3 \rightarrow 2H_2O + 2N_2O_5$$

$$2N_2O_5 \rightarrow 4NO_2 + O_2$$

nitrogen pentoxide which decomposes → nitrogen dioxide and dinitrogen tetroxide

During distillation the proportion of residual water rises as decomposition proceeds. When the temperature reaches $123°C$, a constant boiling-point mixture, or azeotropic† mixture, containing 69.8% nitric acid is obtained which, as the name suggests, distils unchanged. This is ordinary concentrated nitric acid, density 1.42 g cm^{-3}, and usually yellow in colour. Resonance structures for nitric acid and the nitrate ion are:

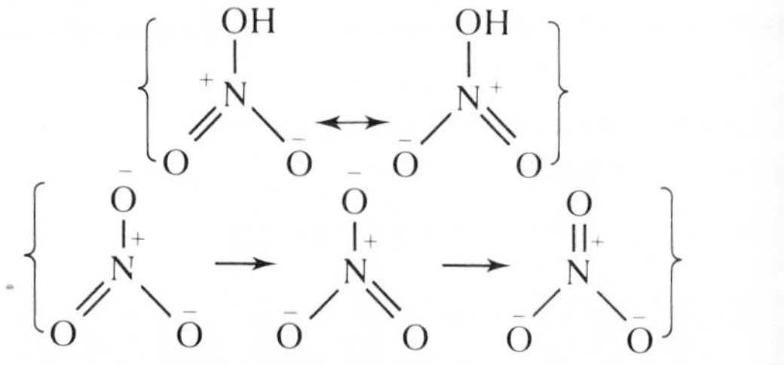

trigonal planar

Nitric acid is extremely corrosive to the skin.

Fuming nitric acid is prepared by heating concentrated sulphuric acid with a dried anhydrous metal nitrate, the acid with the lower boiling-point being displaced. Nitric acid is extremely corrosive and an apparatus made entirely of glass, without rubber stoppers, rubber connections, or corks should be used. This may be a retort but a proprietary glass-jointed distillation assembly is preferred, the nitric acid vapour condensing in a water-cooled condenser. It is convenient to use either sodium nitrate or potassium nitrate:

$$KNO_3 + H_2SO_4 \rightarrow KHSO_4 + HNO_3$$

The manufacture of nitric acid and nitrates by the catalytic oxidation of ammonia is described in Chapter 12, p. 196.

Nitric acid ionizes in a number of ways. Dissolved in water, it is a strong acid:

$$HNO_3 + H_2O \rightleftharpoons H_3O^+ + NO_3^-$$

† Greek, a = without, zeo = to boil, trepo = to change

Anhydrous nitric acid undergoes ionization and is a conducting liquid:

$$2HNO_3 \rightleftharpoons H_2NO_3^+ + NO_3^-$$

and

$$H_2NO_3^+ \rightleftharpoons H_2O + NO_2^+$$

which becomes

$$2HNO_3 \rightleftharpoons NO_2^+ + NO_3^- + H_2O$$

This occurs when a very strongly ionized acid is *not* present. If (say) perchloric acid (chloric(VII) acid) is added,

$$HNO_3 + HClO_4 \rightleftharpoons NO_2^+ + H_2O + ClO_4^-$$

and

$$H_2O + HClO_4 \rightleftharpoons H_3O^+ + ClO_4^-$$

$$\therefore HNO_3 + 2HClO_4 \rightleftharpoons NO_2^+ + 2ClO_4^- + H_3O^+$$

The concept of acids has been developed in Chapter 15, pp. 235–39. Similarly, for sulphuric acid in nitric acid:

$$HNO_3 + 2H_2SO_4 \rightleftharpoons NO_2^+ + 2HSO_4^- + H_3O^+$$

The nitrating action of nitric acid in the presence of sulphuric acid depends on the presence of the nitryl cation, formerly the nitronium ion, NO_2^+. Nitrobenzene is formed by the action of a mixture of concentrated nitric acid and sulphuric acid on benzene, with the temperature being kept below 60°C to avoid further nitration:

$$C_6H_6 + HNO_3 \rightarrow C_6H_5.NO_2 + H_2O$$

In aqueous solution nitric acid is a strongly ionized acid. It affects indicators, forms nitrates with basic oxides and hydroxides, and liberates carbon dioxide from carbonates and hydrogen-carbonates:

$$CuO + 2H^+ \rightarrow H_2O + Cu^{2+}$$

$$Mg(OH)_2 + 2H^+ \rightarrow Mg^{2+} + 2H_2O$$

$$CaCO_3 + 2H^+ \rightarrow Ca^{2+} + H_2O + CO_2$$

Freshly prepared nitric acid is without action on marble chippings. Water must be present for the acidic character to appear. While nitric acid oxidizes ethanol with violence it may be used with sulphuric acid to prepare esters from alcohols. The esters are covalent and do not contain nitrate ions. Glycerol (propane-1,2,3-triol) forms glyceryl trinitrate, or nitroglycerine, which explodes violently when struck or heated, being used in dynamite,

cordite and blasting gelatine:

$$
\begin{array}{ll}
CH_2.OH & CH_2.ONO_2 \\
| & | \\
CH.OH \; + 3HNO_3 \rightarrow & CH.ONO_2 \; + 3H_2O \\
| & | \\
CH_2.OH & CH_2.ONO_2
\end{array}
$$

Nitric acid is a powerful oxidizing agent and hydrogen cannot be collected by the action of laboratory concentrations of acid on metals. However, magnesium and manganese will displace hydrogen from 1–2% nitric acid.

A wide range of nitrogenous products result from the reduction of nitric acid and equations must be regarded as approximate, showing what is believed to be the principal reaction. The final products depend on the concentration of acid, the nature of the reducing agent and the temperature. Ion-electron equations may be written, assuming that water is present to permit ionization as in the commercial concentrated acid:

$$4H^+ + 2NO_3^- + 2e^- \rightleftharpoons 2H_2O + 2NO_2$$

$$4H^+ + NO_3^- + 3e^- \rightleftharpoons 2H_2O + NO$$

$$10H^+ + 2NO_3^- + 8e^- \rightleftharpoons 5H_2O + N_2O$$

$$7H^+ + NO_3^- + 6e^- \rightleftharpoons 2H_2O + NH_2OH$$

$$10H^+ + NO_3^- + 8e^- \rightleftharpoons 3H_2O + NH_4^+$$

Nearly all metals are attacked by concentrated or dilute nitric acid: platinum, rhodium, iridium, tantalum and gold being the exceptions. Weak metals, or those of high valency, able to combine at the higher oxidation states, yield oxides or oxoacids: these include tin, tungsten, molybdenum and antimony. Aluminium, chromium, iron, cobalt and nickel are rendered passive by the action of concentrated nitric acid in forming a protective oxide layer. With a given metal, the final product of chemical action will depend on the concentration of acid and the temperature. Copper reduces concentrated nitric acid to nitrogen dioxide (mainly), moderately dilute acid to nitrogen oxide (mainly) and dilute acid to nitrogen and dinitrogen oxide:

concentrated:
$$4H^+ + 2NO_3^- + Cu \rightarrow Cu^{2+} + 2H_2O + 2NO_2$$

50% v/v:
$$8H^+ + 2NO_3^- + 3Cu \rightarrow 3Cu^{2+} + 4H_2O + 2NO$$

Zinc, having a greater tendency to lose electrons, yields dinitrogen oxide, except in dilute nitric acid, where reduction proceeds to ammonia, or hydroxylamine:

$$10H^+ + 2NO_3^- + 4Zn \rightarrow$$
$$4Zn^{2+} + 5H_2O + N_2O$$

$$10H^+ + NO_3^- + 4Zn \rightarrow$$
$$4Zn^{2+} + 3H_2O + NH_4^+$$

With dilute nitric acid, tin yields hydroxylamine and ammonia.

Many non-metallic elements are oxidized by nitric acid to the highest oxide or oxoacid. Carbon is oxidized by hot nitric acid to carbon dioxide:

$$C + 4HNO_3 \rightarrow CO_2 + 4NO_2 + 2H_2O$$

Red phosphorus, arsenic, sulphur and iodine are oxidized to the acid by warm concentrated nitric acid:

$$P + 5HNO_3 \rightarrow H_3PO_4 + 5NO_2 + H_2O$$
$$\text{orthophosphoric}$$
$$\text{acid}$$

$$As + 5HNO_3 \rightarrow H_3AsO_4 + 5NO_2 + H_2O$$
$$\text{orthoarsenic}$$
$$\text{acid}$$

$$S + 6HNO_3 \rightarrow H_2SO_4 + 6NO_2 + 2H_2O$$
$$\text{sulphuric acid}$$

$$I_2 + 10HNO_3 \rightarrow 2HIO_3 + 10NO_2 + 4H_2O$$
$$\text{iodic acid}$$

A mixture of nitric acid vapour and hydrogen, passed over heated platinized asbestos, yields ammonia:

$$HNO_3 + 4H_2 \rightarrow 3H_2O + NH_3$$

Nitric acid oxidizes many compounds. Warm turpentine inflames, forming torrents of brown fumes of nitrogen dioxide. Tin(II) chloride reduces nitric acid to hydroxylamine:

$$7H^+ + NO_3^- + 3Sn^{2+} \rightarrow$$
$$3Sn^{4+} + 2H_2O + NH_2OH$$

Iron(II) ions are oxidized to iron(III), the colour changing from pale green to deep brown, due to the complex ions $Fe^{2+}.NO$, which decomposes on warming to a yellow-brown solution. Brown fumes of nitrogen dioxide appear:

$$4H^+ + NO_3^- + 3Fe^{2+} \rightarrow 3Fe^{3+} + 2H_2O + NO$$

$$2NO + O_2 \rightarrow 2NO_2$$

Sulphites and sulphurous acid are oxidized to sulphates and sulphuric acid respectively:

$$SO_3^{2-} + 2H^+ + 2NO_3^- \rightarrow$$
$$SO_4^{2-} + H_2O + 2NO_2$$

Aqua regia is a mixture of concentrated hydrochloric acid (3 volumes) and concentrated nitric acid (1 volume) used to dissolve noble metals, which are not attacked by either acid acting alone. The action is due to chlorine liberated:

$$3HCl + HNO_3 \rightarrow Cl_2 + NOCl + 2H_2O$$

Complex ions, such as $AuCl_4^-$ and $PtCl_6^{2-}$, are formed:

$$Au + HNO_3 + 4HCl \rightarrow$$
$$HAuCl_4 + NO + 2H_2O$$

$$3Pt + 4HNO_3 + 18HCl \rightarrow$$
$$3H_2PtCl_6 + 4NO + 8H_2O$$

All nitrates are soluble in water and all are decomposed on heating. The nitrates of Group I decompose to the nitrite on strong heating:

$$2KNO_3 \rightarrow 2KNO_2 + O_2$$

Other nitrates break down further, to the oxide, nitrogen dioxide and oxygen:

$$2Cu(NO_3)_2 \rightarrow 2CuO + 4NO_2 + O_2$$

Further decomposition occurs where the oxide dissociates:

$$2Hg(NO_3)_2 \rightarrow 2HgO + 4NO_2 + O_2$$

$$2HgO \rightarrow 2Hg(s) + O_2(g)$$

The nitrates of metals exerting higher valencies and of some weaker metals tend to be hydrolysed in the absence of an excess of nitric acid, precipitating basic nitrates.

The detection of the elements in pure nitric acid is outlined below, but the experimental work requires skill and safety precautions. By the action of water-free nitric acid on dry powdered sodium chloride, hydrogen chloride gas is displaced and when passed over heated sodium reacts to form sodium chloride and hydrogen:

$$HNO_3 + NaCl \rightarrow HCl + NaNO_3$$

$$2HCl + 2Na \rightarrow 2NaCl + H_2$$

(*Note*. The action of very, very dilute nitric acid on magnesium yields hydrogen, but both acid and water contain combined hydrogen.) The thermal decomposition of nitric acid in a silica tube yields oxygen, nitrogen dioxide and steam: oxygen may be collected over water, into which excess acid and the other products disperse. Nitric acid vapour is reduced to ammonia by hydrogen in the presence of a heated platinum catalyst. Passage of ammonia over heated copper(II) oxide yields nitrogen.

In qualitative analysis, nitrates must be distinguished from nitrites (see p. 371). Dilute sulphuric acid has no action on nitrates but warm concentrated sulphuric acid produces nitric acid vapour. Addition of copper turnings to the reaction mixture gives brown fumes of nitrogen dioxide and a blue-green solution containing copper(II) ions. The thermal decomposition of nitrates is characteristic. The Brown Ring test may be used to detect nitrates in solution, in the absence of bromides, iodides and chromates which also give brown colorations. The solution is acidified with dilute sulphuric acid and mixed with a freshly prepared solution of iron(II) sulphate. Concentrated sulphuric acid is poured cautiously down the inside of the tube to form a lower dense layer. Nitrates form a brown layer at the interface containing the complex ion, $Fe^{2+}.NO$ (p. 368).

The presence of nitrate, or other nitrogenous compounds, may be shown by reduction with aluminium or Devarda's Alloy (Cu 50, Al 45, Zn 5), and hot alkali, ammonia being evolved:

$$3NO_3^- + 5OH^- + 2H_2O + 8Al \rightarrow \\ 8AlO_2^- + 3NH_3$$

If ammonium salts are present, ammonia from this source should be expelled initially by the action of hot alkali alone. The liquid should be cooled before addition of Devarda's Alloy. In the titrimetric (volumetric) determination of nitrates, ammonia may be generated in this way from a known mass of sample, absorbed in a known volume of standard acid, the excess of which is determined by titration against a standard base.

Important oxides and oxoacids of phosphorus

Structures of the oxides are given in Fig, 21.4 (p. 353).

Hypophosphorous acid, H_3PO_2, HPH_2O_2, and hypophosphites

Hypophosphites are formed by the action of hot alkali on white phosphorus, phosphine being evolved. All hypophosphites are soluble. Using aqueous barium hydroxide, a solution of barium hypophosphite is formed. Carbon dioxide is passed in to remove excess alkali as a precipitate of barium carbonate. The solution is evaporated to crystallization. Hypophosphorous acid is isolated by addition of a calculated volume of sulphuric acid to a solution containing a known mass of barium hypophosphite, when barium sulphate is precipitated:

$$4P + 3OH^- + 3H_2O \rightarrow PH_3(g) + 3H_2PO_2 - \\ \text{isolated as} \\ Ba(H_2PO_2)_2 \\ \text{barium hypophosphite}$$

$$Ba^{2+} + 2H_2PO_2^- + 2H^+ + SO_4^{2-} \rightarrow \\ BaSO_4(s) + 2H_3PO_2$$

Evaporation, with care to avoid decomposition, yields a colourless crystalline solid, m.p. 26.5°C, freely soluble in water. It is a monobasic acid. The sodium and calcium salts are used in pharmaceutical preparations.

Hypophosphorous acid and its salts are decomposed by heat; phosphine being evolved during disproportionation:

$$4NaH_2PO_2 \rightarrow Na_4P_2O_7 + 2PH_3 + H_2O \\ \text{sodium} \\ \text{diphosphate} \\ \text{(pyrophosphate)}$$

The acid and its salts are chiefly important as very powerful reducing agents, being oxidized to orthophosphoric acid and phosphates. Potassium permanganate is reduced immediately in the cold, concentrated sulphuric acid reduced to sulphur and many heavy metals are precipitated from solutions of their salts. Mercury(II) chloride is reduced to a white precipitate of dimercury(I) dichloride in the cold, being further reduced to grey mercury on warming:

$$H_2PO_2^- + 4HgCl_2 + 2H_2O \rightarrow \\ 2Hg_2Cl_2(s) + 4Cl^- + H_3PO_4 + 3H^+$$

$$H_2PO_2^- + 2Hg_2Cl_2(s) + 2H_2O \rightarrow \\ 4Hg(s) + H_3PO_4 + 4Cl^- + 3H^+$$

Warm copper(II) sulphate solution yields a brown precipitate, possibly copper(I) hydride, on reduction with a hypophosphite,

$$3H_2PO_2^- + 4Cu^{2+} + 6H_2O \rightarrow \\ 4CuH(s) + 3H_3PO_4 + 5H^+$$

Diphosphorus trioxide, phosphorus trioxide (phosphorous oxide), P_4O_6, and the phosphorous acids

Phosphorus trioxide is prepared by burning white phosphorus in a slow current (limited supply) of air or in oxygen under reduced pressure. It is contaminated with phosphorus pentoxide. On passing the mixture of vapours through a brass tube maintained at 50–60°C, the less volatile pentoxide is conveniently trapped in a glass wool plug. Phosphorus trioxide passes on to be condensed in a U-tube cooled in a freezing-mixture. The white waxy solid (m.p. 23.8°C; b.p. 173°C) is melted in the U-tube using warm water and run off. The relative molecular mass in the vapour state and when dissolved in organic solvents, such as benzene and carbon disulphide, indicates a molecular formula P_4O_6, confirmed by electron diffraction measurements.

When gently heated, phosphorus trioxide burns in air to the pentoxide, the oxide of the principal valency state,

$$P_4O_6 + 2O_2 \rightarrow P_4O_{10}$$

On heating phosphorus trioxide vapour in the absence of air above 210°C, phosphorus and phosphorus tetroxide, probably a polymerized form of P_4O_8 or $(PO_2)_n$, are formed. This reacts with water as a mixed anhydride of phosphorous and phosphoric acids. Phosphorus trioxide acts as phosphorous anhydride, reacting with cold water to form phosphorous acid:

$$P_4O_6 + 6H_2O \rightarrow 4H_3PO_3$$

Use of hot water causes decomposition of the acid.

Orthophosphorous acid is best prepared by the hydrolysis of phosphorus trichloride. Air is drawn through phosphorus trichloride at 60°C and the vapour passed into water at 0°C when a crystalline mass is formed:

$$PCl_3 + 3H_2O \rightarrow H_3PO_3 + 3HCl$$

Being very soluble in water, orthophosphorous acid is dried in a vacuum desiccator (p. 94). It is a colourless crystalline solid, melting-point 74°C. Pyro- (diphosphorous acid) and meta-forms of the acid ($H_4P_2O_5$ and HPO_2 respectively) have been prepared but are not important, so that the prefix ortho- is not usually added. Phosphorous acid is dibasic in inorganic compounds and is believed to contain two structurally different forms in equilibrium:

$$
\begin{array}{ccc}
\text{OH} & & \text{O} \\
| & & \| \\
\text{HO—P—OH} & \rightleftharpoons & \text{HO—P—OH} \\
& & | \\
& & \text{H}
\end{array}
$$

Salts containing the $H_2PO_3^-$ and HPO_3^{2-} ions, but not PO_3^{3-}, are known. When heated, phosphorous acid disproportionates to phosphine and orthophosphoric acid:

$$4H_3PO_3 \rightarrow 3H_3PO_4 + PH_3$$

Phosphorous acid and phosphites are powerful reducing agents being oxidized to orthophosphoric acid and phosphates but are not as strong as hypophosphorous acid in this respect. Acidified potassium permanganate is decolorized on warming and warm concentrated sulphuric acid, reduced to sulphur dioxide. Warm mercury(II) chloride solution is reduced to a white precipitate of dimercury(I) dichloride and then to grey mercury while warm silver nitrate solution is reduced to silver:

$$HPO_3^{2-} + 2HgCl_2 + H_2O \rightarrow \\ Hg_2Cl_2(s) + H_3PO_4 + 2Cl^-$$

$$HPO_3^{2-} + Hg_2Cl_2(s) + H_2O \rightarrow \\ 2Hg(s) + H_3PO_4 + 2Cl^-$$

$$HPO_3^{2-} + 2Ag^+ + H_2O \rightarrow 2Ag(s) + H_3PO_4$$

Diphosphorus pentoxide, phosphorus pentoxide (phosphoric oxide), P_4O_{10}, and the phosphoric acids

Phosphorus pentoxide is formed as a white solid when phosphorus burns in an excess of air or oxygen:

$$4P + 5O_2 \rightarrow P_4O_{10}$$

It is polymorphic, existing in three forms. Phosphorus pentoxide is the most stable oxide of phosphorus. Electron diffraction experiments and vapour density determinations indicate that the formula is P_4O_{10} in the vapour. It has a strong affinity for water, being the most efficient drying agent known. It absorbs moisture from the air to form the meta-phosphoric acid,

$$P_4O_{10} + 2H_2O \rightarrow 4HPO_3$$

and then the pyro- and ortho-acids.

Phosphorus pentoxide reacts violently with ice-cold water to form pure metaphosphoric acid, which on boiling gives orthophosphoric acid:

$$HPO_3 + H_2O \rightarrow H_3PO_4$$

Phosphorus pentoxide may be used to remove the elements of water from compounds. On warming, ethanamide (acetamide) is dehydrated to ethanonitrile, cyanomethane (acetonitrile):

$$CH_3.C\overset{\displaystyle O}{\underset{\displaystyle NH_2}{\diagup}} \ - H_2O \rightarrow \ CH_3.C\equiv N$$

ethanamide	ethanonitrile
acetamide	cyanomethane
	acetonitrile

Nitric acid is dehydrated to nitrogen pentoxide, and sulphuric acid to sulphur trioxide:

$$2HNO_3 - H_2O \rightarrow N_2O_5$$

$$H_2SO_4 - H_2O \rightarrow SO_3$$

In the absence of the true ortho-acid, which would be $P(OH)_5$, the first dehydration product H_3PO_4 is called orthophosphoric acid. It is tribasic. On heating to above 200°C, pyrophosphoric acid, diphosphoric acid, $H_4P_2O_7$, is formed; this gives only two series of salts, containing the $H_2P_2O_7{}^{2-}$ and $P_2O_7{}^{4-}$ ions. At red heat more water is lost and metaphosphoric acid, HPO_3, remains until volatilized:

orthophosphoric acid

$$\underset{O}{\overset{OH}{HO-P-OH}} + \underset{O}{\overset{OH}{HO-P-OH}}$$

$$\downarrow \text{200–300°C}$$

pyrophosphoric acid
diphosphoric acid

$$\underset{O}{\overset{OH}{HO-P}}-O-\underset{O}{\overset{OH}{P-OH}} + H_2O$$

$$\downarrow \text{above 300°C}$$

metaphosphoric acid
(polymerized)

$$\left[\underset{O}{\overset{O}{HO-P}} + \underset{O}{\overset{O}{P-OH}} \right]_n + 2H_2O$$

The changes are reversed by water. Orthophosphoric and pyrophosphoric acids are colourless crystalline solids while metaphosphoric acid is a glassy, translucent substance. The use of Greek prefixes has been explained on p. 353.

Orthophosphoric acid is usually prepared by the oxidation of red phosphorus by nitric acid in the presence of a little iodine as catalyst:

$$P + 5HNO_3 \rightarrow H_3PO_4 + 5NO_2 + H_2O$$

The acid is diluted to density $1.2\,g\,cm^{-3}$ to avoid undue violence and the mixture boiled under reflux until all phosphorus has dissolved. After repeated evaporation with nitric acid to oxidize lower acids of phosphorus, the final evaporation to 150°C is done in a platinum basin, when a thick syrup, which would attack porcelain, is obtained. The liquid is cooled in a freezing-mixture, inoculated if necessary, and allowed to crystallize. The crystals are dried over phosphorus pentachloride in a small desiccator. Commercially, phosphoric acid is made by the action of sulphuric acid on calcium phosphate, the other product, calcium sulphate, being insoluble. Orthophosphoric acid may be isolated as a colourless crystalline solid, m.p. 42°C, but is usually used in the laboratory as a syrupy liquid (density $1.75\,g\,cm^{-3}$). Phosphoric acid has no powers of oxidation and having a very low volatility may be used to displace acids from their salts, notably hydrogen iodide from potassium iodide:

$$KI + H_3PO_4 \rightarrow HI + KH_2PO_4$$

Only the orthophosphates formed by ammonia and the alkali metals except lithium are soluble in water. However, insoluble orthophosphates dissolve in acids which are highly ionized, soluble hydrogenphosphates being formed. Important phosphates have been described under the metals: sodium (p. 278), magnesium and calcium (p. 296). Phosphoric acid may be estimated volumetrically (titrimetric analysis) by titration with standard alkali using the appropriate indicator (p. 279).

Pyrophosphoric acid, diphosphoric acid, is a colourless crystalline solid (m.p. 61°C), forming stable salts in two series, of which $Na_4P_2O_7$ and $Na_2H_2P_2O_7$ are examples. Sodium pyrophosphate (tetrasodium diphosphate) is formed by heating disodium hydrogenphosphate to expel water:

$$2Na_2HPO_4 \rightarrow Na_4P_2O_7 + H_2O$$

Heating sodium dihydrogenphosphate up to 200°C, avoiding further decomposition to the metaphosphate, yields disodium pyrophosphate (disodium diphosphate):

$$2NaH_2PO_4 \rightarrow Na_2H_2P_2O_7 + H_2O$$

Metaphosphoric acid is the final product of heating orthophosphoric acid and the first on adding phosphorus pentoxide to ice-cold water. Sodium metaphosphate, $(NaPO_3)_n$, is a collective name for many metaphosphates formed by sodium in which n assumes different values, most of the compounds being polymers and some, high polymers. Calgon, *sodium hexametaphosphate*, is used in laundry work and the textile industry: it softens water by removal of calcium and magnesium ions by absorption into a complex anion aggregate. The metal ions are said to be *sequestered*.

Nitric acid and ammonium molybdate form a bright yellow precipitate of ammonium phosphomolybdate with solutions of orthophosphates on warming to 60°C. Ammonium phosphomolybdate has the complex formula: $(NH_4)_3[PO_4(MoO_3)_{12}]$. A pale buff precipitate which may be seen is probably hydrated molybdic acid, $H_2MoO_4 . H_2O$. Arsenates give a similar yellow precipitate to phosphates, ammonium arsenomolybdate being formed but only on boiling. However, the two salts may be distinguished by their precipitation reactions with neutral silver nitrate solution: silver phosphate is yellow and silver arsenate, brown. In the gravimetric estimation of phosphates, precipitation of ammonium phosphomolybdate is used as a preliminary step to avoid interference in the next stage from other ions present. The precipitate is dissolved in ammonia, and on addition of solutions of ammonium chloride and magnesium chloride, magnesium ammonium phosphate is precipitated. The precipitate is filtered, washed, dried and ignited to constant mass at 1100°C to magnesium pyrophosphate and weighed:

$$2MgNH_4PO_4 . 6H_2O \rightarrow$$
$$Mg_2P_2O_7 + 2NH_3 + 13H_2O$$

Diarsenic trioxide, arsenic trioxide (arsenious oxide), As_4O_6, and arsenites

Arsenic trioxide, known as white arsenic or merely as arsenic, is a colourless crystalline solid formed when arsenic or a sulphide of arsenic burns in air. X-ray analysis of the solid state, electron diffraction studies of the vapour and vapour density: (relative to H = 1) determinations show that the molecule is As_4O_6, having the same structure as P_4O_6. However, unlike the oxide of phosphorus, arsenic trioxide is the principal oxide of arsenic. Arsenic trioxide sublimes on heating. Up to 800°C the molecules are As_4O_6 but by 1800°C dissociation to As_2O_3 is complete. Sublimation gives a very pure product, which is used as a primary standard in titrimetric analysis.

Arsenic trioxide is very poisonous and the lethal dose, which varies, is about 0.1 g. It is slightly soluble in water (solubility = 1.8 g $100 g^{-1}$ at 20°C) becoming hydrated to a very weak acid, either $HAsO_2$ or H_3AsO_3, but only the oxide can be isolated from solution:

$$As_4O_6 + 2H_2O \rightleftharpoons 4HAsO_2$$
meta-arsenious
acid

$$As_4O_6 + 6H_2O \rightleftharpoons 4H_3AsO_3$$
ortho-arsenious
acid

With alkalis, arsenites are formed, some with condensed complex formulae (i.e. by elimination of water between anions to form $-O-$ bridges), but most are derivatives of the meta-acid: e.g. $NaAsO_2$, sodium arsenite, and Scheele's Green, $Cu(AsO_2)_2$, copper(II) arsenite. Paris Green, a mixed arsenite and ethanoate (acetate) of copper,

$$(CH_3COO)_2Cu . 3Cu(AsO_2)_2,$$

has been used in horticultural poisons. Arsenic trioxide dissolves in concentrated hydrochloric acid to form the chloride:

$$As_4O_6 + 12HCl \rightleftharpoons 4AsCl_3 + 6H_2O$$

No salts are formed with sulphuric acid and nitric acid. As^{3+} ions have not been detected so the formation of arsenic trichloride by the trioxide is not a basic property. Arsenic trioxide shows acidic properties, as already indicated, but it is not amphoteric although it reacts with hydrochloric acid.

Numerous organic derivatives have been made.

Arsenic trioxide may be reduced to arsenic by heating with charcoal or potassium cyanide, yielding a black mirror of arsenic on the cool part of the tube:

$$As_4O_6 + 6C \rightarrow 4As(g) + 6CO(g)$$

$$As_4O_6 + 6KCN \rightarrow 4As(g) + 6KOCN$$

Arsenites would be mixed with anhydrous sodium carbonate for this method of reduction. In the following reductions, arsenites may be used but on acidification form arsenic trioxide. The oxide is therefore shown in the equations. The oxide is reduced to arsenic by tin(II) chloride in concentrated hydrochloric acid and by boiling with copper in concentrated hydrochloric acid:

$$As_4O_6 + 12HCl + 6SnCl_2 \rightarrow$$
$$4As(s) + 6SnCl_4 + 6H_2O$$

$$As_4O_6 + 12HCl + 12Cu \rightarrow$$
$$4As + 12CuCl + 6H_2O$$

It reduces Fehling's solution to copper(I) oxide. In common with other solutions containing arsenic, arsenites are reduced to arsine by hydrogen generated *in situ*.

Arsenic trioxide is oxidized to arsenic pentoxide by hypochlorites (chlorates(I)), nitric acid, permanganates and dichromates. When reacting with iodine under the mildly alkaline conditions afforded by sodium hydrogencarbonate solution, arsenic trioxide (as sodium arsenite) is oxidized to arsenic pentoxide (as sodium arsenate). The uses of arsenic trioxide and arsenites in titrimetric (volumetric) analysis are described more fully under iodine (p. 432) and hypochlorites (p. 437):

$$AsO_2^- + 2H_2O - 2e^- \rightleftharpoons AsO_4^{3-} + 4H^+$$
arsenite   arsenite

$$AsO_2^- + 2H_2O + I_2 \rightarrow AsO_4^{3-} + 4H^+ + 2I^-$$
removed by
HCO_3^-

$$AsO_2^- + H_2O + ClO^- \rightarrow Cl^- + AsO_4^{3-} + 2H^+$$
hypochlorite
chlorate(I)

In qualitative analysis, arsenites are detected by the pale yellow precipitate of silver arsenite given by neutral silver nitrate solution, and the presence of arsenic itself confirmed by any of the usual tests.

Diarsenic pentoxide, arsenic pentoxide (arsenic oxide), As_2O_5, and arsenates

Arsenic pentoxide is a white amorphous deliquescent solid of unknown relative molecular mass, and expressed As_2O_5. It cannot be formed by direct synthesis from the elements. Arsenic trioxide is oxidized by concentrated nitric acid to a solution of orthoarsenic acid:

$$As_4O_6 + 4HNO_3 + 4H_2O \rightarrow 4H_3AsO_4 + 2'N_2O_3'$$

On evaporation and cooling in ice, crystals of hydrated acid ($H_3AsO_4.H_2O$) separate from the syrupy liquid. Water is lost on heating to above 200°C, and arsenic pentoxide remains.

On strong heating, arsenic pentoxide decomposes to the trioxide and oxygen:

$$2As_2O_5 \rightarrow As_4O_6 + 2O_2$$

Arsenic pentoxide is an acidic oxide, with no trace of basic behaviour. It is acidic to a greater degree than the trioxide. It dissolves in acid to give the moderately strong (on the first ionization) ortho-arsenic acid:

$$H_3AsO_4 \rightleftharpoons H^+ + H_2AsO_4^-$$

$$K = 5 \times 10^{-3} \text{ mol dm}^{-3} \text{ at } 20°C$$

In addition to the ortho-acid, pyroarsenic acid, (diarsenic acid) $H_4As_2O_7$, and meta-arsenic acid, $HAsO_3$, have been claimed. These are parallel to the acids of phosphorus. Salts containing the ions, AsO_4^{3-}, $HAsO_4^{2-}$ and $H_2AsO_4^-$, are well known and much more important than the acids. Arsenates resemble the phosphates in many ways, often being isomorphous with them. Disodium hydrogenarsenate, $Na_2HAsO_4.12H_2O$, is used in calico printing. Lead(II) arsenate has been used for spraying fruit trees but the use of this and other horticultural arsenical products is declining because of their toxicity. The molybdate test for phosphates and arsenates has been given under phosphates already. Arsenates may be distinguished from arsenites by the chocolate-brown precipitate of silver arsenate given with neutral silver nitrate solution. Arsenates may be estimated volumetrically (titrimetrically) as arsenites after reduction by sulphur dioxide and titration with iodine or potassium bromate. Alternatively, as with phosphates, addition of ammonia, ammonium chloride and magnesium chloride gives a crystalline precipitate of magnesium ammonium arsenate which is ignited to magnesium pyroarsenate (diarsenate) and weighed:

$$2MgNH_4AsO_4.6H_2O \rightarrow$$
$$Mg_2As_2O_7 + 2NH_3 + 13H_2O$$

Arsenates may be reduced to arsine by hydrogen formed *in situ*.

Diantimony trioxide, antimony trioxide, antimony(III) oxide, Sb_4O_6, and antimonates(III)

Antimony trioxide is an amphoteric oxide, forming salts with alkalis and with acids. This illustrates the emergence of metallic properties with antimony. The Stock Notation, not used for compounds between non-metals, may therefore be used. Antimony trioxide is a white solid, which sublimes on heating and exists as two solid forms, below 570°C in a structure similar to P_4O_6 and As_4O_6, and above, as a macromolecular structure.

Antimony trioxide is formed when antimony burns in air or oxygen, and by the complete hydrolysis of antimony trichloride with excess boiling water:

$$4Sb + 3O_2 \rightarrow Sb_4O_6$$

$$SbCl_3 + H_2O \rightarrow SbOCl(s) + 2HCl$$
antimony
oxide chloride

$$4SbOCl(s) + 2H_2O \rightarrow Sb_4O_6(s) + 4HCl$$

Antimony trioxide is insoluble in water and dilute acids but reacts with concentrated hydrochloric acid to form the chloride and with concentrated sulphuric acid to form the sulphate:

$$Sb_4O_6 + 12HCl \rightarrow 4SbCl_3 + 6H_2O$$

$$Sb_4O_6 + 6H_2SO_4 \rightarrow 2Sb_2(SO_4)_3 + 6H_2O$$

With alkali, the antimonate(III) is formed:

$$Sb_4O_6 + 4OH^- \rightarrow 4SbO_2^- + 2H_2O$$

Sodium meta-antimonate(III) precipitates heavy metal antimonates(III) from solutions and reduces silver nitrate to metallic silver. Free antimonic(III) acid (antimonous acid) has not been isolated but salts are known of various *condensed* forms.

Antimony dioxide, antimony tetroxide, Sb_4O_8 or $(SbO_2)_n$

This is obtained on heating antimony, its sulphide or oxides to above 300°C in air. It decomposes above 900°C to the trioxide. The structure is unknown but is possibly $Sb(SbO_4)$.

Diantimony pentoxide, antimony pentoxide, antimony(V) oxide, Sb_2O_5, and antimonates(V)

Antimony pentoxide is formed by the action of concentrated nitric acid on antimony, or by oxidation of the trioxide with nitric acid. The hydrated oxide formed is carefully dehydrated. It may also be formed by hydrolysis of the pentachloride. Antimony pentoxide is a yellow solid. It is an acidic oxide but only slightly soluble in water. However, the solution is acidic. On fusion with alkali the antimonate(V) is formed. X-ray analysis has shown that sodium 'pyroantimonate', formerly written $Na_2H_2Sb_2O_7.5H_2O$, is $Na[Sb(OH)_6]$, sodium hexahydroxoantimonate(V), a complex form of the meta-antimonate(V), $NaSbO_3$. Potassium pyroantimonate,

$$K[Sb(OH)_6].\tfrac{1}{2}H_2O$$

is used in solution to precipitate the less soluble sodium salt in qualitative analysis.

Bismuth trioxide, bismuth(III) oxide, Bi_2O_3

Bismuth trioxide is a yellow basic oxide, with no acidic character, prepared by the standard methods for metal oxides. Its relative molecular mass is unknown. A crystalline hydroxide $Bi(OH)_3$ may be precipitated and a wide range of salts is known. These are readily hydrolysed to oxide salts. $(BiO)_2CO_3$, dibismuth dioxide carbonate, formerly bismuthyl carbonate, is the only carbonate formed in this group: a measure of the basic nature of the hydroxide.

Sodium bismuthate, approximately 'NaBiO₃'

Bismuth trioxide may be oxidized but no pure higher oxides have been isolated. However, a series of impure bismuthates has been prepared. When bismuth trioxide is fused with caustic soda and sodium peroxide, a substance of variable unknown composition, formulated $NaBiO_3$ and named sodium bismuthate, is formed. It is a light brown powder, used in the laboratory as an oxidizing agent: manganese(II) ions are oxidized to permanganate by sodium bismuthate in the presence of concentrated nitric acid:

$$2Mn^{2+} + 5NaBiO_3 + 14H^+ \rightarrow$$
$$2MnO_4^- + 5Bi^{3+} + 7H_2O + 5Na^+$$

Sulphides

The sulphides of arsenic, antimony and bismuth have characteristic colours and are well known in the laboratory. (Tetra)phosphorus trisulphide, P_4S_3, is present with potassium chlorate in some brands of 'strike-anywhere' matches with additives to ensure smooth striking and burning. The several nitrogen sulphides are unimportant but nitrogen sulphide N_4S_4 (a cyclic structure) is mentioned to show the unusual behaviour of nitrogen. The trisulphides of arsenic and antimony are acidic, forming salts with yellow ammonium sulphide and alkali, while that of bismuth is typical of a metal.

Nitrogen sulphide, N_4S_4

Methods of preparation are based on the reversible reaction between sulphur and anhydrous ammonia:

$$10S + 4NH_3 \rightleftharpoons 6H_2S + N_4S_4$$

with removal of hydrogen sulphide to obtain complete conversion to N_4S_4. Nitrogen sulphide forms golden-yellow crystals, soluble in organic solvents. It has a covalent cyclic structure.

Sulphides of phosphorus, P_4S_3, P_4S_7, P_4S_{10}, etc.

The sulphides P_4S_3 (trisulphide), P_4S_7 (heptasulphide), and P_4S_{10} (pentasulphide) are yellow crystalline solids formed by heating red phosphorus with sulphur in an inert atmosphere. They are hydrolysed by water to hydrogen sulphide and phosphorous and phosphoric acids.

Sulphides of arsenic, As_4S_4, As_4S_6, As_2S_5, thioarsenites and thioarsenates

Arsenic sulphide, As_4S_4, similar in structure to N_4S_4, is the red-orange mineral realgar, and may be formed by fusing arsenic trioxide with sulphur in the correct proportions out of contact with air or by heating the elements together:

$$As_4O_6 + 7S \rightarrow As_4S_4 + 3SO_2$$

$$4As + 4S \rightarrow As_4S_4$$

It inflames readily with oxidizing agents and is used in *Bengal Fire* flares in pyrotechny, giving arsenic trioxide and sulphur dioxide. With alkalis and sulphide solutions it forms thioarsenites with separation of arsenic.

Arsenic trisulphide, As_4S_6, has been shown to have a similar structure to the trioxide, As_4O_6. It is found as the mineral orpiment (corruption of Latin, *auri pigmentum* = gold paint) which is golden-yellow in colour. Arsenic trisulphide may be formed by heating the trioxide with sulphur in the correct proportions out of contact with air or by passing hydrogen sulphide into a solution of arsenic trichloride or arsenic trioxide in hydrochloric acid, or an arsenite:

$$As_4O_6 + 9S \rightarrow As_4S_6 + 3SO_2$$

$$4AsCl_3 + 6H_2S \rightleftharpoons As_4S_6 + 12HCl$$

Although arsenic trisulphide will not react with concentrated hydrochloric acid while hydrogen sulphide is present, the reaction is reversed by prolonged boiling when both H_2S and $AsCl_3$ are volatilized away. Arsenic trisulphide burns to the trioxide and sulphur dioxide on heating in air. It is an acidic sulphide. Arsenic trisulphide is insoluble in water but dissolves readily in alkalis to form the arsenite ($LiAsO_2$) and thioarsenite ($LiAsS_2$) and mixed compounds ($LiAsOS$):

$$As_4S_6 + 4OH^- \rightarrow AsO_2^- + 3AsS_2^- + 2H_2O$$

Colourless ammonium sulphide yields a solution of ammonium thioarsenite, a similar reaction occurring with alkali metal sulphides:

$$As_4S_6 + 2S^{2-} \rightarrow 4AsS_2^-$$

Acidification precipitates arsenic trisulphide:

$$AsO_2^- + 3AsS_2^- + 4H^+ \rightarrow As_4S_6(s) + 2H_2O$$

$$4AsS_2^- + 4H^+ \rightarrow As_4S_6(s) + 2H_2S$$

Yellow ammonium (poly)sulphide causes oxidation to ammonium thioarsenate, which on acidification deposits the pentasulphide, also yellow:

$$As_4S_6 + (6S^{2-} + 4S) \rightarrow 4AsS_4^{3-}$$
$$\text{thioarsenate}$$

$$2AsS_4^{3-} + 6H^+ \rightarrow As_2S_5(s) + 3H_2S$$

Antimony trisulphide behaves similarly† but arsenic trisulphide is more acidic, dissolving in ammonium carbonate solution or alkali carbonate solutions:

$$As_4S_6 + 2CO_3^{2-} \rightarrow AsO_2^- + 3AsS_2^- + 2CO_2$$
$$\text{arsenite} \quad \text{thioarsenite}$$

Arsenic pentasulphide, As_2S_5, may be prepared by heating the trisulphide with sulphur in the calculated proportions, by heating the elements out of contact with air, or by acidification of ammonium thioarsenate. It is a yellow solid. Passage of hydrogen sulphide into a hot 5M hydrochloric acid containing quinquevalent arsenic appears to give reduction to the tervalent state, with precipitation of sulphur and arsenic trisulphide:

$$H_3AsO_4 + 2HCl \rightarrow HAsO_2 + 2H_2O + Cl_2$$

$$H_3AsO_4 + H_2S \rightarrow HAsO_2 + 2H_2O + S(s)$$

$$4HAsO_2 + 6H_2S \rightarrow As_4S_6(s) + 8H_2O$$

Arsenic pentasulphide is acidic, dissolving in aqueous alkali and carbonates to form mixed arsenates and thioarsenates and in solutions of sulphides to form thioarsenates; with ammonium sulphide:

$$As_2S_5 + 3S^{2-} \rightarrow 2AsS_4^{3-}$$
$$\text{(tetra)thioarsenate}$$

Sulphides of antimony: Sb_2S_3, (Sb_2S_4, Sb_2S_5)

Antimony trisulphide exists in two forms: orange precipitated by passage of hydrogen sulphide through a solution of antimony(III) chloride, and black, the naturally occurring stibnite, obtained by heating the elements in the correct proportion out of contact with air:

$$2SbCl_3 + 3H_2S \rightleftharpoons Sb_2S_3(s) + 6HCl$$

$$2Sb + 3S \rightarrow Sb_2S_3$$

† But thioantimonite ≡ thioantimonate(III)

In the first case, prolonged boiling in the solution used for its preparation converts the orange form to black. On heating, antimony trisulphide burns in air or oxygen to the oxide and sulphur dioxide, the combusion finding use in match heads and fireworks. It dissolves in concentrated hydrochloric acid to give the trichloride and hydrogen sulphide. Antimony trisulphide will not dissolve in ammonium carbonate solution but is similar to arsenic trisulphide in its reaction with ammonium sulphide and alkalis: parallel equations may be written, showing the formation of thioantimonate(III) and (V).

There is some doubt about the chemical individuality of Sb_2S_4 and Sb_2S_5. 'Antimony Pentasulphide' is made commercially by acting on boiling alkali with sulphur and antimony trisulphide: sodium antimonate(v) is precipitated and thioantimonate(v) in the filtrate is decomposed by acid:

$$2SbS_4^{3-} + 6H^+ \rightarrow Sb_2S_5(s) + 3H_2S$$

It is used to produce 'red rubber'. Antimony 'pentasulphide' is golden yellow in colour.

Bismuth trisulphide, bismuth(III) sulphide, Bi_2S_3

Found as the mineral bismuthinite, bismuth trisulphide may be synthesized by heating the elements or precipitated from solution by hydrogen sulphide. It is a dark brown solid with the usual properties of metal sulphides, insoluble in alkali and sulphide solutions.

Halides, halides of oxoradicals and oxide halides

These are listed in Table 21.6. Oxide halides were formerly called oxyhalides.

The anomalous halides of nitrogen

Nitrogen trichloride is formed as a very explosive yellow oil by the action of chlorine or sodium hypochlorite (chlorate(I)) solution on ammonium chloride:

$$NH_4Cl + 3Cl_2 \rightarrow NCl_3 + 4HCl$$

There is no nitrogen tribromide, NBr_3, and although NI_3 can be isolated, iodine and ammonia react to form the explosive nitrogen iodide, $NI_3 \cdot NH_3$, a black solid. NCl_3 and NI_3 are slowly hydrolysed by water and rapidly by alkali to ammonia and the oxoacid or oxosalt:

$$NCl_3 + 3H_2O \rightarrow NH_3 + 3HClO$$

$$NI_3 + 3H_2O \rightarrow NH_3 + 3HIO$$

This should be compared with the behaviour of the other halides. Nitrogen trifluoride, NF_3, is an inert gas formed by electrolysis of fused ammonium hydrogen (di)fluoride, $NH_4F \cdot HF$, at 125°C. It is not hydrolysed.

The halides of phosphorus

Phosphorus trifluoride is a stable gas formed by a reaction between phosphorus trichloride and arsenic trifluoride or other suitable fluorides. It is colourless, does not fume in air and is only slowly hydrolysed by water. The other trihalides are formed by union of the elements. They are readily hydrolysed to the hydrogen halide and phosphorous acid:

$$PBr_3 + 3HOH \rightarrow H_3PO_3 + 3HBr$$

Phosphorus pentafluoride is a fuming gas, the most stable pentafluoride formed here and hydrolysed to phosphoric acid. It forms a strong complex acid and salts: hexafluorophosphoric acid, $H^+(PF_6^-)$, and hexafluorophosphates, $Na^+(PF_6^-)$. Phosphorus pentafluoride has five covalent bonds. Phosphorus pentachloride has five covalent bonds in the vapour state, but X-ray analysis has shown that the solid consists of $(PCl_4^+)(PCl_6^-)$, the only example of the PCl_6^- ion. Hydrolysis (see below) gives phosphoric acid. Phosphorus pentabromide is less stable structurally, consisting of $(PBr_4^+)Br^-$ while no penta-iodide has been discovered.

Table 21.6 Halides and related compounds of the Group V elements

NF_3		PF_3, PF_5	AsF_3, AsF_5	SbF_3, SbF_5	BiF_3
NCl_3	(P_2Cl_4),	PCl_3, PCl_5	$AsCl_3$	$SbCl_3$, $SbCl_5$	$BiCl_3$
—		PBr_3, PBr_5	$AsBr_3$	$SbBr_3$	$BiBr_3$
NI_3	(P_2I_4),	PI_3	(As_2I_4), AsI_3	SbI_3	BiI_3
$NOCl$		$POCl$	$AsOCl$	$SbOCl$	$BiOCl$
—		$POCl_3$	—	—	—

Phosphorus trichloride, PCl₃

In the synthesis of this compound the apparatus used (Fig. 20.9, p. 338) is similar to that for the preparation of tin(IV) chloride. Dry white phosphorus is placed in the retort from which air has been displaced by carbon dioxide. On passing chlorine, phosphorus burns spontaneously. Phosphorus trichloride distils over.

The reaction may be controlled by the rate at which chlorine is passed and the position of the delivery tube relative to the phosphorus. Too vigorous a combusion causes phosphorus to vaporize. Phosphorus trichloride is redistilled and collected as a colourless fuming liquid, b.p. 76°C;

$$2P + 3Cl_2 \rightarrow 2PCl_3$$

Phosphorus trichloride is violently hydrolysed by water to phosphorous acid and reacts with compounds containing the −OH group generally:

$$PCl_3 + 3HOH \longrightarrow H_3PO_3 + 3HCl$$

$$PCl_3 + 3CH_3.C\overset{\displaystyle O}{\underset{\displaystyle OH}{<}} \longrightarrow H_3PO_3 + 3CH_3.C\overset{\displaystyle O}{\underset{\displaystyle Cl}{<}}$$

ethanoic acid   ethanoyl chloride
(acetic acid)   (acetyl chloride)

Phosphorus trichloride absorbs oxygen from air to form phosphoryl(v) chloride, $POCl_3$, and reacts with sulphur to give thiophosphoryl(v) chloride, $PSCl_3$. Phosphorus trichloride reacts with chlorine to form phosphorus pentachloride which is a yellow solid, dissociating on heating:

$$PCl_3 + Cl_2 \rightarrow PCl_5$$

Phosphorus pentachloride, PCl₅

When phosphorus trichloride is dripped into an atmosphere of chlorine, phosphorus pentachloride is formed. The flask may be cooled in ice water as in Fig. 21.7.

Phosphorus pentachloride sublimes and dissociates on heating into chlorine and phosphorus trichloride: at 160°C, 13.5% has dissociated at atmospheric pressure and 100% by 300°C. The degree of dissociation may be determined by vapour density methods (as for $N_2O_4 \rightleftharpoons 2NO_2$, p. 369). Phosphorus pentachloride is hydrolysed violently by water and reacts with compounds containing an −OH group generally, with evolution of hydrogen chloride gas:

$$PCl_5 + H_2O \rightarrow \underset{\substack{\text{phosphoryl(v)}\\\text{chloride}}}{POCl_3} + 2HCl$$

$$POCl_3 + 3H_2O \rightarrow H_3PO_4 + 3HCl$$

With acids containing −OH groups (i.e. all except HCl, etc., H₂S):

$$CH_3.C\overset{\displaystyle O}{\underset{\displaystyle OH}{<}} + PCl_5 \longrightarrow$$

ethanoic acid
(acetic acid)

$$CH_3.C\overset{\displaystyle O}{\underset{\displaystyle Cl}{<}} + POCl_3 + HCl$$

ethanoyl chloride
(acetyl chloride)

$$\underset{\displaystyle O}{\overset{\displaystyle O}{>}}S\overset{\displaystyle OH}{\underset{\displaystyle OH}{<}} + 2PCl_5 \longrightarrow \underset{\displaystyle O}{\overset{\displaystyle O}{>}}S\overset{\displaystyle Cl}{\underset{\displaystyle Cl}{<}} + 2POCl_3 + 2HCl$$

sulphuric acid   sulphuryl chloride

$$\overset{\displaystyle O}{\underset{\displaystyle ^-O}{>}}\overset{+}{N}{-}OH + PCl_5 \longrightarrow \overset{\displaystyle O}{\underset{\displaystyle ^-O}{>}}\overset{+}{N}{-}Cl + POCl_3 + HCl$$

nitric acid   nitryl chloride,
                               formerly nitroxyl chloride

$$\overset{\displaystyle C\overset{O}{\diagup}}{\underset{\displaystyle C\underset{O}{\diagdown}}{\mid}}\overset{\displaystyle OH}{\underset{\displaystyle OH}{<}} + PCl_5 \longrightarrow CO + CO_2 + POCl_3 + 2HCl$$

ethanedioic acid
(oxalic acid)

Phosphoryl(v) chloride, phosphorus oxychloride, POCl₃

Phosphoryl(v) chloride may be prepared by heating a mixture of phosphorus trichloride and finely powdered potassium chlorate (chlorate(v)):

$$3PCl_3 + KClO_3 \rightarrow 3POCl_3 + KCl$$

followed by distillation. It is a colourless, fuming liquid, b.p. 108°C. On hydrolysis, orthophosphoric acid and hydrochloric acid are formed:

$$POCl_3 + 3H_2O \rightarrow H_3PO_4 + 3HCl$$

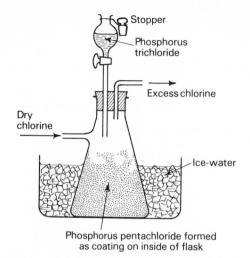

Phosphorus pentachloride formed
as coating on inside of flask

Fig. 21.7 The preparation of phosphorus pentachloride

Halides of arsenic

Arsenic forms all trihalides but in the quinquevalent state, only arsenic pentafluoride.

Arsenic Trichloride, $AsCl_3$

The recommended preparation involves refluxing arsenic trioxide with disulphur dichloride in a stream of chlorine:

$$2As_4O_6 + 3S_2Cl_2 + 9Cl_2 \rightarrow 8AsCl_3 + 6SO_2$$

Arsenic trichloride may be distilled over. Alternatively, arsenic is heated in a stream of chlorine. A solution of arsenic trichloride results from boiling arsenic trioxide with concentrated hydrochloric acid:

$$As_4O_6 + 12HCl \rightleftharpoons 4AsCl_3 + 6H_2O$$

Arsenic trichloride is a volatile liquid (b.p. 130°C) which fumes in air and is very poisonous. Because of its volatile nature, care must be exercised in using it. It has the same pyramidal structure as phosphorus trichloride but differs in being only partially (reversibly) hydrolysed:

$$4AsCl_3 + 6H_2O \rightleftharpoons As_4O_6 + 12HCl$$

Halides of antimony

Both pentafluoride and pentachloride are formed in addition to all trihalides. Antimony pentachloride catalyses the chlorination of ethyne (acetylene) to *Westron*, $C_2H_2Cl_4$.

Antimony trichloride, antimony(III) chloride, $SbCl_3$

Finely powdered antimony burns spontaneously in chlorine to form the trichloride. However, the compound is usually prepared by heating antimony trisulphide with concentrated hydrochloric acid:

$$Sb_2S_3 + 6HCl \rightarrow 2SbCl_3 + 3H_2S$$

The solution is evaporated to crystallization. Antimony trichloride is a soft white solid ('butter of antimony'), m.p. 73°C and deliquescent. It is hydrolysed reversibly by cold water to a white precipitate of antimony oxide chloride:

$$SbCl_3 + H_2O \rightleftharpoons SbOCl(s) + 2HCl$$

It is unlikely that simple Sb^{3+} ions exist in solution, considerable hydrolysis taking place. Well defined oxide chlorides, SbOCl and $Sb_4O_5Cl_2$, which re-dissolve in excess acid, have been identified. On boiling in water, antimony trioxide is formed but it will re-dissolve in concentrated hydrochloric acid to form antimony trichloride:

$$4SbOCl + 2H_2O \rightarrow Sb_4O_6 + 4HCl$$

the whole hydrolysis being

$$4SbCl_3 + 6H_2O \rightleftharpoons Sb_4O_6 + 12HCl$$

Solutions of antimony trichloride contain added hydrochloric acid to reduce hydrolysis. Antimony oxide chloride, formerly antimonyl or antimony oxychloride, is only slightly soluble in water but dissolves in benzene, carbon disulphide and chloroform, indicating that it has some covalent nature.

Halides of bismuth

Bismuth forms all trihalides but no pentahalides.

Bismuth trichloride, bismuth(III) chloride, $BiCl_3$

The normal methods for the preparation of salts of metals may be used. Union of the elements and the action of hydrochloric acid on the trioxide are methods similar to those described above:

$$2Bi + 3Cl_2 \rightarrow 2BiCl_3$$

$$Bi_2O_3 + 6HCl \rightarrow 2BiCl_3 + 3H_2O$$

Bismuth trichloride is a colourless crystalline solid, m.p. 232°C and deliquescent. It is hydrolysed reversibly by water with precipitation of white bismuth oxide chloride, re-dissolving in concentrated hydrochloric acid:

$$BiCl_3 + H_2O \rightleftharpoons BiOCl(s) + 2HCl$$

Hydrolysis does not go further even on boiling with a large excess of water. Solutions of bismuth trichloride contain added hydrochloric acid to reduce hydrolysis.

Salts of the oxoacids

Salts of the oxoacids are formed by elements with electropositive (metallic) character. Weak metallic character, bringing weakness of the base, results in easily hydrolysed salts. Carbonates, being salts of a very weak acid, are not usually formed by very weak metals. The valency in the middle groups of the Periodic Table drops as two electrons become inert, tervalency becoming more important than quinquevalency in Group V. Salts of the cations Sb^{3+} and Bi^{3+} are known but the former readily forms complexes, e.g. $Sb(C_2O_4)_2{}^-$, and oxide (and hydroxide) salts are the most stable salts of bismuth. In recent investigations, the cations $Bi(OH)^{2+}$ and BiO^+ have not been detected and the name bismuthyl has accordingly lapsed.

The metallic state is not evident with arsenic although there is a report of the formation of *arsenic (III) sulphate*, $As_2(SO_4)_3$, by the action of sulphur trioxide on arsenic trioxide at 100°C. Antimony forms Sb^{3+} with reluctance. The trioxide will react with concentrated sulphuric acid to form a deliquescent *sulphate*, $Sb_2(SO_4)_3$, water causing hydration and then hydrolysis to *antimony oxide sulphate*, $(SbO)_2SO_4$. Concentrated nitric acid yields the *nitrate*, $Sb(NO_3)_3$, which is completely hydrolysed to the hydrated pentoxide on addition of water. The *nitrate* and *sulphate*, $Bi(NO_3)_3 \cdot 5H_2O$ and $Bi_2(SO_4)_3$, formed by the action of the appropriate acid on the trioxide, are readily hydrolysed to $BiONO_3$ and $(BiO)_2SO_4$ respectively.

The most important oxosalts in this family are the bismuth oxide salts. *Bismuth oxide nitrate*, $BiONO_3$, is the subnitrate of bismuth used in pharmacy. (*Di*)*bismuth* (*di*)*oxide carbonate*, $(BiO)_2CO_3$, the only carbonate formed by the elements of this group, may be prepared by mixing solutions of bismuth nitrate and ammonium carbonate, the white precipitate being dried at 100°C.

Qualitative and quantitative analysis for arsenic, antimony and bismuth in their compounds

Tests for nitrites, nitrates and ammonia, and for phosphates have been included in the appropriate sections. The other elements are precipitated in Group II of the Qualitative Analysis Tables, although arsenic may ultimately be found to be present as arsenite or arsenate.

The separation of arsenic, antimony and bismuth in qualitative analysis

The sulphides are precipitated from hot acidic solution by passing hydrogen sulphide. Quinquevalent arsenic requires at least 5M hydrochloric acid at about 80°C for reduction to the tervalent state but the final acidity should be adjusted to about 0.25M (use Congo Red paper or methyl violet indicator) to ensure complete precipitation of all sulphides. Bismuth sulphide is brown, arsenic trisulphide yellow, and antimony sulphide orange. Initial separation is achieved either by warming with yellow ammonium sulphide or boiling with lithium hydroxide reagent when the sulphides of arsenic and antimony (Analysis Group IIb) dissolve and bismuth (Analysis Group IIa) remains.

Bismuth sulphide is dissolved in nitric acid and precipitated as the white hydroxide by ammonia. Freshly prepared sodium stannate(II) solution reduces this to black bismuth. The sulphides of arsenic and antimony are re-precipitated by adding dilute hydrochloric acid dropwise to the Analysis Group IIb extract. Separation is achieved by boiling with concentrated hydrochloric acid, when the sulphide of antimony dissolves but that of arsenic does not. Arsenic trisulphide is sufficiently acidic to dissolve in a cold solution of ammonium carbonate from which it is deposited by hydrogen sulphide as a yellow precipitate. Passage of hydrogen sulphide into the solution of antimony chloride gives an orange precipitate of antimony trisulphide. If tin is present (it makes up Table IIb with arsenic and antimony), oxalic acid is added to form a stable complex which is not decomposed by hydrogen sulphide.

Special tests for small quantities of arsenic and antimony

These are of forensic interest because of the poisonous nature of their compounds. *Attention is drawn to the extremely poisonous nature of arsine and stibine.*

Fleitmann's test

Compounds of arsenic, except arsenates, but not compounds of antimony are reduced on boiling

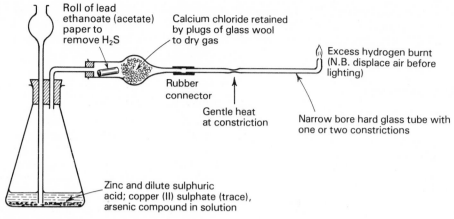

Fig. 21.8   Marsh's test for arsenic and antimony (in fume cupboard)

with sodium hydroxide solution and zinc (As-free grade), which generate hydrogen. Arsine is produced. A filter paper, moistened with silver nitrate solution, shows a black stain of silver produced by the reducing action of arsine.

Reinsch's test

Tervalent arsenic compounds, on boiling with a strip of clean metallic copper and hydrochloric acid, give a grey film of arsenic on the copper. The strip is washed, dried and heated in air when arsenic is oxidized and sublimes as white arsenic trioxide. Antimony compounds form a violet deposit oxidized on heating in air to the trioxide which, however, does not sublime. It may be dissolved in 2-3-dihydroxybutanedioic acid (tartaric acid) solution and orange antimony sulphide precipitated by hydrogen sulphide.

Gutzeit's test (the most suitable)

A solution containing arsenic is reduced to arsine by warming with zinc (As-free grade) and dilute sulphuric acid containing a little copper(II) sulphate solution to ensure a brisk generation of hydrogen. The gas is detected by a filter paper moistened in silver nitrate, black silver being produced. Alternatively, mercury(II) chloride goes yellow then brown and black with increasing quantities of arsine. If a trial experiment without the solution under test reveals a trace of arsenic in the reagents, then the two stains obtained under comparable conditions are compared. Any hydrogen sulphide may be removed from the gases evolved by inserting a loose plug of cotton wool, soaked in lead(II)

ethanoate (acetate) solution, just inside and below the mouth of the tube. Positive results are obtained with antimony salts.

Marsh's test

This is based on the same principle but the hydrides are dissociated by heat, care being taken to ensure that all air has been displaced from the apparatus, which is shown in Fig. 21.8. A mirror of arsenic is formed just past the point of heating while frequently, antimony appears just before that point, because stibine is less stable than arsine. The arsenic mirror is soluble in sodium hypochlorite (chlorate(I)) while that of antimony is not. Alternatively, the mirrors may be obtained on a cold porcelain dish held in the flame where excess hydrogen and hydride burn: incomplete combustion yields the element whereas complete combustion yields the oxide:

$$4AsH_3 + 3O_2 \rightarrow 4As + 6H_2O$$

$$4AsH_3 + 6O_2 \rightarrow As_4O_6 + 6H_2O$$

Arsenic compounds have been found in natural gas pipe lines in California, following the discovery of a white solid (43% by mass combined As) choking an inlet pipe for a monitoring device. Wet analytical techniques involving oxidation to As(v) by nitric acid enabled concentrations of 20 mg dm^{-3} to 1 μg dm^{-3} to be determined while proton †NMR techniques were used to identify the compounds. 50–80% of the arsenic content was found to be trimethylarsine, $(CH_3)_3As$. In this investigation

† Nuclear Magnetic Resonance

Marsh's test has proved to be the best analytical method, over 150 years after its introduction into early Victorian forensic science.

The neutral silver nitrate test

The following coloured precipitates of silver salts are formed by neutral silver nitrate:

silver arsenite	pale yellow
silver arsenate	chocolate brown
silver phosphate	pale yellow

The molybdate test may be applied for phosphate and arsenate.

The titrimetric (volumetric) estimation of arsenic, antimony and bismuth

Arsenic (as arsenites and arsenates) and antimony may be determined by use of iodine, potassium bromate or potassium iodate (pp. 432, 441, 442).

Bismuth may be precipitated as the sulphide which is agitated with a measured volume of standard silver nitrate, acidified with dilute nitric acid and filtered. The less soluble silver sulphide is removed:

$$Bi_2S_3(s) + 6Ag^+ \rightarrow 2Bi^{3+} + 3Ag_2S(s)$$

The excess silver is determined by titration under acidic conditions against thiocyanate.

22

Group VI: sulphur

The unusual nature of sulphur

8	
O	
2, 6	
16	
S	
2, 8, 6	
34	
Se	
2, 8, 18, 6	
52	
Te	
2, 8, 18, 18, 6	
84	
Po	
2, 8, 18, 32, 18, 6	

Sulphur is the second element of Group VI. Oxygen (Chapter 16) shows anomalous properties as the first member of the family. The gap between oxygen and sulphur is very much greater than that between any other successive pair of elements in this family, being paralleled in Group VII by the behaviour of fluorine.

Sulphur has been known since early times, being used by the Egyptians and the Greeks. Large deposits of natural sulphur occur in Texas and Louisiana in the USA and in Sicily. Sulphur occurs combined in mineral sulphides, notably zinc blende, ZnS, galena, PbS, pyrites, FeS_2, and copper pyrites, $CuFeS_2$. The sulphur content of these minerals is used for the production of sulphur dioxide and sulphuric acid. The paramount importance of sulphuric acid as an industrial chemical and limitations on the export of sulphur from the USA have stimulated the production of sulphur compounds

from gypsum, $CaSO_4.2H_2O$, and anhydrite, $CaSO_4$, in the UK. Sulphur is an important by-product in the refining of oil from certain regions. The commercial use of bacteria to oxidize sulphides to sulphates has recently been introduced.

The allotropy of sulphur has been described separately in the section, *Allotropy, polymorphism and isomerism* on pp. 107–9 of Chapter 6. Some physical data for sulphur are given in Tables 22.1 and 22.2.

Table 22.2 Group VI: sulphur: physical data

	Sulphur
Atomic number	16
Electronic configuration	2, 8, 6
[Core]outer electrons	$[Ne](3s)^2(3p)^4$
Relative atomic mass	32.064
Atomic radius/nm	0.104
Ionic radius/nm (S^{2-})	0.184
Standard electrode potential (H scale:	
298 k)/V $H_2S \rightleftharpoons S + 2e^- + 2H^+$	+0.14
Electronegativity	2.5
Co-ordination number	2, 4, 6
Abundance (% by mass)	0.05

Oxygen and sulphur in relation to other elements of Group VI (*p* block): oxygen, sulphur, selenium and tellurium

The Group VI elements are highly electronegative with feeble metallic properties becoming only just perceptible as the atomic number increases. The elements are the first to form stable, monatomic anions. Many differences are apparent between oxygen and sulphur, although sulphur can replace oxygen in compounds and many analogous compounds are formed. The valency of sulphur may rise to six, while that of oxygen remains at two. That oxygen differs markedly from the other elements within the family in its range and

Table 22.1 A comparison of physical data for rhombic and monoclinic sulphur

	Rhombic ⇌ Monoclinic	
	95.5°C	
Density/g cm^{-3}	2.07	1.96
Melting-point/°C	114	119
Boiling-point/°C		444.6
Atomic volume/cm^3 mol^{-1}*	15.56	16.35

*Traditional term. It is the volume of 1 mole of atoms as solid element

Table 22.3   Group VI (oxygen group): O, S, Se, Te, Po Comparison of selected physical data p block elements* $(ns)^2 (np)^4$

	O	S	Se	Te
1st ionization energy/ kJ mol^{-1} (273 K)	1312	1003	941	719
Atomic radius/nm	0.066	0.104	0.117	0.137
Ionic radius/nm (X^{2-})	0.140	0.184	0.198	0.221
van der Waals' radius/nm	0.140	0.185	0.20	0.22
Electronegativities	3.5	2.5	2.4	2.1

*$n \neq 1$

variety of properties is revealed by a comparison of the data shown in Table 22.3 and the Figures which follow. Its 1st ionization energy (Fig. 22.1) is much higher than that of sulphur, the atom and ion are smaller (Fig. 22.2) and the electronegativity (Fig. 22.3), higher. On the other hand, sulphur allies itself to selenium and tellurium.

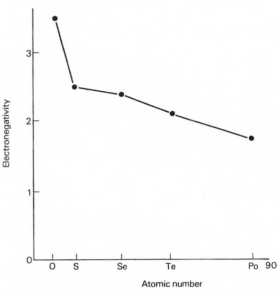

Fig. 22.3   Electronegativity plotted against atomic number for Group VI elements

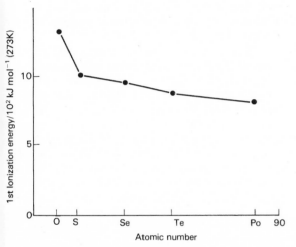

Fig. 22.1   1st ionization energy of Group VI elements plotted against atomic number

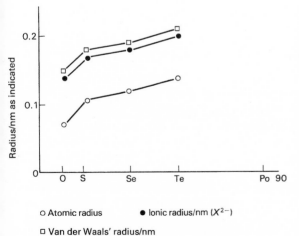

○ Atomic radius   ● Ionic radius/nm (X^{2-})

□ Van der Waals' radius/nm

Fig. 22.2   Atomic, ionic and van der Waals' radii plotted against atomic number for elements of Group VI. Note that ionic and van der Waals' radii are almost the same

While oxygen is characterized by its wide range of ionic oxides, the others tend towards covalency, and the anions S^{2-}, Se^{2-}, Te^{2-} are found only in association with the more electropositive (basic) cations. Oxygen and sulphur are clearly non-metals, but some metallic character is discernible in the others yet remains latent until it emerges in the fifth member of the family, polonium.

Catenation, chain-forming between atoms of an element, is illustrated profusely by sulphur compounds, in covalent molecules and in anions. Sulphur forms stable compounds with two, three or more† sulphur atoms linked in a chain. The element itself is S_8 up to 800°C. The limit for the chain-forming ability of oxygen is reached with three atoms in the highly reactive allotrope ozone, O_3. Peroxides and peroxy compounds, which contain $-O-O-$, are also unstable. Of the congeners, selenium forms halides with chains of

† This plays havoc with the calculation of oxidation numbers rendering them meaningless. S(I) in S_2Cl_2 could be added to the list on p. 124

atoms ($Se_n\dot{C}l_2$, including Se_2Cl_2) but no chain anions while the tendency is not apparent in tellurium compounds. Hydrogen bonding is less marked with sulphur compounds, hydrogen sulphide being more acidic than water and gaseous under laboratory conditions. Unlike those formed by sulphur, the chlorides of oxygen are explosive.

A general survey of bonding in sulphur compounds

The elements of Group VI are separated from the inert gases by only one group, the halogens. Sulphur may attain the electronic configuration of argon by acquiring two electrons from strong metals to form anions of double negative charge or by sharing two pairs of electrons to give a double covalency. Of course, sulphur retains the proton number (atomic number) of $+16$.

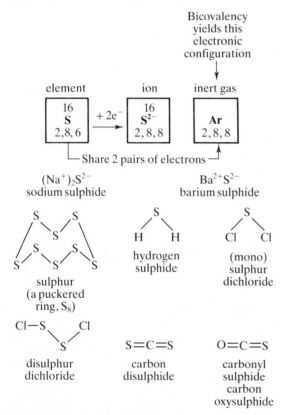

In these compounds, sulphur shows a valency of two. By use of the energy levels of the $3d$ orbitals, sulphur may assume higher oxidation states in which the orbit is exceeded, showing valencies of 4 and 6.

Quadricovalency is shown in sulphur tetrafluoride and the unstable tetrachloride:

while in the very stable gas, sulphur hexafluoride, sulphur exhibits sexacovalency. In the structures assumed by zinc sulphide, zinc blende and wurtzite, each sulphur atom is covalently bonded in a tetrahedral configuration to four zinc atoms, and each zinc atom similarly to four sulphur atoms (Fig. 6.6, p. 100). The atoms assume formal charges of two: $Zn(2-)$, $S(2+)$.

In the oxoacids and oxoanions of sulphur classical bond diagrams require expansion of the valency octet of sulphur. To avoid this, co-ionic bonding can be used, following the suggestion of Lewis (1916). However, the comparison of the measured interatomic distances in these compounds with the bond lengths of S–O known has shown the co-ionic structures to be unsatisfactory. Therefore, classical bond diagrams are preferred, with the recognition that where appropriate, resonance brings equalization of bond lengths. The alternative bond diagrams may be illustrated by the sulphate ion, the electronic configurations of sulphur being shown:

S	S^{2-}		
sulphur	sulphide ion	sulphate ion	
2, 8, 6	2, 8, 8	2, 8, 8 co-ionic structure	2, 8, 12 classical structure

The classical diagram must be fused with resonating structures such as those following to give the final resonance hybrid:

See Chapter 7, p. 124 for an account of the oxidation numbers shown by sulphur in its compounds. A range of structures is shown below with the valency of sulphur and the shapes of the resulting molecules indicated:

$$\underset{\underset{\text{bent (V-shaped)}}{\underset{\text{(4)}}{\underset{\text{dioxide}}{\text{sulphur}}}}}{O\diagdown\!\!\diagdown S\diagup\!\!\diagup O}$$

sulphur dioxide (4) bent (V-shaped)

thionyl chloride (4)

sulphurous acid (4)

trigonal planar

sulphur trioxide (6) trigonal planar

sulphuryl chloride (6)

sulphuric acid (6)

tetrahedral

Sulphur

The allotropy of sulphur and a detailed description of the action of heat on the element is given in Chapter 6, pp. 107–9.

Sulphur is a pale yellow solid non-metallic element, without taste or odour, a poor conductor of heat and a non-conductor of electricity. On slow heating it melts at 119°C (monoclinic-sulphur) while rapid heating causes melting at 114°C (rhombic-sulphur). The boiling-point, 444.6°C at atmospheric pressure, is a fixed point in thermometry. Sulphur is insoluble in water, moderately soluble in benzene and very soluble in carbon disulphide.

Sulphur combines directly with all elements except nitrogen, tellurium, iodine, gold, platinum and iridium. Lithium, sodium and potassium react in the cold, as do copper, silver and mercury. The others require heating. The products are called sulphides except where combination is with a more electronegative element. Molten sulphur burns in air with a blue flame and very vigorously in oxygen with a violet-blue flame to form sulphur dioxide, SO_2. In industry, carbon disulphide is produced by a reaction between sulphur and carbon (coke) in an electric furnace. More detail about this compound appears under carbon in Chapter 20, p. 335.

Non-metallic elements do not react with dilute mineral acids (i.e. with H^+ ions) but may be oxidized to acidic oxides or oxoacids by oxidizing acids. Hot concentrated sulphuric acid oxidizes sulphur to sulphur dioxide while hot concentrated nitric acid forms sulphuric acid. Both oxidations are catalysed by addition of bromine:

$$S + 2H_2SO_4 \rightarrow 2H_2O + 3SO_2$$

$$S + 6HNO_3 \rightarrow 2H_2O + H_2SO_4 + 6NO_2$$

With hot, concentrated alkali solution a complex series of reactions takes place with the formation of thiosulphate and polysulphides:

$$6OH^- + 4S \rightarrow 2S^{2-} + \underset{\text{thiosulphate}}{S_2O_3^{2-}} + 3H_2O$$

$$S^{2-} + 4S \rightarrow \underset{\text{pentasulphide}}{S_5^{2-}}$$

Hydride: hydrogen sulphide, sulphuretted hydrogen, H_2S

Hydrogen sulphide is usually prepared by the action of somewhat diluted hydrochloric acid (1 volume acid : 1 volume water) on commercial sticks of iron(II) sulphide:

$$FeS + 2HCl \rightarrow FeCl_2 + H_2S$$

Acid spray is removed by passage through water and the gas collected by upward displacement of air. Alternatively, it may be collected over hot water, hydrogen sulphide being fairly soluble in cold water. The gas may be dried by phosphorus pentoxide although the reaction with calcium chloride is very slow and this drying agent could be used. Concentrated sulphuric acid would be reduced to sulphur:

$$H_2SO_4 + 3H_2S \rightarrow 4H_2O + 4S(s)$$

The main impurity is hydrogen resulting from the action of acid on uncombined iron in the sticks of iron(II) sulphide. However, iron(II) sulphide can be prepared specially so as to contain very little free iron. Pure hydrogen sulphide results from the action of hot concentrated hydrochloric acid on antimony trisulphide:

$$Sb_2S_3 + 6HCl \rightarrow 2SbCl_3 + 3H_2S$$

Hydrogen sulphide is a colourless gas, b.p. −60.4°C, with a sweet, highly offensive odour of rotting eggs. It dissolves in water, 2.6 volumes of gas dissolving in 1 volume of water at 20°C, giving a solution with a small electrical conductance due to feeble ionization:

$$H_2S \rightleftharpoons H^+ + HS^-$$

$$HS^- \rightleftharpoons H^+ + S^{2-}$$

On heating above 400°C, it dissociates into the elements. Hydrogen sulphide burns with a blue flame in excess air or oxygen to sulphur dioxide and in a restricted supply to sulphur, steam being the other product:

$$2H_2S + 3O_2 \rightarrow 2H_2O + 2SO_2$$

$$2H_2S + O_2 \rightarrow 2H_2O + 2S$$

A concentration of 0.1% by volume of hydrogen sulphide rapidly proves fatal to man, but the danger is not great because at a concentration of one-tenth of this, the air becomes most unpleasant. Solutions of caustic alkalis absorb hydrogen sulphide to form the sulphide:

$$2OH^- + H_2S \rightarrow S^{2-} + 2H_2O$$

Excess hydrogen sulphide gives the hydrogen-sulphide, formerly hydrosulphide:

$$S^{2-} + H_2S \rightarrow 2SH^-$$

Solutions of the hydroxides of calcium, strontium and barium form the metal hydrogen-sulphides with hydrogen sulphide.

Hydrogen sulphide is a weak dibasic acid.

Sulphides may also be formed by heating the metal in hydrogen sulphide: sodium gives sodium sulphide and tin, tin(II) sulphide:

$$2Na + H_2S \rightarrow Na_2S + H_2$$

$$Sn + H_2S \rightarrow SnS + H_2$$

Insoluble sulphides may be precipitated by the action of hydrogen sulphide on a solution of a metal salt:

$$HgCl_2 + H_2S \rightarrow HgS(s) + 2H^+ + 2Cl^-$$

$$Cu^{2+} + H_2S \rightarrow CuS(s) + 2H^+$$

$$Mn^{2+} + H_2S \rightarrow MnS(s) + 2H^+$$

In qualitative analysis, the black stain produced on a piece of filter paper impregnated with lead(II) ethanoate (acetate) solution is used as a test for hydrogen sulphide, usually already suspected because of its odour.

Selective precipitation of sulphides may be achieved by control of pH (p. 232).

$$H_2S \rightleftharpoons H^+ + HS^-$$

$$K_1 = \frac{[H^+][HS^-]}{[H_2S]}$$

$$= 9.1 \times 10^{-8} \text{ mol dm}^{-3} \text{ at } 25°C$$

$$HS^- \rightleftharpoons H^+ + S^{2-}$$

$$K_2 = \frac{[H^+][S^{2-}]}{[HS^-]}$$

$$= 1.2 \times 10^{-15} \text{ mol dm}^{-3} \text{ at } 25°C$$

By multiplying,

$$K_1K_2 = \frac{[H^+]^2[S^{2-}]}{[H_2S]}$$

$$= 1.1 \times 10^{-22} \text{ mol}^2 \text{ dm}^{-6} \text{ at } 25°C$$

The solubility of hydrogen sulphide, $[H_2S]$, is approximately 0.1 M.

$$\therefore [H^+]^2[S^{2-}] = 1.1 \times 10^{-23} \text{ mol}^3 \text{ dm}^{-9} \text{ at } 25°C$$

In qualitative analysis, by controlling the concentration of hydrogen ions, the concentration of sulphide ion in saturated hydrogen sulphide may be adjusted to achieve precipitation of the sulphides of Analysis Group II elements (mercury(II), lead, bismuth, copper, cadmium, arsenic, antimony and tin). Solubility products are exceeded momentarily. The hydrogen ion concentration should be 0.25M and, by calculation, the sulphide ion concentration, $1.8 \times 10^{-22} \text{ mol dm}^{-3}$. In Analysis Group IV, the precipitation of the comparatively more soluble sulphides of nickel, cobalt, manganese and zinc is arranged by addition of ammonia (and ammonium chloride) which raises the pH, i.e. lowers the hydrogen ion concentration.

Hydrogen sulphide is a powerful reducing agent and is readily oxidized to sulphur. The reaction is essentially

$$S^{2-} - 2e^- \rightleftharpoons S(s)$$

but hydrogen sulphide is a weak electrolyte,

$$\therefore H_2S - 2e^- \rightleftharpoons 2H^+ + S(s)$$

Sulphur dioxide is also a powerful reducing agent, but differs from hydrogen sulphide in that sulphur is not precipitated except in their interaction when moist:

$$2H_2S + SO_2 \rightarrow 3S + 2H_2O$$

A magenta-coloured acidified solution of potassium permanganate is decolorized by hydrogen sulphide, an acidified orange solution of potassium dichromate turned green and a pale yellow solution of iron(III) chloride turned pale green, all with separation of sulphur:

$$2MnO_4^- + 6H^+ + 5H_2S \rightarrow$$
$$2Mn^{2+} + 8H_2O + 5S(s)$$

$$Cr_2O_7^{2-} + 8H^+ + 3H_2S \rightarrow$$
$$2Cr^{3+} + 7H_2O + 3S(s)$$

$$2Fe^{3+} + H_2S \rightarrow 2Fe^{2+} + 2H^+ + S(s)$$

Halogens in the presence of water form hydracids with hydrogen sulphide, red bromine water being decolorized:

$$Br_2 + H_2S \rightarrow 2HBr + S(s)$$

Hydrogen sulphide reduces concentrated nitric acid, the liquid becoming hot and much nitrogen dioxide being evolved:

$$2HNO_3 + H_2S \rightarrow 2NO_2 + 2H_2O + S(s)$$

Oxidation of the sulphur to sulphuric acid also occurs. Nitrous acid, i.e. acidified sodium nitrite solution, oxidizes hydrogen sulphide to sulphur and is reduced to nitrogen (nitric) oxide:

$$2HNO_2 + H_2S \rightarrow 2NO + S(s) + 2H_2O$$

In order to obtain the molecular formula of hydrogen sulphide, the density relative to hydrogen must be known in addition to determining the volume proportion of hydrogen in the gas. In a bent tube over mercury, tin is heated with hydrogen sulphide to give hydrogen with no change in volume: the apparatus is that used for dinitrogen oxide (Fig. 21.6, p. 367).

1 volume of hydrogen sulphide + tin yields
1 volume of hydrogen + solid tin(II) sulphide

Applying Avogadro's Hypothesis,

∴ 1 molecule of hydrogen sulphide + tin yields
1 molecule of hydrogen + tin(II) sulphide

But hydrogen may be shown to be H_2,

∴ hydrogen sulphide is H_2S_x

where x is unknown.

However, the relative density is 17 (H = 1), so that the relative molecular mass is 34,

$$\therefore H_2S_x \equiv 34$$

$$2 + 32x = 34$$

$$\therefore x = 1$$

Hydrogen sulphide is H_2S

Sulphides of the metals

The term sulphide is applied to those binary compounds of sulphur in which the other element is less electronegative (or relatively electropositive). Sulphides may be synthesized from the elements, formed by reaction of hydrogen sulphide. Certain sulphides (e.g. Na_2S) may be prepared by reduction of the sulphates. Metal sulphides may be ionic or covalent. Many occur natively and some are termed *glances*, e.g. PbS, NiS, Ag_2S, ZnS, where the reflection of light is similar to that obtained from a metal, this type also showing electrical conductance.

Solutions containing sulphides of the alkali metals and alkaline earths tend to dissolve sulphur with the formation of polysulphides. For sodium, Na_2S_2, Na_2S_4 and Na_2S_5 have been prepared by various methods. Ammonium sulphide solution (colourless) will react with flowers of sulphur to form the polysulphide, known as yellow ammonium sulphide, $(NH_4)_2S \cdot S_x$.

Soluble sulphides are hydrolysed to varying extents. Those of Group I give alkaline solutions:

$$S^{2-} + H_2O \rightleftharpoons SH^- + OH^-$$

A purple coloration is given by sulphide solution on addition of a little disodium pentacyanonitrosylferrate:[†]

$$Fe(CN)_5NO^{2-} + S^{2-} \rightarrow Fe(CN)_5NOS^{4-}$$

The sulphides of calcium, strontium and barium dissolve slowly:

$$CaS + H_2O \rightleftharpoons Ca^{2+} + HS^- + OH^-$$

$$HS^- + H_2O \rightleftharpoons H_2S + OH^-$$

Magnesium sulphide rapidly evolves hydrogen sulphide. Aluminium sulphide, synthesized from the elements, is instantly hydrolysed:

$$Al_2S_3 + 6H_2O \rightarrow 2Al(OH)_3 + 3H_2S$$

Difficulty is experienced in bringing the very insoluble sulphides into solution. Concentrated nitric acid or bromine water may be used to oxidize them to soluble products. In qualitative analysis, mercury(II) sulphide can be dissolved in concentrated hydrochloric acid to which either bromine water or concentrated nitric acid, the latter making aqua regia, has been added:

$$3HgS + 2HNO_3 + 6HCl \rightarrow$$
$$3HgCl_2 + 2NO + 3S + 4H_2O$$

† Formerly, sodium nitroprusside

The other sulphides react with non-oxidizing acids with liberation of hydrogen sulphide:

$$ZnS + 2H^+ \rightarrow Zn^{2+} + H_2S$$

On heating in air, a number of sulphides burn to the oxide, while a few give the sulphate or the metal:

$$2ZnS + 3O_2 \rightarrow 2ZnO + 2SO_2$$

$$2Sb_2S_3 + 9O_2 \rightarrow Sb_4O_6 + 6SO_2$$

$$BaS + 2O_2 \rightarrow BaSO_4$$

$$HgS + O_2 \rightarrow Hg(g) + SO_2(g)$$

Oxides, oxoacids and their salts, and acid chlorides

While five oxides of sulphur, SO, SO_2, S_2O_3, SO_3 and SO_4, have been isolated, only SO_2 and SO_3 are of any importance and stability under ordinary conditions. At least twelve oxoacids of sulphur have been described of which only H_2SO_3 and H_2SO_4, corresponding to SO_2 and SO_3 respectively, are of importance together with their acid chlorides. In addition two peroxoacids will be described and thiosulphuric acid mentioned:

Sulphur dioxide
SO_2

Sulphurous* acid
H_2SO_3

Thionyl chloride
$SOCl_2$

Sulphur trioxide
SO_3

Sulphuric* acid
H_2SO_4

Thiosulphuric* acid
$H_2S_2O_3$

Chlorosulphonic acid
Chlorosulphuric acid
$HO.SO_2.Cl$

Sulphuryl chloride
Sulphonyl chloride
SO_2Cl_2

Caro's acid
Peroxo(mono)sulphuric acid
Sulphomonoperacid
H_2SO_5

Peroxodisulphuric acid
Sulphodiperacid
$H_2S_2O_8$

Sulphur dioxide, SO_2, sulphurous acid (trioxosulphuric acid), H_2SO_3, and its salts

Sulphur dioxide is a colourless gas, with the pungent acrid odour associated with burning sulphur, leaving a characteristic taste in the mouth. The gas is poisonous and has been used for fumigation purposes. It is denser than air, relative (vapour) density 32 (H = 1, air = 14.4) and easily liquefied (b.p. $-10.0°C$, critical temperature 157°C) under 250 kPa pressure at room temperature or by cooling in a freezing-mixture at atmospheric pressure. Sulphur dioxide is rarely prepared in the laboratory but obtained from siphons of liquefied gas. Liquid sulphur dioxide is an excellent solvent, with a very small electrical conductance due to slight ionization,

$$2SO_2 \rightleftharpoons SO^{2+} + SO_3^{2-}$$
acidic ion   basic ion

The concept of acids and bases (pp. 237–8) has been extended to non-protonic solvents, SO^{2+} being the acidic ion and SO_3^{2-} the basic ion, so that thionyl chloride, $SOCl_2$, is an acid and sodium sulphite, Na_2SO_3, a base and on neutralization, sodium chloride is precipitated:

$$SO^{2+} + 2Cl^- + 2Na^+ + SO_3^{2-} \rightarrow$$
$$2NaCl(s) + 2SO_2$$

Liquid sulphur dioxide has a rather high dielectric constant ($= 13.5$ at 15°C) and the power to solvate ions. Other examples of non-protonic solvents include $(CH_3CO)_2O$ (ethanoic anhydride), I_2, ICl_3, BrF_3 and IF_5. In this context, an acid gives the cation of the parent solvent and a base, the anion. Physical measurements by electron diffraction and dipole moments indicate that sulphur dioxide is not a linear molecule.

Sulphur dioxide may be prepared by the combustion of sulphur or certain sulphides, the reduction of concentrated sulphuric acid or by the action of acids on sulphites and hydrogensulphites. Industrially, combustion of pyrites, zinc blende, etc., or sulphur may be used. In the laboratory, hot concentrated sulphuric acid is reduced by a metal: at 130°C, copper reduces sulphuric acid in a complex process producing sulphides of copper, copper(II) sulphate and sulphur dioxide. The principal

Full IUPAC names:
H_2SO_3 trioxosulphuric acid
H_2SO_4 tetraoxosulphuric acid
$H_2S_2O_3$ trioxothiosulphuric acid

reaction is:

$$Cu + 2H_2SO_4 \rightarrow CuSO_4 + SO_2 + 2H_2O$$

Otherwise, a moderately concentrated solution of an acid is added to sodium sulphite or hydrogen-sulphite (or sodium disulphite, $Na_2S_2O_5$). Sulphuric acid, being involatile, is preferred to hydrochloric acid.

$$SO_3^{2-} + 2H^+ \rightarrow SO_2 + H_2O$$

The gas is dried with concentrated sulphuric acid and collected by upward displacement of air. Sulphur dioxide is very soluble in water, 1 volume dissolving 80 volumes of gas at $0°C$ and 39 volumes at $20°C$, to form a colourless acidic solution containing sulphurous acid, although this acid has never been isolated pure:

$$SO_2 + H_2O \rightleftharpoons H_2SO_3$$

$$H_2SO_3 \rightleftharpoons H^+ + HSO_3^-$$

$$HSO_3^- \rightleftharpoons H^+ + SO_3^{2-}$$

The solution smells strongly of free sulphur dioxide. It is seen that sulphur dioxide is sulphurous anhydride and an acidic oxide.

The molecular formula of sulphur dioxide may be deduced by the same method as that used for carbon dioxide. Sulphur burns in oxygen to give sulphur dioxide without change in volume of the gas.

By Avogadro's Hypothesis,

1 molecule of oxygen + sulphur yields
1 molecule of sulphur dioxide

and since oxygen may be shown to be O_2, the formula of sulphur dioxide is S_xO_2 where x is unknown. But the density relative to hydrogen is 32, equal to half of the relative molecular mass;

$$\therefore S_xO_2 \equiv 2 \times 32$$

$$32x + 32 = 2 \times 32$$

$$\therefore x = 1$$

Sulphur dioxide is SO_2

Sulphur dioxide may be reduced by hydrogen on passing the gases together over warm platinum black, sulphur and steam being formed:

$$SO_2 + 2H_2 \rightarrow S + 2H_2O$$

Hydrogen sulphide reduces sulphur dioxide solution to sulphur, to which it is itself oxidized:

$$2H_2S + SO_2 \rightarrow 3S(s) + 2H_2$$

Sulphur dioxide will support the combustion of burning magnesium, forming magnesium oxide and sulphur, and also magnesium sulphite and thiosulphate:

$$2Mg + SO_2 \rightarrow 2MgO + S$$

Sulphur dioxide is chiefly notable for its reducing power, usually in solution but the gas itself shows an affinity for oxygen and chlorine. Sulphur dioxide reacts reversibly with oxygen to form sulphur trioxide over a catalyst of platinum black at $400°C$:

$$2SO_2 + O_2 \rightleftharpoons 2SO_3$$

When sulphur dioxide is bubbled through hydrogen peroxide, oxidation to sulphuric acid occurs and the liquid becomes hot:

$$H_2O_2 + SO_2 \rightarrow H_2SO_4$$

Sodium peroxide becomes incandescent in the gas, forming the sulphate:

$$Na_2O_2 + SO_2 \rightarrow Na_2SO_4$$

The puce-coloured lead(IV) oxide also becomes incandescent when warmed to form white lead(II) sulphate:

$$PbO_2 + SO_2 \rightarrow PbSO_4$$

Sulphur dioxide reacts with chlorine in the presence of granules of active charcoal (blood charcoal), ethene (ethylene), or camphor to form sulphuryl chloride:

$$SO_2 + Cl_2 \rightarrow SO_2Cl_2$$

A solution of sulphur dioxide contains sulphurous acid, which acts as a powerful reducing agent, being readily oxidized to sulphuric acid. This happens when the solution is left exposed to the air. Sulphites are also readily oxidized to sulphates:

$$SO_3^{2-} + H_2O - 2e^- \rightleftharpoons SO_4^{2-} + 2H^+$$

No precipitate of sulphur is formed, distinguishing the oxidation of sulphurous acid from that of hydrogen sulphide. Sulphurous acid is fairly strong. The colour change of an aqueous oxidizing agent may be used to confirm the presence of sulphur dioxide. The red of bromine water is discharged:

$$Br_2 + H_2O + SO_3^{2-} \rightarrow 2Br^- + 2H^+ + SO_4^{2-}$$

Chlorine water (pale greenish-yellow) reacts similarly. The magenta colour of acidified potassium permanganate disappears:

$$2MnO_4^- + 6H^+ + 5SO_3^{2-} \rightarrow$$
$$5SO_4^{2-} + 2Mn^{2+} + 3H_2O$$

Orange aqueous potassium dichromate forms green chromium(III) ions:

$$Cr_2O_7^{2-} + 8H^+ + 3SO_3^{2-} \rightarrow$$
$$2Cr^{3+} + 4H_2O + 3SO_4^{2-}$$

The yellow colour of acidified iron(III) salt solutions turns to the pale green of the iron(II) salts in reaction with sulphur dioxide although a red complex intermediate may be seen:

$$2Fe^{3+} + SO_3^{2-} + H_2O \rightarrow 2Fe^{2+} + SO_4^{2-} + 2H^+$$

Moist sulphur dioxide is used industrially as a bleach for materials which would be destroyed by chlorine. Unlike chlorine, sulphur dioxide bleaches by reduction, the process being slowly reversed by air in sunlight. The gas is used to bleach silk and wool, sponges and straw. As calcium hydrogensulphite, it is used in the processing of wood pulp in the manufacture of paper. However, oxidation to sulphur trioxide for conversion into sulphuric acid is the main application.

Therein lies a major air pollution problem. Sulphur dioxide emissions in the waste combustion gases from electric power stations cause major environmental damage, due to the aptly-called 'acid rain'. Methods for removal of the gas include the use of limestone, which is converted to gypsum, calcium sulphate, and as a possible supplementary method, the injection of magnesium hydroxide into flue gases. The level of sulphur dioxide emission has to conform to the EC large plant directive (1985), for which the emission targets are given below:

	Percentage reduction on 1980 UK Power Station emission level	SO₂ emissions/ Mt
1980	—	2.87
1993	20	2.30
1998	40	1.72
2003	60	1.15

That acid rain has had apparently less effect in the UK cultivated farm land than in Scandinavia may well be due to the traditional practice of 'liming' the soil, adding crushed limestone or chalk, which neutralizes acid fall-out from power station chimney gases. However, a grim picture emerges from a report (1990) published jointly by the Royal Society and its sister societies in Norway and Sweden. The report is based on the work of 30 scientific teams over the period 1983–1988 and cost £5 million (paid for by the Central Electricity Generating Board and British Coal). It concludes that wholesale destruction of aquatic life in thousands of lakes and rivers has been caused by acid rain. Contamination has been building up for 150 years or more in some areas. In the UK 70% of sulphur dioxide emissions originate from our comparatively highly sulphurous coal used in electricity production. A de-sulphurization plant has been installed in the huge Drax power station but these plants are costly and it may prove cheaper to burn imported coal of lower grade and minimal sulphur content or to use natural gas.

The Norwegians claim that 20% of their acidification in the south originated in Britain. This is probably true but the spread of radioactivity from the Chernobyl nuclear power station disaster turned up in some unexpected places. Poisoning of fish occurs when aluminium compounds, leached out of the soil and rock, are deposited in the gills. Calcium and magnesium carbonates slow down or neutralize this process which means that in areas where the soil is thin or the rocks are acidic (e.g. granite) the problem is more acute. Trout and salmon in up to a third of the rivers and lakes in Galloway, Gwynedd, the Lake District and the Pennines have been wiped out or severely affected. Magazine articles about salmon and trout fishing over the years have been carrying a similar message.

Sulphurous acid is a dibasic acid forming normal and acid salts, viz. sulphites, hydrogensulphite-(bisulphites) and disulphites(metabisulphites),

H_2SO_3 sulphurous acid	$(Na^+)_2SO_3^{2-}$ sodium sulphite
$Na^+HSO_3^-$ sodium hydrogensulphite (bisulphite)	$(Na^+)_2S_2O_5^{2-}$ sodium disulphite (metabisulphite)

From a well-cooled solution of sodium hydrogensulphite, sodium disulphite, sodium metabisulphite or pyrosulphite is deposited, elimination of water probably giving a pyro form, an example of *condensation*:

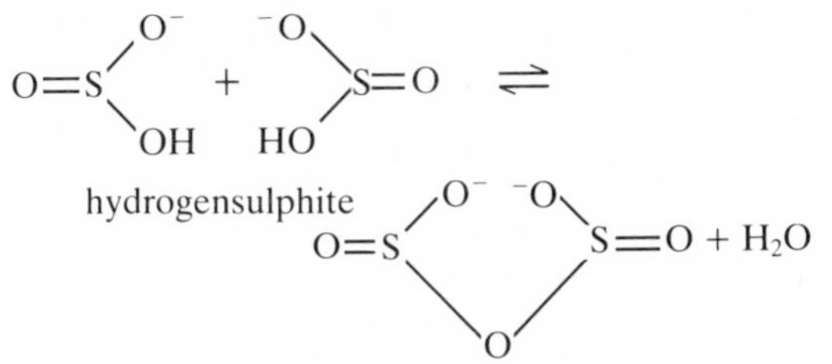

hydrogensulphite

disulphite
(pyrosulphite)

Sodium disulphite readily dissolves in water to regenerate hydrogensulphite. Important sulphites and hydrogensulphites belong to the alkali metals and calcium, and are described with those metals. Sulphites in solution are somewhat alkaline due to hydrolysis but unlike hydrogencarbonates, hydrogensulphites are acidic due to ionization of the hydrogensulphite ion:

$$SO_3^{2-} + H_2O \rightleftharpoons HSO_3^- + OH^-$$

$$HSO_3^- \rightleftharpoons H^+ + SO_3^{2-}$$

Acidification yields sulphur dioxide:

$$SO_3^{2-} + H^+ \rightleftharpoons HSO_3^-$$

$$HSO_3^- + H^+ \rightleftharpoons H_2SO_3 \rightleftharpoons H_2O + SO_2$$

The gas, because of its high solubility is hardly evolved from dilute solutions.

Powdered roll sulphur reacts with a boiling solution of a sulphite to form a thiosulphate:

$$SO_3^{2-} + S \rightarrow S_2O_3^{2-}$$

Sodium disulphite is used in photography, calcium hydrogensulphite in extracting lignin from wood-pulp and in the laboratory, sodium hydrogensulphite solution is used to prepare derivatives of aldehydes and ketones:

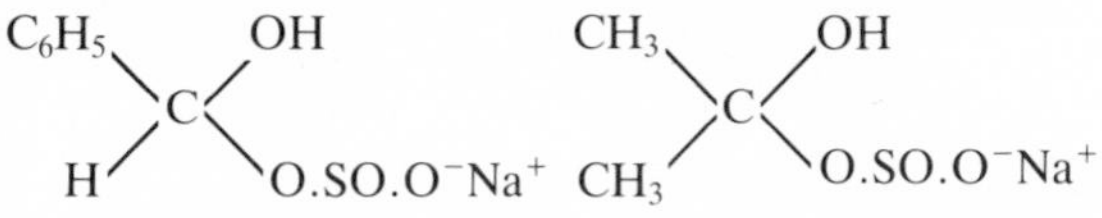

benzaldehyde
sodium hydrogensulphite

propanone (acetone)
sodium hydrogensulphite

Thionyl chloride, sulphinyl chloride, SOCl₂

The acid chloride of sulphurous acid is thionyl chloride, just as ethanoyl chloride (acetyl chloride) is the acid chloride of ethanoic acid (acetic acid):

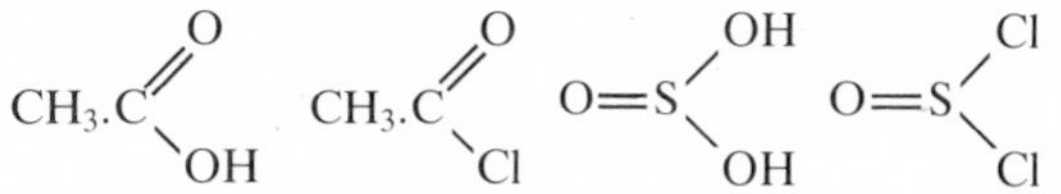

ethanoic acid (acetic acid)   ethanoyl chloride (acetyl chloride)   sulphurous acid   thionyl chloride

$>S=O$ is the thionyl or sulphinyl group.

Thionyl chloride is usually made by the action of phosphorus pentachloride on sulphur dioxide or a sulphite. Dry sulphur dioxide is passed into solid phosphorus pentachloride with shaking until the whole is liquid:

$$PCl_5 + SO_2 \rightarrow SOCl_2 + POCl_3$$

Thionyl chloride (b.p. 76°C) is separated by fractional distillation from phosphorus oxychloride, phosphoryl(v) chloride (b.p. 108°C).

Thionyl chloride is a colourless liquid which fumes in moist air and is hydrolysed by water to sulphurous acid and hydrochloric acid:

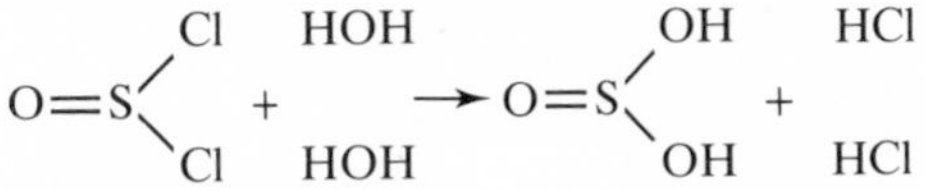

Thionyl chloride may be used in the preparation of acid chlorides, both acyl and aryl, by acting on the anhydrous acid:

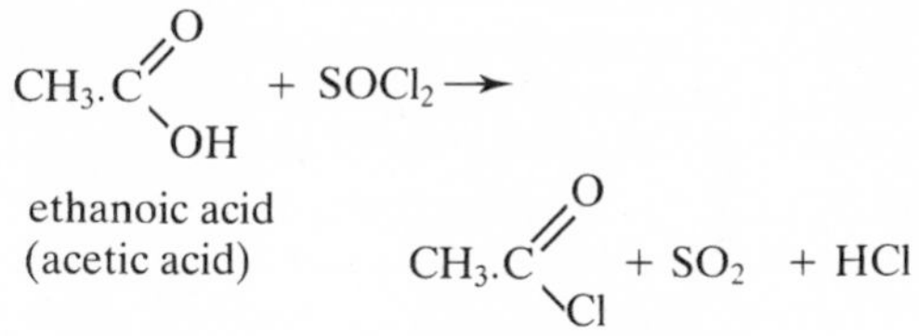

ethanoic acid (acetic acid)

ethanoyl chloride (acetyl chloride)

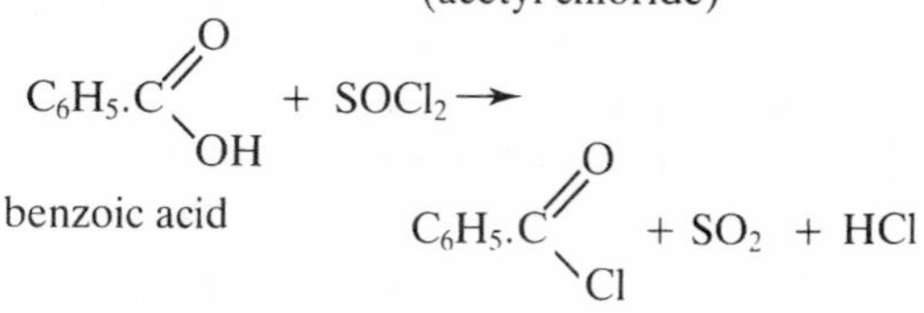

benzoic acid

benzoyl chloride

Note that the other products of the reactions are volatile.

The stability of the thionyl halides falls off with increasing atomic number of the halogen, thionyl fluoride being the most stable whereas there is no thionyl iodide.

Sulphur trioxide, SO_3, sulphuric acid (tetraoxosulphuric acid), H_2SO_4, and its salts

The commercial production of sulphuric acid is described in Chapter 13 (p. 198).

Sulphur trioxide is prepared by the reversible reaction between sulphur dioxide and oxygen in the presence of a heated catalyst of platinized asbestos or vanadium(v) oxide, the fundamental reaction of the Contact process for the production of sulphuric acid. Sulphur trioxide is condensed as a silky, felted mass of needles by cooling in crushed ice, excess sulphur dioxide passing on at this temperature. The reaction is exothermic and the optimum working temperature 400–450°C. 98–99% conversion may take place:

$$2SO_2 + O_2 \rightleftharpoons 2SO_3; \quad \Delta H = -2 \times 94.6 \text{ kJ mol}^{-1}$$

Sulphur trioxide is formed by the thermal decomposition of certain sulphates:

$$\underset{\substack{\text{iron(III)} \\ \text{sulphate}}}{Fe_2(SO_4)_3} \rightarrow \underset{\substack{\text{iron(III)} \\ \text{oxide}}}{Fe_2O_3} + 3SO_3(g)$$

$$\underset{\substack{\text{sodium hydrogen} \\ \text{sulphate}}}{2NaHSO_4} \rightarrow H_2O(g) + \underset{\substack{\text{sodium disulphate} \\ \text{(pyrosulphate)}}}{Na_2S_2O_7} \rightarrow$$

$$H_2O(g) + Na_2SO_4 + SO_3(g)$$

The structure of solid sulphur trioxide is not fully understood. The vapour condenses to a liquid, b.p. 44.5°C, which freezes at 16.8°C to ice-like crystals of the α-form. However, the liquid usually forms long silky needles in a felted mass of the β modification which is rather like asbestos. This change is catalysed by the presence of a trace of sulphuric acid, formed by moisture present. The β (asbestos) form is apparently more stable than the α (ice) form.

Sulphur trioxide is an acidic oxide, fuming in moist air and reacting explosively with water to form sulphuric acid: it is sulphuric anhydride,

$$SO_3 + H_2O \rightarrow H_2SO_4$$

Sulphuric acid may be dehydrated by phosphorus pentoxide to yield sulphur trioxide. This will form sulphates with basic oxides, barium oxide becoming incandescent:

$$BaO + SO_3 \rightarrow BaSO_4$$

Sulphur trioxide reacts with sulphuric acid to form pyrosulphuric acid (or disulphuric acid), also called oleum:

$$SO_3 + H_2SO_4 \rightarrow H_2S_2O_7$$

Chlorosulphuric acid may be made by the action of dry hydrogen chloride on sulphur trioxide, or sulphur trioxide in concentrated sulphuric acid:

$$HCl + SO_3 \rightarrow Cl.SO_2.OH$$

On heating sulphur trioxide to 1000°C, decomposition into sulphur dioxide and oxygen is virtually complete:

$$2SO_3 \rightleftharpoons 2SO_2 + O_2$$

Sulphur trioxide is a powerful oxidizing agent: hydrogen bromide is oxidized to bromine, a reversal of the action of sulphur dioxide on bromine water:

$$SO_3 + 2HBr \rightarrow H_2O + Br_2 + SO_2$$

Its oxidizing power is usually seen in the reactions of concentrated sulphuric acid.

Pure sulphuric acid is a colourless, odourless, oily liquid, density 1.84 g cm^{-3} at 15°C. The laboratory acid usually contains about 2% water by mass and is called concentrated sulphuric acid. It is extremely destructive of skin and clothing. Addition of sulphur trioxide, as oleum, to the concentrated acid yields the 100% acid which freezes at 10.37°C. A constant boiling-point mixture contains 98.5% sulphuric acid and boils at 317°C under atmospheric pressure with some decomposition into sulphur trioxide and steam, giving white acrid poisonous vapours. Concentrations higher than this mixture fume in air and contain free sulphur trioxide. A number of solid hydrates have been isolated.

The 100% acid has a definite electrical conductance and a number of involved ionizations, starting with auto-ionization, are believed to occur:

$$2H_2SO_4 \rightleftharpoons H_3SO_4^+ + HSO_4^-$$

and

$$H_2SO_4 \rightleftharpoons H_2O + SO_3$$

giving

$$H_2SO_4 + H_2O \rightleftharpoons H_3O^+ + HSO_4^-$$

$$H_2SO_4 + SO_3 \rightleftharpoons H_2S_2O_7$$

$$H_2SO_4 + H_2S_2O_7 \rightleftharpoons HS_2O_7^- + H_3O^+ + SO_3$$

Of importance in the nitration of aromatic hydrocarbons, nitric acid ionizes in sulphuric acid:

$$2H_2SO_4 + HNO_3 \rightleftharpoons NO_2^+ + 2HSO_4^- + H_3O^+$$

the nitryl, formerly nitronium, ion, NO_2^+, being the nitrating agent. Sulphuric acid in dilute aqueous solution is a strongly ionized dibasic acid, although weaker than hydrochloric acid and nitric acid,

$$H_2SO_4 + H_2O \rightleftharpoons H_3O^+ + HSO_4^-$$

$$HSO_4^- + H_2O \rightleftharpoons H_3O^+ + SO_4^{2-}$$

Theories about the way in which acids function have been outlined in Chapter 15, pp. 235–9.

Sulphuric acid has a powerful affinity for water. On dilution with water much heat (908 kJ mol^{-1}) is evolved. Acid must be cautiously added to water with stirring and cooling: otherwise, the less dense water floats without mixing on the surface of the acid. Boiling occurs at the interface and liquid may be dispersed violently. Sulphuric acid is used as a drying agent for gases and in vacuum desiccation. Concentrated sulphuric acid will remove the water of crystallization from blue copper(II) sulphate pentahydrate leaving a white anhydrous salt:

$$CuSO_4.5H_2O - 5H_2O \rightarrow CuSO_4$$

Concentrated sulphuric acid will remove the elements of water from many organic compounds. Sucrose forms sugar carbon:

$$C_{12}H_{22}O_{11} - 11H_2O \rightarrow 12C$$

methanoic (formic) acid and methanoates (formates) yield carbon monoxide, ethanedioic (oxalic) acid and ethanedioates (oxalates) on heating, carbon monoxide and dioxide:

$$H.C\overset{\displaystyle O}{\underset{\displaystyle OH}{<}} \quad -H_2O \rightarrow CO$$

$$\begin{array}{c} C\overset{O}{\underset{OH}{<}} \\ | \\ C\underset{O}{\overset{OH}{<}} \end{array} \quad .2H_2O - 3H_2O \rightarrow CO + CO_2$$

Excess concentrated sulphuric acid at 180–200°C dehydrates ethanol to ethene (ethylene):

$$\begin{array}{c} CH_3 \\ | \\ CH_2OH \end{array} \quad - H_2O \rightarrow \begin{array}{c} CH_2 \\ \| \\ CH_2 \end{array}$$

However, concentrated sulphuric acid surrenders its own water to phosphorus pentoxide:

$$2H_2SO_4 + P_4O_{10} \rightarrow 4HPO_3 + 2SO_3$$

Concentrated sulphuric acid will combine with aromatic hydrocarbons to form sulphonic acids with elimination of water. This is called sulphonation. Benzenesulphonic acid is best prepared by adding benzene in small portions to fuming sulphuric acid, with agitation and cooling. The sulphonation is exothermic:

$$C_6H_6 + HO.SO_2.OH \rightarrow C_6H_5.SO_2.OH + H_2O$$

The sulphur atom is attached to the carbon of the aromatic ring.

Hot concentrated sulphuric acid is an oxidizing agent. The action is complex and equations indicate the probable main reaction. Sulphur dioxide is the first reduction product of sulphuric acid (regard it as sulphur trioxide solution) but sulphur and hydrogen sulphide may be formed by more powerful reduction:

$$\text{SO}_3 \text{ forms SO}_2, \text{ S}, \text{ H}_2\text{S}$$

Oxidation numbers: $\quad +6 \qquad +4 \quad 0 \quad -2$

Hot concentrated sulphuric acid oxidizes silver, mercury and aluminium to sulphates with liberation of sulphur dioxide:

$$2Ag + 2H_2SO_4 \rightarrow Ag_2SO_4 + 2H_2O + SO_2$$

$$Hg + 2H_2SO_4 \rightarrow HgSO_4 + 2H_2O + SO_2$$

$$2Al + 6H_2SO_4 \rightarrow Al_2(SO_4)_3 + 6H_2O + 3SO_2$$

Vigorous reduction of hot concentrated sulphuric acid occurs with zinc, sulphur dioxide and much sulphur being formed:

$$Zn + 2H_2SO_4 \rightarrow ZnSO_4 + 2H_2O + SO_2$$

$$3Zn + 4H_2SO_4 \rightarrow 3ZnSO_4 + 4H_2O + S$$

Non-metallic elements do not form sulphates: hot concentrated sulphuric acid forms the acidic oxide or oxoacid. Carbon and sulphur form the dioxide, arsenic the trioxide and phosphorus, orthophosphoric acid:

$$C + 2H_2SO_4 \rightarrow CO_2 + 2H_2O + 2SO_2$$

$$S + 2H_2SO_4 \rightarrow 3SO_2 + 2H_2O$$

$$4As + 6H_2SO_4 \rightarrow As_4O_6 + 6H_2O + 6SO_2$$

$$4P + 8H_2SO_4 \rightarrow 4H_3PO_4 + 2H_2O + 7SO_2 + S$$

Oxidation of compounds also occurs. Notable examples include the halogen hydrides:

$$2HBr + H_2SO_4 \rightarrow 2H_2O + SO_2 + Br_2$$

$$6HI + H_2SO_4 \rightarrow 4H_2O + S + 3I_2$$

$$8HI + H_2SO_4 \rightarrow 4H_2O + H_2S + 4I_2$$

The oxidizing characteristics of hot concentrated sulphuric acid limit its use in the displacement of other more volatile acids from their salts. From the ionic standpoint, the displaced acids may be stronger than sulphuric acid, but only volatility matters here. Nitrates yield nitric acid on warming, and chlorides, hydrogen chloride:

$$KNO_3 + H_2SO_4 \rightarrow KHSO_4 + HNO_3$$

$$NaCl + H_2SO_4 \rightarrow NaHSO_4 + HCl$$

However, hydrogen bromide is partially oxidized to bromine and hydrogen iodide virtually wholly to iodine during displacement from their salts by concentrated sulphuric acid.

Aqueous solutions of sulphuric acid have all the usual properties of acids, due to the presence of hydrogen ions. The acid is neutralized by bases, liberates carbon dioxide from carbonates and hydrogencarbonates, and generates hydrogen with metals placed above hydrogen in the electrochemical series. The concentration of a solution may be determined by titration against a standard base.

Sulphuric acid is a dibasic acid giving two series of salts: sulphates and hydrogensulphates:

$$H_2SO_4 \qquad NaHSO_4 \qquad Na_2SO_4$$
sodium hydrogen sulphate, sodium bisulphate — sodium sulphate

Sulphates and hydrogensulphates of importance are described under the metal. Sulphates may be detected in solution by the formation of a white precipitate of barium sulphate on addition of hydrochloric acid and barium chloride solution. Only fluorosilicates also do this. In the absence of hydrochloric acid many other precipitates could be formed, e.g. sulphite, carbonate or any insoluble salt of a weakly ionized acid, all of which dissolve in a strongly ionized acid. Sulphates may be estimated gravimetrically by precipitation as barium sulphate, weighed as such and the mass of sulphate ion calculated:

$$\frac{\text{Mass sulphate ion}}{\text{Mass barium sulphate}} = \frac{\text{Formula mass } SO_4{}^{2-}}{\text{Formula mass } BaSO_4}$$

Most sulphates are soluble, the exceptions being those of lead, calcium, strontium and barium. Sulphates are notable for forming isomorphous series, examples from three such series being:

$$FeSO_4 . 7H_2O \qquad FeSO_4 . (NH_4)_2SO_4 . 6H_2O$$
Green vitriol — Mohr's salt
iron(II) sulphate heptahydrate — ammonium iron(II) sulphate

$$(NH_4)_2SO_4 . Fe_2(SO_4)_3 . 24H_2O$$
$$(NH_4)Fe(SO_4)_2 . 12H_2O$$
iron(III) alum
ammonium iron(III) sulphate

A comparative study of the thermal decomposition of anhydrous metallic sulphates is given in Table 22.4.

Table 22.4 Decomposition of anhydrous metallic sulphates*

Metallic sulphate	Temperature at beginning of decomposition /°C	Temperature of energetic decomposition /°C	Products of decomposition /°C
$FeSO_4$	167	480	Fe_2O_3, $2SO_2$
Fe_2O_3, $2SO_3$	492	560	Fe_2O_3
$Al_2(SO_4)_3$	590	639	Al_2O_3
$PbSO_4$	637	705	$6PbO$, $5SO_3$
$CuSO_4$	653	670	$2CuO$, SO_3
$MnSO_4$	699	790	Mn_3O_4
$ZnSO_4$	702	720	$3ZnO$, $2SO_3$
$2CuO$, SO_3	702	736	CuO
$NiSO_4$	702	764	NiO
$CoSO_4$	720	770	CoO
$3ZnO$, $2SO_3$	755	767	ZnO
$MgSO_4$	890	972	MgO
Ag_2SO_4	917	925	Ag
$6PbO$, $5SO_3$	952	962	$2PbO$, SO_3 (?)
$CaSO_4$	1200	—	CaO

* Decomposition under moving current of air
(Reproduced from: Stephens, F. M. jun. *Chem. Engng Progress*, 1953, **49**, 455)

Sulphuric acid forms two series of organic esters. Linkage of the groups is through oxygen, not sulphur as in the sulphonic acids:

ethyl hydrogen sulphate — dimethyl sulphate

The thermal decomposition of sulphates has been described. Pyrosulphates (disulphates) are formed as intermediates from hydrogensulphates and subsequently decompose to sulphur trioxide and the sulphate. The term 'oil of vitriol' for sulphuric acid arises from its formation by distillation of hydrated iron(II) sulphate, green vitriol, the overall change being:

$$2FeSO_4.7H_2O \rightarrow Fe_2O_3 + SO_2 + SO_3 + 14H_2O$$

$$H_2O + SO_3 \rightarrow H_2SO_4$$

Glauber obtained sulphuric acid by this method in 1648.

The detection of the elements united in sulphuric acid is not easy. Oxygen is evolved on thermal decomposition of sulphuric acid in a silica tube: the gas may be collected over water, into which the other products disperse:

$$H_2SO_4 \rightarrow H_2O + SO_3$$

$$2SO_3 \rightleftharpoons 2SO_2 + O_2$$

Zinc reduces hot concentrated sulphuric acid to a large proportion of sulphur:

$$3Zn + 4H_2SO_4 \rightarrow 3ZnSO_4 + 4H_2O + S$$

The presence of hydrogen must be shown indirectly, dilute acid not being used because of the presence of hydrogen already in water. Hydrogen chloride is displaced from dried sodium chloride of known composition by the action of concentrated sulphuric acid and is passed over heated sodium:

$$NaCl + H_2SO_4 \rightarrow NaHSO_4 + HCl$$

$$2HCl + 2Na \rightarrow H_2 + 2NaCl$$

Sodium chloride is re-formed and hydrogen may be collected and identified. The presence of hydroxyl groups in sulphuric acid is deduced from the action of phosphorus pentachloride and other reactions in which chlorosulphuric (chlorosulphonic) acid and sulphuryl chloride are formed:

$Cl.SO_2.OH$	$Cl.SO_2.Cl$
chlorosulphuric acid	sulphuryl chloride

The structure of sulphuryl chloride may be shown by X-ray methods to be an irregular tetrahedron, due to the two double and two single bonds, and sulphuric acid is formed by reaction between sulphur trioxide and water. Structures are therefore:

| sulphuryl chloride | sulphuric acid |

The use of co-ionic bonding and resonance in this structure has been discussed.

Chlorosulphuric acid, chlorosulphonic acid, $HO.SO_2.Cl$

The first acid chloride of sulphuric acid is chlorosulphuric acid. It may be prepared by the action of dry hydrogen chloride on sulphur trioxide, dissolved in sulphuric acid, as oleum, $H_2S_2O_7$:

$$HCl + SO_3 \rightarrow Cl.SO_2.OH$$

Alternatively, the calculated quantity of phosphorus pentachloride is added to a mixture of concentrated and fuming sulphuric acids. The mixture becomes very hot and hydrogen chloride is evolved. Further quantitites of hydrogen chloride are expelled on warming and the liquid distilled using an air condenser, the distillate up to 161°C being collected and fractionated further:

$$HO.SO_2.OH + PCl_5 \rightarrow$$
$$\text{sulphuric acid} \quad Cl.SO_2.OH + POCl_3 + HCl$$

$$2HO.SO_2.OH + POCl_3 \rightarrow$$
$$2Cl.SO_2.OH + HCl + HPO_3$$

Chlorosulphuric acid is a colourless, fuming liquid, b.p. 151°C, which reacts violently with water to form sulphuric and hydrochloric acids:

$$HO.SO_2.Cl + H_2O \rightarrow HO.SO_2.OH + HCl$$

With the calculated proportion of pure anhydrous hydrogen peroxide, the peroxo acids, peroxo-(mono)sulphuric acid and peroxodisulphuric acid, are prepared. These reactions establish their structures:

peroxo (mono) sulphuric acid (Caro's acid)

peroxodisulphuric acid

Sulphuryl chloride, sulphonyl chloride, SO_2Cl_2

This is the second acid chloride of sulphuric acid. It is prepared by the reaction between chlorine and sulphur dioxide in the presence of granules of active charcoal (blood charcoal, etc.), camphor, ethanoic anhydride (acetic anhydride) or ethene as catalysts:

$$SO_2 + Cl_2 \rightarrow Cl.SO_2.Cl$$

Sulphuryl chloride is a colourless liquid, b.p. 69°C, which is slowly hydrolysed in water, although large quantities will hydrolyse suddenly with violence, sulphuric acid and hydrochloric acid being formed on complete hydrolysis, with chlorosulphuric acid as the intermediate step:

partial:
$$Cl.SO_2.Cl + H_2O \rightarrow HO.SO_2.Cl + HCl$$

complete:
$$HO.SO_2.Cl + H_2O \rightarrow HO.SO_2.OH + HCl$$

Amidosulphuric acid, sulphamidic acid (sulphamic acid), NH$_2$.SO$_2$.OH

Prepared by reacting urea with 100% sulphuric acid, this half amide of sulphuric acid is itself very stable and a strong monobasic acid:

$$NH_2.CO.NH_2 + 2H_2SO_4 \rightarrow$$
$$2NH_2.SO_2.OH + CO_2 + H_2O$$

It is used as a primary standard in titrimetric (volumetric) analysis. All known simple metallic salts are soluble in water.

Amidosulphuric acid is a colourless, crystalline solid, m.p. 205°C, which reacts rapidly and quantitatively with sodium nitrite solution:

$$NH_2.SO_2.OH + NaNO_2 \rightarrow$$
$$N_2 + H_2O + NaHSO_4$$

Compare this reaction with similar reactions of primary amines (aminoalkanes) and amides. Slow hydrolysis of an acidified solution occurs:

$$NH_2.SO_2.OH + H_2O \xrightarrow{H_3O^+} NH_4HSO_4$$

Sulphuric diamide, sulphonyl diamide (sulphamide), SO$_2$(NH$_2$)$_2$

Melting with decomposition at 93°C, this colourless crystalline solid is formed by interaction of ammonia and sulphuryl chloride in an inert solvent trichloromethane (chloroform):

$$Cl.SO_2.Cl + 4NH_3 \rightarrow NH_2.SO_2.NH_2 + 2NH_4Cl$$

Again, there are parallel reactions in Organic Chemistry. It is readily soluble in water, the solution being hydrolysed on boiling:

$$NH_2.SO_2.NH_2 + 2H_2O \rightarrow (NH_4)_2SO_4$$

Peroxo(mono)sulphuric acid, H$_2$SO$_5$, peroxodisulphuric acid, H$_2$S$_2$O$_8$, and potassium peroxodisulphate(persulphate), K$_2$S$_2$O$_8$

Peroxomonosulphuric acid is also known as sulphomonoperacid but is usually called Caro's acid. Peroxodisulphuric acid, formerly persulphuric acid, is also called sulphodiperacid. The first acid forms no salts but the sparingly soluble potassium persulphate and ammonium persulphate are in common use.

Peroxodisulphuric acid may be prepared by the electrolysis of ice-cold 50% sulphuric acid with a high current density at the platinum anode, virtually no oxygen being evolved:

$$2HSO_4^- - 2e^- \rightarrow H_2S_2O_8$$

Usually sulphuric acid, density 1.3 g cm^{-3}, saturated with potassium sulphate at 0–4°C, is used with an anode density of 1×10^4 A m^{-2}. Hydrogen is evolved at the cathode which is partitioned from the anode to avoid interaction of the products. Colourless potassium peroxodisulphate, potassium persulphate, crystallizes out.

Caro's acid is prepared by prolonged triturating of potassium peroxodisulphate with moderately concentrated sulphuric acid cooled in ice, followed by dilution with crushed ice, partial hydrolysis of peroxodisulphuric acid occurring:

$$H_2S_2O_8 + H_2O \rightarrow H_2SO_5 + H_2SO_4$$

Alternatively, Caro's acid is prepared by mixing anhydrous peroxodisulphuric acid and anhydrous hydrogen peroxide:

$$H_2S_2O_8 + H_2O_2 \rightarrow 2H_2SO_5$$

Both acids may be prepared by the action of anhydrous hydrogen peroxide on chlorosulphuric acid in the correct proportions. These reactions, fixing the structure of the acids, have been given on p. 40. Peroxodisulphuric acid is a colourless hygroscopic crystalline solid, m.p. 65°C, melting with decomposition. Caro's acid forms large colourless hygroscopic crystals, m.p. 45°C. Both are hydrolysed by water.

The structural diagrams show:

$$\underset{\text{Caro's acid}}{\begin{array}{c} \text{HO} \diagdown \underset{O \diagup}{S} \diagup \overset{O-O}{\diagdown} \underset{\diagdown O}{S} \diagup \text{OH} \end{array}} + H_2O \rightarrow$$

$$\underset{\text{Caro's acid}}{\begin{array}{c} \text{HO} \diagdown \underset{O \diagup}{S} \diagup \overset{O-O-H}{} \end{array}} + \underset{\text{sulphuric acid}}{\begin{array}{c} \text{HO} \diagdown \underset{O \diagup}{S} \diagdown \overset{OH}{\diagup O} \end{array}}$$

$$\underset{\text{Caro's acid}}{\begin{array}{c} \text{HO} \diagdown \underset{O \diagup}{S} \diagup \overset{O-O-H}{} \end{array}} + H_2O \rightarrow$$

$$\underset{\text{sulphuric acid}}{\begin{array}{c} \text{HO} \diagdown \underset{O \diagup}{S} \diagdown \overset{OH}{\diagup O} \end{array}} + \underset{\text{hydrogen peroxide}}{\begin{array}{c} H-O \\ | \\ H-O \end{array}}$$

$$2H_2O_2 \rightarrow 2H_2O + O_2$$

They are very powerful oxidizing agents. Comparison of the structures with that of hydrogen peroxide shows the $-O-O-$ 'bridge' and the fact that Caro's acid resembles hydrogen peroxide in having the $-O-O-H$ grouping. A comparison of the reactions of the three compounds shows that peroxodisulphuric acid is sometimes slower in acting, probably due to hydrolysis being a necessary first stage. Thus, peroxodisulphates (persulphates) liberate iodine slowly from potassium iodide solution while Caro's acid and hydrogen peroxide react immediately. The potassium and ammonium salts are the usual laboratory, persulphate, reagents, for which the ion-electron equation is:

$$S_2O_8^{2-} + 2e^- \rightleftharpoons 2SO_4^{2-}$$

Iron(II) salts are oxidized to the iron(III) state, iodides to iodine and metallic zinc or copper to their sulphates:

$$2Fe^{2+} + S_2O_8^{2-} \rightarrow 2Fe^{3+} + 2SO_4^{2-}$$

$$2I^- + S_2O_8^{2-} \rightarrow I_2 + 2SO_4^{2-}$$

$$Zn + S_2O_8^{2-} \rightarrow Zn^{2+} + 2SO_4^{2-}$$

Aqueous manganese(II) salts are oxidized to manganese(IV) oxide unless silver nitrate is present, when oxidation goes to the permanganate stage (Marshall's reaction):

$$Mn^{2+} + 2H_2O + S_2O_8^{2-} \rightarrow \\ MnO_2(s) + 4H^+ + 2SO_4^{2-}$$

$$2Mn^{2+} + 8H_2O + 5S_2O_8^{2-} \rightarrow \\ 2MnO_4^- + 16H^+ + 10SO_4^{2-}$$

Caro's acid, peroxodisulphuric acid and its salts may be distinguished from hydrogen peroxide because pure samples neither decolorize acidified potassium permanganate solution, nor give a blue coloration of CrO_5 with acidified potassium dichromate solution.

Thiosulphuric acid(trioxothiosulphuric acid), $H_2S_2O_3$, and thiosulphates

Sodium thiosulphate, $Na_2S_2O_3.5H_2O$, has been described under sodium, p. 281. It is used as photographic fixer because of the ready formation of complexes with silver ions and as an antichlor, reacting with traces of chlorine after bleaching. In titrimetry, sodium thiosulphate is used in the iodometric determination of oxidizing agents. Acidification of sodium thiosulphate solution gives sulphur dioxide and sulphur:

$$S_2O_3^{2-} + 2H^+ \rightarrow H_2S_2O_3$$

$$H_2S_2O_3 \rightarrow H_2SO_3 + S(s)$$

$$H_2SO_3 \rightleftharpoons H_2O + SO_2$$

In dilute solution there is a considerable latent period before sulphur appears, interpreted as either due to the slow coagulation of sulphur particles or the slow decomposition of thiosulphuric acid. The decomposition is very complex and other products may also be obtained.

Thionic acids, $H_2S_nO_6$ ($n = 2, 3, 4, 5, 6$ and much higher values)

From our point of view, the thionic acids are only important as illustrations of the ready way in which sulphur-sulphur bonds are formed, examples of *catenation*. The structures of dithionates and trithionates have been confirmed by X-ray analysis:

$$\underset{\substack{H_2S_2O_6 \\ \text{dithionic acid}}}{\begin{array}{c} O\ \ O \\ \| \ \ \| \\ HO-S-S-OH \\ \| \ \ \| \\ O\ \ O \end{array}} \qquad \underset{\substack{H_2S_3O_6 \\ \text{trithionic acid}}}{\begin{array}{c} O \quad\ S\ \ O \\ \| \diagup\ \diagdown \| \\ HO-S \qquad S-OH \\ \| \qquad\quad \| \\ O \qquad\quad O \end{array}}$$

$$\underset{\text{dithionate ion}}{\begin{array}{c} O\ \ O \\ \| \ \ \| \\ \bar{O}-S-S-\bar{O} \\ \| \ \ \| \\ O\ \ O \end{array}} \qquad \underset{\text{trithionate ion}}{\begin{array}{c} O \quad\ S\ \ O \\ \| \diagup\ \diagdown \| \\ \bar{O}-S \qquad S-\bar{O} \\ \| \qquad\quad \| \\ O \qquad\quad O \end{array}}$$

Dithionic acid is a strongly ionized acid, being somewhat distinct from the others. Sodium tetrathionate is formed in the reaction between sodium thiosulphate solution and iodine, utilized in titrimetric (volumetric) analysis:

$$2S_2O_3^{2-} + I_2 \rightarrow 2I^- + S_4O_6^{2-}$$
$$\text{tetrathionate}$$

The halides of sulphur

Both fluorine and chlorine form a range of compounds with sulphur, although the chemical identity of some of them is doubtful. Only one (unstable) compound of bromine has been isolated and none of iodine. Only important compounds are described. Those which appear definite are:

$$S_2F_2 \qquad\qquad S_2F_{10} \quad SF_6$$
$$S_2Cl_2 \quad SCl_2 \quad SCl_4$$
$$S_2Br_2$$

Sulphur hexafluoride and *(di)sulphur decafluoride* are inert, the first a gas at room temperature, subliming at −64°C from the solid; the second, a liquid boiling at 29°C under atmospheric pressure. The maximum covalency of six is exerted.

Disulphur dichloride, a pale yellow fuming liquid, b.p. 138°C, with a pungent odour, is an excellent solvent for sulphur used in vulcanization of rubber. Disulphur dichloride is made by action of chlorine on molten sulphur, as in Fig. 20.9, p. 338, or by refluxing sulphuryl chloride and sulphur at 30–70°C with aluminium chloride as catalyst, followed by re-distillation using an air condenser:

$$2S + Cl_2 \rightarrow S_2Cl_2$$

$$SO_2Cl_2 + 2S \rightarrow S_2Cl_2 + SO_2$$

The hydrolysis of disulphur dichloride is complex, giving sulphur, sulphite and thiosulphate, hydrogen sulphide and thionic acids.

The action of chlorine on ice-cold disulphur dichloride yields *sulphur dichloride*, SCl_2, a garnet-red liquid, b.p. 59°C. Iodine trichloride and tin(IV) chloride are catalysts.

Replacing the silicon chip

Computers are based on the *silicon chip* which is not more than 1×10^{-5} m thick, a hundred times smaller than the transistor which replaced the valve, itself about 10 cm long. The next development will almost certainly be in the direction of molecular dimensions. Metals conduct electricity because electrons can be promoted between energy levels very easily. Semiconductors (in the middle of the Periodic Table, e.g. Si) have an energy barrier which prevents the free flow of electrons until some of these receive enough energy (become excited) to reach energy levels through which they can move freely. Chemists are attempting to design molecules which can be stacked up in a crystal in such a way that the metal atoms are lined up in a single direction in the crystal. The electrons can then flow along this line of atoms. Sulphur and selenium have proved ideal ligands for this scheme. The performance of such one dimensional metals depends on the architectural details of the molecular structure, the interatomic distances and alignment of the framework. Such a one dimensional metal conductor structure is shown in Fig. 22.4 and underwrites the elegance of the original concept.

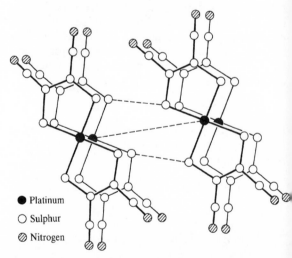

● Platinum
○ Sulphur
⊘ Nitrogen

Fig. 22.4   The structure of one of the 'molecular metals' as shown by X-rays. The flat molecules are stacked in layers so the electrons can 'hop' between the molecules (by courtesy of SERC)

23

Group VII: the halogens

Fluorine, chlorine, bromine and iodine

9
F
2, 7
17
Cl
2, 8, 7
35
Br
2, 8, 18, 7
53
I
2, 8, 18, 18, 7
85
At
2, 8, 18, 32, 18, 7

The name halogen (Greek, *hals genon* = sea-salt producing) was used by Berzelius because he wished to indicate that chlorine, bromine and iodine occurred in the sea as salts. Sodium chloride is sea-salt. The halogens are a distinctive family of diatomic non-metallic elements showing a progressive gradation in reactivity. Fluorine shows many anomalies as the first member of the family, and indeed, it could be argued that fluorine is so distinctive that it is not a halogen, in the family sense, at all. The extreme non-metallic properties of fluorine are toned down with successive members of the family until some slight metallic characteristics appear with iodine. The radioactive element, astatine, is not described here.

As the atomic number increases the elements darken in colour, becoming less volatile and less reactive. Fluorine is a pale yellow gas at room temperature, lighter in colour than chlorine which is a greenish-yellow gas. Bromine is a dark red liquid which gives off a dense red vapour, while iodine is a shiny grey crystalline solid, giving a violet vapour on heating. Fluorine, chlorine and the vapour of bromine are extremely irritant to nose and throat passages and are very poisonous.

Chlorine was discovered by Scheele in 1774 but not recognized as an element. The gas was formed in a reaction between hydrochloric acid, or muriatic acid as it was known (Latin, *muria* = brine), and manganese(IV) oxide (pyrolusite). Because aqueous solutions of the new substance evolved oxygen in sunlight and because it was formed by oxidation of muriatic acid, Berthollet called the gas oxymuriatic acid in 1785. In a later series of experiments, Davy failed to show the presence of oxygen, concluding that the gas was an element. In 1810, he called it chlorine (Greek, *chloros* = greenish-yellow). The oxygen had originated from water. The action of concentrated sulphuric acid on calcium fluoride (fluorspar), generating hydrogen fluoride, had been known since about 1720. In 1810, Ampère put forward the hypothesis that this substance was a compound of hydrogen and a hitherto unknown element analogous to chlorine which he proposed to call fluorine. This element was isolated by the electrolysis of potassium hydrogenfluoride containing anhydrous hydrogen fluoride by Moissan in 1886. Courtois discovered iodine (Greek, *ioïdes* = violet in colour) in 1812; after extraction of the ashes of seaweed with water and crystallization of sodium carbonate, he added sulphuric acid to the mother liquor and on heating, obtained a violet vapour condensing to a dark solid. Bromine (Greek, *bromos* = stench) was isolated by Balard in 1826 by the action of chlorine on the bittern or residual liquor remaining after crystallization of salt from sea water.

Physical data for the halogens are listed in Table 23.1.

A general survey of bonding involving the halogens of the *p* block

The main characteristic of the halogens is their marked univalency in both ionic and covalent bonding. Each halogen is positioned immediately before an inert gas in the Periodic Table. Acceptance of an electron or the sharing of a single pair of electrons, to which it has contributed one, results in the halogen acquiring an inert gas electronic configuration.

Table 23.1   Group VII (halogens): physical data p block elements $(ns)^2(np)^5$ $n \neq 1$

	F	Cl	Br	I
Atomic number	9	17	35	53
Electronic configuration	2, 7	2, 8, 7	2, 8, 18, 7	2, 8, 18, 18, 7
[Core]outer electrons	$[He](2s)^2(2p)^6$	$[Ne](3s)^2(3p)^6$	$[Ar,3d^{10}](4s)^2(4p)^6$	$[Kr,4d^{10}](5s)^2(5p)^6$
Relative atomic mass	18.9984	35.453	79.904	126.9044
Atomic radius/nm	0.064	0.099	0.114	0.133
Ionic radius/nm (X^-)	0.136	0.181	0.195	0.216
van der Waals' radius/nm	0.135	0.180	0.195	0.215
Atomic volume/cm^3 mol^{-1}*	17.2	23.5	27.1	34.2
Standard electrode potential (H scale: 298 K)/V	+3.06	+1.36	+1.06	+0.536
Boiling-point/°C	−188	−34.1	+58.8	+187
Melting-point/°C	−220	−101	−7.3	+114
Electronegativity	4.0	3.0	2.8	2.4
Electron affinity/kJ mol^{-1} (273 K)	−349	−364	−343	−314
Co-ordination number	1	1–4	1–5, not 4	1–7
Abundance (% by mass)	0.027	0.048	c.0.003	3×10^{-5}

*Traditional term. It is the volume of one mole of atoms as solid element

Ionization energies for the halogens are shown in Table 23.2. There is some tendency with iodine (see values in bold typeface) to form positively charged ions but by far the most important chemistry from our viewpoint lies in the covalently bound atoms and the uninegatively charged ions of the elements in this group.

First ionization energies are plotted against atomic number in Fig. 23.1 and electron affinities similarly in Fig. 23.2. Somewhat surprisingly, chlorine has the highest electron affinity in the group.

Atomic and ionic radii are plotted against atomic number in Fig. 23.3, which also shows van der Waals' radii, values of which are virtually the same as those for the radii of X^- ions, and electronegativities similarly in Fig. 23.4. These values which illustrate the gradation of properties within the group must be compared with those of elements

Table 23.2   Group VII (the halogens): ionization energies/ kJ mol^{-1} (273 K)

	F	Cl	Br	I
1st $M(g) \rightarrow M(g)^+ + e^-$	1697	1254	1139	**1003**
2nd $M^+(g) \rightarrow M^{2+}(g) + e^-$	3377	2296	2100	**1843**
3rd $M^{2+}(g) \rightarrow M^{3+}(g) + e^-$	6040	3850	2480	**3039**
4th $M^{3+}(g) \rightarrow M^{4+}(g) + e^-$	8413	5162	4853	—

Electronic configurations

3 **Li** 2, 1						9 **F** 2, 7	10 **Ne** 2, 8
11 **Na** 2, 8, 1	12 **Mg** 2, 8, 2	13 **Al** 2, 8, 3	14 **Si** 2, 8, 4	15 **P** 2, 8, 5	16 **S** 2, 8, 6	17 **Cl** 2, 8, 7	18 **Ar** 2, 8, 8
19 **K** 2, 8, 8, 1						35 **Br** 2, 8, 18, 7	36 **Kr** 2, 8, 18, 8
37 **Rb** 2, 8, 18, 8, 1						53 **I** 2, 8, 18, 18, 7	54 **Xe** 2, 8, 18, 18, 8
55 **Cs** 2, 8, 18, 18, 8, 1							

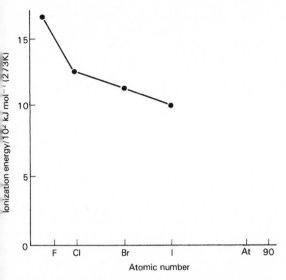

Fig. 23.1 First ionization potential plotted against atomic number for Group VII elements

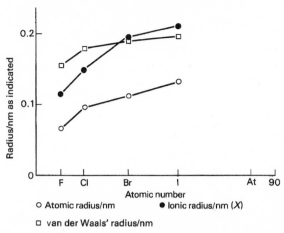

Fig. 23.3 Atomic, ionic and van der Waals' radii plotted against atomic number for Group VII elements (Note that ionic and van der Waals' radii are almost the same)

in other groups, and with the periodic treatment in Chapters 3 and 5.

The halides of the stronger metals are ionic while those formed by the non-metallic elements and weak metals are covalent. Fluorine, being the most electronegative element, enters into ionic combination with certain metals, e.g. mercury(II), tin(IV), bismuth(III), which otherwise form covalent halides, the roman numerals in the examples signifying final oxidation states. Fluorine is also more electronegative than oxygen, displacing it from water. As the atomic number increases, the forces of attraction between positive nucleus and valence electrons decrease because of the increased atomic radius and the screening effect of intermediate shells. Fluorine has the greatest tendency to form

anions in solution and iodine the smallest, astatine not being considered here. This is shown by their mutual displacements: the halogens are displaced from simple halide salts by a halogen of lower atomic number.

Standard electrode potentials are plotted against atomic number in Fig. 23.5. The positive values show that atoms of the elements of lower atomic number have a very high tendency to acquire an

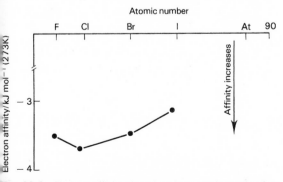

Fig. 23.2 Electron affinity plotted against atomic number for Group VII elements: (caution: see page 44 for definition and sign convention)

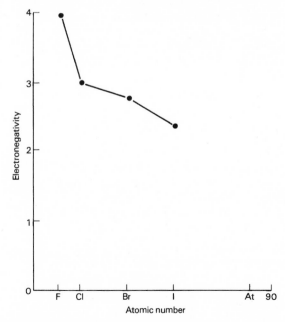

Fig. 23.4 Electronegativity plotted against atomic number for Group VII elements

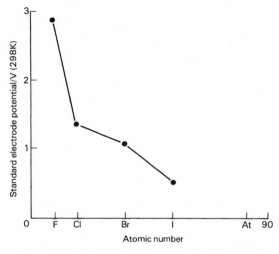

Fig. 23.5 Standard electrode potentials plotted against atomic number for Group VII elements

electron each to form the anion of single charge:

$$\tfrac{1}{2}Cl_2 + e^- \rightleftharpoons Cl^-$$

at the electrode Cl^-, $Cl_2(Pt)$.

These electrode potentials are the potential differences for the cells made respectively with the standard hydrogen electrode, e.g.

$$Pt, H_2 \mid H^+ \mid Cl^- \mid Cl_2, Pt$$

implying the reaction

$$\tfrac{1}{2}H_2 + \tfrac{1}{2}Cl_2 \rightarrow H^+ + Cl^-$$

Consider the simple anions of unit negative charge: with increase of size the extra charge will tend to be dragged away by small cations with the formation of covalent bonds. *Fajans' Rules* (p. 82) are well illustrated by the silver halides of which silver iodide is covalent, the others being more ionic as the atomic number of the halogen decreases. The distortion of electron shells, to some extent balanced by pulls in all directions in the solid lattice, becomes much more directional in the vapour state and covalency sometimes results. Aluminium chloride is probably ionic in the solid state but covalent as a vapour, while the bromide and iodide remain covalent under both conditions. The covalent structure for the chloride is shown below.

	Sodium fluoride		Potassium chloride		Rubidium bromide		Caesium iodide	
	Na^+	F^-	K^+	Cl^-	Rb^+	Br^-	Cs^+	I^-
Electronic configurations:	2,8	2,8	2,8,8	2,8,8	2,8,18,8	2,8,18,8	2,8,18,18,8	2,8,18,18,8
		Ne		Ar		Kr		Xe

The halogens and the alkali metals are separated by the family of inert gases.

Sodium chloride	Magnesium chloride	Aluminium chloride	Silicon tetrachloride	Phosphorus trichloride	Sulphur dichloride	Chlorine
		Cl | Cl—Al—Cl	Cl | Cl—Si—Cl | Cl	Cl | P Cl⁄ ⁀Cl	Cl⁄S⁀Cl	
Na^+ Cl^-	$Mg^{2+}(Cl^-)_2$					Cl—Cl
2, 8	2, 8	2, 8, 6	2, 8, 8	2, 8, 8	2, 8, 8	2, 8, 8
Ne	Ne	*not* inert gas configuration *but* (Al_2Cl_6) $[Al(OH_2)_6]^{3+}(Cl^-)_3$ are formed	Ar	Ar	Ar	(in all examples) Ar
ionic	*ionic* with trace of covalent character	*intermediate* anhydrous: covalent vapour (dimerized), probably ionic solid; hydrate: ionic (H₂O shown electrostatically bound and not by co-ionic links, for simplicity)	*covalent*	*covalent*	*covalent*	*covalent*
dissolves in water	*dissolves:* hydrolysis on evaporation	*some hydrolysis*	*hydrolysed*	*hydrolysed*	*hydrolysed*	*partial hydrolysis*

Electronic configurations are given for the element forming the halide shown.

Bond diagrams for some halides of Group I (alkali) metals and some chlorides formed by elements of period 3 are given opposite. The chlorine atom acquires the electronic configuration of argon.

Fluorine brings out the highest covalencies in elements with which it combines. To illustrate this point, highest covalencies shown to fluorine and chlorine by arsenic, sulphur and iodine may be compared:

$$AsF_5 \quad\quad SF_6 \quad\quad IF_7$$
$$AsCl_3 \quad\quad SCl_4 \quad\quad ICl_3$$

Notice a similarity between fluorine and hydrogen here. Both have a single valency and both, when combined, are satellites rather than central in covalent molecules, remaining necessarily on the periphery of molecules. On the other hand, the relatively large size of the iodine atom limits the number of such atoms which may be grouped around a central atom, as in the sequence, PF_5, PCl_5, PBr_5 *but* PI_3, of the highest halides of phosphorus.

Higher covalencies than one are shown by all of the halogens except fluorine. As in other groups, with increasing atomic number there is a regular shift towards lower valency and emergence of electropositive characteristics. Chlorine has valencies of 1–7 inclusive except 2, the extreme values bringing the most stability to the compounds formed.

Expressed another way, in combination fluorine shows only the oxidation state $= -1$, while the others have principal oxidation states $= -1$, $+1$, $+5$, $+7$ although the last for bromine is only probable while chlorine forms additionally $HClO_2$ in which it has an oxidation number $= +3$; ClO_2, $+4$; $ClO_3(Cl_2O_6)$, $+6$. Fluorine does not have $3d$ orbitals available for bonding and this limits its oxidation number in combination to -1. The others have such $3d$ energy levels available and so can form a range of compounds, in which the useful, but limited, notion of the octet of electrons has to be set aside. The extension of the octet rule for covalent compounds has been outlined in Chapter 4, p. 54, using fluorides of silicon, phosphorus and sulphur (SiF_6^{2-}, PF_6^- and SF_6) as examples. The limitations of the Theory of Resonance have been indicated already when depicting the structures of oxoacids of phosphorus and sulphur. In passing note that such compounds as the fluorides (as molecules and anions) mentioned above have entities in which many atoms are crowded together, causing further electrical interaction and splitting of energy levels. Further, it has been noted generally that in a sequence of oxoanions of general formulae, XO^{n-}, XO_2^{n-}, XO_3^{n-}, XO_4^{n-}, reaction rates fall off with increasing proportion of oxygen. This may be due to a steric hindrance factor.

Structures of oxides, oxoacids and oxoanions of chlorine merit closer scrutiny.

Cl_2O Chlorine monoxide	ClO_2 Chlorine dioxide	$Cl_2O_6 \rightleftharpoons 2ClO_3$ Chlorine hexoxide (dissociates)	Cl_2O_7 Chlorine heptoxide

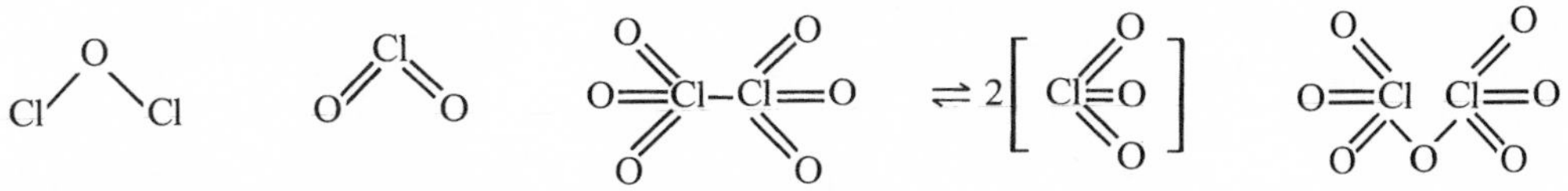

bent (V-shaped)	bent (V-shaped)		—	tetrahedral about each Cl atom
2,8,8 (Ar)	2,8,11 odd electron molecule	2,8,14	2,8,13 odd electron molecule	2,8,14

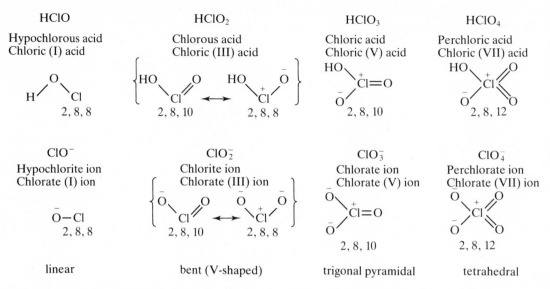

The electron configurations are those acquired by chlorine.

Oxidation states

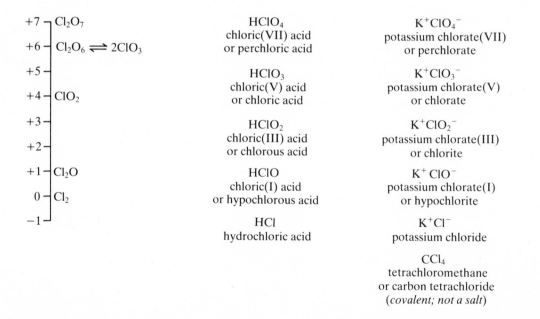

Notice that ClO_2 and ClO_3 are odd-electron molecules. Co-ionic structures may be written but seem to have no special advantages in being a closer representation of the bonding as shown by bond length measurements. Cl_2O_7 is the only exothermic oxide formed by chlorine and the only one which is thermodynamically stable. Oxoacids of chlorine and their anions have similarly been written with co-ionic bonds using resonance to satisfy the octet rule, but double bond structures are preferred because the measured bond lengths indicate much double bond character. It is interesting to note that the shape of the perchlorate ion is an irregular tetrahedron, one pair of $Cl-O$ bonds being shorter than the other pair. This might have been expected from the ideas propounded in discussing the shapes by simple inorganic molecules in Chapter 5, p. 71.

Oxidation states displayed by chlorine have been shown on the 'ladder' below the structures. Although the IUPAC rules specify that the *Stock Notation* 'should preferably *not* be applied to compounds between non-metals'. it can be applied to both cations and anions. Its use for anions leads naturally to its use for the corresponding acids. Such names are listed with the more usual names and have been adopted by the *Nuffield* scheme for oxoacids and their salts in the periodic family.

Oxygen compounds of the halogens are very irregular in behaviour, probably due to the very high electronegativity of oxygen.

Univalent halogen atoms are incorporated into many complex anions. Usually the chloride complex is more stable than the bromide, which in turn is more stable than the iodide while the fluoride complexes follow no general rule. Beryllium forms the fluoroberyllate ion, BeF_4^{2-}, and boron the fluoroborate ion, BF_4^-, neither having analogues with the other halogens. On the other hand, bivalent lead forms complex ions of formula $PbCl_3^-$, $PbCl_4^{2-}$ and $PbCl_6^{4-}$, and similar ions for bromine and iodine but not fluorine. Tin(IV) forms the complex ions SnF_6^{2-} and $SnCl_6^{2-}$ of comparable stability but greater than that of the corresponding bromide and iodide.

Slight electropositive behaviour emerges with iodine. The hydride is notable for being an endothermic compound at room temperature, readily undergoing thermal dissociation. Liquid iodine monochloride, ICl, conducts electricity, as will solutions in ethanoic (acetic) acid and nitrobenzene, with deposition of iodine on the cathode. Compounds of general formula IX_3, yellow in colour and less stable than the basic salt, $(IO)X$, have been prepared. The formation of a basic ion is typical of weakly electropositive character. Compounds of tervalent iodine include the chloride, ICl_3; nitrates, $I(NO_3)_3$, $IO(NO_3)$; iodate, $I(IO_3)_3$; ethanoate (acetate), $I(O.OC.CH_3)_3$; perchlorate (chlorate(VII)), $I(ClO_4)_3.2H_2O$, and a basic sulphate, $(IO)_2SO_4$. X-ray studies have not so far revealed the presence of positively charged iodide ions. Iodine trichloride seems to ionize:

$$2ICl_3 \rightleftharpoons ICl_2^+ + ICl_4^-$$

In oxoacids and salts, iodic acid is more stable than bromic acid, which in turn is more stable than chloric acid. There is no fluoric acid. Iodine displaces the other halogens from chlorates and bromates, a reversal of the displacement order from simple halides.

A laboratory introduction to the halogens

Concentrated sulphuric acid on metal halides

The sodium halides are colourless crystalline solids. On addition of concentrated sulphuric acid with warming, if desired, sharp differences appear. The halides of other metals react similarly.

Hydrogen fluoride is evolved from fluorides: it is a colourless acidic gas (b.p. 19.6°C) with a pungent odour and etches the walls of the test-tube, causing a greasy appearance as droplets of liquid collect:

$$NaF + H_2SO_4 \rightarrow \underset{\substack{\text{sodium} \\ \text{hydrogensulphate}}}{NaHSO_4} + HF$$

$$\underset{\text{from glass}}{SiO_2} + 4HF \rightarrow \underset{\substack{\text{silicon} \\ \text{tetrafluoride}}}{SiF_4} + 2H_2O$$

Hydrogen chloride (solution = 'spirits of salt') is formed from chlorides; it is a colourless gas which has an acrid odour and fumes in moist air giving droplets of hydrochloric acid:

$$NaCl + H_2SO_4 \rightarrow NaHSO_4 + HCl$$

Bromides yield hydrogen bromide, similar in appearance to hydrogen chloride, and a reddish vapour due to some free bromine, sulphuric acid ($= SO_3 + H_2O$) being reduced to sulphur dioxide and water:

$$KBr + H_2SO_4 \rightarrow KHSO_4 + HBr$$

$$2HBr + H_2SO_4 \rightarrow Br_2 + SO_2 + 2H_2O$$

Iodides give a complex series of reactions in which violet iodine vapour is evolved, some condensing to a dark solid, with the formation of hydrogen sulphide and free sulphur. Some hydrogen iodide, similar in appearance to hydrogen chloride, is evolved and initially, there may also be sulphur dioxide. Various reactions probably occur:

$$KI + H_2SO_4 \rightarrow KHSO_4 + HI$$

$$H_2SO_4 + 2HI \rightarrow SO_2 + I_2 + 2H_2O$$

$$H_2SO_4 + 6HI \rightarrow S + 3I_2 + 4H_2O$$

$$H_2SO_4 + 8HI \rightarrow H_2S + 4I_2 + 4H_2O$$

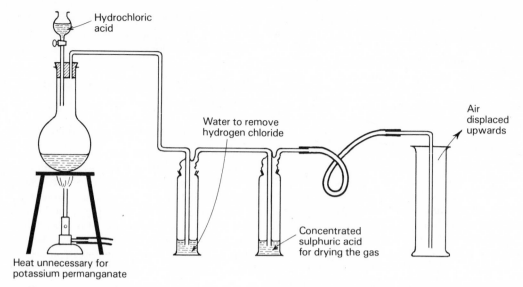

Fig. 23.6   The preparation of chlorine (in fume cupboard). Alternatively, a proprietary brand of glass-jointed apparatus may be used for the preparation

This illustrates the increasing liability of the hydrides to oxidation as the atomic number increases.

Concentrated sulphuric acid with manganese(IV) oxide (dioxide), as the oxidizing agent, on the metal halides†

Except for fluorides, the free halogen is evolved in each case:

$$MnO_2 + 3H_2SO_4 + 2NaBr \rightarrow$$
$$MnSO_4 + 2NaHSO_4 + Br_2 + 2H_2O$$

Elusive fluorine

Chemical oxidation methods fail to liberate fluorine from fluorides. Fluorine is the most electronegative element and strongest oxidizing agent. Fluoride ions will only lose electrons by electrolysis of the fused salt.

The laboratory preparations of chlorine, bromine and iodine

Chlorine may be generated by the oxidation of concentrated hydrochloric acid. Granular manganese(IV) oxide is frequently used as the oxidizing agent, the reaction being speeded by warming. Alternatively, concentrated hydrochloric acid†† may be allowed to drip on to solid potassium

permanganate (manganate(VII)) in the cold. Washing the gaseous product with water removes hydrogen chloride vapour. The gas is dried by concentrated sulphuric acid and collected by upward displacement of air (relative densities: chlorine 35.5, air 14.4). Alternatively, chlorine may be collected, without drying, over brine.

$$MnO_2 + 4H^+ + 2Cl^- \rightarrow$$
$$Cl_2 + Mn^{2+} + 2H_2O$$

$$2MnO_4^- + 16H^+ + 10Cl^- \rightarrow$$
$$5Cl_2 + 2Mn^{2+} + 8H_2O$$

Other oxidizing agents may be used, including lead(IV) oxide and red lead. Using an adaptation of the method outlined in the previous section, concentrated sulphuric acid may be run on to a mixture of sodium chloride and manganese(IV) oxide, but a more controllable evolution of gas results from the use of diluted acid (e.g. 11 acid: 8 water, by volume) with warming. Suitable apparatus is shown in Fig. 23.6. Another very effective way of generating chlorine is by the action of dilute acid on bleaching powder, a mixture of calcium

† To avoid repetition, where two or more halogens undergo similar reactions only one equation will be written. The others follow by substituting the appropriate symbol

†† *Warning*: make sure that concentrated sulphuric acid is not picked up by mistake, because a dangerous explosion could result

hypochlorite and basic calcium chloride, $Ca(OCl)_2$ + $CaCl_2 \cdot Ca(OH)_2 \cdot H_2O$:

$$OCl^- + Cl^- + 2H^+ \rightarrow H_2O + Cl_2$$

Chlorine is a dense, yellowish-green gas of characteristic choking odour, turning moist litmus paper red and bleaching it.

Both bromine and iodine are prepared by the reactions between sulphuric acid, preferably diluted as above, sodium bromide or iodide and manganese(IV) oxide (manganese dioxide). Bromine is evolved as a dark red, very dense vapour (relative densities: bromine 80, air 14.4) and has an extremely unpleasant irritating action on the nose and throat passages. It condenses to a dark red-brown liquid which can cause severe burns on the skin. It is extremely corrosive and either a retort or a convenient glass-jointed distillation apparatus is used for the preparation. The traditional retort method is shown in Fig. 23.7.

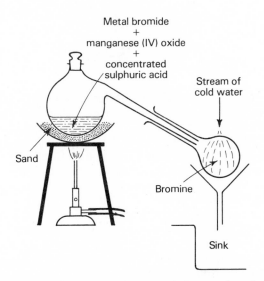

Fig. 23.7 The preparation of bromine (in fume cupboard). A proprietary form of glass-jointed distillation apparatus may be preferred

Iodine is evolved as violet vapour, condensing to shiny grey crystals which may be purified by sublimation. It may be conveniently prepared on the small scale (Fig. 23.8) in a beaker, the crystals subliming on to the bottom of an evaporating dish filled with cold water. Impure iodine may be freed from other halogens by sublimation after mixing with potassium iodide.

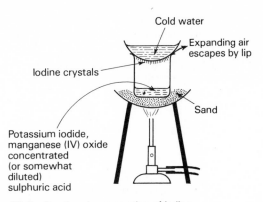

Fig. 23.8 Small-scale preparation of iodine

A comparison of the reactions of chlorine, bromine and iodine showing the progressive gradation of properties

In common with the first member of other families, fluorine shows some anomalous properties, and as the study of fluorine in the laboratory presents special problems, it may with advantage be postponed until the broad trends in halogen chemistry have been appreciated.

Generally, the following reactions may be classified as oxidation by halogens. Ionic oxidation requires the loss of electrons while covalent oxidation may appear as the gain of halogen or oxygen, or the loss of hydrogen or an electropositive element. When halogen molecules become anions, reduction has occurred:

$$Cl_2 + 2e^- \rightleftharpoons 2Cl^-$$

The halogens are especially notable for their powerful affinity for hydrogen.

1 Synthesis of hydrides of the halogens

Chlorine and hydrogen readily combine to form hydrogen chloride. Mixtures of the gases explode in sunlight, in the glare of burning magnesium or when a lighted taper is applied. Quiet combination occurs over a catalyst of activated charcoal, or in diffuse daylight. A jet of one gas will burn in an atmosphere of the other with a silvery-grey flame. Hydrogen and bromine vapour combine with a feeble flame when ignited or passed over a heated platinized asbestos catalyst at 200–300°C to form hydrogen bromide. In both cases the colour of the halogen is discharged and steamy fumes of the hydride remain. Hydrogen and iodine vapour combine reversibly to give an equilibrium mixture when heated with a catalyst of platinized asbestos at about 400°C. Note that a catalyst speeds up both forward and reverse reactions but does not affect the final position of equilibrium because it is unchanged finally in energy content. The syntheses may be expressed:

$$H_2 + Cl_2 \rightarrow 2HCl$$

$$H_2 + Br_2 \rightarrow 2HBr$$

$$H_2 + I_2 \rightleftharpoons 2HI$$

Much hydrogen chloride is synthesized in this way from the by-products of the industrial electrolysis of brine. The affinity of the halogen for hydrogen falls off with increase in atomic number.

2   Combination with non-metallic elements

Hydrogen forms hydrides (see above).

Consider the non-metals of Periods 2 and 3 of the Periodic Table. Chlorine, bromine and iodine do not combine directly with carbon, nitrogen or oxygen. Interhalogen compounds are studied briefly later. When heated, boron reacts with chlorine and bromine to form the trihalides (BCl_3, BBr_3), while the iodide has to be prepared indirectly. Both chloride and bromide are liquids.

Silicon unites with chlorine, bromine and iodine when heated. Tetrahalides ($SiCl_4$, etc.) are formed: the chloride and bromide are liquids, while the iodide is a solid.

Phosphorus reacts readily with the halogens, the white allotrope being more reactive than the red. White phosphorus ignites spontaneously, while red phosphorus may need warming, to form the trichloride. Excess chlorine converts the liquid trichloride to the solid pentachloride:

$$2P + 3Cl_2 \rightarrow 2PCl_3$$

$$PCl_3 + Cl_2 \rightarrow PCl_5$$

Bromine reacts violently with red phosphorus and explosively with white phosphorus. Excess bromine converts the liquid tribromide to solid pentabromide just as for chlorine. White phosphorus inflames with solid iodine while red phosphorus reacts on warming. The solid tri-iodide is formed. There is no penta-iodide.

Heated sulphur combines with chlorine and bromine but not iodine to form the disulphur dihalides, formerly monohalides, both of which are red liquids:

$$2S + Br_2 \rightarrow S_2Br_2$$
disulphur dibromide

Except for the phosphorus pentahalides, having ionic structures in the solid state, $(PCl_4^+)(PCl_6^-)$ and $(PBr_4^+)Br^-$, the halides being considered are covalent and readily hydrolysed by water. The hydrolysis of chlorides is studied in detail on p. 420.

3   Combination with metallic elements

Most metals unite with halogens. Reaction may bring incandescence or mere surface corrosion. Chlorine is more reactive than bromine while iodine is the least reactive and requires heating to higher temperatures for reactions to occur. The halide of the higher valency state, if a number of halides exist under the conditions of the reaction, is formed. The higher valency state represents a higher state of oxidation. In the case of mercury, the relative proportions of mercury and iodine ground together with a little alcohol determines whether mercury(II) or dimercury(I) di-iodide is formed:

$$Hg + I_2 \rightarrow HgI_2$$
mercury(II) iodide
(red)

$$2Hg + I_2 \rightarrow Hg_2I_2$$
dimercury(I) di-iodide
(green)

Selecting examples from different groups: union of the elements yields the chlorides, bromides and iodides of potassium (e.g. KCl), magnesium (e.g. $MgBr_2$), aluminium (e.g. AlI_3), tin(IV) (e.g. $SnCl_4$) and bismuth (e.g. $BiBr_3$). The reactions will vary in intensity according to the electropositive nature of the metal. Potassium will act with violence. Aluminium iodide is synthesized in an atmosphere of hydrogen as it inflames on formation in air. Iron reacts with chlorine to give iron(III) chloride ($FeCl_3$) with incandescence, with bromine vapour to give first iron(II) bromide ($FeBr_2$) and then the rather unstable iron(III) bromide ($FeBr_3$), while iron(II) iodide (FeI_2), there being no iron(III) iodide, is the final product with iodine.

4   Order of displacement

Chlorine displaces bromine from bromides and iodine from iodides while bromine displaces iodine from iodides. This order is reversed with displacement from some of the oxosalts. Bromine appears as a reddish-brown coloration. Iodine forms the brown complex ion, I_3^-, and then a dark precipitate of the element, when there is insufficient iodide ion to form the complex:

$$Br_2 + 2I^- \rightarrow I_2 + 2Br^-$$

These are examples of ionic oxidation. Addition of the immiscible dense liquid tetrachloromethane (carbon tetrachloride) results in a light brown lower layer of bromine or a violet layer of iodine dissolved in the organic liquid.

5   Reactions with hydrides

(a)   Hydrides of carbon

Taking one example of substitution from Organic Chemistry: chlorine, bromine but not iodine will bring about substitution of hydrogen atoms in (saturated) paraffin hydrocarbons in diffuse daylight, hydrogen chloride (HCl) also being formed at each stage:

$$CH_4$$
methane
$$\downarrow +Cl_2$$
$$CH_3Cl \xrightarrow{+Cl_2} CH_2Cl_2 \xrightarrow{+Cl_2} CHCl_3$$
monochloro-   dichloro-   trichloro-
methane   methane   methane
$$\downarrow +Cl_2$$
$$CCl_4$$
tetrachloromethane

In direct sunlight, chlorine and methane explode:

$$CH_4 + 2Cl_2 \rightarrow 4HCl + C$$

Chlorine, bromine and, in some cases, iodine, will form addition products with unsaturated hydrocarbons. The reactivity falls off in this order. Ethene (ethylene) will react with all three halogens:

$$\begin{array}{c}CH_2 \\ \parallel \\ CH_2\end{array} + \begin{array}{c}Br \\ | \\ Br\end{array} \rightarrow \begin{array}{c}CH_2Br \\ | \\ CH_2Br\end{array}$$
ethene   1:2 dibromoethane
(ethylene)   (ethylene dibromide)

Anomalous reactions may occur. Ethyne (acetylene), rising from the action of water on lumps of calcium carbide (ethynide, acetylide), inflames spontaneously in an atmosphere of chlorine, yielding hydrogen chloride and carbon:

$$C_2H_2 + Cl_2 \rightarrow 2C + 2HCl$$

Bromine reacts by addition. Warm turpentine, introduced into chlorine on glass wool, may inflame and a burning wax taper will continue to burn but with a red flame, emitting clouds of carbon and hydrogen chloride:

$$C_{10}H_{16} + 8Cl_2 \rightarrow 10C + 16HCl$$
turpentine

$$C_xH_y + \frac{y}{2}Cl_2 \rightarrow xC + yHCl$$
wax: a
mixture of solid
hydrocarbons

(b)   Ammonia

Concentrated aqueous ammonia dripped into chlorine reacts vigorously to produce nitrogen and ammonium chloride. Red flashes may be observed in a darkened room:

$$8NH_3 + 3Cl_2 \rightarrow N_2 + 6NH_4Cl$$

Bromine reacts similarly. However, when excess chlorine is passed into aqueous ammonia, the dangerously explosive nitrogen trichloride separates as an oil,

$$2NH_3 + 6Cl_2 \rightarrow 2NCl_3(l) + 6HCl$$

Bromine forms ammonium perbromide, NH_4Br_3, when in excess while iodine gives a brown precipitate of 'nitrogen iodide', which is explosive when dry:

$$5NH_3 + 3I_2 \rightarrow 3NH_4I + NH_3.NI_3$$
'nitrogen
iodide'

(c)   Hydrogen sulphide

When hydrogen sulphide is mixed with chlorine or bromine vapour, or passed through water containing bromine or iodine, sulphur is deposited and the colour of the halogen disappears. All three halogens react similarly at first:

$$H_2S + Br_2 \rightarrow 2HBr + S$$

Sulphur may react further to form disulphur dihalides with chlorine and bromine (S_2Br_2), which are then hydrolysed:

$$S + 3Br_2 + 4H_2O \rightarrow H_2SO_4 + 6HBr$$

(d)   Salt-like hydrides of metals

The hydrides of the Group I metals, except lithium, react vigorously with chlorine:

$$KH + Cl_2 \rightarrow KCl + HCl$$

6 Reactions with hydrides (ctd): water and bleaching

Chlorine is moderately soluble in water: 3.04 volumes dissolving in 1 volume of water at 0°C and 2.37 volumes at 15°C or about 0.8 g in 100 g water at atmospheric pressure. A yellowish-green solution results. Some chlorine reacts with the water to form a mixture of hydrochloric and hypochlorous acids, the latter being responsible for the powerful bleaching and germicidal power of wet chlorine:

$$Cl_2 + H_2O \rightleftharpoons HCl + HClO$$

$$HClO + dyestuff \rightarrow HCl + oxidized\ dyestuff$$

On cooling to below 0°C, hydrates of chlorine, $Cl_2.xH_2O$, where x has been assigned values 6, 7, 8, 10, separate out: on warming the hydrates dissociate to chlorine. Hypochlorous acid is unstable and chlorine water slowly liberates oxygen in sunlight:

$$2HClO \rightarrow 2HCl + O_2$$

or $$2Cl_2 + 2H_2O \rightarrow 4HCl + O_2$$

Chlorine and steam, passed through a red-hot silica tube, also react to give oxygen and hydrogen chloride.

Bromine is slightly more soluble in water, 3.6 g dissolving in 100 g water at 20°C. No reaction similar to that for chlorine has been detected although both the expected products, hydrobromic acid and hypobromous acid, exist. Bromine water is a good oxidizing agent, not as powerful as chlorine water but more useful in the laboratory because the discharge of the reddish-orange colour is readily seen. It could be used for bleaching. It is more stable than chlorine water because the bromine remains uncombined, but in sunlight it suffers a similar hydrolysis:

$$2Br_2 + 2H_2O \rightarrow 4HBr + O_2$$

On cooling, red hydrates separate out. Iodine is only very slightly soluble in water and its oxidizing and bleaching powers are feeble.

Chlorine is used for the commercial bleaching of dyed cotton and linen materials. It is not used for wool and silk, which contain proteins, the $>$NH groups of which would react with chlorine. Bromine is not used for commercial bleaching.

7 Reactions with hydroxide (hydroxyl) ions in aqueous alkali

With excess cold dilute sodium hydroxide, chlorine forms a solution containing sodium chloride and sodium hypochlorite, otherwise sodium chlorate(I)

$$Cl_2 + 2OH^- \rightarrow Cl^- + ClO^- + H_2O$$
$$\text{hypochlorite}$$
$$\text{chlorate(I)}$$

Bromine forms sodium bromide and hypobromite; iodine, sodium iodide and hypoiodite.

When hot concentrated alkali is used, sodium chloride and chlorate, bromide and bromate, iodide and iodate are formed respectively. For chlorine:

$$3Cl_2 + 6OH^- \rightarrow 5Cl^- + ClO_3^- + 3H_2O$$
$$\text{chlorate}$$
$$\text{chlorate(v)}$$

Similar reactions occur in solution for other hydroxides of Group I (alkali) metals and for calcium, strontium and barium of Group II. Hypobromites are less stable than hypochlorites but very much more stable than hypoiodites which quickly disproportionate to iodide and iodate. In one investigation, potassium hypoiodite solution was 30% decomposed in 1 hour while sodium hypochlorite solution was 50% decomposed in 3 years. In hot strong alkali, conditions favour the decomposition of the hypochlorite ion and its analogues.

With solid lime, chlorine forms bleaching powder which is essentially a mixture of basic calcium hypochlorite and basic calcium chloride. Bromine forms a red analogue.

8 Oxidation of compounds of metals

Chlorine and bromine water will oxidize solutions of iron(II) salts to the iron(III) state; the pale green of the former changing to the yellow of iron(III)

$$2Fe^{2+} + Cl_2 \rightarrow 2Fe^{3+} + 2Cl^-$$

Potassium hexacyanoferrate(II) solution may be oxidized to the hexacyanoferrate(III) by chlorine or bromine,

$$2Fe(CN)_6^{4-} + Cl_2 \rightarrow 2Fe(CN)_6^{3-} + 2Cl^-$$

Chlorine, bromine and iodine will oxidize a tin(II) compound in solution to the tin(IV) condition,

$$Sn^{2+} + I_2 \rightarrow Sn^{4+} + 2I^-$$

9   Oxidation of sulphites and thiosulphates

Sulphurous acid and sulphites are oxidized by chlorine water and bromine water to sulphuric acid and sulphates respectively. For the reaction to go to completion, iodine requires the presence of sodium hydrogencarbonate as a buffer solution to avoid the accumulation and atmospheric oxidation of hydriodic acid. For bromine:

$$SO_3^{2-} + Br_2 + H_2O \rightarrow 2H^+ + SO_4^{2-} + 2Br^-$$

Iodine oxidizes sodium thiosulphate solution to sodium tetrathionate, a solution of iodine in potassium iodide being used in titrimetric (volumetric) analysis:

$$I_2 + 2S_2O_3^{2-} \rightarrow S_4O_6^{2-} + 2I^-$$

Quite different is the reaction with chlorine, the final product being sulphate and chloride ions. Bromine reacts similarly. Sodium thiosulphate is used as an anti-chlor after bleaching with chlorine:

$$S_2O_3^{2-} + 4Cl_2 + 5H_2O \rightarrow$$
$$2SO_4^{2-} + 8Cl^- + 10H^+$$

The oxidative power diminishes from chlorine to iodine.

The laboratory preparation of hydrogen chloride, bromide and iodide

All are colourless gases with an acrid odour, fuming in moist air and turning damp litmus paper red. They are extremely soluble in water, extinguish a lighted splint, and are denser than air. They are increasingly prone to oxidation with increase in atomic number, a fact which brings a variety of preparative methods.

1   By displacement with an acid of higher boiling-point

Hydrogen chloride, a solution of which is called 'spirits of salt', is evolved by the action of concentrated sulphuric acid on common salt, sodium chloride. To moderate the action, rock salt may be substituted for powdered salt, with a consequent reduction in surface area for a given mass and a slower and more controllable reaction. Alternatively, the acid may be diluted (11 acid : 8 water, by volume). The rate of reaction may then be controlled by heating. Sulphuric acid is used because it is cheap and readily available. The gas may be collected by downward delivery (relative densities: hydrogen chloride 18.25, air 14.4) as in Fig. 23.9, or over mercury,

$$NaCl + H_2SO_4 \rightarrow NaHSO_4 + HCl$$

Sulphuric acid tends to oxidize hydrogen bromide and reacts with hydrogen iodide avidly. Syrupy phosphoric acid, which has no oxidizing power, may be used to displace hydrogen bromide and hydrogen iodide from their salts. The potassium salt is covered with syrupy phosphoric acid and the mass carefully warmed to fusion. While heat is maintained, slightly moist gas is evolved which may be dried by passage through the appropriate

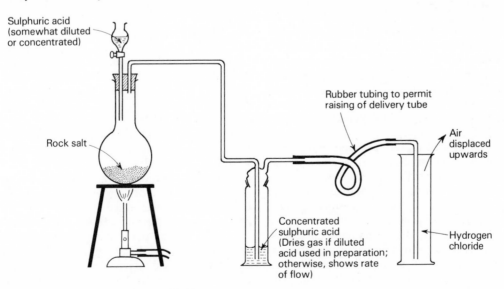

Sulphuric acid (somewhat diluted or concentrated)

Rubber tubing to permit raising of delivery tube

Air displaced upwards

Rock salt

Concentrated sulphuric acid (Dries gas if diluted acid used in preparation; otherwise, shows rate of flow)

Hydrogen chloride

Fig. 23.9   The preparation of hydrogen chloride

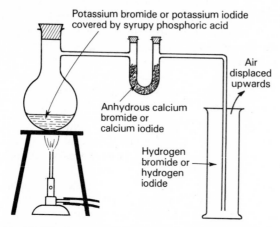

Potassium bromide or potassium iodide covered by syrupy phosphoric acid

Air displaced upwards

Anhydrous calcium bromide or calcium iodide

Hydrogen bromide or hydrogen iodide

Fig. 23.10   A preparation of hydrogen bromide and hydrogen iodide

anhydrous calcium salt, bromide or iodide, as in Fig. 23.10.

$$KI + H_3PO_4 \rightarrow KH_2PO_4 + HI$$

monopotassium
dihydrogen
orthophosphate
or
potassium
dihydrogen
phosphate

The gases are collected by downward delivery (relative densities: hydrogen bromide 40.5, hydrogen iodide 64, air 14.4).

2   By hydrolysis of phosphorus halides

The halides are synthesized by union of the elements and hydrolysed by the minimum quantity of water, excess being avoided because it would dissolve the gases. Hydrogen bromide and iodide are prepared in this way. Volatilized bromine or iodine may be removed by reaction with red phosphorus sprinkled on glass wool in a Drechsel bottle. The apparatus is that shown in Fig. 23.9. In the preparation of hydrogen iodide, red phosphorus and iodine are warmed until they fuse together in a flask which is then connected to the Drechsel bottle as described and drops of water added. Flashes of light are observed initially, possibly due to phosphine igniting in air, and gas is evolved copiously on warming. Being very dense, hydrogen iodide is easily collected by upward displacement of air. Alternatively, iodine may be dissolved in a little potassium iodide solution. The reaction with bromine is more vigorous: drops of bromine are

allowed to fall on to sand covering a paste of red phosphorus, sand and water; flashing is again observed:

$$2P + 3Br_2 \rightarrow 2PBr_3$$

phosphorus
tribromide

$$PBr_3 + 3H_2O \rightarrow 3HBr + H_3PO_3$$

phosphorous
acid

A comparison of the reactions of hydrogen chloride, bromide and iodide

These compounds become progressively less stable with increasing atomic number of the halogen.

1   The nature of their aqueous solutions: hydrochloric, hydrobromic and hydriodic acids (the halogen hydracids)

The dry gases are covalent and remain covalent when dissolved in non-ionizing solvents, such as toluene and other hydrocarbons, trichloromethane (chloroform) and 1,2-dibromoethane (ethylene dibromide). However, the gases are extremely soluble in water: at 10°C, 14 mol dm^{-3} hydrogen chloride, 15 mol dm^{-3} hydrogen bromide and 12 mol dm^{-3} hydrogen iodide form hydrochloric, hydrobromic and hydriodic acids respectively. Interaction with water occurs bringing about ionization:

$$HI + H_2O \rightleftharpoons H_3O^+ + I^-$$

The hydrated hydrogen ion, H_3O^+ (hydronium, hydroxonium or oxonium ion), is the usual form of the proton in aqueous solution. Usually it is written H^+. Various devices have been employed for obtaining solutions of very soluble gases without suck-back occurring, and two of these are illustrated in Fig. 23.11.

Because of the reaction with water, solutions of these gases have a lower vapour pressure than would be expected from two miscible but inert components. Accordingly, negative deviations from Raoult's Law occur (consult texts on Physical Chemistry). Concentrated solutions lose gas on heating until the liquid reaches a constant maximum boiling-point at a given pressure when the whole of the residual liquid distils unchanged. Dilute solutions similarly lose water until the temperature has risen to the same temperature again, when residual liquid of the same composition

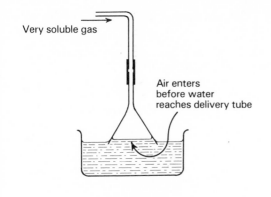

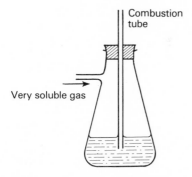

Fig. 23.11   Dissolving a very soluble gas in water (avoiding suck-back)

distils over. This is known as a *maximum constant boiling-point mixture or azeotropic* (Greek† = to boil unchanged) *mixture*. The composition of this mixture depends on the pressure and the substances involved. At 101 kPa (atmospheric pressure), constant boiling hydrochloric acid contains 20.2% HCl and boils at 110°C, hydrobromic acid 47.5% HBr boiling at 126°C, and hydriodic acid 57% HI also boiling at 126°C.

In dilute solution these acids, together with perchloric (chloric(VII)) acid, are the most strongly ionized and therefore, on the Ionic Theory, the strongest acids. Salts are formed with soluble and insoluble bases, carbon dioxide displaced from carbonates and metals above hydrogen in the electrochemical series cause hydrogen to be displaced:

$$OH^- + H^+ \rightarrow H_2O \qquad \text{e.g. KBr}$$

$$NH_3 + H^+ \rightarrow NH_4^+ \qquad \text{e.g. } NH_4I$$

$$MgO + 2H^+ \rightarrow H_2O + Mg^{2+} \text{ e.g. } MgI_2$$

$$CO_3^{2-} + 2H^+ \rightarrow H_2O + CO_2 \text{ e.g. NaCl}$$

$$Fe + 2H^+ \rightarrow Fe^{2+} + H_2 \qquad \text{e.g. } FeCl_2$$

Where the simple hydrated metal cation forms a complex, the metals below hydrogen in the electrochemical series may react. Thus, concentrated hydrochloric acid attacks copper in air:

$$2Cu + 4H^+ + 8Cl^- + O_2 \rightarrow 2CuCl_4^{2-} + 2H_2O$$

Salts may be formed from the covalent gases, too. Anhydrous aluminium chloride is formed by heating aluminium in a stream of hydrogen chloride:

$$2Al + 6HCl \rightarrow 2AlCl_3 \text{ (as } Al_2Cl_6 \text{ vapour)} + 3H_2$$

Where two (or more) chlorides exist under the conditions of the experiment, that belonging to the lowest oxidation state is formed; e.g. iron reacts to give anhydrous iron(II) chloride:

$$Fe + 2HCl \rightarrow FeCl_2 + H_2$$

2   Thermal dissociation

The thermal dissociation of hydrogen iodide may be readily demonstrated. On heating the gas in a loosely corked boiling tube, or on plunging a red-hot glass rod or iron wire into the gas, copious violet fumes of iodine are observed:

$$2HI \rightleftharpoons H_2 + I_2$$

Strong heating of hydrogen bromide gas in a loosely corked boiling tube gives some brown coloration due to bromine. Data in Table 23.3 show that the iodide dissociates more readily than the bromide while the chloride is very little dissociated. Compare with the affinity shown by the elements towards hydrogen.

Table 23.3   The thermal dissociation of the hydrides of the halogens

Hydrogen chloride	temp/°C	727	1727	2727
	% dissociation	1.3×10^{-3}	0.41	1.30
Hydrogen bromide	temp/°C	1024	1108	1222
	% dissociation	0.5	0.73	1.08
Hydrogen iodide	temp/°C	327	527	1217
	% dissociation	19.1	24.9	35.6

3   Reactions with non-metallic elements

Of the non-metallic elements, only oxygen and the halogens themselves react with the halogen hydrides which thereby are oxidized. Chlorine displaces bromine from bromides and iodine from

† *a* = without, *zeo* = to boil, *trepo* = to change

iodides, while bromine displaces iodine from iodides. Excess chlorine on the displaced iodine forms brown solid iodine monochloride and yellow crystals of the trichloride:

$$2HI + Cl_2 \rightarrow 2HCl + I_2$$

$$I_2 + Cl_2 \rightarrow 2ICl; \quad ICl + Cl_2 \rightarrow ICl_3$$

Hydriodic acid is readily oxidized by the air in sunlight with liberation of iodine, seen as a brown coloration:

$$4HI + O_2 \rightarrow 2I_2 + 2H_2O$$

In sunlight, hydrobromic acid, but not hydrochloric acid, is oxidized similarly, being tinted brown by the liberated bromine. In the presence of copper(II) chloride as catalyst, hydrogen chloride may be oxidized to chlorine by air. The optimum temperature for the reaction is 400–450°C and about 60% of hydrogen chloride is oxidized: the product is chlorine greatly diluted by nitrogen of the air, the residual hydrogen chloride being removed by washing. This is the now obsolete Deacon process:

$$4HCl + O_2 \rightleftharpoons 2H_2O + Cl_2$$

4   Relative ease of oxidation

Hydrogen iodide surrenders iodine very readily. While bromine may be obtained from hydrogen bromide fairly easily, hydrogen chloride is affected by the stronger oxidizing agents only. Concentrated nitric acid liberates iodine from hydrogen iodide faster than bromine from hydrogen bromide, nitrogen dioxide being also formed:

$$2HNO_3 + 2HI \rightarrow 2H_2O + 2NO_2 + I_2$$

Concentrated sulphuric acid rapidly attacks hydrogen iodide with the formation of iodine, sulphur and hydrogen sulphide, while hydrogen bromide is slowly attacked to give sulphur dioxide and bromine. Hydrogen peroxide reacts with hydrogen bromide and hydrogen iodide only:

$$H_2O_2 + 2HBr \rightarrow 2H_2O + Br_2$$

Almost any oxidizing agent will liberate iodine from hydrogen iodide or acidified iodides. Nitrous acid (i.e. sodium nitrite solution and dilute hydrochloric acid, or here just sodium nitrite solution) and iron(III) chloride will act on hydrogen iodide but not the others:

$$2HNO_2 + 2H^+ + 2I^- \rightarrow 2H_2O + 2NO + I_2$$

$$2Fe^{3+} + 2I^- \rightarrow 2Fe^{2+} + I_2$$

5   The solubility of salts

The chlorides, bromides and iodides of the typical metals of Group I (alkali) and Group II (alkaline-earth) metals follow the same pattern. The solubility (expressed in grams dissolving in 100 g water) increases in the order chloride, bromide and iodide except for the halides of caesium where it is reversed. In Table 23.4, the figures refer to anhydrous salts in 100 g water at 25°C unless another temperature is stated. The second column represents the solubility in moles of anhydrous salt in 100 g water under the same conditions. Note the run of values for the magnesium halides. Common insoluble halides include those of copper(I), silver, lead(II) and dimercury(I). Mercury(II) bromide is sparingly soluble and the iodide, insoluble.

Table 23.4 Solubilities of halides at 25°C (unless stated otherwise)

	Anhydrous salt			Anhydrous salt	
	g 100 g^{-1} water	mol 100 g^{-1} water		g 100 g^{-1} water	mol 100 g^{-1} water
NaCl	35.9	0.614	MgCl$_2$	54.1 (20°C)	0.568
NaBr	94.1	0.914	MgBr$_2$	102.5	0.557
NaI	184.1	1.23	MgI$_2$	140.0 (20°C)	0.552
KCl	34.4	0.461	CaCl$_2$	81.5 (20°C)	0.734
KBr	65.0	0.546	CaBr$_2$	140	0.876
KI	144.6	0.871	CaI$_2$	204 (20°C)	1.02

6   Addition to unsaturated hydrocarbons

Hydrogen chloride adds on slowly to unsaturated hydrocarbons, hydrogen bromide slowly except in the presence of oxygen or a peroxide and hydrogen iodide always rapidly. The stronger the affinity of the halogen for hydrogen, the slower the addition reaction which requires rupture of the halide bonding:

$$\begin{array}{ccc} CH_2 & H & CH_3 \\ \parallel & + \ | & \rightarrow \ | \\ CH_2 & Br & CH_2Br \end{array}$$

ethene   bromoethane
(ethylene)   (ethyl bromide)

The action of water on halides

Most of the examples quoted here will be chlorides but the action of water on bromides and iodides is very similar. Fluorides will be considered in detail later. Individual halides are described in detail under the appropriate element.

The chlorides of the strongly electropositive metals are ionic while those of the non-metals are covalent. Ionic compounds conduct electricity when fused or dissolved in an ionizing solvent, are ionic even in the solid state and usually have high melting- and boiling-points, although halides generally are more volatile than other salts, and they are soluble in the relatively few solvents of high dielectric constant. On the other hand, covalent compounds are non-conductors of electricity unless combination with an ionizing medium occurs, are easier to melt, boil or sublime and are soluble in organic solvents. Some halides are intermediate in behaviour and bonding. This is well brought out by the action of water on the halides.

The typically covalent hydrogen halides react with water to give electrically conducting solutions:

$$HCl + H_2O \rightleftharpoons H_3O^+ + Cl^-$$

Ionic halides which dissociate into ions when dissolved in water are formed by the strongly electropositive metals, found in Groups I and II (except beryllium).

Most covalent halides are hydrolysed by water and a theory has been developed to account for different products of hydrolysis and to explain why some covalent halides are inert. Covalent halides are formed by non-metals and the less electropositive metals. Consider the chlorides of the non-metals of Period 2, except interhalogen compounds:

$$(LiCl) \quad (BeCl_2) \quad BCl_3 \quad CCl_4 \quad NCl_3$$

$$OCl_2 \quad (FCl)$$

Boron trichloride provides a good example. The boron atom has only six electrons in the valency shell and a co-ionic bond is formed by a lone pair of electrons from the oxygen of water acting as a donor:

$$BCl_3 + 3H_2O \longrightarrow H_3BO_3 + 3HCl$$
boron trichloride → orthoboric acid

$$NCl_3 + 3H_2O \longrightarrow NH_3 + 3HClO$$
nitrogen trichloride → ammonia + hypochlorous acid / chloric (I) acid

$$Cl_2O + H_2O \longrightarrow 2HClO$$
chlorine monoxide

$$(Cl_2 + H_2O \longrightarrow HCl + HClO)$$

There was room for two extra electrons in the valency shell of boron but this does not hold for the remaining atoms of Period 2 when combined in halides. Further, for the Period 2 elements the octet cannot be expanded as in later periods: there are no energy levels corresponding to $3d$ orbitals. Tetrachloromethane (carbon tetrachloride) is inert to water, carbon having reached the covalency maximum. Nitrogen and oxygen, in combination, differ from carbon in having lone pairs of electrons. Nitrogen trichloride and chlorine monoxide are hydrolysed by a mechanism peculiar to themselves, although the hydrolysis of chlorine itself follows a similar pattern. Possibly hydrogen bonding occurs with the lone pair of nitrogen or oxygen, the hydrolyses proceeding as shown under the scheme for the hydrolysis of boron trichloride, with which they should be compared.

Now, consider the chlorides formed by the non-metals of Period 3:

(NaCl) (MgCl$_2$) (AlCl$_3$) SiCl$_4$ PCl$_3$ SCl$_2$ Cl$_2$

With expansion of the valency octet possible, now that $3d$ orbitals are available, the chlorides of all elements outside Period 2 are considered to be hydrolysed by a mechanism similar to that for boron trichloride. Stepwise hydrolysis will account for the hydrolysis products of silicon tetrachloride and phosphorus trichloride:

$$SiCl_4 + 4H_2O \rightarrow SiO_2.H_2O + 4HCl$$
hydrated silica

$$PCl_3 + 3H_2O \rightarrow H_3PO_3 + 3HCl$$
phosphorous
acid

Phosphorus pentachloride is hydrolysed similarly. The hydrolysis of sulphur dichloride is similar to that of disulphur dichloride, giving a complete mixture of products. However, it is interesting to note that sulphur hexafluoride (fluorine bringing out the maximum valency) is inert, sulphur having reached its covalency maximum (12). Similarly, selenium hexafluoride, SeF$_6$, and hexafluorophosphoric acid, HPF$_6$, are not hydrolysed.

In any periodic family, increase of atomic number means a movement away from the electronegative, or comparatively electronegative, towards the electropositive. This is shown in the middle groups very well. The principal valency drops by 2 and ionic chlorides tend to appear. Hydrolysis becomes more difficult. Thus in Group V, bismuth has no chloride in the quiniquevalent state and the trichloride is hydrolysed reversibly to bismuth oxide chloride but no further, whereas with antimony trichloride hydrolysis in boiling water yields the oxide:

$$BiCl_3 + H_2O \rightleftharpoons BiOCl(s) + 2HCl$$

$$SbCl_3 + H_2O \rightleftharpoons SbOCl(s) + 2HCl$$

$$4SbOCl(s) + 2H_2O \rightleftharpoons Sb_4O_6(s) + 4HCl$$

Lead in Group IV has a stable dichloride which is moderately soluble in hot water but not hydrolysed, and an unstable covalent tetrachloride which is hydrolysed in a similar manner to silicon tetrachloride:

$$PbCl_4 + 2H_2O \rightarrow PbO_2(s) + 4HCl$$

Along the diagonal boundary separating metals from non-metals in the Periodic Table interesting reactions with water occur. The hydrolysis of antimony trichloride is quite different from that of arsenic trichloride which behaves as a chloride of a non-metal but reversibly:

$$4AsCl_3 + 6H_2O \rightleftharpoons As_4O_6 + 12HCl$$

Beryllium chloride is covalent. When fused about one molecule in a thousand is ionized. With water there is a large evolution of heat and a tetrahydrate, BeCl$_2$.4H$_2$O, is formed. The tetrahedral structure, formed by [sp^3] hybridization is shown on p. 289. The bromide behaves similarly while the iodide is hydrolysed completely. Hydrated ions are presumably present in solutions of the chloride and bromide, which are strongly hydrolysed. Aluminium chloride is probably ionic in the solid state and differs from the bromide and iodide in this respect. It fumes in air and reacts with water with evolution of heat to give the hexahydrate, AlCl$_3$.6H$_2$O or [Al(H$_2$O)$_6$]Cl$_3$. The octahedral structure, formed by [sp^3d^2] hybridization is shown on p. 305.

Anhydrous chlorides of metals cannot always be prepared by evaporation of solutions or by heating the hydrated crystals. An oxide chloride (formerly oxychloride), hydroxide chloride or an oxide may result and hydrogen chloride escapes:

$$MgCl_2.6H_2O \rightarrow Mg(OH)Cl + HCl + 5H_2O$$

$$2AlCl_3.6H_2O \rightarrow Al_2O_3^- + 6HCl + 9H_2O$$

$$SnCl_2.2H_2O \rightarrow Sn(OH)Cl + HCl + H_2O$$

Dehydration may be effected by heating in a stream of hydrogen chloride gas.

Solutions of the halides of the less electropositive metals are hydrolysed in solution and have acidic solutions. The hydrated metallic ion loses a proton to the solvent and the acidic nature results:

$$Zn(H_2O)_4{}^{2+} + H_2O \rightleftharpoons Zn(H_2O)_3OH^+ + H_3O^+$$

The common insoluble halides have already been mentioned (p. 420).

The distinctive character of fluorine, hydrogen fluoride and fluorides

Fluorine is the most electronegative element. Elements have been assigned values for their electronegativities by Pauling. Values for the halogens and oxygen are given below and show that oxygen is second only to fluorine in electronegativity:

F (4.0)   O (3.5)   Cl (3.0)   Br (2.8)   I (2.4)

The formation of ionic and covalent fluorides is accompanied by the evolution of large heats of reaction. Fluorine cannot be liberated from its compounds by chemical oxidation methods. It is prepared by electrolysis of a concentrated solution of potassium fluoride (about 60%) in anhydrous hydrofluoric acid just below 100°C. Anodes are made of ungraphitized carbon because, although carbon is relatively unattacked, graphite swells up and disintegrates as the layers of atoms are penetrated by the small fluorine atoms. The cathode is of steel and the tank containing the electrolyte of Monel metal. Anhydrous hydrofluoric acid (hydrogen fluoride) must be replenished as necessary. Hydrogen is discharged at the cathode and fluorine at the anode. Volatilized hydrogen fluoride is trapped by refrigeration at −80°C (b.p. +19.6°C, m.p. −83.7°C) and the remaining 1–2% by granular sodium fluoride, forming sodium hydrogenfluoride:

$$HF + NaF \rightarrow NaHF_2$$

Fluorine and fluorides are highly poisonous.

Fluorine is the most powerful oxidizing agent. It behaves as the most active member of the halogens: the order of reactivity diminishing in the series F > Cl > Br > I. However, in common with some of the other first members of periodic families it shows somewhat anomalous behaviour. This is all the more distinctive because of the extreme electronegativity. Fluorine is the most reactive element. Anomalous behaviour is probably related to the fact that valency electrons of the elements placed in the second period are incompletely screened from the positive nucleus by

the duplet of electrons. This increases the affinity for electrons.

Consult Tables 23.1 and 23.2 for a comparison of 1st ionization energies and of electron affinities respectively for the successive halogens. Electronegativities are compared in Fig. 23.4. For a wider perspective consult Fig. 3.6, p. 43 (ionization energies) and Fig. 5.7, p. 79 (electronegativities).

The extreme reactivity of fluorine

The sequence of reactions which follow are arranged in the same order as those given for chlorine, bromine and iodine, to facilitate comparison at each stage.

1   Synthesis of the hydride

Fluorine explodes with hydrogen in the dark, even at −252°C, the boiling-point of hydrogen, at which fluorine is a solid. Hydrogen fluoride is formed. A jet of fluorine may be burned in hydrogen:

$$H_2 + F_2 \rightarrow 2HF$$

2   Combination with non-metallic elements

Fluorine combines directly with all non-metals except nitrogen, helium, neon and argon. Usually where combustion occurs and where several valency states are possible, the product displays the highest covalency of the element. Fluorine cannot itself exceed unicovalency. Sometimes, covalent fluorides occur in series. Boron and silicon burn to the trifluoride and tetrafluoride respectively, charcoal to the tetrafluoride (distinction from other halogens) while phosphorus and arsenic both form the trifluoride and with excess fluorine, the pentafluoride. Sulphur, selenium and tellurium burn to form the hexafluorides. Fluorine and oxygen combine directly under the influence of an electric discharge when cooled in liquid air to give dioxygen difluoride (fluorine dioxide), O_2F_2. In calculating oxidation numbers (rules, p. 124), note that fluorine is always in the −1 oxidation state even though this means giving oxygen a positive oxidation number. Fluorine, under special conditions, will react with the other halogens, the primary products having the capacity to take up more fluorine. Chlorine forms ClF and ClF_3, bromine forms BrF, BrF_3 and BrF_5, and iodine IF_5 and IF_7. XeF_2, XeF_4, XeF_6 and very unstable KrF_2 and KrF_4 have been reported from the reaction of fluorine with noble gases. By extending the basic ideas underlying Fajans' Rules (p. 82) to the attack of pairs of fluorine atoms on the relatively large atoms of the higher noble gases, one can see, at a preliminary stage, what has happened.

3   Combination with metallic elements

Most metals take fire in fluorine, especially when heated. Alkali metals and alkaline-earths usually inflame immediately. Aluminium, tin, lead, zinc, chromium, silver and nickel are among those which react with incandescence, when heated. Copper, gold, and the platinum family metals require strong heating. The fluoride formed belongs to the higher valency state where more than one fluoride is stable under the conditions of the experiment: as an illustration iron forms iron(III) fluoride. At ordinary temperatures, some metals are protected from further attack by a surface layer of the fluoride. Nickel and copper are protected in this way and an alloy, Monel metal, may be used for the construction of storage vessels for fluorine. Fluorine is more reactive than the other halogens in reactions with metals.

4   Order of displacement

Fluorine displaces other halogens from ionic and covalent compounds. Cold solid potassium chloride yields potassium fluoride and chlorine, while potassium bromide burns to the fluoride and bromine trifluoride, BF_3, and potassium iodide to the fluoride and iodine pentafluoride, IF_5:

$$(2K^+) 2Cl^- + F_2 \rightarrow 2F^- + Cl_2 (+2K^+)$$

$$(2K^+) 2Br^- + 4F_2 \rightarrow 2F^- + 2BrF_3 (+2K^+)$$

$$(2K^+) 2I^- + 6F_2 \rightarrow 2F^- + 2IF_5 (+2K^+)$$

Fluorine displaces chlorine from tetrachloromethane (carbon tetrachloride),

$$CCl_4 + 2F_2 \rightarrow CF_4 + 2Cl_2$$

5   Reactions with hydrides

Solid methane explodes violently with liquid fluorine. Gaseous fluorine reacts with ammonia to give nitrogen and some nitrogen trifluoride, which is inert. Hydrogen sulphide burns in the gas to form sulphur hexafluoride and hydrogen fluoride. Chlorine and bromine go no further than the disulphur dihalide (S_2Cl_2 and S_2Br_2) when in excess.

6   Reactions with hydrides (ctd): water

In moist air, fluorine forms hydrogen fluoride and oxygen, and with liquid water, hydrofluoric acid and ozonized oxygen, and some hydrogen peroxide:

$$2F_2 + 2H_2O \rightarrow 4HF + O_2$$

Notice that this is a displacement reaction. Chlorine and bromine react slowly with water in sunlight.

7   Reactions with hydroxide (hydroxyl) ions in aqueous alkali

Unlike the other halogens, fluorine does not form oxosalts with alkalis and there are no oxoacids. With dilute alkali (2% sodium hydroxide) oxygen difluoride, formerly fluorine monoxide, and (sodium) fluoride result; with concentrated alkali, oxygen and the fluoride:

$$2OH^- + 2F_2 \rightarrow OF_2 + 2F^- + H_2O$$

$$4OH^- + 2F_2 \rightarrow 4F^- + O_2 + 2H_2O$$

8   Oxidation of compounds of metals

Fluorine will oxidize chlorate and iodate ions to perchlorate (chlorate(VII)) and per-iodate (iodate(VII)), hydrogensulphate to peroxodisulphate and chromium(III) ions to dichromate. There are of course many more examples.

9   Attempts at organic substitution reactions

Direct fluorination of organic compounds is difficult. Fluorine reacts violently and the products include hydrogen fluoride and carbon tetrafluoride. Compare the reactions of fluorine and the other halogens with methane noting what happens with chlorine in both diffuse light and sunlight. Fluorination methods include the use of fluorine diluted with nitrogen in the presence of a catalyst of copper gauze and replacement of halogen atoms by fluorine from the fluoride of a weakly electropositive metal, e.g. HgF_2.

Xenon tetrafluoride, XeF_4

Experimental work on platinum chemistry by Neil Bartlett at the University of British Columbia in 1962 led to the discovery of a compound with an unexpected cation, that of oxygen, in $O_2^+PtF_6^-$.

On account of the closeness of the 1st ionization energies of xenon, Xe $(1167 \, kJ \, mol^{-1})$ and molecular oxygen, O $(1180 \, kJ \, mol^{-1})$, both being of comparable molecular diameter (c. 0.40 nm), he postulated a reaction between the gases xenon and platinum hexafluoride:

if $$PtF_6 + O_2 \rightarrow O_2{}^+PtF_6{}^-$$

then $$PtF_6 + Xe \rightarrow Xe^+PtF_6{}^-$$

On mixing the gases he obtained a red crystalline solid of the predicted formula, an elegant illustration of scientific method which, as it were, opened a new window on chemistry, and led to the *inert* gases becoming known as the *noble* gases.

Later that year, the preparation of xenon tetrafluoride was reported from Canada by Bartlett and from England. J. H. Holloway and R. D. Peacock of the University of Birmingham described the preparation in *The Proceedings of the Chemical Society*, December 1962.

Xenon and fluorine combine at dull red heat to form xenon tetrafluoride in a yield of 30–50% in a nickel tube packed with nickel sheet. Xenon tetrafluoride, appreciably volatile at 20°C, is collected in a trap cooled to −72°C by a mixture of propanone (acetone) and solid carbon dioxide. Xenon tetrafluoride forms white crystals which volatilize without decomposition in a stream of nitrogen.

The compound does not attack glass at 20°C. It has been stored for several weeks at −25°C without change. Xenon tetrafluoride dissolves quietly in potassium iodide solution, fluoride ions and xenon being formed quantitatively,

$$XeF_4 + 4I^- \rightarrow Xe + 2I_2 + 4F^-$$

Also white volatile solids, xenon difluoride and xenon hexafluoride have been synthesized. XeF_6 has a low vapour pressure of 4 kPa (30 mm Hg) at 25°C and can be stored indefinitely in a nickel vessel. An oxofluoride $XeOF_4$, an oxide XeO_3 (a white, dangerously explosive solid) and perxenates (e.g. $Na_4XeO_6 . 8H_2O$) have now been isolated.

The structures of XeF_2 and XeF_4 are described in Table 5.4, p. 73. They are respectively linear and square planar. XeF_2 has ten electrons in five electron-pairs around the Xe atom, and XeF_4, six pairs, its shape being based on the octahedron with two opposite points occupied by lone pairs. XeF_6 has seven electron-pairs and on the supposition that it might be related structurally to IF_7 (which has seven pairs of electrons) a distorted octahedron

has been predicted. Recently this structure has been confirmed.

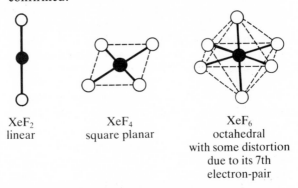

XeF$_2$	XeF$_4$	XeF$_6$
linear	square planar	octahedral with some distortion due to its 7th electron-pair

The Xe−F bonds are surprisingly stable to heat. Early experimental work had been conducted at low temperatures on the contrary supposition.

Hydrogen fluoride: HF

Hydrogen fluoride is liberated when concentrated sulphuric acid is warmed with calcium fluoride:

$$CaF_2 + H_2SO_4 \rightarrow CaSO_4 + 2HF$$

In industry, 98% sulphuric acid acts on high-grade fluorspar, calcium fluoride, in a rotating kiln at 200°C. Principal impurities are sulphur dioxide from sulphuric acid and silicon tetrafluoride from the action of hydrogen fluoride on the silica ($<1\%$) remaining after concentration of the fluorspar. After scrubbing with sulphuric acid to remove dust and acid spray, hydrogen fluoride is absorbed in water in absorption towers to give 80% aqueous acid as the main product, the residual vapour being absorbed to give 60% and 30% aqueous acid. Fractionation of the 80% acid yields anhydrous hydrogen fluoride (b.p. 19.6°C) and 60% acid (b.p. 88°C). The maximum constant boiling-point acid contains 38% hydrogen fluoride and boils at 112°C, but corrosion problems prevent fractionation into 100% and 38% hydrogen fluoride. Dry hydrogen fluoride is known as anhydrous hydrofluoric acid, or AHF. It is the starting material for most industrial fluorine compounds. AHF is transported in mild steel cylinders or tank wagons.

Hydrogen fluoride is liberated when potassium (or sodium) hydrogenfluoride is heated. A copper or platinum apparatus may be used and the vapour condensed using a freezing mixture,

$$KHF_2 \rightarrow KF + HF$$

Hydrogen fluoride is a colourless, volatile, fuming liquid which has a highly toxic vapour and causes severe burns to the skin. It is more highly associated than any other known substance. Linear polymers are formed by what is believed to be electrostatic attraction, called *hydrogen bonding* (p. 220), between the fluorine atom of one molecule and the hydrogen atom of the next. Comparison of the boiling-points of the hydrides shows the anomalous behaviour of hydrogen fluoride which, having apparently the lowest relative molecular mass of the series, might reasonably be expected to be gaseous:

	HF	HCl	HBr	HI
b.p. °C	19.6	−85	−67	−36

The degree of association may be determined by density methods. At the boiling-point under atmospheric pressure the vapour has the relative molecular mass 63 corresponding to $(HF)_{3.1}$; at 100°C and the same pressure, it is virtually HF. It has been suggested that $(HF)_6$ polymers are present:

$$6HF \rightleftharpoons (HF)_6; \quad \Delta H = -170 \, kJ \, mol^{-1}$$

There is no direct evidence for the structure of this polymer but it is apparently linear, and not a ring, the actual shape being zig-zag. In being associated hydrogen fluoride resembles ammonia and water, hydrides of the first elements in the previous two groups. Hydrogen bonding is characteristic of elements of the highest electronegativity.

AHF is an ionizing solvent and resembles water and liquid ammonia in this property; being quite different from the other hydrides of the group:

$$2HF \rightleftharpoons H_2F^+ + F^-$$

The simple fluoride ion then forms the HF_2^- ion,

$$HF + F^- \rightleftharpoons HF_2^-$$

AHF is outstanding as a solvent. Potassium fluoride dissolves to form K^+ and F^- ions, potassium chloride gives the fluoride and hydrogen chloride while potassium nitrate, a salt of a strong oxoacid, gives a type of oxonium ion:

$$K^+F^- = K^+ + F^-$$

$$K^+Cl^- + HF = K^+ + F^- + HCl$$

$$K^+NO_3^- + 2HF = K^+ + H_2NO_3^+ + 2F^-$$

Usually the only anion is F^- but some salts of very strong acids retain their anions; examples include potassium perchlorate (chlorate(vii)), $KClO_4$, and potassium fluoroborate, KBF_4.

In aqueous solution, hydrofluoric acid is about 10% ionized in decimolar solution, having about half the strength of monochloroethanoic acid (monochloracetic acid). Conductivity measurements are explained in terms of two equilibria:

$$HF \rightleftharpoons H^+ + F^-$$

$$(HF + H_2O \rightleftharpoons H_3O^+ + F^-)$$

$$F^- + HF \rightleftharpoons HF_2^-$$

$H^+[F_2H]$ is completely ionized while HF is only partially ionized, unlike HCl, HBr and HI. The H—F bond is very strong indeed and ionization proves difficult. The normal acidic properties are shown but evaporation of a solution containing molar proportions of potassium fluoride and hydrofluoric acid yields an acid salt, known as potassium hydrogenfluoride or acid potassium fluoride:

$$K^+ + F^- + HF \rightarrow KHF_2$$

Aqueous hydrofluoric acid attacks all metals except gold and the platinum family of metals, the formation of complexes accounting for the action with the more noble metals. A 5% (w/w) solution of hydrofluoric acid is an excellent remover of iron mould on fabrics by complex formation. Moist hydrogen fluoride attacks silica while AHF does not:

$$SiO_2 + 4HF \rightleftharpoons SiF_4 + 2H_2O$$

$$SiF_4 + 2HF \rightleftharpoons H_2SiF_6$$

Silicon tetrafluoride and hexafluorosilicic acid are formed. Silicates form hexafluorosilicates:

$$SiO_3^{2-} + 6HF \rightarrow SiF_6^{2-} + 3H_2O$$

Glass contains sodium and calcium silicates so this reaction may be used for etching. Frosting of glass is achieved by immersion in a solution of potassium hydrogenfluoride containing hydrochloric acid. Experimental work with aqueous solutions may be done using polythene ware.

Hydrogen fluoride vapour reacts vigorously with solid sodium and potassium fluorides to give the corresponding acid fluorides. It reacts with metals to form anhydrous fluorides. Iron forms iron(ii) fluoride with hydrogen fluoride whereas fluorine gives the iron(iii) compound.

The use of AHF as a fluorinating agent in Organic Chemistry has been widely developed. In the presence of antimony dichlorotrifluoride, 1 : 1 : 1-trichloroethane is converted into 1 : 1 : 1-trifluoroethane:

$$CH_3.CCl_3 + 3HF \xrightarrow{SbCl_2F_3} CH_3.CF_3 + 3HCl$$

Hydrogen fluoride will form addition products with unsaturated hydrocarbons in a similar manner to hydrogen iodide and the others,

$$
\begin{array}{c}
CH_2 \\
\parallel \\
CH_2
\end{array}
+
\begin{array}{c}
H \\
\mid \\
F
\end{array}
\rightarrow
\begin{array}{c}
CH_3 \\
\mid \\
CH_2F
\end{array}
$$

ethene   fluoroethane
(ethylene)   (ethyl fluoride)

As an industrial catalyst AHF has applications in alkylation and acylation. The reaction of 2-methylpropane (isobutane) with 2-methylpropene (butylene) to form 2,2,4-trimethylpentane iso-octane, for high octane aviation spirit is catalysed by AHF.

$$C_4H_8 + C_4H_{10} \rightarrow C_8H_{18}$$

$$
\begin{array}{c}
CH_2 \\
\parallel \\
CH_3-C \\
\mid \\
CH_3
\end{array}
+
\begin{array}{c}
CH_3.CH.CH_3 \\
\mid \\
CH_3
\end{array}
\rightarrow
$$

$$
\begin{array}{ccc}
& CH_3 & \\
& \mid & \\
CH_3.C.CH_2.CH.CH_3 & & \\
\mid & \mid & \\
CH_3 & CH_3 &
\end{array}
$$

For acylations, the action of AHF is similar to that of anhydrous aluminium chloride but has advantages where the carboxylic acids are used,

$$RH + R_1.COOH \rightarrow R.CO.R_1 + H_2O$$

where R and R_1 are alkyl groups. Aromatic hydrocarbons can be nitrated and sulphonated effectively using solutions of potassium nitrate or potassium sulphate in AHF. Dissociation in this medium has been described already.

Fluorides

Solubility data for the fluorides of metals often reveal striking differences from other halides. In Group I (alkali) metals, lithium fluoride and sodium fluoride are much less soluble than the chlorides but the solubilities increase in the order, fluoride to iodide. Lead(II) fluoride is sparingly soluble like the other halides of bivalent lead. However, the fluorides of the Group II (alkaline-earth) metals are sparingly soluble (except for the first member, beryllium) in contrast to the other halides which are very soluble, the solubilities increasing from chloride to iodide. Solubilities of the fluorides and chlorides of these metals are

Table 23.5 Solubilities of fluorides and chlorides of Group II (alkaline-earth) metals at the temperature stated/g (anhydrous salt) 100 g^{-1} water

	Mg	Ca	Sr	Ba
MF$_2$	0.012 (18°C)	2.53×10^{-3} (18°C)	0.0117 (18°C)	0.209 (18°C)
MCl$_2$	54.1 (20°C)	81.5 (20°C)	55.5 (20°C)	37.2 (20°C)

contrasted in Table 23.5. In contrast, silver fluoride is very soluble (180 g dry salt in 100 g water at 25°C) while the other silver halides are 'insoluble'.

Fluorides may be classed as ionic or covalent. The fluorides of the Group I (alkali) metals are ionic and dissolve in water while those of Group II (alkaline-earth) metals except beryllium, are ionic but sparingly soluble. Only able to achieve unicovalency, fluorine brings out the highest covalency in other elements. In consequence with a maximum covalency there may be no room for electron-pairs donated by the oxygen of water and as there are no energy levels corresponding to $3d$ orbitals in fluorine, so hydrolysis of these compounds does not occur. Covalent fluorides which are inert include tetrafluoromethane (carbon tetrafluoride), nitrogen trifluoride, and the hexafluorides of sulphur and selenium. Maximum covalency is also seen in stable complex ions, BF_4^-, AlF_6^{3-} and SiF_6^{2-}. Covalent fluorides, which are hydrolysed similarly to the corresponding chlorides, but to a lesser extent, include boron trifluoride, silicon tetrafluoride, phosphorus trifluoride and arsenic trifluoride. The first two are partially hydrolysed and give the complex tetrafluoroboric and hexafluorosilicic acids:

$$BF_3 + HF \rightleftharpoons HBF_4$$

$$SiF_4 + 2HF \rightleftharpoons H_2SiF_6$$

Beryllium fluoride is covalent, readily soluble in water and hydrolysed on evaporation.

Of the halogens, fluorine has the greater tendency to form the ionic bond so that some elements, which otherwise have covalent halides, form fluorides with ionic characteristics. The transition to ionic fluorides is shown by tin(IV) fluoride, antimony trifluoride and mercury(II) fluoride. The first is readily soluble in water and hydrolysed. Antimony trifluoride is much less hydrolysed than the other halides. Mercury(II) fluoride is one of the few strong electrolytes of mercury, being hydrolysed with excess water. Bismuth, the next congener of antimony in Group V, forms a trifluoride which is insoluble in water. In Group IV, lead, unlike tin, does not form a tetrafluoride, the lead(IV) ion being too unstable presumably to form what would probably be an ionic compound, reflecting the increase in electropositive nature of the elements as the atomic number increases. Aluminium fluoride has been considered to be ionic in all physical states while the chloride is ionic in the solid state, and otherwise covalent as are the bromide and iodide: aluminium fluoride is attacked by water very slowly at 400°C whereas the other halides are hydrolysed. Pauling considers that aluminium fluoride is a covalent macromolecular structure and this is likely to be so.

Finally, the oxides and fluorides of an element are very often physically alike. The association of hydrogen fluoride and water has been mentioned already. Similar acids are formed with boron, HBO_2 and HBF_4, silicon, H_2SiO_3 and H_2SiF_6, etc. Silicon tetrafluoride and carbon dioxide have similar physical constants. However, fluorine (4.0) is more electronegative than oxygen (3.5) and induces the highest oxidation state in any element with which it combines, itself always remaining in oxidation state −1; thus, oxygen in F_2O has an oxidation number of +2. So the formula becomes OF_2, and the name, oxygen difluoride.

Fluorine and its compounds are so distinctive that it can be maintained that the element is not a halogen in the family sense but an element quite unique in its chemical reactivity.

Interhalogen compounds principally of fluorine

ClF	BrF	BrCl	ICl
ClF_3	BrF_3		ICl_3
	BrF_5		
		IF_6	
		IF_7	

The interhalogen compounds are volatile and covalent, being prepared by synthesis from the elements or by further halogenation, as in the fluorination of BrF_3 and IF_5 to form BrF_5 and IF_7 respectively. The less electronegative element is placed on the left in formulae and the naming is then straightforward. In each case the smaller atom is present in larger atomic proportion: as the atomic volume increases, the capacity for covalency increases. The compounds are very reactive, some reactions being extremely violent. Chlorine trifluoride and bromine trifluoride react violently with organic

material and set fire to asbestos: asbestos wool burns with incandescence. These compounds are important in the preparation of uranium hexafluoride in atomic energy work.

Some of the interhalogens ionize slightly (i.e. undergo auto-ionization) and are ionizing solvents so that *non-protonic acid–base systems* may be developed. Here, an acid gives the cation of the parent solvent and the base, the anion:

$$2ICl_3 \rightleftharpoons ICl_2^+ + ICl_4^-$$
$$2BrF_3 \rightleftharpoons BrF_2^+ + BrF_4^-$$
$$2IF_5 \rightleftharpoons IF_4^+ + IF_6^-$$
$$ \uparrow \qquad \uparrow$$
$$ \text{acids} \quad \text{bases}$$

The following bases and acids have been prepared for the system, $2BrF_3 \rightleftharpoons BrF_2^+ + BrF_4^-$

bases: $KBrF_4$, $AgBrF_4$, $Ba(BrF_4)_2$

acids: $BrF_2.SbF_6$, $(BrF_2)_2SnF_6$, $BrF_2.AuF_4$, etc.

To which compounds do these correspond in the $2H_2O \rightleftharpoons H_3O^+ + OH^-$ system? Many non-protonic systems do not have high dielectric constants showing that this is not an essential characteristic of an ionizing medium. Acid-base theory was developed on pages 235–9.

Fluorine in atomic energy

Concentration of the fissionable isotope of uranium, $^{235}_{92}U$, is achieved by a diffusion process based on uranium hexafluoride. The source of fluorine is anhydrous hydrofluoric acid and various intermediates made from it, elemental fluorine and halogen fluorides, chlorine trifluoride, ClF_3 (b.p. 12°C), bromine trifluoride, BrF_3 (b.p. 126°C), and the pentafluoride, BrF_5 (b.p. 40°C), which are made from the elements at 200–300°C. Special corrosion problems arise and a range of chemically resistant lubricants and liquids, the fluorocarbons, have been developed. A number of methods for the production of uranium hexafluoride have been used but the principles are illustrated in the following outline:

$$U_3O_8 \xrightarrow[\text{H}_2]{\text{Reduce}} UO_2 \xrightarrow{\text{AHF}}$$

triuranium octaoxide 70% concentrate   uranium dioxide

$$UF_4 \xrightarrow[\text{F}_2]{\text{ClF}_3 \text{ or}} UF_6$$

uranium tetrafluoride   uranium hexafluoride

Purification based on fractionation under reduced pressure

Uranium hexafluoride sublimes without melting at 56°C and atmospheric pressure but may be melted under pressure. After concentration of $^{235}_{92}UF_6$, the tetrafluoride is formed by reduction with hydrogen at 550°C and is reduced by magnesium in a steel bomb heated to 600–650°C in an electric furnace. The uranium (m.p. 1660°C) is melted by the heat of the reaction. A slag of magnesium fluoride is formed. The uranium ingot may be machined and processed for use as fuel elements in atomic reactors.

Fluorine products of high chemical and thermal stability

The principal products are fluorocarbons, freons and fluorinated polymers, all of which are carbon compounds, and sulphur hexafluoride.

Fluorocarbons contain carbon and fluorine only. In straight and branched chains, fluorocarbons up to C_{12} analogues of aliphatic hydrocarbons and some ring compounds have been prepared. The main characteristics of fluorocarbons include low toxicity and non-flammability in addition to high chemical stability. They are used as heat transfer fluids, hydraulic fluids and special lubricants. Because of their low dielectric constants, there are electrical applications.

Freons are saturated aliphatic chlorofluorocarbons or chlorofluorohydrocarbons (CFCs). Freon 12, dichlorodifluoromethane, CF_2Cl_2 (b.p. −29.8°C at atmospheric pressure), has been widely used in refrigeration on account of its chemical stability and low toxicity. Freon 22, chlorodifluoromethane, $CHClF_2$ (b.p. −40.8°C at atmospheric pressure) is used in deep-freezing of foodstuffs. Such a freon is known as a CFC, on account of the successive elements in its name, chlorine-fluorine-carbon. Until recently aerosols for the dispersal of insecticides, paints and lacquers, and various toilet preparations contained a *freon propellant*. Bromotrifluoromethane, $CBrF_3$, and dibromodifluoromethane, CBr_2F_2, have been used as fire-extinguishers in aircraft; being less toxic and more efficient than tetrachloromethane (carbon tetrachloride) or bromomethane (methyl bromide) but too expensive for general use.

By international agreement the use of CFCs is being discontinued because the compounds are believed to cause the depletion of ozone in the upper reaches of the atmosphere, the so-called 'ozone hole', or at least to be a major culprit. In view of the otherwise exceptionally useful properties of these compounds, such a turn of affairs is a major setback, and serves to remind us that we cannot go on mass producing chemicals indiscriminately year after year, especially as more and more countries develop their chemical industries. Pollution problems no longer belong to the 'backyard' but to the whole world, and some of them may prove to be irreversible.

Fluorine-containing polymers are extremely resistant to chemical attack. Polytetrafluoroethylene ($-CF_2-CF_2-$)$_n$ known as PTFE, Teflon or Fluon, and polychlorotrifluoroethylene, ($-CF_2-CFCl-$)$_n$ known as Kel-F, Fluorothene or Hostoflon have a chemical and thermal stability which far exceeds that of other organic polymers. Both are used widely in the chemical and electrical industries. Teflon has an extremely low coefficient of resistance and is used for self-lubricating light duty bearings. Domestically, non-stick ovenware is a great boon in our kitchens and most of us have used a non-stick cake tin or frying-pan, amongst other utensils.

Sulphur hexafluoride, SF_6, made by burning sulphur in fluorine, followed by purification from lower fluorides, is characterized by chemical inertness, non-corrosive and non-toxic properties. Sulphur hexafluoride has a relatively high volatility for a gas of such high relative molecular mass (146) and condenses to a solid at −64°C under atmospheric pressure. It has a high dielectric constant (2.4–2.5), and has electrical uses as insulating material in transformers, condensers, X-ray equipment and co-axial cables.

Halogen and halide aspects of qualitative and quantitative analysis

Laboratory tests for halogen anions

Reactions are summarized in Tables 23.6 and 23.7. Mixtures of anions cause more difficulty, although fluorides may readily be distinguished from the other halides by the differing solubility of silver and calcium salts.

Chloride, bromide and iodide all present

Carbon tetrachloride is added to form an immiscible layer beneath the aqueous solution. Addition of AR† sodium nitrite (solid) and ethanoic acid results in a violet solution of iodine in tetrachloromethane

† Analytical reagent i.e. *very* pure

Table 23.6   Tests on single solid metal halides using concentrated sulphuric acid

	F⁻	Cl⁻	Br⁻	I⁻
Warm solid halide with concentrated sulphuric acid:				
1 alone	Hydrogen fluoride evolved ('greasy' tube)	Hydrogen chloride evolved	Hydrogen bromide and some bromine evolved	Iodine, hydrogen sulphide, a little hydrogen iodide and sulphur formed
2 and manganese dioxide	as 1	Chlorine evolved	Bromine evolved	Iodine evolved
3 and powdered silica, holding a drop of water in a looped platinum wire in the vapour	Silicon tetrafluoride evolved; hydrolysed to gelatinous silicic acid	as 1	as 1	as 1

Table 23.7   Tests on solutions of anions

	F⁻	Cl⁻	Br⁻	I⁻
1 Addition of dilute nitric acid and silver nitrate solution followed by concentrated aqueous ammonia	—	Curdy white precipitate of silver chloride turning violet in light and readily soluble in ammonia to form $Ag(NH_3)_2^+$	Curdy cream precipitate of silver bromide sparingly soluble in ammonia to form $Ag(NH_3)_2^+$	Curdy yellow precipitate of silver iodide insoluble in ammonia
2 Addition of calcium chloride solution	Slimy white precipitate of calcium fluoride	—	—	—
3 Addition of lead ethanoate solution	White precipitate of lead fluoride	White precipitate of lead chloride soluble in hot water	White precipitate of lead bromide soluble in hot water	Yellow precipitate of lead iodide soluble in hot water to give an almost colourless solution
4 Addition drop by drop of chlorine water or chlorine in tetrachloromethane or acid and very dilute sodium hypochlorite solution	—	—	Bromine liberated giving a brown coloration dissolving in tetrachloromethane to give a brown layer	Iodine liberated to give a brown coloration dissolving in tetrachloromethane to give a violet layer disappearing as iodine is oxidized to iodic acid with excess oxidant
5 Addition of dilute ethanoic acid and sodium nitrite solution (use AR quality)	—	—	—	Iodine liberated giving a violet solution in a layer of tetrachloromethane

(carbon tetrachloride):

$$2HNO_2 + 2I^- + 2H^+ \rightarrow 2H_2O + 2NO + I_2$$

Excess sodium nitrite is added, the aqueous solution decanted off and evaporated to expel traces of iodine. Addition of tetrachloromethane (carbon tetrachloride), acidification with nitric acid and the addition drop by drop of dilute potassium permanganate solution liberates bromine which dissolves to give a brown solution in tetrachloromethane:

$$2MnO_4^- + 10Br^- + 16H^+ \rightarrow$$
$$2Mn^{2+} + 5Br_2 + 8H_2O$$

The aqueous layer is decanted, evaporated somewhat to boil off traces of bromine and excess permanganate reduced by addition of sodium nitrite:

$$2MnO_4^- + 5NO_2^- + 6H^+ \rightarrow$$
$$2Mn^{2+} + 5NO_3^- + 3H_2O$$

The silver nitrate test will now show the presence of chloride (hence the use of dilute *nitric* acid).

Chloride in the presence of bromide and/or iodide

This test depends on the formation and hydrolysis of chromyl chloride. There is no corresponding bromide or iodide. A fluoride does exist. The solid powder is mixed intimately with powdered potassium dichromate and warmed with concentrated sulphuric acid. A reddish-brown vapour, chromyl chloride, is evolved and is condensed into water to form a yellow solution, also coloured by products from other halides:

$$Cr_2O_7^{2-} + 4Cl^- + 6H^+ \longrightarrow 2CrO_2Cl_2 + 3H_2O$$
$$4KCl + K_2Cr_2O_7 + 3H_2SO_4 \longrightarrow$$
$$3K_2SO_4 + 3H_2O + 2CrO_2Cl_2$$

a chromic acid

Addition of ether and hydrogen peroxide gives an upper blue ethereal layer containing CrO_5, thus proving the presence of chromium which must have arrived as chromyl chloride.

Bromide and/or iodide in the presence of chloride

Iodide and chloride may be distinguished in the presence of each other by the silver nitrate test. Silver chloride may be extracted with aqueous ammonia from the mixture of insoluble silver chloride and iodide, being then re-precipitated with nitric acid which reacts with free ammonia, destroying the ammine complex:

$$Ag(NH_3)_2^+ + 2H^+ \rightarrow Ag^+ + 2NH_4^+$$
$$Ag^+ + Cl^- \rightarrow AgCl(s)$$

Otherwise, iodine and then bromine may be displaced by chlorine water, chlorine in tetrachloromethane or acidified dilute sodium hypochlorite. Iodine is liberated first and gives a violet layer in added tetrachloromethane. This gives way to brown bromine as iodine is oxidized to iodate by excess oxidant:

$$Cl_2 + 2I^- \rightarrow I_2 + 2Cl^-$$
$$I_2 + 5Cl_2 + 6H_2O \rightarrow 2HIO_3 + 10HCl$$
$$Cl_2 + 2Br^- \rightarrow Br_2 + 2Cl^-$$

The volumetric (titrimetric) assay of chlorides, bromides and iodides

This is achieved by titration against silver nitrate in neutral solution with potassium chromate as indicator or by the addition of excess silver nitrate and titration of the excess in acid solution against a standard ammonium or potassium thiocyanate solution using ferric alum indicator. Both methods are given in detail under silver nitrate (p. 492).

Iodide is more usually determined by reaction with potassium iodate (iodate(v)), either by addition of an excess of an iodate solution of any sufficient strength and titration of the liberated iodine against standard sodium thiosulphate solution or by titration against potassium iodate in a solution containing an excess (>6M) hydrochloric acid. Both methods of titrimetric analysis are given under potassium iodate (p. 442).

Iodine in volumetric (titrimetric) analysis

Iodine is only slightly soluble but dissolves in potassium iodide solution to give a complex ion,

$$I_2 + I^- \rightleftharpoons I_3^-$$

This readily surrenders iodine as a reaction proceeds:

$$I_2 + 2e^- \rightleftharpoons 2I^-$$

Two types of determination are used. In the first, iodine is liberated by the action of an oxidizing agent on an excess of acidified potassium iodide, and titrated against sodium thiosulphate solution (Iodometry). In the other, a standard solution of iodine (in potassium iodide) is used in the titration as the oxidizing agent (Iodimetry). Both sodium thiosulphate and iodine solutions are usually standardized and not made up by direct weighing because the exact degree of hydration of sodium thiosulphate, $Na_2S_2O_3 . 5H_2O$, is uncertain and iodine is difficult to dry completely. Neither solution is completely stable and periodic standardization is desirable.

Sodium thiosulphate may be standardized by the first method: a measured volume of the standard solution of oxidizing agent is pipetted into an excess of strongly acidified potassium iodide. Sodium thiosulphate solution is run in until the deep burgundy-red of the iodine solution lightens to a pale straw colour. Some freshly prepared starch solution is added and a blue-black coloration appears due to action with the iodine. Titration is continued cautiously until this colour is discharged. Potassium iodate, bromate and dichromate are all suitable primary standards: potassium permanganate is also simple to use:

$$IO_3^- + 5I^- + 6H^+ \rightarrow 3I_2 + 3H_2O$$

$$BrO_3^- + 6I^- + 6H^+ \rightarrow Br^- + 3H_2O + 3I_2$$

$$Cr_2O_7^{2-} + 14H^+ + 6I^- \rightarrow 2Cr^{3+} + 7H_2O + 3I_2$$

$$2MnO_4^- + 16H^+ + 10I^- \rightarrow$$
$$2Mn^{2+} + 8H_2O + 5I_2$$

In reacting with iodine, sodium thiosulphate forms sodium tetrathionate, the other product being sodium iodide:

$$I_2 + 2e^- \rightleftharpoons 2I^-$$

$$2S_2O_3^{2-} - 2e^- \rightleftharpoons S_4O_6^{2-}$$

$$\therefore I_2 + 2S_2O_3^{2-} \rightarrow 2I^- + S_4O_6^{2-}$$
$$\text{tetrathionate}$$

The calculation is straightforward, the essential relationship coming from the relevant pair of equations. Thus for the use of iodate, 1 mol of KIO_3 yields 3 mol I_2 which requires 6 mol of $Na_2S_2O_3 . 5H_2O$.

Many oxidizing agents may be estimated by this method.

A solution of iodine can be standardized against thiosulphte or, better, against a solution of sodium arsenite as a primary standard, prepared usually from arsenic trioxide. This last compound may be obtained in a high degree of purity, being finally sublimed and dried over sulphuric acid in a desiccator. A weighed quantity of arsenic trioxide is dissolved in alkali (for speed) and neutralized with 1.0M hydrochloric acid using phenolphthalein as indicator. Sodium hydrogencarbonate is added to neutralize hydriodic acid as formed in order to avoid reduction of arsenate by this powerful reducing agent. Sodium carbonate or hydroxide would react with iodine. Arsenite (care!—it is very poisonous) is run into a known volume of iodine solution using starch as indicator near to the endpoint as before:

$$As_4O_6 + 4H_2O - 8e^- \rightleftharpoons 2As_2O_5 + 8H^+,$$

$$I_2 + 2e^- \rightleftharpoons 2I^-$$

giving

$$As_4O_6 + 4H_2O + 4I_2 \rightarrow 2As_2O_5 + 8I^- + 8H^+$$
$$\text{removed by}$$
$$\text{NaHCO}_3$$

Standard iodine solutions are used for the estimation of arsenic and antimony, sulphides and sulphites, and tin(II) compounds in strong acid.

The principal oxides of the halogens, their oxoacids and salts

The oxides of the halogens prove individual in behaviour and do not display a progressive pattern. No oxoacids of fluorine have been isolated, but the following oxoacids of the other halogens are known:

Hypochlorous acid	HClO	chloric(VII) acid
Chlorous acid	HClO₂	chloric(II) acid
Chloric acid	HClO₃	chloric(V) acid
Perchloric acid	HClO₄	chloric(VII) acid
Hypobromous acid	HBrO	bromic(I) acid
—		
Bromic acid	HBrO₃	bromic(V) acid
—		
Hypoiodous acid	HIO	iodic(I) acid
—		
Iodic acid	HIO₃	iodic(V) acid
Per-iodic acids	H₅IO₆, HIO₄ and H₄I₂O₉	

It is conventional to write formulae with symbols in the order shown although each hydrogen atom is part of an —OH group, and not directly attached to a halogen atom. The suffix -ic refers to a higher valency state of the halogen than -ous. A *hypo* acid contains a smaller proportion of oxygen while a *per* acid contains a larger proportion of oxygen than the acid with what is otherwise the same name. In the series HXO_3, iodic acid is the most stable; iodine will displace bromine from bromic acid and bromates, and chlorine from chloric acid and chlorates. Bromine will also displace chlorine at least partially from chloric acid and chlorates. The series HXO is unstable and readily forms HXO_3, the salts behaving similarly: hypochlorous acid and hypochlorites are more stable than the bromine counterparts which are in turn more stable than those of iodine. In the chlorine series, with the increasing proportion of oxygen, the oxidizing power decreases but strength and stability of the acids increase.

Oxygen fluorides (oxides of fluorine)

Four oxygen fluorides have been described with the quite engaging structural formulae FOF, FOOF, FOOOF and FOOOOF but no oxoacids or their salts have been claimed.

Oxygen difluoride (fluorine monoxide), OF_2

A slow stream of fluorine, passed through 2% aqueous sodium hydroxide, yields oxygen difluoride, formerly† fluorine monoxide, in 70% yield:

$$2F_2 + 2OH^- \rightarrow 2F^- + OF_2 + H_2O$$

Little oxygen difluoride is formed with water alone while moderately concentrated alkali yields ozonized oxygen. Oxygen difluoride is a colourless gas (b.p. $-145°C$, m.p. $-224°C$), more poisonous though chemically less reactive than fluorine, and a powerful oxidizing agent. On heating it dissociates into the elements.

Dioxygen difluoride (fluorine dioxide), O_2F_2

Dioxygen difluoride, O_2F_2, is formed when fluorine and oxygen are subjected to an electric discharge when cooled in liquid air. It is a brown gas which decomposes above $-100°C$.

Oxides of chlorine

Chlorine forms a range of oxides which, in general, are endothermic, explosive compounds (coloured except for the heptoxide).

Chlorine monoxide, Cl_2O

When dry chlorine is passed over freshly precipitated and dried mercury(II) oxide, which is kept water-cooled, chlorine monoxide and brown mercury(II) oxide chloride are formed:

$$2Cl_2 + 2HgO \rightarrow Cl_2O + HgCl_2.HgO$$

Chlorine monoxide is a brownish-yellow gas (m.p. $-11.6°C$, b.p. $+2°C$) which may be condensed in a freezing mixture while excess chlorine passes on. On heating, the gas explodes to the elements:

$$2Cl_2O \rightarrow 2Cl_2 + O_2$$

It detonates violently with rubber and other organic matter and reacts with most metals to form a mixture of chlorides and oxides. It is a powerful oxidizing agent. Chlorine monoxide dissolves in water, with the formation of hypochlorous acid. On heating, the oxide is recovered; it is hypochlorous anhydride:

$$Cl_2O + H_2O \rightleftharpoons 2HClO$$

Chlorine dioxide, ClO_2

Chlorine dioxide is an orange-yellow gas (m.p. $-59°C$, b.p. $+11°C$) with a distinctive odour and will explode violently to form the elements.

When small additions of dried, powdered potassium chlorate are made to concentrated sulphuric acid, a series of sharp explosions is experienced due to the decomposition of chlorine dioxide by heat generated in the reaction. This experiment can be extremely dangerous:

$$KClO_3 + H_2SO_4 \rightarrow KHSO_4 + HClO_3$$
chloric acid
chloric(v) acid

$$3HClO_3 \rightarrow HClO_4 + 2ClO_2 + H_2O$$
perchloric acid
chloric(vii) acid

Chlorine dioxide is a powerful oxidizing agent, exploding on heating and with organic matter.

It dissolves in water to form a fairly stable solution which gradually forms a mixture of oxoacids, and ultimately chloric and hydrochloric acids respectively of oxidation states $+5$ and -1:

$$6ClO_2 + 3H_2O \rightarrow 5HClO_3 + HCl$$

With aqueous alkalis, a mixture of chlorite and chlorate, otherwise chlorate(III) and chlorate(v)

† Oxygen difluoride because fluorine is more electronegative than oxygen

result:

$$2ClO_2 + 2OH^- \rightarrow ClO_2^- + ClO_3^- + H_2O$$

Chlorine dioxide is a mixed anhydride.

Chlorine dioxide is made commercially by the action of nitrogen dioxide on sodium chlorate solution:

$$ClO_3^- + NO_2 \rightarrow NO_3^- + ClO_2$$

Its technical applications include the bleaching of flour, the bleaching of wood-pulp for paper-making and the sterilization of water supplies.

Chlorine hexoxide, $Cl_2O_6 \rightleftharpoons 2ClO_3$

Very unstable but less explosive than the dioxide, chlorine hexoxide has been prepared by the action of chlorine dioxide on ozonized oxygen. It is a dark red liquid (m.p. +3.5°C) which dissociates on heating and breaks down eventually to the elements. It reacts violently with water but with cooled water vapour, perchloric acid monohydrate (oxonium perchlorate) and chloric acid result:

$$Cl_2O_6 + 2H_2O \rightarrow (H_3O^+)(ClO_4^-) + HClO_3$$

oxonium perchlorate   chloric acid
or perchloric acid   chloric(v) acid
monohydrate
$HClO_4 . H_2O$

It is a mixed anhydride; aqueous alkali forms a mixture of chlorate and perchlorate,

$$Cl_2O_6 + 2OH^- \rightarrow ClO_3^- + ClO_4^- + H_2O$$

chlorate   perchlorate
chlorate(v)   chlorate(vii)

Chlorine heptoxide, Cl_2O_7

The dehydration of perchloric acid, chloric(vii) acid, yields chlorine heptoxide. The anhydrous or highly concentrated acid is cautiously added to phosphorus pentoxide, cooled in a freezing-mixture. Chlorine heptoxide may be distilled off under reduced pressure,

$$4HClO_4 + P_4O_{10} \rightarrow 2Cl_2O_7 + 4HPO_3$$

metaphosphoric
acid

Chlorine heptoxide is a colourless, oily liquid (m.p. −91.5°C, b.p. 80°C). Although the most stable oxide of chlorine, it is easily detonated by heat or shock. It slowly decomposes on standing and with

water slowly forms perchloric acid,

$$Cl_2O_7 + H_2O \rightarrow 2HClO_4$$

perchloric acid
chloric(vii) acid

It is perchloric anhydride.

The oxoacids of chlorine and their salts

Chlorine forms the widest range of oxoacids in the halogen family.

Hypochlorous acid and hypochlorites, chloric(i) acid and its salts

Hypochlorous acid is formed together with hydrochloric acid when chlorine reacts with water,

$$Cl_2 + H_2O \rightleftharpoons HClO + HCl$$

Hydrochloric acid is completely ionized at this concentration, and hypochlorous acid, feebly ionized. The removal of chloride ions, as a covalent chloride, disturbs the equilibrium and results in more hypochlorous acid being formed. Hypochlorous acid may be prepared by the action of chlorine water on a suspension of yellow precipitated mercury(II) oxide, when the basic mercury(II) chloride formed, mercury(II) oxide chloride, may be filtered off and hypochlorous acid distilled off under reduced pressure:

$$2Cl_2 + H_2O + 2HgO(s) \rightarrow$$
$$HgCl_2 . HgO(s) + 2HClO$$

During distillation, hypochlorous acid may break down into the anhydride and water which reunite on condensation,

$$2HClO \rightleftharpoons H_2O + Cl_2O$$

The anhydrous acid is unknown. The pale yellow aqueous solution decomposes in sunlight in a number of ways:

$$2HClO \rightarrow 2HCl + O_2$$

$$HCl + HClO \rightleftharpoons Cl_2 + H_2O$$

$$3HClO \rightarrow HClO_3 + 2HCl$$

The decomposition is sensitive to catalysis by platinum black, manganese(IV) oxide and cobalt(II) oxide. Hypochlorous acid is weaker than carbonic acid and may be displaced by it from hypochlorites. With alkalis, hypochlorous acid forms salts, which are largely hydrolysed:

$$HClO + OH^- \rightleftharpoons H_2O + ClO^-$$

The more electropositive metals will react with hypochlorous acid to give hydrogen and the hypochlorite. Hypochlorous acid is a powerful oxidizing agent and bleach, being responsible for the bleaching action of damp chlorine.

Sodium hypochlorite, sodium chlorate(I), NaClO

When chlorine is passed into a cold solution of sodium hydroxide or carbonate, a mixture of sodium hypochlorite and chloride remains in solution:

$$2OH^- + Cl_2 \rightarrow ClO^- + Cl^- + H_2O$$

Alternatively, the electrolysis of brine may be arranged so that the products, chlorine and sodium hydroxide, unite and hydrogen escapes. One of many products of the alkali industry, described on p. 213 of Chapter 14.

Solid hydrates have been reported but sodium hypochlorite is usually found as the aqueous solution. It is a powerful oxidizing agent and has strong germicidal properties. Its applications include use as a disinfectant, as a domestic bleach and in industry for the bleaching of the vegetable fibres, cotton and linen, and of wood-pulp in the production of rayon. Sodium hypochlorite is losing favour as a domestic (lavatory) bleach in some European countries. In Germany, for example, hydrogen peroxide seems to be the major constituent.

The oxidizing action of hypochlorites may be expressed by the ion-electron equations:

in alkali:

$$ClO^- + H_2O + 2e^- \rightleftharpoons Cl^- + 2OH^-$$

in acid:

$$ClO^- + 2H^+ + 2e^- \rightleftharpoons Cl^- + H_2O$$

or because hypochlorous acid is a weak acid

$$HClO + H^+ + 2e^- \rightleftharpoons Cl^- + H_2O$$

The oxidation potential is higher in acid solution. Hypochlorite solutions also contain chloride ions so dilute acids liberate chlorine:

$$HClO + H^+ + Cl^- \rightarrow Cl_2 + H_2O$$

Iodine is liberated from acidified solutions of iodides:

$$HClO + H^+ + 2I^- \rightarrow Cl^- + H_2O + I_2$$

and iron(II) ions oxidized to the iron(III) state in acid solution:

$$HClO + H^+ + 2Fe^{2+} \rightarrow Cl^- + H_2O + 2Fe^{3+}$$

Solutions of chromium(III) salts are oxidized to chromate, seen by the colour change from green to yellow, on boiling with sodium hypochlorite solution and alkali:

$$2Cr(OH)_3 + 3ClO^- + 4OH^- \rightarrow$$
initial precipitate
of chromium(III)
hydroxide
$$2CrO_4^{2-} + 3Cl^- + 5H_2O$$

Similarly, manganese(II) salts are oxidized to manganese(IV) oxide, and in alkaline solution, to the manganate(VII) condition as sodium permanganate:

$$Mn(OH)_2(s) + ClO^- \rightarrow$$
initial precipitate
of manganese(II)
hydroxide
$$MnO_2(s) + Cl^- + H_2O$$

$$2MnO_2(s) + 3ClO^- + 2OH^- \rightarrow$$
$$2MnO_4^- + 3Cl^- + H_2O$$

Solutions of lead(II) nitrate or lead(II) ethanoate (acetate), the common soluble salts of bivalent lead, are oxidized to lead(IV) oxide on boiling with sodium hypochlorite solution, the yellow precipitate formed initially darkening to chocolate brown:

$$PbO(s) + ClO^- \rightarrow PbO_2(s) + Cl^-$$

Ammonia is oxidized to nitrogen and hydrogen peroxide to oxygen:

$$2NH_3 + 3ClO^- \rightarrow N_2 + 3Cl^- + 3H_2O$$

$$H_2O_2 + ClO^- \rightarrow O_2 + Cl^- + H_2O$$

Oxygen may be catalytically generated from alkaline solutions of hypochlorites by the action of cobalt(II) oxide (formed by addition of cobalt nitrate solution), manganese(IV) oxide or platinum black:

$$2OCl^- \rightarrow 2Cl^- + O_2$$

The oxidizing capacity of hypochlorites is often expressed in terms of the chlorine equivalent of the hypochlorite ion. This is the *available chlorine* liberated by excess dilute (hydrochloric) acid:

$$ClO^- + 2H^+ + Cl^- \rightarrow Cl_2 + H_2O$$

so that $NaClO \equiv Cl_2$ or 71 g chlorine is liberated by 74.5 g sodium hypochlorite. The available chlorine is expressed as the mass/g of chlorine liberated by

100 cm^3 of hypochlorite solution. Sodium hypochlorite commonly contains 14% available chlorine on a mass per volume of solution basis.

On heating solutions of hypochlorite, sodium chlorate and sodium chloride result:

$$3ClO^- \rightarrow ClO_3^- + 2Cl^-$$

Calcium hypochlorite, calcium chlorate(I), Ca(ClO)₂.4H₂O

The passage of chlorine into a cold suspension of calcium hydroxide (milk of lime) yields a mixture of calcium hypochlorite and chloride, from which hydrated crystals of calcium hypochlorite may be isolated on evaporation *in vacuo*. Anhydrous calcium hypochlorite, stabilized with added lime, may also be prepared. Both are available commercially. Calculation shows that anhydrous calcium hypochlorite on treatment with hydrochloric acid yields 99.4% available chlorine as a theoretical maximum.

Bleaching powder

Nearly dry slaked lime readily absorbs chlorine to form bleaching powder. In commercial production, electrolytic chlorine, diluted with air, passes up a rotating inclined cylinder down which slaked lime falls, being constantly stirred mechanically to expose a fresh uncombined surface. The temperature is regulated by water-cooling. The product, bleaching powder or 'chloride of lime', is an off-white solid which smells of chlorine. It is used for bleaching and as a disinfectant but is giving way to sodium hypochlorite solution. The development of bleaching powder as a source of chlorine by Tennant and Macintosh in patents of 1799 and 1800 revolutionized bleaching processes in the textile industry and led to a rapid expansion in production of cotton and linen, the bleached product being of more uniformly high quality. Furthermore, the bleach fields used in those days for exposing treated fabrics to sunlight were released for agricultural purposes.

Elucidating the chemical constitution of bleaching powder created many problems, which were finally resolved by X-ray analysis in 1935. The results were interpreted as revealing that fully chlorinated bleaching powder is a complex mixture of a probably basic calcium hypochlorite and a non-deliquescent basic calcium chloride:

$$3Ca(ClO)_2 . 2Ca(OH)_2 . 2H_2O$$

or

$$Ca(ClO)_2 . 2Ca(OH)_2$$

with

$$CaCl_2 . Ca(OH)_2 . H_2O$$

which is usually written:

$$Ca(ClO)_2 + CaCl_2 . Ca(OH)_2 . H_2O$$

In addition, unchanged slaked lime, hydrated calcium chloride, $CaCl_2 . 4H_2O$, and calcium chlorate are present. Bleaching powder takes up moisture from the atmosphere but to nothing like the extent to which it would if the calcium chloride were present as the deliquescent, normal, rather than the basic, salt. Furthermore, although normal calcium chloride is freely soluble in ethanol, very little may be extracted by this solvent from bleaching powder.

Bleaching powder is partially soluble in water, the hypochlorite dissolving and being responsible for the bleaching and oxidizing powers. Examples of oxidation include ammonia to nitrogen, acidified potassium iodide solution to iodine, iron(II) ions to iron(III) and the preparation of lead(IV) oxide. Equations involving the ClO^- ion may be written as for sodium hypochlorite. On boiling the aqueous solution, calcium chlorate and chloride are formed. Oxygen may be prepared conveniently from the catalytic decomposition of bleaching powder by warming with a solution containing cobalt(II) ions, the nitrate or chloride. A suggested mechanism involves the formation and decomposition of a dioxide of cobalt:

$$\underset{\text{from lime}}{Co^{2+} + 2OH^-} \rightarrow \underset{\substack{\text{cobalt(II)} \\ \text{oxide}}}{CoO} + H_2O$$

$$CoO + \underset{\substack{\text{from calcium} \\ \text{hypochlorite}}}{ClO^-} \rightarrow CoO_2 + Cl^-$$

$$CoO_2 + ClO^- \rightarrow CoO + O_2 + Cl^-$$

One method of assaying bleaching powder requires the measurement of the volume of oxygen liberated from a known mass of bleaching powder acting in suspension on approximately '5 volume' hydrogen peroxide:

$$OCl^- + H_2O_2 \rightarrow Cl^- + H_2O + O_2$$

whence $Ca(ClO)_2 \equiv 2Cl_2$ (available chlorine, see

above also) $\equiv 2H_2O_2 \equiv 2O_2$ and calculation shows:

1 cm^3 of oxygen at STP (273 K, 101 kPa)
$$\equiv 0.00317 \text{ g available chlorine}$$

Available chlorine is a measure of the oxidizing capacity of the hypochlorite ion and may be liberated by the action of dilute acids, the available chlorine being expressed as a percentage weight/weight. Calculation reveals that

$$Ca(ClO)_2 + CaCl_2.Ca(OH)_2.H_2O$$

ought to contain 41% available chlorine but in practice a good sample may contain up to 38%. The available chlorine in bleaching powder is usually determined by titrimetric (volumetric) analysis. Acidification of bleaching powder leads to the following changes:

$$\begin{aligned}Ca(ClO)_2 \\ CaCl_2.Ca(OH)_2.H_2O\end{aligned} + \begin{aligned}2H^+ \\ 4H^+\end{aligned} \rightarrow$$

$$\begin{aligned}Ca^{2+} \quad\quad + \boxed{2HClO} \\ 2Ca^{2+} + 3H_2O + \boxed{2H^+ + 2Cl^-}\end{aligned}$$

$$\downarrow$$
$$2H_2O + 2Cl_2$$

$\therefore Ca(ClO)_2 + CaCl_2.Ca(OH)_2.H_2O$

$$+ \begin{Bmatrix}3H_2SO_4, \\ 6HCl \ or \\ 6HNO_3\end{Bmatrix} \rightarrow 3\begin{Bmatrix}CaSO_4, \\ CaCl_2 \ or \\ Ca(NO_3)_2\end{Bmatrix} + 5H_2O + 2Cl_2$$

Solid bleaching powder is attacked by carbonic acid formed by moist air, giving a mixture of calcium carbonate and calcium chloride, and the decomposition products of hypochlorous acid, chlorine and oxygen:

$$Ca(ClO)_2 + H_2O + CO_2 \rightarrow$$
$$CaCO_3 + 2HClO$$

$$CaCl_2.Ca(OH)_2.H_2O + CO_2 \rightarrow$$
$$CaCl_2 + CaCO_3 + 2H_2O$$

$\therefore Ca(ClO)_2 + CaCl_2.Ca(OH)_2.H_2O + 2CO_2 \rightarrow$
$$2CaCO_3 + CaCl_2 + 2HClO + H_2O$$

followed by

$$2HClO \rightarrow 2HCl + O_2$$

and $\quad\quad HCl + HClO \rightarrow Cl_2 + H_2O$

Bleaching powder may be used for bleaching cotton and linen. The fabric is soaked in a weak aqueous suspension of bleaching powder and exposed to the air when hypochlorous acid is liberated and causes bleaching by oxidation. The process is completed and excess bleaching powder decomposed by addition of dilute hydrochloric or sulphuric acid. The fabric is treated with an anti-chlor, such as sodium thiosulphate, and washed.

Titrimetric (volumetric) assay of hypochlorites

Hypochlorites are used as oxidizing agents, and their capacity for oxidation, usually expressed in terms of 'available' chlorine liberated by dilute acids, must be determined. Available chlorine is expressed as a percentage of mass/volume of solution (for solutions) or mass/mass (for bleaching powder).

For estimation, sodium hypochlorite solution is diluted accurately to a convenient strength for titration while a weighed sample of bleaching powder (2–3 g) is ground with water and swirled into a graduated flask. The suspension is made up to the mark (250 cm^3) and well shaken just before use. [2.5 g bleaching powder in 250 cm^3, or 10 g dm^{-3}, with about 36% chlorine (i.e. c 0.5 mol in 100 g) available gives a roughly 0.05M solution of available chlorine on acidification.]

A measured volume (25 cm^3) of hypochlorite is added to an excess of a known volume of standard sodium (meta)arsenite solution (50 cm^3, 0.05 M— *very poisonous, no pipettes!*):

$$ClO^- + H_2O + 2e^- \rightleftharpoons Cl^- + 2OH^-$$

$$AsO_2^- + 2H_2O - 2e^- \rightleftharpoons AsO_4^{3-} + 4H^+$$

$\therefore ClO^- + AsO_2^- + H_2O \rightarrow$
$$Cl^- + AsO_4^{3-} + 2H^+$$

or $\quad 4ClO^- + As_4O_6 \rightarrow 4Cl^- + 2As_2O_5$

where arsenic trioxide is used to make up the solution. After five minutes or so, sodium hydrogencarbonate (2 g) is added with shaking and the liquid titrated against iodine (p. 432). The arsenite is titrated against iodine alone and the oxidizing capacity of the sample of hypochlorite related to arsenite only:

$4ClO^- \equiv As_4O_6 \equiv 8$ faradays $\equiv 8Cl$ or $4 Cl_2$
$$\text{(as available chlorine)}$$

Alternatively, the hypochlorite (25 cm^3) is pipetted into an excess of potassium iodide solution (2 g solid) and acidified with a weakly ionized acid, ethanoic (acetic) acid, to avoid the risk of oxidation by chlorate impurities, liable to occur in the presence of a strong acid. Iodine is liberated and titrated against standard (0.05M) sodium thiosulphate solution using starch as indicator,

$$ClO^- + 2H^+ + 2I^- \rightarrow I_2 + Cl^- + H_2O$$

$$I_2 + 2S_2O_3^{2-} \rightarrow S_4O_6^{2-} + 2I^-$$

$ClO^- \equiv I_2 \equiv 2Na_2S_2O_3.5H_2O$
$\equiv 2$ faradays $\equiv 2Cl$ or Cl_2 (as available chlorine)

Chlorous acid and chlorites, chloric(III) acid and its salts

These are much less important than the other oxoacids and their salts. They are mild oxidizing agents and bleaches.

Chloric acid and chlorates, chloric(V) acid and its salts

The salts are more important than the acid. Neither the potassium nor the sodium salt form hydrates.

Potassium chlorate is formed by chlorination of hot strong aqueous caustic potash until the appropriate change in mass has taken place:

$$3Cl_2 + 6OH^- \rightarrow ClO_3^- + 5Cl^- + 3H_2O$$

On cooling, potassium chlorate separates out (solubility = 7.4 g in 100 g water at 20°C), is filtered at the pump, washed in cold water and re-crystallized. It is readily soluble in hot water (Fig. 6.1, p. 92). Sodium chlorate is much more soluble and may be prepared by the same method, but sodium chloride is crystallized out first from boiling solution. After filtration of the hot liquor, sodium chlorate crystallizes out on cooling. It is filtered, washed and re-crystallized.

The electrolytic production of chlorates has been described briefly in Chapter 14 (p. 213).

On heating, chlorates melt and decompose, oxygen being evolved and the final solid product being the chloride. Potassium chlorate (m.p. 359°C), when heated to just above the melting-point, forms the perchlorate (chlorate(VII)), which on further heating yields the chloride, decomposition being complete by 445°C. The decomposition is catalysed by oxides of copper(II), iron(II), cobalt(II) and nickel(II) but principally by manganese(IV) oxide when evolution of oxygen commences at 200°C:

$$4KClO_3 \rightarrow 3KClO_4 + KCl$$

$$KClO_4 \rightarrow KCl + 2O_2$$

or $\qquad 2KClO_3 \rightarrow 2KCl + 3O_2$

Chlorates are oxidizing agents, acidic conditions proving favourable:

$$ClO_3^- + 6H^+ + 6e^- \rightleftharpoons Cl^- + 3H_2O$$

However, the reactions do not move quickly to completion. Iodine is liberated from iodides, hot sulphite solution oxidized to sulphate and iron(II) ions to iron(III):

$$ClO_3^- + 6H^+ + 6I^- \rightarrow 3I_2 + Cl^- + 3H_2O$$

$$ClO_3^- + 3SO_3^{2-} \rightarrow 3SO_4^{2-} + Cl^-$$

$$ClO_3^- + 6H^+ + 6Fe^{2+} \rightarrow 6Fe^{3+} + Cl^- + 3H_2O$$

On heating with concentrated hydrochloric acid, a mixture of chlorine and chlorine dioxide in varying proportions is formed. This is called euchlorine (Davy). The reaction may be expressed:

$$8ClO_3^- + 24H^+ + 16Cl^- \rightarrow$$
$$9Cl_2 + 6ClO_2 + 12H_2O$$

The dangerously explosive action of concentrated sulphuric acid on potassium chlorate has already been described (p. 433).

Chlorates form highly dangerous mixtures with charcoal, sulphur, phosphorus and easily oxidized organic matter, such as sugar, which explode very easily and with violence:

$$2KClO_3 + 3S \rightarrow 2KCl + 3SO_2$$

Potassium chlorate has been used in the manufacture of matches and fireworks. Its sale is restricted in areas where political unrest may lead to bomb-making, although there have been many sad cases of mutilation arising from youngsters attempting to make fireworks from sodium chlorate mixtures with combustible material. Sodium chlorate is sold as a weed-killer which destroys all organic matter, including soil bacteria; a fire hazard to fencing and clothing, etc. results from accidental spillage of chlorate solution during the application to gravel paths and drives. Sodium chlorate is used in the commercial oxidation of aniline to aniline black.

Chloric acid is known only in solution. It may be prepared by addition of dilute sulphuric acid to well-cooled barium chlorate solution when barium sulphate is precipitated:

$$Ba(ClO_3)_2 + H_2SO_4 \rightarrow BaSO_4 + 2HClO_3$$

Chloric acid forms a yellow solution, similar to nitric acid in odour and moderately stable in the dark. By evaporation in a vacuum desiccator, the concentration may rise to 40% when decomposition occurs:

$$3HClO_3 \rightarrow HClO_4 + Cl_2 + 2O_2 + H_2O$$

Concentrated chloric acid is fairly strong, zinc displacing hydrogen, but iron reduces it to hydrochloric acid. Organic matter, e.g. wood and paper, inflames in contact with concentrated chloric acid.

Titrimetric (volumetric) assay of chlorates

An acidified solution of a chlorate is reduced to chloride which is determined by Volhard's Method, using silver nitrate and ammonium thiocyanate. Reduction is effected by zinc amalgam in the presence of dilute sulphuric acid:

$$ClO_3^- + 6H^+ + 3Zn \rightarrow Cl^- + 3H_2O + 3Zn^{2+}$$
$$ClO_3^- \equiv Cl^- \equiv Ag^+ \equiv SCN^-$$

Note that the problem of slowness in reacting is resolved by converting the chlorate ion into its reduction product, chloride ion, and determining that by a proven titrimetric method.

Perchloric acid and perchlorates, chloric(VII) acid and its salts

Perchloric acid is usually prepared from potassium perchlorate.

Potassium perchlorate is formed by heating potassium chlorate to just above the melting-point:

$$4KClO_3 \rightarrow KCl + 3KClO_4$$

As the temperature rises, the perchlorate decomposes until at 445°C it has completely disappeared:

$$KClO_4 \rightarrow KCl + 2O_2$$

This stage is very susceptible to catalysis so clean apparatus and a carefully controlled temperature are important. Potassium chlorate is just-melted in a silica basin and stirred until the mass becomes pasty and almost solid. When samples evolve a minimal quantity of chlorine on warming with concentrated hydrochloric acid, the conversion of chlorate to perchlorate has occurred and the mass cooled. Perchlorates, as distinct from chlorates, do not oxidize hydrochloric acid. After grinding the solid with cold water to extract the more soluble potassium chloride, the perchlorate is recrystallized from residual chlorate using hot water. It is unusual for a potassium salt to be sparingly

soluble in water, only 1.67 g dissolving in 100 g water at 25°C. There is no hydrate.

On the industrial scale, electrolytic oxidation at the anode is utilized. Concentrated sodium chlorate solution is electrolysed with iron cathodes and platinum anodes, these having a high oxygen overpotential:

$$ClO_3^- + H_2O - 2e^- \rightarrow ClO_4^- + 2H^+$$

Addition of potassium chloride precipitates sparingly soluble potassium perchlorate.

Potassium perchlorate is used in explosives.

Perchlorates decompose to the chloride and oxygen on heating, do not oxidize hydrochloric acid and are reduced by powerful reducing agents only.

Perchloric acid is the most stable oxoacid of chlorine and the only one isolated in the pure state. It is prepared by the distillation of potassium perchlorate with 96–97.5% sulphuric acid under reduced pressure:

$$KClO_4 + H_2SO_4 \rightarrow KHSO_4 + HClO_4$$

It is a colourless mobile liquid which may explode on heating. Under normal atmospheric pressure it boils with decomposition at 90°C, but without decomposition at 19°C under 1.5 kPa (11 mm Hg) pressure.

Only anhydrous perchloric acid and concentrated solutions are oxidizing agents; paper, wood and other suitable organic matter inflaming on contact with the anhydrous perchloric acid while on warming, ethanol explodes. The oxidation potential is very much reduced in cold dilute solution, only the most powerful reducing agents, e.g. titanium(III) chloride, causing reduction. Dehydration with phosphorus pentoxide yields chlorine heptoxide. Anhydrous perchloric acid fumes in moist air and is very hygroscopic. A solid monohydrate, shown by X-ray analysis to be oxonium perchlorate, $(H_3O^+)(ClO_4^-)$ analogous to ammonium perchlorate $(NH_4^+)(ClO_4^-)$ and isomorphous with it, is formed by absorption of moisture. A maximum constant boiling-point mixture containing 71.6% acid, and boiling at 203°C under atmospheric pressure, is formed with water. Anhydrous perchloric acid, itself covalent, hisses when added to water, forming the most strongly ionized acid known. It yields hydrogen with metals without being reduced, a distinction from chloric acid. The salts are even weaker oxidizing agents than the acid.

Titrimetric (volumetric) assay of perchlorates

In dilute aqueous solution perchlorates are reduced by only the most powerful reducing agents. Titanium(III) sulphate is oxidized to the titanium(IV) state:

$$ClO_4^- + 8H^+ + 8e^- \rightleftharpoons Cl^- + 4H_2O$$

$$Ti^{3+} - e^- \rightleftharpoons Ti^{4+}$$

$$\therefore 8Ti^{3+} + ClO_4^- + 8H^+ \rightarrow$$
$$8Ti^{4+} + Cl^- + 4H_2O$$

A roughly 0.013M (M/80) solution is prepared. A measured volume of the perchlorate solution (25 cm^3) is boiled with a known volume of excess standard titanium(III) sulphate solution (50 cm^3 : 0.05M) and dilute sulphuric acid for 30 minutes. The flask is fitted with a narrow-bore tube, about 0.5 m long, to restrict access of air which would otherwise oxidize the reducing agent.

Excess iron(III) alum solution is added to oxidize residual titanium(III) sulphate,

$$Fe^{3+} + Ti^{3+} \rightarrow Fe^{2+} + Ti^{4+}$$

The chloride formed is estimated by Volhard's Method, using silver nitrate and ammonium thiocyanate,

$$ClO_4^- \equiv Cl^- \equiv Ag^+ \equiv SCN^-$$

Note that the titrimetric analysis is not based on the redox process but on a proven method which determines the amount of reduction product, in this case chloride ion.

Some oxides, oxoacids and oxosalts of bromine

The oxides of bromine are of little importance and decompose below room temperature. Some (4%) bromine monoxide, Br_2O, is formed when bromine vapour passes over precipitated and dried mercury(II) oxide at 50–100°C. This is similar to the preparation of chlorine monoxide. A tenfold increase in yield results from shaking a solution of bromine in tetrachloromethane (carbon tetrachloride) with mercury(II) oxide. Bromine monoxide is a brown solid (m.p. −17.5°C) which begins to dissociate into the elements at −16°C. Bromine dioxide, BrO_2, is formed when bromine vapour and oxygen are exposed to an electric discharge at the temperature of liquid air. It is a yellow solid which begins to dissociate above −40°C.

Only two oxoacids exist. Hypobromous acid also called bromic(I) acid is similar in preparation and reactions to hypochlorous acid but is weaker and less stable, but more stable and stronger than hypoiodous acid. No solid hypobromites have been isolated but an orange analogue of bleaching powder containing bromine has been made, although no iodine compound has been isolated. One standard method for the estimation of urea depends on measuring the volume of nitrogen produced by decomposition with alkaline sodium hypobromite. The yield is 95.7% of that calculated for the reaction represented:

$$O{=}C\begin{array}{l} \diagup NH_2 \\ \diagdown NH_2 \end{array} + 3NaBrO \longrightarrow$$
$$3NaBr + N_2 + CO_2 + 2H_2O$$
$$\downarrow 2NaOH$$
$$Na_2CO_3 + H_2O$$

No analogues of chlorous acid and the chlorites have been isolated for bromine and iodine, although there is some evidence for the existence of the bromite and perbromate ions. Iodine forms the per-iodic acids and per-iodates. Bromic acid, $HBrO_3$, of Br(V), is very similar to chloric acid and known only in solution.

Bromine reacts with hot concentrated aqueous alkali to form a mixture of bromate and bromide:

$$3Br_2 + 6OH^- \rightarrow BrO_3^- + 5Br^- + 3H_2O$$

Acidification reverses this reaction and is useful analytically.

Potassim bromate (bromate(V)) is a versatile oxidizing agent in acid solution for titrimetric (volumetric) analysis. It is readily prepared and purified by re-crystallization, may be dried without decomposition at 180°C and forms a perfectly stable solution. The only draw-back is the low reacting mass which demands high accuracy in weighing:

$$BrO_3^- + 6H^+ + 6e^- \rightleftharpoons Br^- + 3H_2O$$

When oxidation of the reducing agent being determined is complete, free bromine is liberated by further addition of bromate ion, acting on the bromide ion which has been formed in the main reaction:

$$BrO_3^- + 5Br^- + 6H^+ \rightarrow 3Br_2 + 3H_2O$$

Addition of methyl orange or methyl red provides an irreversible method of detecting the end-point, the colour being destroyed by the liberated bromine. Hydrochloric acid is usually added to cause instant bleaching by chlorine, liberated by slight excess of bromate at the end-point. To avoid premature bleaching of the (irreversible) indicator, potassium bromate solution must be run in slowly.

Arsenic trioxide may be determined in the presence of concentrated hydrochloric acid at about 60°C:

$$As_4O_6 + 4H_2O - 8e^- \rightleftharpoons 2As_2O_5 + 8H^+$$

$$3As_4O_6 + 4BrO_3^- \rightarrow 4Br^- + 6As_2O_5$$

Antimony trioxide, Sb_4O_6, may be determined similarly. Note that the bromide ion obtained in the end-product does not correspond to the product obtained with potassium iodate in the presence of concentrated hydrochloric acid (see later).

Oxides and oxoacids of iodine

Iodine is virtually insoluble in water but dissolves in aqueous alkali to form a mixture of the iodide and hypoiodite, the latter being considerably hydrolysed:

$$I_2 + 2OH^- \rightarrow IO^- + I^- + H_2O$$

$$IO^- + H_2O \rightleftharpoons HIO + OH^-$$

Hypoidous acid, iodic(I) acid is much weaker than hypochlorous acid and hypobromous acid. Hypoiodites and the acid readily form iodate:

$$3IO^- \rightarrow IO_3^- + 2I^-$$

Both acid and salts are oxidizing agents. The only important oxide is iodine pentoxide, I_2O_5, which is a solid and the anhydride of iodic acid, HIO_3. Of the other oxides, I_4O_9 appears to be an iodate, $I(IO_3)_3$, and I_2O_4, or $(IO_2)_n$, possibly $IO(IO_3)$. Iodous acid and iodites do not appear to exist.

Iodine pentoxide, iodic acid (iodic(V) acid) and its salts

Iodine reacts with hot strong aqueous alkali to form a mixture of the iodate and iodide:

$$3I_2 + 6OH^- \rightarrow IO_3^- + 5I^- + 3H_2O$$
iodate
iodate(v)

Addition of acid to this mixture re-forms iodine, a reaction of considerable importance in titrimetry.

Iodic acid may be prepared by the action of fuming nitric acid on iodine at 85–90°C when rapid oxidation occurs; oxides of nitrogen are evolved and iodic acid separates as a white solid,

$$I_2 + 10HNO_3 \rightarrow 2HIO_3 + 10NO_2 + 4H_2O$$

The liquid is decanted and the iodic acid re-crystallized from 50% nitric acid, in which it has a reduced solubility and with which it will not react, drained of liquid and heated gently to drive off traces of nitric acid. It is finally re-crystallized from the minimum quantity of hot water by evaporation and filtered at the pump on sintered glass. It is very soluble.

Iodine pentoxide is formed by heating iodic acid to constant mass at 180–200°C:

$$2HIO_3 \rightarrow I_2O_5 + H_2O$$

It is a colourless solid, a strong oxidizing agent and deliquescent. It is the anhydride of iodic acid which it forms with water. On heating, it is decomposed into the elements. It has been used for the quantitative oxidation of carbon monoxide in gas analysis, forming carbon dioxide which is absorbed in baryta and weighed,

$$I_2O_5 + 5CO \rightarrow I_2 + 5CO_2$$

A better preparation of iodic acid employs the double decomposition of barium iodate with dilute sulphuric acid so the general preparation of iodates may be conveniently studied first.

Potassium iodate, potassium iodate(V), KIO_3

Powdered iodine is gently warmed to about 50°C with a solution of potassium chlorate containing a little nitric acid, chlorine being evolved. A number of reactions occur:

$$I_2 + 2ClO_3^- \rightarrow 2IO_3^- + Cl_2$$

$$I_2 + Cl_2 \rightarrow 2ICl$$

$$ClO_3^- + ICl \rightarrow IO_3^- + Cl_2$$

$$I_2 + 5Cl_2 + 6H_2O \rightarrow$$
$$2H^+ + 2IO_3^- + 10H^+ + 10Cl^-$$

On cooling, a mixture of potassium iodate, KIO_3, and potassium biniodate, $KIO_3 \cdot HIO_3$, crystallizes out, is filtered and washed. The solid is dissolved in boiling water and neutralized with caustic potash solution using litmus as an external indicator. The solution is then just-acidified with nitric acid, filtered and cooled. Colourless crystals of potassium iodate separate out, are filtered at the pump on to a sintered glass funnel, washed in a little cold water, drained by suction and dried in a vacuum desiccator over concentrated sulphuric acid. Potassium iodate is very much more soluble in hot water (32.3 g 100 g^{-1} water) than cold water (8.1 g 100 g^{-1} water). There are no hydrates.

Barium iodate, barium iodate(v), $Ba(IO_3)_2.H_2O$

Barium iodate is conveniently prepared by mixing sodium iodate solution, prepared in a way similar to the first stage above, and barium nitrate solution when sparingly soluble barium iodate monohydrate is deposited:

$$Ba^{2+} + 2IO_3^- + H_2O \rightarrow Ba(IO_3)_2.H_2O(s)$$

Iodic acid, Iodic(v) acid HIO_3

Addition of barium iodate to an excess of boiling dilute sulphuric acid results in the precipitation of the less soluble barium sulphate, which is filtered off. The solution is evaporated until very viscous and concentrated nitric acid added. This precipitates iodic acid by reducing the solubility. After decantation, the solid is washed with glacial ethanoic (acetic) acid on a sintered glass Buchner filter, drained and dried in a vacuum desiccator. Iodic acid is deliquescent and extremely soluble (76.3 g 100 g^{-1} solution at 25°C and 86.0 g 100 g^{-1} solution at 110°C, i.e. 76.3% and 86.0% of saturated solution respectively). Because of its powerful oxidizing nature, easily oxidized organic solvents may not be used in the purification.

Iodic acid is a strong acid and oxidizing agent,

$$IO_3^- + 6H^+ + 6e^- \rightleftharpoons I^- + 3H_2O$$

turning litmus red and bleaching it. It is the most stable acid in the series HXO_3. Iodic acid will set fire to the easily combusted elements, sulphur and phosphorus and to organic materials. Iodic acid and iodates in acid solution liberate iodine from solutions containing iodine ions:

$$IO_3^- + 5I^- + 6H^+ \rightarrow 3I_2 + 3H_2O$$

Iodic acid oxidizes sulphurous acid and initiates further reactions, acting on the iodide product with liberation of iodine which then reacts with more sulphurous acid. With excess iodic acid, iodine appears only when all of the sulphurous acid has

reacted, a striking example of a time-reaction:

$$IO_3^- + 3SO_3^{2-} \rightarrow I^- + 3SO_4^{2-}$$

$$IO_3^- + 5I^- + 6H^+ \rightarrow 3I_2 + 3H_2O$$

$$SO_3^{2-} + I_2 + H_2O \rightarrow$$
$$SO_4^{2-} + 2H^+ + 2I^-$$

$$\therefore 5SO_3^{2-} + 2H^+ + 2IO_3^- \rightarrow$$
$$5SO_4^{2-} + H_2O + I_2$$

Hydrogen sulphide is oxidized to sulphur:

$$5H_2S + 2IO_3^- + 2H^+ \rightarrow I_2 + 5S + 6H_2O$$

On heating, potassium iodate is decomposed to oxygen and potassium iodide, but less readily than the bromate and chlorate. Because potassium iodate is so easily purified, it is used in the preparation of pure potassium iodide. On heating:

$$2KIO_3 \rightarrow 2KI + 3O_2$$

Acid salts are readily crystallized: potassium biniodate, $KIO_3.HIO_3$, and the tri-iodate, $KIO_3.2HIO_3$, have no parallels with chlorates and bromates.

Iodates in titrimetric (volumetric) analysis

Potassium iodate is readily prepared in a high degree of purity by re-crystallization. It may be dried safely at 120°C. However, the small value of the reacting mass requires high accuracy in weighing for titrimetric purposes.

A standard solution of potassium iodate may be used to standardize sodium thiosulphate solution which is titrated against the iodine liberated by the action of a measured volume of iodate on acidified potassium iodide solution. Conversely, iodate or strongly ionized acids may be assayed by this reaction, especially in very dilute solution. A suitable excess of two reagents is maintained during the application involving the third:

$$IO_3^- + 5I^- + 6H^+ \rightarrow 3I_2 + 3H_2O$$

$$2S_2O_3^{2-} + I_2 \rightarrow S_4O_6^{2-} + 2I^-$$

When potassium iodate is used for quantitative oxidation, a further reaction occurs between excess iodate and the iodide formed initially from it, a reaction similar to that occurring with potassium bromate and bromine (p. 440). However, in the presence of an excess of hydrochloric acid the liberated iodine is oxidized by iodate to iodine monochloride, ICl. There is no corresponding reaction for bromine–bromate. Iodine monochloride is stabilized in concentrated hydrochloric acid as the complex ion, ICl_2^-:

$$IO_3^- + 6H^+ + 6e^- \rightleftharpoons I^- + 3H_2O$$

$$IO_3^- + 5I^- + 6H^+ \rightarrow 3I_2 + 3H_2O$$

$$2I_2 + IO_3^- + 6H^+ + 10Cl^- \rightarrow 5ICl_2^- + 3H_2O$$

In ICl_2^-, iodine may be regarded as I^+, so that the iodide ion has undergone the oxidation state changes:

$$I^-(-1) \xrightarrow{-e^-} I(0) \xrightarrow{-e^-} I^+(+1)$$

The equations may be combined into one ion-electron equation:

$$IO_3^- + 6H^+ + 2Cl^- + 4e^- \rightleftharpoons ICl_2^- + 3H_2O$$

Reactions involving the formation of iodine monochloride have been developed by Andrews (1903) and Jamieson (1924) for the determination of reducing agents. Initially, iodine is liberated; the concentration increases as the titration proceeds and then decreases as iodine is further oxidized until at the end-point only iodine monochloride remains. Addition of tetrachloromethane (carbon tetrachloride) results in a lower immiscible layer assuming a purple colour while iodine is present. The disappearance of the last violet tinge in the lower organic layer shows the completion of the process. The reaction mixture is shaken vigorously in a stoppered bottle between additions.

Three examples of estimation using this method follow:

arsenic trioxide (arsenites):

$$As_4O_6 + 4H_2O - 8e^- \rightleftharpoons 2As_2O_5 + 8H^+$$

$$2IO_3^- + As_4O_6 + 4H^+ + 4Cl^- \rightarrow 2As_2O_5 + 2ICl_2^- + 2H_2O$$

tin(II) chloride:

$$Sn^{2+} - 2e^- \rightleftharpoons Sn^{4+}$$

$$2Sn^{2+} + IO_3^- + 6H^+ + 2Cl^- \rightarrow 2Sn^{4+} + ICl_2^- + 3H_2O$$

and iodides:

$$I^- + 2Cl^- - 2e^- \rightleftharpoons ICl_2^-$$

$$2I^- + IO_3^- + 6H^+ + 6Cl^- \rightarrow 3ICl_2^- + 3H_2O$$

When calculating results you will see that the reacting mass for each faraday of charge transferred in the redox process is 0.25 mol of KIO_3 while in the thiosulphate standardization described earlier it was 0.167 mol of KIO_3, i.e. respectively $KIO_3/4$ and $KIO_3/6$.

Per-iodic acids and per-iodates

An heptoxide of iodine, I_2O_7, has not been isolated but a number of per-iodic acids, which may be regarded as derived from it by the addition of water, are known. The common per-iodic acid obtained on crystallization is the 'ortho' acid, H_5IO_6 (orthoper-iodic acid), which on heating to 80°C under 1.6 kPa (12 mm Hg) pressure gives a 'pyro' form (diper-iodic acid), $H_4I_2O_9$, and at 100°C under the same low pressure, the 'meta' acid, HIO_4 (tetraoxoiodic(VII) acid). Orthoperiodic acid, H_5IO_6, is much weaker than chloric acid and in common with many weak acids with an excessive number of hydroxyl groups, loses water to form condensation products. Salts of these acids and certain hypothetical condensed acids are known of general formula M_5IO_6, $M_4I_2O_9$, M_3IO_5, $M_8I_2O_{11}$, and MIO_4 where M is univalent. Because of hydrolysis, salts prepared in aqueous solution have only two hydrogen atoms of the acid replaced. The per-iodates are usually only sparingly soluble in water. They are powerful oxidizing agents.

Thermal decomposition of the iodates of Group II (alkaline-earths) yield per-iodates. Barium iodate decomposes at red heat iodine and oxygen being expelled:

$$5Ba(IO_3)_2 \rightarrow Ba_5(IO_6)_2 + 4I_2(g) + 9O_2(g)$$

Sodium iodate behaves similarly, but the potassium salt yields potassium iodide and oxygen. Compare this behaviour with that of the chlorates (p. 438).

Addition of barium per-iodate to dilute sulphuric acid precipitates barium sulphate, leaving a solution of orthoper-iodic acid:

$$Ba_5(IO_6)_2 + 5H_2SO_4 \rightarrow 2H_5IO_6 + 5BaSO_4(s)$$

24

Period 4: transition and associated elements

The first transition series

Transition elements are characterized by atoms in which an inner electron shell (or level) is in process of expansion but is not yet complete. The penultimate shell in the isolated atoms contains more than eight but less than eighteen electrons. The extra electrons may be used in bonding. From the definition, the series, scandium–nickel inclusive, are transition elements: metallic physically, having a tendency to combine in several oxidation states, forming coloured ions and complexes, and behaving catalytically. Scandium shows only one oxidation state, namely 3, and on forming the Sc^{3+} ion achieves an inert gas electronic configuration. Its claim to belong to the series on consideration of properties is therefore tenuous. Scandium resembles aluminium. Copper and zinc are sometimes included in the series but are not, by definition, transition elements themselves. But the copper(II) ion has an incomplete electron shell and resembles the cations of other transition elements while the relatively simple chemistry of zinc results from the completed penultimate electron shell. This explains why the term 'associated' is introduced.

The electronic configurations of the free atoms, together with the various oxidation states and other information, are listed in Table 24.1. The terms coordination number, oxidation state (or number) and valency will be met in the text. All describe an atom in a compound, but with an increasing level of sophistication. The *coordination number* equals the number of nearest neighbours to the atom, no matter what the bonding. *Oxidation state* (or *number*) is a straightforward arbitrarily assigned number (usually a Roman numeral) with a positive or negative sign, decided by the application of a set of rules (hence, arbitrary) and ignoring the nature of the bonding. On the other hand, *valency*

addresses this problem, ranging from the apparently elementary ideas of the early chemists (though the idea represented a major step forward) to the intricacies of atomic and molecular orbital theories and again is a theoretical term. Often valency can be summarized by the oxidation state number but the state of affairs in complexes necessarily requires a much longer description, i.e. a complex one. However, a number assigned by an internationally agreed set of rules ensures that compounds can be named and their chemistry investigated while the arguments about bonding and valency wax and wane. It is difficult to give an adequate picture of the rarer valency states and selected examples are given only where compounds are later described. These rarer valency states are enclosed in parentheses.

The number of oxidation states assumed is at a maximum near the middle of the series, manganese representing the peak. Elements before manganese have fewer electrons for combining purposes so that all elements up to and including manganese are able to use all electons in excess of the penultimate octet in covalent (and ionic) bond formation. The oxides of the higher valency states are covalent and acidic, while those of the lower states are ionic and basic. The oxides of copper and zinc form covalent macromolecular structures. Intermediate oxides tend to be amphoteric.

In the Mendeléeff Periodic Table, the transition and associated elements were placed with the typical elements according to their maximum (group) valency, now unambiguously the oxidation state, then their only resemblance to each other, except for Fe Co Ni (a Döbereiner Triad, p. 12) assigned to a collective Group VIII: Sc(III), Ti(IV), V(V), Cr(VI), Mn(VII), Fe Co Ni, Cu(I) and Zn(II). To accommodate these elements, subgroups were formed. The expected decrease in atomic radius with increasing atomic number, or proton number

Table 24.1 Some physical properties of transition and associated elements $(1s)^2(2s)^2(2p)^6(3s)^2(3p)^6(3d)^{1-10}(4s)^{2(1)}$

	21 Sc	22 Ti	23 V	24 Cr	25 Mn	26 Fe	27 Co	28 Ni	29 Cu	30 Zn
	2	2	2	2	2	2	2	2	2	2
	8	8	8	8	8	8	8	8	8	8
	9	10	11	13	13	14	15	16	18	18
	2	2	2	1	2	2	2	2	1	2
[Ar core], outer electrons	$(3d)^1(4s)^2$	$(3d)^2(4s)^2$	$(3d)^3(4s)^2$	$(3d)^5(4s)^1$	$(3d)^5(4s)^2$	$(3d)^6(4s)^2$	$(3d)^7(4s)^2$	$(3d)^8(4s)^2$	$(3d)^{10}(4s)^1$	$(3d)^{10}(4s)^2$
Relative atomic mass	44.956	47.90	50.942	51.996	54.9380	55.847	58.9332	58.71	63.546	65.37

Oxidation states
Valency states

	Sc	Ti	V	Cr	Mn	Fe	Co	Ni	Cu	Zn
					7					
				6	6	(6)				
			5							
		4	4		4	(4)				
	3	3	3	3	3	3	(3)	(3)	(3)	2
		(2)	2	2	2	2	2	2	2	
								(1)	1	
				(0)		(0)		(0)		

Metallic radius/nm	0.162	0.147	0.134	0.127	0.126	0.126	0.125	0.124	0.128	0.138
Atomic radius/nm	0.144	0.132	0.122	0.117	0.117	0.117	0.116	0.115	0.135	0.131

Ionic radius/nm

	Sc	Ti	V	Cr	Mn	Fe	Co	Ni	Cu	Zn
M^+	—	—	—	—	—	—	—	—	0.096	—
M^{2+}	—	—	0.088	0.074	0.080	0.074	0.072	0.069	0.072	0.074
M^{3+}	0.081	0.076	0.074	0.063	0.066	0.064	0.063	—	—	—
M^{4+}	—	0.068	—	—	—	—	—	—	—	—

Standard electrode potential $(M^{2+}; M,$ except $Cr^{3+}), E^{\ominus}_{298}/V$	—*	—*	−1.18	−0.74*	−1.19	−0.440	−0.227	−0.250	+0.337†	−0.763
Electronegativity	1.2	1.3	1.45	1.55	1.6	1.65	1.7	1.75	1.75	1.65
Atomic volume/ cm^3 mol^{-1}	14.6	10.8	8.39	7.24	7.62	7.10	6.77	6.59	7.12	9.17
Boiling-point/ °C	2831	3287	3380	2260	1900	2750	2870	2730	2627	907
Melting-point/ °C	1541	1660	1720	1830	1242	1527	1490	1455	1083	419
Density/g cm^{-3}	2.99	4.50	5.96	7.20	7.20	7.86	8.90	8.90	8.92	7.14

*No Sc^{2+}, Ti^{2+} and Cr^{2+} rapidly oxidized in air
†Note change of sign

i.e. nuclear charge, which is seen in traversing a period as an electron shell is filled is seen in this period and is illustrated in Fig. 24.1 which shows M^{2+} ionic radii and metallic radii as well. This shrinkage with rise in atomic number is naturally accompanied by an increase in relative atomic mass, so the density also increases from scandium to copper, shown in Fig. 24.2, as the $3d$ electron shell is formed.

When atomic volume is plotted against atomic number as in Fig. 24.3 the relative compactness of the transition metal structures is again seen.

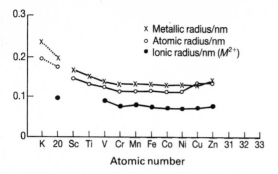

Fig. 24.1 Atomic, ionic (M^{2+}) and metallic radii plotted against atomic number for elements in the K—Kr sequence

Another consequence of increasing nuclear charge as the $3d$ level is filled, besides that of causing shrinkage of the atom and its ions allied to it, is the rise in ionization energies, and especially the higher ones, showing that electrons are more firmly controlled by the nucleus. Values are shown in Table 24.2 and the trend is illustrated in Fig. 24.4 for the first and second ionization energies. Values vary slightly in the literature. Comparison of the successive ionization energies of a transition element (e.g. V) with those of a typical metal of high oxidation state (or valency) e.g. Al, shows that the latter has a sharply

increased 4th ionization energy while the former has a more equable series of energies. When the graphical illustrations are matched with the electronic configurations of the elements it is seen that in the filling of the $3d$ level, the extra stability (exchange energy, p. 46) associated with a half filled shell, $(3d)^5 (4s)^1$, and a completed shell, $(3d)^{10} (4s)^1$, for chromium and copper respectively, is evident from the minor peaks. The rise in first ionization energy seen in comparing chromium and manganese $(3d)^5 (4s)^2$

$$\underset{(3d)^5(4s)^1}{\text{Cr}} \longrightarrow \underset{(3d)^5}{\text{Cr}^+} + e^-; \quad \Delta H = 652\,\text{kJ mol}^{-1}$$

$$\underset{(3d)^5(4s)^2}{\text{Mn}} \longrightarrow \underset{(3d)^5(4s)^1}{\text{Mn}^+} + e^-; \quad \Delta H = 717\,\text{kJ mol}^{-1}$$

is reversed when the second ionization energies are compared:

$$\underset{(3d)^5}{\text{Cr}^+} \longrightarrow \underset{(3d)^4}{\text{Cr}^{2+}} + e^-; \quad \Delta H = 1592\,\text{kJ mol}^{-1}$$

$$\underset{(3d)^5(4s)^1}{\text{Mn}^+} \longrightarrow \underset{(3d)^5}{\text{Mn}^{2+}} + e^-; \quad \Delta H = 1505\,\text{kJ mol}^{-1}$$

This is also seen for the $(3d)^{10}$ configuration. The density and atomic volume graphs should also be inspected.

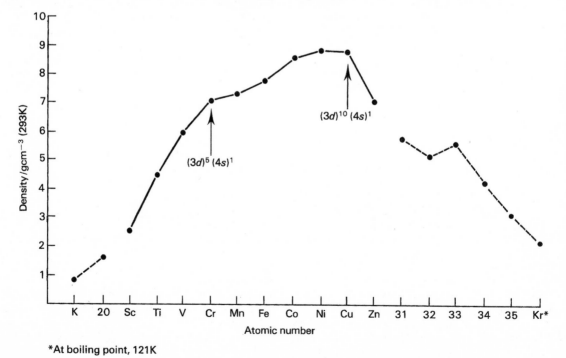

*At boiling point, 121K

Fig. 24.2 Density plotted against atomic number for elements of the K—Kr sequence which are solid at 293K

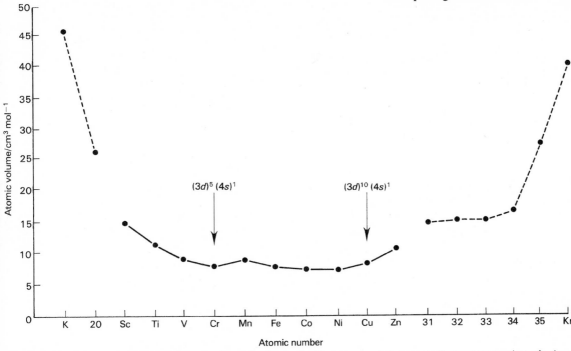

Fig. 24.3   Atomic volume plotted against atomic number for the K—Kr sequence of elements to illustrate comparison of values for transition elements with others in the same period

Finally, increasing atomic number brings a rise in electronegativity but this is much less pronounced than the trend seen across a period of typical elements, as shown in Fig. 24.5 although the values are high when compared with those shown by the s-block elements.

Comparing the 3d elements

A general survey of the transition and associated elements of Period 4 may be based on the more important oxidation and valency states.

Oxidation state 0, M(0)

Carbonyls are formed by the combination of carbon monoxide and a metal. Only simple carbonyls will be considered here; they are formed by chromium, iron and nickel: $Cr(CO)_6$, $Fe(CO)_5$, $Ni(CO)_4$. In complexes, if the ligand is a neutral molecule (here, CO) the metal is assigned an oxidation number equal to the charge (in this case, zero). All bonding electrons come from carbon monoxide, so that the metal is assigned zero valency Ni^0, although the structural picture is quite complicated. While the bonding is essentially co-ionic from carbon, it has been shown by physical methods that there is some double bond character, too.

$$M + \bar{C}{\equiv}\overset{+}{O} \rightarrow \bar{M}{-}C{\equiv}\overset{+}{O}$$

The pair of electrons is donated by carbon and the final number of electrons in the proximity of the metal atoms may be calculated:

$$
\begin{array}{ll}
Cr(CO)_6: & 24 \text{ from } Cr + (6 \times 2) = 36 \\
Fe(CO)_5: & 26 \text{ from } Fe + (5 \times 2) = 36 \\
Ni(CO)_4: & 28 \text{ from } Ni + (4 \times 2) = 36
\end{array} \right\} 2, 8, 18, 8
$$

Table 24.2   Period 4: Transition and associated elements: ionization energies/kJ mol^{-1} (273 K)

Ionization		Sc	Ti	V	Cr	Mn	Fe	Co	Ni	Cu	Zn
1st	$M(g) \rightarrow M^+(g) + e^-$	633	659	652	650	717	762	758	736	745	906
2nd	$M^+(g) \rightarrow M^{2+}(g) + e^-$	1235	1312	1418	1592	1505	1563	1650	1756	1959	1737
3rd	$M^{2+}(g) \rightarrow M^{3+}(g) + e^-$	2393	2653	2827	2991	3251	2952	3232	3396	3550	3830
4th	$M^{3+}(g) \rightarrow M^{4+}(g) + e^-$	7130	4168	4507	4785	4940	5290	4950	5300	5330	5730

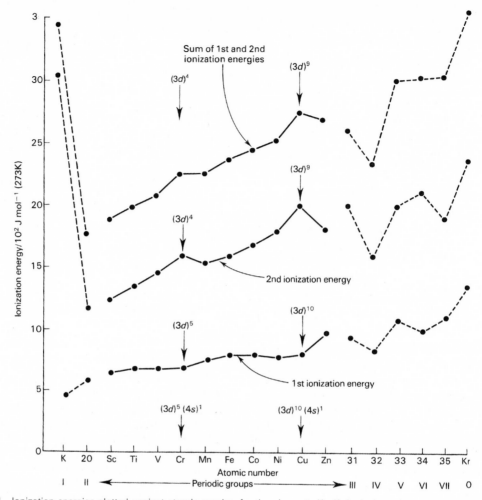

Fig. 24.4   Ionization energies plotted against atomic number for the elements K—Kr inclusive

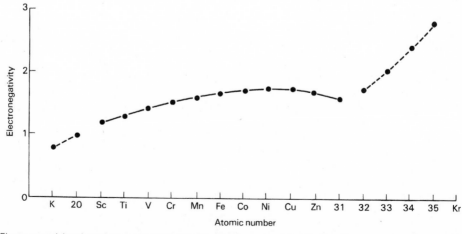

Fig. 24.5   Electronegativity plotted against atomic number for elements of the K—Kr sequence values for transition elements

They acquire the electronic configuration of the inert gas, krypton.

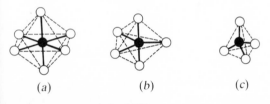

in shape the molecules are:

(a) octahedral   (b) trigonal bipyramidal

(c) tetrahedral

(a)   (b)   (c)

Oxidation state +1, $M(\text{I})$

Although there is some evidence for compounds of this state being formed by manganese and nickel, it is important for copper only. In the copper(I) condition, the 18 electron group is complete and not transitional in character. Complex ions of copper(I) include the ammine, $Cu(NH_3)_2^+$, a cation with neutral ligands so that the oxidation number equals the ionic charge (+1) and the sequence of chlorides, $CuCl_2^-$, $CuCl_3^{2-}$, and $CuCl_4^{3-}$ anions in which the role of the metal is to balance the negative contributions of the chlorine to leave the ions negatively charged as shown, and in which the oxidation number is again +1. There is an interesting relationship between the relative stabilities of the Cu(I) and Cu(II) states in the presence of water, described later.

Oxidation state +2, $M(\text{II})$

This oxidation state becomes increasingly important as the atomic number rises. Scandium forms no compounds in this state. For titanium, it is confined to the oxide, sulphide and halides, which are readily oxidized in air while the ion cannot exist in the presence of water, titanium and a titanium(IV) compound resulting:

$$2Ti^{2+} \rightarrow Ti + Ti^{4+}$$

Because of their ease of oxidation, salts of bivalent vanadium and chromium are difficult to isolate. Bivalent manganese forms very stable compounds. The elements, vanadium–zinc, form sulphates which are isomorphous heptahydrates, although for copper(II) and manganese(II) sulphates the pentahydrates are more usual. Except for manganese and scandium, double sulphates with ammonium sulphate are formed. Examples of sulphates include:

$$FeSO_4 . 7H_2O \qquad FeSO_4 . (NH_4)_2SO_4 . 6H_2O$$

iron(II) sulphate   ammonium iron(II) sulphate

Ionic in nature, these salts contain the hexaquoiron(II) cation, $Fe(H_2O)_6^{2+}$, in which the oxidation number of iron = +2, the cationic charge, because the water ligands are neutral.

The colours of the hydrated cations†, $M(H_2O)_6^{2+}$, in this oxidation state are various, and are associated with incomplete electron shells:

V^{2+}	Cr^{2+}	Mn^{2+}	Fe^{2+}
violet	bright blue	pale pink	pale green
Co^{2+}	Ni^{2+}	Cu^{2+}	Zn^{2+}
pink	green	blue	colourless

The simple ions show the metals in the bivalent state and this is also true in the complexes (including the hydrated ions) but the net electron contributions must be counted to decide this. The earlier transition metals form oxides and hydroxides (or hydrated oxides) in this oxidation state which are readily oxidized. Those of titanium and vanadium have not been isolated, while yellow chromium(II) hydroxide, $Cr(OH)_2$, is rapidly changed by water into Cr_3O_4. Pink covalent manganese(II) hydroxide, $Mn(OH)_2$, is soon oxidized to Mn_2O_3. White iron(II) hydroxide, $Fe(OH)_2$, rapidly darkens to green then black and forms hydrated iron(III) oxide, $Fe_2O_3 . H_2O$. Cobalt(II) hydroxide, $Co(OH)_2$, appears in blue and pink forms, readily oxidized to hydrated cobalt(III) oxide, $Co_2O_3 . H_2O$. Nickel(II) hydroxide, $Ni(OH)_2$, is green, copper(II) hydroxide,

† *Formerly* vanadous, chromous, manganous, ferrous, cobaltous, nickelous, cupric (and zinc) ions

$Cu(OH)_2$, light blue, and zinc hydroxide, $Zn(OH)_2$, colourless; none being oxidized in air. All are precipitated from aqueous solution of salts in the gelatinous condition.

The complex cyanides† form well-defined series the first members being readily oxidized and the stability rising with atomic number of the transition metal:

$K_4[V(CN)_6]$
potassium
hexacyanovanadate(II)

$K_4[Cr(CN)_6]$
potassium
hexacyanochromate(II)

$K_4[Mn(CN)_6]$
potassium
hexacyanomanganate(II)

$K_4[Fe(CN)_6]$
potassium
hexacyanoferrate(II)

$K_4[Co(CN)_6]$
potassium
hexacyanocobaltate(II)

$K_2[Ni(CN)_4]$
potassium
tetracyanoniccolate(II)

$K_2[Cu(CN)_4]$
potassium
tetracyanocuprate(II)

$K_2[Zn(CN)_4]$
potassium
tetracyanozincate

Many complexes with ammonia are formed. All give ammines of the type $MCl_2.6NH_3$, but for copper and zinc, tetrammines are the most stable: $Cu(NH_3)_4^{2+}$ and $Zn(NH_3)_4^{2+}$. Examples of the shapes of these complexes may be written:

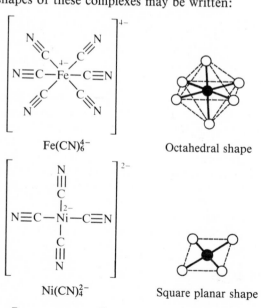

Fe(CN)$_6^{4-}$ Octahedral shape

Ni(CN)$_4^{2-}$ Square planar shape

Cu(NH$_3$)$_4^{2+}$ Square planar shape

Electronic configurations of metal

Fe 2,8,14,2 Fe^{2+} 2,8,14   Fe(CN)$_6^{4-}$ 2,8,18,8 (Kr)

Ni 2,8,16,2 Ni^{2+} 2,8,16   Ni(CN)$_4^{2-}$ 2,8,18,6

Cu 2,8,18,1 Cu^{2+} 2,8,17   Cu(NH$_3$)$_4^{2+}$ 2,8,18,7

Notice the relationship between the square planar structure and the octahedral shape from which it may be derived by replacing two ligands with electron-pairs. For Ni(CN)$_4^{2-}$ refer to *ligand field theory*, p. 65.

Oxidation state +3, M(III)

By measuring ionization potentials (p. 42), the relative ease of formation of the M^{3+} cations from metals of the transition series being studied may be determined. Thus, in Table 24.3, scandium requires the least amount of energy to shed three electrons from each atom, to form Sc^{3+}.

Table 24.3 Period 4: Transition and associated elements: sum of 1st, 2nd and 3rd ionization energies/kJ mol^{-1} (273K)

	Sc	Ti	V	Cr	Mn	Fe	Co	Ni	Cu	Zn
$M_{(g)} \rightarrow$ $M_{(g)}^{3+} + 3e^-$	4261	4624	4895	5235	5473	5277	5640	5888	6254	6473

While what determines the stability of the various oxidation states is incompletely understood, the thermal stability of the halides (Cl, Br, I) of the transition elements at 298 K has been studied with this in mind by rigorous application of thermodynamics

$$MX_3(s) \rightarrow MX_2(s) + \tfrac{1}{2}X_2(g)$$

Lattice energies, ionization energies and entropy change were among the factors considered but the most important single factor turned out to be the 3rd ionization energy (Table 24.2):

$$M^{2+} \rightarrow M^{3+} + e^-$$

Tervalency is the most stable valency state for scandium, but only just perceptible in nickel. Thus in decreasing ease of formation of M^{3+} from M^{2+}, the order generally given is:

Sc   Ti   V   Fe   Cr   Co   Mn   Ni

† *Formerly* potassium vanadocyanide, potassium chromocyanide, potassium manganocyanide, potassium ferrocyanide, potassium cobaltocyanide, potassium nickelicyanide, potassium cupricyanide, potassium zinc cyanide

which corresponds to the order of increase of the 3rd ionization energies. The salts formed by cobalt and manganese in this valency state are very unstable, but the stability is greater in complexes. There are no salts of nickel, copper and zinc.

The colours of the hydrated cations†, $M(H_2O)_6^{3+}$, are characteristic:

Sc^{3+}	Ti^{3+}	V^{3+}	Cr^{3+}
colourless	violet	green	violet or green

Mn^{3+}	Fe^{3+}	Co^{3+}
red	violet (solutions yellow/brown)	blue

From vanadium onwards the sulphates form alums, which are double salts, not complexes:

$$KV(SO_4)_2 . 12H_2O \quad \text{or} \quad K_2SO_4 . V_2(SO)_3 . 24H_2O$$
potassium vanadium(III) sulphate

Of the many complexes, the cyanides†† form a notable series where stability increases with atomic number:

$K_3[V(CN)_6]$	$K_3[Cr(CN)_6]$
potassium hexacyanovanadate(III)	potassium hexacyanochromate(III)

$K_3[Mn(CN)_6]$	$K_3[Fe(CN)_6]$
potassium hexacyanomanganate(III)	potassium hexacyanoferrate(III)

$$K_3[Co(CN)_6]$$
potassium hexacyanocobaltate(III)

Simple ions:

$$Fe^{3+} \qquad \bar{C}{\equiv}N \qquad Co^{3+}$$

Electronic configurations:

$$2,8,13 \qquad\qquad 2,8,14$$

Complex ions:

Electronic configurations:

Fe now 2,8,18,7   Co now 2,8,18,8 (Kr)
octahedral structures

The highest valency states of vanadium, chromium, manganese and iron (respective oxidation states +5, +6, +7 and +6)

Important compounds are formed by vanadium, chromium and manganese in their highest valency states. The oxides are covalent and acidic; on reaction with alkali, the transition metal appears in the anion of the salt. In what follows the right superscript position is used (by international convention) to indicate the state of valency. The names of compounds incorporate the oxidation number of the transition metal oxidation state, assigned according to the rules (p. 124).

V^V *Vanadium pentoxide, vanadium(v) oxide*, V_2O_5, forms vanadates which are in some ways analogous to phosphates. Ammonium metavanadate(v), NH_4VO_3, is useful as a source of vanadium in the laboratory. More complicated structures are formed by *condensation*, the elimination of water between anions. Vanadates are readily reduced.

Cr^{VI} *Chromium trioxide, chromium(vi) oxide*, CrO_3, which forms bright red needle-like crystals, is a powerful oxidizing agent, giving chromates, which are yellow (and analogous to sulphates), and dichromates, which are orange. Potassium chromate, K_2CrO_4, is isomorphous with the sulphate, K_2SO_4. In addition to potassium dichromate, $K_2Cr_2O_7$, salts containing a higher proportion of CrO_3 are known.

Mn^{VII} *Manganese heptoxide, manganese(vii) oxide*, Mn_2O_7, is a dark green, volatile liquid which is very liable to decompose explosively and is an extremely powerful oxidizing agent. The corresponding salts, permanganates (manganates(vii)),

† *Formerly* scandium, titanous, vanadic, chromic, manganic, ferric, cobaltic ions
†† *Formerly* potassium vanadicyanide, potassium chromicyanide, potassium manganicyanide, potassium ferricyanide, potassium cobalticyanide

are formed by fusion of a manganese(II) salt or manganese(IV) oxide with alkali and an oxidizing agent. Potassium permanganate, $KMnO_4$, is important in the laboratory and is isomorphous with the perchlorate (chlorate(VII)), $KClO_4$. Permanganates are purple in colour.

Vanadic acid, or *hydrated vanadium pentoxide* ($V_2O_5.H_2O = 2HVO_3$, $2V_2O_5.H_2O = H_2V_4O_{11}$), is precipitated by addition of acid to a solution of a vanadate. Neither the chromic acids nor permanganic acid have been isolated from solution. For vanadium, chromium and manganese, the valencies 5, 6, and 7 exerted in these compounds are the group valencies of the subgroups of the Mendeléeff Periodic Table.

Fe^{VI}   Finally, unstable ferrates corresponding to an unknown oxide, FeO_3, of sexavalent iron may be made. Potassium ferrate(VI), K_2FeO_4, may be prepared by oxidizing freshly precipitated iron(III) hydroxide electrolytically in alkali. It is deep red in colour and isomorphous with potassium sulphate(VI) and potassium chromate(VI).

The many colours of transition metal ions

The colour of a compound corresponds to what is *not* absorbed from the radiation which falls upon it from the visible range of the electromagnetic spectrum. Light which is absorbed causes electrons to be promoted to higher energy levels ($\Delta E = hv$, v being the frequency of the radiation and h, Planck's constant). The electron may return in a series of intermediate energy jumps corresponding to an energy, and radiation, release outside the visible spectrum. In addition, energy may be used in the various vibrational energy levels in the molecule or ion. Thus, energy corresponding to light of certain visible frequencies is absorbed and redistributed so that absorption of frequencies near to red, for example, leaves the characteristic blue of the hydrated copper(II) ion. However, by another mechanism in complex ions (and molecules) electrons may be snatched from the orbitals of one atom to another, usually though not always, from ligand to central atom. Such *electron transfer spectra* are seen with certain oxoanions, e.g. CrO_4^{2-} and MnO_4^-.

Addition of ammonia solution darkens the shade of blue of the hydrated ion of copper(II) as water molecules in the aquo-complex are displaced by ammonia, the new ligand. For the aquo-nickel(II) ion the colour changes progressively from green to blue to violet. This *blue-shift* is seen whatever the metal ion present. Maximum absorption of light, as would be expected, corresponds to the energy difference between the modified d orbitals of the central atom of the ion, which have been separated into higher and lower levels (p. 66) by the approach of ligands with their lone pair electron-clouds, and promoted electrons move between these levels. Ammonia induces an increase in the splitting (ΔE higher: v higher) while Cl^-, when added to an aquo-complex, e.g. $Ni(H_2O)_6^{2+}$ causes a colour change from green to yellow-green, a *red-shift*, corresponding to a decrease in the splitting (ΔE lower: v lower). Ligands may be listed in the *spectro-chemical* series according to their capacity to increase the *ligand field splitting* of d orbitals:

$$I^- < Br^- < Cl^- < F^- < H_2O < NH_3 < CN^-$$

Thus, the cyanide ion is most effective. It has the strongest ligand field and gives the greatest energy separation of the split d orbitals.†

The colours of complex ions require vacant d orbitals to be available and depend also on the nature of the ligand. The intensity of the colour, dependent on the ease with which electrons are promoted, increases with the degree of covalency in the ligand connection and is enhanced when the octahedral configuration gives way to the tetrahedral. The theory is being developed but from what has been seen here, it will be realized that colour changes in complex ion chemistry are not as haphazard as they may have seemed at one time.

Influence of ligands on the relative stability of oxidation states

The relative stability of oxidation states of transition metals in solution depends to a marked degree on the nature of the ligand. The effect may be illustrated by comparing redox potentials and requires some appreciation of *ligand field theory* (p. 65).

Consider the iron(III) and iron(II) equilibria shown in Table 24.4. The cyanide complexes are of the low spin (inner orbital) type while the hydrated ions are both high spin (outer orbital) in type. This is caused by the strong electric field exerted by the former compared with the weak field exerted by the

† The *ligand field stabilization energy*, or *crystal field stabilization energy* (CFSE), is the name given to the energy effect of the crystal environment on d orbitals, usually in octahedral field, and is included in calculations of lattice energy, solvation energy etc. The ligands create an electrostatic field around the central atom of the complex as well as undergoing bonding of the covalent (co-ionic) type

Table 24.4 Ligands and redox systems

System	Standard electrode potential (298 K)/Volt
$Fe(H_2O)_6^{3+} + e^- \rightleftharpoons Fe(H_2O)_6^{2+}$	+0.771
$Fe(CN)_6^{3-} + e^- \rightleftharpoons Fe(CN)_6^{4-}$	+0.466
$Co(H_2O)_6^{3+} + e^- \rightleftharpoons Co(H_2O)_6^{2+}$	+1.84
$Co(NH_3)_6^{3+} + e^- \rightleftharpoons Co(NH_3)_6^{2+}$	+0.1
$Co(CN)_6^{3-} + e^- \rightleftharpoons Co(CN)_6^{4-}$	−0.8

latter. There is but a small gain in going from Fe(III) to Fe(II) in crystal field stabilization energy for the cyanides, due to the extra electron (Fe(III)d^5; Fe(II)d^6). The Fe(III) complex cyanide acts as a mild oxidizing agent (by electron loss), and is susceptible to replacement of cyanide groups by water and other ligands (e.g. $Fe(CN)_5(H_2O)^{2-}$). The oxidizing potential of Fe^{3+} is seen to be lowered by combination with CN^- on formation of the hexacyanoferrate(III) complex ion, reflecting the wider splitting of 3*d* energy levels of Fe(III) with CN^- rather than with H_2O. In the case of the cobalt complexes, the relative effects of H_2O, NH_3 and CN^- as ligands may be compared, the oxidizing potential of Co^{3+} being lowered by the successive combinations, correlating with the magnitude of the electric fields exerted by the ligands, and sharing the enhanced stability of the cyanide complex.

The stereoisomerism of complex ions: ligands and chelate ligands

The complex ions of transitional elements are of great importance. Most of the *ligands*, the name given to the atoms attached to the central metal ion, are attached by only one donor atom, and are called *monodentate*† ligands (i.e. they bite the central atom only once). Some ligands have two donor atoms in each molecule and are called *bidentate ligands* (e.g. ethylenediamine, ethane-1,2-diamine, $NH_2CH_2CH_2NH_2$) while those with more than two such atoms per molecule are called *polydentate ligands*. Ligands which 'bite' on to a central atom in more than one position simultaneously are termed *chelate*†† ligands. The stereochemistry of complex ions is outlined on p. 70 and specific examples appear in the text.

Catalytic activity

Many examples of the catalytic activity of the metals of this series and their compounds may be collected.

1 Titanium esters catalyse ester exchange reactions and reactions of the Aldol condensation (more accurately, polymerization) type;

2 Vanadium pentoxide:

$$2SO_2 + O_2 \rightleftharpoons 2SO_3$$
Contact process

3 Basic zinc chromate(VI):

$$CO + 2H_2 \rightarrow CH_3OH$$
methanol production

4 Manganese(II) ethanoate:

ethanal(acetaldehyde) oxidized to ethanoic (acetic) acid

5 Finely divided iron:

$$N_2 + 3H_2 \rightleftharpoons 2NH_3$$
Haber synthesis

6 Cobalt on charcoal:

polymerization of ethene (ethylene) to butadiene

7 Nickel:

general hydrogenations

8 Copper:

$$2CH_3OH + O_2 \rightarrow 2HCHO + 2H_2O$$
oxidation of methanol to methanal (formaldehyde)

Always important, catalytic applications of the transition metals and their compounds to industrial processes continue to proliferate. Two examples of carbido-metal cluster catalysts are shown on p. 346.

† Latin, dens, dentis = tooth
†† Greek, khele = claw

Scandium

Scandium and its compounds are unimportant. Nominally a transition metal, all three valency electrons are always utilized. Mendeléeff predicted the properties of eka-boron, later isolated by Nilson in 1879 and named scandium.

Titanium

Pure titanium has been available commercially since 1949. It is a grey metal which combines with oxygen at red heat, fluorine and chlorine on warming and nitrogen on strong heating to form respectively the dioxide, TiO_2, tetrahalides, TiF_4 and $TiCl_4$, and a nitride, TiN, respectively. The very unstable Ti^{2+} state has been mentioned. The other valencies are 3 (formerly titanous) and 4 (formerly titanic). Titanium(III) salts are coloured, having an odd electron in excess of the octet and are readily oxidized to the titanium(IV) state, which is colourless. Titanium(III) solutions will evolve hydrogen in the presence of platinum:

$$2Ti^{3+} + 2H^+ \rightarrow 2Ti^{4+} + H_2$$
$$\phantom{2Ti^{3+}}\text{2, 8, 9}\text{2, 8, 8}$$

In the short (Mendeléeff) form of the Periodic Table, titanium was placed in a subgroup of Group IV. It resembles tin of that group in reacting with hot concentrated nitric acid to form the dioxide, titanium(IV) oxide. The titanium(III) state is more basic than the titanium(IV) state, as is usual. It resembles the corresponding state of vanadium (V^{3+}) and chromium (Cr^{3+}). A solution of titanium(IV) chloride, titanium tetrachloride, in excess hydrochloric acid, or titanium(IV) sulphate in sulphuric acid, may be reduced by zinc to the titanium(III) condition, revealed by its violet colour:

$$2Ti^{4+} + Zn \rightarrow 2Ti^{3+} + Zn^{2+}$$

Reduction may be accomplished electrolytically, or by use of tin and acid in place of zinc, but not by hydrogen sulphide or sulphur dioxide.

Titanium salts, in the presence of dilute acid, or titanium dioxide in sulphuric acid, are coloured yellow by hydrogen peroxide which forms a peroxoacid, or hydrated peroxide, $TiO_3.2H_2O$. As little as 0.001% Ti may be detected in this way. Higher concentrations produce an orange-red coloration.

Ti^{IV}: titanium(IV) oxide, titanium dioxide

Naturally occurring iron(II) titanium(IV) oxide, sometimes called iron(II) titanate, ilmenite, $FeO.TiO_2$, is reacted with sulphuric acid, the liquid purified and hydrolysed on boiling to precipitate titanium dioxide. The product is ground and processed. Under ordinary conditions, titanium dioxide, TiO_2, is completely inert, of excellent whiteness, brightness and opacity, and ideal as a white pigment. It is used in paints, printing inks, in the production of linoleum and textiles, vitreous enamel, paper-making and in the rubber and plastics industry.

Ti^{IV}: titanium(IV) chloride, titanium tetrachloride

This is made from the elements, or by passing chlorine over a heated mixture of the dioxide and carbon, as was the case with aluminium chloride, which it resembles (e.g. reactions with ethers, alcohols, etc. in benzene solution; see p. 312). It is a colourless, volatile liquid (b.p. 136°C) which fumes strongly in moist air and is a non-conductor of electricity.

Titanium compounds in titrimetric analysis

Titanium(III) salts are very powerful reducing agents, being more powerful than those of tin(II):

$$Ti^{3+} - e^- \rightleftharpoons Ti^{4+}$$

Titanium(III) chloride solution is rapidly oxidized by air and an inert atmosphere is required in contact with the solution when it is used. However, a solution of titanium(III) sulphate in 2M sulphuric acid, prepared similarly, is much more stable, will remain standard for about 12 hours and may be used in an open burette. Alternatively, titanium(IV) sulphate in sulphuric acid is reduced by zinc amalgam as required for titrimetric work.

Titanium(III) sulphate solution, which is violet in colour, may be standardized by titration against potassium permanganate at 50–60°C,

$$5Ti^{3+} + MnO_4^- + 8H^+ \rightarrow$$
$$5Ti^{4+} + Mn^{2+} + 4H_2O$$

or against iron(III) alum with a trace of ammonium thiocyanate as indicator,

$$Ti^{3+} + Fe^{3+} \rightarrow Ti^{4+} + Fe^{2+}$$

It may be used for the determination of chlorates and perchlorates, i.e. chlorates(v) and (vii), (directly or by back titration, using iron(iii) alum) and various nitro compounds and their derivatives.

Vanadium

Vanadium forms compounds in which it exerts the valencies 2, 3, 4 and 5 which are associated respectively with the colours lavender-violet, green and blue, and colourless in solution, the highest number representing the most stable oxidation state. In the Mendeléeff Classification, vanadium was placed in a subgroup of Group V, following his maxim that the highest valency state is equal to the group valency.

Vanadium is a hard silvery-white lustrous metal. Stable in air at ordinary temperatures, it is oxidized to oxides of successive oxidation states on heating strongly in air or oxygen, giving V_2O_3, VO_2 and V_2O_5, no oxide of formula VO having been isolated. The metal reacts with nitrogen and carbon on heating and is attacked by chlorine to yield the tetrachloride, VCl_4. Halides are formed in all valency states. Vanadium resists the action of acids, except under oxidizing conditions when both concentrated nitric acid and concentrated sulphuric acid form the pentoxide.

The colours associated with the different valency states of vanadium are seen during reduction of ammonium metavanadate(v) solution. Reduction with acid and zinc or zinc amalgam converts the colourless solution ($NH_4^+VO_3^-$) to blue (VO^{2+}, vanadyl(iv)), then green (V^{3+}) and finally lavender (V^{2+}). Reduction of the warm acidified solution with hydrogen sulphide yields the blue vanadyl(iv) solution (VO^{2+}) with separation of sulphur.

Addition of hydrogen peroxide and a little sulphuric acid to a vanadate(v) solution gives a yellow or orange-yellow coloration due to the formation of a peroxovanadic acid. The free acid has not been isolated but peroxovanadates, of which KVO_4 is an example, have been obtained. Titanium compounds behave similarly.

V^V: Vanadium pentoxide, vanadium(v) oxide

V_2O_5 is the most important. It is a yellow-brown solid, amphoteric but predominantly acidic, with alkali giving vanadates of the type M_3VO_4, $M_4V_2O_7$ and MVO_3, which are usually hydrated. In the precipitate formed by addition of acid to solutions of vanadates, two hydrated forms, $V_2O_5.H_2O$ and $2V_2O_5.H_2O$, corresponding to 'metavanadic acid', HVO_3, and 'pyrovanadic acid', $H_2V_4O_{11}$, have been detected. Rapid addition of excess acid to a concentrated vanadate(v) solution, so that the oxide is not precipitated, yields a solution containing the VO^{3+} ion, although no salts have been isolated. Various complexes are formed. Vanadium pentoxide is a versatile catalyst, used in the oxidation of sulphur dioxide to trioxide (Contact process), of sucrose by concentrated nitric acid to ethanedioic (oxalic) acid, of ethanol by air and in various hydrogenation reactions of aromatic hydrocarbons.

V^{IV}: Vanadium dioxide, vanadium(iv) oxide

VO_2 is amphoteric but predominantly basic. It is a dark blue solid, which forms an unstable vanadate(iv) with alkali. There are no salts of the ion V^{4+} but those of the vanadyl(iv) cation, VO^{2+}, are well known; $(VO)SO_4$ forms grey-green crystals. Vanadium dioxide is readily oxidized by heating in air, or by concentrated nitric acid, to form the pentoxide. Many complexes are formed.

V^{III}: Vanadium sesquioxide, vanadium(iii) oxide

V_2O_3 is a black solid, which is slowly oxidized by air at ordinary temperatures and inflames on heating in air. It is basic, giving the pale-green hydrated sulphate, $V_2(SO_4)_3.3H_2O$, alums and also various complex sulphates. The complex hexacyanovanadate(iii), $K_3[V(CN)_6]$, corresponding to potassium hexacyanoferrate(iii), is the least stable transition metal complex cyanide of this type.

V^{II}: Compounds

No oxide VO has been isolated. The addition of alkali to a vanadium(ii) solution gives a brown precipitate which probably contains $V(OH)_2$ but it is rapidly oxidized. This is a wholly basic state. The sulphate, a vitriol, $VSO_4.7H_2O$, which is violet in colour is very readily oxidized. Few complexes occur in this oxidation state. Vanadium(ii) sulphate absorbs nitrogen oxide as does iron(ii) sulphate, which forms $Fe^{2+}.NO$. The hexacyanovanadate(ii), $K_4[V(CN)_6].3H_2O$, is a yellow-brown solid and is similar to potassium hexacyanoferrate(ii), $K_4[Fe(CN)_6].3H_2O$, except that it is much more readily oxidized, even by the air.

Chromium

The principal valencies exerted by chromium are 3, 6 and 2 in decreasing order of stability, these also being the numbers describing the oxidation states. In the hexacarbonyl, $Cr(CO)_6$, zero valency is assigned to chromium because all bonding electrons come from carbon monoxide in the co-ionic structure: similarly the oxidation number of chromium here is zero (although it is combined) because the complex is neutral and the ligand uncharged. In the Mendeléeff Classification, chromium was placed in a subgroup of Group VI, the group valency being the maximum exerted.

Chromium is a bluish-white, hard, shiny metal which is very resistant to chemical attack and not tarnished in air. Although it is unaffected by dry air at ordinary temperatures, some oxidation occurs on strong heating. Chromium unites with the halogens on heating, chlorine forming chromium(III) chloride, $CrCl_3$. It reacts with sulphur to form chromium(III) sulphide, Cr_2S_3, on heating and with boron, carbon and silicon, nitrogen and phosphorus. Such information is readily summarized on an outline copy of the Periodic Table. Dilute acids attack chromium slowly in the cold, but rapidly when hot, to form chromium(II) (Cr^{2+}) salts which are oxidized by air to the chromium(III) (Cr^{3+}) condition. Hot concentrated sulphuric acid oxidizes chromium to chromium(III) sulphate and is reduced to sulphur dioxide,

$$6H_2SO_4 + 2Cr \rightarrow Cr_2(SO_4)_3 + 3SO_2 + 6H_2O$$

but concentrated nitric acid renders the metal very passive, an oxide film protecting the metal from further action. Chromium may be prepared by electrolysis or by the Goldschmidt Thermite Reduction, in which chromium(III) oxide is reduced with aluminium:

$$Cr_2O_3 + 2Al \rightarrow Al_2O_3 + 2Cr$$

$$\Delta H = -456 \text{ kJ mol}^{-1}$$

Cr^{VI}: chromium(VI) oxide, chromium trioxide, and derivatives of 'chromic acid'

chromium(VI) oxide
chromium trioxide
chromic anhydride

the simplest chromic acid
(not isolated)
giving chromates, e.g.
$(K^+)_2CrO_4^{2-}$

a chromic acid formed
by condensation (not isolated)
giving dichromates, e.g.
$(K^+)_2Cr_2O_7^{2-}$

chromyl chloride
an acid chloride

potassium chlorochromate

The electronic configuration of chromium becomes 2, 8, 18, 2. The compounds are analogous to certain compounds formed by sulphur:

$$SO_3 \qquad H_2SO_4 \qquad H_2S_2O_7$$
sulphur trioxide   sulphuric acid   di- or pyro-sulphuric acid

$$SO_2Cl_2 \qquad Cl.SO_2.\overset{-}{O}\overset{+}{K}$$
sulphuryl chloride   potassium chlorosulphate*

No chromic acid has been isolated pure, and there is no chlorochromic acid corresponding to chlorosulphuric acid.

Cr^{VI}: chromium (VI) oxide and 'chromic acid'; chromates and dichromates

Careful addition of concentrated sulphuric acid to a well-cooled concentrated aqueous solution of sodium dichromate with stirring leads to deposition of red needle-like crystals of chromium(VI) oxide, CrO_3, (m.p. 196°C):

$$2H^+ + Cr_2O_7^{2-} \rightarrow 2CrO_3 + H_2O$$

The mixture is filtered through asbestos or sintered glass, washed with concentrated nitric acid, dried at 120°C and cooled in a desiccator. Chromium(VI) oxide is an extremely powerful oxidizing agent, especially to organic matter: ethanol is immediately oxidized and usually catches fire. It is an acidic oxide, chromic anhydride, and very soluble in water to form 'chromic acid', a powerful oxidizing agent and cleansing fluid for glassware. Although salts of simple chromic acid are well known, condensation reactions with elimination of water

* *Formerly* potassium chlorosulphonate

occur:

$$H_2O + CrO_3 \rightarrow H_2CrO_4$$
simple chromic
acid

$$H_2CrO_4 \rightleftharpoons H^+ + HCrO_4^- \rightleftharpoons 2H^+ + CrO_4^{2-}$$
hydrogenchromate   chromate
(bichromate)

$$2HCrO_4^- \rightleftharpoons H_2O + Cr_2O_7^{2-}$$
dichromate

Besides salts corresponding to these ions, e.g. K_2CrO_4 and $K_2Cr_2O_7$, salts of the acids, $H_2Cr_3O_{10}$ and $H_2Cr_4O_{13}$ formed by elimination of water between H_2CrO_4 molecules, are known. Spectroscopic examination of solutions of dichromates has shown that they contain largely the dichromate ion, $Cr_2O_7^{2-}$, in concentrated solution and the hydrogenchromate ion, $HCrO_4^-$, produced by hydrolysis, in increasing proportion as the concentration drops. Soluble chromates are yellow and dichromates redorange. Important soluble chromates are potassium chromate, K_2CrO_4, and sodium chromate, $Na_2CrO_4.10H_2O$. Potassium chromate solution is used as a laboratory reagent to detect metals which form characteristic insoluble chromates. Important insoluble chromates include the pigment, chrome yellow, lead(II) chromate, $PbCrO_4$, brick-red silver chromate, Ag_2CrO_4, and yellow barium chromate, $BaCrO_4$. Sodium dichromate, $Na_2Cr_2O_7.2H_2O$, is deliquescent and used when strong solutions of dichromate are required for oxidations in Organic Chemistry, as in the preparation of ethanal (acetaldehyde) from ethanol. Potassium dichromate, $K_2Cr_2O_7$, is not hydrated, less soluble (10 g in 100 g water at 15°C), easily dried by gentle fusion and conveniently used as a primary standard in titrimetric (volumetric) analysis.

Chromates and dichromates in solution are interconvertible. Even weak acids cause the change from yellow chromate to orange dichromate, while this is reversed by addition of alkali or a carbonate solution:

$$2CrO_4^{2-} + 2H^+ \rightarrow$$
$$2HCrO_4^- \rightleftharpoons H_2O + Cr_2O_7^{2-}$$

$$HCrO_4^- + OH^- \rightarrow H_2O + CrO_4^{2-}$$

$$2HCrO_4^- + CO_3^{2-} \rightarrow 2CrO_4^{2-} + H_2O + CO_2$$

The commercial production and laboratory preparation of chromates and dichromates

In industrial production, finely powdered chromite ore ($FeO.Cr_2O_3$) is heated with anhydrous sodium carbonate and a flux of lime in air in a reverbatory furnace:

$$4(FeO.Cr_2O_3) + 8Na_2CO_3 + 7O_2 \rightarrow$$
$$8Na_2CrO_4 + 2Fe_2O_3 + 8CO_2$$

Sodium chromate is extracted with water, and concentrated sulphuric acid added to produce sodium dichromate while hydrated sodium sulphate crystallizes out. On further concentration, sodium dichromate crystallizes as the dihydrate, $Na_2Cr_2O_7.2H_2O$. Addition of potassium chloride to a concentrated solution of sodium dichromate precipitates the less soluble potassium dichromate, $K_2Cr_2O_7$. The corresponding chromates may be prepared by addition of alkali. Sodium dichromate is used in leather tanning, and as an oxidizing agent for organic compounds in the laboratory and in industry.

In the laboratory, chromates may be prepared by oxidation of chromium(III) hydroxide with a solution of sodium peroxide, or hydrogen peroxide in alkali:

$$2Cr(OH)_3 + 3Na_2O_2 \rightarrow$$
$$2Na_2CrO_4 + 2NaOH + 2H_2O$$

The green gelatinous precipitate of chromium(III) hydroxide, formed by addition of alkali to a solution of a chromium(III) salt, dissolves as yellow chromate. Solid chromium(III) salts may be fused with sodium carbonate (or fusion mixture: mixed potassium carbonate and sodium carbonate) and an oxidizing agent, sodium peroxide or potassium nitrate, in a reaction similar to that used industrially.

Cr^{VI}: potassium dichromate (potassium μ-oxobistrioxochromate(VI)) as a laboratory reagent

Potassium dichromate is used as a primary standard in titrimetric (volumetric) analysis (see p. 122). The solubility of the salt increases markedly with rise in temperature: 10 g in 100 g water at 0°C and 94 g in 100 g water at 100°C, make re-crystallization easy. Potassium dichromate may be dried to known composition, not decomposing even on fusion. Very strong heating causes decomposition:

$$4K_2Cr_2O_7 \rightarrow 4K_2CrO_4 + 2Cr_2O_3 + 3O_2$$

The solution is stable and is especially useful in the presence of hydrochloric acid, which reacts with potassium permanganate, another widely used oxidant.

Potassium dichromate solution may be used in making qualitative tests for sulphur dioxide and hydrogen sulphide. Both cause the colour change from orange to green, but hydrogen sulphide gives a precipitate of sulphur:

$$Cr_2O_7^{2-} + 2H^+ + 3SO_2 \rightarrow$$
$$2Cr^{3+} + H_2O + 3SO_4^{2-}$$
$$Cr_2O_7^{2-} + 8H^+ + 3H_2S \rightarrow 2Cr^{3+} + 7H_2O + 3S$$

Chromates in analysis

Potassium chromate, potassium tetraoxochromate(VI), may be estimated by conversion to dichromate on acidification and titration against ammonium iron(II) sulphate. It is used in analysis as an indicator for determination of chloride or bromide by silver nitrate in neutral solution, brick-red silver chromate not being precipitated by silver nitrate until all halide has been precipitated:

$$2Ag^+ + CrO_4^{2-} \rightleftharpoons Ag_2CrO_4(s)$$

This precipitation may be used as a qualitative test for chromate ions in neutral solution. Details of the test are given under silver nitrate (p. 491). Lead(II) ethanoate (acetate) solution gives a precipitate of yellow lead(II) chromate with solutions of chromates. But hydrogen peroxide affords the most sensitive test: an intense blue coloration appears when hydrogen peroxide is added to an acidified chromate (or dichromate) solution. This is probably due to a peroxide, possibly CrO_5, which may be extracted by ether as a blue immiscible layer.

Cr^{VI}: chromyl chloride and potassium chlorochromate

Chromyl chloride, CrO_2Cl_2, is the acid chloride of simple chromic acid, itself not isolated, and potassium chlorochromate (trioxochlorochromate), a salt of chlorochromic acid, also not isolated. Structures are given on p. 456. Chromyl fluoride may also be prepared but there is no corresponding bromide or iodide.

Chromyl chloride is prepared by addition of concentrated sulphuric acid to an intimate mixture of sodium chloride and potassium dichromate, followed by gentle distillation in an all-glass apparatus:

$$K_2Cr_2O_7 + 4NaCl + 6H_2SO_4 \rightarrow$$
$$2CrO_2Cl_2 + 2KHSO_4 + 4NaHSO_4 + 3H_2O$$

It is a dark red liquid, b.p. 116°C under atmospheric pressure, dense and rather like bromine in appearance. It is immediately hydrolysed by water to give a yellow solution containing hydrochloric acid and chromic acid:

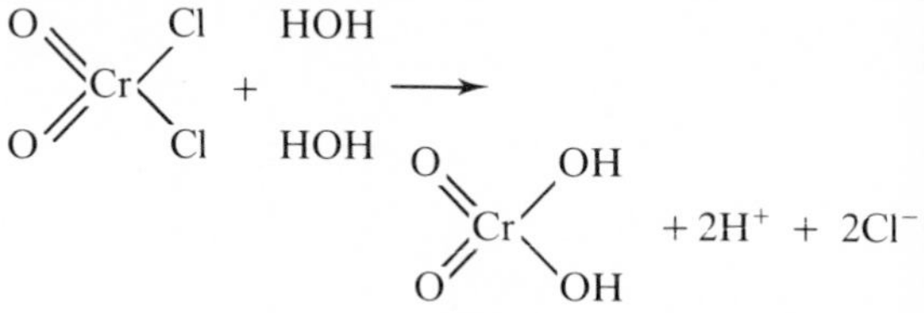

This reaction may be adopted as a qualitative test for chromates or chloride (or fluoride) in the presence of bromide and/or iodide. The presence of chromic acid may be shown by the intense blue coloration given with hydrogen peroxide.

Potassium chlorochromate, $KO.CrO_2.Cl$, may be prepared by repeated evaporation of finely divided potassium dichromate with concentrated hydrochloric acid on a water bath at 60–70°C. Red crystals of potassium chlorochromate separate. It is hydrolysed by water, but may be re-crystallized from hydrochloric acid. Alternatively, chromyl chloride will react with saturated potassium chloride solution to give potassium chlorochromate. On gentle heating, decomposition occurs and chlorine is evolved with oxygen:

$$4(KO.CrO_2.Cl) \rightarrow$$
$$K_2Cr_2O_7 + Cr_2O_3 + 2KCl + O_2 + Cl_2$$

Qualitative tests for chromates and dichromates

Soluble chromates are yellow in solution giving orange dichromates on acidification; the change being reversed by addition of caustic alkali or carbonate solution.

Characteristic precipitates of chromates may be obtained: barium chloride gives yellow barium chromate, lead(II) ethanoate (acetate) solution gives yellow lead(II) chromate and silver nitrate, brick-red silver chromate. Insoluble chromates dissolve in strongly ionized acids (e.g. nitric acid) but are insoluble in those which are feebly ionized (e.g. ethanoic acid).

Oxidation by hydrogen peroxide in acid yields the intensely blue CrO_5, soluble in ether.

Reduction to the chromium(III) state may be effected by means of sulphur dioxide or hydrogen sulphide on the solution or heating the solid with concentrated hydrochloric acid (above the temperature for potassium chlorochromate):

$$K_2Cr_2O_7 + 14HCl \rightarrow$$
$$2KCl + 2CrCl_3 + 3Cl_2 + 7H_2O$$

Cr^{III}: chromium (III) oxide, 'chromium(III) hydroxide' and chromium(III) salts (formerly chromic compounds)

Chromium(III) oxide, chromium sesquioxide, Cr_2O_3, is a bulky green powder obtained by the thermal decomposition of ammonium dichromate, a reaction which is self-propagating when started, or by heating a mixture of potassium dichromate and ammonium chloride:

$$(NH_4)_2Cr_2O_7(s) \rightarrow Cr_2O_3(s) + N_2(g) + 4H_2O(g)$$

Chromium(III) oxide is an ionic structure, of the corundum (Al_2O_3) type

Electronic configurations: $\underset{2,\,8,\,11}{(Cr^{3+})_2}$ $\underset{2,\,8}{(O^{2-})_3}$

It is the most stable oxide of chromium, unaffected by heat, not reduced by hydrogen, reacting slowly with acids to form chromium(III) salts and when fused with alkali in air, or with an oxidizing agent, such as potassium nitrate, forming the chromate(VI):

$$2Cr_2O_3 + 8KOH + 3O_2 \rightarrow 4K_2CrO_4 + 4H_2O$$

Addition of caustic soda solution to a solution of a chromium(III) salt gives a pale green gelatinous precipitate, called chromium(III) hydroxide, nominally $Cr(OH)_3$, although it seems to have no definite composition. Acids form chromium(III) salts while the precipitate passes into the colloidal state with excess alkali, although it is possible that a chromate(III) is formed. Chromium(III) hydroxide loses water on heating, to give chromium(III) oxide. It is oxidized by hydrogen peroxide in sodium hydroxide solution or by a solution of sodium peroxide to sodium chromate(VI):

$$2Cr(OH)_3(s) + 3Na_2O_2 \rightarrow$$
$$2Na_2CrO_4 + 2NaOH + 2H_2O$$

There is no chromium(III) carbonate and chromium(III) sulphide is immediately hydrolysed. Addition of aqueous sodium carbonate or ammonium sulphide to a solution of a chromium(III) salt precipitates the hydroxide. The behaviour of chromium(III) salts in solution is complex and will be considered separately. Anhydrous chromium(III) chloride, $CrCl_3$, may be prepared by heating chromium in chlorine to a high temperature, or passing the gas over a heated mixture of chromium(III) oxide and carbon:

$$Cr_2O_3 + 3C + 3Cl_2 \rightarrow 3CO + 2CrCl_3$$

It sublimes in chlorine to a reddish-violet solid. Heated in air, it forms chromium(III) oxide and chlorine, while in the absence of air, chromium(II) chloride, $CrCl_2$, and chlorine are formed. Potassium chromium(III) sulphate, or chrome alum, is the most important salt and is readily obtained by reduction of potassium dichromate solution containing sulphuric acid using sulphur dioxide or ethanol. The temperature is kept below 60°C to avoid the formation of the more soluble green modification of chromium(III) sulphate. On standing, deep violet octahedral crystals, isomorphous with other alums, separate. Reduction of $(K^+)_2Cr_2O_7^{2-}$ liberates Cr^{3+} ions equal in ionic concentration amount to the K^+ ions, which crystallize to give the double salt, sulphuric acid supplying SO_4^{2-} ions:

$$K_2SO_4.Cr_2(SO_4)_3.24H_2O \quad \text{or} \quad KCr(SO_4)_2.12H_2O$$

$$Cr_2O_7^{2-} + 2H^+ + 3SO_2 \rightarrow$$
$$2Cr^{3+} + H_2O + 3SO_4^{2-}$$

$$Cr_2O_7^{2-} + 8H^+ + 3CH_3CH_2OH \rightarrow$$
$$2Cr^{3+} + 3CH_3CHO + 7H_2O$$
$$\underset{\text{(acetaldehyde)}}{\text{ethanal}}$$

Qualitative tests for chromium(III) ions

Addition of alkali to chromium (III) solutions precipitates green chromium(III) hydroxide which dissolves in excess to form a green solution and with sodium peroxide is oxidized to yellow sodium chromate.

On fusion with sodium carbonate and potassium nitrate in a loop of platinum wire, chromium(III) salts form a yellow bead of chromate.

Cr^{II}: chromium(II) (formerly chromous) compounds

Chromium(II) salts are very readily oxidized to the chromium(III) state and air must be excluded during their preparation.

Reduction of potassium dichromate solution with zinc and hydrochloric acid yields a green solution of chromium(III) chloride which is further reduced to the chromium(II) salt, seen as a bright blue solution:

$$2Cr^{3+} + Zn(s) \rightarrow 2Cr^{2+} + Zn^{2+}$$

Filtration of this mixture into saturated sodium ethanoate solution gives a red crystalline precipitate of chromium(II) ethanoate, $[Cr(CH_3COO)_2]_22H_2O$, a diamagnetic dimerized compound. This is washed in cold water saturated with carbon dioxide. Addition of hydrochloric acid, under an atmosphere of carbon dioxide, yields chromium(II) chloride solution or with sulphuric acid, chromium(II) sulphate, which crystallizes as $CrSO_4.7H_2O$, a vitriol, isomorphous with iron(II) sulphate, $FeSO_4.7H_2O$. Chromium(II) salts are powerful reducing agents, being oxidized in air and by water in the presence of platinum (as a catalyst) or hydrochloric acid:

$$2Cr^{2+} + 2H^+ \rightarrow 2Cr^{3+} + H_2$$

Anhydrous chromium(II) chloride may be prepared as a white powder by heating chromium(III) chloride in dry hydrogen,

$$2CrCl_3 + H_2 \rightarrow 2CrCl_2 + 2HCl$$

or at white heat by the action of hydrogen chloride on the metal,

$$Cr + 2HCl \rightarrow CrCl_2 + H_2$$

Complexes of Cr^0 and Cr^{III}

Cr^0: chromium hexacarbonyl, $Cr(CO)_6$

This preparation uses a Grignard reagent. An ethereal solution of ethyl magnesium bromide is slowly added to an ice-cold ethereal suspension of chromium (III) chloride in the presence of carbon monoxide (under pressure, preferably). The product is hydrolysed by dilute acid. Distillation of the separated, dried ethereal layer yields the carbonyl as a white powder, which may be purified by sublimation.

Chromium hexacarbonyl melts at 149°C. It may be recrystallized from organic solvents. For a carbonyl, it is remarkably stable, not being attacked by concentrated sulphuric acid, concentrated hydrochloric acid or bromine. Thermal decomposition commences apparently above 120°C in the vapour phase. Such compounds of carbon monoxide which readily dissociate are, of course, lethal.

Cr^{III}: a wide variety and range of complexes

An exceedingly large number of complexes based on the chromium(III) ion, Cr^{3+}, may be prepared. The largest group contain ammonia or amines, which are derived from ammonia. Salts containing the hexammine complex ion, $Cr(NH_3)_6^{3+}$, may be prepared. Nitrogen supplies the electrons for each

co-ionic bond to chromium. The ion has an octahedral structure:

Cr now has the electronic configuration 2, 8, 18, 5. The salts are yellow or orange.

Replacing pairs of ammonia molecules by ethane-1,2-diamine (ethylenediamine) leads to the appearance of optical isomerism; the ions of the isomers being related as object to image in a plane mirror:

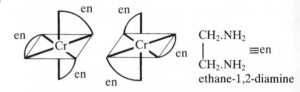

The possibility of co-ionic bond formation, with its electric charge distribution, has been ignored, to avoid unnecessary complications. In the complex ion $[Cr(en)_2Cl_2]^+$, *cis-trans* geometrical isomerism occurs, and the *cis*-isomer may be resolved further into optical isomers:

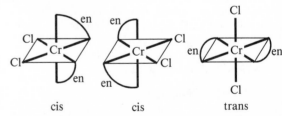

The stereoisomerism of the ethanedioato (oxalato)-complex, $Cr(C_2O_4)_3^{3-}$, of chromium(III) can be explained similarly.

Analogous to potassium hexacyanoferrate(III), potassium hexacyanochromate(III), $K_3[Cr(CN)_6]$, has already been mentioned: it is a yellow solid, giving a yellow solution and is much less stable than the ion complex.

Finally, the complex hydrated salts require mentioning. Four isomers of formula $CrCl_3.6H_2O$ may be prepared and the structures are given below, only the ionic chloride being precipitated by addition of silver nitrate. Similar complexes of chromium(III) sulphate are formed.

$$CrCl_3 . 6H_2O$$

$[Cr(OH_2)_6]^{3+}(Cl^-)_3$ $[Cr(OH_2)_5Cl]^{2+}(Cl^-)_2 . H_2O$
violet   pale green
All Cl^- precipitated $\frac{2}{3}$ Cl precipitated by
by $AgNO_3$ aq. $AgNO_3$ aq.

$$[Cr(OH_2)_4Cl_2]^+Cl^- . 2H_2O$$
deep green
$\frac{1}{3}$ Cl precipitated by
$AgNO_3$ aq.

In addition, $[Cr(H_2O)_3Cl_3]$, which is covalent, has been prepared in ethereal solution.

Manganese

In its compounds manganese may exert valencies of 1–7 inclusive, except 5. The manganese(II) ion, Mn^{2+}, is the most stable cation formed while in covalent combination, septavalent manganese, as in the permanganate ion, MnO_4^-, is most stable. In the Mendeléeff Periodic Table manganese was allocated to a subgroup of Group VII, which also included the halogens. In oxidation state +7, potassium perchlorate, $KClO_4$, and potassium permanganate, $KMnO_4$, otherwise called potassium chlorate (VII) and manganate (VII) are isomorphous. In a few complex cyanides, manganese appears as oxidation state +1 in $K_5[Mn(CN)_6]$ and as oxidation state +2 in $K_3[Mn(CN)_5NO]$ in which it is considered univalent and tervalent respectively.

Manganese is a hard silver-white metal which is highly reactive, especially if impure. It is used in the steel industry for the removal of oxygen and sulphur from molten steel. Manganese will liberate hydrogen from cold water slowly, but rapidly from hot:

$$Mn + 2H_2O \rightarrow Mn(OH)_2(s) + H_2$$

Dilute acids form manganese(II) salts, with displacement of hydrogen. Alkali is without action. Manganese does not display passivity after immersion in concentrated nitric acid, and the usual reactions occur with acids under oxidizing conditions. Manganese reacts with many non-metals. In oxygen or air, trimanganese tetroxide, Mn_3O_4, is the final product above 940°C. Chlorine forms manganese(II) chloride. On very strong heating, manganese reacts with nitrogen to form the nitride, probably Mn_3N_2. It also reacts with phosphorus and sulphur.

Manganese may be prepared by the Goldschmidt Thermite process. Because the reduction of the manganese(IV) oxide ore, pyrolusite, is violent, it is usual to convert this to trimanganese tetroxide by heating to redness:

$$3MnO_2 \rightarrow Mn_3O_4 + O_2$$

The product is mixed with coarse aluminium powder in a fire-clay crucible and ignited with a mixture of barium peroxide and aluminium powder, itself lit by a magnesium ribbon fuse. The temperature may exceed 2500°C and fused manganese results:

$$8Al + 3Mn_3O_4 \rightarrow 4Al_2O_3 + 9Mn$$

Mn^{VII}: potassium permanganate (potassium tetraoxomanganate(VII))

By analogy to chlorine heptoxide, perchloric acid and the perchlorate ion, the following structures may be given:

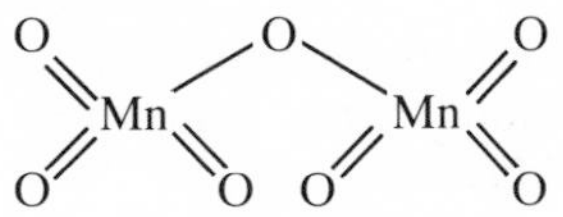

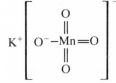

manganese(VII) oxide
manganese heptoxide

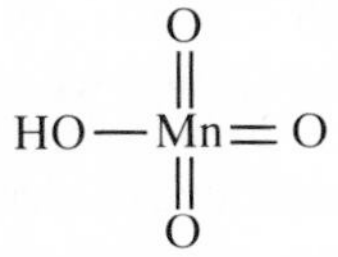

permanganic acid
(not isolated)

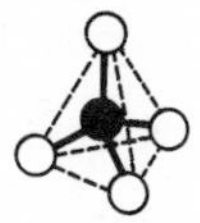

tetrahedral resonance hybrid
potassium permanganate
potassium manganate(VII)

Mn now has the electronic configuration 2, 8, 18, 4, all electrons in excess of 2, 8, 8 having been used. Potassium perchlorate, $KClO_4$, and permanganate, $KMnO_4$, form a continuous series of solid solutions.

Manganese(VII) oxide (permanganic anhydride), Mn_2O_7, is a volatile dark oil, freezing to a green solid, which is liable to explode violently. It is wholly acidic. Permanganic acid has not been isolated from solution. The most important salt is potassium permanganate, used in the laboratory and as an antiseptic. It is prepared by the oxidation of a lower manganese compound under alkaline conditions.

When finely divided manganese(IV) oxide, manganese dioxide, is stirred into a fused mass of caustic potash and an oxidizing agent, potassium chlorate, and the mixture heated strongly, potassium manganate(VI) is formed:

$$3MnO_2 + 6KOH + KClO_3 \rightarrow$$
$$3K_2MnO_4 + KCl + 3H_2O$$

After cooling, the solid is boiled with water when the green solution of potassium manganate(VI) is hydrolysed to a magenta solution of potassium permanganate, the action being assisted by passing carbon dioxide to neutralize the alkali formed:

$$3MnO_4^{2-} + 2H_2O \rightarrow$$
$$2MnO_4^- + MnO_2(s) + 4OH^-$$

Disproportionation has occurred:

$$3Mn^{VI} \rightarrow 2Mn^{VII} + Mn^{IV}$$

The liquid is filtered through glass wool, or sintered glass, and evaporated until crystallization occurs. Potassium permanganate forms small crystals which are almost black in appearance. In the commercial process, oxidation of manganate is effected by chlorine, avoiding undue loss of manganese as the dioxide, or electrolytically:

$$2MnO_4^{2-} + Cl_2 \rightarrow 2MnO_4^- + 2Cl^-$$

$$MnO_4^{2-} - e^- \rightarrow MnO_4^-$$

Potassium permanganate is a powerful oxidizing agent. Oxygen is evolved on heating:

$$2KMnO_4 \rightarrow K_2MnO_4 + MnO_2 + O_2$$

Hydrochloric acid is oxidized to chlorine and hydrogen peroxide to oxygen:

$$2MnO_4^- + 16H^+ + 10Cl^- \rightarrow$$
$$2Mn^{2+} + 8H_2O + 5Cl_2$$

$$2MnO_4^- + 6H^+ + 5H_2O_2 \rightarrow$$
$$2Mn^{2+} + 8H_2O + 5O_2$$

Potassium permanganate is used for the titrimetric (volumetric) estimation of reducing agents (p. 121).

The decolorization of potassium permanganate solution may be used as a qualitative test for sulphur dioxide and hydrogen sulphide, the latter giving a precipitate of sulphur:

$$2MnO_4^- + 5SO_2 + 2H_2O \rightarrow$$
$$2Mn^{2+} + 5SO_4^{2-} + 4H^+$$

$$2MnO_4^- + 5H_2S + 6H^+ \rightarrow$$
$$2Mn^{2+} + 5S + 8H_2O$$

If insufficient acid is present in the redox systems outlined above, a brown coloration due to manganese(IV) oxide is observed.

Mn^{VI}: potassium manganate(VI)

The formation of potassium (tetraoxo)manganate(VI), K_2MnO_4, in preparing the permanganate has been described. Manganates(VI) are usually brown but form green solutions. They are powerful oxidizing agents. Manganates are extremely difficult to purify. Manganic acid, H_2MnO_4, has not been isolated and it is considered improbable that manganese(VI) oxide, the trioxide, exists.

Mn^{IV}: manganese(IV) oxide (manganese dioxide)

This is probably the only simple compound formed by manganese in this oxidation state. Manganese(IV) oxide, MnO_2, occurs naturally as pyrolusite, psilomelane and braunstein: it is used as a depolarizer in electric dry cells and in the decolorizing of glass, where it oxidizes green iron(II) silicate to the paler iron(III) silicate while the pinkish hue compensates for any greenish tint left. Hydrated manganese(IV) oxide reddens litmus and reacts with alkali to form an unstable product called a manganate(IV). Salts with mineral acids have not been isolated although complex hexafluorides and hexachlorides have been prepared: K_2MnF_6 is golden-yellow and K_2MnCl_6, dark red. Manganese(IV) oxide resembles lead(IV) oxide in preparation and oxidizing properties.

Manganese(IV) oxide may be prepared by the alkaline oxidation of manganese(II) salts by sodium hypochlorite (chlorate(I)) or bleaching powder; it is hydrated:

$$Mn^{2+} + 2OH^- \rightarrow Mn(OH)_2(s)$$

$$Mn(OH)_2(s) + ClO^- \rightarrow MnO_2(s) + H_2O + Cl^-$$

Excess hypochlorite results in oxidation to permanganate. Anhydrous manganese(IV) oxide is formed by the thermal decomposition of manganese(II) nitrate at 200°C, no oxygen being evolved:

$$Mn(NO_3)_2 \rightarrow MnO_2 + 2NO_2$$

It is a dark brown, almost black, powder.

Manganese(IV) oxide will catalytically decompose hydrogen peroxide and catalyse the thermal decomposition of potassium chlorate.

Manganese(IV) oxide is an oxidizing agent:

$$MnO_2(s) + 4H^+ + 2e^- \rightleftharpoons Mn^{2+} + 2H_2O$$

Warm concentrated hydrochloric acid is oxidized to chlorine:

$$MnO_2(s) + 4H^+ + 2Cl^- \rightarrow Mn^{2+} + 2H_2O + Cl_2$$

In the presence of hot dilute sulphuric acid, iron(II) sulphate is oxidized to the iron(III) state and ethanedioic acid to carbon dioxide:

$$MnO_2(s) + 4H^+ + 2Fe^{2+} \rightarrow$$
$$Mn^{2+} + 2H_2O + 2Fe^{3+}$$

$$MnO_2(s) + 2H^+ + \begin{array}{c} COOH \\ | \\ COOH \end{array} \rightarrow$$
$$Mn^{2+} + 2H_2O + 2CO_2$$

Either of these reactions may be used for the estimation of manganese(IV) oxide in its minerals. A weighed sample is heated with a measured volume of standard iron(II) (ammonium) sulphate solution and dilute sulphuric acid until no black particles remain, and excess iron(II) ion determined by titration against potassium permanganate solution. Alternatively, pyrolusite may be boiled with ethanedioic (oxalic) acid or sodium ethanedioate (oxalate) and sulphuric acid; the salt being a primary standard and not liable to atmospheric oxidation is to be preferred. When manganese(IV) oxide is heated with concentrated sulphuric acid, oxygen is evolved:

$$2MnO_2 + 2H_2SO_4 \rightarrow 2MnSO_4 + 2H_2O + O_2$$

On strong heating, manganese(IV) oxide forms trimanganese tetroxide:

$$3MnO_2 \rightarrow Mn_3O_4 + O_2$$

Mn^{III}: manganese(III) oxide (manganese sesquioxide)

Anhydrous manganese(III) oxide is found as the mineral braunite, Mn_2O_3, while a hydrated form, $Mn_2O_3 . H_2O$, which is chemically more reactive, occurs as manganite. Manganese forms both anions and cations in oxidation state +3. Simple salts of tervalent manganese are not stable. However, complexes are more stable, the hexacyano-manganates(III) being an example: $K_3Mn(CN)_6$ is a red crystalline solid, isomorphous with potassium hexacyanoferrate(III) and slowly hydrolysed.

Manganese(III) oxide is formed as a grey solid by heating manganese(IV) oxide to 700°C for several hours:

$$4MnO_2 \rightarrow 2Mn_2O_3 + O_2$$

It is converted into trimanganese tetroxide when heated above 940°C in air or reduced by hydrogen above 230°C. Hot concentrated hydrochloric acid is oxidized to chlorine by manganese(III) oxide:

$$Mn_2O_3(s) + 6H^+ + 2e^- \rightleftharpoons 2Mn^{2+} + 3H_2O$$

$$Mn_2O_3(s) + 6H^+ + 2Cl^- \rightarrow$$
$$2Mn^{2+} + 3H_2O + Cl_2$$

Mn^{II} and Mn^{III}: trimanganese tetroxide (formerly mangano-manganic oxide)

Mn_3O_4 is also known as red manganese oxide and occurs native as hausmannite. It is the most stable oxide and is eventually formed by heating manganese, other oxides of the metal or salts with volatile anions to about 1000°C,

$$6MnO + O_2 \rightarrow 2Mn_3O_4$$

It behaves chemically as a mixed oxide, $Mn_2O_3 . MnO$, and oxidizes hot concentrated hydrochloric acid to chlorine:

$$Mn_3O_4(s) + 8H^+ + 2Cl^- \rightarrow$$
$$3Mn^{2+} + 4H_2O + Cl_2$$

Mn^{II}: manganese(II) (formerly manganous) compounds

Manganese(II) salts are the only stable simple salts. The oxide and hydroxide are entirely basic, but are readily oxidized. Manganese(II) oxide is a greenish-grey solid formed by the thermal decomposition of manganese(II) ethanedioate or carbonate in the absence of air:

$$\begin{array}{c} COO \\ | \\ COO \end{array} Mn \rightarrow MnO + CO + CO_2$$

$$MnCO_3 \rightarrow MnO + CO_2$$

When finely divided, it is oxidized by air at ordinary temperatures to trimanganese tetroxide and manganese(III) oxide. In preparation and oxidation, it resembles iron(II) oxide. Manganese(II) hydroxide resembles iron(II) hydroxide. Addition of alkali to a solution of a manganese(II) salt yields a white precipitate of manganese(II) hydroxide which rapidly darkens in air:

$$Mn^{2+} + 2OH^- \rightarrow Mn(OH)_2(s)$$

$$2Mn(OH)_2 + O_2 \rightarrow 2MnO_2 . H_2O(s)$$

Salts are prepared by the usual methods. When hydrated, manganese(II) salts are pink. Solutions are pale pink. A wide range of oxosalts are formed. Manganese(II) carbonate is precipitated as a pink precipitate by addition of sodium hydrogencarbonate solution to an aqueous manganese(II) salt. The thermal decomposition of the nitrate has been described: instead of manganese(II) oxide and oxygen, manganese(IV) oxide is formed with nitrogen dioxide.

$$Mn(NO_3)_2 \rightarrow MnO_2 + 2NO_2$$

Manganese(II) sulphate, used in the dyestuffs industry, is the most important salt, and is the most stable transition metal sulphate. It is the final product of heating almost any manganese compound with concentrated sulphuric acid. Industrially, manganese(IV) oxide is used:

$$2MnO_2 + 2H_2SO_4 \rightarrow 2MnSO_4 + 2H_2O + O_2$$

Several hydrates exist, among them $MnSO_4 . 5H_2O$, isomorphous with $CuSO_4 . 5H_2O$ and $MnSO_4 . 7H_2O$, isomorphous with the other vitriols, e.g. $ZnSO_4 . 7H_2O$.

The tendency to form complexes is less in the manganese(II) state. Notable are the hexacyanomanganates(II), $K_4[Mn(CN)_6] . 3H_2O$, analogous to the hexacyanoferrates(II), $K_4[Fe(CN)_6] . 3H_2O$.

Qualitative tests for manganese(II) salts

In qualitative anlysis, ammonium sulphide or hydrogen sulphide in the presence of ammonia and ammonium chloride precipitates flesh-pink manganese(II) sulphide, MnS.

Aqueous ammonia precipitates manganese(II) hydroxide, which rapidly darkens owing to oxidation to hydrated manganese(IV) oxide. Although ammonium chloride will suppress the hydroxide ion concentration sufficiently to prevent precipitation of manganese(II) hydroxide by aqueous ammonia, the alkaline solution formed, slowly deposits hydrated manganese(IV) oxide.

Manganese(II) salts may be oxidized to the permanganate condition, in the absence of chloride ions, by sodium bismuthate and nitric acid:

$$2Mn^{2+} + 5NaBiO_3 + 14H^+ \rightarrow$$
$$2MnO_4^- + 5Na^+ + 5Bi^{3+} + 7H_2O$$

This oxidation may be effected by boiling with red lead, or lead(IV) oxide, in concentrated nitric acid:

$$Pb_3O_4 + 4H^+ \rightarrow$$
$$2Pb^{2+} + 2H_2O + PbO_2$$
$$2Mn^{2+} + 5PbO_2 + 4H^+ \rightarrow$$
$$2MnO_4^- + 5Pb^{2+} + 2H_2O$$

Fusion with caustic alkali or fusion mixture and potassium chlorate or nitrate causes oxidation to the manganate(VI) state:

$$MnSO_4 + Na_2CO_3 \rightarrow MnCO_3 + Na_2SO_4$$
$$MnCO_3 \rightarrow MnO + CO_2$$
$$MnO + Na_2CO_3 + O_2 \rightarrow Na_2MnO_4 + CO_2$$

Compounds of manganese will give an amethyst borax bead after heating in the oxidizing flame and cooling.

Iron

The principal valency states of iron are 2 and 3, and are of comparable stability, these being the main oxidation states. In the simple carbonyl, $Fe(CO)_5$, iron is assigned zero valency because bonding electrons are donated by carbon monoxide in the co-ionic structure: similarly, with the oxidation number. Oxidation of iron with potassium nitrate in alkaline solution yields potassium ferrate(VI), K_2FeO_4 but this red, easily decomposed solid will not be considered further. In the Mendeléeff Table, iron was placed with cobalt and nickel in a separate triad.

The metallurgy of iron is complex. When pure, it is a soft, white, ductile and malleable metal, but its properties are profoundly modified by traces of other metals and non-metals, but principally carbon, to form steels. Alloys of iron have been described separately. Iron is *ferromagnetic* up to 770°C.

When iron is heated strongly in air or oxygen, tri-iron tetroxide is formed:

$$3Fe + 2O_2 \rightarrow Fe_3O_4$$

Iron wire burns when heated in oxygen or the oxy-coal gas flame. When steam is passed over iron at red heat, the same oxide is formed in a reversible reaction:

$$3Fe + 4H_2O \rightleftharpoons Fe_3O_4 + 4H_2$$

Hydrogen is swept away by the steam, the equilibrium is disturbed and the reaction proceeds to the right. Passing hydrogen over tri-iron tetroxide, for the same reason, causes reduction to iron. In moist air and traces of electrolytes, rusting occurs. Rust is hydrated iron(III) oxide, $Fe_2O_3.xH_2O$. Electrolytic corrosion and the use of zinc and tin in protecting iron (as steel) has been discussed on pages 128–31. Iron displaces hydrogen from dilute hydrochloric and sulphuric acids, and metals lower in the electrochemical series from their salt solutions:

$$Fe + 2H^+ \rightarrow Fe^{2+} + H_2$$

$$Fe + Cu^{2+} \rightarrow Fe^{2+} + Cu$$

Hot concentrated sulphuric acid is reduced to sulphur dioxide:

$$2Fe + 6H_2SO_4 \rightarrow Fe_2(SO_4)_3 + 3SO_2 + 6H_2O$$

Dilute nitric acid yields ammonium nitrate and iron(II) nitrate,

$$4Fe + 10HNO_3 \rightarrow$$
$$4Fe(NO_3)_2 + NH_4NO_3 + 3H_2O$$

but after an initial reaction, concentrated nitric acid renders iron passive. This passivity is due to a thin continuous film of tri-iron tetroxide and brings reduced reactivity: copper is not displaced from copper(II) sulphate solution and the effect of dilute mineral acids is greatly reduced or even inhibited. Passivity goes when the oxide film is destroyed by scratching. Grease-free iron filings when mixed with sulphur and heated, react with incandescence to form iron(II) sulphide, FeS. Iron wire or steel wool burns with incandescence when heated in a stream of chlorine to form a sublimate of iron(III) chloride, $FeCl_3$.

Very pure iron may be prepared by reduction of a pure oxide by heating in hydrogen,

$$Fe_2O_3 + 3H_2 \rightarrow 2Fe + 3H_2O$$

by the electrolysis of an iron(II) salt, or the thermal dissociation of iron pentacarbonyl on heating,

$$Fe(CO)_5 \rightleftharpoons Fe + 5CO$$

The redox system: $Fe^{3+} + e^- \rightleftharpoons Fe^{2+}$

The interconversion of iron(II) and iron(III) compounds is important not only in the estimation of iron and its compounds but also in the investigation of oxidizing and reducing agents generally.

Iron(II) and iron(III) ions are of comparable stability, but solutions of iron(II) salts are oxidized by air unless in acid solution. Covalent iron(III) compounds are more stable than those of iron(II), a general trend in the transition series and in harmony with Fajan's Rules (p. 82).

We may explore this briefly by reference to the redox expression for the Fe^{3+}; Fe^{2+} equilibrium on p. 119.

$$E/V = +0.771 + 0.0591 \log \left(\frac{m_{Fe^{3+}}/\text{mol dm}^{-3}}{m_{Fe^{2+}}/\text{mol dm}^{-3}} \right)$$

The electrical potential is that experienced by a platinum wire immersed in a solution containing both ions, but action only follows if the wire is connected to another immersed party in another suitable solution, with a conduction bridge (electrolytic agar jelly). A positive potential with respect to the solution means that the wire has to surrender electrons to the equilibrium system because of the tendency (when compared with the hydrogen electrode system) for Fe^{3+} ions to form Fe^{2+}. If the electrodes were to be coupled to one which has the I^-; I_2(Pt) system

$$\tfrac{1}{2}I_2 + e^- \rightleftharpoons I^-$$

for which $E_{298}^{\ominus}$ is +0.536V, less than that for the above system, the half cell reactions would be

$$Fe^{3+} + e^- \rightarrow Fe^{2+}$$

$$I^- - e^- \rightarrow \tfrac{1}{2}I_2$$

and the overall reaction, whether in the cell or more conventionally

$$Fe^{3+} + I^- \rightarrow Fe^{2+} + \tfrac{1}{2}I_2$$

In the original expression, the value of E depends on the logarithm of the ratio of the molar concentrations of the ions $Fe^{3+} : Fe^{2+}$

If $Fe^{3+}/Fe^{2+} > 1$, then $E > +0.771V$ meaning higher oxidating power

If $Fe^{3+}/Fe^{2+} < 1$, then $E < +0.771V$ meaning lower oxidating power

Suppose sodium hydroxide solution is added to the mixed solution of Fe^{3+} and Fe^{2+} ions, so that the supernatant liquid above the precipitated hydroxides is 1.0M for hydroxide ions. Now the solubility products at laboratory temperature are

$$Fe(OH)_3[Fe^{3+}][OH^-]^3 = 1.1 \times 10^{-36} \text{ mol}^4 \text{ dm}^{-12}$$

$$Fe(OH)_2[Fe^{2+}][OH^-]^2 = 1.6 \times 10^{-14} \text{ mol}^3 \text{ dm}^{-9}$$

If $[OH^-] = 1.0 \text{ mol dm}^{-3}$

then $\dfrac{[Fe^{3+}]}{[Fe^{2+}]} = \dfrac{1.1 \times 10^{-36}}{1.6 \times 10^{-14}}$

$$= 6.87 \times 10^{-23}$$

whence $E = -0.54V$

This is a complete change from the value $E = +0.771V$, the value for neutral or acid conditions (in which the hydrogen ions can be considered not to interfere).

The change

$$Fe^{2+} \rightarrow Fe^{3+} + e^- \text{ (to the electrode)}$$

must have the dominant tendency, in spite of the low concentration of the (instantly replenishable) Fe^{2+} ions. An example illustrates the change.

Iron(III) salts will oxidize copper and copper(I) compounds to the copper(II) state. In the presence of aqueous ammonia iron(II) salts will reduce copper(II) salts to the copper(I) state while in the presence of strong alkali e.g. potassium hydroxide, the reduction goes one oxidation state further to metallic copper(0).

Fe^{III}: iron(III) oxide and iron(III) salts (formerly ferric compounds)

Iron(III) oxide, Fe_2O_3, occurs abundantly in nature as haematite and as limonite, $2Fe_2O_3.3H_2O$. In the combustion of iron pyrites to form sulphur dioxide, it is formed as the main by-product:

$$4FeS_2 + 11O_2 \rightarrow 2Fe_2O_3 + 8SO_2$$

Iron(III) oxide may be prepared by heating green vitriol, iron(II) sulphate heptahydrate, or indeed the metal, iron(II) oxide or any salt of a volatile acid:

$$FeSO_4.7H_2O \rightarrow FeSO_4 + 7H_2O$$

$$2FeSO_4 \rightarrow Fe_2O_3 + SO_3 + SO_2$$

A fine red powder is formed which may be used as a polishing powder, being known as jewellers' rouge, and in many pigments. Iron(III) oxide may be used as a catalyst in the production of sulphur trioxide by the Contact process. The oxide is regarded as basic, salts being formed with acids, but there is a perceptible acidic character seen in the formation of ferrates(III) by fusion of metal oxides, or alkali carbonates, with iron(III) oxide:

$$Fe_2O_3 + 6H^+ \rightarrow 2Fe^{3+} + 3H_2O$$

$$Fe_2O_3 + Na_2CO_3 \rightarrow 2NaFeO_2 + CO_2$$

Iron(III) oxide is readily reduced on heating with carbon or in a stream of carbon monoxide, hydrogen or coal gas:

$$Fe_2O_3 + 3C \rightarrow 2Fe + 3CO$$

$$Fe_2O_3 + 3CO \rightarrow 2Fe + 3CO_2$$

$$Fe_2O_3 + 3H_2 \rightarrow 2Fe + 3H_2O$$

Hydrated iron(III) oxide, or iron(III) hydroxide, of which the formula is usually written $Fe(OH)_3$ although this composition is not reached, is precipitated by the action of aqueous alkali or carbonate on a solution of an iron(III) salt. A foxy-red gelatinous precipitate, readily soluble in acids to form iron(III) salts, insoluble in alkali and readily passing over to the colloidal state, is formed. No iron(III) carbonate has been isolated.

Fe^{III}: iron (III) sulphate and ammonium iron(III) alum

The iron(III) ion, Fe^{3+}, has the electronic configuration 2, 8, 13. In a range of hydrated salts it is usually amethyst in colour, and probably associated with six molecules of water: $Fe(H_2O)_6{}^{3+}$. Aqueous solutions are acidic in reaction due to hydrolysis:

$$Fe(OH_2)_6{}^{3+} + H_2O \rightleftharpoons Fe(H_2O)_5(OH)^{2+} + H_3O^+$$

Unless acid is added, solutions turn brown in colour with separation of basic salts. Ammonium iron(III) sulphate, or simply iron(III) alum, is a double salt, not a complex:

$$\underset{\text{iron(III) sulphate}}{(Fe^{3+})_2(SO_4{}^{2-})_3} \quad (NH_4)_2SO_4.Fe_2(SO_4)_3.24H_2O$$

$$\underset{\text{iron(III) alum}}{\text{or } NH_4{}^+Fe^{3+}(SO_4{}^{2-})_2.12H_2O}$$

Iron(III) sulphate is prepared by oxidation of iron(II) sulphate. This is dissolved in dilute sulphuric acid, heated, and an oxidizing agent producing volatile reduction products is added. With concentrated nitric acid, brown fumes of nitrogen dioxide appear as the complex $Fe^{2+}.NO$ ion breaks down:

$$6Fe^{2+} + 8H^+ + 2NO_3^- \rightarrow 6Fe^{3+} + 2NO + 4H_2O$$

The anhydrous salt is white. On heating, iron(III) oxide and sulphur trioxide are formed:

$$Fe_2(SO_4)_3 \rightarrow Fe_2O_3 + 3SO_3$$

When a pure iron(III) salt is required, iron(III) alum is usually chosen. It is readily crystallized. Concentrated solutions of the constituent salts may be mixed and set aside for crystallization, but more usually, the calculated proportions of iron(II) sulphate heptahydrate and ammonium sulphate are dissolved in water containing sulphuric acid and concentrated nitric acid added to cause oxidation as before. Evaporation to half bulk drives off volatile products and on cooling, amethyst octahedral crystals of hydrated ammonium iron(III) sulphate are formed.

Fe^{III}: the halides

The fluoride, chloride and bromide may be synthesized from the elements, although iron(II) bromide is the first product with the last-named halogen. Iron(III) iodide has not been isolated; the elements react to form iron(II) iodide. Iron(III) bromide is similar to the chloride but is less stable. Both are extremely soluble in water, differing from iron(III) fluoride which is only slightly soluble.

Iron(III) chloride is prepared by passing dry chlorine over heated iron wire or steel wool. The reaction proceeds with incandescence when started and iron(III) chloride sublimes as almost black iridescent scales. The apparatus is the same as that used for aluminium chloride (p. 313). Iron(III) chloride is rapidly hydrolysed in moist air. It is deliquescent and very soluble in water (solubility = 92 g in 100 g water at 20°C). The solution is ionized and hydrolysed, being acidic (degree of hydrolysis = 0.47 in 0.003M solution at room temperature):

$$Fe(H_2O)_6{}^{3+} + H_2O \rightleftharpoons Fe(H_2O)_5(OH)^{2+} + H_3O^+$$

Hydrolysis is suppressed by addition of hydrochloric acid, which gives a yellowish solution; otherwise, brown basic salts are deposited. Iron(III) chloride forms four hydrates, of which the golden-brown hexahydrate, $FeCl_3.6H_2O$, is the most common.

Iron(III) chloride dissolves in many organic solvents, forming solutions of very small electrical conductivity. In ethanol, propanone (acetone), ethoxyethane (ether) and pyridine the molecular weight corresponds to $FeCl_3$ but is higher in other solvents, corresponding to Fe_2Cl_6 in ethyl ethanoate (acetate). The vapour is dimerized: at 400°C the (relative) vapour density corresponds to Fe_2Cl_6. With addition of chlorine to suppress thermal dissociation:

$$2FeCl_3 \rightleftharpoons 2FeCl_2 + Cl_2$$

it corresponds to $FeCl_3$ at 750°C. Without added chlorine, dissociation into iron(II) chloride and chlorine commences at 500°C. The volatilization of iron(III) chloride occurs at about 300°C. Iron(III) chloride is most probably covalent and dimerized:

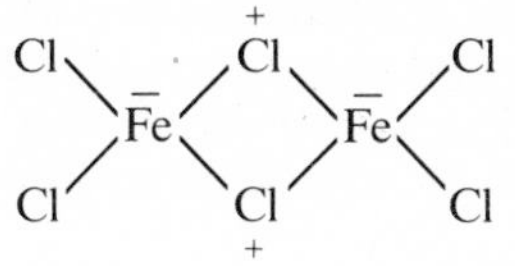

In many ways iron(III) chloride resembles aluminium chloride, which it may replace in many Friedel–Crafts' reactions in Aromatic Organic Chemistry.

Qualitative tests for iron(III) salts

Addition of aqueous alkali or sodium carbonate solution precipitates foxy-red gelatinous iron(III) hydroxide.

In acid solution, reduction to the iron(II) state is brought about by hydrogen sulphide,

$$2Fe^{3+} + H_2S \rightarrow 2Fe^{2+} + 2H^+ + S(s)$$

but in the presence of ammonia, a black precipitate said to be iron(III) sulphide, is formed:

$$2Fe^{3+} + 3H_2S \rightarrow Fe_2S_3(s) + 6H^+$$

Addition of acid decomposes the black precipitate to the iron(II) state and sulphur:

$$Fe_2S_3 + 4H^+ \rightarrow 2Fe^{2+} + 2H_2S + S(s)$$

Many other reducing agents may be used: tin(II) chloride is commonly used to reduce iron(III) chloride to iron(II) chloride:

$$2Fe^{3+} + Sn^{2+} \rightarrow 2Fe^{2+} + Sn^{4+}$$

Traces of iron(III) may be detected in solution by addition of potassium thiocyanate or ammonium thiocyanate solution, when a deep red coloration is produced. Traces of iron(III) in an iron(II) salt, which gives no coloration when pure, may be detected in this way. A complex thiocyanato-iron(III) cation ion is formed:

$$Fe^{3+} + SCN^- \rightleftharpoons Fe(SCN)^{2+}$$

probably an octohedral structure $Fe(SCN)(H_2O)_5^{2+}$. The colour disappears on addition of a solution containing fluoride ions due to the formation of a stable complex fluoride, FeF_6^{3-}.

Potassium hexacyanoferrate(II) solution when added to Fe(III) cation solutions gives a dark blue precipitate, called Prussian Blue, usually written as iron(III) hexacyanoferrate(II):

$$4Fe^{3+} + 3Fe(CN)_6{}^{4-} \rightarrow Fe_4[Fe(CN)_6]_3(s)$$

The composition of this precipitate is considered later. No precipitate is given with potassium hexacyanoferrate(III) solution, only a brown or green coloration.

Fe^{II} and Fe^{III}: tri-iron tetroxide (formerly ferroso-ferric oxide)

Fe_3O_4 has also been known as magnetic oxide of iron, and triferric tetroxide. It occurs naturally as magnetite and lodestone. Tri-iron tetroxide is formed when iron is heated in oxygen, steam or carbon dioxide:

$$3Fe + 2O_2 \rightarrow Fe_3O_4$$

$$3Fe + 4H_2O \rightleftharpoons Fe_3O_4 + 4H_2$$

$$3Fe + 4CO_2 \rightarrow Fe_3O_4 + 4CO$$

It is chemically very resistant but dissolves in hydrochloric acid to give a mixture of iron(II) and iron(III) chlorides:

$$Fe_3O_4 + 8HCl \rightarrow FeCl_2 + 2FeCl_3 + 4H_2O$$

Tri-iron tetroxide behaves as iron(II) ferrate(III) $Fe^{II}(Fe^{III}O_2)_2$.

Fe^{II}: iron (II) oxide and iron(II) salts (formerly ferrous compounds)

Iron(II) oxide, FeO, is basic. It is prepared as a black powder by reduction of iron(III) oxide using hydrogen below 300°C,

$$Fe_2O_3 + H_2 \rightarrow 2FeO + H_2O$$

and as a product of the thermal decomposition of iron(II) ethanedioate (oxalate), a yellow solid which usually contains an indefinite amount of hydration:

$$\begin{matrix} COO \\ | \\ COO \end{matrix} Fe \rightarrow CO + CO_2 + FeO \quad \text{plus water vapour}$$

On exposure to air, iron(II) oxide is oxidized to the iron(III) state, often with incandescence.

Iron(II) hydroxide, $Fe(OH)_2$, may be prepared in the absence of air (free and dissolved) as a white precipitate by the addition of alkali to a solution of an iron(II) salt. Usually a green gelatinous precipitate appears and rapidly darkens as iron(III) hydroxide or hydrated iron(III) oxide is formed. The iron(II) state is very prone to oxidation under alkaline conditions. Iron(II) hydroxide reacts with acids to form salts.

Iron(II) forms a carbonate, which occurs naturally as spathic iron ore. In the absence of air, iron(II) carbonate is formed as a white precipitate on addition of sodium carbonate solution to iron(II) sulphate solution.

Fe^{II}: hydrated iron(II) sulphate (green vitriol) and ammonium iron(II) sulphate (Mohr's Salt)

Iron(II) sulphate heptahydrate, green vitriol, $FeSO_4 \cdot 7H_2O$, is the common iron(II) salt. It is used in the manufacture of ink and iron compounds. On the commercial scale, iron(II) sulphate may be made by oxidation of iron pyrites with air and water, scrap iron being added to neutralize acid liberated:

$$2FeS_2 + 7O_2 + 2H_2O \rightarrow 2FeSO_4 + 2H_2SO_4$$

In the laboratory, iron is dissolved in dilute sulphuric acid and green vitriol crystallized from acid solution as pale green crystals, which could be named hexaquoiron(II) sulphate monohydrate, $[Fe(H_2O)_6]SO_4 \cdot H_2O$. On heating, the white anhydrous salt is formed, decomposing on stronger heating to give red iron(III) oxide, the usual preparation of this oxide. Iron(II) sulphate solution forms a dark brown additive complex with nitrogen oxide, $Fe^{2+} \cdot NO$. Solutions of iron(II) sulphate are readily oxidized, the salt being frequently used during the investigation of oxidizing agents. Addition of sulphuric acid to the solution renders it less liable to oxidation by air. Iron(II) sulphate reduces silver and gold salts to the metal and mercury(II) chloride to dimercury(I) dichloride in light:

$$Ag^+ + Fe^{2+} \rightarrow Ag(s) + Fe^{3+}$$

$$Au^{3+} + 3Fe^{2+} \rightarrow Au(s) + 3Fe^{3+}$$

$$2HgCl_2 + 2Fe^{2+} \rightarrow \underset{\text{covalent}}{Hg_2Cl_2(s)} + 2Fe^{3+} + 2Cl^-$$

Crystals of iron(II) sulphate effloresce and may become oxidized during storage. Where a non-oxidized salt is required, ammonium iron(II) sulphate is very suitable. This salt is popular as a cheap primary standard in volumetric analysis, although for very accurate titrimetric work with permanganate, sodium ethanedioate(oxalate), which is anhydrous, may be preferred.

Ammonium iron(II) sulphate, Mohr's salt $FeSO_4 \cdot (NH_4)_2SO_4 \cdot 6H_2O$, is a double salt, not a complex. Calculated quantities of green vitriol and ammonium sulphate are dissolved in hot water and set aside for crystallization. Alternatively, iron and sulphuric acid may be used in place of iron(II) sulphate. Unlike green vitriol, this hydrate does not lose water by efflorescence nor is it oxidized in air.

Fe^{II}: the halides

Iron(II) fluoride and iron(II) chloride may be prepared in the anhydrous condition by the action of the hydrogen halide on the heated metal,

$$Fe + 2HCl \rightarrow FeCl_2 + H_2$$

Iron(II) bromide and iodide may be synthesized from the elements. Iron(II) chloride may also be made by reduction of heated anhydrous iron(III) chloride in hydrogen,

$$2FeCl_3 + H_2 \rightarrow 2FeCl_2 + 2HCl$$

By the action of hydrochloric acid on iron a solution of iron(II) chloride is formed from which, preferably in the presence of acid, greenish-blue crystals of the tetrahydrate, $FeCl_2.4H_2O$, separate. Iron(II) chloride is very soluble; at 20°C, 69.0 g anhydrous salt saturates 100 g water. It is more readily oxidized than the sulphate. Unlike iron(III) chloride, it is only slightly hydrolysed.

The theoretical aspects of the redox equilibrium

$$Fe^{3+} + e^- \rightleftharpoons Fe^{2+}$$

have been discussed under redox processes in Chapter 7.

Qualitative tests for iron(II) salts

Addition of alkali to a solution of an iron(II) salt precipitates iron(II) hydroxide, which usually appears as a green gelatinous precipitate which rapidly darkens.

Potassium hexacyanoferrate(III) solution gives a dark blue precipitate, Turnbull's Blue, usually written as iron(II) hexacyanoferrate(III), when added to a solution containing iron(II) ions:

$$3Fe^{2+} + 2Fe(CN)_6^{3-} \rightarrow Fe_3[Fe(CN)_6]_2(s)$$

The constitution of Turnbull's Blue is considered later. Potassium hexacyanoferrate(II) solution gives a white precipitate of iron(II) hexacyanoferrate(II), which is usually rendered pale blue by oxidation.

Iron(II) salts will reduce many oxidizing agents in solution. Potassium permanganate and bromine water are separately decolorized and potassium dichromate turned from orange to green:

$$5Fe^{2+} + MnO_4^- + 8H^+ \rightarrow$$
$$5Fe^{3+} + Mn^{2+} + 4H_2O$$

$$2Fe^{2+} + Br_2 \rightarrow 2Fe^{3+} + 2Br^-$$

$$6Fe^{2+} + Cr_2O_7^{2-} + 14H^+ \rightarrow$$
$$6Fe^{3+} + 2Cr^{3+} + 7H_2O$$

The interconversion of iron(II) and iron(III) compounds

Solutions of iron(III) salts are yellow, although hydrolysis may produce a brownish hue. Iron(III) salts, which are mild oxidizing agents, are formed by the ready oxidation of iron(II) salts but in turn are relatively easy to reduce back to the iron(II) condition. Iron(II) solutions are pale green in colour. The course of redox reactions may be followed by the colour changes observed, but samples should be tested by qualitative tests given above to follow the course of the interconversion. The choice of method will depend on whether the anions and cations introduced interfere with subsequent use of the solution.

Solutions of iron(II) may be most conveniently oxidized by compounds which give volatile reduction products: the appropriate acid being added to provide the extra anions required on increase of the valency of iron, otherwise basic salts result. Hydrogen peroxide or concentrated nitric acid may be used:

$$2Fe^{2+} + 2H^+ + H_2O_2 \rightarrow 2Fe^{3+} + 2H_2O$$

$$6Fe^{2+} + 8H^+ + 2NO_3^- \rightarrow 6Fe^{3+} + 4H_2O + 2NO$$

Chlorine may be used for aqueous or anhydrous iron(II) chloride:

$$2Fe^{2+} + Cl_2 \rightarrow 2Fe^{3+} + 2Cl^-$$

$$2FeCl_2 + Cl_2 \rightarrow 2FeCl_3$$

Solutions of iron(III) may be reduced by addition of the corresponding acid and zinc, or by zinc amalgam:

$$2Fe^{3+} + Zn(s) \rightarrow Zn^{2+} + 2Fe^{2+}$$

The use of hydrogen sulphide and tin(II) chloride has already been mentioned. Reduction may be conveniently effected by sulphurous acid, or sulphur dioxide:

$$2Fe^{3+} + SO_3^{2-} + H_2O \rightarrow 2Fe^{2+} + 2H^+ + SO_4^{2-}$$

Anhydrous iron(III) chloride may be reduced by heating in a stream of hydrogen:

$$2FeCl_3 + H_2 \rightarrow 2FeCl_2 + 2HCl$$

Iron(II) chloride is formed.

The titrimetric (volumetric) estimation of iron and its salts

Iron may be dissolved in dilute sulphuric acid to give iron(II) sulphate. Iron(II) may then be determined in acid solution by titration against standard potassium dichromate, or potassium permanganate (p. 121).

In the estimation of iron(II) salts, reduction to the iron(II) state may be used followed by titration as above; zinc amalgam (with dilute sulphuric acid) and tin(II) chloride are suitable reducing agents, followed by titration against dichromate.

In addition to the direct determination of iron, other oxidizing and reducing agents may be determined indirectly, especially when insoluble. Thus, zinc reduces iron(III) sulphate solution,

$$Zn(s) + 2Fe^{3+} \rightarrow Zn^{2+} + 2Fe^{2+}$$

and manganese(IV) oxide will oxidize warm acidified iron(II) sulphate.

$$MnO_2(s) + 2Fe^{2+} + 4H^+ \rightarrow$$
$$Mn^{2+} + 2Fe^{3+} + 2H_2O$$

A weighed sample is added to a measured volume (excess) of either iron(II) sulphate (for an oxidizing agent) or iron(III) alum solution (for a reducing agent), and after reaction the iron(II) content is determined by titration against permanganate or dichromate in the presence of dilute sulphuric acid.

Complexes of Fe^0, Fe^{II} and Fe^{III}

Fe^0: iron pentacarbonyl

The action of carbon monoxide on iron powder at 100–200°C under pressure yields iron pentacarbonyl, condensing to a pale yellow liquid, b.p. 103°C:

$$Fe + 5CO \rightarrow Fe(CO)_5$$

It dissociates upon further heating and in light. Because both the final complex and the ligand are uncharged, iron is assigned to oxidation state 0 even though it is clearly in chemical combination. Similarly, since the central atom makes no contribution to the bonding electrons (in the co-ionic bonding diagram) iron is considered to be zerovalent in this compound. Iron pentacarbonyl has been used as an 'anti-knock' in petrol but is now used for producing high-grade iron(III) oxide on combustion.

Fe^{II}: potassium hexacyanoferrate(II) (formerly potassium ferrocyanide)

Potassium hexacyanoferrate(II) results from the action of aqueous potassium cyanide on iron(II)

sulphate solution. To avoid excess cyanide, a freshly prepared and not acidified solution of iron(II) sulphate is added to potassium cyanide solution until a small permanent precipitate of iron(II) cyanide is formed hitherto the complex ion having been formed:

$$Fe^{2+} + 6CN^- \rightarrow Fe(CN)_6^{4-}$$

The solution is boiled, filtered and evaporated. Pale lemon crystals of the trihydrate, $K_4[Fe(CN)_6].3H_2O$, separate on cooling, and should be dried below 100°C, at which temperature they lose water. It is readily soluble, 20 g dissolving in 100 g water at 15°C and at 100°C, 100 g.

In qualitative organic analysis, the formation of sodium hexacyanoferrate(II) is used in the reactions associated with the Lassaigne sodium test for nitrogen:

$$\underset{\text{added}}{Na} + \underset{\substack{\text{from organic} \\ \text{compound}}}{C} + \underset{\substack{\text{from organic} \\ \text{compound}}}{N} \overset{\text{fuse}}{\longrightarrow} NaCN$$

The cyanide is then converted into hexacyanoferrate(II) in alkaline solution and iron(III) chloride in hydrochloric acid added to give the characteristic dark blue-green coloration (Prussian Blue).

Potassium hexacyanoferrate(II) is obtained commercially from the 'spent-oxide' of coal gas production, where this is still made, hydrogen cyanide being absorbed in moist iron(III) oxide.

When acidified, aqueous potassium hexacyanoferrate(II) gives a white precipitate of hexacyanoferric(II) acid or hydrogen hexacyanoferrate(II), $H_4Fe(CN)_6$. With copper(II) sulphate solution, either neutral or only slightly acidic, chocolate brown copper(II) hexacyanoferrate(II) is precipitated:

$$2Cu^{2+} + Fe(CN)_6^{4-} \rightarrow Cu_2[Fe(CN)_6](s)$$

When potassium hexacyanoferrate(II) is heated with dilute acids, hydrogen cyanide (b.p. 26°C) is liberated;

$$6H^+ + Fe(CN)_6^{4-} \rightarrow Fe^{2+} + 6HCN(g)$$

while concentrated sulphuric acid and a little water yields pure carbon monoxide on heating:

$$12H^+ + Fe(CN)_6^{4-} + 6H_2O \rightarrow$$
$$Fe^{2+} + 6NH_4^+ + 6CO$$

Fe^{III}: potassium hexacyanoferrate(III) (formerly potassium ferricyanide)

Potassium hexacyanoferrate(III) cannot be prepared by a reaction similar to that for the hexacyanoferrate(II), iron(III) hydroxide being precipitated on mixing iron(III) chloride and potassium cyanide solutions, its solubility product $(1.1 \times 10^{-36}\,mol^4\,dm^{-12}$ at laboratory temperature) being exceedingly small and the cyanide alkaline by salt hydrolysis (p. 241). Instead, potassium hexacyanoferrate(II) solution is oxidized with chlorine,

$$2Fe(CN)_6^{4-} + Cl_2 \rightarrow 2Fe(CN)_6^{3-} + 2Cl^-$$

After evaporation, red crystals of potassium hexacyanoferrate(III), $K_3[Fe(CN)_6]$, separate on cooling.

Prussian Blue: hexacyanoferrates(II) and (III) of iron

The use of potassium hexacyanoferrate(II) in the detection of iron(III) ions and potassium hexacyanoferrate(III) for iron(II) ions has been described; dark blue precipitates are formed which are given nominal formulae:

$$3Fe(CN)_6^{4-} + 4Fe^{3+} \rightarrow Fe_4[Fe(CN)_6]_3$$

hexacyano-   iron(III)   iron(III)
ferrate(II)   hexacyanoferrate(II)
(Prussian Blue)

$$2Fe(CN)_6^{3-} + 3Fe^{2+} \rightarrow Fe_3[Fe(CN)_6]_2$$

hexacyano-   iron(II)   iron(II)
ferrate(III)   hexacyanoferrate(III)
(Turnbull's Blue)

But this is too simple; mutual oxidation–reduction occurs:

$$Fe^{2+} + Fe(CN)_6^{3-} \rightleftharpoons Fe^{3+} + Fe(CN)_6^{4-}$$

The bulk of the evidence suggests that both blue precipitates contain iron(III) and hexacyanoferrate(II) ions, with potassium ions. The composition may be given simply as $KFe^{3+}[Fe(CN)_6]^{4-}$. Such mixed valency compounds, in which different numbers of electrons are associated with different atoms of the same element, usually metals, within the same compound are characterized by their intense colour. Examples are found in enzymes and magnetic solids, natural and synthetic pigments. One interesting development is based on these mixed valency compounds with a linear chain structure, where the movement of electrons along the chain means that the compounds are potentially one-dimensional (i.e. linear) semiconductors. The 'hopping' of electrons is promoted by visible radiation (the compounds are coloured) and may be studied† to give an understanding of what happens within the molecule, and to assess probable performance as a semiconductor.

Fe^{III}: disodium pentacyanonitrosylferrate(III) (formerly sodium nitroprusside)

When potassium hexacyanoferrate(II) solution is boiled with 30% nitric acid a complex series of reactions take place, and after neutralization with sodium carbonate solution, ruby-red crystals of disodium pentacyanonitrosylferrate,

$$Na_2[Fe(CN)_5NO].2H_2O,$$

may be separated. Iron is present in the iron(III) state. In computing the oxidation number of the metal atom in an ion, the groups NO, NS, CO and CS, when linked directly to the metal, are treated as neutral (*IUPAC* rules). Its alternative systematic name is therefore sodium pentacyanonitrosylferrate(III). A purple coloration is given with soluble sulphides, a useful analytical test.

Cobalt

There are two oxidation states of cobalt, $+2$ and $+3$. The principal valency of cobalt in simple salts is two, (Co^{2+}), but in complexes, the higher oxidation state of three is the most stable with presumably a valency of three, although even fluorine does not produce CoF_3 entirely. The division is sharp; cobalt(II) complexes sometimes decomposing water to give hydrogen, while cobalt(III) ions, Co^{3+}, react to give oxygen. Cobalt was placed with iron and nickel in a triad of Mendeléeff's Table, but such a triad could be formed from any three adjacent principal transition metals. The chemistry of cobalt is much simpler than that of iron because there is only one series of simple salts.

Cobalt is a bright, bluish-white, very hard metal which is ferromagnetic to above 1000°C. It is not attacked by water and is stable in air, but very slowly oxidized at red heat to tricobalt tetroxide, Co_3O_4. However, the finely divided metal, formed by reduction of cobalt(II) oxide in hydrogen, is pyrophoric. The dilute mineral acids attack cobalt less readily than iron: dilute nitric acid is the most

† By resonance Raman spectroscopy

reactive, while hydrochloric acid and sulphuric acid slowly yield cobalt(II) salts and hydrogen. Cobalt is oxidized by hot concentrated sulphuric acid but rendered passive by concentrated nitric acid. There is virtually no action with alkalis. Cobalt forms cobalt(II) halides with chlorine and bromine but a mixture of cobalt(II) and cobalt(III) fluorides with fluorine.

Co^{II}: cobalt(II) oxide and cobalt(II) salts (formerly cobaltous compounds)

Cobalt(II) oxide, CoO, is basic; it is formed as a green powder by the thermal decomposition of the carbonate, hydroxide or nitrate in an inert atmosphere or *in vacuo*. In air it reacts with oxygen to form Co_3O_4, tricobalt tetroxide. Cobalt(II) hydroxide, $Co(OH)_2$, may be precipitated from cobalt(II) salt solutions; addition of alkali to a cold solution gives a blue form which gradually changes to pink, a form which is given immediately on addition of the cobalt(II) solution to alkali. Cobalt(II) hydroxide is basic and resembles the corresponding hydroxides of the earlier members of the series in being oxidized by air; the product is a hydrated form of cobalt(III) oxide, $Co_2O_3 . H_2O$.

The cobalt(II) state is associated with brilliant colour. Cobalt blue is a well known artists' pigment. The blue of cobalt compounds is used in china and pottery decoration; cobalt silicate, or smalt, being formed by heating the oxide, Co_3O_4, with silica and potassium carbonate after which the fused mass is cooled and ground to a fine powder. Various coloured compounds are formed with the oxides of certain metals, and are used in their identification. When magnesium oxide is moistened with cobalt(II) nitrate solution and strongly heated on the charcoal block, a pale pink mass results; with zinc oxide, Rinmann's Green is formed, and with aluminium oxide, Thénard's Blue. The oxides are produced in each case by heating a suitable compound with sodium carbonate, or fusion mixture, on the charcoal block prior to treatment.

Cobalt(II) compounds are prepared by the standard methods. A pink basic oxide is formed by addition of aqueous carbonate or hydrogencarbonate to a solution of a cobalt(II) salt. Anhydrous cobalt(II) chloride is formed by the action of chlorine on the heated metal and in appearance is a pale blue solid. It is deliquescent and forms a series of hydrates, with 1, 2 and 6 molecules of water of crystallization, with the colour becoming increasingly pink as the degree of hydration rises. Cobalt(II) sulphate forms a vitriol, $CoSO_4 . 7H_2O$, or $[Co(H_2O)_6]SO_4 . H_2O$, which is carmine in colour, from which double sulphates of the type, $K_2SO_4 . CoSO_4 . 6H_2O$, may be prepared. Cobalt(II) nitrate, $Co(NO_3)_2 . 6H_2O$, is a pink, crystalline, deliquescent solid, which has not been dehydrated and is used as a laboratory reagent.

The colour changes exhibited by solutions of cobalt(II) compounds defy a simple unified explanation. The behaviour of cobalt(II) chloride solution warrants special study. The pink colour changes to an intense blue on concentrating the solution and on addition of hydrochloric acid or potassium chloride solution. During electrolysis, the blue coloration moves towards the anode. These changes are interpreted in terms of the formation of complex chlorides as anions. Furthermore, an alcoholic solution of cobalt(II) chloride is blue but addition of mercury(II) chloride or zinc chloride restores the pink colour. Both mercury(II) chloride and zinc chloride will take up chloride ions to form complex chlorides, thereby, it is supposed, releasing pink cobalt(II) ions from their blue complex.

Suggested reactions may be written:

Addition of Cl^- or concentration:

$$\underset{\text{pink}}{Co(H_2O)_6{}^{2+}} + \underset{\text{colourless}}{4Cl^-} \rightleftharpoons \underset{\text{blue}}{CoCl_4{}^{2-}} + 6H_2O$$

Addition of $HgCl_2$:

$$\underset{\substack{\text{blue}}}{CoCl_4{}^{2-}} + \underset{\substack{\text{covalent}\\\text{colourless}}}{2HgCl_2} \rightleftharpoons \underset{\substack{\text{pink}\\\textit{(hydrated}\\\textit{as above)}}}{Co^{2+}} + \underset{\substack{\text{colourless}}}{2HgCl_4{}^{2-}}$$

Cobalt(II) chloride solution may be used as 'invisible' ink: the pale pink colour is not apparent but on warming, blue writing becomes clearly visible.

Co^{II} and Co^{III}: tricobalt tetroxide (formerly cobalto–cobaltic oxide)

This is the usual oxide of cobalt and is formed on heating cobalt(II) nitrate strongly:

$$3Co(NO_3)_2 \rightarrow Co_3O_4 + 6NO_2 + O_2$$

It is also formed on heating cobalt(II) oxide in air and reverts to this oxide on very strong heating. Tricobalt tetroxide is a black powder, reduced by hydrogen to the metal when heated and reacting with hydrochloric acid to give cobalt(II) chloride and chlorine:

$$Co_3O_4 + 4H_2 \rightarrow 3Co + 4H_2O$$

$$Co_3O_4 + 8HCl \rightarrow 3CoCl_2 + Cl_2 + 4H_2O$$

Tri-iron tetroxide gave a mixture of iron(II) and iron(III) chlorides with hydrochloric acid.

Qualitative tests for cobalt(II) salts

Addition of ammonium sulphide solution or passage of hydrogen sulphide in the presence of added aqueous ammonia and ammonium chloride precipitates the black sulphide, CoS, from solutions of cobalt(II) salts. Cobalt(II) sulphide, insoluble in dilute hydrochloric acid, reacts slowly with hot concentrated hydrochloric acid but is readily oxidized if concentrated nitric acid (forming aqua regia) or a crystal of potassium chlorate is also present, depositing sulphur.

Aqueous caustic alkali precipitates a blue form of the hydroxide which changes to pink in excess of reagent on warming. Oxidation to dark brown $Co_2O_3.H_2O$ occurs on standing in air. Addition of ammonium chloride prevents precipitation by caustic alkali. Aqueous ammonia gives a blue precipitate probably of a basic salt, soluble in excess or in ammonium chloride solution. The ammoniacal solution goes red on standing in air as various complexes are formed through oxidation. Addition of hydrogen peroxide hastens this.

Potassium nitrite solution in the presence of ethanoic (acetic) acid and with shaking in air, precipitates yellow crystalline potassium hexanitrocobaltate(III), $K_3[Co(NO_2)_6]$, from cobalt salt solutions. Potassium cyanide solution precipitates a reddish-brown cobalt(II) cyanide, soluble in excess of the reagent, to form potassium hexacyanocobaltate(II), $K_4[Co(CN)_6]$,

$$Co^{2+} + 2CN^- \rightarrow Co(CN)_2(s)$$

$$Co(CN)_2(s) + 4CN^- \rightarrow [Co(CN)_6]^{4-}$$

Addition of a fairly concentrated solution of potassium thiocyanate or ammonium thiocyanate gives potassium or ammonium tetrathiocyanatocobaltate(II), blue in colour and extracted from the aqueous layer on addition of pentan-1-ol (amyl alcohol):

$$Co^{2+} + 4CNS^- \rightarrow Co(SCN)_4{}^{2-}$$

Cobalt compounds form a blue borax bead.

Complexes of Co^{II} and Co^{III}

Complex cyanides similar to those of vanadium, chromium, manganese and iron are formed. While potassium hexacyanocobaltate(II), $K_4[Co(CN)_6]$, is readily oxidized in air, the hexacyanocobaltate(III) $K_3[Co(CN)_6]$, is very stable. As was the case with potassium hexacyanoferrate(III), the hexacyanocobaltate(III) cannot be prepared direct, the instability of Co^{3+} being an additional problem, but by oxidation of potassium hexacyanocobaltate(II) with chlorine. Although complexes are formed in the cobalt(II) state (see above), their formation is most pronounced in the cobalt(III) state. Sodium hexanitrocobaltate(III) is the most important complex in the cobalt(III) state.

Co^{III}: sodium hexanitrocobaltate(III) (formerly sodium cobaltinitrite)

A current of air drawn through a warm solution containing cobalt(II) nitrate, sodium nitrite and ethanoic (acetic) acid causes oxidation to the cobalt(III) state and the formation of the hexanitrocobaltate(III) complex ion:

$$Co^{2+} + NO_2{}^- + 2H^+ \rightarrow Co^{3+} + NO + H_2O$$

$$Co^{3+} + 6NO_2{}^- \rightarrow [Co(NO_2)_6]^{3-}$$

Sodium hexanitrocobaltate(III), $Na_3[Co(NO_2)_6]$, is a yellow crystalline solid and may conveniently be precipitated from solution by addition of alcohol.

Co^{III}: aquo, chloro and ammino complexes of cobalt(III)

$[Co(H_2O)_6]^{3+}(Cl^-)_3$ $[Co(NH_3)_6]^{3+}(Cl^-)_3$
hexa-aquocobalt(III)   hexa-amminecobalt(III)
chloride   chloride

$[Co(NH_3)_5(H_2O)]^{3+}(Cl^-)_3$ $[Co(NH_3)_5Cl]^{2+}(Cl^-)_2$
aquopenta-amminecobalt(III)   chloropenta-amminecobalt(III)
chloride   chloride

$[Co(NO_2)_3(NH_3)_3]$
trinitrotriamminecobalt(III)

In naming a complex cation (or neutral complex) the ligand order adopted is: (negative ligand) (neutral ligand) (principal atom) (oxidation number). For a complex anion the order is the same but the suffix -ate is added before the oxidation number. Anionic ligands have names ending in -o.

$[Co(NH_3)_6][Cr(C_2O_4)_3]$
hexa-amminecobalt(III) triethanedioatochromate(III)

Ligands with cumbersome names are placed in parentheses and prefixed by *bis-*, *tris-*, *tetrakis-*, *pentakis-*, and *hexakis-* according to their number.

$$[Co(NH_2CH_2CH_2NH_2)_3]^{3+}(Cl^-)_3$$
tris(ethane-1,2-diamine)cobalt(III) chloride
tris(ethylenediamine)cobalt(III) chloride

$$[CoCl_2(NH_3)_2((CH_3)_2NH)_2]^+Cl^-$$
or $$[CoCl_2(NH_3)_2(Me_2NH)_2]^+Cl^-$$
dichlorodiamminebis(dimethylamine)cobalt(III) chloride

The stereochemistry of some of these complexes has been described in Chapter 5, pages 70–1.

The structural configuration of cobalt(III) with six ligands which mutually repel each other is basically the octahedron, and this allows the possibility of both geometrical and optical isomerism. In the structures which follow, no special significance should be afforded to the obliquely viewed squares because these have no chemical significance. They are drawn solely to enhance the three dimensional effect by showing the equatorial plane, containing the central atom and four ligands (notionally) at right angles to the plane of the paper.

$$[Co(NH_3)_4Cl_2]^+Cl^-$$
dichlorotetra-ammine cobalt(III) chloride

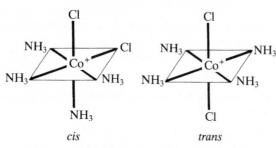

cis   *trans*

geometrical isomers

$$[Co(en)_2Cl_2]^+Cl^-$$
$$[Co(NH_2CH_2CH_2NH_2)_2Cl_2]^+Cl^-$$
dichlorobis(ethane-1,2-diamine)cobalt(III) chloride
dichlorobis(ethylenediamine)cobalt(III) chloride

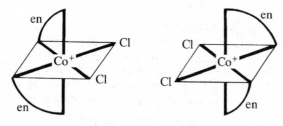

optical isomers

$$[Co(en)_3]^{3+}(Cl^-)_3$$
$$[Co(NH_2CH_2CH_2NH_2)_3]^{3+}(Cl^-)_3$$
tris(ethane-1,2-diamine)cobalt(III) chloride
tris(ethylaminediamine)cobalt(III) chloride

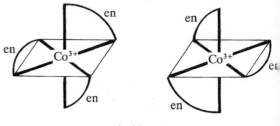

optical isomers

In all of these structures the nature of the Co-N linkage has been ignored, being depicted as simply covalent, rather than co-ionic, which would convey a different impression of electric charge distribution, $Co^- \!-\! N^+$. Reference to the hexammine of chromium (p. 64) will make this point clear.

Nickel

In nearly all of its compounds, nickel has a valency of two. Zero valency is assigned to nickel in the tetracarbonyl, $Ni(CO)_4$, and it is also in oxidation state zero: the former because no electrons from nickel are used in the co-ionic bond diagrams and the latter by application of the rules (p. 124). In a curious series of complex cyanides and their derivatives, some of which are very unstable, nickel exhibits the oxidation states 0, +1 and +2 respectively: $K_4[Ni(CN)_4]$, $K_2[Ni(CN)_3]$, $K_2[Ni(CN)_4]$. Only the last complex cyanide is important. Iron, cobalt and nickel were grouped as a triad in Mendeléeff's Table. It is of interest to recall that while nickel has a higher atomic number than cobalt, it has the lower relative atomic mass.

Nickel is a hard, white metal, which is malleable and ductile, takes a polish readily and is well suited to electro-plating. It is ferromagnetic up to 340°C. Nickel is only slowly tarnished in air. On strong heating, the oxide, NiO, is formed. Nickel is not attacked by water but at red heat in steam, the oxide, NiO, and hydrogen are formed. Of the dilute mineral acids, only nitric acid attacks the metal appreciably and fuming nitric acid renders it passive. Nickel is not attacked by aqueous or fused caustic alkali. On heating in chlorine, nickel chloride, $NiCl_2$, is formed.

Finely divided nickel, formed by reduction of the heated oxide in hydrogen, is an extremely efficient catalyst in hydrogenations:

$$\begin{array}{ccccc} CH_2 & H & & CH_3 \\ \| & + & | & \xrightarrow[140°C]{Ni} & | \\ CH_2 & H & & CH_3 \end{array}$$

Commercially, hydrogenation with nickel as catalyst is used to convert inedible oils into solid fat for the production of margarine.

Ni^II: nickel(II) (formerly nickelous) compounds

This is the only series of salts. Nickel oxide, NiO, is a green solid formed by thermal decomposition of the hydroxide, carbonate or nitrate in the absence of air. It is a basic oxide. The hydroxide, $Ni(OH)_2$, is formed as an apple-green precipitate by addition of caustic alkali to aqueous nickel salts and is insoluble in excess of caustic alkali but forms a lavender ammine with ammonia, $Ni(NH_3)_6^{2+}$. Both the oxide and hydroxide take up oxygen from the air but no compounds of a higher valency state have been isolated. Salts are prepared by the standard methods.

Sodium carbonate solution produces a light green precipitate of a basic carbonate of variable composition from solutions of nickel salts. The normal carbonate is said to be formed when sodium hydrogencarbonate is used.

Nickel sulphate, $NiSO_4.7H_2O$, $[Ni(H_2O)_6]SO_4$. H_2O, a green crystalline vitriol, is the most common salt. Double sulphates of the type $K_2SO_4.NiSO_4.6H_2O$ may be prepared.

Anhydrous nickel chloride may be prepared as a yellow solid by the action of chlorine on the hot metal. Nickel chloride will form a double salt with ammonium chloride, $NiCl_2.NH_4Cl.6H_2O$, obtained by crystallization from a solution containing both salts in equi-molar proportions.

Qualitative tests for nickel salts

Hydrogen sulphide in the presence of aqueous ammonia and ammonium chloride, or addition of ammonium sulphide solution, precipitates black nickel sulphide. NiS, from solutions of nickel salts. Nickel sulphide dissolves slowly in hot concentrated hydrochloric acid, but rapidly in the presence of an oxidizing agent, concentrated nitric acid (forming aqua regia) or a crystal of potassium chlorate.

Caustic alkali precipitates apple-green nickel hydroxide, insoluble in excess alkali, while aqueous ammonia precipitates a green basic salt soluble in

excess to give the lavender ammine, $Ni(NH_3)_6^{2+}$, when added to a solution containing nickel ions. Similarly, a complex cyanide, potassium tetracyanoniccolate(II),† $K_2Ni(CN)_4$, is formed by addition of excess potassium cyanide solution to the green precipitate of nickel cyanide formed initially:

$$Ni(CN)_2 + 2CN^- \rightarrow Ni(CN)_4^{2-}$$

In the presence of dilute ammonia, solutions of nickel salts form a red precipitate with a 1% alcoholic solution of dimethyl glyoxime.

Complexes of Ni⁰ and Ni^II

Ni⁰: nickel tetracarbonyl, Ni(CO)₄

Nickel tetracarbonyl, $Ni(CO)_4$, is formed by passing carbon monoxide over finely divided nickel below 100°C:

$$Ni + 4CO \rightleftharpoons Ni(CO)_4$$

It is a colourless liquid, b.p. 43°C, which readily volatilizes to a lethal vapour. On heating it dissociates into nickel and carbon monoxide.

Ni^II: complexes met in laboratory tests

Of the organic complexes, the dioximes are the most stable. Dimethyl glyoxime is mentioned above, nickel ions forming a red insoluble complex:

$$\begin{array}{cccc} CH_2OH & CHO & CH_3.C=O & CH_3.C=NOH \\ | & | & | & | \\ CH_2OH & CHO & CH_3.C=O & CH_3.C=NOH \end{array}$$

glycol   glyoxal   dimethyl   dimethyl
(the   (the   glyoxal   glyoxime
alcohol)   aldehyde)   (α:α-diketone)   (the oxime)

nickel (II) dimethyl glyoxime complex:
bis (dimethylglyoximato) nickel (II)
nickel(II) butane -2,3-dionedioxime

The complex tetracyanoniccolate(II) ion, $Ni(CN)_4^{2-}$ of square planar coordination, departs from the octahedral type $M(CN)_6^{4-}$ seen hitherto. The solids form yellow or yellow-orange hydrates:

$$Na_2[Ni(CN)_4].3H_2O \quad K_2[Ni(CN)_4].H_2O$$

† IUPAC: nickel being a trivial name, the Latin derivation niccolate is preferred to nickelate

Many ammines of nickel have been prepared, but the most stable complex salts contain $Ni(NH_3)_6^{2+}$, the hexamminenickel(II) cation. Ethane-1,2-diamine, $NH_2CH_2CH_2NH_2(= en)$ acts as a bidentate ligand in nickel(II) complex cations, $[Ni(en)_3]^{2+}$.

Copper

Copper is not a transition metal according to the definition because it has a completed penultimate shell of electrons. Its oxidation states are +1 and +2. Two series of compounds are formed, the metal showing valencies of 1 and 2. In the latter state, the copper(II) ion, Cu^{2+}, has the electronic configuration 2, 8, 17 and with its incomplete shell of 17 electrons shows the characteristics associated with the transition metals: in solution the copper(II) ion is blue and complex ions are readily formed.

Copper is a tough, malleable, ductile metal which resists chemical attack. Freshly prepared, it is light salmon-pink in colour but becomes reddish-brown in air through slight tarnishing with the formation of oxide and sulphide. Copper is not attacked by water or steam. Strong heating in air yields copper(II) oxide with some copper(I) oxide. Moist air produces a splendid green patina, or layer, seen on copper roofs and on bronze statues. In industrial areas this is a basic sulphate, $CuSO_4 . 3Cu(OH)_2$, while near the sea, in marine air, a basic chloride is formed, $CuCl_2 . 3Cu(OH)_2$. On heating, copper reacts with chlorine to form copper(II) chloride, $CuCl_2$, and with sulphur vapour to form copper(I) sulphide, Cu_2S, principally.

Copper is not attacked by caustic alkali, even when fused. Only under oxidizing conditions is it attacked by acids. It is not rendered passive by concentrated nitric acid. Oxidations are complex and simplified full equations may be given. Concentrated nitric acid gives nitrogen dioxide while moderately dilute (50% v/v) nitric acid forms nitrogen oxide, in addition to copper(II) nitrate and water:

$$Cu + 4HNO_3 \rightarrow Cu(NO_3)_2 + 2H_2O + 2NO_2$$

$$3Cu + 8HNO_3 \rightarrow 3Cu(NO_3)_2 + 4H_2O + 2NO$$

Hot concentrated sulphuric acid is reduced to sulphur dioxide and water, and oxidizes copper to copper(II) sulphate, although copper(I) sulphide is also formed:

$$Cu + 2H_2SO_4 \rightarrow CuSO_4 + 2H_2O + SO_2$$

In air, dilute sulphuric acid attacks copper slowly to form copper(II) sulphate:

$$2Cu + 2H_2SO_4 + O_2 \rightarrow 2CuSO_4 + 2H_2O$$

Hydrogen is not liberated by copper from dilute mineral acids. In air, it dissolves in concentrated hydrochloric acid to give a dark green solution of chlorocuprate(II) complexes:

$$2Cu + 4HCl + O_2 \rightarrow 2CuCl_2 + 2H_2O$$

$$CuCl_2 + Cl^- \rightarrow CuCl_3^-$$

$$CuCl_2 + 2Cl^- \rightarrow CuCl_4^{2-}$$

In air, copper will also react with aqueous ammonia, and potassium cyanide solution, the resulting copper ions forming complexes, $Cu(NH_3)_4^{2+}$ and $Cu(CN)_3^{2-}$ respectively.

Copper is positioned low in the electrochemical series. It is below hydrogen. Active metals displace it from solution:

$$Cu^{2+} + Zn(s) \rightarrow Cu(s) + Zn^{2+}$$

while it will displace less active metals:

$$Cu(s) + 2Ag^+ \rightarrow Cu^{2+} + 2Ag(s)$$

It may be conveniently prepared by electrolysis.

The relative stability of copper(I) and copper(II) compounds

The relationship between the compounds of the two series of compounds formed by copper is most unusual. Their relative stability depends on the conditions and type of compound and its structure.

Covalent compounds of copper(II) are less stable than their copper(I) counterparts. Copper(II) iodide is very unstable, if it exists by itself at all, breaking down to form copper(I) iodide and iodine; on heating, copper(II) bromide dissociates into the copper(I) halide and bromine while at a higher temperature copper(II) chloride behaves similarly:

$$2CuI_2 \rightarrow 2CuI + I_2$$

$$2CuBr_2 \rightarrow 2CuBr + Br_2$$

$$2CuCl_2 \rightarrow 2CuCl + Cl_2$$

On warming, copper(II) cyanide decomposes similarly to the copper(I) compound and cyanogen:

$$2Cu(CN)_2 \rightarrow 2CuCN + (CN)_2(g)$$

Strong heating converts copper(II) oxide to copper(I) oxide and oxygen, and copper(II) sulphide to copper(I) sulphide and sulphur:

$$4CuO \rightarrow 2Cu_2O + O_2$$

$$2CuS \rightarrow Cu_2S + S$$

Note that both copper(I) oxide and copper(I) sulphide occur in nature.

The copper(I) salts of oxoacids are decomposed by water forming the copper(II) salt and copper:

$$2Cu^+ \rightarrow Cu^{2+} + Cu(s)$$

the presumably hydrated copper(I) ion being unstable and readily forming the familiar blue hydrated copper(II) ion. Copper(I) salts of the oxoacids are prepared in non-aqueous media. The simple copper(I) ion, Cu^+, is present in copper(I) sulphate, Cu_2SO_4, which is colourless. With water, copper(II) sulphate and copper are formed:

$$Cu_2SO_4 \rightarrow CuSO_4 + Cu(s)$$

Any salts of copper(I) may be stabilized towards water by the formation of suitable complexes: those of thiourea, $NH_2.CS.NH_2$ (abbreviated = tu), being easily prepared, e.g. $(Cu.3tu)_2SO_4.2H_2O$.

Copper(II) compounds are usually the starting point for the preparation of copper(I) compounds. In general, the conditions necessary for the formation of a copper(I) compound from copper(II) require the formation of either a very insoluble compound or a little-ionized complex ion. It may be noted that the tendency to complex formation is greater with copper(II).

The disproportionation $2Cu^+ \rightleftharpoons Cu + Cu^{2+}$

The relative stabilities of the Cu(I) and Cu(II) states in aqueous solution are reflected in a comparison of the standard electrode (redox) potentials, E:

$$\tfrac{1}{2}Cu^{2+} + e^- \rightleftharpoons \tfrac{1}{2}Cu \qquad +0.337V$$

$$Cu^+ + e^- \rightleftharpoons Cu \qquad +0.520V$$

$$Cu^{2+} + e^- \rightleftharpoons Cu^+ \qquad +0.153V$$

and by eliminating e^- in the last two equations,

$$2Cu^+ \rightleftharpoons Cu + Cu^{2+} \qquad +0.367V$$

From p. 120 we see

$$E = E_{298}^{\ominus} + \frac{RT}{nF} \ln \frac{\text{[Oxidized form]}}{\text{[Reduced form]}}$$

For $Cu^+ + e^- \rightleftharpoons Cu$, $n = 1$ and so

$$E/V = +0.520 + \frac{0.0591}{1} \log \left(\frac{m_{Cu^+}/\text{mol dm}^{-3}}{m_{Cu}/\text{mol dm}^{-3}} \right)$$

in which m_{Cu^+}/m_{Cu} is the ratio of the respective concentrations/mol dm^{-3} or here gdm^{-3}.

For $Cu^{2+} + e^- \rightleftharpoons Cu^+$, $n = 1$ and similarly

$$E/V = +0.153 + \frac{0.0591}{1} \log \left(\frac{m_{Cu^{2+}}/\text{mol dm}^{-3}}{m_{Cu^+}/\text{mol dm}^{-3}} \right)$$

At equilibrium, for $2Cu^+ \rightleftharpoons Cu + Cu^{2+}$ the redox potentials of both component systems must be equal,

$$\therefore +0.520 + 0.0591 \log \left(\frac{m_{Cu^+}}{m_{Cu}} \right)$$

$$= +0.153 + 0.0591 \log \left(\frac{m_{Cu^{2+}}}{m_{Cu^+}} \right)$$

Rearranging,

$$\log \left(\frac{m_{Cu^{2+}}.m_{Cu}}{m^2_{Cu^+}} \right) = 6.21$$

But the expression in brackets is the equilibrium constant for the reaction

$$2Cu^+ \rightleftharpoons Cu + Cu^{2+}$$

$$K_c = \frac{m_{Cu^{2+}}.m_{Cu}}{m^2_{Cu^+}} = 1.6 \times 10^6 \text{ (no units)}$$

This huge value shows that the equilibrium lies far to the right so that in aqueous solution Cu^+ is unstable and readily disproportionates. This is unexpected. The d^{10} configuration has a high exchange energy and is expected to be very stable. However, Cu^+ is large compared with Cu^{2+} and with its single charge has a lower density of charge, and is less attractive to lone pairs of electrons on donor molecules, e.g. water, ammonia and ions, cyanide, etc. The hydrated Cu^+ ion is, of course, unknown but it is interesting to note that in the complex ammines and cyanides Cu^+ has a co-ordination number of 2 instead of 4, as with Cu^{2+}: $Cu(NH_3)_2^+$ and $Cu(CN)_2^-$.

Cu^{II}: copper(II) oxide and copper(II) salts (formerly cupric compounds)

Copper(II)oxide is an almost black basic oxide with a macromolecular covalent structure. It may be prepared by the usual methods: heating the nitrate, a basic carbonate or the hydroxide:

$$2Cu(NO_3)_2 \rightarrow 2CuO + 4NO_2 + O_2$$

$$CuCO_3.Cu(OH)_2 \rightarrow 2CuO + CO_2 + H_2O$$

$$Cu(OH)_2 \rightarrow CuO + H_2O$$

With acids, copper(II) salts are formed:

$$CuO + 2H^+ \rightarrow Cu^{2+} + H_2O$$

Copper(II) oxide is reduced to the metal by heating in hydrogen or carbon monoxide, or mixed with charcoal,

$$CuO + H_2 \rightarrow Cu + H_2O$$

$$CuO + CO \rightarrow Cu + CO_2$$

$$CuO + C \rightarrow Cu + CO$$

Copper(II) hydroxide is thrown down as a gelatinous light blue precipitate on addition of caustic alkali to a solution of a copper(II) salt, and may be dried to the formula $Cu(OH)_2$. Slight amphoteric behaviour has been detected but it is virtually completely basic. Boiling results in the formation of hydrated oxide, when the precipitate becomes black. Aqueous ammonia similarly precipitates basic salts, blue or greenish-blue in colour, which dissolve in excess to form the tetramminecopper(II) salt, sometimes known as the cuprammonium salt:

$$NH_3 + H_2O \rightleftharpoons NH_4^+ + OH^-$$

$$2Cu^{2+} + SO_4^{2-} + 2OH^- \rightarrow$$
$$CuSO_4.Cu(OH)_2(s)$$

$$CuSO_4.Cu(OH)_2(s) + 8NH_3 \rightarrow$$
$$2Cu(NH_3)_4^{2+} + SO_4^{2-} + 2OH^-$$

Tetramminecopper(II) solutions may be used as a solvent for cellulose in the production of rayon.

The normal carbonate is unknown, both sodium carbonate and hydrogencarbonate solutions precipitating basic carbonates from copper(II) solutions. Certain basic carbonates occur as minerals: the green malachite, $Cu(OH)_2.CuCO_3$, and the blue azurite, $Cu(OH)_2.2CuCO_3$, are examples. Copper(II) nitrate, $Cu(NO_3)_2.3H_2O$, forms dark blue deliquescent crystals and is prepared by the normal methods: its thermal decomposition has been mentioned. Copper(II) ethanoate (acetate), $(CH_3COO)_2Cu.H_2O$, forms dark greenish-blue crystals. Basic copper(II) ethanoate, formed by the action of vinegar on sheet copper, has the approximate composition: $(CH_3COO)_2Cu.Cu(OH)_2$, and is the important green pigment, verdigris. Paris Green, no longer used as a pigment but with horticultural applications, because of its poisonous nature, is an insoluble double salt of copper(II) ethanoate and arsenite: $(CH_3COO)_2Cu.3Cu(AsO_2)_2$.

Cu^{II}: copper(II) sulphate, ammonium copper(II) sulphate and tetramminecopper(II) sulphate

Copper(II) sulphate is the most important copper salt. White when anhydrous, it forms a pentahydrate, $CuSO_4.5H_2O$, of a characteristic blue colour. Lower hydrates, $CuSO_4.3H_2O$ and $CuSO_4.H_2O$, may be prepared, and a heptahydrate, $CuSO_4.7H_2O$, isomorphous with $FeSO_4.7H_2O$, may also be isolated. The pentahydrate is known as blue vitriol. On heating, the pentahydrate forms the monohydrate at about 110°C, becoming anhydrous at about 250°C. Stronger heating yields a basic sulphate, and eventually above 700°C, copper(II) oxide. The formation of the blue pentahydrate is utilized as a sensitive test for water; the reaction is also exothermic. The crystal structure of blue vitriol is complex, but it is considered that each copper(II) ion has four molecules of water associated with it: $Cu(H_2O)_4^{2+}$ forming the tetraquocopper(II) ion although in aqueous solution the hexa-aquo ion is assumed, of somewhat irregular structure, $Cu(H_2O)_6^{2+}$.

Copper(II) sulphate may be prepared by the usual laboratory methods. Commercially, scrap copper is sprayed with hot dilute sulphuric acid in a current of air:

$$2Cu + 4H^+ + O_2 \rightarrow 2Cu^{2+} + 2H_2O$$

From the resulting liquor, the pentahydrate may be crystallized. Copper(II) sulphate has fungicidal properties and with milk of lime is used as Bordeaux mixture for spraying vines, etc. Other applications of copper(II) sulphate include timber preservation, electroplating, the production of pigments containing the basic carbonates, and in calico printing.

When the required proportions of copper(II) sulphate and ammonium sulphate are dissolved in hot water, a double salt, ammonium copper(II) sulphate, $CuSO_4.(NH_4)_2SO_4.6H_2O$, which is pale blue in colour, crystallizes. Alternatively, half of a given volume of sulphuric acid may be neutralized with ammonia and mixed with the remainder which is then neutralized with copper(II) oxide or basic copper(II) carbonate and filtered. The indicator, used during the neutralization of sulphuric acid by ammonia, is removed by boiling the solution with addition of animal charcoal. The indicator is adsorbed. On filtration it is removed with the animal charcoal. Ammonium copper(II) sulphate contains the simple (hydrated) copper(II) ion.

The complex tetramminecopper(II) sulphate, or

cuprammonium sulphate, containing the complex $Cu(NH_3)_4^{2+}$, is prepared by addition of excess 0.880 aqueous ammonia to copper(II) sulphate solution. Initially, a light bluish-green basic sulphate is precipitated and dissolves to give a deep blue solution. Addition of alcohol causes the precipitation of dark blue crystals which must be filtered rapidly at the pump and dried in a desiccator over quicklime. Tetramminecopper(II) sulphate monohydrate, $[Cu(NH_3)_4]SO_4.H_2O$, is a brilliant dark indigo-blue crystalline solid; the formula corresponds to that of blue vitriol, $[Cu(OH_2)_4]SO_4.H_2O$.

Passing dry ammonia over anhydrous copper(II) sulphate leads to the formation of a dark indigo-blue ammine, with the evolution of heat. On gentle heating, the ammine evolves torrents of ammonia, eventually leaving the white anhydrous copper(II) sulphate. Ammonia is absorbed until the pentammine, $CuSO_4.5NH_3$, is formed. This decomposes on warming to form the tetrammine, $CuSO_4.4NH_3$, the anhydrous form of the complex prepared above. Further heating expels ammonia.

Cu^{II}: the halides

Copper(II) *fluoride* is a white crystalline solid, probably of ionic structure, formed by passing hydrogen fluoride over heated copper(II) oxide. Both anhydrous copper(II) chloride and copper(II) bromide form flat, covalent macromolecular chains in which adjacent copper atoms are linked by two separate halogen atoms:

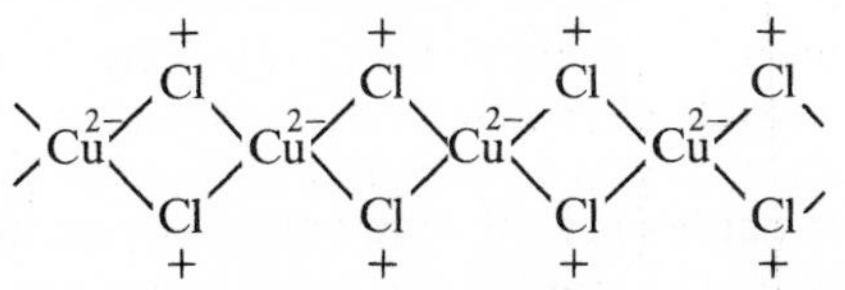

All three halides form hydrates. Addition of potassium iodide to a solution of copper(II) ions, results in the precipitation of white copper(I) iodide and the liberation of iodine:

$$2Cu^{2+} + 4I^- \rightarrow 2CuI(s) + I_2(s)$$

Copper(II) *chloride* may be prepared by reacting copper(II) oxide or carbonate with hydrochloric acid and evaporation to crystallization. With care, blue crystals of the dihydrate, $CuCl_2.2H_2O$, which appear green when moist, separate. Further heating, preferably in a current of dry hydrogen chloride, yields dark brown anhydrous copper(II) chloride. Alternatively, copper may be heated in

chlorine to give the anhydrous compound. It is deliquescent. The structure of the anhydrous compound has been given; that of the dihydrate is also covalent:

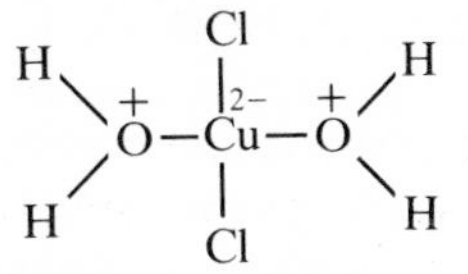

Copper(II) chloride is very soluble in water, 76.0 g dissolving in 100 g water at 25°C.

Concentrated solutions of copper(II) chloride are dark green-brown in colour, while dilute solutions have the blue colour associated with the hydrated copper(II) ion. Addition of alkali-metal chlorides or passing in hydrogen chloride changes the colour to brownish-yellow. The phenomenon is linked with the formation of complexes and differing degrees of hydration and has probably not been fully elucidated. The blue colour is due to $Cu(H_2O)_6^{2+}$ (in solution) and the dark colour to complex anions $CuCl_3^-$ and $CuCl_4^{2-}$, mixed with unionized molecules summed up in an almost wholly fictitious equation:

$$2CuCl_2 + 6H_2O \rightarrow Cu(H_2O)_6^{2+} + CuCl_4^{2-}$$

Various complexes may be isolated, including $Li(CuCl_3).2H_2O$ and $K(CuCl_3)$ in addition to the acids $HCuCl_3.3H_2O$ and $H_2CuCl_4.5H_2O$, all of which form dark red crystals. The anion is an extended structure

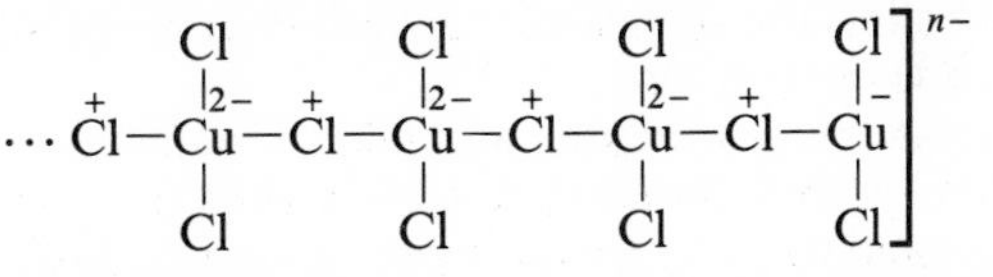

$$(CuCl_3)_n{}^{n-}$$

called the *catena* μ†-chloro-dichlorocuprate(II) ion. In addition, potassium chloride and copper(II) chloride will crystallize as the double salt, $2KCl.CuCl_2.2H_2O$.

At red heat, copper(II) chloride loses chlorine to form copper(I) chloride:

$$2CuCl_2 \rightarrow 2CuCl + Cl_2$$

Copper(II) chloride was used as a catalyst in the oxidation of hydrogen chloride to chlorine by air (Deacon's process).

† μ indicates that a bridging group follows

Qualitative tests for copper(II) salts

Hydrogen sulphide in neutral or acid solution precipitates black copper(II) sulphide, CuS, which is insoluble in hot dilute sulphuric acid but readily soluble in hot nitric acid with deposition of sulphur.

Caustic alkali precipitates gelatinous light blue copper(II) hydroxide, and ammonia, the green-blue basic salt from solutions containing copper(II) ions. The precipitates dissolve in excess ammonia to form the deep blue tetramminecopper(II) ion. Hydrogen sulphide will decompose this ammine, precipitating black copper(II) sulphide.

Potassium cyanide solution precipitates yellow copper(II) cyanide, $Cu(CN)_2$, from aqueous copper salts. The precipitate rapidly turns white, forming copper(I) cyanide, CuCN, and liberating cyanogen, $(CN)_2$ (*extemely poisonous*):

$$2Cu(CN)_2 \rightarrow 2CuCN + (CN)_2(g)$$

With excess potassium cyanide a colourless solution containing complex potassium tetracyanocuprate(I), $K_3[Cu(CN)_4]$, is produced:

$$CuCN + 3CN^- \rightarrow Cu(CN)_4^{3-}$$

The complex tetracyanocuprate(I) also results from addition of potassium cyanide solution to a tetramminecopper(II) solution. The complex ammine is sufficiently dissociated to allow this, the dark blue colour being discharged as the reaction proceeds. The complex cyanide is very stable and so little dissociated that hydrogen sulphide will not precipitate a sulphide from it. This reaction may be used to allow the detection of cadmium in the presence of copper(II) ions.

The yellow cadmium sulphide is effectively masked by black copper(II) sulphide when they are precipitated by the action of hydrogen sulphide on a solution containing their ammines, $Cu(NH_3)_4^{2+}$ and $Cd(NH_3)_4^{2+}$, both of oxidation state +2. However, addition of potassium cyanide solution produces the complex cyanide ions: $Cu(CN)_4^{3-}$ (valency change) and $Cd(CN)_4^{2-}$ (no valency change) now of respective oxidation states +1 and +2. The latter is sufficiently dissociated to allow precipitation of cadmium sulphide upon passing hydrogen sulphide.

Potassium iodide solution reduces a solution of a copper(II) salt to a precipitate of copper(I) iodide, which is white, and brownish-yellow iodine.

Potassium hexacyanoferrate(II) solution forms a chocolate-brown precipitate of copper(II) hexacyanoferrate(II), $Cu_2[Fe(CN)_6]$, from copper(II) salts in neutral solution, or in the presence of ethanoic (acetic) acid but not of a strongly ionized acid.

Copper may be obtained from solutions of copper(II) salts by displacement with metals higher in the electrochemical series (e.g. iron) or as red scales by blow-pipe reduction on a charcoal block of a mixture of a copper salt and fusion mixture. A greenish coloration appears in the flame test, and the borax bead obtained in an oxidizing flame is green when hot and blue when cold.

Cu^I: copper(I) oxide and copper(I) salts (formerly cuprous compounds)

The limitation on the existence of certain copper(I) oxosalts in the presence of water has been mentioned.

Copper(I) oxide occurs as the mineral cuprite. It appears as a yellowish-orange, brick-red or dark brown insoluble powder according to the conditions of its preparation. Copper(I) oxide has a covalent macromolecular structure: it is prepared by partial reduction of copper(II) ions and precipitation in alkaline solution. Copper(II) sulphate solution is run into a boiling aqueous solution containing the reducing agent, sodium sulphite, and a source of chloride ions, sodium chloride. A colourless solution containing a complex chloride of copper(I) results:

$$2Cu^{2+} + SO_3^{2-} + H_2O \rightarrow 2Cu^+ + SO_4^{2-} + 2H^+$$

$$Cu^+ + 3Cl^- \rightleftharpoons CuCl_3^{2-}$$

The solution is slowly added to a boiling solution of borax when copper(I) oxide is precipitated. As the reaction proceeds, the finely divided yellow copper(I) oxide becomes brick-red and crystalline. It is filtered at the pump, washed in water and propanone, being finally dried in warm air. Borax solution is slightly alkaline because of hydrolysis:

$$2CuCl_3^{2-} + 2OH^- \rightarrow Cu_2O(s) + H_2O + 6Cl^-$$

Alternatively, sodium hydroxide solution is added to a boiling solution containing copper(II) sulphate and glucose (an aldehyde) when a blue, green and finally red precipitate of copper(I) oxide appears.

Copper(I) oxide is used in the manufacture of rectifiers and the production of red glass. On heating in hydrogen or carbon monoxide, or mixed with charcoal, it is reduced to copper. It is a basic oxide but unusual secondary reactions occur with the mineral acids. Dilute sulphuric acid forms copper(II) sulphate solution leaving a residue of metallic copper, the copper(I) ion being unstable in the presence of water:

$$Cu_2O + 2H^+ \rightarrow Cu(s) + Cu^{2+} + H_2O$$

Dilute nitric acid forms copper(II) nitrate and oxides of nitrogen, due presumably to a further reaction between the acid and precipitated copper. Hydrochloric acid does not liberate copper but the covalent copper(I) chloride forms a complex anion:

$$Cu_2O + 2H^+ + 6Cl^- \rightarrow$$
$$2CuCl_3^{2-} \text{ (and } CuCl_4^{3-}) + H_2O$$

Copper(I) oxide dissolves in a solution of ammonia and ammonium sulphate to form a colourless solution of $[Cu(NH_3)_2^+]_2SO_4^{2-}$ which is oxidized to the blue tetramminecopper(II) complex by air. The notable copper(I) ammine cation is $Cu(NH_3)_2^+$, which is analogous to that formed by silver ions, $Ag(NH_3)_2^+$.

Copper(I) sulphate may be prepared under anhydrous conditions by the action of dimethyl sulphate on copper(I) oxide:

$$Cu_2O + (CH_3)_2SO_4 \rightarrow Cu_2SO_4 + (CH_3)_2O$$

the other product being methoxymethane (dimethyl ether). Copper(I) sulphate is a grey crystalline solid which disproportionates in water to copper(II) sulphate and copper:

$$Cu_2SO_4 + 6H_2O \rightarrow Cu(H_2O)_6^{2+} + SO_4^{2-} + Cu(s)$$

Thiourea, $NH_2.CS.NH_2$ ($= tu$), and its derivatives may be used for the preparation of copper(I) complexes of oxosalts which are water-soluble but do not disproportionate. Tri(thiourea)-copper(I) sulphate, $(Cu.3tu)_2SO_4.2H_2O$, may be prepared from copper(II) sulphate and thiourea, the latter acting as a reducing agent as well as forming a complex. Copper(I) nitrate is only known in complexes, of which penta(thiourea)-dicopper(I) nitrate, $(Cu_25tu)(NO_3)_2.3H_2O$, is an example and is prepared from thiourea and copper(II) nitrate solution.

Cu^I: the halides

The existence of pure copper(I) fluoride is considered doubtful. Copper(I) chloride is a colourless solid (m.p. 430°C) which is oxidized rapidly in moist air to the green basic chloride, atacamite, $CuCl_2.3Cu(OH)_2$. Copper(I) bromide is a pale greenish-yellow solid (m.p. 483°C) very similar to the chloride, but prepared by heating copper(II) bromide strongly in air,

$$2CuBr_2 \rightarrow 2CuBr + Br_2$$

Copper(II) chloride also dissociates in this way. White copper(I) iodide is precipitated on mixing solutions of potassium iodide and copper(II) sulphate,

$$2Cu^{2+} + 4I^- \rightarrow 2CuI + I_2$$

Iodine, liberated simultaneously during the redox process, may be removed by reaction with sodium thiosulphate solution. In the vapour state, copper(I) chloride is dimerized even at 1700°C, copper(I) bromide at least to 1100°C but copper(I) iodide is not associated in the vapour; the molecular formulae in the vapour state are therefore $(CuCl)_2$, $(CuBr)_2$ and CuI respectively. The solid halides form the zinc blende structure in which each atom is covalently bound to four equidistant atoms of the other element spaced tetrahedrally; co-ionic bonding holds the structure together. The macromolecule formed is called an *inner complex compound*.

Copper(I) chloride is prepared by boiling a solution containing copper(II) chloride (usually copper sulphate and sodium chloride) with concentrated hydrochloric acid and copper turnings until the initial colour of greenish-blue lightens to straw-colour and clear. At this stage reduction to the copper(I) state has occurred with the formation of a complex chloride, tri- and tetrachlorocuprate(I):

$$CuCl_2 + Cu + 4Cl^- \rightarrow 2CuCl_3^{2-} \text{ (and } CuCl_4^{3-})$$

On pouring this solution into an excess of water, in which a little sodium disulphite has been dissolved to provide a non-oxidizing medium, a white precipitate of copper(I) chloride is formed:

$$CuCl_3^{2-} \rightleftharpoons CuCl + 2Cl^-$$

To avoid atmospheric oxidation, the precipitate is filtered at the pump taking care not to expose the solid to the air and washed repeatedly in glacial ethanoic acid. It is dried at 120°C in an air oven. Final washings may be with ethanol and acetone to facilitate drying. Alternatively, copper(II) chloride (as before) may be reduced by passing sulphur dioxide:

$$2Cu^{2+} + 2Cl^- + SO_2 + 2H_2O \rightarrow 2CuCl + SO_4^{2-} + 4H^+$$

Copper(I) chloride is insoluble in water. Soluble complexes are formed with concentrated hydrochloric acid and solutions of alkali metal chlorides, containing the anions: $CuCl_2^-$, $CuCl_3^{2-}$ and $CuCl_4^{3-}$. Aqueous ammonia forms the soluble ammine which is ionized: $[Cu(NH_3)_2^+]Cl^-$. Both of these colourless complexes may be used in gas analysis for the absorption of carbon monoxide, water being necessary to form the carbon monoxide complex, $[Cu(H_2O)(CO)Cl]_2$. Diammine-copper(I) chloride is oxidized in air, becoming blue in colour due to the formation of ammines of copper(II). Ethyne precipitates red copper(I) ethymide from diamminecopper(I) chloride,

$$2Cu(NH_3)_2Cl + C_2H_2 \rightarrow Cu_2C_2 + 2NH_4Cl$$

Copper(I) ethynide, which is explosive when dried, reacts with acid to liberate pure ethyne.

Copper(I) chloride is used as a catalyst in the Sandmeyer Reaction in Aromatic Organic Chemistry.

The volumetric (titrimetric) estimation of copper and its salts

Copper may be estimated electrolytically by deposition on a rotating platinum gauze cathode from ammoniacal solution. It may also be determined iodometrically as follows:

To a measured volume of a solution of a copper(II) salt, solid potassium iodide is added and the liberated iodine is titrated against a standard solution of sodium thiosulphate using starch as indicator:

$$2Cu^{2+} + 4I^- \rightarrow 2CuI + I_2$$

For satisfactory titrations, the acidity must be controlled to give a pH of 4.0–5.5. Free mineral acid is neutralized by addition of 0.880 aqueous ammonia drop by drop to render the solution deep blue and then dilute ethanoic acid added until the pale blue-green is restored, followed by an equal portion of the acid.

Copper(I) compounds may be determined volumetrically by oxidizing weighed samples with acidified iron(III) sulphate solution followed by titration of the liberated iron(II) sulphate against a standard oxidant:

$$Cu_2O + 2H^+ + 2Fe^{3+} \rightarrow 2Cu^{2+} + 2Fe^{2+} + H_2O$$

Complexes of Cu^I and Cu^{II}

The tendency to form complexes is stronger with copper(II) than copper(I). In quite different structures (complex but not in the complex ion sense)

associated with unusual electrical conductance, described in the following section, Cu^{3+} appears, as well as Cu^+ and Cu^{2+}.

Cu^I:

$$Cu(NH_3)_2^+$$
diamminecopper(I) ion

$$[Cu(NH_3)_2]_2SO_4 . H_2O$$
diamminecopper(I) sulphate monohydrate

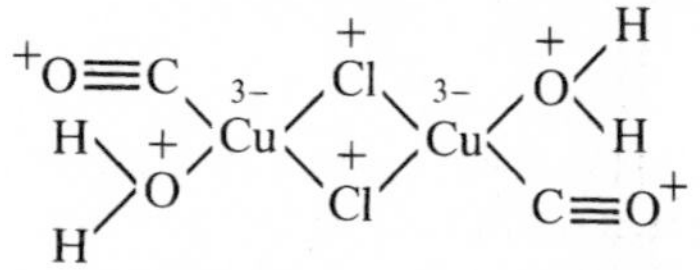

carbon monoxide complex of ammoniacal copper(I) chloride Cu 2, 8, 18, 8 (Kr)

$$Cu(CN)_2^- \qquad Cu(CN)_3^{2-} \qquad Cu(CN)_4^{3-}$$
complex cyanocuprate(I) ions

$$CuCl_2^- \qquad CuCl_3^{2-} \qquad CuCl_4^{3-}$$
complex chlorocuprate(I) ions

$$K_2[Cu(C_2H)_3]$$
potassium triethynylcuprate(I)

Cu^{II}:

$$[CuCl_2(CH_3NH_2)_2] \qquad\qquad Cu(H_2O)_6^{2+}$$
dichlorobis (methylamine)   hexa-aquo copper(II) ion
copper(II)   'copper(II) ion'

$$Cu(NH_3)_4^{2+}[Cu(NH_3)_4(H_2O)_2]^{2+}$$
tetramminecopper(II) ion
cuprammonium ion

$$CuCl_3^- \qquad CuCl_4^{2-}$$
tri- and tetra-chlorocuprate(II) ions

$$K^+(CuCl_3^-) \quad (K^+)_2(CuCl_4^{2-}).2H_2O$$
brown-red   blue
potassium trichloro- dipotassium tetrachloro-
cuprate(II) (p. 479) cuprate(II) dihydrate

$$[CuCl_2(CH_3NH_2)_2]$$
dichlorobis(methylamine) copper(II)

Superconductivity: $YBa_2Cu_3O_7$ (Cu^{2+} and Cu^{3+})

Superconductors are materials that, in theory, carry an electric current for ever without losing energy. Originally discovered in studies with mercury at very low temperatures, until recently all experiments were made on materials at temperatures below 77K (the boiling point of liquid nitrogen). Recently, ceramic oxides of the systems

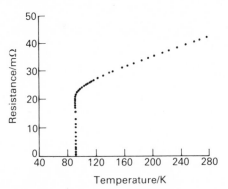

Fig. 24.6 The temperature dependence of the electrical resistance of a sample of $YBa_2Cu_3O_4$—a zero resistance state achieved at 90K (P. P. Edwards, M. R. Harrison and R. Jones, *Chem. Br.*, 1987, **23**, 962)

Ba—La—Cu—O and Y—Ba—Cu—O have been investigated, the latter reaching zero resistance at 90K (Fig. 24.6).

Two 'compounds' are of interest here, both being oxygen deficient versions of what is known as a perovskite structure, of ideal stoichiometry $YBa_2Cu_3O_9$ (Perovskite, $CaTiO_3$, gives its name to a pseudo-cubic unit cell structure in which the valency of the cations is unimportant: any pair of ions of radii appropriate for the coordination number and aggregate charge (oxidation state) of $+6$ appearing). These compounds are $YBa_2Cu_3O_7$ and $YBa_2Cu_3O_6$, the former with a layered structure (see below) being the superconductor and the latter, a semiconductor.

Such $1:2:3$ (Y : Ba : Cu) structures are made by sintering and quenching/annealing appropriate materials, to give a stoichiometric composition somewhere between the ideal formulae given in regard to their oxygen content. If the oxidation states are calculated for copper in each, assuming the normal oxidation states for the other elements (Y, $+3$; Ba, $+2$: O, -2)

Oxidation state for copper	$YBa_2Cu_3O_7$	$YBa_2Cu_3O_6$
	$+7/3$	$+5/3$
	Superconductor	*Semiconductor*

The structures are shown below in Fig. 24.7. Analysis suggests that the former has one Cu^{3+} and two Cu^{2+} ions (average oxidation number = 7/3) and the latter, one Cu^+ and two Cu^{2+} ions (average oxidation number = 5/3). The numbers alongside the symbols in what follows are coordinated numbers: local electrical neutrality requirements favour Cu^{3+} ions in the Cu(1) plane of the superconductor while Cu^{2+} ions are expected to dominate in the square pyramidal Cu(2) locality.

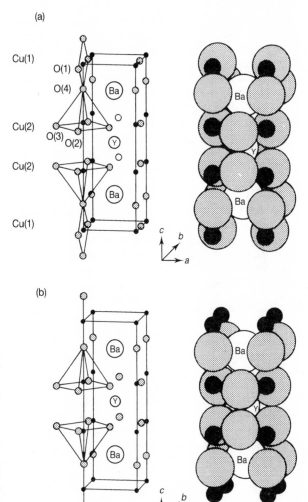

Fig. 24.7 Idealized structures of (a) $YBa_2Cu_3O_7$, a 90K superconductor, and (b) $YBa_2Cu_3O_6$, a semiconductor (Ibid)

There are many theories of superconductivity, the resurgence in the subject appearing in late 1986.

Zinc

Zinc is not a transition metal. The penultimate shell is complete both in the metal and in the ion, Zn^{2+}. Zinc is bivalent only and its ions are colourless. In compounds, zinc is, of course, in oxidation state $+2$.

Pure zinc is a bluish-white metal of moderate

484 Period 4: transition and associated elements

strength which has excellent resistance to atmospheric corrosion. It is widely used as a surface coating on iron and steel to prevent rusting. Over the range of temperature 110–150°C zinc is both malleable and ductile. In damp air zinc is quickly covered by a coherent layer of basic carbonate. At red heat, zinc burns vigorously in air to form the fibrous zinc oxide, ZnO, once called philosopher's wool. Its combustion in the oxy-coal gas flame is spectacular. Zinc is not attacked by cold water but decomposes steam at red heat,

$$Zn + H_2O \rightarrow ZnO + H_2$$

Although zinc is high in the electrochemical series, there is no action with cold water because of the high overpotential of hydrogen at a zinc surface and the protective layer of insoluble zinc hydroxide formed. Because of the hydrogen overpotential, pure zinc is only very slowly attacked by the dilute acids under non-oxidizing conditions. However, the reactivity is altered profoundly by the presence of specks of more noble impurities, such as copper, which allow the discharge of hydrogen from their surfaces with consequent solution of zinc:

$$Zn + 2H^+ \rightarrow Zn^{2+} + H_2$$

Copper(II) sulphate is frequently added to the dilute sulphuric acid used to dissolve zinc, numerous galvanic couples being formed between zinc metal and copper displaced by zinc from solution:

$$Cu^{2+} + Zn(s) \rightarrow Zn^{2+} + Cu(s)$$

Impure zinc reacts rapidly with hydrochloric acid.

Under oxidizing conditions, zinc reacts vigorously. Hot concentrated sulphuric acid forms much sulphur and sulphur dioxide while hydrogen sulphide may be given off with slightly diluted acid:

$$Zn + 2H_2SO_4 \rightarrow ZnSO_4 + 2H_2O + SO_2$$

$$3Zn + 4H_2SO_4 \rightarrow 3ZnSO_4 + 4H_2O + S$$

$$4Zn + 5H_2SO_4 \rightarrow 4ZnSO_4 + 4H_2O + H_2S$$

The products with nitric acid depend on the concentration. The gaseous product is mainly dinitrogen oxide (nitrous oxide) with fairly concentrated acid:

$$10H^+ + 2NO_3^- + 4Zn \rightarrow 4Zn^{2+} + 5H_2O + N_2O$$

while ammonia, and by neutralization, ammonium nitrate, is formed with dilute acid:

$$10H^+ + NO_3^- + 4Zn \rightarrow 4Zn^{2+} + 3H_2O + NH_4^+$$

Under alkaline conditions, nitrates are reduced by zinc to ammonia:

$$7OH^- + NO_3^- + 4Zn \rightarrow$$
$$4ZnO_2^{2-} + 2H_2O + NH_3$$

or

$$7OH^- + NO_3^- + 4Zn + 6H_2O \rightarrow$$
$$4Zn(OH)_4^{2-} + NH_3$$

Using potash, potassium zincate remains in solution. It is also formed with hydrogen when caustic potash, aqueous or fused, attacks zinc:

$$Zn + 2OH^- \rightarrow ZnO_2^{2-} + H_2$$

A mixture of finely powdered sulphur and zinc powder, not zinc dust because it contains oxide, burns furiously when ignited with a taper, to form a voluminous mass of the sulphide, ZnS, as cream flakes. Zinc burns in chlorine on heating to form the chloride ZnCl_2.

Zn^{II}: zinc(II) oxide and zinc(II) compounds

There is only one series of salts. Zinc oxide is a covalent amphoteric oxide. White when cold, on heating to above 250°C, the solid becomes lemon in colour. It may be prepared by the standard methods: heating the carbonate, nitrate, etc. Industrially it is made by burning zinc vapour in air. Zinc oxide is used as the pigment, zinc white, which does not darken in air (unlike lead white pigments), as a filler for rubber and for its antiseptic qualities in zinc ointment. At red heat, zinc oxide is reduced by carbon, as in the extraction:

$$ZnO + C \rightarrow Zn + CO$$

With acids, zinc salts are formed, while caustic alkalis form zincates:

$$ZnO(s) + 2H^+ \rightarrow Zn^{2+} + H_2O$$
$$ZnO(s) + 2OH^- \rightarrow ZnO_2^{2-} + H_2O$$

or

$$ZnO(s) + 2OH^- + H_2O \rightarrow Zn(OH)_4^{2-}$$

illustrating its amphoteric behaviour.

Addition of the calculated quantity of alkali to a solution of a zinc salt produces a gelatinous white precipitate of zinc hydroxide, Zn(OH)_2. The precipitate dissolves in excess caustic alkali to form the zincate, in aqueous ammonia to form ammines, Zn(NH_3)_6^{2+}, and in acids to form salts. Zinc hydroxide is decomposed by heat:

$$Zn(OH)_2 \rightarrow ZnO + H_2O$$

Zinc carbonate occurs naturally as calamine, and is used to soothe skin irritation as calamine lotion. Addition of sodium hydrogencarbonate solution to a zinc salt solution precipitates the white normal carbonate, while use of sodium carbonate precipitates a basic zinc carbonate, also white:

$$Zn^{2+} + 2HCO_3^- \rightarrow ZnCO_3(s) + CO_2 + H_2O$$

$$3Zn^{2+} + CO_3^{2-} + 4OH^- + 2H_2O \rightarrow$$
$$ZnCO_3.2Zn(OH)_2.2H_2O(s)$$

Zinc sulphate is formed by the standard methods: it forms colourless crystals of the heptahydrate, $ZnSO_4.7H_2O$, white vitriol, isomorphous with the corresponding hydrates of the vanadium(II), chromium(II), manganese(II), iron(II), nickel and cobalt sulphates, $[M(H_2O)_6]SO_4.H_2O$. Double sulphates of the type $K_2SO_4.ZnSO_4.6H_2O$ are formed. Zinc sulphide, occurring naturally as zinc blende and wurtzite, may be synthesized from the elements or precipitated by passing hydrogen sulphide into a solution of a zinc salt, but not in the presence of a mineral acid: it is white. The commercial production and uses of lithopone, a mixture of zinc sulphide and barium sulphate, have been described (p. 299).

Zn^{II}: zinc (II) halides

Anhydrous zinc chloride may be prepared by heating zinc in chlorine or hydrogen chloride, zinc oxide to 700°C in chlorine when oxygen is displaced, or the hydrate in hydrogen chloride. The action of hydrochloric acid on zinc gives a solution which becomes syrupy on evaporation and sets on cooling to a colourless solid. Evaporation to dryness causes hydrolysis to a basic zinc chloride, $ZnCl_2.nZnO$, or $Zn(OH)Cl$, formerly called zinc oxychloride, now zinc oxide chloride. Anhydrous zinc chloride is a colourless deliquescent solid which melts at 275°C and is very soluble in water, 420 g dissolving in 100 g water at 25°C. It is also soluble in alcohols, ethers, esters and ketones. Mixed with zinc oxide, zinc chloride is used as a dental cement, setting to a hard mass of zinc oxide chloride, and acting as a preservative. A solution of zinc chloride, named 'killed spirits' because of its preparation by addition of zinc to hydrochloric acid, is used as a flux in soldering. In Organic Chemistry, zinc chloride is used in condensation reactions as a catalyst, and for dehydrations because of its affinity for water.

The other halides are similar, except that the anhydrous fluoride may be prepared by evaporation of the aqueous solution.

Qualitative tests for zinc salts

Addition of caustic alkali to a solution containing zinc ions precipitates the white gelatinous hydroxide, which dissolves in excess alkali to form the zincate. Aqueous ammonia also precipitates the hydroxide, which dissolves in excess to give ammines. In the presence of ammonium chloride, suppression of the hydroxide ion concentration in aqueous ammonia prevents precipitation.

From neutral or alkaline solutions hydrogen sulphide precipitates white (often discoloured) zinc sulphide, insoluble in weakly ionized ethanoic (acetic) acid but readily soluble in mineral acids.

From aqueous solutions of zinc salts, a solution containing disodium hydrogenphosphate and ammonium chloride precipitates white zinc ammonium phosphate,

$$HPO_4^{2-} \rightleftharpoons H^+ + PO_4^{3-}$$

$$Zn^{2+} + NH_4^+ + PO_4^{3-} \rightarrow ZnNH_4PO_4(s)$$

White zinc hexacyanoferrate(II) is precipitated by addition of potassium hexacyanoferrate(II) solution to aqueous zinc salts:

$$2Zn^{2+} + Fe(CN)_6^{4-} \rightarrow Zn_2[Fe(CN)_6](s)$$

Excess of potassium hexacyanoferrate(II) forms the double salt, zinc potassium hexacyanoferrate(II),

$$Zn_3K_2[Fe(CN)_6]_2.$$

When zinc compounds, mixed with fusion mixture, made from equal amounts of sodium carbonate and potassium carbonate, are heated on the charcoal block, a yellow incrustation of the oxide, which becomes white as it cools, is formed. Addition of a little cobalt nitrate solution and further heating forms a green mass, Rinmann's Green, a solid solution of cobalt oxide in zinc oxide.

Complexes of Zn^{II}

Of the complexes formed, only the ammines and cyanides need be mentioned. Complex cyanides $K[Zn(CN)_3]$ and $K_2[Zn(CN)_4]$ are formed, containing the anions, $Zn(CN)_3^-$ and $Zn(CN)_4^{2-}$, the tricyanozincate(II) and tetracyanozincate(II) ions, respectively . The ammines may contain up to six molecules of ammonia for each zinc cation, $Zn(NH_3)_6^{2+}$. Passage of gaseous ammonia into a hot saturated solution of zinc chloride in water until the precipitate formed initially dissolves gives the tetrammine, which crystallizes on cooling as $[Zn(NH_3)_4]Cl_2$. On evaporation and cooling, the diammine is obtained: $[Zn(NH_3)_2]Cl_2$. Passing dry ammonia over zinc chloride at $-15°C$ yields a voluminous mass of the hexammine, $[Zn(NH_3)_6]Cl_2$. The zincate ion, ZnO_2^{2-} or $Zn(OH)_4^{2-}$, may be regarded as complex.

25

Period 5: silver

The element, its character and uses

Silver's electronic configuration is 2, 8, 18, 18, 1. The penultimate shell of electrons being complete, silver is not a transition metal. The principal oxidation state is +1, and it moves to oxidation state +2 only very rarely, and to +3 extremely so. However, the metal may be regarded as being associated with the transition elements of Period 5, although the $4d$ electrons remain inviolate in the principal compounds. Silver may also be studied in relation to copper and gold, the so-called 'coinage' group of metals, placed in the B subgroup of Group I in the Mendeléeff Table. All show univalency, the only resemblance to the alkali metals, the typical elements of Group I. However, it is only in their physical characteristics that copper, silver and gold are similar. The chemistry of silver is relatively simple, because it is virtually univalent only, while copper exhibits valencies of one and two, and gold of one and three. The only simple compound in which silver shows a valency of two is silver(II) fluoride, AgF_2.

Some characteristics of copper, silver and gold are listed in Table 25.1. Certain assumptions are made in calculating some of the data, e.g. metallic radii depend on the coordination number used, in this case 12.

Silver is a lustrous white metal of distinctive appearance, malleable and ductile, and the best conductor of heat and electricity among the common elements. It is commercially valuable. Its use in currency has been discontinued in the UK (see p. 175). Besides ornamentation, silver finds use in electronic engineering. Silver is tarnished in air by sulphurous compounds which form a brown or

Table 25.1 Some physical characteristics of copper, silver and gold

	Cu	Ag	Au
Atomic number	29	47	79
Electron shells (energy levels) filled	2, 8, 18, 1	2, 8, 18, 18, 1	2, 8, 18, 32, 18, 1
[Core], outer electrons	$[Ar](3d)^{10}(4s)^1$	$[Kr](4d)^{10}(5s)^1$	$[Xe](5d)^{10}(6s)^1$
Relative atomic mass	63.546	107.868	196.967
1st Ionization energy/kJ mol^{-1}	745	730	890
Oxidation states*	+1, +2	+1, (+2), ((+3))	+1, +3
Principal valency states*	+1, +2	+1	+1, +3
Metallic radius/nm	0.128	0.144	0.146
Atomic radius/nm	0.135	0.153	0.150
Ionic radius (M^+)/nm	0.096	0.126	0.137
Standard electrode potential, (H scale: 298 K)/V	$Cu^+; Cu$ +0.52 $Cu^{2+}; Cu$ +0.337	$Ag^+; Ag$ +0.799	—
Electronegativity	1.75	1.4	1.4
Atomic volume**/cm^{3}mol^{-1}	7.12	10.3	10.2
Boiling-point/°C	2567	2152	2360
Melting-point/°C	1083	960.5	1063
Density/g cm^{-3}	8.93	10.5	19.3

*Oxidation states assigned by rules (p. 124). Valency states not always straightforward, e.g. in complexes, macromolecules, new compounds
**Traditional term. It is the volume of one mole of atoms as solid element

black sulphide surface layer. It is not attacked by oxygen although molten silver will absorb about twenty times its own volume of oxygen from the atmosphere, but on cooling, the gas is expelled ('spitting'). Silver reacts slowly with the halogens at red heat to form halides (e.g. AgCl), and with sulphur to form silver sulphide (Ag_2S).

Silver is very low in the metals section of the electrochemical series. It is not attacked by water or steam, or by dilute non-oxidizing acids. Hot concentrated sulphuric acid forms silver sulphate and sulphur dioxide:

$$2Ag + 2H_2SO_4 \rightarrow Ag_2SO_4 + SO_2 + 2H_2O$$

Silver reacts with nitric acid in various concentrations giving oxides of nitrogen and silver nitrate:

$$3Ag + 4H^+ + NO_3^- \rightarrow 3Ag^+ + 2H_2O + NO$$

Silver does not dissolve in aqua regia, because silver chloride is insoluble. Silver does not react with alkali, either aqueous or fused, or with ammonia. Silver dishes may be used in the laboratory for molten alkali. In the presence of oxygen, silver will react with potassium cyanide solution:

$$4Ag + 8CN^- + O_2 + 2H_2O \rightarrow$$
$$4Ag(CN)_2^- + 4OH^-$$

Silver may be displaced from silver nitrate solution by metals higher in the electrochemical series:

$$Cu(s) + 2Ag^+ \rightarrow Cu^{2+} + 2Ag(s)$$

Silver may be prepared by electrochemical displacement, electrolysis or reduction of ammoniacal solutions. In electroplating, a solution of potassium dicyanoargentate(I) (argentocyanide), $K[Ag(CN)_2]$, ensures a fine deposition of silver by maintaining a steady low concentration of silver ions through dissociation of the $Ag(CN)_2^-$ complex ion. Silver may be purified by fusion with sodium carbonate in a crucible lined with bone ash or calcined dolomite.

Silver salts are sensitive to light. Few are soluble; these include the nitrate, chlorate (chlorate(V)), perchlorate (chlorate(VII)) and fluoride, all of which contain highly electronegative anions. The fluoride forms a hydrate, $AgF.2H_2O$. Silver salts may be brought into solution by formation of complex ions, such as an ammine, $Ag(NH_3)_2^+$, or a cyanide, $Ag(CN)_2^-$. Although the silver ion is colourless, it will be observed that the salts formed by silver are frequently coloured and this colour is not due to the anion involved. The formation of a coloured compound has been cited as indicating the

formation of covalent bonds but the exact connection has not been made clear. The silver ion has the electronic configuration 2, 8, 18, 18. It readily gains an electron in being reduced to the metal and also readily passes into the covalent state.

Metallic silver will catalyse oxidation with oxygen or air in Organic Chemistry, usually being heated: methanol is oxidized to methanal (formaldehyde),

$$2CH_3OH + O_2 \rightarrow 2H.CHO + 2H_2O$$

and ethene (ethylene) to ethene (ethylene) oxide,

$$2\,\underset{CH_2}{\overset{CH_2}{\|}} + O_2 \longrightarrow 2\,\underset{CH_2}{\overset{CH_2}{|}}\!\!\diagdown\!\!\underset{\diagup}{O}$$

both products being important intermediates in the chemical industry.

Ag^I: silver(I) compounds (formerly sometimes called argentous compounds)

Ag^I: silver oxide, Ag_2O

Silver oxide is a basic oxide with a covalent macromolecular structure. In the absence of a hydroxide of silver, it is precipitated as a dark brown, almost black, solid by addition of caustic alkali to silver nitrate solution,

$$2Ag^+ + 2OH^- \rightarrow Ag_2O(s) + H_2O$$

After filtration, silver oxide may be washed and dried at about 80°C although it is extremely difficult to achieve complete dehydration.

A suspension of silver oxide in water is slightly alkaline, reacting blue to litmus. This basic tendency is shown by moist silver oxide when absorbing carbon dioxide from the air to form the carbonate,

$$Ag_2O + CO_2 \rightarrow Ag_2CO_3$$

Because of the insolubility of silver chloride, bromide and iodide, it is sometimes convenient to use moist silver oxide for the mild hydrolysis of organic halides:

$$CH_3.CH_2.I + \text{'AgOH'}(s) \rightarrow$$
$$\text{iodoethane}$$
$$CH_3.CH_2.OH + AgI(s)$$
$$\text{ethanol}$$

On the other hand, dry silver oxide yields an ether:

$$2CH_3.CH_2.I + Ag_2O \rightarrow$$
$$CH_3.CH_2.O.CH_2.CH_3 + 2AgI$$
ethoxyethane (diethyl ether)

Silver oxide is decomposed on heating to 160–250°C:

$$2Ag_2O \rightarrow 4Ag + O_2$$

It is reduced by hydrogen below 100°C,

$$Ag_2O + H_2 \rightarrow 2Ag + H_2O$$

and is reduced to the metal in reacting with hydrogen peroxide,

$$Ag_2O + H_2O_2 \rightarrow 2Ag + H_2O + O_2$$

Silver oxide may be precipitated from silver nitrate solution by aqueous ammonia, but the precipitate dissolves in excess to form the diamminesilver(I) cation, $Ag(NH_3)_2^+$. This solution may be used to detect reducing agents but should be freshly prepared, because on standing in air the very explosive fulminating silver, usually described as silver nitride, Ag_3N, is precipitated. In an alternative preparation, silver oxide is precipitated by the addition of dilute caustic soda to silver nitrate solution and the precipitate of silver oxide just dissolved by dropwise addition of aqueous ammonia. The solution is called ammoniacal silver nitrate or ammoniacal silver oxide. By slow reduction, a silver mirror may be deposited on the walls of the containing vessel, while rapid reduction gives a coarse precipitate of silver. The reaction vessel must be clean to form a good mirror. Ethanal (acetaldehyde) is oxidized to ammonium ethanoate:

$$2Ag(NH_3)_2^+ + CH_3CHO + H_2O \rightarrow$$
$$2Ag + CH_3COO^- + 3NH_4^+ + NH_3$$

Aldehydes, methanoates (formates), 2-hydroxy-propanoates (lactates) and 2,3-dihydroxybutane-dioates (tartrates) and certain amines are among the organic reducing agents which cause reduction to silver. The carbohydrate, glucose, is also an aldehyde and is often used in the laboratory for the production of a silver mirror. Commercially, silvering is hardened by use of tin(II) chloride.

The passage of ethyne through ammoniacal silver oxide precipitates silver ethynide (acetylide):

$$2Ag(NH_3)_2^+ + C_2H_2 \rightarrow Ag_2C_2 + 2NH_3 + 2NH_4^+$$

This is a dangerously explosive compound when dry.

Ag^I: the silver halides

Silver fluoride differs from the other halides and is unusual in forming hydrates. The anhydrous salt, formed by union of the elements, is deep yellow in colour, ionic (with the rock salt lattice) and forming two colourless hydrates, $AgF.4H_2O$ stable below 18.7°C and $AgF.2H_2O$ stable up to 39.5°C. Silver fluoride is very soluble in water (179 g saturating 100 g water at 25°C) and very deliquescent.

The other silver halides (AgCl, AgBr, AgI) are very insoluble in water and made by mixing solutions containing the appropriate ions. Curdy and partly colloidal precipitates are formed. Silver chloride is white, 3.0×10^{-3} g dissolving in 100 g water at 25°C, and ionic in structure. Silver bromide is cream, 5.5×10^{-4} g dissolving in 100 g water at 25°C and also ionic. Silver iodide is pale yellow, and has a complex covalent macromolecular structure with 5.6×10^{-6} g dissolving in 100 g water at 25°C.

Although silver chloride is precipitated as white curds, on heating it fuses at 449°C and solidifies on cooling to a translucent horny mass. Horn-silver is a naturally occurring form of silver chloride. Silver chloride is sensitive to light, and rapidly turns violet after precipitation in sunlight. It is used in printing-out papers in photography. While very insoluble in water, silver chloride will dissolve in reagents which form silver complexes: it is readily soluble in aqueous ammonia to give the diamminesilver(I) cation,

$$AgCl + 2NH_3 \rightleftharpoons [Ag(NH_3)_2]^+ + Cl^-$$

while potassium cyanide solution forms the dicyanoargentate(I) anion, $[Ag(CN)_2]^-$,

$$AgCl + 2CN^- \rightleftharpoons [Ag(CN)_2]^- + Cl^-$$

and aqueous sodium thiosulphate, a series of thiosulphatoargentate(I) anions,

$$[Ag(S_2O_3)]^-, \quad [Ag(S_2O_3)_2]^{3-}, \quad [Ag(S_2O_3)_3]^{5-}$$

the formation of which is used to remove unchanged silver halide after development in photography (p. 494). Silver chloride is about one hundred times more soluble (c 3.0 gdm^{-3}) in 1.0M hydrochloric acid than in water, due to the formation of chloroargentate(I) complexes, $AgCl_2^-$ and $AgCl_3^{2-}$. They are unimportant.

Much silver chloride (and thiocyanate) may be accumulated as silver residues in the laboratory after titrimetry. Because of the value placed on silver, recovery becomes worth while. Silver chloride may be fused with sodium carbonate, or fusion mixture, when silver results from the breakdown of carbonate and oxide,

$$2Na_2CO_3 + 4AgCl \rightarrow$$
$$4Ag + O_2 + 2CO_2 + 4NaCl$$

Silver chloride is reduced by zinc or aluminium in dilute acid, or by heating in hydrogen:

$$Zn(s) + 2AgCl \rightarrow Zn^{2+} + 2Cl^- + 2Ag(s)$$

$$2AgCl + H_2 \rightarrow 2Ag + 2HCl$$

Silver bromide is similar to silver chloride, but is even less soluble in water. It dissolves in the complex-forming reagents, but is only sparingly soluble in aqueous ammonia. Silver bromide is more photosensitive than either silver chloride or silver iodide. Silver bromide is used in various photographic emulsions. Silver iodide is the most insoluble halide, virtually insoluble in ammonia and only moderately soluble in sodium thiosulphate solution.

Ag^I: silver nitrate, AgNO$_3$

Silver nitrate is the common salt of silver made commercially. Silver nitrate may be prepared by the action of warm moderately strong nitric acid on silver:

$$3Ag + 4H^+ + NO_3^- \rightarrow 3Ag^+ + 2H_2O + NO$$

After concentration by evaporation and cooling, large transparent crystals separate. Silver nitrate is very soluble in water, 257 g saturating 100 g water at 25°C, and various organic solvents, including alcohols, amines and nitriles. The crystals are anhydrous and on melting (m.p. 209°C) silver nitrate may be cast into sticks.

Silver nitrate is decomposed at red heat into the metal, nitrogen dioxide and oxygen:

$$2AgNO_3 \rightarrow 2Ag + 2NO_2 + O_2$$

Silver nitrate is readily reduced to silver; with iron(II) sulphate:

$$Ag^+ + Fe^{2+} \rightarrow Ag + Fe^{3+}$$

Silver nitrate is used as a constituent of marking ink for linen, in the fibres of which silver is deposited by the reducing power of organic matter, and in hair-dyes. Black stains of silver produced on the skin are formed by reduction. Silver nitrate is used in the photographic supplies industry for the preparation of photosensitive emulsions. The reduction of ammoniacal silver nitrate (or oxide) has been described. Silver nitrate is important in inorganic and organic analysis and these applications will be studied in detail later in the chapter.

Ag(I): other salts of interest

Silver carbonate, Ag$_2$CO$_3$

Silver carbonate is conveniently prepared as a yellow precipitate formed on mixing solutions of silver nitrate and sodium carbonate. It may also be precipitated by passing carbon dioxide into ammoniacal silver nitrate solution. Silver carbonate darkens in light, and on heating decomposes to give as final products, silver, oxygen and carbon dioxide:

$$2Ag_2CO_3 \rightarrow 4Ag + 2CO_2 + O_2$$

Silver cyanide, AgCN

Silver cyanide is formed as a white precipitate on mixing silver nitrate solution and potassium cyanide and is soluble in an excess of the latter to form the complex dicyanoargentate(I) anion etc.,

$$Ag^+ + CN^- \rightarrow AgCN$$

$$AgCN + CN^- \rightarrow Ag(CN)_2^- \text{ (and } Ag(CN)_3^{2-})$$

Silver cyanide has a linear macromolecular structure of repeating AgCN units.

Silver ethynide, silver acetylide, Ag$_2$C$_2$

What happens when ethyne (acetylene) is passed through an ammoniacal solution of silver oxide (or silver nitrate) has been described earlier. Ethynides, formerly acetylides, have also been called dicarbides. Since the hydrogen atoms in ethyne are potentially acidic (differing from those in ethene), silver ethynide is a salt. This acidity appears in all ethynes with a terminal hydrogen atom and is caused by the triple covalent bond's near monopoly of valency electrons, weakening the bond which dissociates to the ions:

$$R.C \equiv CH + Ag^+ \text{ (as } Ag(NH_3)_2^+) \rightarrow$$
$$[R.C \equiv C]^- Ag^+ + H^+ \text{ (as } NH_4^+)$$

where R is an alkyl (C_nH_{2n+1}) group.

Silver ethynide is a white solid, explosive when dry and easily detonated by a blow or by heating to 120–140°C. The explosion products are carbon and silver only, although the ethynide is never entirely free of impurities, notably water and oxygen. Since there are apparently no gaseous products to propagate the explosion, it must be caused by expansion of the surrounding air by an intense release of heat from the bonding of this endothermic compound.

With dilute nitric or sulphuric acid, pure ethyne (acetylene) is evolved:

$$Ag_2C_2 + 2H^+ \rightarrow 2Ag^+ + \begin{matrix} CH \\ \mathrm{|||} \\ CH \end{matrix}$$

Alternatively, warm potassium cyanide forms the complex dicyanoargentate(I), liberating ethyne:

$$Ag_2C_2 + 4CN^- + 2H_2O \rightarrow$$
$$2Ag(CN)_2^- + 2OH^- + C_2H_2$$

A similar compound of copper(I) may be prepared, Cu_2C_2, a reddish-brown solid.

Silver sulphate, Ag₂SO₄

Silver sulphate, Ag_2SO_4, is precipitated from a fairly concentrated solution of silver nitrate by adding a sulphate solution, or by action of hot concentrated sulphuric acid on silver. A saturated solution of silver sulphate, containing 0.83 g in 100 g water at 25°C, is sometimes used in qualitative analysis in place of silver nitrate to avoid the possibility of slight precipitation of silver sulphate from high sulphate ion concentrations during the test for halides. Silver sulphate solution is sensitive to light and is stored in a dark bottle.

Silver sulphide, Ag₂S

Silver sulphide, Ag_2S, which is formed as the brown tarnish on silver, may be precipitated by addition of a sulphide solution to aqueous silver nitrate. On heating in air, silver and sulphur dioxide are formed:

$$Ag_2S + O_2 \rightarrow 2Ag + SO_2$$

Qualitative tests for silver salts

Addition of hydrochloric acid to solutions of silver salts, of which very few are soluble in neutral solution, precipitates white, curdy silver chloride which goes violet in light. It is soluble in aqueous ammonia, sodium thiosulphate and potassium cyanide to form complex ions.

In neutral solution, coloured salts may be precipitated: iodides give yellow silver iodide, chromates dark red silver chromate, and sulphides black silver sulphide.

Ammoniacal solutions may be reduced to silver (as a mirror) with glucose, 2, 3-dihydroxybutanedioates (tartrates) and other reducing agents.

Complexes of Ag^I, Ag^II and Ag^III

The following important complex ions have been mentioned:

$$N\equiv C-\overset{-}{Ag}-C\equiv N \qquad \underset{H}{\overset{H}{N}}\!-\!\overset{+}{Ag}\!-\!\underset{H}{\overset{H}{N}}$$

dicyanoargentate(I)   diamminesilver(I)
(argentocyanide)   cation
anion
also $Ag(CN)_3^{2-}$

and certain complex thiosulphates have been isolated:

$Na[Ag(S_2O_3)]$	$K[Ag(S_2O_3)]$
$Na_3[Ag(S_2O_3)_2]$	$K_3[Ag(S_2O_3)_2].H_2O$
$Na_5[Ag_3(S_2O_3)_4]$	$K_5[Ag_3(S_2O_3)_4]$

Names are straightforward but cumbersome:

e.g. $K_3[Ag(S_2O_3)_2].H_2O$
tripotassium bis(thiosulphato)argentate(I)
monohydrate

where the Latin *bis* is preferred to the Greek *di* to avoid possible ambiguity in depicting whole groups of atoms.

Solid silver chloride will absorb ammonia up to the composition indicated by $AgCl.3NH_3$. Ag(II) forms organic complexes with compounds of the type $[AgB_4]^{2+}$ where B represents one molecule (one mole, strictly) of organic amine; somewhat analogous to the copper(II) ammines. Ag(III) occurs in $K[AgF_4]$, potassium tetrafluoroargentate(III) a tribute to the extremely electronegative nature of fluorine, behaviour met elsewhere.

The laboratory importance of silver nitrate

Silver nitrate in inorganic qualitative analysis

(a)   *Tests for halides*

The sodium carbonate extract (or original solution, if suitable) is acidified with nitric acid, warmed to expel carbon dioxide and silver nitrate solution added. A white precipitate easily soluble in aqueous ammonia shows a chloride, a cream precipitate soluble with difficulty, a bromide, and a pale yellow precipitate insoluble in ammonia, an iodide.

(b) Tests in neutral solution

To the liquid obtained above, silver nitrate solution is added to ensure complete precipitation of halides and to give an excess. Dilute ammonia solution is added drop by drop so that a neutral interface between the lower acidic and the upper alkaline layers is formed: at this interface, insoluble silver salts of feebly ionized acids separate and have characteristic colours:

arsenate	chocolate-brown
arsenite	pale yellow
chromate	dark red
ethanedioate	white
phosphate	pale yellow

Silver salts are soluble in ammonia due to ammine formation, $Ag(NH_3)_2^+$, and silver salts of feebly ionized acids are soluble in fully ionized acids because of the removal of anions as un-ionized acid. Silver phosphate dissolves in nitric acid:

$$\underset{\text{insoluble}}{(Ag^+)_3PO_4^{3-}} \rightleftharpoons \underset{\text{in solution}}{3Ag^+ + PO_4^{3-}}$$

$$H^+ + PO_4^{3-} \rightleftharpoons HPO_4^{2-}$$

$$H^+ + HPO_4^{2-} \rightleftharpoons H_2PO_4^-$$

$$\underset{\text{excess}}{H^+} + H_2PO_4^- \rightleftharpoons H_3PO_4$$

Removal of phosphate ions disturbs the solubility equilibrium and silver phosphate dissolves.

Silver nitrate in titrimetric (volumetric) analysis

Silver nitrate may be obtained in a very high degree of purity. It is a primary standard. Solutions are photosensitive and must be stored in dark bottles.

(a) In neutral solution

Silver nitrate may be used to estimate chloride, bromide and iodide in neutral solution using potassium chromate as indicator (Mohr's Method). Standard silver nitrate solution is run into a measured volume of the halide solution, neutralized if necessary, containing potassium chromate. The white or cream silver halide is precipitated until at the end-point, dark red silver chromate appears in the precipitate. To achieve an accurate indication of the end-point about $1\,cm^3$ of 5% w/v potassium chromate solution should be used for a final volume of $50\,cm^3$, if the reactants are decimolar. Conditions must be neutral; alkali precipitates silver oxide and acid prevents precipitation of silver chromate. Acidic solutions may be neutralized by addition of precipitated chalk, while alkaline solutions may be acidified with excess dilute nitric acid and then neutralized. Silver nitrate is always added to the halide and not the other way, otherwise silver nitrate would give a precipitate of silver chromate, becoming less soluble as it ages and re-dissolving only very slowly in the low concentrations near the end-point. The precipitated silver halide adsorbs halide or silver ions, whichever is in excess, so titrations must be conducted slowly with constant agitation. Iodides are usually determined by a redox method because silver iodide forms an adsorption complex with silver chromate making the end-point difficult to see.

Silver chromate is precipitated after the silver halide because it is more soluble. In separate saturated solutions of silver chloride and silver chromate at 18°C, the following concentrations have been measured and solubility products calculated:

AgCl
solubility $= 1.3 \times 10^{-3}\,g\,dm^{-3}$
i.e. $9.0 \times 10^{-6}\,mol\,dm^{-3}$
  $[Ag^+] = [Cl^-] = 9.0 \times 10^{-6}\,mol\,dm^{-3}$
$\therefore [Ag^+][Cl^-] = 8.1 \times 10^{-11}\,mol^2\,dm^{-6}$

Ag_2CrO_4
solubility $= 2.5 \times 10^{-2}\,g\,dm^{-3}$
i.e. $= 7.5 \times 10^{-5}\,mol\,dm^{-3}$
  $[CrO_4^{2-}] = 7.5 \times 10^{-5}\,mol\,dm^{-3}$
  $[Ag^+] = 2 \times 7.5 \times 10^{-5}\,mol\,dm^{-3}$
$\therefore [Ag^+]^2[CrO_4^{2-}] = 1.7 \times 10^{-12}\,mol^3\,dm^{-9}$

During the addition of silver nitrate solution to the chloride solution containing potassium chromate, silver chloride is precipitated until at the stoichiometric end-point, a precipitate of dark red silver chromate appears. The solution is saturated with silver chromate as well as silver chloride. At the stoichiometric point for the mixture,

$$[Ag^+] = [Cl^-] = 9.0 \times 10^{-6}\,mol\,dm^{-3}$$

and for silver chromate,

$$(9.0 \times 10^{-6})^2[CrO_4^{2-}] = 1.7 \times 10^{-12}$$

$$\therefore [CrO_4^{2-}] = 2.1 \times 10^{-2} \, mol \, dm^{-3}$$

or about 0.02M

i.e. the concentration of potassium chromate = $2.1 \times 10^{-2} \times 194 \, g \, dm^{-3}$ which is $\simeq 0.4\%$ w/v. In practice this concentration proves excessive, giving a strong yellow tint to the solution and making it difficult to observe the first appearance of silver chromate. Usually 1 cm^3 of 5% w/v potassium chromate solution is added for a final volume of 50 cm^3, i.e. $[CrO_4^{2-}] \simeq 0.1\%$ w/v, and the concentration of silver nitrate required for its precipitation is given by the rough calculation:

$$[Ag^+]^2[CrO_4^{2-}] = 1.7 \times 10^{-12} \, mol^3 \, dm^{-9}$$

$$[Ag^+]^2 = \frac{1.7 \times 10^{-12}}{2.1 \times 10^{-2} \times \frac{1}{4}}$$

$$\therefore [Ag^+] = 1.8 \times 10^{-5}$$

$$= 18 \times 10^{-6} \, mol \, dm^{-3}$$

But at equivalence

$$[Ag^+] = 9.0 \times 10^{-6} \, mol \, dm^{-3}$$

$\therefore$ extra $[Ag^+] = 9.0 \times 10^{-6} \, mol \, dm^{-3}$ required (and 1 cm^3 0.1M silver nitrate contains 1×10^{-4} mol silver ions).

$\therefore$ Required volume of decimolar silver nitrate to be added to a final volume of roughly 100 cm^3

$$= 0.01 \, cm^3 \text{ of 0.1M silver nitrate}$$

One drop of silver nitrate solution is roughly 0.05 cm^3, so that the error is negligible for decimolar solutions. For use with centimolar solutions, as in the determination of the chloride content of boiler water, a blank titration against a measured quantity of indicator in distilled water of the same total volume as the halide solution in the main titration must be done. In the above calculations, the figures should be regarded as approximate.

(b)   In acid solution

When a solution containing thiocyanate ions, usually either the ammonium or potassium salt, is run into silver nitrate solution, white silver thiocyanate is precipitated. The equivalence point may be revealed by addition of iron(III) alum solution, and nitric acid to avoid hydrolysis, when the slightest excess of thiocyanate forms the red thiocyanatoiron(III) complex:

$$Ag^+ + SCN^- \rightarrow AgSCN(s)$$

$$Fe^{3+} + SCN^- \rightarrow Fe(SCN)^{2+}$$

Without nitric acid, iron(III) alum would be hydrolysed to a brown insoluble basic salt. Usually, about 2 cm^3 of 10% w/v iron(III) alum solution is used for a 50 cm^3 final volume. Both ammonium and potassium thiocyanate are colourless deliquescent solids which readily dissolve in water to form solutions which are standardized by titration against silver nitrate.

Chlorides may be estimated by Volhard's Method. A measured volume of the chloride solution is added to a known volume (excess) of standard silver nitrate solution. The precipitated silver chloride is removed by filtration and the filtrate, with washings, is titrated against thiocyanate, which may be standardized against fresh portions of the silver nitrate solution. Precipitated silver chloride must be removed because it would react with thiocyanate to give the less soluble silver thiocyanate at the stoichiometric end-point instead of allowing it to react with iron(III) ions:

$$\underset{1.3 \times 10^{-3} \, g \, dm^{-3}}{AgCl(s)} + SCN^- \rightleftharpoons \underset{0.2 \times 10^{-3} \, g \, dm^{-3}}{AgSCN(s)} + Cl^-$$

Solubilities at 18°C

The precipitated silver chloride adsorbs silver ions which are lost on filtration from solution giving some inaccuracy. As silver bromide is less soluble ($0.14 \times 10^{-3} \, g \, dm^{-3}$ at 18°C) than silver thiocyanate, filtration is unnecessary in estimating bromides. The method may be adapted for the determination of acids giving insoluble silver salts in neutral solution, such as phosphoric acid or phosphates: after filtration of silver phosphate, the filtrate containing excess silver nitrate is acidified and titrated against thiocyanate.

Silver nitrate in gravimetric analysis

By precipitation with silver nitrate solution in the presence of dilute nitric acid, certain anions may be precipitated as silver salts. Halides, thiocyanates and cyanides may be determined in this way. Conditions are controlled to produce a precipitate which is readily filtered, washed and dried for weighing. Determinations must be done in subdued light to avoid decomposition.

Silver nitrate and the relative molecular masses of organic acids

Most carboxylic acids form insoluble silver salts which rarely contain water of crystallization and may be readily decomposed on gentle heating to leave a residue of metallic silver. As silver salts of weak acids usually dissolve in strongly ionized acids, and alkali reacts with silver nitrate, carboxylic acids are neutralized by addition of aqueous ammonia, any excess of which is volatilized by boiling. The silver salt is precipitated under neutral conditions, filtered, washed and dried in a steam oven or vacuum desiccator. A weighed quantity is ignited to constant mass.

Consider an organic carboxylic acid of relative molecular mass M (molar mass $= M$ g mol^{-1}) containing $n -COOH$ groups per molecule. M/n g of acid will form $(M/n - 1 + 107.9)$ g of silver salt. On ignition,

$$\frac{\text{mass of silver salt}}{\text{mass of residual silver}} = \frac{M/n - 1 + 107.9}{107.9}$$

This is just one of the complementary classical methods of investigating relative molecular masses, and eventually structures, in Organic Chemistry. It may be used to calculate either M or n if the other is known, or if it is not, as one of several converging methods in an investigation.

The estimation of silver

This may be achieved titrimetrically or gravimetrically by adapting the methods described above.

Photography and silver compounds

Taking a photograph depends on the light-sensitive nature of silver compounds, and especially that of silver bromide. The precise nature of all the changes which occur are not fully known.

Using hot solutions of silver nitrate, potassium bromide and gelatin, under suitably controlled conditions, a photosensitive emulsion of silver bromide in gelatin is applied to glass plates and celluloid films. By making suitable additions of dyes and minute traces of sulphur compounds, as well as controlling the grain size of silver bromide, the photographic plate or film may be manufactured to the required sensitivity, i.e. speed and range of wave-lengths.

By means of a focused lens and a shutter, light falls on to the emulsion, usually for a small fraction of a second. Particles of silver are formed where silver bromide has decomposed in proportion to the intensity of the light at that point, but are not visible to the eye:

$$2Ag^+Br^- \xrightarrow{\text{light}} 2Ag + Br_2$$

The medium, gelatin and impurities, such as sulphide ions, are essential to the working of this process.

The latent image may be developed by means of a reducing agent, with the formation of areas of silver around the particles formed on exposure to light. Quinol, an organic reducing agent, is well known as a developer. With suitable development, a negative is produced showing an image which is darkest where most light had fallen:

$$Ag^+Br^- + e^- \rightarrow Ag + Br^-$$

Unreduced silver bromide must now be removed before the film is taken from the developing tank into the light.

After thorough washing, the film is treated with a fairly concentrated solution of sodium thiosulphate ('hypo' to the photographer) in order to 'fix' the negative, and prevent fogging on further exposure to light. Silver ions dissolve as complex thiosulphatoargentate(I) anions and are removed:

$$Ag^+Br^- + S_2O_3^{2-} \rightarrow Ag(S_2O_3)^- + Br^-$$

$$Ag(S_2O_3)^- + S_2O_3^{2-} \rightarrow Ag(S_2O_3)_2^{3-}$$

$$Ag(S_2O_3)_2^{3-} + S_2O_3^{2-} \rightarrow Ag(S_2O_3)_3^{5-}$$

The negative is thoroughly washed and dried. It consists of deposits of silver of various intensities in gelatin and is no longer photosensitive.

The required photograph, or positive, may be obtained by placing the negative over a photosensitive emulsion containing silver chloride or silver bromide on a piece of suitable paper and giving it a controlled exposure. The paper is developed, fixed and washed as before. By contact printing, dark areas on the negative become light on the positive and vice versa.

Colour photography requires the incorporation of various organic compounds which, according to the patented processes, act as dyes and screening agents.

However, various methods of obtaining coloured prints by electronic photography are being marketed. Silver halide photography is expensive, each film has to be processed separately irrespective of the number of exposures used and irrespective of the quality of the photographs, and there is a considerable delay between taking photographs and actually seeing the prints. The polaroid process has overcome this disadvantage but at a cost, and the quality of the print is not as high.

In electronic photography an electronic camera replaces the conventional silver halide camera and a magnetic disc, the film. At present the disc allows 50 pictures to be taken and at any convenient time these may be viewed on insertion into a printer linked to television. Satisfactory pictures may then be selected and printed in a process which takes about one minute. The disc is re-usable. The process is clean, dry, instant and the picture can be modified when on the screen. The process takes coloured pictures, and, of course, was evolved initially as part of the video industry. There are several patented methods of preparing the prints. Titanium dioxide is used as a white pigment to promote the photofading of various organic dyes used in the dyesheets.

Period 6: mercury

The element and its oxidation states

The electronic configuration of mercury is as follows: 2, 8, 18, 32, 18, 2. The penultimate shell of electrons is complete. Although mercury is not a transition metal, it occupies, like silver in the previous chapter, the tail-end of a series resulting from expansion of such a shell. It is sometimes studied in conjunction with zinc and cadmium, which occupy the corresponding position in the previous two periods, and are similar to each other, but mercury is so different from other metals that it is best studied alone. In the Mendeléeff Periodic Classification, zinc, cadmium and mercury occupied the B subgroup of Group II (alkaline earths). Certain similarities concern zinc and cadmium on the one side and magnesium in particular on the other, and there is some semblance to the adjacent elements of Group III, the boron group but mercury is quite different. Table 26.1 shows physical data. Remember that certain assumptions are made in calculating some of these values (e.g. coordination number for metallic radius) and it is not unusual to find that quite impeccable sources differ. The table offers a ready comparison. Notice how the standard electrode potentials of mercury have a positive sign, indicating its noble, unreactive nature, and the high values of its ionization energies compared with those for adjacent elements of the Periodic Table as well as congeners.

Mercury forms two series of compounds which are well established. One of them illustrates the way in which the use of oxidation state, or number, and valency differ. The number assigned to the

Table 26.1 Some physical characteristics of zinc, cadmium and mercury

	Zn	Cd	Hg
Atomic number	30	48	80
Electron shells (energy levels) filled	2, 8, 18, 2	2, 8, 18, 18, 2	2, 8, 18, 32, 18, 2
[Core], outer electrons	$[Ar](3d)^{10}(4s)^2$	$[Kr](4d)^{10}(5s)^2$	$[Xe](5d)^{10}(6s)^2$
Relative atomic mass	65.37	112.40	200.59
1st Ionization energy/kJ mol^{-1}	906	867	1003
2nd Ionization energy/kJ mol^{-1}	1737	1621	1804
1st + 2nd Ionization energies/kJ mol^{-1}	2643	2488	2807
Oxidation states*	2	2	1, 2
Principal valency states*	2	2	$Hg_2 2, Hg 2$
Metallic radius/nm	0.138	0.154	0.157
Atomic radius/nm	0.131	0.148	0.148
Ionic radius (M^{2+})/nm	0.074	0.097	0.110
Standard electrode potential,			$\frac{1}{2}Hg^{2+}$; Hg +0.789
(H Scale: 298K)/V	Zn^{2+}; Zn −0.763	Cd^{2+}; Cd −0.402	Hg^{2+}; Hg +0.854
Electronegativity	1.65	1.45	1.45
Atomic volume/cm^{3}mol^{-1}	9.17	13.0	14.8
Boiling-point/°C	907	764	357
Melting-point/°C	419	321	−38.9
Density/g cm^{-3}	7.1	8.6	13.6

*Oxidation states assigned by rules (p. 124). Valency states not always straightforward, e.g. complexes (Hg_2^{2+}), macromolecules, new compounds

oxidation state is calculated according to the rules on p. 124 and the position here is akin to that of rule 5: 'The oxidation number of oxygen is −2 except in peroxides (O_2^{-2}) where it is −1, the ionic charge being shared between two atoms'. In other words, the structure, no matter what bonding holds it together, is completely dissociated into ions (in theory) to allocate the oxidation number.

Now, consider what were formerly called mercurous and mercuric compounds, and the combined state of their mercury atoms:

Mercuric compounds

e.g. $HgCl_2$, mercuric chloride
Hg^{II} is largely covalent in compounds.
valency state, Hg   2
oxidation state, Hg   +2
Hg^{II} represents the bivalent mercury(II) atom.
For the mercury(II) state,

$HgCl_2$ is *either* mercury(II) chloride
or mercury(II) dichloride.

Mercurous compounds

e.g. Hg_2Cl_2, mercurous chloride
Hg_2^{2+} is largely ionic in compounds.
valency state, Hg_2 2
oxidation state, Hg   +1
Hg_2^{2+} represents the doubly charged di[mercury(I)] cation.
For the mercury(I) state,

Hg_2Cl_2 is *either* dimercury(I) chloride
or dimercury(I) dichloride
or mercury(I) chloride, which is
only valid if the formula of the elementary unit is specified.

The dimercury cation is a dimeric complex and is unique of its type. Any ambiguity which could arise over abandoning the [ ] brackets is removed by declaring the stoichiometric proportions in the compound, and in calculations involving the mole, (see p. 8 for definition), the formula or symbol for the entity. To avoid any possible misunderstanding, in this book:

Hg_2Cl_2 is dimercury(I) dichloride
$HgCl_2$ is mercury(II) chloride

(If the Hg_2^{2+} cation were to be treated as the oxidation number entity, the former would be called dimercury(II) chloride.)

Mercury is a very dense silvery-grey metal with a high surface tension (4.7×10^{-2} N m^{-1} at room temperature), not wetting surfaces but forming small globules and accounting for its old name of quicksilver. Mercury is a relatively poor conductor of heat and electricity compared with silver but as a liquid metal it is used in thermometers and for the collection and manipulation of gases under anhydrous conditions. Mercury is the only liquid metal at ordinary temperature although caesium (m.p. 29°C) and gallium (m.p. 30°C) are very readily melted. The vapour pressure of mercury is very low (0.027 Pa at 0°C), which makes it useful in barometers, manometers and high-vacuum apparatus, but the vapour is very poisonous. Stringent safety precautions are taken in laboratories over the use of mercury for this reason. A cubic metre of air, saturated with mercury vapour, contains 14 mg at 20°C. The vapour pressure rises rapidly on heating, the volatility of mercury being without parallel among the metals, illustrated by boiling-points, mercury (357°C) caesium (670°C) and gallium (2403°C). The vapour is almost wholly monatomic, a state otherwise found only in the inert gases.

Mercury is notable for the formation of alloys with other metals, some of which are compounds. The alloys are called *amalgams*, and are solid or liquid depending on the relative proportions of the constituent metals. Iron, cobalt and nickel are not affected by mercury, which may be transported in iron flasks. Small quantities are contained in polythene or stoneware bottles. Tiny pieces of sodium pressed under mercury, using a pestle and mortar, dissolve with a flash and hiss to form sodium amalgam, useful as a reducing agent and forming hydrogen and sodium hydroxide in water, especially if a metal, such as iron, without a hydrogen overpotential is placed in contact with it,

$$2Na(Hg) + 2H_2O \rightarrow 2Na^+ + 2OH^- + H_2 + 2Hg$$

Zinc, heated with mercury and dilute sulphuric acid over a boiling water bath, forms zinc amalgam, useful as a reducing agent. An amalgam containing silver and tin rapidly sets to a hard mass and is used as a dental filling†. Amalgamated aluminium, called an aluminium–mercury couple, may be used for the reduction of alkyl halides to paraffin hydrocarbons in the presence of aqueous alcohol.

Mercury is unaffected by air at room temperature. Heated almost to boiling it slowly reacts with oxygen to form mercury(II) oxide, which dissociates on stronger heating, a most useful characteristic

---

† The author sometimes wonders why dentists, dental nurses and patients do not *apparently* show signs of mercury poisoning.

associated with the discoveries of Priestley and the work on combustion by Lavoisier:

$$2Hg + O_2 \rightleftharpoons 2HgO$$

Mercury(II) oxide prepared in this way is red in colour, but darkens to brown on heating. Ozone causes mercury to 'tail', possibly due to oxide formation. The halogens attack mercury, putting another limitation on its use in manipulating gases. Chlorine forms mercury(II) chloride, $HgCl_2$. Sulphur and mercury react on grinding to form mercury(II) sulphide, HgS, at first black and then vermilion in colour. Mercury is not attacked by water or steam, alkalis and the dilute mineral acids under non-oxidizing conditions. Hot concentrated sulphuric acid is reduced to sulphur dioxide and forms mercury(II) sulphate, $HgSO_4$, if the acid is in excess, and dimercury(I) sulphate, Hg_2SO_4, if mercury is in excess:

$$Hg + 2H_2SO_4 \rightarrow HgSO_4 + SO_2 + 2H_2O$$

$$2Hg + 2H_2SO_4 \rightarrow Hg_2SO_4 + SO_2 + 2H_2O$$

Warm dilute nitric acid with excess mercury forms dimercury(I) dinitrate, while excess concentrated acid gives the mercury(II) *oxosalt* (i.e. ionic) and oxides of nitrogen;

$$6Hg + 8HNO_3 \rightarrow 3Hg_2(NO_3)_2 + 4H_2O + 2NO$$

$$Hg + 4HNO_3 \rightarrow Hg(NO_3)_2 + 2H_2O + 2NO_2$$

The dimercury(I) and mercury(II) (formerly mercurous and mercuric) states

Mercury, in both the dimercury(I) and mercury(II) states, may be assigned the valency two. All common dimercury(I) compounds are ionic, being based on the ion: $[Hg-Hg]^{2+}$, while mercury(II) compounds, except those formed with exceptionally electronegative anions, are much more covalent than similar compounds of other metals. Mercury forms many more complexes in the mercury(II) state than in the dimercury(I) state, and is notable for forming organic compounds with very stable $Hg-C$ covalent bonding, as in $Hg(CH_3)_2$. Stable bonds with nitrogen are also formed. An isolated mercury atom has the electronic configuration 2, 8, 18, 32, 18, 2, the rare mercury(II) ion, Hg^{2+}, 2, 8, 18, 32, 18, and the covalently bound atom, 2, 8, 18, 32, 18, 4. In the dimercury(I) ion, each mercury atom, if bound covalently in the dimer as shown, has the configuration, 2, 8, 18, 32, 18, 2, i.e. that of the

original metal. There is little tendency to acquire the octet of electrons.

The degree to which mercury(II) compounds are ionized depends very much on the anion. The oxoanions, nitrate, sulphate, chlorate (chlorate(v)) and perchlorate (chlorate(vII)) form ionized salts which are easily soluble in water, are strong electrolytes and somewhat hydrolysed. Mercury(II) oxosalts are rarely hydrated, the mercury(II) cation having little electrical affinity for oxygen. Mercury(II) fluoride is also ionized, but is largely hydrolysed in solution to mercury(II) oxide and hydrofluoric acid. The other halides are only moderately soluble in water but soluble in organic solvents and readily volatilize. Mercury(II) chloride is dissociated to the order of less than 1% in 0.005M aqueous solution:

$$HgCl_2 \rightleftharpoons HgCl^+ + Cl^-$$

Mercury(II) thiocyanate, $Hg(SCN)_2$, is also very little ionized and mercury(II) cyanide, $Hg(CN)_2$, is completely covalent and unhydrolysed in solution.

Dimercury(I) compounds are based on the unusual Hg_2^{2+} ion, not on Hg^+. This leads to apparent doubling of formulae. The crystal structure of dimercury(I) dichloride, determined by X-ray analysis, corresponds to Hg_2Cl_2. Dimercury(I) salts are usually ionic and the dimercury(I) ion has been shown to be Hg_2^{2+} by various methods. Nomenclature has been discussed earlier. Discrete Hg_2 units have been shown in crystal structures and the Raman spectrum of the nitrate solution suggests an absorption band due to Hg–Hg. Further all Hg(I) compounds are diamagnetic while the monatomic Hg would have one unpaired electron rendering it paramagnetic: this is coupled in Hg–Hg. Further evidence for the presence of the Hg_2^{2+} ion comes from electrochemistry.

The EMF of a concentration cell in each half of which is a mercury pole (and a suitable conducting head) immersed in a dimercury(I) dinitrate solution of known concentration, is given by

$$E = \frac{0.0591}{z} \log_{10} c_1/c_2$$

where z is the charge on the ion, and c_1/c_2 are the concentrations. When the ratio of the concentrations is 10/1, $E/V = 0.0591/z$. But for decimolar and centimolar solutions (calculated on '$HgNO_3$'†), the measured value is 0.0274 V at

---

† The formula does not affect the result, because a ratio of concentrations is used.

25°C, so that $z = 2$ and the dimercury(I) ion, hitherto unnamed, Hg_2^{2+}.

The variation of equivalent electrical conduction with $\sqrt{}$concentration and values obtained for the depression of the freezing-point of solutions may also be explained by the presence of the Hg_2^{2+} ion. Except for the very soluble oxosalts, dimercury(I) salts are less soluble than the corresponding mercury(II) compounds. That they are also largely ionized in comparison with the covalent mercury(II) compounds may be explained by Fajans' approach to valency: the double charge of the dimercury(I) ion is more diffuse, being spread over two atoms, and has less polarizing action than that of the mercury(II) ion.

When liquid mercury is shaken with a solution of an ionized mercury(II) salt (nitrate), dimercury(I) ions are produced according to the equilibrium:

$$Hg(l) + Hg^{2+} \rightleftharpoons Hg_2^{2+}$$

and taking the active mass of liquid mercury as constant, at a given temperature we have

$$\frac{[Hg_2^{2+}]}{[Hg^{2+}]} = \text{constant, } K, \text{ where [ ] refer to molar concentrations}$$

The value of K may be deduced from redox potentials. Using the electrochemical symbols of Chapter 7, two expressions for the (standard) Gibbs' Free Energy (Chapter 8) may be equated

$$-zFE_{298}^{\ominus} = -RT\ln K$$

where K is the equilibrium constant deduced in the original thermodynamical arguments for ideal gases reacting. This expression may be rearranged and simplified,

$$E_{298}^{\ominus} = \frac{RT}{zF}\ln K_{298} = 0.0591 \log K_{298}$$

The standard electrode potential, redox potential for the disproportionation may be deduced:

$$\tfrac{1}{2}Hg^{2+} + e^- \rightleftharpoons \tfrac{1}{2}Hg(l)$$
$$E_{298}^{\ominus} = 0.789 \text{ V}$$

$$Hg^{2+} + e^- \rightleftharpoons \tfrac{1}{2}Hg_2^{2+}$$
$$E_{298}^{\ominus} = 0.920 \text{ V}$$

$$Hg(l) + Hg^{2+} \rightleftharpoons Hg_2^{2+}$$

for which $\therefore E_{298}^{\ominus} = (0.920\text{-}0.789) \text{ V}$

$$E_{298}^{\ominus} = 0.131 \text{ V}$$

substituting $\quad 0.131 = 0.0591 \log K_{298}$

whence $\quad\quad K_{298} = 165$

$$\therefore [Hg_2^{2+}]/[Hg^{2+}] = 165/1$$

Thus an aqueous solution of a dimercury(I), Hg_2^{2+}, compound contains 0.6% of mercury(II), Hg^{2+}, although the numbers should be treated as very approximate and you will recall any dissolved mercury metal has been ignored. Thus, any reactant which forms an insoluble mercury(II) compound or a mercury(II) complex, thereby removing mercury(II) ions from the equilibrium, will cause more dimercury(I) to dissociate and the disproportionation goes to completion. Clearly there will be no dimercury(I) analogues to these compounds. On the other hand, if dimercury(I) can be removed, it should be possible to reduce ionized mercury(II) salts with mercury, but mercury(II) compounds are on the whole unionized.

Dimercury(I) dicyanide has not been isolated because it decomposes immediately to a solution of covalent mercury(II) cyanide and mercury:

$$Hg_2^{2+} + 2CN^- \rightarrow Hg(CN)_2 + Hg(l)$$

Addition of alkali to dimercury(I) dinitrate solution yields a black precipitate of mercury and mercury(II) oxide, and hydrogen sulphide yields mercury and mercury(II) sulphide, also black in appearance:

$$Hg_2^{2+} + 2OH^- \rightarrow HgO(s) + Hg(l) + H_2O$$

$$Hg_2^{2+} + H_2S \rightarrow HgS(s) + Hg(l) + 2H^+$$

Thus the oxide, hydroxide, cyanide and sulphide of the dimercury(I) state do not appear to exist.

Hg(I): dimercury(I) compounds and their reactions

Dimercury(I) oxosalts

Dimercury(I)

Addition of sodium hydrogencarbonate solution to dimercury(I) dinitrate solution precipitates yellow dimercury(I) carbonate, which loses carbon dioxide and disproportionates on heating:

$$Hg_2^{2+} + 2HCO_3^- \rightarrow Hg_2CO_3(s) + H_2O + CO_2$$

$$Hg_2CO_3 \rightarrow CO_2 + HgO + Hg$$

Dimercury(I) dinitrate, $Hg_2(NO_3)_2$

This is formed by the action of cold dilute nitric acid on mercury, and purified by re-crystallization to give the colourless dihydrate, $Hg_2(NO_3)_2 \cdot 2H_2O$. It is freely soluble in water, easily hydrolysed to basic salts and decomposed on heating to mercury(II) oxide, which dissociates into the elements on further heating, no residue remaining:

$$Hg_2(NO_3)_2 \rightarrow 2HgO + 2NO_2(g)$$

$$2HgO \rightleftharpoons 2Hg(g) + O_2(g)$$

Dimercury(I) sulphate, Hg₂SO₄

A white precipitate of dimercury(I) sulphate is formed by interaction of dimercury(I) dinitrate and a sulphate (sulphuric acid) in solution. Alternatively, concentrated sulphuric acid is heated with excess mercury, already described, or mercury(II) sulphate, $HgSO_4$, is ground with mercury.

$$Hg_2^{2+} + SO_4^{2-} \rightarrow Hg_2SO_4(s)$$

$$2Hg + 2H_2SO_4 \rightarrow Hg_2SO_4 + SO_2 + 2H_2O$$

$$Hg + HgSO_4 \rightarrow Hg_2SO_4$$

The dimercury(I) halides

Dimercury(I) difluoride is very soluble in water and hydrolysed to mercury, mercury(II) oxide and hydrofluoric acid:

$$Hg_2F_2 + H_2O(s) \rightarrow HgO + Hg(l) + 2HF$$

The other halides are very insoluble in water.

Dimercury(I) dichloride, Hg_2Cl_2, may be precipitated by addition of a soluble chloride (dilute hydrochloric acid) to dimercury(I) dinitrate, $Hg_2(NO_3)_2$, in solution. On the large scale it is produced by heating mercury with mercury(II) chloride (or a mixture of mercury(II) sulphate and sodium chloride) when dimercury(I) dichloride sublimes and is collected:

$$HgCl_2 + Hg \rightarrow Hg_2Cl_2(g)$$

Density determinations reveal that dimercury(I) dichloride vapour is dissociated. Spectroscopic data are interpreted as showing that dissociation yields mercury(II) chloride and mercury, not HgCl molecules:

$$Hg_2Cl_2(g) \rightleftharpoons HgCl_2(g) + Hg(g)$$

Dimercury(I) dichloride, as calomel, is used as a purgative in medicine and in mercury ointments. The name, calomel ('beautiful black'), refers to the black precipitates formed by addition of ammonia and alkali (p. 362). Aqueous ammonia converts dimercury(I) dichloride from a white insoluble compound into a black chloroamidodimercury(I), formerly dimercury(I) aminochloride which changes to the mercury(II) compound with separation of mercury:

$$Hg_2Cl_2 + 2NH_3 \rightarrow$$
$$H_2N \cdot Hg \cdot Hg \cdot Cl + NH_4^+ + Cl^-$$

$$H_2N \cdot Hg \cdot Hg \cdot Cl \rightarrow Hg \Big\langle \begin{matrix} NH_2 \\ Cl \end{matrix} + Hg$$

'mercury(II) aminochloride'
chloroamidomercury(II)

Dimercury(I) dichloride is reduced by tin(II) chloride solution to mercury, a grey precipitate,

$$Hg_2Cl_2 + SnCl_2 \rightarrow 2Hg(l) + SnCl_4$$

Dimercury(I) di-iodide, greenish-yellow in colour, is formed by precipitation with potassium iodide solution and dimercury(I) dinitrate solution. It quickly disproportionates to mercury(II) iodide and mercury. Excess potassium iodide yields a solution of potassium tetraiodomercurate(II) and a residue of mercury:

$$Hg_2I_2 + 2I^- \rightarrow HgI_4^{2-} + Hg(l)$$

Qualitative tests for dimercury(I) salts

Addition of hydrochloric acid to solutions of dimercury(I) salts precipitates white dimercury(I) dichloride, blackened by aqueous ammonia to form chloroamidomercury(II) and mercury. Dimercury(I) dichloride is oxidized to the soluble mercury(II) halide on boiling with aqua regia. Tin(II) chloride reduces it to mercury.

Hydrogen sulphide precipitates a mixture of mercury(II) sulphide and mercury, which is black, from dimercury(I) solutions, while caustic alkali gives a mixture of mercury(II) oxide and mercury, again a black residue.

On heating with an excess of solid sodium carbonate, all mercury compounds yield mercury. In a test-tube a grey mirror consisting of fine droplets of mercury condenses on the cool part.

Hg(II): mercury(II) compounds and their reactions

Mercury(II) oxide, HgO

This oxide, is basic and dimorphous. A red form results from heating mercury in air or oxygen just below the boiling-point (357°C). It reverts to mercury on stronger heating, the colour darkening,

$$2Hg + O_2 \rightleftharpoons 2HgO$$

The red form is also produced by thermal decomposition of mercury(II) nitrate, obtained by the action of nitric acid on mercury,

$$2Hg(NO_3)_2 \rightarrow 2HgO + 4NO_2 + O_2$$

or on the commercial scale by gentle ignition of mercury and mercury(II) nitrate,

$$Hg(NO_3)_2 + Hg \rightarrow 2HgO + 2NO_2$$

Mixing boiling solutions of mercury(II) chloride with sodium carbonate with continued heating of the precipitated carbonate in the liquid also produces the red variety,

$$HgCl_2 + CO_3^{2-} \rightarrow HgO(s) + CO_2 + 2Cl^-$$

The yellow modification is precipitated when mercury(II) chloride solution is poured into an excess of aqueous caustic soda, there being no mercury(II) hydroxide,

$$HgCl_2 + 2OH^- \rightarrow HgO(s) + H_2O + 2Cl^-$$

The yellow form is slightly metastable with respect to the red: the difference being due to particle size. A basic chloride results from the addition of alkali to mercury(II) chloride solution.

Mercury(II) oxide does not react with alkali, but forms salts with acids. It will dissolve in alkali-metal halide solutions on warming. Sodium chloride forms un-ionized mercury(II) chloride, while potassium iodide forms the complex potassium tetraiodomercurate(II):

$$HgO + H_2O + 2Cl^- \rightarrow HgCl_2 + 2OH^-$$

$$HgO + H_2O + 4I^- \rightarrow HgI_4^{2-} + 2OH^-$$

Mercury(II) oxosalts

Only a basic *mercury(II) carbonate* is known but the other oxosalts are formed. The preparation and thermal decomposition of *mercury(II) nitrate* have been described: the hydrated salt, $Hg(NO_3)_2 . H_2O$, is very soluble in water, being hydrolysed, and ionic. Its use in titrimetric (volumetric) analysis is described later. *Mercury(II) sulphate* may be prepared by the action of hot concentrated sulphuric acid on mercury or mercury(II) compounds: it forms a monohydrate, $HgSO_4 . H_2O$, and is hydrolysed to basic salts by excess water. Mercury(II) sulphate is used as a catalyst (with

iron(III) sulphate) for the industrial production of ethanal (acetaldehyde) from ethyne (acetylene) which is passed into warm dilute sulphuric acid containing the catalyst.

$$HC\equiv CH + H_2O \rightarrow [H_2C=CHOH]$$
$$\rightarrow CH_3 . CHO$$

Mercury(II) sulphide, HgS

This, as the mineral cinnabar, is the important naturally occurring source of mercury, which volatilizes on combustion of the ore in air:

$$HgS + O_2 \rightarrow Hg(g) + SO_2(g)$$

Mercury(II) sulphide may be prepared as a black 'mud' on grinding mercury with flowers of sulphur, moistened with ammonium sulphide. On prolonged heating to 40–50°C in caustic potash solution, a red crystalline variety, the pigment vermilion, is formed. Mercuric(II) sulphide is dimorphic, the black form metastable below 386°C changing spontaneously to the red, stable form.

Black mercury(II) sulphide may be precipitated by passage of hydrogen sulphide into a solution of a mercury(II) salt. It dissolves in aqua regia. The red variety may be prepared conveniently by subliming the black form which is prepared as above, and then washed and dried.

The mercury(II) halides

Anhydrous mercury(II) fluoride is a colourless, crystalline salt of ionic structure which forms a yellow dihydrate, $HgF_2 . 2H_2O$, also indicative of ionic bonding. It is hydrolysed by excess water to basic salts and mercury(II) oxide. The other halides are covalent, little hydrolysed, sublime readily and are readily soluble in organic solvents. Their structures are covalent; that of mercury(II) iodide being macromolecular. They have low solubilities in water, mercury(II) chloride being much more soluble in hot water than cold (5.6 g saturating 100 g water at 10°C and 54 g at 100°C), mercury(II) bromide being much less soluble than the chloride, and the iodide, insoluble (solubility $4.8 \times 10^{-3} g\,dm^{-3}$ at 25°C). Electrical conductance measurements show that mercury(II) chloride and bromide are little ionized in solution, the degree being less than 0.01 (i.e. 1%) in 0.005M solution:

$$HgCl_2 \rightleftharpoons HgCl^+ + Cl^-$$

Mercury(II) chloride, also called corrosive sublimate, is very poisonous. Very dilute solutions have been used as an antiseptic because it kills bacteria. Mercury(II) chloride may be prepared by the commercial methods: heating mercury in chlorine or heating a mixture of mercury(II) sulphate and sodium chloride, with a little manganese(IV) oxide to oxidize dimercury(I) compounds to the mercury(II) state, when mercury(II) chloride sublimes on to a suitably cooled surface:

$$Hg + Cl_2 \rightarrow HgCl_2$$

$$HgSO_4 + 2NaCl \rightarrow HgCl_2 + Na_2SO_4$$

The sublimate may be purified by dissolving in hot water and cooling, when colourless needle-like crystals separate. Mercury(II) chloride is very soluble in ethanol, ethoxyethane (ether) and ethyl ethanoate (acetate). It is not affected by hot concentrated nitric acid and volatilizes away when treated with boiling concentrated sulphuric acid. There can be little doubt that mercury(II) chloride is covalent.

Mercury(II) chloride is readily reduced to dimercury(I) dichloride, and then to mercury. In solution, tin(II) chloride precipitates white dimercury(I) dichloride, rapidly turning dark grey as mercury is formed:

$$2HgCl_2 + SnCl_2 \rightarrow Hg_2Cl_2(s) + SnCl_4$$

$$Hg_2Cl_2(s) + SnCl_2 \rightarrow 2Hg(l) + SnCl_4$$

On warming, excess methanoic (formic) acid brings about a similar reduction:

$$2HgCl_2 + H.COOH \rightarrow$$
$$Hg_2Cl_2(s) + CO_2 + 2HCl$$

$$Hg_2Cl_2(s) + H.COOH \rightarrow 2Hg(l) + CO_2 + 2HCl$$

Mercury(II) chloride forms complexes with ammonia which are referred to by the descriptions, fusible white precipitate and infusible white precipitate, the latter volatilizing before fusion. Fusible white precipitate is the diammine, obtained by passing dry ammonia into a solution of mercury(II) chloride in ethyl ethanoate (acetate),

$$HgCl_2 + 2NH_3 \rightarrow HgCl_2.2NH_3$$

This diammine is hydrolysed by water to

$$HgO.HgCl_2.2NH_3,$$

infusible white precipitate, which is sometimes given as $Hg(NH_2)Cl.$†

The dimorphism of mercury(II) iodide has been described in Chapter 6 (p. 106). The scarlet form changes to the yellow modification at 126°C. On precipitation of mercury(II) iodide, by addition of potassium iodide to mercury(II) chloride in solution, a transient yellow coloration of the form which is metastable at room temperature is observed:

$$HgCl_2 + 2I^- \rightarrow HgI_2(s) + 2Cl^-$$

Mercury(II) iodide dissolves in excess potassium iodide solution to form colourless potassium tetraiodomercurate(II):

$$HgI_2(s) + 2I^- \rightleftharpoons HgI_4^{2-}$$

Rendered alkaline, the solution is known as Nessler's Reagent and is a very sensitive test for ammonia, or ammonium compounds, which form a yellow coloration in very small quantity or a brown precipitate.

Mercury(II) complexes

A complex compound, *mercury(II) tetrathiocyanatocobaltate(II)*, $Hg[Co(SCN)_4]$, is formed as a blue precipitate when a solution containing cobalt(II) chloride and ammonium thiocyanate is added to a solution of mercury(II) chloride. A range of complex cyanides, ammines, sulphides and halides are known. Those with ammonia are mentioned above.

Mercury(II) nitrate in titrimetric (volumetric) analysis

In an attempt to avoid the lengthy Volhard titrimetric method for the estimation of chloride using silver nitrate and potassium thiocyanate solutions, a direct determination involving mercury(II) nitrate solution, containing nitric acid to suppress hydrolysis, was developed. A 10% solution of sodium hexanitrocobaltate(III) is used as indicator, when the stoichiometric end-point is revealed by a slight turbidity which is permanent after shaking. The end-point is sharp. Mercury(II) nitrate solution is run into a solution containing chloride ions and indicator: covalent mercury(II) chloride is formed until all chloride ions have been removed, when excess mercury(II) ions cause precipitation of mercury(II) hexanitrocobaltate(III) at the end-point. The solution of mercury(II) nitrate is standardized by titration against a prepared standard solution of potassium chloride.

† i.e. deduct H_2O from $HgO.HgCl_2.2NH_3$ and halve the result.

Qualitative tests for mercury(II) compounds

Passage of hydrogen sulphide through solutions containing mercury(II) compounds yields finally a black precipitate of mercury(II) sulphide, but lighter coloured intermediates are also seen: white, yellow, and brown, the first of which has the composition, $HgCl_2.2HgS$. Mercury(II) sulphide, insoluble in most reagents, dissolves readily in concentrated hydrochloric acid containing nitric acid (aqua regia) or potassium chlorate as oxidizing agents.

Reduction to mercury may be effected by tin(II) chloride solution, which precipitates white dichloride solution, which precipitates white di-

mercury(I) dichloride initially, or by copper foil, previously cleaned in nitric acid:

$$Cu(s) + HgCl_2 \rightarrow Hg(l) + CuCl_2$$

Potassium iodide solution precipitates red (transient yellow seen) mercury(II) iodide, soluble in excess to give a colourless solution of the tetraiodomercurate(II) complex, alkaline solutions of which give a brown coloration with traces of ammonia.

All mercury compounds yield the metal on heating with a large excess of sodium carbonate. In a test-tube a film of mercury condenses on the cool upper points as a grey mirror.

Examination questions

The level of each question is indicated after its number, as follows.

A   Advanced level. In some countries this is equivalent to the seventh year of a secondary course.

S   Scholarship or Special level. In some countries this is equivalent to the eighth year of a secondary course *or* the first year of a university or technical college course.

OS The Open Scholarship examinations were abolished in 1984. They were replaced by college entrance examinations which are taken in the fourth term of the VI form and are for pre-'A' level candidates only.

Questions have been included from all of the Boards examining at Advanced and Scholarship levels, and from the old Open Scholarship examinations of the Universities of Oxford and Cambridge.

The author wishes to thank the undermentioned authorities for granting permission for the reproduction of examination questions from past papers:

The Local Examination Syndicate of the University of Cambridge International Examinations (CL)

The Senate of the University of London (L)

The Joint Matriculation Board of the Northern Universities (N)

University of Oxford Delegacy of Local Examinations (OL)

The Oxford and Cambridge Schools Examination Board (O&C)

The Southern Universities' Joint Board for School Examinations (S)

The Northern Ireland Schools Examinations Council (NI)

The Welsh Joint Education Committee (W)

The Syndics of the Cambridge University Press (C) for Open Scholarship Questions

The Oxford Colleges' Admissions Office (O).

Acknowledgement is made after each question, the source being indicated by one of the abbreviations given above. The nomenclature used is that given in the original question as set in the papers of the Board indicated. Unfashionable terms are explained by a footnote.

The style of examinations in the United Kingdom and Northern Ireland has changed somewhat in recent years. Generally, there are three theoretical papers: the first, multiple choice, the second comprises structured questions (with information to be interpreted, graphs and diagrams), while the third consists of the more traditional type of question (each built up in graded degrees of difficulty or even left open-ended where there is no rigid answer scheme). The questions selected here are intended for use as written work to consolidate knowledge and to focus attention on salient features of the chemistry studied. They will prove useful even as mental exercises to test what information has been retained, and what can be adapted to novel situations after reading. Limitations on space preclude the use of structured questions, while it would be unethical to include material from the pre-tested multiple choice banks of questions.

Chapter 1

The development of fundamental ideas in 19th century chemistry

1 *S* (a) 'Although in the development of chemical theory, Dalton's work stands pre-eminent, almost equal prominence must be given to the combined contribution of Avogadro and Cannizzaro since this provided the first correct methods for determining

 (i) the molecular masses of gaseous substances,

 (ii) the relative atomic masses of gaseous elements,

(iii) the relative atomic masses of certain non-gaseous elements,

(iv) the empirical formulae of many compounds and the molecular formulae of many of these.'

 Elaborate and justify this statement.

(b) A mixture of methane, ethene and ethyne has a vapour density of 11.3. When $10.0 \, \text{cm}^3$ of this mixture and 30.0 cm^3 of oxygen are sparked together over aqueous caustic potash the volume contracts to 5.5 cm^3 and then disappears when pyrogallol is introduced. All volumes are measured under the same conditions of temperature, pressure and humidity. Calculate the composition of the original mixture. (H = 1, C = 12.)   (N)

2 *A*   Describe fully, using diagrams and equations, an experiment which you could carry out in the laboratory to verify the Law of Constant Proportions.

Explain what is meant by isotopes. To what extent does the existence of isotopes affect the exactness of this law as originally stated?   (S)

3 *OS*   Write an account of your reasons for believing in the existence of atoms, molecules and ions.   (C)

4 *OS*   Explain clearly why you believe in the existence of atoms.   (O)

5 *S*   What do you understand by the terms 'hypothesis' and 'theory' in science? Give ONE good example of each.

Write a critical account of the development from hypotheses of any TWO theories which are of importance in chemistry, which are quite distinct from each other and which belong to different aspects of the subject.

THEORIES WHICH APPEAR SUBSTANTIALLY IN ANY OTHER QUESTION ATTEMPTED, OR TO BE ATTEMPTED, IN ANY SECTION OF THIS PAPER ARE DISQUALIFIED FROM THE DISCUSSION.   (S)

6 *A*   The following chemists made contributions to the phlogiston theory and its displacement by the oxygen theory of combustion. Discuss the work of each of these chemists in relation to the phlogiston theory.
(a) Johann Joachim Becher
(b) Georg Ernst Stahl
(c) Joseph Priestley
(d) Henry Cavendish
(e) Joseph Black
(f) Antoine Laurent Lavoisier
   This requires further reading in a text on the history of chemistry.   (NI)

7 *A*   Partington in his *Short History of Chemistry* states:
   'Although the theory of phlogiston had the advantage of co-ordinating a large number of facts into a system, it retarded the progress of chemistry and prevented a number of the best investigators from seeing the correct explanation of the facts they brought to light.'
   Discuss this statement in relation to the contributions of Priestley, Cavendish, Black and Lavoisier to the theory of phlogiston.
   Requires further reading about a discarded theory of combustion.   (NI)

Chapter 2

The properties of elements and the structure of their atoms

1 *A*   What do you understand by the terms *electron, proton, neutron, atomic number, isotope*?
   What is the atomic structure of the chlorine isotopes? Explain why determinations of the relative atomic mass of chlorine give a value of 35.5.   (OL)

2 *A*   From a consideration of the properties of the elements and of those of their principal compounds, justify the positions of the following elements in the Periodic Classification: (a) iron, (b) silicon, (c) hydrogen.   (L)

3 *A*   Name the three fundamental particles of which an atom is composed and give their relative charges and masses.
   What bearing do these particles have on (a) the atomic number (*Z*) and (b) the relative atomic mass (*M*) of an element? Explain what is meant by an 'isotope'.
   Write electronic structures for the following compounds, indicating clearly the type of bond present: (i) carbon dioxide, (ii) magnesium oxide, (iii) phosphorus trichloride.   (N)

4 *A*   Explain the terms (a) *atomic number,* (b) *relative atomic mass,* (c) *isotope*, and show that a given element occupies a single fixed position in the Periodic Table.
   Discuss the position of argon and potassium in the Periodic Table.   (CL)

5 *A*   From a consideration of their properties, and those of their compounds, justify the positions of the following elements in the Periodic Table: (a) barium, (b) phosphorus, (c) iodine.   (CL)

6 *A*   What is (a) a proton, (b) an electron, (c) a neutron? Show how these particles contribute to the relative atomic structure of magnesium (atomic number 12, relative atomic mass 24), and chlorine (atomic number 17, relative atomic mass 35). Explain how the proportions of the particles change when the atoms of these elements are converted into ions.
   How do you account for the existence of chlorine atoms of relative atomic mass 37?   (CL)

7 *OS*   Discuss critically the classification of elements into A and B subgroups of the Periodic Table.   (O)

8 *OS*   Discuss the so-called diagonal relationships in the Periodic Table, with particular reference to the elements lithium and magnesium, boron and silicon, oxygen and chlorine.   (O)

9 *OS*   Evaluate the importance of the discovery of the electron to the progress of chemistry.   (O)

10 *A*   (a) What do you understand by the term *relative atomic mass*?
(b) Outline, with the aid of a labelled diagram, the use of the mass spectrometer in the determination of relative atomic masses.

Naturally occurring gallium, Ga, is a mixture of two isotopes, gallium-69 and gallium-71. Use this information, together with the relative atomic mass of gallium in your *Data Booklet*, to calculate the percentage abundance of each isotope.

(c) The mass spectrum of chlorine, $Cl_2(g)$, consists of peaks at m/e values of 70, 72 and 74 of relative abundance $9:6:1$. Explain these observations as fully as you can.   (CL)

11 *A*   The Sun uses a series of nuclear fusion reactions to produce heat. Two of these reactions are shown below.

$$^1_1H + {}^1_1H \rightarrow X + {}^0_{+1}e$$

$$^2_1H + Y \rightarrow {}^3_2He + \gamma$$

(a) (i) Complete the *two* names of the following: $_{-1}^0e$ and γ.
(ii) Identify the nuclei X and Y, giving their atomic symbols, atomic numbers and mass numbers.
(b) The stable nucleus $^{12}_6C$ is formed in the Sun from the fusion of helium nuclei. On Earth $^{14}_6C$ is formed in the atmosphere from cosmic rays bombarding nitrogen atoms.
(i) The isotope $^{14}_6C$ is used in carbon dating. It has a half-life of 5 730 years. What is meant by half-life?
(ii) A piece of cloth from 'The Turin Shroud' was recently dated using carbon dating. Explain carbon dating.
(c) The surface of the Earth is warmed by sunlight. The energy of sunlight is mostly in the form of yellow light.
(i) Calculate the energy associated with one photon of yellow light of frequency 6×10^{14} Hz (Planck's constant, $h = 6.6 \times 10^{-34}$ Js).
(ii) With the aid of an energy level diagram explain what happens when an atom absorbs a photon.
(d) Ultra-violet radiation is more energetic than yellow light and is capable of breaking bonds and initiating reactions such as that between methane and chlorine.

$$CH_4 + Cl_2 \rightarrow CH_3Cl + HCl$$

(i) Write the balanced equation for the step which initiates this reaction.
(ii) Write balanced equations for the *two* propagation steps.
(iii) Write a balanced equation for a termination step in this reaction.
(iv) Explain why only a flash of ultra-violet light, rather than continuous ultra-violet light, is needed for this reaction.   (NI)

12 *A*   (a) Define (i) *atomic number*, (ii) *mass number*.
(b) What is the nature of the three main types of radiation emitted by radioactive elements? How are they affected by a magnetic field and how do they differ in penetrating power?

(c) Complete the following nuclear equation, adding all the mass numbers, atomic numbers and symbols:

$$^{226}Ra \rightarrow \alpha +$$

The α-particles emitted in the radioactive decay of radium-226 can be counted by means of a Geiger counter. Each α-particle gains electrons to form helium gas. It is found that 1.82×10^{17} α-particles give 6.75×10^{-3} cm³ of helium, measured at s.t.p. By using these data, together with any other data required in the *Data Booklet*, obtain a value for the Avogadro constant, L.   (CL)

13 *A*   Uranium is extracted from pitchblende and other ores of uranium. The uranium-235 isotope is purified and used in nuclear reactors.
(a) The extraction of uranium from its ore can be represented by the following reaction scheme.

$$Ore \xrightarrow{HNO_3} UO_2(NO_3)_2 \xrightarrow{Heat} UO_3 \xrightarrow{H_2} UO_2$$

$$\xrightarrow{HF} UF_4 \xrightarrow{Mg} U$$

(i) Give the oxidation number of uranium in the following compounds.

$$UO_2(NO_3)_2, \; UO_3 \text{ and } UF_4.$$

(ii) Write balanced equations for the following conversions.

1. $UF_4 \xrightarrow{Mg} U$

2. $UO_2 \xrightarrow{HF} UF_4$

(b) One of the nuclear reactions taking place in a reactor is

$$^{235}_{92}U + {}^1_0n \rightarrow {}^{147}_{60}Nd + {}^{87}_{32}Ge + 2{}^1_0n + \gamma$$

(i) State two of the properties of the radiation represented by the symbol.
(ii) By what means is it possible to determine accurately the number 87 in the species $^{87}_{32}Ge$?
(iii) Write a balanced equation for the breakdown of an atom of uranium-235 to an atom of thorium-231 and an α-particle.
(c) Uranium metal has properties typical of a metal.
   It has a body-centred cubic structure. Its electrical conductivity is similar to that of lead. Its van der Waals' radius is 0.16 nm.
(i) Sketch a body-centred cubic structure.
(ii) Use the concept of delocalised bonding in uranium to explain its electrical conductivity.
(iii) Explain what is meant by the van der Waals' radius.   (NI)

14 *S*   (a) A week after the Chernobyl accident in April 1986, a sample of London rainwater showed an activity of 81 ± 10 counts per 10 seconds when measured in a liquid counter of capacity 7.0 cm³. It was estimated that

$\frac{5}{8}$ of the activity was due to ^{131}I, a radioisotope which undergoes β^- decay with a half-life of 8.07 days.
(i) Write an equation for the radioactive decay of ^{131}I.
(ii) Explain the significance of the ± 10 after the count rate.
(iii) Estimate the concentration of ^{131}I in atoms dm^{-3} in the rainwater sample. State what assumptions you have made in making this estimate.
(iv) On the following day the level of activity in the rainwater fell almost to zero, but the activity became high again after the bottle containing the rainwater was shaken. What can you deduce from this about the chemical and physical form in which the ^{131}I is held in the rainwater?
(b) Samples of fresh milk were also tested for ^{131}I activity, and the following results were recorded at a farm in the South of England.

Date when milk was obtained	Activity/counts $s^{-1} dm^{-3}$
May 8	41 ± 5
May 10	24 ± 5
May 12	21 ± 5
May 14	11 ± 4
May 16	8 ± 4
May 18	7 ± 3
May 20	5 ± 3

(i) Explain briefly how ^{131}I can come to be present in milk.
(ii) Suggest a reason why the activity in the milk falls off more rapidly than the half-life of ^{131}I.   (CL)

Chapter 3

The properties of elements and the electronic configuration of their atoms

1 S   The atomic numbers of the inert gases are 2, 10, 18, 36, 54 and 86. Make use of this information in deducing
(a) the properties of the elements having atomic numbers 15 and 56;
(b) the salient characteristics of the newly discovered elements astatine (atomic number 85) and francium (atomic number 87);
(c) the probable position in the Periodic Classification of a metallic element which has a specific heat of 0.057 and an atomic number between 47 and 52, and which forms two chlorides containing respectively 23.7 and 48.1% of chlorine.   (N)

2 S   Explain how it is possible to predict many of the properties of an element from a knowledge of its atomic number.
On the basis of your knowledge of the Periodic Classification of the elements give the electronic structures of the elements having atomic numbers 21 and 34 and deduce the main properties of these elements.   (N)

3 S   On the basis of current views on atomic structure explain the meaning and significance of the following statements with, where possible, illustrative examples:

(a) there is no precise arithmetical relationship between atomic number and relative atomic mass;
(b) the reactivity of an atom is determined by its electronic configuration;
(c) a monatomic ion has the same atomic number and atomic weight as the neutral atom;
(d) isotopes of an element have the same atomic number.   (N)

4 A   Describe the modern theory of the structure of atoms, defining any particles you mention. Explain how this theory accounts for (a) the periodicity of elements, (b) the variations in valency within a short period, and (c) a characteristic valency in a group of the Periodic Table.
Chlorine has atomic number 17. Sketch the structures of its isotopes of relative atomic mass 35 and 37.   (S)

5 OS   Describe how the chemical properties of the following elements depend on their positions in the Periodic Table:
He 2,   N 7,   O 8,   Na 11,   Br 35.   (C)

6 A   Explain the formation of the line spectra of hydrogen and show how the line spectra are related to the electronic structure and ionization energy of the atom. What evidence concerning atomic structure is provided by X-ray spectra?   (CL)

7 A   Explain concisely what is meant by the Periodic Classification of the elements, indicating, with one example of each, the significance of the terms 'group', 'short period' and 'transition element'.
How is the electronic structure of an element related to (a) its position in the Periodic Classification, (b) its valency?
Give the electronic structures of the following atoms: argon, carbon, chlorine, sodium.
How is it that iron (atomic number 26) can have two valencies?   (N)

8 S   Write short accounts of (a) **either** (i) the role of radioactivity in the development of chemical theory, **or** (ii) the role of the inert gases in the development of chemical theory, **and** (b) **either** (iii) the allotropy of sulphur and of phosphorus, **or** (iv) theories of catalysis.   (N)

9 A   Write an account of electronic energy levels in atoms. Include in your account reference to
(a) spectroscopic evidence
(b) the use of the equation $E = h\nu$
(c) the flame colorations of group I chlorides
(d) ionisation energies.   (NI)

Chapter 4

The electronic theory of valency and the periodic classification

1 A   Give a *brief* explanation of the difference between an electrovalent and a covalent link.

What is a co-ordinate link (or dative covalency)?

What kind of valency bonds do you think exist in ammonium chloride, calcium sulphate, copper sulphate pentahydrate? Give your reasons. (O&C)

2 *S* What are the principal physical and chemical properties by which metals are distinguished from non-metals?

Explain how metallic properties in an element are related to its atomic structure and to its position in the Periodic Table. (L)

3 *A* Name **two** general properties of covalent compounds and **two** general properties of electrovalent compounds. Explain carefully, using carbon tetrachloride as an example, the nature of the covalent link.

Classify the following substances as electrovalent or covalent and give in each case their electronic structures: (a) hydrogen chloride, (b) calcium sulphide, (c) magnesium chloride, (d) ammonia. (N)

4 *S* Write an essay on 'Modern Views on Valency'. (L)

5 *OS* 'Metallic character of the elements in the Periodic Classification decreases in progressing from left to right in the table but increases in progressing from top to bottom.' Summarize clearly and concisely in note form what you feel are the key points in favour of or against such a statement, illustrating your answer by carefully chosen examples where possible. (C)

6 *A* Give an account of the way in which a knowledge of the properties and electronic structures of the rare gases has been of use in advancing our understanding of valency and the periodic system of the elements.(O&C)

7 *A* Give an account of the types of valency found in the following substances, describing the structure of each: (a) magnesium chloride, (b) sulphuric acid, (c) carbon tetrachloride, (d) diamond, (e) potassium ferrocyanide.

Show *briefly* how the *physical* properties of these substances depend upon the type of valency present.(S)

8 *S* Give a brief account of the bonding in the following: (a) sodium chloride, (b) carbon tetrachloride, (c) nitrogen, (d) hydrogen chloride gas, (e) diamond.

How does the nature of the bonding affect the properties of these substances? (CL)

9 *A* Indicate the relative masses and charges of (a) a proton, (b) an electron, (c) a neutron. In terms of these particles, describe the composition of the nucleus of a chlorine atom having an atomic number of 17 and a mass number of 35, and explain why the relative atomic mass of chlorine as determined by experiment is not a whole number. Account for the differences in chemical and physical properties of sodium chloride, carbon tetrachloride and ammonium chloride. (CL)

10 *A* Give the electronic formulae of hydrogen chloride, ammonium chloride, sodium chloride, and phosphorus trichloride. Discuss the relationship between the formula and the properties of the compound in each case. (CL)

11 *A* Give the electronic structures of the atoms of carbon, chlorine, argon and potassium.

Show how these structures account for the following: (a) both potassium and chlorine have valency 1; (b) the formation of a compound between potassium and chlorine which is an electrolyte; (c) the formation of a compound between carbon and chlorine which is a non-electrolyte. (CL)

12 *OS* In Mendeléeff's Periodic Table, the top right-hand corner is occupied by non-metals and the lower left-hand part by metals; the region between contains a number of elements of which some of the properties are metallic, some non-metallic. Give as many examples of these 'metalloid' elements as you can, pointing out in each case the properties that lead us to class the elements as metalloid. What explanations can you advance for the occurrence of metals, non-metals and metalloids in the three regions of the table? (C)

13 *OS* Explain the characteristics of the different types of bonds which occur in chemical compounds. Discuss the bonding in the following: (a) sodium nitrate, (b) the $NH_3.BF_3$ complex, (c) carbon monoxide, (d) ethanoic acid in water, and in benzene solution. (C)

14 *OS* What are the principal physical and chemical properties of the metallic state? Explain how metallic properties are related to position in the Periodic Table. (C)

15 *OS* Define the terms *electrovalency* (ionic valency) and *covalency*. Compare and contrast the properties of electrovalent compounds with those of covalent compounds. In which of the compounds $AlCl_3$, $NaCl$, $SnCl_4$, $SnCl_2$, CaO, do you think electrovalencies are present? Give reasons for your opinions. (C)

16 *OS* Some elements in the divalent state behave very like calcium, whereas others behave more like cadmium or mercuric mercury(II). Into which group do the following elements fall: magnesium, iron(II) (ferrous), lead? Give reasons for your answers. (O)

17 *A* Describe the electronic structures (main shells only) of the atoms of the elements of atomic numbers, 3, 9, 12 and 15, and indicate the types of chemical bonds you would expect these elements to form.

Taking your examples from the short period which includes the elements of atomic numbers 11 to 18, write electronic formulae (valence electrons only) for the following: (a) the hydride of **one** metal, (b) the simplest hydrides of **two** non-metals, (c) the simplest chlorides of **two** non-metals. (W)

18 *A* (a) State **three** ways in which the members of each of the following two groups of elements resemble one another in chemical properties:

(i) lithium, sodium, potassium;

(ii) magnesium, calcium strontium.

What similarities in electronic structure are responsible for these group resemblances?

(b) Write electronic formulae (valence electrons only) for

(i) the hydride of an alkali metal,

(ii) the chloride of element number (i.e. atomic number) 14,

(iii) the free elements numbers 17 and 18,

(iv) the ions formed in aqueous solution by the hydride of element number 17,

(v) phosphine,

(vi) phosphonium iodide. (W)

19 A The octet rule was introduced by G. N. Lewis in 1916. Since then it has been used to explain many aspects of theoretical chemistry.

(a) Write an account of the applications of the octet rule to:

(i) ionic compounds,

(ii) covalent compounds with single bonds,

(iii) covalent compounds with multiple bonds.

(b) Discuss the limitations of the octet rule. (NI)

Chapter 5

Bonding and the structures displayed by elements and their compounds

1 S From knowledge of their physical and chemical properties, what can you infer concerning the nature of the valency forces in the substances of empirical composition KBr, HCl, PbO_2, $K_4Fe(CN)_6$, and $HgCl_2$?(CL)

2 OS Give a short account of the stable electronic configurations of the elements as ions and as atoms in simple molecules, and discuss the view that electron-pairing is a more important concept in chemistry than the rule of eight (inert gas structure concept). (C)

3 OS Distinguish between atomic number and relative atomic mass, and discuss the chemical significance of these quantities. Write a short essay on the valency of nitrogen and discuss, in the course of this, the electronic structures of CH_3CN, NH_4Cl, N_2O, HNO_2. (C)

4 OS 'The valency of elements in the first row can usually be described by the octet rule.' Give examples of the success and failure of this rule. Does the rule apply to the compounds of the second row elements? (O)

5 OS According to Fajans' theory, the tendency towards covalency is dependent upon ionic size and charge. Discuss the usefulness of this theory, giving examples. (O)

6 OS Discuss the order of topics which should be adopted in teaching fundamental ideas about valency and chemical combination. (O)

7 A Describe the structures of **four** of the following substances (**one** from each pair) with the aid of electronic formulae showing valence electron shells only. State the type(s) of valence encountered in each substance you select and indicate briefly in each case how the *physical* properties of the substance (e.g. solubility, volatility, electrical conductance) are related to its valence type.

(a) **either** diamond **or** quartz;

(b) **either** ethane **or** carbon tetrachloride;

(c) **either** magnesium chloride **or** sodium sulphide;

(d) **either** sulphuric acid **or** orthophosphoric acid. (W)

8 A Discuss the variation in the boiling points of the hydrides of carbon, nitrogen, oxygen and fluorine.

Describe the electronic arrangement of the ammonia molecule. The H–N–H angle in ammonia is approximately 107°, whereas in a perfectly tetrahedral molecule it would be $109\frac{1}{2}°$ while the F–N–F angle in NF_3 is approximately $102\frac{1}{2}°$. Suggest reasons for these observations.

By writing equations and stating conditions, outline the reactions which occur between ammonia and (a) sodium, (b) copper(II) oxide (cupric oxide), (c) chlorine, (d) sodium hypochlorite. (S)

9 S Explain the various types of valency from the standpoint of atomic structure.

Describe the bonding and geometrical shape of the molecules of each of the following compounds:

(a) ammonia;

(b) sodium chloride;

(c) potassium ferrocyanide;

(d) sulphur hexafluoride.

Explain the fact the chlorine trifluoride has a T-shaped molecule. (S)

10 S (a) What geometric shapes would you expect the molecules of each of the following compounds in the vapour state to have: carbon tetrachloride, nitrogen trifluoride, oxygen difluoride and hydrogen fluoride? Give your reasons.

(b) Describe and discuss the probable geometric changes in the reaction sequence:

$$P_4 \xrightarrow{O_2} P_4O_6 \xrightarrow{O_2} P_4O_{10}$$ (S)

11 A (a) In elementary nitrogen, N_2, the bonding is covalent but the bonding between potassium and chlorine in potassium chloride is electrovalent. Explain why this is so and construct suitable diagrams to illustrate the bonding in each case. Show the situation before and after bond formation.

(b) Say why the covalent bonds in hydrogen chloride and in water are polarised and draw suitable diagrams for both compounds to show the bond polarity. It has been suggested that hydrogen chloride reacts with water:

$$\begin{array}{c} H \\ \diagdown \\ O + H - Cl \rightarrow \\ \diagup \\ H \end{array} \left[\begin{array}{c} H \\ \diagup \\ H - \bar{O} \\ \diagdown \\ H \end{array} \right]^{+} + Cl^{-}$$

Explain why this hypothesis seems feasible.

(c) Using similar principles as in (b), explain, with diagrams, which of the two hydrolysis reactions below would be preferred for bromine(I) chloride:
(i) $BrCl + H_2O \rightarrow BrOH + H^+ + Cl^-$
(ii) $BrCl + H_2O \rightarrow ClOH + H^+ + Br^-$.
(d) How can you account for the abnormally high boiling point of water compared with hydrogen sulphide?   (S)

12 *A*   The electronegativity scale was formulated by Pauling in 1931. Part of the scale is shown below.

Element	Si	H	C	S	Br	Cl	O
Electronegativity value	1.8	2.1	2.5	2.5	2.8	3.0	3.5

Make use of electronegativity values in your answers to the following.
(a) Explain the meaning of electronegativity using the hydrogen bromide molecule as an example.
(b) Give a simple explanation, in terms of electronegativity only, of the relative acid strengths of ethanoic acid, the three chloroethanoic acids and monobromoethanoic acid.
(c) Using diagrams discuss the polar or non-polar nature of the following molecules, taking into account their bonding and where appropriate their shape.

$$Cl_2; \quad HCl; \quad SiCl_4; \quad SCl_2.$$ (NI)

13 *A*   Give the formulae of the chlorides (if any) of the elements in the third period of the Periodic Table (sodium to argon). Describe and account for the bonding in these chlorides and their reaction (if any) with water.

Draw a 'dot-and-cross' diagram to show the electronic structure of a molecule of boron trifluoride, BF_3. Use the electron pair repulsion theory to predict the shape of this molecule.

State and explain how you would expect boron trifluoride to react with ammonia. Draw a dot-and-cross diagram to show the electronic structure of the product of this reaction.   (CL)

14 *A*   Lithium chloride is formed from lithium (atomic number 3) and chlorine (atomic number 17). Its lattice energy may be calculated from a Born-Haber cycle using the following experimental data.

	$\Delta H/kJ \, mol^{-1}$
First ionisation energy of lithium	$= +520$
Heat of atomisation of lithium	$= +159$
Heat of formation of lithium chloride	$= -409$
Heat of atomisation of chlorine per mole of chlorine atoms	$= +122$
Electron affinity of chlorine atoms	$= -349$

(a) Using the usual chemical symbols, the state symbols (s), (l), (g) and the symbol for an electron, e^-, write an equation in each case to define the following terms.
(i) The first ionisation energy of lithium.
(ii) The second ionisation energy of lithium.
(iii) The heat of formation of lithium chloride.

(iv) The electron affinity of chlorine.
(v) The lattice energy of lithium chloride.
(b) (i) Construct a labelled Born-Haber cycle for the formation of lithium chloride.
(ii) Using the constructed Born-Haber cycle, or any other method, calculate the lattice energy of lithium chloride.
(c) Using the s, p, and d notation for electrons write the electronic configurations of lithium and chlorine, and their ions.
(d) Lithium chloride reacts separately with silver nitrate solution and with concentrated sulphuric acid.

Write a balanced equation in each case to represent these reactions.   (NI)

15 *S*   Suggest explanations for **five** of the following observations.
(a) SiF_4 is tetrahedral but XeF_4 is planar.
(b) The NO_2^+ ion is linear but the NO_2 molecule is bent.
(c) $1.0 \, mol \, dm^{-3}$ solutions of $NaNO_3$, $Mg(NO_3)_2$ and $Al(NO_3)_3$ have pH values of 7, 6 and 3 respectively.
(d) CaCl(s) does not exist though its standard enthalphy change of formation is estimated as $-180 \, kJ \, mol^{-1}$.
(e) $LiAlH_4$ and $LiBH_4$ are both powerful reducing agents; the former must be used as a solution in dry ether, but the latter can be used in aqueous solution.
(f) The standard electrode potentials of the systems

$$MnO_4^-(aq) + 5e^- + 8H^+(aq) \rightarrow Mn^{2+}(aq) + 4H_2O(l)$$
$$E^O = +1.51 \, V$$
and

$$Mn^{3+}(aq) + e^- \rightarrow Mn^{2+}(aq) \qquad E^O = +1.49 \, V$$

are nearly the same, but whereas a solution of potassium manganate(VII) can be kept for some weeks, a solution of manganese(III) sulphate is rapidly decomposed. (CL)

16 *A*   State and explain how **each** of the following properties varies across the third period of the Periodic Table from sodium to argon:
(a) the first ionisation energy of the element,
(b) the boiling point of the element,
(c) the acid/base behaviour of the oxide.   (CL)

Chapter 6

Crystallization and the crystalline State

1 *A*   Give an account of the element sulphur including a description of its main allotropic forms. Rhombic and monoclinic sulphur behave similarly in chemical reactions, but oxygen and ozone and red and white phosphorus behave differently. What explanation can you suggest for these facts?   (O&C)

2 *A*   Explain clearly, with examples where possible, what you understand by **five** of the following: polymer, enantiotrope, enantiomorph, transition point, transition element, hydrate.   (O&C)

3 *S*   Write an essay on 'The Allotropy of Sulphur'.(S)

4 A Describe the differences in properties of yellow phosphorus and red phosphorus, and explain how the yellow form may be converted into the red form. Explain what happens when phosphorus is heated (a) in dry air, (b) with nitric acid. How may the product of reaction (a) be changed into the product of reaction (b)? (CL)

5 S What is meant by the term *allotropy*? Using sulphur and phosphorus as illustrative examples, distinguish carefully between enantiotropy and monotropy.

How would you determine the transition temperature of rhombic and prismatic (monoclinic) sulphur? (CL)

6 S (a) Describe, as far as you can, the crystal structures and the bonding in
(i) sodium chloride;
(ii) silica;
(iii) copper metal.
Comment on and explain appropriate properties, characteristic of these materials, which are a direct consequence of their structure.
(b) By choosing FOUR different examples, explain how the shapes of molecules are a function of electron (shared and unshared) pairs bonded to the central atom. (S)

7 OS 'Elements can often exist in a number of different forms.' Discuss this statement, illustrating your answer with as wide a variety of examples as you can. (O)

8 S '*Allotropy* is a very general term which includes several different phenomena.' Discuss this statement. (W)

9 A Discuss how the physical properties of SEVEN of the following substances are related to their structure and bonding.
(a) Iodine
(b) Graphite
(c) Polythene
(d) Sodium chloride
(e) Diamond
(f) Quartz
(g) Iron
(h) White phosphorus
(i) Ice (NI)

Chapter 7

Oxidation, reduction and electrochemical processes

1 A Give **four** processes to which you could apply the term *reduction*, with a reaction to illustrate each. Explain what connection there is between the processes you mention which justifies the application of the term *reduction* to them. (O&C)

2 A Explain fully the meaning of the term *reducing agent*.
Give **one** example of the reducing action of each of the following with a different reagent in each case; (a) ozone; (b) hydrogen sulphide; (c) carbon monoxide;

(d) nitrous acid; (e) stannous chloride; (f) potassium ferrocyanide. Write equations for the reactions. (OL)

3 A Define oxidation and reduction in terms of electron transfer and illustrate your answer by reference to the reactions which take place between the following pairs of substances in aqueous solution: (a) metallic zinc and copper sulphate; (b) bromine and ferrous sulphate; (c) iodine and hydrogen sulphide; (d) copper sulphate and potassium iodide. (L)

4 S (a) Two elements A and B form insoluble carbonates. Their electrode potentials are -2.76 and -0.76 volts respectively, that of copper being $+0.34$ volts. Suggest two different methods, one for A and one for B, by which these elements could be isolated starting from the carbonate in each case.
(b) In some forms of the Periodic Table the long periods are displayed with two elements in each group. Select one such pair of elements and, by means of a table, compare and contrast the chemical and physical properties of the elements, their oxides and their chlorides.

Discuss briefly whether the pairing of these elements is justified. (N)

5 A Account for the following experimental observations, naming the substances formed, and writing ionic equations wherever possible. By finding the change in oxidation number, if any, of zinc, chromium, mercury, and sulphur in the examples, classify as oxidation, reduction, neutralization, disproportionation the TYPE of reaction THESE ELEMENTS undergo in each case.
(a) A white precipitate forms when sodium hydroxide solution is added to zinc chloride solution but this precipitate dissolves when excess of the alkali is added.
(b) A white precipitate forms when tin(II) chloride solution is added to a mercury(II) chloride solution and, on warming, the precipitate darkens.
(c) A green precipitate forms when an excess of sodium hydroxide solution is added to chromium(III) chloride solution but this precipitate dissolves to form a yellow solution when the mixture is warmed with aqueous hydrogen peroxide.
(d) When sodium thiosulphate ($Na_2S_2O_3$) solution is acidified, a colloidal solution appears and a gas is evolved which decolorises acidified potassium manganate(VII) (potassium permanganate) solution. (S)

6 A Arrange the metals calcium, copper, magnesium, potassium, tin and zinc in order of their chemical activity and justify your arrangement by reference to the reactions of these metals with (a) oxygen, (b) water and (c) dilute hydrochloric acid. Equations are not required.

Name and define a physical property the magnitude of which determines the position of a metal in the electrochemical series. (N)

7 OS What is an oxidizing agent? Illustrate your answer by referring to oxygen, hydrogen peroxide, iodine and potassium dichromate. (O)

8 *A*   Represent the electronic structures of (a) the atoms, (b) the ions, of the elements potassium and sulphur and state how the structures of the ions differ from those of the atoms.

Define oxidation and reduction in terms of electron transfer.

Write ionic equations for the following reactions and state which atoms or ions are oxidized and which are reduced, giving your reasons.

(i) Chlorine is passed into an aqueous solution of ferrous chloride.

(ii) Hydrogen sulphide is passed into an aqueous solution of ferric chloride.

(iii) An iodine solution is mixed with an aqueous solution of sodium thiosulphate.

(iv) Aqueous solutions of copper sulphate and potassium iodide are mixed.   (N)

9 *A*   The following (represented in molecular terms) are oxidation–reduction reactions occurring in aqueous solution. By considering each in electronic terms, show clearly what is oxidized and what is reduced, giving reasons.

(a) $2FeCl_2 + Cl_2 = 2FeCl_3$
(b) $I_2 + 2Na_2S_2O_3 = 2NaI + Na_2S_4O_6$

State what colour changes would be observed in both and account for them in chemical terms.

By what reactions could you (c) convert ferric(III) chloride to ferrous(II) chloride, (d) obtain sulphur from hydrogen sulphide? Write the corresponding equations.

The salt $X_2M_2O_7$ (where X and M are metallic elements of relative atomic mass $X = 39$, $M = 52$) can be used in acidic solution for reactions in which the overall reaction of the salt is $M_2O_7^{2-} + 14H^+ + 6e^- \rightarrow 2M^{3+} + 7H_2O$.

Deduce the concentration in grams per dm^{-3} of a decinormal† solution of the salt. Write *ionic* equations for the reactions which occur when the acidic solution of the salt is mixed with (i) potassium iodide solution, (ii) ferrous sulphate solution.   (N)

10 *S*   Give with examples a summary of the various types of chemical processes to which the term oxidation might be applied.

The equations representing the oxidizing action of potassium iodate in (a) dilute and (b) concentrated hydrochloric acid are as follows:

(a) $KIO_3 + 6HCl = KCl + 3H_2O + (I + 5Cl)$
(b) $KIO_3 + 6HCl = KCl + 3H_2O + ICl + (4Cl)$

If a solution of potassium iodate made up as decinormal† for use in dilute acid were used by mistake for the second reaction what normality would it appear to have?   (O&C)

11 *A*   Give, with one example of each, **three** different ways, other than by the addition of oxygen, in which you could recognize that oxidation had taken place in a reaction.

Sulphur dioxide, manganese dioxide and hydrogen peroxide can, in different reactions, be oxidizing or reducing agents. Explain why this is so and give **one** example of each type of reaction for each substance.   (O&C)

12 *A*   Arrange the metals sodium, zinc, silver, calcium and copper in their correct order in the electrochemical series. Describe the properties and reactions of these metals, their oxides, hydroxides and carbonates, to illustrate the relation between electrochemical character and Periodic Classification.   (S)

13 *A*   Describe simple experiments you would make to place the metals zinc, magnesium, calcium and copper in their correct order in the electrochemical series.

Often the more electropositive elements have the more basic oxides. Discuss this statement in connection with the above four metals.   (CL)

14 *OS*   Give an account of the electrochemical series of the elements, and discuss its value.

Supposing that a sample of a newly discovered metallic element became available, what experiments would you perform in order to find its place in the electrochemical series?   (O)

15 *S*   Before there was any electronic theory the following changes were described as oxidation:

(a) ferrous salts to ferric salts;
(b) potassium ferrocyanide to potassium ferricyanide **or** potassium manganate to potassium permanganate;
(c) sulphur dioxide to sulphur trioxide;
(d) lead sulphide to lead sulphate;
(e) hydrogen iodide to water and iodine.

Discuss these examples from the electronic point of view, asking yourself precisely what has been oxidized, and seeking to find a description of oxidation from this point of view which is generally applicable. (If you prefer to discuss some other example of oxidation instead of one of those given you may do so, but most credit will be given for a discussion of the examples above.)   (S)

16 *A*   Give an account of the processes of oxidation and reduction, illustrating your answers by reference to the reactions between the following:

(a) hydrogen sulphide and oxygen,
(b) stannous chloride and mercuric chloride,
(c) iodine and sodium thiosulphate,
(d) zinc and copper sulphate,
(e) ferrous, permanganate and hydrogen ions.   (S)

17 *OS*   Discuss the use of electrolytic methods in the preparation or manufacture of chemical compounds, with special reference to the production of either hydrogen peroxide or chlorates.   (C)

18 *OS*   Define the terms *electrode potential* and *electrochemical series*. Discuss the reactions which occur at platinum electrodes when aqueous solutions of (a) lithium fluoride, (b) magnesium iodide, (c) cobalt nitrate, (d) potassium methanoate are electrolysed.

† redox normal ≡ one Faraday of electric charge transferred.

An aqueous solution of a metallic salt is electrolysed for an hour with a current of 0.290 amp., 0.334 g of metal being deposited at the cathode. An oxide of the metal contains 11.2% of oxygen. What deductions can you make concerning the properties of the metal?
[1 Faraday = 96 500 coulombs per g equivalent]. (C)

19 *OS*   Define the terms 'oxidation' and 'reduction'. Discuss the reactions of (a) hydrogen peroxide, and (b) the halogens and their oxy-acids, in the light of your definitions.
Explain what is meant by the statement that potassium dichromate is a stronger oxidizing agent than ferric sulphate. (C)

20 *OS*   What factors determine the nature of the reaction (if any) that takes place when a metal is immersed in an aqueous solution of an acid?
Suggest explanations of the following observations.
(a) Impure commercial zinc dissolves more rapidly in dilute sulphuric acid than high purity zinc.
(b) Iron is not attacked by concentrated nitric acid at room temperature.
(c) Amalgamated aluminium is rapidly attacked by cold water, whereas the unamalgamated metal is almost unattacked. (C)

21 *OS*   Define oxidizing agent. Comment on the oxidizing properties (if any) of the groups NO_3, PO_4, SO_4 and ClO_4 when present (a) in the molecules of anhydrous acids, (b) in aqueous solutions as ions. (C)

22 *OS*   In what ways is the chemistry of a metal and its compounds related to the position of the metal in the electrochemical series? (O)

23 *A*   Explain briefly what is meant by the *electrochemical series* and place hydrogen, iron, copper and calcium in their correct sequence in this series. Discuss the relevance of this sequence to the methods used to extract the three metals from their compounds. (W)

24 *A*   Give a brief account of the interpretation of *oxidation* and *reduction* in electronic terms. Illustrate your answer by means of **one** distinct example each of oxidation–reduction reactions selected from the chemistry of the compounds of any **four** of the following: (a) hydrogen, (b) halogens, (c) iron, (d) chromium, (e) manganese, (f) tin. (W)

25 *A*   Potassium dichromate(VI) is a useful reagent in the laboratory. It may be used in the preparation of other chromium compounds or as an oxidising agent.
(a) Chrome alum may be prepared by the reduction of potassium dichromate(VI).
(i) State the formula and colour of chrome alum.
(ii) Name a reducing agent that is commonly used in the preparation of chrome alum.
(iii) Describe the colour change that takes place during the reduction.

(b) Potassium dichromate(VI) is a powerful oxidizing agent and will oxidize tin(II) to tin(IV). The half equations are:

$$Cr_2O_7^{2-} + 14H^+ + 6e^- \rightarrow 2Cr^{3+} + 7H_2O$$

$$Sn^{2+} \rightarrow Sn^{4+} + 2e^-$$

(i) Write a balanced equation for the reaction of tin(II) ions with dichromate(VI) ions.
(ii) 20 cm^3 of a solution of tin(II) ions required 18.4 cm^3 of a 0.1 mol dm^{-3} potassium dichromate(VI) solution for complete oxidation.
Calculate the mass of tin ions contained in 1 dm^3 of the solution. (Sn = 119)
(c) Potassium dichromate(VI) is used in the oxidation of alcohols. The product of the oxidation depends on the structure of the alcohol. Draw the structures of the oxidation products, if any, when the following alcohols are gently heated with acidified potassium dichromate(VI) solution. If you consider that the alcohol does not undergo oxidation write 'no reaction'

$$CH_3CH_2CH_2OH, \ CH_3CH_2CH(OH)CH_3, \ (CH_3)_3COH.$$

(d) In solution the dichromate(VI) ion is in equilibrium with the chromate(VI) ion

$$2CrO_4^{2-} + 2H^+ \rightleftharpoons Cr_2O_7^{2-} + H_2O.$$

(i) Name a reagent that you would add to the equilibrium mixture in order to move the equilibrium to the left, or to the right.
(ii) Describe the expected colour change when the equilibrium moves to the left. (NI)

26 *S*   (a) Write an ion-electron equation for the oxidation of
(i) ethanedioate ion, $C_2O_4^{2-}$, to carbon dioxide,
(ii) thiosulphate ion, $S_2O_3^{2-}$, to sulphate ion.
(b) the oxidation of thiosulphate ion by chlorine follows a different course from the oxidation of thiosulphate ion by iodine.
In an investigation of the reaction, chlorine was passed into aqueous sodium thiosulphate containing 15.8 g dm^{-3} of $Na_2S_2O_3$. After removal of the excess dissolved chlorine, separate 25.0 cm^3 portions of the resulting solution were found to require
(i) 25.0 cm^3 of 1.00 mol dm^{-3} sodium hydroxide for neutralisation,
(ii) 20.0 cm^3 of 1.00 mol dm^{-3} silver nitrate for complete precipitation of the chloride ions present.
A further 25.0 cm^3 portion of the resulting solution gave 1.165 g of barium sulphate precipitate when an excess of aqueous barium chloride was added.
Deduce an equation for the reaction of aqueous sodium thiosulphate with chlorine.
Suggest a reason for the formation of different sulphur-containing products from chlorine and iodine in this reaction.

(c) Ethanedioic acid forms a salt of formula $K_xH_y(C_2O_4)_z \cdot nH_2O$, which is used to standardise bases and oxidising agents, as it can be obtained pure and does not deteriorate on standing.

Separate 25.0 cm^3 portions of a solution containing 5.08 g dm^{-3} of this salt were found to require
(i) 15.0 cm^3 of 0.100 mol dm^{-3} sodium hydroxide for neutralisation,
(ii) 20.0 cm^3 of 0.0200 mol dm^{-3} acidified potassium manganate(VII) for complete oxidation.
Deduce values for x, y, z and n and hence the formula of the salt. (CL)

27 S (a) What do you understand by the term *standard electrode potential*?
(b) Describe, with the aid of a fully labelled diagram of the apparatus, how you would determine the standard potential of an iron(II)/iron electrode in the laboratory.
(c) Discuss the use of standard electrode potential data in predicting the course of a chemical reaction, including any limitations in their use for this purpose.

Use the *Data Booklet*, to predict the outcome of the reaction between iron(II) sulphate and hydrogen peroxide in aqueous solution.
(d) Explain the variation in the electrode potential, E, of an iron(II)/iron electrode with pH shown in the following diagram as fully as you can. [Solubility product, K_{sp}, of iron(II) hydroxide $= 6.0 \times 10^{-15}$ mol^3 dm^{-9} at 25°C.]

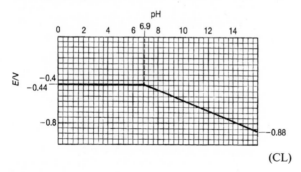

(CL)

Chapter 8

The principles governing the extraction of elements

1 S Discuss and illustrate each of the following statements.
(a) The methods employed in the extraction of metals from their natural sources are determined largely by the electrode potentials of the metals.
(b) Physical properties show that the bonds formed by a metal and a halogen are covalent in some compounds and electrovalent in others.
(c) A metal atom can be part of an anion. (N)

2 A What general methods are available for the extraction of a metal from one of its naturally occurring compounds? Illustrate by examples how the choice of method is governed by the position of the metal in the electrochemical series. (L)

3 S Without giving technical details explain the various reactions which occur between iron, carbon and their oxides at different levels in a blast furnace.

To what extent do these reactions bear out the following observations?
(a) At moderate temperatures (below 700°C) carbon monoxide is a more effective reducing agent than carbon.
(b) At high temperatures carbon is a more effective reducing agent than carbon monoxide.
(c) Iron will reduce carbon dioxide at high temperatures but not at low ones. (O&C)

4 S Discuss the corrosion of metals, with special reference to the methods used for their protection in the processes of isolation from their ores, and under ordinary conditions of exposure. (S)

5 A Arrange the metals, copper, iron, zinc, mercury, lead, magnesium and tin in order of decreasing reactivity. Summarize the methods of extraction of these metals to show that the method used depends on the position of the metal in the electrochemical series.

Outline the chemistry of the extraction from their ores of **two** of the metals: magnesium, zinc, tin.

What reaction, if any, takes place when a concentrated solution of sodium hydroxide is heated with each of these three metals? (S)

6 OS Explain what is meant by the electrochemical series, and illustrate how the methods used for the isolation of elements from their compounds can be related to the positions of the elements in the series. (C)

7 OS Describe **two** general methods of extracting metals from their ores. In what way is the choice of the method in practice related to the position of the metal in the electrochemical series? (C)

8 OS Give an account, with examples, of the methods commonly used to isolate metals from (a) their oxides, (b) their sulphides, (c) their chlorides. (C)

9 S What is meant by the *electrochemical series*? Show how the methods used for the isolation of (a) sodium, (b) aluminium, (c) zinc, (d) mercury, are related to the positions of these metals in the series. (CL)

10 OS Give an account of the principal methods used for the extraction of metals. Indicate for representative examples, what determines the choice of method in particular instances. (O)

11 A Bromine, in the form of bromide ions, is found to the extent of 65 mg dm^{-3} in sea-water, and accounts for 0.19% of the dissolved solids. Compounds of bromine such as dibromoindigo, extracted as Tyrian purple, are formed from bromide ion by molluscs (oysters, mussels and whelks). This accounts for the variable concentration of bromide ion in some regions of the Mediterranean Sea.

The residence time of bromine, in sea-water is about 7×10^3 years.

(a) Explain what is meant by residence time. Suggest a possible reason why the concentration of bromide ion might be lower in surface waters rather than in the depths of the ocean.

(b) Write an account of the essential chemistry which is associated with the production of bromine from sea-water.

(c) Give TWO examples of the use of bromine and explain their economic importance. (NI)

12 *A* Discuss the following topics which are some aspects of chemistry associated with the sea.

(a) A sea is an immense body of water. Describe how you would obtain a sample of *local* sea-water for analysis. Explain why you could not regard the results of your analysis as applying to all parts of the sea from which your sample has been extracted.

(b) Write an account of the essential chemistry which is associated with the production of magnesium from sea-water. Give TWO examples of the use of magnesium and explain their economic importance.

(c) Explain what a polychlorobiphenyl is and why it is regarded as a pollutant if found in the sea. (NI)

Chapter 9

The extraction and uses of typical elements

(Further examples will be seen under each element or group in following chapters.)

1 *A* Give the names and formulae of two important minerals which contain calcium. Describe the preparation of the metal from one of them.

Show how the chief chemical properties of calcium accord with the fact that it occupies a position near the top of the electrochemical series. (L)

2 *A* What are the chief sources of magnesium? Describe one method by which the metal is obtained on the large scale. What are the major differences between the compounds of magnesium and the corresponding compounds of calcium? (L)

3 *A* Describe the preparation of white phosphorus and its conversion to the red variety, both on an industrial scale. Give an account of the allotropy of phosphorus. How, and under what conditions, does phosphorus react with (a) dry chlorine, (b) wet iodine? (L)

4 *A* Describe briefly, with special attention to the reactions and principles involved, the methods available for the extraction of the metals aluminium and zinc from their crude ores. Show how the methods used are determined by the position of the metal in the electrochemical series.

Explain the advantages of *galvanizing* over *tin plating* as a method of protecting iron from rusting. (S)

5 *A* Describe, with equations, the *chemical reactions* employed in the extraction of lead from its ores. Name the chief impurities in the crude metal, and describe **one** method for their removal.

How and under what conditions does a solution of a lead salt react with: (a) dilute hydrochloric acid, (b) potassium iodide, and (c) sodium hydroxide? (S)

6 *A* Outline the essential principles and chemical reactions involved in the commercial production of any **three** of the following from appropriate raw materials: aluminium, copper, hydrogen, iodine, sodium carbonate. (Diagrams and descriptions of industrial plant are not required.) (W)

7 *OS* Give a brief account of the methods by which aluminium and magnesium are made from their compounds. Comment on the general principles involved. (C)

8 *S* Briefly describe methods involving the use of electricity for the following industrial processes: (a) the extraction of aluminium from bauxite, (b) the purification of crude copper, (c) the manufacture of calcium carbide. State **one** important use for each of the products. (CL)

9 *A* Describe the electrothermal production of phosphorus. How may white phosphorus be converted into red phosphorus?

Give one method of preparing orthophosphoric acid from white phosphorus. Write down equations for the neutralization of phosphoric acid in solution with sodium hydroxide to the end-points indicated by (a) methyl orange, (b) phenolphthalein. (S)

10 *A* The following processes find application in the isolation of elements from their compounds. Describe concisely **one** example in each case illustrating the use for this purpose of any **four** of these methods:

(a) electrolysis of a fused electrolyte;
(b) electrolysis of an aqueous electrolyte;
(c) complex ion formation;
(d) reduction with carbon;
(e) reduction with a metal. (W)

11 *A* 1986 was the 100th anniversary of the Hall-Héroult process for the manufacture of aluminium. The essential chemistry of the process has changed little over the past 100 years.

(a) Write an account of the chemistry associated with the conversion of bauxite to aluminium.

(b) Giving practical details, explain how you would determine the amount of iron present in a sample of aluminium foil.

(c) Former periods of history were described as the stone age, the bronze age and the iron age. It is said that we are now living in the aluminium age. Discuss the truth of this statement. (NI)

Chapter 10

The extraction and uses of transition and associated metals

1 *A* Describe the manufacture of pig iron from haematite and its conversion into steel. What reactions occur between iron and nitric acid under various conditions?   (L)

2 *A* How does zinc occur in nature? Give an account of the electrolytic extraction of zinc.

Outline important uses of zinc that depend on (a) its resistance to oxidation and (b) its solubility in acids or in ammonium chloride.   (L)

3 *A* Discuss the chemical reactions and principles involved in the extraction of copper from its sulphide ores. To what properties is the usefulness of copper due? Comment briefly on the action of acids on this metal.(L)

4 *A* Describe the manufacture of pig iron from haematite (Fe_2O_3), paying particular attention to the chemical reactions which occur in the process. (Details of industrial plant are **not** required.)

State concisely how you would prepare from iron (a) ferric oxide, (b) anhydrous ferric chloride.   (N)

5 *A* Describe, without giving details of the manufacture, how pure copper may be obtained from a copper sulphide ore. State the colour, solubility in water and behaviour on heating in air, of the following compounds: cuprous oxide, copper carbonate, hydrated copper nitrate, hydrated copper sulphate.   (O&C)

6 *A* What are the principal sources of zinc and how is it obtained from them? (Chemical processes only are required, without technical detail.)

Describe the principal features (chemical and physical) and uses of (a) metallic zinc, (b) zinc oxide, and (c) zinc chloride.   (O&C)

7 *A* Describe the manufacture of pig iron from haematite, and outline briefly **two** methods for converting pig iron into steel.

How, starting from iron filings, would you prepare specimens of (a) ferric oxide, (b) ferrous ammonium sulphate?   (CL)

8 *A* State the approximate compositions of pig iron, wrought iron and steel. Give **two** properties of pig iron which make it suitable for casting. Explain the fundamental chemical reactions involved in the conversion of (a) haematite (Fe_2O_3) into pig iron, and (b) pig iron into steel.

Discuss the relative advantages and disadvantages of the Bessemer and Open Hearth processes of steel-making.   (S)

9 *OS* Give the chemical reactions on which the extraction of zinc from its ores depends, and outline one of the processes used. In what essential way do these processes differ from the production of iron in the blast furnace? What suggestions can you advance for making zinc by a process similar to that used for iron?   (C)

(NB: The first part refers to the retort distillation method.)

Chapter 11

Industrial alloys

1 *A* An alloy is suspected to contain lead, copper, aluminium, nickel and magnesium. Devise a scheme of qualitative analysis which would enable the presence of these metals to be detected and confirmed in a specimen of the alloy.   (L)

2 *S* How is Devarda's Alloy used in the estimation of nitrates?

Suppose you have carried out a qualitative analysis of Devarda's Alloy (Cu 50%, Al 45%, Zn 5%). Prepare a report in which you describe your method and the identification and confirmatory tests you employed and the results you obtained.

Explain, with the aid of equations, the *tests* and the principles underlying the various *separations* which you performed.   (S)

3 *OS* 'Light metals' are now used in many industries. Give examples of such metals and outline the methods used in their production from natural sources. What merits or disadvantages have these metals in comparison with steel?   (O)

4 *A* Describe the reactions involved in, and the principles underlying, the determination of the qualitative composition of any **two** of the following alloys by standard analytical procedures: (a) bronze (Cu, Sn), (b) solder (Sn, Pb), (c) type metal (Pb, Sb), (d) Raney iron (Al, Fe).   (W)

Chapter 12

Commercial processes based on air, water, petroleum and coal

1 *S* Outline the inorganic industrial processes which could be carried out employing only limestone, anhydrite ($CaSO_4$), silica, alumina, coke, air and water as raw materials. You may assume that pulverized coal and electrical power for heating purposes are available, together with any essential catalysts.   (OL)

2 *A* The Haber process for the preparation of ammonia makes use of the reaction represented by the equation $N_2 + 3H_2 \rightleftharpoons 2NH_3$, 22 000 calories being given out in the formation of 2 moles of ammonia.

From this information deduce, giving reasons, the experimental conditions necessary for the production of ammonia.

What additional conditions are necessary, in practice, to secure the best yield of ammonia?

Describe how ammonia is converted to nitric acid on a large scale.   (N)

3 *A*   Give an account of the action of (a) air, (b) steam on carbon, indicating the industrial applications of these reactions.

Explain, with diagrams, the difference in structure between a candle flame and a bunsen flame.   (L)

4 *A*   Describe briefly, with a clear statement of the operating conditions and the underlying principles involved, (a) the manufacture of water gas, (b) the manufacture of hydrogen from water gas, and (c) the use of hydrogen in the manufacture of ammonia.

Comment on the properties of hydrogen which cause difficulties in assigning it to a place in the Periodic Table.   (N)

5 *A*   Describe how coal gas is manufactured. What are its main constituents and how does it differ from producer gas and water gas? Describe how methanol is manufactured on an industrial scale. Name one other product that could be manufactured in the same plant.   (O&C)

6 *A*   How is oxygen obtained from the air on a large scale?

What products can be obtained, and under what conditions, by the action of gaseous oxygen on the following: phosphorus, sodium, methanol, barium oxide?   (O&C)

7 *A*   Describe, with the aid of a simplified diagram, the manufacture of coal gas. Explain carefully (a) the nature of the crude products, and (b) their separation in the raw state from the coal gas.

State the approximate percentages of any **three** gases present in coal gas. Give **two** reasons why hydrogen sulphide is always removed from coal gas and explain fully the chemical reactions involved in its removal.

What is the effect on the product of operating the gas retorts at temperatures about 300°C below their traditional temperatures?   (S)

8 *S*   'The fixation of nitrogen is vital to the progress of civilized humanity.' (Sir W. Crookes, 1898.)

Give an account of the methods which have been employed to meet this necessity, and of the problems involved in nitrogen fixation.

Explain *in brief outline* why a practical solution to this problem was of vital importance.   (S)

9 *A*   Describe briefly (no diagrams required) the methods used to produce three fuel gases of industrial importance. How may hydrogen be obtained from one of these fuels?

How, and under what conditions, does hydrogen react with (a) chlorine, (b) compounds of arsenic, (c) ethene?   (CL)

10 *A*   Outline the commercial preparation of (a) water gas, and (b) producer gas. How would you prepare a sample of carbon monoxide in the laboratory? How and under what conditions does carbon monoxide react with (i) nickel, (ii) chlorine, (iii) sodium hydroxide?   (CL)

11 *A*   Describe the industrial synthesis of ammonia and emphasize the fundamental principles that are involved. How may ammonia be dried? Briefly explain how the reactants are prepared and indicate how ammonia is converted into nitric acid on the large scale.   (S)

12 *A*   The diagram below outlines the steps in the production of an NPK compound fertiliser.

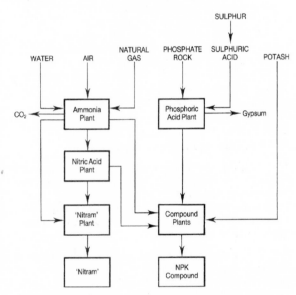

(a) Explain what is meant by an NPK fertiliser, and name the components of a typical NPK fertiliser.

(b) Discuss the chemistry associated with the production of nitric acid from ammonia placing emphasis on chemical reactions and physicochemical principles.

(c) Discuss the chemistry associated with the production of ammonium nitrate from nitric acid omitting technical detail.

(d) Explain the advantages of locating the production of an NPK fertiliser, as in the above flow scheme, on one industrial site   (NI)

13 *A*   On Christmas Eve, 1923, ICI at Billingham in the North East of England, made its first ammonia. ICI continues to make ammonia, and supplies Northern Ireland with 100 000 tonnes of liquid ammonia per year.

The chemistry of ammonia production continues to change although the essential Haber synthesis is still the same. In 1923 coal was the source of hydrogen. Today it is methane.

(a) Describe the production of ammonia under the following headings.

(i) steam reforming of the hydrocarbon feedstock
(ii) purification and compression of synthesis gas. Synthesis gas is a mixture of carbon monoxide and hydrogen.
(iii) synthesis of ammonia
(b) Suggest reasons why natural gas replaced coal in ammonia manufacture.
(c) Carbon dioxide is made in large quantities during ammonia manufacture. Explain *two* major uses of carbon dioxide. (NI)

14 *A* Describe the industrial process for the manufacture of nitric acid from ammonia, referring to the application of the principles of kinetics and equilibrium to the process where appropriate.

Give **two** large-scale uses of nitric acid.

Explain **each** of the following observations.
(a) Benzene is nitrated by concentrated nitric acid in the presence of concentrated sulphuric acid.
(b) The pH of aqueous ammonium nitrate is less than 7. (CL)

Chapter 13

The industrial production of sulphuric acid

1 *A* Outline the production from zinc blende of (a) zinc, (b) sulphuric acid by the Contact process (no diagrams are required).

How does concentrated sulphuric acid react with (i) ethanol at 180°C, (ii) carbon, (iii) calcium phosphate, (iv) potassium ferrocyanide? (OL)

2 *S* Liebig declared that a nation's industrial pre-eminence may be measured by its consumption of sulphuric acid. Justify this statement by
(a) discussing briefly the acid nature, acid strength, high boiling-point, affinity for water and oxidizing power of sulphuric acid,
(b) describing its use (i) in the fertilizer industry, (ii) in any **two** of the following: plastics, dyes, rayon, galvanizing, chemicals, woollen textiles. (N)

3 *A* Outline an industrial method for the preparation of sulphuric acid, laying emphasis on the physico-chemical principles involved.

How would you prepare from sulphuric acid, using no other compounds of sulphur, specimens of (a) hydrogen sulphide, (b) sulphuryl chloride, (c) pure sulphur dioxide? (L)

4 *A* Outline the essential principles and chemical reactions on which the Contact process for the manufacture of sulphuric acid is based.

'The uses of sulphuric acid are determined by its chemical properties as an *acid, a dehydrating agent* and an *oxidizing agent*.' Justify this statement. (W)

5 *A* Describe the manufacture of sulphuric acid by the Contact process. Technical details are not required but

a **full** explanation of the principles and reactions involved should be given, including the sources of sulphur dioxide. What are the relative advantages of the Contact and Lead Chamber processes? (S)

6 *A* Outline the chemistry of the reactions that occur in the Contact process for the manufacture of sulphuric acid from sulphur, paying particular attention to the choice of reaction conditions.

Sulphuric acid is also manufactured from anhydrite, $CaSO_4$, by heating it with sand, ashes (containing Al_2O_3) and coke in a furnace:

$$2CaSO_4 + C \rightarrow 2CaO + 2SO_2 + CO_2$$

Suggest reasons for
(a) the use of sand and ashes in this method,
(b) the increasing importance in recent years of the use of anhydrite in the Contact process.

When radioactive sulphur, ^{35}S, is heated with aqueous sodium sulphite a colourless solution of a compound **C** is obtained. Acidification of this solution with dilute hydrochloric acid gives a precipitate of a solid which is the only radioactive product of the reaction; the remaining solution contains no radioactivity. What conclusions can you draw from these observations about the structure of compound **C**? Explain the reactions involved as far as you can. (CL)

Chapter 14

Chemicals from salt: the alkali industry

1 *S* Outline the production of some compounds of commercial importance using only common salt, limestone, coke, air and water as raw materials. It may be assumed that necessary catalysts are available but not electrical power. (OL)

2 *A* Give a concise account of the manufacture of washing soda from sodium chloride by the Solvay process. Explain the chemistry of this process, drawing attention to physico-chemical principles involved.

Briefly describe how to determine the weights of sodium bicarbonate and sodium carbonate in one dm^3 of a solution containing both these substances by titration with a standard acid, indicating how the result is calculated. (OL)

3 *A* Name the products obtained during the electrolysis of a concentrated sodium chloride solution with the electrodes well separated and explain the processes which occur at the electrodes. What happens if the solution is stirred during the process or the electrodes are close together, the solution being kept cold?

Indicate briefly the commercial importance of all the products.

Describe **one** industrial method for the preparation of metallic sodium. (N)

4 S How may the following be prepared from sodium chloride: (a) metallic sodium, (b) sodium peroxide, (c) constant boiling hydrochloric acid, (d) chlorine monoxide, (e) bleaching powder?

Describe the principal chemical and physical properties of each. (N)

5 A How are (a) sodium hypochlorite, (b) sodium bicarbonate prepared industrially from brine?

What is meant by the 'strength' of a solution of sodium hypochlorite and how is it determined in the laboratory? (L)

6 A With the aid of equations and *brief* explanatory notes, outline the processes by which (a) sodium, (b) sodium carbonate, (c) sodium hydroxide are obtained from common salt.

Under what conditions does sodium react with (i) hydrogen, (ii) oxygen, (iii) ammonia? What products are formed and what is their action on water? (S)

7 A Describe the manufacture of sodium carbonate by the Ammonia-Soda process. Give an account of the reactions and show the flow of all the materials involved.

Explain (a) why sodium carbonate is alkaline in solution and (b) one reason why the Ammonia-Soda process cannot be adapted to the manufacture of potassium carbonate. (S)

8 OS If an island contained only sodium chloride and calcium carbonate as its main resources, and if hydroelectric power (but no coal) were available, describe the possible types of chemical industry that could be carried on. (C)

9 OS Describe the industrial production of substances using common salt, air, water and limestone (or chalk) as raw materials. (O)

Chapter 15

Hydrogen, the hydrides and water

1 A Give details of (a) the manufacture of hydrogen from water and coke, (b) the hydrogenation of oils to give edible fats, (c) the use of hydrogen in the production of methanol.

What is heavy water? State briefly for what it is used. (OL)

2 S Write an essay on 'Water: its occurrence and its chemical and physical properties'.

[The same credit will be given for treating two or three aspects fully as for covering a larger number more briefly.] (N)

3 A How is hydrogen obtained on the large scale? What are its principal uses? How and under what conditions does it react with (a) halogens, (b) olefins, (c) sulphur? (L)

4 A Give a concise account of the different types of natural waters. What is meant by the hardness of water and how is it estimated?

Describe **two** methods employed for the large-scale softening of hard water. (L)

5 S Discuss the differences shown by the members of the series methane, ammonia, water and hydrogen fluoride. (L)

6 A **Either**, How, and under what conditions, does water react wtih carbon, chlorine, fluorine and iron? What light is thrown on the chemical character of a metal by its reaction with water?

Or, Describe, with examples, **three** general types of reaction in which water can take part. Compare the reactions of water with sulphuryl chloride, bismuth trichloride and sodium carbonate. (O&C)

7 A Describe one method for the manufacture of hydrogen.

Give an account of the reactions that can take place between hydrogen and (a) chlorine, (b) nitrogen, paying particular attention to the effect of varying the conditions of reaction. (O&C)

8 A Describe the reactions which take place between water or steam and each of the following: carbon, iron, sodium carbonate, plaster of Paris, phosphorus pentachloride, calcium carbide, aluminium sulphide.

The sulphate of a monovalent metal contained 55.96% water of crystallization. When treated with barium chloride, 1.61 g of the crystals precipitated 1.16 g of barium sulphate. How many molecules of water of crystallization are contained in the formula? [H = 1, O = 16, S = 32, Ba = 137.] (S)

9 S What is an isotope? Explain the existence of two isotopes of chlorine, how they resemble and how they differ from each other. Given a supply of heavy water, D_2O, how would you prepare: (a) ND_3, (b) DI, (c) C_2D_2, (d) $Al(OD)_3$? (S)

10 S Compare and contrast the properties of the simplest hydride of each of the following elements: carbon, nitrogen, oxygen, sulphur and chlorine. (S)

11 S Give a *comparative* account of the properties of the hydrides of the common non-metallic elements.(CL)

12 A State the conditions, name the products formed, and write equations for the reactions of water (or steam) with the following: (a) carbon, (b) sodium peroxide, (c) ferric chloride, (d) nitrogen dioxide, (e) calcium ethynide.

Write brief notes on the industrial importance of the products obtained in reactions (a) and (e). (CL)

13 OS Compare the methods of preparation and the properties of phosphine, hydrogen sulphide and hydrogen iodide. (C)

14 *A* (a) Describe the reactions of calcium, copper, silver, sodium, strontium and tin with water or steam and arrange the metals in order of decreasing activity.
(b) Write equations **only** for the reactions of (i) hydrogen with chromic and stannous oxides, (ii) heat on the nitrates of stannous tin and silver.
(c) Describe briefly the action of water on sodium chloride and on the lower chlorides of arsenic and phosphorus. State the nature of the electronic bonds involved in each chloride.   (S)

15 *OS* Discuss the methods of formation and compare the properties of hydrides of phosphorus, nitrogen, arsenic and calcium.   (C)

16 *OS* Give an account of the properties of the hydrides of nitrogen, oxygen and fluorine. In what ways and for what reasons, do these compounds differ from one another and from the corresponding compounds of phosphorus, sulphur and chlorine?   (C)

17 *OS* Explain why the chemical properties of the isotopes of a given element are closely similar to one another. In what physical properties would you expect differences between isotopes to be shown?
Given a supply of heavy water, D_2O, how would you prepare samples of (a) DI, (b) ND_3, (c) CaD_2, (d) CH_3OD?   (C)

18 *OS* Outline the chemistry of the hydrides. How far can they be grouped into types?   (O)

19 *OS* Consider regularities and irregularities in two series of compounds:
(a) CH_4, NH_3, OH_2 and FH, and
(b) CH_3CH_3, NH_2NH_2 and HOOH.   (O)

20 *OS* Describe and discuss the reactions of water with **five** of the following: CCl_4, $SiCl_4$, $ZnCl_2$, CO_2, Br_2, LiH, H_2S.   (O)

21 *OS* It is a peculiarity of the Periodic Table that fluorine has one hydride, oxygen two and nitrogen at least three. Discuss these hydrides and explain the observation.   (O)

22 *OS* Certain elements react with water. Give examples of such reactions and attempt to classify them.   (C)

23 *S* Define the term *isotope*. Explain the existence of isotopes of hydrogen, indicating points of similarity and difference in their behaviour. Outline how you would separate isotopes of hydrogen from a mixture.
By writing equations, naming reactants and products, and stating conditions, outline reaction schemes for the preparation of the following from heavy water, D_2O: (a) DI, (b) C_2D_2, (c) CD_4, (d) PD_3, (e) D_2O_2, (f) HD.   (S)

24 *S* '*An acid is a compound which ionizes to give the cation of the parent solvent.*' Write an essay tracing the development of ideas about *acids*. It may be helpful to think about contributions made by Boyle and/or contemporaries, Lavoisier, Davy, Arrhenius, Bronsted and Lowry (and others) up to the present day although a treatment in depth of modern theory will be equally acceptable.   (S)

25 *S* Outline the principal features in the chemistry of ionic and molecular hydrides.
Starting with heavy water, D_2O, as the only source of deuterium write reaction schemes for the preparation of the following: NaD, ND_3, $CH_2D.CH_2Br$, PD_4I, NaD_2PO_2.
Suggest why the compound $LiAlH_4$ is considerably more soluble than $NaBH_4$ in ether.   (O&C)

Chapter 16

Oxygen, oxides and peroxides

1 *A* Classify and define the different kinds of oxides, giving **one** example of each class.
Discuss the classification of (a) red lead; (b) nitrogen dioxide (NO_2).
Briefly describe how you would prepare (i) cuprous oxide from cupric sulphate, (ii) manganese dioxide from manganous sulphate.   (OL)

2 *A* Describe (a) the laboratory preparation of a solution of hydrogen peroxide from hydrated barium peroxide, (b) **two** distinctive tests for hydrogen peroxide. Give **two** reactions in which hydrogen peroxide acts as an oxidizing agent.
$25 \ cm^3$ of hydrogen peroxide solution were diluted to $200 \ cm^3$, then $25 \ cm^3$ of the diluted solution, acidified with dilute sulphuric acid, were found to react with $21.5 \ cm^3$ of 0.11N potassium permanganate†. Calculate the weight of hydrogen peroxide in one dm^3 of the original solution. [H = 1; O = 16.]   (OL)

3 *A* Classify oxides according to their chemical behaviour and constitution. Give **two** examples of each class with illustrative reactions and equations.   (L)

4 *A* Outline (a) an industrial method for obtaining oxygen from the atmosphere, (b) a laboratory method for preparing pure oxygen.
On what grounds is the formula O_3 assigned to the molecule of ozone?   (L)

5 *A* Outline, with essential experimental conditions, **one** method for the preparation of a reasonably pure sample of each of the following: (a) phosphoric oxide P_2O_5 from red phosphorus, (b) lead oxide PbO from lead, (c) sodium peroxide Na_2O_2 from sodium, (d) nitrous oxide (N_2O).
How would you classify each of these oxides? Give reasons to justify your classification.   (N)

† redox normal = one Faraday of electric charge transferred.

6 *A* Describe **one** method of preparing in the laboratory a dilute aqueous solution of hydrogen peroxide.

Explain what is meant by a '20 volume' solution of hydrogen peroxide.

What will be the result of adding hydrogen peroxide solution to (a) lead sulphide, (b) an acidified aqueous solution of potassium iodide, (c) silver oxide? What is the function of the hydrogen peroxide (i) in (a), (ii) in (c)? (N)

7 *A* Classify the following oxides: magnesium oxide, water, nitrogen dioxide, lead dioxide, ferrosoferric oxide and stannous oxide.

Define each class of oxide mentioned and, if possible, give for each example the necessary reactions and equations to justify your grouping. (S)

8 *A* How may a concentrated aqueous solution of hydrogen peroxide be prepared?

Describe **two** reactions in each case in which hydrogen peroxide reacts as (a) an oxidizing agent, (b) a reducing agent.

Calculate the concentration in grams per dm^3 of a 10-volume solution of hydrogen peroxide. [H = 1; O = 16]. (CL)

9 *A* Describe, with the aid of a labelled diagram, the preparation of ozonized oxygen. What reactions take place between ozone and (a) aqueous hydrogen peroxide, (b) acidified potassium iodide solution, (c) lead sulphide, (d) ethene?

When oxygen is heated to 4000°C and cooled rapidly some ozone is obtained, but when the product is heated to about 300°C all the ozone is decomposed. How do you account for this? (CL)

10 *A* Calculate the concentration in grams per dm^3 of a '20-volume' solution of hydrogen peroxide, and describe, with essential practical details, **one** method by which you would determine this concentration in the laboratory. Write the equations for the reactions of an alternative method of estimating hydrogen peroxide.

Explain the reactions which occur when chromic hydroxide reacts with sodium hydroxide solution containing hydrogen peroxide. (S)

11 *OS* How may (a) SO_2, (b) BaO_2, (c) PbO_2, (d) SiO_2 be made in the laboratory?

From your knowledge of the physical properties of these substances what can you infer concerning the nature of the chemical bonding in them? (C)

12 *OS* Cite what evidence you can in support of **three** of the following statements.
(a) Many elements attain higher valencies in oxides than in halides.
(b) The nature of the chemical bond in oxides of formula XO_2 varies widely with variation in the nature of X.

(c) Water molecules in the liquid state exert considerable attractive forces on one another.
(d) Where a metal exhibits more than one valency, increase in valency is accompanied by increase in acidic properties. (C)

13 *OS* Give a short general account of the properties of the metallic oxides.

Explain why it is usual to place PbO_2 in a different class of oxide from BaO_2. Mention other examples of these two classes of oxide, and suggest methods whereby an unknown oxide of empirical formula MO_2 could be assigned to its correct class. (C)

14 *OS* Some elements exhibit more than one valency in their compounds with oxygen. By reference to **three** elements discuss how the properties of the oxides of an element vary as the valency changes. (O)

15 *OS* How is hydrogen peroxide prepared, and why is it given the molecular formula H_2O_2?

Give an account of its chemical reactions and describe how you would proceed to determine the strength of an aqueous solution of it. (O)

16 *S* You are provided with a supply of $H_2^{18}O$. Outline how you would prepare from this water, **six** of the compounds of the following formula with all oxygen atoms labelled ^{18}O:
(a) $KMnO_4$ (b) H_2O_2 (c) HNO_3 (d) $NaNO_2$ (e) K_2CO_3 (f) O_3 (g) CH_3CHO. (S)

Chapter 17

Group I: the alkali metals: lithium, sodium, potassium, rubidium and caesium

1 *A* Describe the laboratory preparation of (a) potassium bromide from bromine, (b) potassium chlorate from chlorine, (c) potassium persulphate from potassium hydrogen sulphate.

What are the differences in the properties of corresponding compounds of sodium and potassium? (OL)

2 *A* How is potassium carbonate prepared on the industrial scale? How, and under what conditions, does potassium carbonate react with: (a) aluminium sulphate, (b) sulphur dioxide, (c) iodine, (d) zinc sulphate? (L)

3 *A* Outline an electrolytic method for the industrial preparation of sodium hydroxide from brine. How, and under what conditions, does sodium hydroxide react with (a) carbon monoxide, (b) ammonium alum, (c) silicon, (d) phosphorus? (L)

4 *A* Compare, with explanations, the reactions of sodium hydroxide and of ammonium hydroxide (or ammonia) with each of the following substances: nickel sulphate; aluminium sulphate; calcium chloride; hot dilute nitric acid; ethanol. (O&C)

5 *A*   A solution *S* containing *p* grams per dm³ of sodium carbonate (anhydrous) and *q* grams per dm³ of sodium bicarbonate was titrated with 0.1 M hydrochloric acid. When phenolphthalein was used as indicator the titre for 25 cm³ of *S* was 10.3 cm³. When methyl orange was used as an indicator the titre for 25 cm³ of *S* was 24.0 cm³.
(a) State clearly with reasons what information was given by each titration.
(b) Calculate the concentrations *p* and *q* of the two salts.

25 cm³ of the solution *S* were gently boiled, care being taken to avoid any loss by spray. Excess of barium chloride solution was then added, and the resulting precipitate filtered, washed, and dried. Calculate the weight of the dry precipitate. [H = 1, C = 12, O = 16, Na = 23, Ba = 137.4.]   (O&C)

6 *S*   Compare and contrast the electronic structures and properties of potassium and sodium, and of their important compounds, to show how their inclusion in the same group of the Periodic Table is justified.

Why are the salts of potassium and sodium used to a greater extent than those of other metals, both in the laboratory and on the industrial scale?   (S)

7 *A*   (a) Give an account of the chemical reactions involved in the manufacture of sodium carbonate by the Ammonia-Soda process.
(b) Discuss *briefly* the chemistry of sulphurous acid and its sodium salts.
(c) Describe, with equations, the processes which occur when a piece of sodium is exposed for a long time to moist air.   (S)

8 *A*   Describe the chemistry of the manufacture of sodium hydroxide from brine. (Details of apparatus are not required.)
Under what conditions does sodium hydroxide react with (a) bromine, (b) sulphur, (c) phosphorus, (d) aluminium, and what products are formed in each case? What volume of 4 M sodium hydroxide solution would be required to react with 9 g of aluminium? [Al = 27.]   (S)

9 *A*   Give an account of the Ammonia-Soda process for the manufacture of sodium carbonate.
Discuss the reactions between sodium carbonate and (a) calcium hydroxide, (b) ferric chloride solution, (c) 'hard' water, (d) barium sulphate.   (CL)

10 *OS*   What do you know of the reactions of sodium hydroxide with (a) metals, (b) non-metals?
Suggest a method for the quantitive analysis of a solution containing both sodium hydroxide and sodium carbonate. Why is sodium carbonate a particularly important substance in volumetric (titrimetric) analysis?   (O)

11 *S*   The elements of Group Ia in the Periodic Table are in the order lithium, sodium, potassium, rubidium, caesium, and their atomic numbers are 3, 11, 19, 37 and 55 respectively. Deduce their probable electronic structures. Use these structures to explain the relative reactivities of the elements, the type of bond they form and the relative stability of their compounds. Thus predict for lithium (Li) and caesium (Cs) the chemical properties of the elements and of their hydrides, hydroxides, carbonates and chlorides.   (S)

Chapter 18

Group II: the alkaline-earth metals: beryllium, magnesium, calcium, strontium and barium

1 *A*   Show how the characteristics of a group in the Periodic Table are illustrated by the properties of the elements calcium, strontium and barium and their compounds. State briefly how the existence of groups is accounted for by the theory of atomic structure. (O&C)

2 *A*   Name **three** magnesium minerals, and give an account of **one** method of extracting magnesium.
State exactly how you would analyse an alloy which you are told consists only of magnesium and zinc, in order to prove the presence of these two metals.
Compare the properties, physical and chemical, of (a) the oxides, (b) the sulphides, of magnesium and zinc.   (OL)

3 *A*   Excluding forms of calcium carbonate, name **three** important calcium minerals.
State how metallic calcium is manufactured.
Starting with limestone, outline the manufacture of (a) bleaching powder, (b) calcium cyanamide. Explain the bleaching action of (a) and the use of (b) as a fertilizer.   (OL)

4 *A*   For **either** magnesium **or** zinc give a brief account of the extraction and properties of the metal, and of the properties or reactions of **four** of its compounds.   (O&C)

5 *A*   Give in outline the chemistry of the metal calcium and of **four** of its compounds which are of importance in nature or industry.   (O&C)

6 *S*   Compare and contrast the properties of barium and calcium, and their important compounds, to show how their inclusion in the same group of the Periodic Table is justified.
Briefly describe the procedure and reactions you would use to prepare pure crystals of barium and calcium chlorides from a mixture of the carbonates of barium and calcium.   (S)

7 *A*   How is magnesium obtained on the large scale?
Starting from magnesium how would you prepare (a) anhydrous magnesium chloride, (b) magnesium oxide, (c) magnesium hydroxide?   (CL)

8 *A*   Give a comparative account of the physical and chemical properties of beryllium, magnesium and calcium (atomic number 4, 12 and 20 respectively) and some of their principal compounds with particular reference to the atomic structures of the three elements and to their location in the Periodic Classification of the elements. (W)

9 S Discuss the properties of magnesium and its compounds in relation to those of sodium and aluminium and their compounds. How can the similarities and differences be correlated with the atomic structures of the three elements? [Atomic numbers: Na = 11, Mg = 12, Al = 13.] (W)

10 A The following is a simple account of the production of magnesium from sea water:

Sea water is concentrated and calcium removed by the controlled addition of carbonate ions (A). The mixture is filtered and the clear solution treated by controlled addition of hydroxide ions (B). The precipitate is thermally decomposed (C) and the residue converted into anhydrous magnesium chloride (D). Magnesium is then obtained by electrolysis (E).

(a) Write equations, as simply as possible, for reactions A, B, C and D.
(b) What is meant by 'controlled addition'?
(c) Outline the process E, emphasizing what you consider to be the important points (details of plant are NOT required).
(d) Suggest why it is difficult to produce magnesium by heating the residue from the thermal decomposition with carbon.

Mention a metallic ore reducible by carbon.
(e) Write TWO commercial uses for magnesium or its compounds.
(f) Group II metals can only form one oxidation state. Using a suitable electronic configuration diagram (outer shells only), show how they achieve this.
(g) Arrange the atoms of the Group II metals in order of decreasing atomic size (i.e. put the SMALLEST last).

Comment on the stability of the metal cation as the group is ascended.
(h) Suggest why beryllium chloride hydrolyses completely, magnesium chloride hydrolyses partially, but barium chloride does not hydrolyse at all.
(i) State the 'flame' colours for calcium, strontium and barium. Suggest why magnesium compounds fail to produce a visible 'flame' colour. (S)

11 A Explain **each** of the following observations in terms of structure and bonding.
(a) The alkaline earth metals (beryllium, magnesium, calcium, strontium and barium) all have a fixed oxidation number in their compounds.
(b) Beryllium chloride (boiling point 547°C) is much more volatile than magnesium chloride (boiling point 1418°C).
(c) Magnesium sulphate is readily soluble in water at room temperature (33 g per 100 g of water at 20°C), but

barium sulphate is only sparingly soluble (2.4×10^{-4} g per 100 g of water at 20°C).
(d) Beryllium oxide is amphoteric, but the oxides of all the other alkaline earth metals are basic.
(e) Magnesium carbonate undergoes thermal decomposition at a much lower temperature (353°C) than barium carbonate does (1430°C). (CL)

12 A The Dead Sea is an inland lake with no outlet. A comparison between Dead Sea water and an average sample of sea water is shown in the table below.

| | Ion Concentrations/g dm^{-3} | |
Ion	Sea water	Dead Sea water
Cl$^-$	19.4	156.0
Na$^+$	10.7	23.0
Mg^{2+}	1.3	37.0
SO$_4^{2-}$	2.7	0.51
Ca^{2+}	0.42	14.7
K$^+$	0.38	3.9
Br$^-$	0.07	6.2

(a) Why do you think the concentrations of anions and cations are usually much higher in Dead Sea water compared to sea water?
(b) Dead Sea water has been an important economical source of certain chemicals. Give TWO examples and explain why they have economic importance.
(c) Describe how you would determine the concentration of magnesium ions in Irish Sea water.

State what modifications you would make to your method if you were to determine the concentration of magnesium ion in Dead Sea water. (NI)

13 S (a) Explain the following:
(i) Aqueous Be(NO$_3$)$_2$, Mg(NO$_3$)$_2$ and Ba(NO$_3$)$_2$ may be distinguished by their different reaction with aqueous NaOH.
(ii) When concentrated HCl is added to aqueous Pb(NO$_3$)$_2$ a white precipitate is formed which dissolves in an excess of the acid. On pouring this solution into a beaker of cold water, the white precipitate is reformed; it redissolves if the beaker of water is warmed.
(b) When aqueous NaF of concentration 0.0010 mol dm^{-3}, is added to aqueous Ca(NO$_3$)$_2$, a precipitate is formed if [Ca^{2+}(aq)] exceeds 4.0×10^{-5} mol dm^{-3}. If the NaF is replaced by HF of the same concentration, [Ca^{2+}(aq)] must exceed 1.5×10^{-4} mol dm^{-3}, before a precipitate is formed. What quantitative information can be deduced from these observations?
(c) What is the pH of a solution made by mixing equal volumes of 0.0010 mol dm^{-3} HF and 0.10 mol dm^{-3} Ca(NO$_3$)$_2$? (CL)

Chapter 19

Group III: the boron group: boron and aluminium

1 A Give an account of the manufacture of aluminium. Explain applications based on **two** physical and **two**

chemical characteristics of the metal. Give a brief description of the chloride, hydroxide and sulphate of aluminium. (O&C)

2 *A* Give details of (a) the production of pure alumina from bauxite, (b) the extraction of aluminium from alumina.

What is an alum? Name **one** alum and state how you would prepare it.

Describe **one** test which would enable you to identify a colourless, gelatinous precipitate as aluminium hydroxide. (OL)

3 *A* Describe the extraction of metallic aluminium from bauxite, explaining the chemical principles underlying the main steps in the process.

How, and under what conditions, does aluminium react with (a) water, (b) caustic soda, (c) chlorine, (d) sulphuric acid? (L)

4 *A* (a) Write an equation, including state symbols, which represents the first ionization energy of an element *M*.

(b) Describe briefly, with the aid of a diagram, how the first ionization energy of an element can be experimentally determined using a simple electron source, electron accelerator and current detector assembled inside a glass bulb filled with sodium vapour at low pressure.

(c) Typical results for the ionization energies of boron are:

Ionization energy kJ mol^{-1}	1st	2nd	3rd	4th	5th
	$+800$	$+2\,400$	$+3\,700$	$+25\,000$	$+32\,800$

(i) On a piece of graph paper, plot $\log_{10}$ [ionization energy/kJ mol^{-1}] against the number of electrons removed from boron.

(ii) Label the graph at each point with the electronic configuration (s, p, d, or f) of the corresponding electron removed.

(iii) Deduce from the graph, with some explanation, the most likely formula of boron oxide.

(d) Write an equation, including state symbols, for the standard enthalpy change which is known as the lattice energy of boron oxide.

(e) The experimental value of the lattice energy of boron oxide is significantly different from that derived theoretically.

What assumption in the theory must be invalid to account for the discrepancy?

(f) For boron fluoride gas:

(i) draw a bond diagram;

(ii) explain why the structure is electron deficient;

(iii) explain, using a bond diagram, how it forms an addition compound with ammonia. (S)

5 *A* Write an account of the occurrence and extraction of aluminium. Describe the preparation of pure specimens of the following from aluminium: (a) aluminium oxide, (b) anhydrous aluminium chloride, (c) potash alum. (CL)

6 *S* Give a *comparative* account of the chemistry of sodium, magnesium, and aluminium. (CL)

7 *A* Give equations and explain how pure aluminium oxide is separated from a sample containing small quantities of the oxides of iron and silicon.

Using sodium chloride (as the source of chlorine) and aluminium, describe briefly, with the aid of a diagram, how you would make anhydrous aluminium chloride.

How and under what conditions does aluminium react with (a) water, (b) sodium hydroxide, and (c) ferric oxide? (S)

8 *OS* Groups 3 and 3a of the Periodic Table consist of the elements boron, aluminium, gallium, indium and thallium. From your knowledge of the chemistry of aluminium and the trends which occur in chemical properties as a group is descended, describe *what you would deduce to be* the main features of the chemistry of gallium, indium and thallium.

Why is it valid to make deductions of this kind? (C)

Chapter 20

Group IV: the carbon group: carbon, silicon, germanium, tin and lead

1 *A* For **either** tin **or** lead, give an account of the extraction and properties of the metal, and of the properties of its compounds with oxygen and with chlorine. (O&C)

2 *A* What is the chief ore of tin, and how is metallic tin obtained from it?

Briefly describe how you would obtain from tin (a) a solution of stannous chloride, (b) anhydrous stannic chloride.

Compare the chemical properties of stannic oxide and lead dioxide. (OL)

3 *S* Discuss the general chemistry of the group of elements carbon, silicon, tin and lead. (OL)

4 *A* Describe, giving a diagram, the preparation in the laboratory of pure carbon monoxide.

Under what conditions and with what results does carbon monoxide react with (a) sodium hydroxide, (b) lead monoxide?

Name **two** industrial gas mixtures of which carbon monoxide is a constituent, and say briefly how these mixtures are made. Comment briefly on the importance of **one** of these mixtures. (N)

5 *A* Name the most important ore of lead and give the formula of the compound of lead which occurs in it.

How may metallic lead be obtained from this ore?

Starting from lead, describe the preparation, in a reasonably pure state, of (a) red lead, (b) lead dioxide, (c) lead sulphate, (d) normal lead carbonate. (N)

6 *A* How would you prepare (a) lead dioxide from lead, (b) lead from lead dioxide? By a comparison of the general chemistry of tin and lead justify their inclusion in the same group of the Periodic Classification. (L)

7 *S* Give an account of the chemical properties of the element tin and describe four of its principal compounds.

The element germanium (Mendeléeff's *ekasilicon*) lies in Group IV of the Periodic Table below carbon and silicon and above tin and lead. What properties would you predict for this element, for its oxide GeO_2 and for its chloride $GeCl_4$? (O&C)

8 *S* **Either**, Carbon monoxide is said to have a structure like that of nitrogen, and carbon dioxide like that of nitrous oxide. What physical and chemical similarities can you find to justify this view? For each pair, give **two** examples where the two compounds differ in their chemical behaviour, and give what explanations you can for these differences.

Or, Give briefly **two** ways of preparing carbon monoxide and describe **four** of its more important reactions. Explain how **two** of these are applied in industrial operations.

How far can the reactions of carbon monoxide be accounted for by its structure? (O&C)

9 *A* Group IV of the periodic table contains the elements carbon, silicon, germanium, tin and lead in increasing order of atomic number. The most common oxidation states for this group are -4, $+2$ and $+4$.
(a) Write the formula of a hydride of silicon in the -4 oxidation state.
(b) Write the formula of an oxide of lead in the $+2$ oxidation state.
(c) Write the formula of an oxo-acid of tin in the $+4$ oxidation state.
(d) Which of these elements is the end product of three radioactive decay series? What effect has this fact on the relative atomic mass of this element?
(e) Write an equation representing the thermal decomposition of the compound in (a).
(f) Put the hydrides of this group in order of increasing stability to heat (i.e. put the MOST stable LAST).
(g) When magnesium and silica (SiO_2) are heated, magnesium silicide and oxide are the two products.
(i) Write the equation for this reaction.
When dilute acid is added to magnesium silicide, silicon hydride is evolved.
(ii) Write the equation for this reaction.
Silicon hydride burns SPONTANEOUSLY in air forming silica.
(iii) Write the equation for this reaction.
(iv) If the conditions of reaction (iii) are compared with the combustion of methane, in what way does this confirm your answer in (f)?
(h) Place the oxides of the elements in this group (all in the $+4$ oxidation state) in order of increasing acidity (i.e.

put the MOST acidic LAST). Give an explanation of the order listed.
(i) The equation:

$$H_2GeO_3(aq) + 2OH^-(aq) \rightarrow H_2O(l) + GeO_3^{2-}(aq)$$

represents the reaction of germanic(IV) acid with hydroxyl ions.
(i) Which species is donating potons?
(ii) Which species is accepting protons?
(iii) What type of reaction is represented by this equation?
(j) The relationship between the oxidation states of tin and lead may be represented:

$$Sn(II) \rightleftharpoons Sn(IV)$$
$$Pb(II) \rightleftharpoons Pb(IV).$$

What conclusions as to the nature of the redox reactions of tin and lead compounds can be deduced from this? (S)

10 *A* Describe **one** laboratory method of preparing carbon monoxide. How, and under what conditions, does carbon monoxide react with (a) sodium hydroxide, (b) water, (c) nickel?

Briefly describe how the volume of carbon monoxide in a sample of coal gas may be determined. (OL)

11 *S* A substance X dissolved in hot concentrated hydrochloric acid, and on treating the solution with caustic soda, a white precipitate formed which dissolved in excess, giving a solution with strongly reducing properties.

On heating X with sulphur, a brown powder Y was formed, which dissolved in hot concentrated hydrochloric acid. Y dissolved on warming with yellow ammonium sulphide solution, and on adding hydrochloric acid, a yellow precipitate was formed.

When X was strongly heated in air, a white powder Z was obtained which could be dissolved only in concentrated sulphuric acid. When Z was fused with caustic soda, and extracted with hot water, white crystals were obtained.

Identify X, Y and Z, and account for the reactions described, and discuss *briefly* the chemical properties of X shown in them. (S)

12 *S* Explain the reactions involved in the extraction of lead from its principal ore.

Give **one** method by which silver may be extracted from impure lead. Point out any important principle that may underly the method you describe.

Explain, from the standpoint of the position of lead relative to that of hydrogen in the electrochemical series, the action of *dilute* solutions of hydrochloric and sulphuric acids on the metal.

How would you prepare lead dioxide from litharge? (S)

13 *S* Compare the chemistry of tin with that of lead in relation to the stability and properties of (a) their chlorides, (b) their oxides, and (c) their sulphides. (CL)

14 *S* Compare and contrast the properties of the oxides of carbon, silicon, tin and lead. Comment on the nature of the bonds in the oxides of carbon and silicon. (CL)

15 *OS* Describe briefly the preparation of the chlorides of the elements carbon, silicon, tin, lead. Discuss the chemistry of these halides in relation to the position of the four elements in the Periodic Table. (O)

16 *OS* Give a comparative account of the properties of the hydrides, oxides and chlorides of carbon, silicon, tin and lead. To what extent do these properties justify the classification together of these elements? (C)

17 *S* Tabulate **two** differences and **two** similarities in the chemistry of carbon and silicon.

When stannic chloride was allowed to react with an excess of ethylmagnesium bromide (C_2H_5MgBr), a liquid *A* was isolated, the vapour density of which was 117; 0.1935 g of *A* gave 0.1240 g of SnO_2 on repeated evaporation with nitric acid. *A* contained no magnesium and gave no reaction with silver nitrate solution.

On heating 1.41 g of *A* with 0.52 g of stannic chloride in a sealed tube for some time, 1.93 g of a liquid *B* were obtained, 0.2240 g of which gave 0.1332 g of silver chloride when treated with silver nitrate; 0.1865 g of *B* gave 0.1164 g of SnO_2 on treatment with nitric acid. The vapour density of *B* was 121.

(a) What is the formula of *A*?

(b) What is the formula of *B*?

(c) Write an equation for the reaction of *A* with stannic chloride to give *B*.

(d) What would you expect to be the product of the reaction of *B* with silver ethanoate?

[H = 1; C = 12; O = 16; Cl = 35.45; Ag = 107.9; Sn = 118.7.] (O&C)

18 *S* Describe and explain the trends in physical and chemical properties of the Group IV elements (carbon, silicon, germanium, tin and lead) and their compounds.

Explain **each** of the following observations as fully as you can.

(a) The first ionisation energy of lead is higher than that of tin.

(b) The carbonate ion is trigonal planar in shape.

(c) Silicon tetrafluoride forms a complex ion with the formula $[SiF_6]^{2-}$, but carbon tetrafluoride does not form a corresponding complex ion. (CL)

19 *A* The empirical formula of an iodide of tin may be determined by the following method:

A sample of tin foil was cleaned on both sides by wiping it with a piece of tissue paper dampened with propanone (acetone). After drying the foil was cut into pieces and weighed.

A sample of iodine was also weighed. Both the tin and iodine were placed in a flask together with 50 cm³ of 1,1,1-trichloroethane. A condenser was fitted to the flask and the contents refluxed for 30 minutes using a water bath.

When the reaction was finished the solution was decanted, and the remaining tin was washed with fresh solvent, dried and reweighed.

Weighings:

Mass of tin at the start of the experiment = 5.12 g

Mass of tin at the end of the experiment = 1.40 g

Mass of iodine = 15.12 g

(a) Draw the structures of 1,1,1-trichloroethane and propanone.

(b) (i) Using the information given at the start of the question draw a labelled diagram of the apparatus used to react tin with iodine.

(ii) Explain what is meant by refluxing.

(iii) How would you know when the reaction was complete?

(c) (i) Deduce the empirical formula of the tin iodide from the weighings made during the experiment (Sn = 119, I = 127).

(ii) Give the systematic name for this iodide of tin whose empirical formula is the same as the molecular formula.

(iii) Write a balanced equation for the reaction of tin with iodine in this experiment.

(d) State the type of bonding you expect to be present in this tin iodide, and give one reason for your choice.(NI)

Chapter 21

Group V: the nitrogen group: nitrogen, phosphorus, arsenic, antimony and bismuth

1 *A* Describe the two main allotropes of phosphorus. How do the ortho-, meta- and pyrophosphates differ in composition and chemical properties?

Elucidate the statement: sodium dihydrogen phosphate is an acid salt with an acid reaction, but disodium hydrogen phosphate is an acid salt with an alkaline reaction. (O&C)

2 *OS* Give an account of the preparation and properties of the oxides of nitrogen. Discuss their structures.(C)

3 *A* How is nitric acid manufactured from ammonia?

Give **three** examples of the differences in behaviour between nitrites and nitrates and **two** examples of the use of nitrites in organic chemistry. (O&C)

4 *A* Describe, with a diagram of the apparatus, the preparation of phosphorus trichloride from phosphorus, stating the precautions necessary to obtain a reasonably pure product.

How is phosphorus pentachloride obtained from the trichloride? What is the effect of heating phosphorus pentachloride, and how does the pentachloride react with (a) a little water, (b) an alcohol, (c) a ketone, (d) a carboxylic acid? (OL)

5 *A* Outline the preparation in the laboratory of (a) nitric oxide, (b) dinitrogen tetroxide (N_2O_4).

Describe how the formula of nitric oxide is derived, giving your reasoning in full.

Under what conditions and with what results does

(i) nitric oxide react with phosphorus, (ii) nitrogen dioxide react with sodium hydroxide solution?   (N)

6 S   Discuss the physico-chemical principles involved in the production of ammonia from its elements and in its oxidation into nitric acid, and the conditions which these principles dictate for the efficient production of ammonia and its subsequent oxidation.

Illustrate **three** types of reaction displayed by nitric acid.

Discuss the effect of heat on nitrates, suggesting reasons for the differing results.   (N)

7 A   White phosphorus reacts with dilute aqueous solutions of copper(II) ions:

$$\tfrac{1}{2}P_4(s) + 5Cu^{++}(aq) + 6H_2O(l) \rightarrow$$
$$2HPO_3(aq) + 5Cu(s) + 10H^+(aq).$$

(a) Draw a bond diagram of the white phosphorus molecule.

(b) Suggest a safety precaution which should be used when performing this reaction. Give a reason for your answer.

(c) Briefly explain how you would obtain pure dry copper from the mixture when the reaction has finished.

(d) What type of reaction does the equation represent?

(e) Calculate the maximum yield of copper obtainable from 0.124 g of white phosphorus (P = 31; Cu = 64).

(f) If this reaction is to be used for the disposal of white phosphorus after a laboratory experiment, which reagent should be in excess? What other compounds might the elements copper and phosphorus form in this reaction if the conditions were changed?

(g) Explain, with the aid of a simple diagram, what TYPE of bonding is present in copper metal and give a brief account to show how the nature of this bonding affords some explanation for TWO physical properties of the metal.

(h) Name a reagent you would use to re-convert the copper solid into $Cu^{++}(aq)$ and write an equation for the reaction.

Considering the phosphorus-containing product of this reaction:

(i) write the oxidation number of phosphorus in it;

(ii) write the name for the compound;

(iii) draw a bond diagram using outer electron orbitals (shells) only;

(iv) say whether it ionises more (or less) readily than the nitrogen compound of comparable formula under the same conditions.   (S)

8 A   Describe the laboratory preparation of nitrous oxide. What impurities are likely to be present in the gas and how can they be removed? How has the molecular formula of this gas been established?   (L)

9 S   Illustrate from the chemistry of the element nitrogen, its oxides and chlorides, the anomalous character of the first element in a Period Group.   (O&C)

10 A   (a) Draw a bond diagram of the nitrogen molecule and explain why nitrogen gas fails to react with most elements.

(b) To what extent does nitrogen react with hydrogen? State the conditions used in the industrial synthesis of ammonia (the Haber Process) to obtain the maximum economic yield of ammonia. Write an equation.

(c) Complete the equations below which refer to the catalytic oxidation of ammonia:

(i) $4NH_3 + -O_2 \rightarrow -NO_2 + 6H_2O$;

(ii) $-NO_2 + 2H_2O + -O_2 \rightleftarrows -$.

(d) Suggest how the equilibrium in (ii) is shifted in favour of the product. Why is this procedure preferred?

(e) In (c) above, a catalyst is necessary. Explain, using an energy level diagram, how the catalyst accelerates the rate of attainment of equilibrium assuming the reaction is essentially exothermic.

(f) Comment briefly on the commercial significance of these reactions.

(g) The nitrogen(IV) oxide in (c) (i) is said to be an 'odd-electron molecule'.

(i) Explain why this is so and suggest why the molecule readily dimerises.

(ii) State the formula of another odd-electron molecule containing nitrogen.

(h) Dilute nitric acid is added to zinc. After some time the mixture is made alkaline and then warmed. A gas which turns damp red litmus paper blue is evolved.

(i) Write down the formulae of the two cations formed in the reaction while it is still acidic.

(ii) Use ONE of these cations as a reactant in writing a balanced equation which represents the formation of the gas in the second stage.

(iii) State the type of reaction which occurs in the second stage.   (S)

11 S   Describe the preparation of orthophosphoric acid, phosphorous acid and hypophosphorous acid from white phosphorus. Compare the structures and the properties of these acids.

How would you distinguish between (a) a phosphate and an arsenate, (b) an arsenite and an arsenate? (OL)

12 S   Give an account of the chemistry of antimony referring in particular to (a) the colour of its principal compounds, (b) the valency of the metal in its halides, (c) the reaction of antimony chloride with potassium iodide, and of antimony sulphides with yellow ammonium sulphide.

In the light of the facts discussed in the previous part of this question, and given that antimony pentachloride can show marked catalytic activity, discuss whether antimony should be considered as a transition metal.   (O&C)

13 S   Make a comparison of the compounds formed by chlorine with nitrogen, phosphorus, arsenic, antimony, and bismuth.

Comment on any differences shown by these compounds which you would *not* expect to find in members of the same periodic group.   (O&C)

14 S Give a concise comparative account of the *similarities* in the chemistry of the elements arsenic, antimony, and bismuth. To what extent are they shared by (a) nitrogen, (b) phosphorus?   (O&C)

15 S *A* is a reddish-violet crystalline solid, having a slightly moist appearance. It is insoluble in water, in the usual organic solvents and in a solution of sodium hydroxide, but it dissolves freely when warmed with concentrated nitric acid, forming an acid *B*. When heated in either oxygen or chlorine, *A* burns, forming dense white fumes which, in each case, dissolve in water to form acid solutions.

Identify *A* and *B* and account for the reactions described.

Describe, giving equations, **two** uses of each of the substances *A* and *B* in the preparation of inorganic **or** organic compounds.   (S)

16 A What method would you use in the laboratory to prepare a specimen of nitrogen? [No experimental details are required.]

Give an example of (a) a metal, (b) a non-metal, (c) a compound, which combines directly with nitrogen, stating the conditions and products in each case. Describe the action of water on the products obtained in (a), (b) and (c) .   (CL)

17 A Describe the industrial preparation of white (yellow) phosphorus. How would you prepare from it (a) phosphine, (b) phosphorus pentachloride, (c) phosphorus pentoxide?   (CL)

18 A An element *X* forms an oxide X_2O_3 which is amphoteric, and an oxide X_2O_5 which is acidic. Suggest:
(a) a formula for an acid corresponding to each oxide;
(b) a method of preparing a pure specimen of the sulphide X_2S_3, which is insoluble in water;
(c) a method of preparing a pure specimen of the hydride XH_3;
(d) the reaction occurring between *X* and concentrated nitric acid.   (CL)

19 A *X* forms an unstable pale blue solution in water, which rapidly decomposes even in the cold. The solution decolorizes bromine water and an acid solution of potassium permanganate and oxidizes a solution of stannous chloride in dilute hydrochloric acid.

Identify *X* and explain the reactions mentioned above. Describe (a) how you would prepare a solution of *X* and (b) the reaction of *X* with ammonia solution. Briefly explain **one** laboratory use of *X*.   (S)

20 S Explain the following facts:
(a) When disodium hydrogen phosphate solution is added to a solution of calcium nitrate, a white precipitate

forms and the remaining solution is neutral. The precipitate is soluble in dilute hydrochloric acid.
(b) When sodium dihydrogen phosphate solution is added to a solution of calcium nitrate there is no precipitate, and the resulting solution is acid.
(c) When disodium hydrogen phosphate is added to silver nitrate solution there is a yellow precipitate and the resulting solution is acid.
(d) When 50 cm³ of solution containing 3.267 g per dm³ of orthophosphoric acid is titrated with 0.1 M sodium hydroxide solution using methyl orange indicator, 16.67 cm³ of alkali are required for the titration. If phenolphthalein is used, 33.34 cm³ are required. If 10 cm³ of 1 M barium chloride solution is added to the acid before titration, 50 cm³ of alkali are required whichever indicator is used. [H = 1, O = 16, P = 31.](O&C)

21 S Give a comparative account of the chemistry of nitrogen and phosphorus, with special reference to the properties of (a) the elements, (b) their hydrides, (c) their chlorides, (d) their oxides and oxyacids.   (CL)

22 A Compare the methods of preparation and properties of ammonia, phosphine and arsine.

Describe *briefly* how you would determine the concentration of a solution of an ammonium salt, indicating how you would calculate the result in g per dm³. [H = 1, N = 14.]   (S)

23 S Enumerate, with examples, the uses made of phosphorus compounds in chemical synthesis and analysis.

On warming a solution of a crystalline salt *X* in dilute nitric acid with ammonium molybdate solution, a yellow precipitate was formed. 1.0 g of *X* was heated with sodium hydroxide solution and the ammonia evolved neutralized 23.9 cm³ of 0.2 M acid. When *X* was heated to constant weight the loss was 51.2%, ammonia was detected and the residue left was sodium metaphosphate. Deduce what you can about *X* and account for the changes described.

How would you use *X* to estimate the amount of magnesium in a compound? [H = 1, N = 14, O = 16, Na = 23, P = 31.]   (S)

24 OS A gas consisting of nitrogen heavily contaminated with impurities produced the following reactions:
(a) A student who deeply inhaled the gas collapsed and was taken to hospital, but unfortunately died without regaining conciousness.
(b) A sample of the gas when passed through a hot tube gave rise to a black deposit which was soluble in sodium hypochlorite solution.
(c) A sample of the gas passed through a solution of potassium permanganate decolorized it and yielded a pale yellow colloidal precipitate.
(d) A sample of the gas when *completely* burnt in oxygen in a closed vessel which was subsequently cooled, yielded

a white deposit which dissolved in dilute sodium hydroxide solution. The resulting solution, when neutralized, yielded a white precipitate with barium chloride solution, and a yellow precipitate with silver nitrate solution which was soluble both in acids and in alkalis.

Deduce the nature of the impurities in the nitrogen, and write equations for all the reactions described above. (C)

25 *A* Phosphoric acid may be made from phosphate rock.
(a) Write an account of the production of phosphoric acid including reference to the source of phosphate rock and the chemical reactions involved.
(b) Giving equations, state how the following phosphates are made and explain why they are suitable for use as fertilisers:
(i) calcium phosphates
(ii) ammonium phosphates
(c) Apart from making fertilisers, explain TWO uses of phosphoric acid and compare its industrial and economic importance with that of the other mineral acids. (NI)

26 *A* Phosphoric acid and phosphates play a major part in the fertiliser industry. Richardsons in Belfast process up to 50 000 tonnes of phosphate rock each year to produce phosphoric acid, ammonium phosphate and compound fertilisers.
(a) What is meant by a 'fertiliser'?
(b) Describe the production of
(i) phosphoric acid from phosphate rock
(ii) ammonium phosphate from ammonia
(iii) a compound fertiliser from ammonium phosphate.
(c) Why are compounds of phosphorus so essential to agriculture? (NI)

27 *A* Hydrazoic acid, HN_3, is a colourless liquid which boils at 37°C. It is poisonous and violently explosive. The anhydrous acid is a poor conductor of electricity.

The salt of hydrazoic acid, sodium azide, is being used to save lives. It is used to produce nitrogen in the protective bags which some modern cars have to prevent the occupants from hitting the windscreen in accidents.
(a) X-ray diffraction studies have shown that the azide ion, N_3^-, is linear. The azide ion may be written as:

$$\bar{N} = \overset{+}{N} = \bar{N}$$

(i) Suggest a reason why the azide ion is linear.
(ii) Name the type of bonding present in hydrazoic acid and state *two* pieces of evidence to support your answer.
(b) When the protective bag containing sodium azide is heated by a shower of metal sparks, within milliseconds it produces nitrogen which inflates the bag and forms a protective cushion.

$$2NaN_3 \rightarrow 2Na + 3N_2.$$

(i) A typical bag contains 60 g of sodium azide. Calculate the volume of nitrogen gas produced at 25°C when the bag inflates during a car accident.

(Relative atomic masses: N = 14; Na = 23; the volume of one mole of a gas at 25°C is about 24 dm^3.)
(ii) The sodium metal produced during the decomposition of sodium azide is 'scavenged' (reacted) with a fine dust of iron(III) oxide. Write a balanced equation for this reaction.
(c) The azide ion has played an important role in the development of theoretical chemistry. It was investigated as a nucleophile in substitution reactions with alkyl halides. e.g.

$$CH_3Br + N_3^- \rightarrow CH_3N_3 + Br^-$$

(i) Define a nucleophile.
(ii) Explain what is meant by a *substitution* reaction.
(iii) Complete the following reaction of the azide ion in which it acts as a typical nucleophile.

$$CH_3C \underset{Cl}{\overset{\diagup\!\!\diagup O}{\diagdown}} \xrightarrow{N_3^-}$$

(iv) Write a mechanism for the reaction of azide ion with bromomethane. This reaction is similar to the reaction of a primary bromoalkane with hydroxide ion. (NI)

28 *A* Interpret **each** of the following observations as fully as you can.
(a) The first ionisation energy of nitrogen is higher than that of either of the elements immediately preceding or following it in the Periodic Table.
(b) The boiling point of ammonia (−33°C) is higher than that of phosphine, PH_3 (−87°C).
(c) The nitrogen trichloride molecule is pyramidal in shape, but the boron trichloride molecule, BCl_3, is trigonal planar in shape.
(d) The bond energy of the nitrogen−nitrogen bond in the nitrogen molecule is 944 kJ mol^{-1}, whereas the bond energy of the carbon−oxygen bond in the carbon monoxide molecule, which is isoelectronic with nitrogen, is 1074 kJ mol^{-1}.
(e) The equilibrium constant, K_p, for the reaction

$$N_2(g) + 3H_2(g) \rightleftharpoons 2NH_3(g)$$

decreases as the temperature rises, whereas K_p for the reaction

$$N_2(g) + O_2(g) \rightleftharpoons 2NO(g)$$

increases as the temperature rises. (CL)

Chapter 22

Group VI: sulphur (for oxygen, see Chapter 16)

1 *A* Give an account of the properties of metallic sulphides. Make special reference to the behaviour of metallic sulphides when treated with dilute and concentrated solutions of (a) hydrochloric acid and (b) nitric acid. (O&C)

2 *A*   (a) State the THREE most stable oxidation states common to sulphur, selenium, and tellurium in Group VI of the Periodic Table.
(b) Write the formula of one simple compound of each element (i.e. containing one other element only) in each of these oxidation states.
(c) Using bond diagrams, or otherwise, comment on or explain:
(i) sulphurous acid, H_2SO_3 is weaker than sulphuric acid, H_2SO_4;
(ii) sulphurous acid is stronger than selenious acid, H_2SeO_3.
(d) Write an equation for the hydrolysis of a chloride of selenium.
(e) Explain why the structure of hydrogen sulphide is not linear.
(f) You are provided with a short thick-walled sealed glass tube half-filled with liquid sulphur dioxide. How would you use it to determine the critical temperature of sulphur dioxide?   (S)

3 *A*   Give an account of the occurrence of sulphur in nature and of the methods used for its extraction. Describe and explain what happens when rhombic sulphur is (a) heated very slowly, (b) boiled with sodium hydroxide solution, (c) warmed with nitric acid.   (L)

4 *S*   Give an account of the uses of sulphuric acid as a reagent in organic chemistry. Mention as wide a variety of examples as your experience permits, but direct attention in particular to those applications which you consider of most importance.   (N)

5 *OS*   What is the evidence for regarding the formula of sulphuric acid as H_2SO_4?
Suggest a structural formula for this compound giving your reasons for it.   (O)

6 *A*   State what occurs when the following substances are treated with concentrated sulphuric acid, warming if necessary. Where possible give equations for the reactions.
Sulphur, charcoal, manganese, potassium bromide, potassium dichromate, cane sugar, a mixture of sand and calcium fluoride.
Explain how the last reaction is used as a test for a fluoride.   (S)

7 *S*   *Y* was prepared as a fine white precipitate by adding dilute hydrochloric acid to a solution of a solid *X*. *Y* was filtered, washed and dried.
On heating a mixture of *Y* and potassium chlorate an explosion occurred. On warming *Y* with concentrated nitric acid and a little bromine, dense brown fumes were given off, and an acid *Z* was left in solution. *Y* reacted slowly with a hot concentrated solution of sodium hydroxide to give *X* and other products.
Identify *X*, *Y*, *Z* and *fully* explain the above reactions.

Discuss the **nature** of the reactions of (a) *X* with chlorine, (b) *X* with silver bromide, and (c) *Y* with carbon. Mention any commercial uses of these reactions.   (S)

8 *S*   (a) Write a brief account of the chemistry of sulphurous acid, its anhydride and salts.
(b) The oxide of an element *X*, containing 85.2% of *X*, combines with chlorine by direct addition, giving an oxychloride containing 24.7% of chlorine. The oxychloride (i) treated with potassium hydroxide gives a salt isomorphous with potassium manganate, (ii) when boiled with hydrochloric acid and ethanol gives ethanal, water, and a chloride of *X*, from which the original oxide may be obtained by treatment with alkali.
Deduce from these data as much as you can about the element *X*, and the reactions described above. [O = 16, Cl = 35.5.]   (S)

9 *S*   What are the principal differences between sulphurous and sulphuric acids?
Sulphur dioxide and chlorine react in the presence of a catalyst to give a compound of formula SO_2Cl_2, which is completely decomposed by water. How would you attempt to show that the decomposition is represented by the equation:

$$SO_2Cl_2 + 2H_2O = 4H^+ + SO_4^{2-} + 2Cl^-?   (CL)$$

10 *A*   Describe how you would prepare (a) a pure specimen of sulphur dioxide, (b) crystals of sulphur trioxide.
How, and under what conditions, does sulphur dioxide react with (i) ferric chloride, (ii) hydrogen sulphide, (iii) potassium dichromate, (iv) lead dioxide?   (CL)

11 *S*   How would you prepare pure specimens of (a) sodium sulphite, (b) sodium thiosulphate? What happens when aqueous solutions of these salts are treated with (i) dilute aqueous hydrochloric acid, (ii) an aqueous solution of iodine? Give equations.
Under what conditions does gaseous sulphur dioxide react with (a) chlorine, (b) oxygen (c) lead dioxide, (d) hydrogen sulphide? What are the products? Give equations for the reactions that take place.   (CL)

12 *S*   Discuss (a) the *similarities* between oxygen and sulphur which justify classifying them as elements in the same group of the Periodic Table, (b) the *differences* which exist between the chlorides and hydrides of nitrogen and phosphorus in spite of their being in the same group.   (S)

13 *OS*   Chlorine was passed into a solution of sodium thiosulphate containing 15.8 g per dm^3 of the anhydrous salt. After the reaction was completed the dissolved chlorine was removed, and the following results were obtained with the resulting solution:
(a) 20 cm^3 required 20.1 cm^3 of M sodium hydroxide solution for neutralization.
(b) 20 cm^3 required 15.9 cm^3 of M silver nitrate solution.

(c) 5 cm^3 gave 0.2340 g of barium sulphate when excess of barium chloride was added.

Write an equation for the initial reaction. [H = 1, N = 14,   O = 16,   Na = 23,   S = 32,   Cl = 35.5, Ba = 137.].   (C)

14 *OS* 'Sulphur is in many ways chemically similar to oxygen.' Discuss this statement in relation to the properties of the following pairs:
(a) hydrogen sulphide and water;
(b) potassium thiosulphate and potassium sulphate;
(c) carbon disulphide and carbon dioxide.

25 cm^3 of an acidified solution of bromine were added to an excess of an acidified solution of potassium iodide. This solution was then equivalent to 25 cm^3 of a solution of sodium thiosulphate. When the bromine solution reacted directly with the same amount of the thiosulphate solution it was found that 200 cm^3 of the bromine solution were required for complete reaction. Obtain an equation for the reaction of bromine with thiosulphate.
(C)

15 *OS* Compare and contrast the properties of (a) ozone and sulphur dioxide, (b) hydrogen peroxide and hydrogen disulphide (H_2S_2), (c) water and hydrogen sulphide. To what do you attribute the differences in behaviour?

When 1.000 g of the anhydrous potassium salt X of an oxoacid of sulphur was ignited to constant weight sulphur dioxide was given off, and 0.731 g of potassium sulphate remained as residue. Oxidation of 1.000 g of X with nitric acid yielded 1.146 g of potassium hydrogen sulphate as the only product, no compound containing sulphur being lost during oxidation. Calculate the formula of the salt X, assuming that the oxoacid is dibasic. [K = 39, S = 32.]   (C)

16 *OS* How does sulphuric acid react with (a) carbon, (b) phosphorus, (c) hydrogen sulphide, (d) phosphorus pentachloride?

Give an account of the processes involved in the electrolysis of sulphuric acid.   (C)

17 *S* Account for the chemistry of selenium by predicting the properties of the element and its important compounds by reference to the position of the element in the periodic table.   (S)

18 *A* Describe briefly a convenient laboratory reaction for the preparation of sulphur trioxide. What product is formed in its reaction with hydrogen chloride?

Summarize the nett reactions (mechanisms are not required) of the 'lead-chamber' method of preparing sulphuric acid and describe a suitable test for the detection of small amounts of arsenic in the product.

Under what conditions and with what results does sulphuric acid react with any **three** of the following: (a) copper, (b) potassium iodide, (c) ethanol, (d) nitric acid, (e) ethanedioic acid, (f) graphite?   (W)

Chapter 23

Group VII: the halogens: fluorine, chlorine, bromine and iodine

1 *A* By reference to the properties of the halogens and of their compounds with hydrogen, explain the features you would expect to find in a Group in the Periodic Table.   (O&C)

2 *A* Explain why hydrogen bromide is not prepared from sodium bromide and sulphuric acid. Describe, with a diagram of the apparatus, how hydrogen bromide is usually made in the laboratory. Give **one** method of obtaining hydrobromic acid.

Compare the physical and chemical properties of freshly precipitated specimens of silver chloride, bromide and iodide.   (OL)

3 *A* Give an account of the extraction of iodine from the mother liquors remaining after the extraction of sodium nitrate from Chile saltpetre.

Outline the preparation of (a) hydrogen iodide, (b) iodic acid, from iodine.

Compare (i) the effect of heating hydrogen chloride and hydrogen iodide, (ii) the action of oxidizing agents on hydrochloric acid and hydriodic acid.   (OL)

4 *A* Describe with essential practical details how you could obtain in the laboratory reasonably pure specimens of (a) anhydrous magnesium chloride, (b) phosphorus pentachloride starting from phosphorus, (c) anhydrous stannous chloride.

What will be the result of adding an excess of water to each of these chlorides? What light is thereby thrown on the nature of the elements magnesium, phosphorus and tin?   (N)

5 *A* Describe briefly how you would prepare in the laboratory dry samples of (a) hydrogen chloride, (b) chlorine.

Under what conditions and with what results, if any, do these gases react with (i) iron, (ii) excess of ammonia gas, (iii) dry slaked lime, (iv) potassium bromide solution?   (N)

6 *A* 'Chlorine, bromine and iodine form a family of elements remarkable for their general similarity of chemical behaviour and for the gradation of their physical properties.' Discuss and illustrate this statement.   (L)

7 *OS* Discuss the reactions of halogens with sodium hydroxide.   (O)

8 *A* How, starting from iodine, would you prepare (a) a saturated aqueous solution of hydrogen iodide, (b) potassium iodide crystals? Describe **one** method by which you could find the concentration of the solution obtained in (a).   (CL)

9 *A*   Give an account of phosphorus pentoxide and of iodine pentoxide. Comment on any important differences between these two oxides.   (O&C)

10 *A*   Describe briefly what you would observe when concentrated sulphuric acid is warmed with (a) solid sodium chloride, (b) solid sodium bromide. Give equations and account briefly for the formation of the products of these reactions.

How is bromine extracted from a naturally occurring source?

What products are formed when bromine reacts with (i) red phosphorus and a little water, (ii) an aqueous solution of sulphur dioxide?   (N)

11 *A*   Briefly describe the large-scale production of chlorine from sodium chloride solution (**one** method only).

How does chlorine react with (a) ammonia in excess, (b) solid calcium hydroxide, (c) sodium hydroxide solution? Write equations for the reactions in (a) and (c).

What is an acid chloride? State how **one** inorganic acid chloride is prepared.   (OL)

12 *S*   Elaborate the statement: 'Metals are often distinguished from non-metals by the properties of their chlorides, but there are so many exceptions to this method of differentiation that it should be applied only with caution.'

Include phosphorus pentachloride and ferric chloride in your discussion and write electronic formulae for these chlorides and for **three** of the others which you discuss.   (N)

13 *A*   Give a comparative account of the compounds formed by hydrogen with chlorine, bromine and iodine, showing how the properties of these compounds justify the inclusion of the three elements in the same periodic group. Comment on or account for major differences in the behaviour of these compounds.   (O&C)

14 *S*   Give an account of the element bromine with special reference to (a) its manufacture and laboratory preparation and (b) the ways in which it differs in its chemistry from chlorine and iodine.   (O&C)

15 *A*   How may fluorine be prepared, and in what ways does fluorine and its compounds differ from chlorine and its compounds?

Describe, in outline, experiments which you would make to establish the constitution of hydrogen chloride.   (S)

16 *A*   (a) Outline **one** method by which **each** of the following—chlorine, bromine, and iodine—is usually obtained from natural sources.
(b) How is *bleaching powder* prepared?
(c) Describe a method for estimating the *available chlorine* in bleaching powder and explain why such an estimation is necessary.   (S)

17 *A*   Describe what may be seen, and state the chemical reactions which take place, when an aqueous solution of potassium iodide is added to aqueous solutions of each of the following: (a) lead ethanoate, (b) mercuric chloride, (c) copper sulphate, (d) potassium dichromate acidified with sulphuric acid, (e) propanone and sodium hypochlorite.   (CL)

18 *A*   Describe in outline the manufacture of (a) chlorine, and (b) hydrochloric acid.

State and explain **two** important uses for each of these substances.

Relate **four** properties of chlorine, or of its compounds, to its position in the Periodic Table.   (CL)

19 *A*   How may hydrogen iodide be prepared?

How, and under what conditions, does it react with (a) sulphuric acid, (b) ethanol, (c) chlorine? Explain the theoretical basis of **two** methods by which the concentration of an aqueous solution of hydrogen iodide could be determined.   (CL)

20 *A*   Name the main source of bromine, and describe briefly how the element may be extracted from it. Discuss the reactions which take place between bromine and each of the following: (a) hydrogen sulphide, (b) aqueous sulphur dioxide, (c) sodium hydroxide, (d) ethene.(CL)

21 *S*   Compare and contrast the chemistry of chlorine and iodine. Describe **two** tests by which you would distinguish a sample of potassium iodate from one of potassium chlorate.   (CL)

22 *A*   Describe and account for the action of (a) concentrated sulphuric acid, (b) concentrated phosphoric acid on: (i) sodium chloride, (ii) sodium bromide, (iii) sodium iodide.

Explain how you would prepare a 'constant boiling mixture' of hydrochloric acid from sodium chloride. Give reasons for the method you describe.   (S)

23 *A*   Describe the manufacture of iodine from the salt deposits in Chile.

How and under what conditions does iodine react with (a) potassium hydroxide, (b) sodium thiosulphate?

3.810 g of a copper salt were dissolved in water and made up to 250 cm^3. To 25 cm^3 portions of this solution excess of potassium iodide was added and the iodine liberated was found to require 15.15 cm^3 of N†/10 sodium thiosulphate solution. Calculate the percentage of copper in the salt. [Cu = 63.5.]   (N)

24 *OS*   Describe how, starting from calcium fluoride, you would prepare (a) an aqueous solution of hydrogen fluoride, (b) anhydrous hydrogen fluoride, (c) fluorine.

Discuss the separate reactions of fluorine and chlorine with (i) alkalis, (ii) sulphur, (iii) carbon.   (C)

† redox   N   (normal) ≡ one Faraday of electric charge transferred.

25 *OS* Outline the general methods for the preparation of chlorides. Give directions for the preparation of aluminium chloride, ferrous chloride, carbonyl chloride, and chromyl chloride . (C)

26 *OS* Suggest a scheme for classifying, according to their structure and properties, compounds of the general formula A_aX_x, where A is any element and X a halogen. Indicate, with reasons, how the following compounds would fit into your scheme: HF, AgBr, ICl, $HgCl_2$, OF_2, KI_3, BF_3, $AuCl_3$, NCl_3, $SnCl_4$, PCl_5, SF_6. (C)

27 *OS* The compounds NCl_3, $AlCl_3$, PCl_3, $AuCl_3$, $BiCl_3$ and $FeCl_3$ have the same formula type. Do you expect them to have similar properties? (O)

28 *OS* Give a general account of the chemistry of iodine. Do you consider that, in its chemistry, iodine shows signs of metallic character? (O)

29 *A* Describe **three** general methods for the preparation of the anhydrous chlorides of the metals giving **one** illustrative example of each method. Indicate the limitations, if any, of each method.

How may phosphorus pentachloride and chromyl chloride be prepared? Give equations for the reactions of these two compounds with excess of water and explain how you would demonstrate qualitatively that the products shown in your equations have been formed. (W)

30 *A* Describe suitable reactions for the preparation of any **four** of the following chlorides: (a) sulphuryl chloride, (b) phosphorus oxychloride (c) ferrous chloride, (d) ferric chloride, (e) stannous chloride, (f) stannic chloride, (g) aluminium chloride, (h) thionyl chloride.

Comment on the type(s) of valency in the examples you have chosen and describe their reactions with water.(W)

31 *S* Explain briefly how a solution of hypochlorous acid may be prepared and describe its reactions with (a) hydrochloric acid, (b) hydrogen peroxide, and (c) ammonium chloride. (W)

32 *A* (a) Compare and contrast the reactions of the chlorides of the group IV elements with water. Your answer should include reference to the types of bonding in the chlorides.
(b) Sulphur forms a number of chlorides which are completely hydrolysed by water. The concentration of chloride ion in the resulting solution can be determined by titration with silver nitrate solution.

$$Ag^+ + Cl^- \rightarrow AgCl$$

0.590 g of a chloride of sulphur were added to 100 cm³ of water and completely hydrolysed. The hydrochloric acid produced was neutralised with calcium carbonate. 10 cm³ of the solution was titrated with 0.1 mol dm⁻³ silver nitrate solution. An end point was obtained with 11.4 cm³ of the silver nitrate solution. Calculate the simplest formula of the sulphur chloride

(Relative atomic masses: S = 32; Cl = 35.5).

(c) Explain, with examples, the use of phosphorus pentachloride as a reagent in organic chemistry. (NI)

33 *A* Today the production of hydrochloric acid is largely by the reaction of chlorine with hydrocarbons and not from the reaction of salt with sulphuric acid. Other methods of production include the reaction of hydrogen with chlorine.

On the other hand, hydrofluoric acid is still made by the reaction of sulphuric acid with fluorspar.
(a) Discuss the chemistry of the production of hydrochloric acid by the two methods mentioned above.
(b) Discuss the chemistry of the production of hydrofluoric acid.
(c) Discuss the uses and relative importance of hydrochloric acid (hydrogen chloride) and hydrofluoric acid (hydrogen fluoride). (NI)

34 *A* The abundance of halide ions in sea water is shown below:

Ion	Abundance/g dm⁻³
Cl^-	18.98
Br^-	0.065
I^-	0.06
F^-	0.001

(a) Calculate the number of moles of halide ion in one dm³ of sea water

(Relative atomic masses: F = 19; Cl = 35.5; Br = 79; I = 127).

(b) Explain the source of halide ions in sea water.
(c) Describe how you would show, by a titration method, that there are 18.98 g of chloride ion in one dm³ of sea water.
(d) Explain the presence of the chlorine containing compounds DDT, dichlorodifluoromethane and polychlorobiphenyls in sea water. (NI)

35 *A* State what you would observe in **each** of the following experiments. Explain these observations with the aid of the *Data Booklet* where appropriate.
(a) Solid iodine is shaken with aqueous potassium iodide; aqueous sodium thiosulphate is then added.
(b) Aqueous iron(III) sulphate is added to aqueous potassium iodide and a few drops of starch indicator are then added.
(c) Aqueous chlorine is added to aqueous iron(II) sulphate and aqueous sodium hydroxide is then added.
(d) Chlorine is passed into aqueous sodium hydroxide at room temperature and the mixture is then heated. (CL)

36 *S* Suggest explanations for **each** of the following observations.
(a) At room temperature, chlorine is a gas, bromine is a liquid and iodine a solid.
(b) Gaseous iodine and a solution of iodine in CCl_4 are both purple, but a solution of iodine in C_2H_5OH is brown.

(c) Aqueous solutions of HCl, HBr and HI (approximately 1 mol dm^{-3}) are almost completely ionised, but solutions in concentrated ethanoic acid are ionised to approximately 5, 20 and 50% respectively.
(d) When heated with the following concentrated acids, NaCl(s), NaBr(s) and NaI(s) behave as follows:
(i) with phosphoric(v) acid, H_3PO_4, all give the hydrogen halide in good yield.
(ii) with sulphuric(vi) acid, H_2SO_4, only NaCl(s) gives the hydrogen halide in good yield.
(iii) with selenic(vi) acid, H_2SeO_4, all are converted into the free halogens.
(iv) with telluric(vi) acid, H_6TeO_6, there is no reaction.   (CL)

37 *S*   (a) The standard enthalpy changes of formation, $\Delta H_f^{\ominus}$ of the hydrogen halides are shown in the following table.

	$\Delta H_f^{\ominus}/kJ\,mol^{-1}$
HF	-271
HCl	-92.3
HBr	-36.4
HI	$+26.5$

Discuss the trend in these data in relation to the thermal stability of the hydrogen halides.
(b) The acidic strength of the hydrogen halides in aqueous solution may be related to the enthalpy change for the process

$$HX(aq) \rightarrow H^+(aq) + X^-(aq)$$

where X is a halogen.
By constructing a suitable Born–Haber cycle and using the data in the following table, calculate the standard enthalpy change for this process for hydrofluoric acid and hydrochloric acid.

Reaction	$\Delta H^{\ominus}/kJ\,mol^{-1}$	
	X = F	X = Cl
$HX(aq) \rightarrow HX(g)$	$+\ 48$	$+\ 18$
$HX(g) \rightarrow H(g) + X(g)$	$+\ 568$	$+\ 432$
$H(g) \rightarrow H^+(g) + e^-$	$+1312$	$+1312$
$X(g) + e^- \rightarrow X^-(g)$	$-\ 328$	$-\ 349$
$H^+(g) \rightarrow H^+(aq)$	-1091	-1091
$X^-(g) \rightarrow X^-(aq)$	$-\ 515$	$-\ 381$

Comment on your answers in relation to the strength of hydrofluoric acid ($K_a = 6.7 \times 10^{-4}\,mol\,dm^{-3}$) and hydrochloric acid.
(c) Chlorine forms four oxoacids, $HClO_n$, where $n = 1$, 2, 3 or 4. The dissociation constants, K_a, of these acids are shown in the following table.

Acid	$K_a/mol\,dm^{-3}$
HClO	4.0×10^{-8}
$HClO_2$	1.1×10^{-2}
$HClO_3$	strong
$HClO_4$	acids

Comment on the trend in these data and explain it as far as you can in terms of the structures of the different oxoacids.   (CL)

Chapter 24

Period 4: transition and associated elements

1 *A*   State what you understand by the term 'transition metal'. Give **two** examples from the chemistry of each of the metals manganese, iron and copper to illustrate their transitional character.   (O&C)

2 *S*   Give a comparative account of the chemistry of chromium, manganese and iron, and of their more important compounds.   (OL)

3 *S*   Describe (a) the laboratory preparation from manganese dioxide of crystalline potassium permanganate, (b) the preparation and standardization of an approximately 0.02M solution of the permanganate, (c) the various uses of such a solution in volumetric (titrimetric) analysis.   (OL)

4 *S*   What is a complex salt? Describe the laboratory preparation of crystalline specimens of **two** compounds containing complex ions, starting from simple substances normally available in the laboratory. Discuss concisely the valency forces in, and the structures of, the compounds you have chosen.   (L)

5 *S*   Describe the appearance of ferric ammonium alum, $(NH_4)_2SO_4 \cdot Fe_2(SO_4)_3 \cdot 24H_2O$.
Outline (i.e. without details of actual procedure) the qualitative and quantitative experiments you would make to establish its formula.
Calculate the titration volumes of reagents and the weights of final precipitates you would obtain in these experiments from 25.0 cm^3 portions of a solution containing 48.2 g of the compound per dm^3.
[H = 1,   N = 14,   O = 16,   S = 32,   Fe = 56, Ba = 137.5.]   (N)

6 *S*   Using pure potassium dichromate as a standard and *without using iron or an iron compound*, how would you proceed to standardize a solution of potassium permanganate of unknown concentration? Explain fully the reasons for each step.
Given that 25 cm^3 of a solution containing 4.900 g of dried potassium dichromate per dm^3 are equivalent to 24.3 cm^3 of a solution of potassium permanganate, what is the percentage purity of a sample of potassium tetroxalate, $KH_3(C_2O_4)_2 \cdot 2H_2O$, if 0.160 g of the sample in acidified solution requires 23.9 cm^3 of the permanganate solution?
Explain (a) why acid is added in the reaction between ethanedioate and permanganate ions, and (b) why no indicator is required.
[Atomic weights: C = 12, O = 16, K = 39, Cr = 52.]   (N)

7 *S*   Under what conditions is potassium permanganate reduced to (a) potassium manganate, (b) manganese dioxide, (c) manganese sulphate? Give equations.

It is said that manganese sulphate, formed by the initial reaction of acidified potassium permanganate and ethanedioic acid when they are warmed together, catalyses the subsequent stages of the reaction. By what experiments could you attempt to verify the truth of this statement?   (O&C)

8 *S*   Give the formulae for, and explain the constitutions of **four** of the following: hexammine nickel bromide; potassium ferrocyanide; sodium cobaltinitrite; the 'brown-ring' compound (nitrosyl ferrous sulphate); chrome alum.   (O&C)

9 *A*   Give practical details for the preparation of (a) cuprous chloride from cupric oxide, (b) potassium ferricyanide from potassium ferrocyanide, (c) chromium trioxide (CrO_3) from potassium dichromate. For each of the three compounds prepared, state **two** properties of interest.   (OL)

10 *S*   Describe (a) the laboratory preparation and the manufacture of potassium dichromate, (b) the preparation of a standard solution of potassium dichromate, (c) the different uses of such a solution in volumetric (titrimetric) analysis.   (OL)

11 *A*   (a) What valency is shown by chromium in each of the following compounds: (i) $CrCl_3$; (ii) $CrSO_4, 7H_2O$; (iii) Cr_2O_3; (iv) CrO_2Cl_2; (v) $K_2Cr_2O_7$; (vi) K_2CrO_4? [You are not asked to discuss the type of valency.]
(b) How would you convert (i) $CrCl_3$ into $K_2Cr_2O_7$, and (ii) K_2CrO_4 into chrome alum? In (i) give the reactions only; in (ii) give the experimental details.   (O&C)

12 *S*   Give an account of the chemistry of iron illustrating particularly
(a) its transitional character;
(b) the use of its compounds in analytical chemistry.   (O&C)

13 *S*   **Either**, Give an account of the properties, reactions and uses of sodium thiosulphate.
Or, Give examples of the use of (a) complex formation, (b) variation in valency, in the identification of any **three** transition metals.   (O&C)

14 *S*   State briefly with essential practical details how you would prepare cuprous chloride from copper.

Give the reactions by which you could make a sample of barium chloride from barium sulphate and a sample of stannous chloride from tin.

Explain or comment on the following:
(a) Cuprous chloride is insoluble in water and in dilute hydrochloric acid, but dissolves in concentrated hydrochloric acid.
(b) Barium chloride is soluble in water, less so in dilute hydrochloric acid and still less so in concentrated hydrochloric acid.

(c) If stannous chloride is treated with water an insoluble compound is formed which dissolves with difficulty in dilute hydrochloric acid but readily in concentrated hydrochloric acid.   (O&C)

15 *S*   Discuss the following experimental findings.
(a) An orange solution was treated with concentrated sulphuric acid and from the mixture were obtained long scarlet needle-like crystals. These were dissolved in water, a quantity of ammonium hydroxide added, and yellow crystals obtained on crystallization. These crystals when heated gave a green powder which, after fusion with sodium hydroxide in the presence of air, dissolved in water to give a lemon yellow solution. On adding dilute sulphuric acid to this solution the colour changed and on concentration orange-coloured crystals were obtained.
(b) A white solid, insoluble in water, dissolved readily in ammonium hydroxide. On adding to this solution a solution of potassium iodide, a yellow precipitate was obtained. After filtration, this precipitate was dissolved in a solution of sodium cyanide. The resulting solution dissolved granulated zinc, a black powder being precipitated.
(c) Some red crystals were dissolved in water to give an orange-yellow solution. When this solution was treated with dilute sodium hydroxide and hydrogen peroxide it turned pale yellow and liberated a gas which was neutral to litmus. The crystals obtained from the residual solution were treated with concentrated sulphuric acid. Carbon monoxide was liberated leaving a solution from which pale green crystals were obtained on concentration and cooling. These on heating yielded an acid gas and left a reddish-brown powder as residue.   (O&C)

16 *S*   Compare and contrast the nature and behaviour of solutions of double and complex salts. Illustrate your answer by **two** examples of each type.

Write a concise account of the use of complex ion formation in qualitative analysis.   (S)

17 *A*   Describe the preparation of pure crystals of ferrous ammonium sulphate starting with iron.

How would you show that your crystals are free from ferric iron? Outline how you would determine the percentage of iron in the crystals by a volumetric (titrimetric) method.

Explain *briefly* your reasons for believing that ferrous ammonium sulphate is a *double* salt. Contrast **three** of its reactions with those of a *complex* iron salt.   (S)

18 *S*   Describe how (a) sodium cyanide, (b) potassium ferrocyanide, and (c) potassium dichromate may be prepared, and give an account of their properties and reactions. What would you expect to happen when, by using platinum electrodes, an aqueous solution of potassium ferrocyanide is electrolysed?   (CL)

19 *A*   Name **one** common ore of zinc, and describe briefly how the metal may be extracted from it.

With zinc as starting material, how would you prepare a sample of (a) anhydrous zinc chloride, (b) zinc carbonate, (c) brass?

What is the position of zinc in the electrochemical series of the metals? (CL)

20 *A*   Molar solutions of copper sulphate (solution *A*) and potassium sulphate (solution *B*) were mixed in the proportions shown below. The mixed solutions were then allowed to evaporate to dryness and the residues examined.

(a) 10 cm^3 of *A* and 5 cm^3 of *B* gave a mixture of light blue and darker blue crystals.

(b) 10 cm^3 of *A* and 10 cm^3 of *B* gave light blue crystals only.

(c) 5 cm^3 of *A* and 10 cm^3 of *B* gave a mixture of light blue and colourless crystals.

The light blue crystals were found to contain 24.4% of water of crystallization.

Interpret these results as fully as you can and suggest a formula for the light blue crystalline substance. What ions would you expect to be present in its aqueous solution? Give **one** test for the identification of each ion. [H = 1, O = 16, S = 32, K = 39, Cu = 64.] (CL)

21 *A*   Explain why iron and nickel are grouped together in the Period Classification, and describe the properties of these elements and their compounds to show how they differ from each other and from adjacent elements in the Periodic Classification. (S)

22 *A*   Describe, with full practical details, how you would make (a) crystals of ferric chloride, and (b) anhydrous ferric chloride. Write the equation for **one** method of preparation of anhydrous ferrous chloride.

Compare and contrast the properties of ferrous and ferric chlorides. (S)

23 *A*   Explain what is meant by the term 'transition element'.

Give an account of the characteristic properties of transition elements with some physico-chemical explanation of these properties. Choose specific examples from THREE elements to illustrate your answer.

0.5 g of vanadium(v) salt, in aqueous solution, is treated with sulphur dioxide until reaction is complete. In this reaction, vanadium is reduced from vanadium(v) to vanadium(iv). The excess gas is boiled out and the solution found to require 55.0 cm^3 0.02 M potassium manganate(vii) (potassium permanganate) solution to oxidise it back to vanadium(v). Find the percentage weight of vanadium in the sample (V = 51). (S)

24 *A*   A colourless crystalline solid *X*, was dissolved in water. On addition of ammonium sulphide solution a white precipitate was formed. This precipitate was dissolved in the least possible amount of hydrochloric acid and the hydrogen sulphide was boiled off. When a solution of ammonia was added to the resulting solution,

a white precipitate was formed, which was soluble in excess of the ammonia solution.

On heating *X* with concentrated sulphuric acid, a reddish acid gas was given off, part of which condensed to a dark brown liquid.

Identify *X* and explain the reactions involved. Describe and explain what happens when solutions of ammonium chloride and ammonia are together added to a solution of *X*. (S)

25 *OS*   How is potassium dichromate obtained from chromite? Explain how you would make use of potassium dichromate (a) to prepare crystals of chrome alum, (b) to prove that a sample of sodium bromide contained chloride, (c) to make an ethereal solution of the oxide CrO_5. (C)

26 *OS*   Sulphur and chromium have been included in the same group of the Periodic Table; how far can this be justified? (C)

27 *A*   In the preparation of $[Cu(NH_3)_4]SO_4.H_2O$, excess ammonia solution is added to copper(ii) sulphate solution. Ethanol is very carefully added but poured down the side of the reaction vessel to form an upper layer. After standing for twenty-four hours, the crystals of the salt are filtered at the pump, washed with ethanol, washed with ether, and dried by air suction.

(a) Write an ionic equation for the reaction, representing it as an equilibrium.

(b) (i) Why is an excess of ammonia solution used?

(ii) Why is the ethanol added to form a definite upper layer?

(iii) Why is ethanol added rather than evaporating carefully?

(iv) Why are the crystals first washed with ethanol and then with ether?

(c) Name the prepared substance using the I.U.P.A.C. system.

(d) An aqueous solution of this prepared salt and a solution of the salt $(NH_4)_2Cu(SO_4)_2.6H_2O$ are separately tested with (i) hydrogen sulphide solution, (ii) warm sodium hydroxide solution, and (iii) barium chloride solution. Say what differences are observed, if any, tabulating and explaining your answer.

(e) Explain what happens if concentrated hydrochloric acid is added gradually to an aqueous solution of the prepared crystals. (S)

28 *S*   When a green solid *A* is fused with sodium carbonate in the open air, a coloured salt *B* is formed. A solution of pure *B* gives a coloured precipitate with solutions of lead and barium salts. On extracting *B* with water and adding sulphuric acid, a further colour change takes place and crystals of *C* are obtained on evaporation. When warmed with alcohol, an acidified solution of *C* emits a characteristic smell and on evaporation and cooling, deposits dark violet crystals *D*. On warming *C* with ammonium sulphate a colourless gas is produced and *A* is reformed.

Identify A, B, C and D and explain the reactions above. Describe what happens when a mixture of C and sodium chloride is warmed with concentrated sulphuric acid. How is this reaction used to show the presence of the chloride in a mixture of chloride, bromide and iodide? (S)

29 A When a black powder, A, was heated strongly it gave off a colourless non-flammable gas. When A was boiled with excess dilute sulphuric acid and a little ethanedioic acid, a non-flammable odourless gas was evolved. The solution formed gave no precipitate when saturated with hydrogen sulphide, but this saturated solution gave a light-coloured precipitate when ammonia solution was added to it.

A small sample of A was warmed with an excess of concentrated hydrochloric acid and the gas liberated was passed into a concentrated solution of potassium iodide. This solution was made up to 200 cm^3 and 20.0 cm^3 of it was found to require 18.0 cm^3 of decimolar sodium thiosulphate solution for decolorization.

Identify A, explain the reactions involved, and calculate the weight† of A taken in the last experiment. Briefly suggest an alternative estimation of A. (S)

30 OS Outline the preparation of (a) chlorosulphonic acid, (b) chromyl chloride.

When a sample of potassium chlorochromate, $KCrClO_3$, was heated, gas was evolved and a solid residue was obtained. When the gas was shaken with aqueous potassium iodide, iodine was liberated and half the gas disappeared. The remainder proved to be oxygen. When the solid residue was extracted with water, chromic oxide was left behind, and the solution was of a reddish colour and contained chloride. When the solution was treated with dilute sulphuric acid and potassium iodide, three times as much iodine as was liberated by the gas was set free.

Write an equation to describe how the salt decomposed when it was heated. (C)

31 OS Discuss the chemistry of manganese in relation to its position in the Periodic Table. (O)

32 OS Why are iron, cobalt and nickel placed in the same group in the Periodic Table, and in that order? (O)

33 A Describe and explain the reactions used to prepare the following substances from copper sulphate:
(a) copper,
(b) cuprous chloride,
(c) cupric chloride,
(d) copper tetra-ammine sulphate monohydrate.

Give full experimental details of the preparation of any **one** of the above substances. (S)

34 S How would you prepare a pure specimen of (a) cuprous chloride, (b) cuprous oxide? What changes occur when (i) oxygen, (ii) carbon monoxide, (iii) ethyne is bubbled through ammoniacal solutions of cuprous

chloride? What happens when cuprous oxide is treated with (a) concentrated hydrochloric acid, (b) dilute sulphuric acid? Give equations where possible. (CL)

35 A Discuss the features of the electronic configurations of the transition elements of the series starting with Scandium, Sc (no. 21), which distinguish them from the preceding elements.

By reference to examples of your own choice, illustrate the variation of the following properties of the elements and their compounds in this transition series: (a) ionic character, (b) basic properties, (c) stability of oxidation states, (d) complex-forming tendencies. (S)

36 S Give a short account of the effect of complex formation on the chemical properties of metal ions in solution with special reference to **one** of the elements chromium, manganese, nickel or copper.

When a solution of ferrous ions was oxidized to one of ferric ions with an acidified solution of potassium permanganate, 20 cm^3 of the permanganate solution were required to reach the equivalence point. Repeating the titration, but in the presence of a large excess of fluoride ions, however, 25 cm^3 were required. How can you account for this result? (O&C)

37 A (a) Describe how you would prepare a sample of chrome alum crystals using the following ingredients:

> 50 cm^3 water
> 7.0 g potassium dichromate(VI)
> 6 cm^3 concentrated sulphuric acid
> 3 cm^3 ethanol (or sulphur dioxide,
>   as required, from a canister)

Comment on the safety precautions taken in this particular preparation and indicate the expected observations as the preparation proceeds.
(b) Explain the redox changes which take place during this preparation.
(c) The yield of chrome alum crystals obtained in a preparation using the above quantities of reactants was 21.5 g.

Calculate the percentage yield based on the mass of potassium dichromate(VI) used.

(Relative atomic masses: H = 1; O = 16; S = 32;
                         K = 39; Cr = 52);

(d) Suggest TWO reasons why the yield was less than 100%. (NI)

38 A (a) Many transition metal complexes absorb energy in the visible region of the electromagnetic spectrum. Explain how such absorptions arise, illustrating your answer by reference to the complexes $[Cu(NH_3)_4(H_2O)_2]^{2+}$ and $[Cu(NH_3)_2]^+$.
(b) Complexes of copper(II) ions with water, ammonia and ethane-1,2-diamine are respectively pale blue,

† Candidates were given a table of relative atomic masses.

blue-violet and purple. Account for this change in colour in terms of the ligands involved.

(c) Visible spectroscopy may be used in quantitative analysis by using the Beer–Lambert law

$$A = \varepsilon cl$$

where A is the absorbance, ε is a constant, c is the concentration of the species and l is the path length of the cell.

Titanium and vanadium both form coloured complexes with hydrogen peroxide. Separate solutions, each containing 5.00 mg of these materials, were treated with acid and hydrogen peroxide, and diluted to $100\, cm^3$. A third solution was prepared by dissolving 1.00 g of an alloy containing both titanium and vanadium (but no interfering metal) and treating in the same manner as the standard solutions. The absorbance of each of the three solutions was measured at 410 nm and 460 nm in 1 cm cells with the following results.

Solution	Absorbance 410 nm	Absorbance 460 nm
Titanium	0.760	0.515
Vanadium	0.185	0.250
Alloy	0.715	0.657

By assuming the metal ions behave independently and that titanium and vanadium have approximately the same relative atomic mass, calculate the percentage of titanium and of vanadium in the alloy. (CL)

39 S   A deep-red crystalline powder **A** is converted by treatment with acids into a violet substance **B**. If **A** and **B** are separately dissolved in liquid ammonia, and the liquid ammonia allowed to evaporate, the same yellow substance **C** is obtained.

When **A**, **B** and **C** are heated they all liberate ammonia gas to yield the same dark coloured solid **D**, which when strongly heated loses some chlorine. When equal masses of **A**, **B** and **C** are heated it is found that **C** gives the greatest mass of ammonia and **B** gives the least.

0.250 g of **A** was dissolved in water and made up to $100\, cm^3$. $10.0\, cm^3$ portions of this solution were titrated against $0.010\, mol\, dm^{-3}$ silver nitrate and $20.5\, cm^3$ were needed to precipitate the free chloride. Another $10.0\, cm^3$ portion of the original solution was boiled and a green precipitate **E** was formed; the solution above this precipitate needed $30.7\, cm^3$ of the same silver nitrate solution to react with the chloride. The green precipitate **E** was filtered off and found to be readily soluble in aqueous sodium hydroxide; this alkaline solution turned yellow when warmed with hydrogen peroxide, and the yellow solution changed to orange when treated with an excess of sulphuric acid. On warming with ethanol, the orange solution turned green and the acrid smell of a gas **F** detected. On evaporation of the solution, a green powder **G** was formed.

When these experiments were repeated with 0.250 g of **B** the chemical reactions were identical but the values of the titration readings were 11.0 and $33.0\, cm^3$ for the unboiled and boiled solutions respectively. With 0.250 g

of **C**, both the unboiled and boiled solutions gave the same titration reading of $28.7\, cm^3$.

Deduce the likely identity of substances **A** to **G** and draw a reaction scheme to illustrate the reactions given above. (CL)

40 A   Discuss the relative stability of the principal oxidation states of iron in relation to the relevant electronic structures.

Give a concise account of **each** of the following aspects of the chemistry of iron.

(a) The effect of pH and of cyanide ligands on the relative stability of Fe^{2+} and Fe^{3+}.

(b) The catalysis by iron(III) ions of the reaction of iodide ions with peroxodisulphate(VI) ions.

$$2I^-(aq) + S_2O_8^{2-}(aq) \rightarrow I_2(aq) + 2SO_4^{2-}(aq)$$

(c) The corrosion of iron and its prevention, paying particular attention to the electrochemical aspects.

[Standard electrode potential:
$$Fe(OH)_3(s) + e^- \rightarrow Fe(OH)_2(s) + OH^-(aq)$$
$$E^0 = -0.56\, V.]$$ (CL)

41 A   (a) Describe the extraction of pure nickel from nickel(II) oxide.

Give **two** large-scale uses of nickel.

(b) Describe and explain what happens when aqueous ammonia is gradually added to aqueous nickel(II) sulphate until it is present in an excess.

(c) Aqueous nickel(II) sulphate reacts with ethane-1,2-diamine, $NH_2CH_2CH_2NH_2$ (usually denoted by 'en'), to form a complex ion, $[Ni(en)_3]^{2+}$, in which the ethane-1,2,diamine acts as a bidentate ligand.

(i) Explain the meaning of the term *bidentate ligand*.

(ii) What is the co-ordination number of the nickel in this complex ion?

(iii) What is the oxidation number of the nickel in this complex ion?

(iv) Suggest a possible shape for this complex ion.

(v) State and briefly explain whether you would expect this complex ion to be chiral. (CL)

Chapter 25

Period 5: silver

1 A   If a solution of silver nitrate is added to a fairly concentrated neutral solution of a salt, a white precipitate *soluble in dilute nitric acid* may be formed. Name four anions which can give such a precipitate, and give for each anion one other reaction in solution, and one reaction which can be applied to a solid salt, which would confirm its presence. (O&C)

2 A   Give an account of (a) the extraction of silver by the cyanide process, (b) the desilverization of lead by Parkes's process.

Briefly describe (i) how silver nitrate is made from silver, (ii) how silver oxide is obtained from silver nitrate, (iii) **one** use of ammoniacal silver oxide solution.

What silver compound is employed in electroplating? Explain how the metallic silver is liberated during the electrolysis. (OL)

3 *S* Give an account of the reactions of compounds of (a) iron and (b) silver which can be used for (i) the qualitative detection and (ii) the quantitative determination of **other** elements or radicals. (O&C)

4 *A* Copper and silver are classified together in the periodic system. Compare and contrast the chemistry of these elements, and discuss the relation of the group to its neighbours in the Periodic Table. (S)

5 *OS* Illustrate, and explain, the fact that the chemistry of silver differs from that of copper and sodium. (O)

6 *OS* When ammonium persulphate is added to a solution of silver nitrate a black unstable solid is precipitated, which has been described as a higher oxide of silver. Suggest methods for testing the truth of this statement, and for establishing the formula of the new oxide. (O)

7 *OS* Describe and explain what happens when solutions of (a) ammonia, (b) potassium iodide, (c) potassium cyanide, (d) sodium carbonate, are added to (i) copper sulphate solution, (ii) silver nitrate solution.(O)

Chapter 26

Period 6: mercury

1 *A* Give the name and formula of the chief ore of mercury. How is pure mercury obtained from this ore?

Briefly describe the preparation of mercuric chloride from mercury. State how, and under what conditions, mercuric chloride reacts with (a) stannous chloride, (b) potassium iodide, (c) hydrogen sulphide. (OL)

2 *A* Give the name and formula of one ore of mercury. How is the metal (a) extracted from this ore, (b) purified?

Starting from the metal, how would you prepare specimens of (c) mercurous chloride, (d) mercuric chloride?

What deductions have been made from a study of the vapour density of mercurous chloride at different temperatures? (L)

3 *A* Give an account of the occurrence, extraction and purification of mercury.

Describe the reactions which occur when an aqueous solution of mercuric chloride is treated with an excess of (a) an aqueous solution of potassium iodide, (b) an aqueous solution of stannous chloride, (c) an aqueous solution of sodium hydroxide, (d) copper. (L)

4 *S* Discuss, with examples, the difference between double salts and complex salts, and describe the preparation of one typical member of each class.

The freezing-point of a molar solution of potassium iodide is $-3.2°C$. When mercuric iodide (which is insoluble in pure water) is added in excess, the freezing-point of the solution is raised to $-2.4°C$. Comment on this result. (CL)

5 *S* How could the freezing-point of an aqueous solution be determined accurately?

The freezing-point of a molar solution of potassium iodide is $-0.362°C$; when 4.546 g of mercuric iodide are dissolved in 20 cm³ of this solution the freezing-point rises to $-0.272°C$, whereas when 1.083 g of mercuric oxide are dissolved in 20 cm³ of the solution of potassium iodide, the freezing-point rises only to $-0.317°C$. What explanation can you offer for these results? [O = 16.0, K = 39.1, I = 127.0, Hg = 200.6.] (CL)

6 *OS* Outline **one** line of argument in each case, based on experimental evidence, which establishes that zinc is divalent in zinc chloride, and that mercury is not monovalent in mercurous chloride. (C)

7 *OS* Give a brief account of the chemistry of mercury. In a solution of an ionized mercury salt in equilibrium with metallic mercury, the fraction of dissolved mercury present as the mercuric ion is independent of the concentration of the mercurous ion. What does this tell you about the nature of the mercurous ion? (O)

8 *OS* Describe the characteristic features of the chemistry of mercury. How far do the chemical properties of mercury justify its inclusion in the same group of the Periodic Table as zinc? (O)

Miscellaneous Questions

1 *A* In each of the following experiments (a) name the gas evolved, (b) explain the reaction by which it is formed, (c) give one test by which it could be identified.

(i) Magnesium powder is added to a cold aqueous solution of aluminium chloride.

(ii) A mixture of concentrated sulphuric acid and potassium chloride is warmed with an oxidizing agent.

(iii) White phosphorus is boiled with an aqueous solution of potassium hydroxide.

(iv) Solutions of sodium nitrite and ammonium chloride are mixed and heated.

(v) Sodium bisulphate is heated with sodium chloride.

(vi) Concentrated sulphuric acid is dropped on to sodium methanoate. (N)

2 *A* Discuss the chemical reactions which give rise to the following observations.

(a) To a solution of ethanedioic acid, acidified with dilute sulphuric acid and warmed to 60°C, potassium permanganate solution is added. The colour of the potassium permanganate is quickly discharged.

(b) When concentrated sulphuric acid is added to potassium bromide an acid fuming gas is evolved together with brown vapours. A piece of filter paper soaked in potassium dichromate solution and held near the mouth of the tube is turned green.

(c) To an ammoniacal solution containing copper ions and cadmium ions, potassium cyanide solution is added until the blue colour is destroyed. On hydrogen sulphide being passed through this solution a yellow precipitate is obtained.

(d) When hydrogen sulphide is passed into an acid solution containing a high concentration of nitrate ion a white turbidity is seen.   (N)

3 S   Explain the following:
(a) The solubility of alum in water increases slightly with pressure.
(b) A solution of ferrous sulphate gives no precipitate when hydrogen sulphide is passed in but a black one is thrown down on subsequent addition of sodium ethanoate solution.
(c) An element may be electropositive but have a negative electrode potential.
(d) Sodium peroxide is a salt.
(e) On adding hydrochloric acid to a solution containing sodium sulphite, sodium iodide, sodium iodate and a trace of starch no change may be observed for a considerable time. Then the mixture suddenly turns blue.(N)

4 A   (a) For the first three elements of the third period of the Periodic Table, sodium, magnesium, aluminium,
(i) compare their reactivity towards oxygen, water, and chlorine;
(ii) compare the variation in properties of their oxides (basic/acidic nature), chlorides (hydrolysis), and hydrides (nature of structural unit).
(b) Explain why aluminium hydroxide but NOT magnesium hydroxide is precipitated when ammonia solution and ammonium chloride solution are added to a solution containing Mg^{2+} and Al^{3+} ions.
(c) Suggest why aluminium chloride but NOT magnesium chloride can be used as a Friedel–Crafts' catalyst.   (S)

5 S   Consider the reactions which might be possible between pairs of the following reagents: hydrogen peroxide, potassium dichromate ($K_2Cr_2O_7$), potassium iodide, silver nitrate, hydrochloric acid, lead nitrate, sodium hydroxide.

Write equations and conditions for each reaction and mention the TYPE of change which occurs.

In what ways can some of these reactions be used in analytical chemistry?   (S)

6 S   Account for and illustrate the following statements.
(a) Simple cuprous ions do not exist in solution.
(b) Ammonia can act as an acid as well as a base.
(c) Chromium can form both complex anions and complex cations but potassium can do neither.
(d) Dynamic allotropy can be shown by elements only in the gaseous or liquid state.   (O&C)

7 S   An aqueous solution contains one or more of the following cations:

$$Ag^+, Ba^{2+}, Pb^{2+}, Fe^{3+}, Al^{3+}.$$

Devise a scheme by which you could separate and identify the cations present. Sulphur-containing compounds are NOT available as reagents.   (S)

8 A   Give **two** tests in each case by which you could differentiate between solutions of: (a) mercurous and mercuric nitrates, (b) stannous and stannic sulphates, (c) zinc and aluminium chlorides, (d) calcium and barium bicarbonates, (e) sodium bromide and sodium iodide, (f) sodium zincate and sodium aluminate.   (O&C)

9 A   Explain **six** of the following phenomena:
(a) the addition of excess caustic alkali to a solution of aluminium sulphate gives a clear solution;
(b) both methanoic and ethanoic acids give white precipitates with silver nitrate but the methanoic acid precipitate turns black on warming;
(c) iron immersed in a solution of copper sulphate becomes coated with copper;
(d) boiling barium hydroxide solution dissolves white phosphorus;
(e) iodine is liberated from a solution of iodate on adding sodium bisulphite;
(f) an aqueous suspension of chromic hydroxide when boiled with sodium peroxide gives a yellow solution;
(g) when potassium iodide solution is added to mercuric chloride solution a precipitate, which changes colour, is first formed but dissolves in excess of potassium iodide.   (L)

10 S   For each of the following statements write a short critical comment explaining why the statement is justifiable or drawing attention to any inaccuracy in it.
(a) Zinc is oxidized by hydrochloric acid.
(b) Ammonium chloride is readily formed from ammonia and hydrogen chloride and is easily decomposed by heat to these two gases.
(c) Sodium salts are hydrolysed in aqueous solution.
(d) Iodine dissolves in aqueous sodium thiosulphate.
(e) The heat of neutralization of an acid by a base is 13.7 kcal.   (N)

11 A   Describe how you would show experimentally that:
(a) sulphuric acid is dibasic;
(b) the molecular formula of hydrogen chloride is HCl;
(c) manganese dioxide catalyses the decomposition of aqueous hydrogen peroxide.   (CL)

12 S   Identify the compounds A, B, C, D and E, and explain the reactions described.
(a) When a solid A was fused with a mixture of potassium hydroxide and potassium chlorate a green compound formed. On treating a solution of this compound with carbon dioxide, it gave a dark coloured solution which

was decolorized on warming with a mixture of dilute sulphuric and ethanedioic acids.

(b) When a solution of B was slowly added to a concentrated solution of potassium iodide a yellow precipitate immediately formed, which rapidly turned scarlet and then redissolved. When A and B were mixed and warmed with sulphuric acid, a gas of characteristic smell and colour was evolved.

(c) C was a sodium salt whose solution gave a characteristic precipitate with ferrous ammonium sulphate solution. An alkaline solution of hydrogen peroxide reduced a solution of C and pale yellow crystals of D were obtained on evaporation. When D was heated with concentrated sulphuric acid carbon monoxide was evolved.

(d) When sodium bisulphite was added to a solution of E, black crystals, having a vapour of characteristic colour, were formed. This vapour reacted with a solution of sodium hydroxide to give E as one of the products of the reaction. (S)

13 A Identify the substances A, B and C and explain the following reactions.

(a) When nitric acid was added to a bright red powder A and warmed, it changed to a dark brown solid. On filtering and washing, and adding concentrated hydrochloric acid to the brown substance, chlorine was evolved on warming, leaving a yellowish-white residue on cooling.

(b) When dilute sulphuric acid was warmed with a dark red powder B, it gave a blue solution containing a pinkish brown residue. On adding sodium hydroxide solution to the blue solution a pale blue precipitate formed which turned black on heating.

(c) When a mixture of ethanol and dilute sulphuric acid was warmed with the solution of a red crystalline solid C, a green solution was formed, which on cooling deposited purple crystals isomorphous with potash alum. (S)

14 A The part periodic table below is to be used to answer this question.

Periods	I	II	III	IV	V	VI	VII
				Groups			
2	Li	Be	B	C	N	O	F
3	Na	Mg	Al	Si	P	S	Cl
4	K	Ca	Ga	Ge	As	Se	Br
5	Rb	Sr	In	Sn	Sb	Te	I

(a) List, as in the printed table, the formulae of the oxides in Period 3, using, in each case, the highest oxidation number of the element and the oxidation number −2 for oxygen.

(b) List the oxides (a) in order of increasing basic nature (i.e. put the MOST basic LAST).

(c) From the list (a), write the formula of the oxide which has

(i) explosive properties;

(ii) a giant ionic lattice structure;

(iii) a giant molecular lattice structure;

(iv) a discrete molecular structure but is not explosive.

(d) For those elements, as in the printed table, which form IONIC hydrides

(i) write the symbols of any TWO such elements from different groups;

(ii) write the IONIC formulae of the two hydrides of the two elements chosen in (i);

(iii) say what you would need to do experimentally to obtain hydrogen using any ONE of these two hydrides.

(e) For those elements, as in the printed table, whose chlorides hydrolyse RAPIDLY:

(i) write the symbols of any FOUR such elements;

(ii) write an equation for any ONE hydrolysis reaction;

(iii) say what you would observe when the reaction (ii) is performed in the laboratory.

(f) It is often said that the element at the top of each group is NOT typical of the rest of the group. By choosing THREE suitable examples of elements and reactions, briefly illustrate this comment. (S)

15 OS (a) When a gaseous compound containing carbon was heated by a platinum spiral it decomposed without change of volume and deposited sulphur. When exploded at 20°C with an excess of oxygen, the diminution in volume was equal to half the volume of the gas: the diminution in volume of the residue after treatment with a solution of sodium hydroxide was twice the volume of the gas.

Suggest a formula for the compound and indicate very briefly how you would attempt to prepare it.

(b) A volumetric (titrimetric) method for the determination of zinc is based on the fact that when a solution of a zinc salt is titrated with aqueous potassium ferrocyanide an insoluble precipitate is formed. Iron(II) interferes. This interference is decreased by adding an excess of ferricyanide to the solution of the zinc salt before titrating it. The iron reduces some of the ferricyanide and ferric ferrocyanide is very insoluble. If iron(II) is present and this expedient is adopted, would you expect the final result for the zinc to be high or low, and by what amount per g atom of iron? (C)

16 OS Identify the lettered compounds and solutions and comment briefly on the chemistry involved in the following series of reactions:

The black compound G is insoluble in water or aqueous sodium hydroxide, but readily dissolves in hot hydrochloric acid to give a green solution H. When solution H is boiled with copper wire an almost colourless solution I is obtained and, on dilution with water, solution I deposits a white solid J.

J dissolves in aqueous ammonia to give a colourless solution K, but on exposure to air, solution K rapidly turns blue to give solution L. The blue colour is discharged when aqueous potassium cyanide is added to solution L, and solution M is thereby obtained.

Treatment of solution M with zinc powder gives a red-brown precipitate N which is insoluble in dilute bases, but reacts with warm dilute nitric acid to give a blue solution O. Treatment of O with sodium hydroxide solution yields a blue jelly-like precipitate P, which when filtered off and heated vigorously forms the original compound G. (C)

17 *OS* When 1.148 g of the crystals of a compound were heated, ammonia and water were evolved and 0.936 g of an oxide of an element were left. The oxide was unchanged by further heating in oxygen. It was soluble in an aqueous solution of sodium hydroxide, when it was found by the method of back titration that 0.520 g of sodium hydroxide were equivalent to 0.936 g of the oxide. The VD of the fluoride was 105. Suggest a relative atomic mass for the element.

When 1.148 g of the original crystals were dissolved in sodium hydroxide, 0.297 g of the base was neutralized. Write an equation to describe what happens when the compound is heated. [F = 19, N = 14, NaOH = 40.](C)

18 *OS* The element vanadium (V) occurs in the Periodic Table as indicated below:

$$\begin{array}{cccc} & C & N & O & F \\ & Si & P & S & Cl \\ K & Ca & Sc & Ti & V & Cr & Mn \end{array}$$

Discuss the properties you would expect it to have. (O)

19 *OS* What are the general features of the chemistry of the transition metals? Relate these to the chemistry of any one transition metal. (O)

20 *OS* If you were given a metal how would you set about discovering its chief valency by examining its chemical reactions? (O)

21 *OS* Classify and give examples of the reactions which occur between metals and mineral acids under various conditions. (O)

22 *S* Account for **three** of the following observations.
(a) Of the halogens fluorine is the most powerful oxidizing agent and iodine the least powerful.
(b) Elements in the first row of the Periodic Table are often anomalous in their properties compared with the later elements in their groups.
(c) A transition element may have several oxidation states.
(d) Xenon forms compounds more readily than does Krypton. (O&C)

23 *A* (a) Arrange the following metals in descending order of chemical activity according to the electrochemical series: aluminium, copper, iron, lead, magnesium, sodium; and also the following non-metals; bromine, chlorine, fluorine, iodine, oxygen, sulphur.
(b) Explain **briefly** and illustrate the meaning of **four** of the following: complex salt, amphoteric oxide, monotropy, efflorescence, formula weight, equivalent of a reducing reagent.
(c) Identify the following gases and where possible write equations for the reactions which take place.
(i) Burns in air; reacts with a solution of iron(III) (ferric) chloride to give an almost colourless solution and a pale yellow precipitate.

(ii) A colourless gas which turns brown on exposure to the air and forms a blackish-brown solution into aqueous iron(II) (ferrous) sulphate.
(iii) A colourless, pungent smelling gas which dissolves in water to give an aqueous solution with a pH greater than 7. The gas burns readily in oxygen but not in air. (OL)

24 *A* Identify the substances in (a), (b) and (c), and one substance only from each pair in (d) and (e). Explain your reasoning and write equations to represent the chemical reactions which occur.
(a) A is a colourless solid which melts at 44°C; when heated with aqueous sodium hydroxide it gives off a gaseous product which is spontaneously flammable in air. If the solid is exposed to sunlight it reddens very slowly, but the change may be hastened considerably by heating the solid in the absence of air to about 260°C.
(b) B is a colourless solid; it melts when heated and gives off a colourless gas which is a good supporter of combustion. If the heating is continued the whole of the solid disappears. When B is heated with aqueous sodium hydroxide an alkaline gas is evolved.
(c) C is a colourless fuming liquid which turns to a yellowish-white solid when treated with chlorine in the cold. This product, like C, reacts violently with water to form two different acids.
(d) **Either**, D is a solid which dissolves readily in water to give a solution which decolorizes acidified permanganate and liberates iodine from acidified potassium iodide solution. D gives a yellow flame test; and brown fumes are given off when a dilute acid is added to it.

Or, a colourless solid E when heated with concentrated sulphuric acid gives off a colourless gas which burns with a blue flame. When its aqueous solution is added in excess to a solution of ferrous sulphate, the product gives a dark blue coloration when iron(III) (ferric) ions are added. E gives a lilac colour in the flame test. Warning: gas is highly toxic.
(e) **Either**, F is a yellow solid which is only sparingly soluble in hot water to give a solution which reacts with hydrogen sulphide to give a black precipitate. When F is heated with concentrated sulphuric acid it gives off a violet vapour and a salt is formed which is insoluble in water.

Or, G is a colourless fuming liquid; its aqueous solution gives (i) a yellow precipitate with hydrogen sulphide and (ii) a white precipitate with aqueous sodium hydroxide which dissolves when an excess of the alkali is added. (OL)

25 *A* Compare the chemistry of a metal in an A subgroup of the Periodic Table with that of a metal in the corresponding B subgroup, with particular reference to calcium (Group IIA) and zinc (Group IIB). Give reasons for the similarities and differences in chemical behaviour. (Limit your answer to a discussion of the properties of the elements and any three compounds of each element with a common anion.) (O&C)

26 *A* Provide explanations and two illustrative reactions or properties for each of the following statements.
(a) The hydrides NH_3, H_2O, HF are progressively more acidic.
(b) The oxides P_2O_3, As_2O_3, Sb_2O_3, Bi_2O_3 are progressively more basic.
(c) The chlorides NaCl, $MgCl_2$, $AlCl_3$ are progressively more hydrolysed in aqueous solution.   (O&C)

27 *A* Suggest methods of volumetric (titrimetric) analysis which you could use to determine **three** of the following elements. In each case (a) state how you would obtain a suitable solution of the substance to be analysed, (b) outline the principle of the method including essential conditions and the means of detecting the end-point, (c) give equations for the significant reactions.
(i)   Copper in brass.
(ii) Chlorine in carnallite, KCl, $MgCl_2$, $6H_2O$.
(iii) Nitrogen in ammonium nitrate.
(iv) Iron in ferric chloride.
(v) Calcium in limestone.
   Descriptions of experimental procedures are **not** required.   (O&C)

28 *S* Discuss, as fully as you can, the reaction schemes below:

(a)

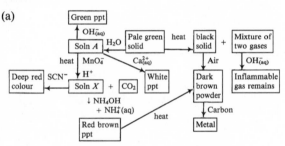

(b)

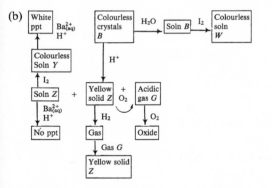

(c)

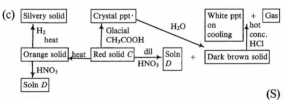

(S)

29 *A* Explain, in detail, the chemistry involved in FOUR of the following reactions. In each case describe what you would expect to observe and give equations where appropriate.
(a) An aqueous solution of sodium hydroxide is gradually added to an aqueous solution of aluminium sulphate until the alkali is in great excess.
(b) A spatula measure of phosphorus pentachloride is added to a beaker containing water to which a few drops of universal indicator solution have been added.
(c) An aqueous solution of chlorine is added to a separating funnel containing a mixture of aqueous potassium iodide and 1,1,1-trichloroethane. The contents of the funnel are then shaken and allowed to settle.
(d) Cobalt(II) carbonate, a purple solid, is suspended in water and concentrated hydrochloric acid is added dropwise until an equal volume of acid has been added.
(e) An aqueous solution of sodium nitrite is added to a solution of phenylamine in hydrochloric acid at 5°C. The mixture is then poured into an alkaline solution of phenol.   (NI)

30 *A* Comment on the similarities *and/or* the differences between the following pairs of compounds.
(i)   tetrachloromethane and silicon tetrachloride,
(ii) sodium hydride and hydrogen chloride,
(iii) carbon dioxide and silicon dioxide,
(iv) sulphur dioxide and sulphur trioxide.
   For each pair consider:
(a) formulae,
(b) physical state,
(c) type of bonding,
(d) reaction, if any, with water.   (NI)

31 *A* Comment on FOUR of the following. Make use of equations and reaction conditions where appropriate.
(a) Aqueous solutions of sodium hydrogencarbonate and sodium hydrogensulphate have pH values above and below seven respectively.
(b) Aluminium hydroxide is soluble in dilute sodium hydroxide solution and hydrochloric acid, but is insoluble in ammonia solution.
(c) Bromine reacts readily with benzene in the presence of a catalyst, but UV light is necessary for its reaction with hexane.
(d) When chlorine is dissolved in cold sodium hydroxide solution, and acidified silver nitrate solution is added only half of the chlorine is precipitated as silver chloride. When the sodium hydroxide solution is heated five sixths of the chlorine may be precipitated as silver chloride.
(e) Phenol is readily soluble in aqueous sodium hydroxide, but does not dissolve in aqueous sodium hydrogencarbonate.
(f) The shapes of the water molecule and the molecule of sulphur hexafluoride illustrate the effect of repulsion between electron pairs.   (NI)

Answers to numerical and problem sections

Chapter 1

1 (b) Volumes: methane $4.0\,cm^3$, ethene $3.0\,cm^3$, ethyne $3.0\,cm^3$

Chapter 2

11 (c) (i) 3.96×10^{-19} J

12 $^{226}_{88}Ra \rightarrow {}^{4}_{2}He(^{4}_{2}\alpha) + {}^{222}_{86}Rn$; $L = 6.04 \times 10^{23}\,mol^{-1}$

14 (a) (i) $^{131}_{53}I \rightarrow {}^{0}_{-1}e\,(\beta^-)^{86} + {}^{131}_{54}Xe$ (+neutrino)
(iii) $(6.4 - 8.2)\,10^8$ atoms dm^{-3}

Chapter 5

14 (b) (ii) -861 kJmol^{-1}

Chapter 7

9 4.9 g dm^{-3}

10 N/15

18 $M^{x+}/x = 31$, $M:O = 127:16$, probably CuO and Cu$_2$O

25 (b) (ii) 32.8 g dm^{-3}

26 (b) $S_2O_3^{2-} + 4Cl_2 + 5H_2O \rightarrow 2SO_4^{2-} + 8Cl^- + 10H^+$ (c) $KHC_2O_4.H_2C_2O_4.2H_2O$; $x = 1$, $y = 3$, $z = 2$, $n = 2$

27 (d) $E = -0.44 + 0.0591/2\,\log_{10}[Fe^{2+}]$ and $[Fe^{2+}][OH^-]^2 = 6.0 \times 10^{-15}\,mol^3\,dm^{-9}$: slope due to variation of $[Fe^{2+}]$ with pH (calculate values).

Chapter 13

6 (b) $C = Na_2S_2O_3$; $S_2O_3^{2-}$ is SO_4^{2-} with one sulphur ($^{35}_{16}S$) atom replacing one oxygen atom.

Chapter 15

8 10

Chapter 16

2 12.9 g dm^{-3}

8 30.4 g dm^{-3}

10 60.7 g dm^{-3}

Chapter 17

5 p = 4.37, $q = 1.14$, 0.237 g

8 $83.3\,cm^3$

Chapter 18

13 (b) $[Ca^{2+}][F^-]^2 = 4.0 \times 10^{-11}\,mol^3\,dm^{-9}$
$[F^-] = 5.2 \times 10^{-4}\,mol\,dm^{-3}$
$K = 5.6 \times 10^{-4}\,mol\,dm^{-3}$
(assume simple ionization)
$\alpha = 0.52$

(c) 3.3

Chapter 20

11 $X = Sn$, $Y = SnS$, $Z = SnO_2$

17 (a) $A = (C_2H_5)_4Sn$; (b) $B = (C_2H_5)_3SnCl$; (c) $3(C_2H_5)_4Sn + SnCl_4 \rightarrow 4(C_2H_5)_3SnCl$; (d) $(C_2H_5)_3SnO.OC.CH_3$

19 (c) (i) SnI_4

Chapter 21

7 0.64 g

15 $A = P$, $B = H_3PO_4$

19 $X = HNO_2$

23 $X = NaNH_4HPO_4.4H_2O$

24 AsH_3

27 (b) (i) $33\,dm^3$

Chapter 22

7 $X = Na_2S_2O_3$, $Y = S$, $Z = H_2SO_4$

8 Group VI, WO_2, WO_2Cl_2

13 $S_2O_3^{2-} + 4Cl_2 + 5H_2O \rightarrow 8Cl^- + 2SO_4^{2-} + 10H^+$

14 $S_2O_3^{2-} + 4Br_2 + 5H_2O \rightarrow 8Br^- + 2SO_4^{2-} + 10H^+$

15 $K_2S_2O_6$

Chapter 23

23 25.3

32 SCl_2

34 (a) 0.536

37 (b) -6 kJmol^{-1} and -59 kJmol^{-1}, respectively

Chapter 24

6 97.6%

15 (a) $Cr_2O_7^{2-}$, (b) AgCl, (c) $K_3Fe(CN)_6$

20 $K_2SO_4.CuSO_4.6H_2O$

23 56.1

24 $ZnBr_2$

28 $A = Cr_2O_3$, $B = Na_2CrO_4$, $C = Na_2Cr_2O_7$, $D = NaCr(SO_4)_2.12H_2O$

29 $A = MnO_2$, 0.78 g

30 $4CrO_2(OK)Cl \rightarrow K_2Cr_2O_7 + Cr_2O_3 + 2KCl + Cl_2 + O_2$

36 $MnO_4^- + 8H^+ + 4e^- \rightarrow Mn^{3+}(as\ MnF_5^{2-}) + 4H_2O$; $MnO_4^- + 8H^+ + 4Fe^{2+} + 29F^- \rightarrow MnF_5^{2-} + FeF_6^{3-} + 4H_2O$

37 (c) 90.5

38 64% Ti, 36% V

39 $A = [Cr(NH_3)_5Cl]Cl_2$; $B = [Cr(NH_3)_4Cl_2]Cl$;
$C = [Cr(NH_3)_6]Cl_3$; $D = CrCl_3$; $E = Cr(OH)_3$;
$F = CH_3CHO$; $G = Cr_2(SO_4)_3$ (amorphous;
zeolitic water)

Chapter 26

4 $2K^+HgI_4{}^{2-}$

5 Formation of $2K^+HgI_4{}^{2-}$ (K_2HgI_4)

Miscellaneous Questions

12 $A = MnO_2$, $B = HgCl_2$, $C = K_3Fe(CN)_6$,
$D = K_4Fe(CN)_6 \cdot 3H_2O$, $E = NaIO_3$

13 $A = Pb_3O_4$, $B = Cu_2O$, $C = (K_2)Cr_2O_7$

15 (a) COS, (b) Too low by $\frac{3}{8}Zn$ ($K_2Zn_3[Fe(CN)_6]_2$
actually precipitated) or $\frac{1}{2}Zn$ if you assumed
$Zn_2[Fe(CN)_6]$ precipitated

16 $G = CuO$, $H = CuCl_2$, $I = CuCl_2{}^-$, $CuCl_3{}^{2-}$,
$CuCl_4{}^{3-}$, $J = CuCl$, $K = Cu(NH_3)_2{}^+$,
$L = Cu(NH_3)_4{}^{2+}$, $M = Cu(CN)_2{}^-$, $Cu(CN)_3{}^{2-}$,
$Cu(CN)_4{}^{3-}$, $N = Cu$, $O = Cu(NO_3)_2$,
$P = Cu(OH)_2$

17 96, $3(NH_4)_2MoO_4 \cdot 4MoO_3 \cdot 4H_2O \rightarrow 7MoO_3$
$+ 6NH_3 + 7H_2O$, i.e.$(NH_4)_6Mo_7O_{24} \cdot 4H_2O$

23 (i) H_2S; (ii) NO; (iii) NH_3

24 $A = P_4$, $B = NH_4NO_3$, $C = PCl_3$, $D = NaNO_2$,
$E = KCN$, $F = PbI_2$, $G = SnCl_4$

28 $A = FeC_2O_4$, $B = Na_2S_2O_3 \cdot 5H_2O$, $C = Pb_3O_4$

Index